AF702180

Translations of

MATHEMATICAL MONOGRAPHS

Volume 22

# Special Functions and the Theory of Group Representations

N. Ja. Vilenkin

AMERICAN MATHEMATICAL SOCIETY
Providence, Rhode Island

СПЕЦИАЛЬНЫЕ ФУНКЦИИ И ТЕОРИЯ
ПРЕДСТАВЛЕНИЙ ГРУПП

Н. Я. ВИЛЕНКИН

Издательство „Наука"
Главная Редакция
Физико-Математической Литературы
Москва 1965

**Translated from the Russian by**
**V. N. Singh**

**AMS ISBN: 978-0-8218-1572-4**

**Library of Congress Card No. 68—19438**

**Copyright © 1968 by the American Mathematical Society**

**Printed in the United States of America**
Reprinted with corrections 1988.
Reprinted by the American Mathematical Society, 2025

*All Rights Reserved*
*No portion of this book may be reproduced*
*without the written permission of the publisher*

The paper used in this book is acid-free and falls within the guidelines
established to ensure permanence and durability. ♾

Information on Copying and Reprinting can be found at the back of this volume.

Visit the AMS home page at URL: http://www.ams.org/

14 13 12 11 10 9 29 28 27 26 25

# PREFACE

In this book we study the theory of special functions from the group-theoretic point of view. At first sight, the theory of special functions appears as a chaotic collection of formulas: apart from the fact that there exists an immense aggregate of the special functions themselves, for each of them there are, at present, all sorts of differential equations, integral representations, recurrence formulas, composition theorems and so on. To establish some kind of order in this chaos of formulas seems to be a completely hopeless task.

However, the development of the theory of group representations has now made it possible to comprehend the theory of the most important classes of special functions from a single point of view. We note that the appraisal of the importance of individual classes of special functions has changed greatly during the last hundred years. In the middle and the second half of the nineteenth century, elliptic and related functions were regarded as the most interesting. However, as F. Klein (see [77]) observed in one of his lectures, there exists another class of special functions, at least as important in view of its numerous applications to astronomy and mathematical physics, namely those associated with the hypergeometric function. The development of mathematics has now confirmed Klein's opinion; the hypergeometric function and its various special and degenerate cases, the functions of Bessel and Legendre, the orthogonal polynomials of Jacobi, Čebyšev, Laguerre and Hermite etc. play a big role in different branches of mathematics and its applications. This class of special functions is often called the special functions of mathematical physics. This is precisely the class of functions which lends itself to a group-theoretic treatment.

The connection between special functions and group representations was first discovered by É. Cartan (see [244]). (However, a connection between the theory of special functions and the theory of invariants, which is one of the aspects of the theory of group representations, was established even earlier). The application of theory of representations to quantum mechanics played a significant part in the investigation part in the investigation of these connections.

Further work in this field was stimulated by the researches of I. M. Gel'fand and M. A. Naĭmark and their students and collaborators in the field of infinite-dimensional representations of groups. In the course of these researches, a connection was established between the theory of group representations and automorphic functions, the theory of special functions over finite fields was constructed, special functions were studied in homogeneous domains, and so on.

The aim of this book is a systematic account of the theory of special functions from the group-theoretic point of view. We have restricted ourselves here to the study of classical special functions and thus of the simplest groups.

The book consists of eleven chapters. In the first chapter basic concepts and facts of the theory of transformation groups and group representations are presented.

In the second chapter, two model examples are analyzed, namely the additive group of real numbers and the group of rotations of the circle. These groups lead, respectively, to the exponential and trigonometric functions. In this chapter certain facts of classical harmonic analysis, the theory of Fourier series and integrals in real and complex domains, are also presented.

The third chapter is devoted to representations of the group of rotations of the three-dimensional euclidean space and the locally isomorphic group $SU(2)$ of unitary unimodular matrices of the second order. The matrix elements of irreducible unitary representations of these groups are expressed by means of Legendre and Jacobi polynomials. The general properties of the matrix elements lead to various relations for these polynomials (orthogonality and completeness relations, recurrence formulas, composition theorem, etc.). At the end of the chapter we consider the discrete analogue of Jacobi polynomials, namely the Clebsch-Gordon coefficients.

In the fourth chapter representations of the group of motions of the euclidean plane are studied. The matrix elements of representations of this group are expressed by means of the Bessel functions, which allows one to derive a number of properties of Bessel functions from group-theoretic considerations. Further properties of Bessel functions and the closely related Hankel and Macdonald functions are studied in Chapter V, devoted to representations of the group of motions of the pseudo-euclidean plane. In this chapter we derive a number of integrals involving the index of cylindrical functions. At the beginning of the chapter the group of linear transformations of the straight line and the theory of $\Gamma$-functions are considered.

The sixth and seventh chapters are devoted to representations of the group of real unimodular matrices of the second order. Representations of this group are connected with certain classes of special functions. In the sixth chapter the conical functions $P_{-\frac{1}{2}+i\lambda}(\operatorname{ch}\theta)$ are studied. In the seventh chapter we consider the realization of representations in the form of integral operators, with the hypergeometric function as the kernel. This leads to a number of properties of the hypergeometric function, some of them well known (the Mellin-Barnes integral representation), others new. In Chapter VIII we similarly study the confluent hypergeometric function and, in greater detail, the closely related Whittaker functions. It is shown that the theory of Whittaker functions is based on the study of representations of the group of third order triangular matrices. Starting from this, we derive various properties of Whittaker functions, e.g. integral representations, recurrence formulas, and continual composition theorems. In the same chapter Laguerre polynomials are considered, and a composition formula is derived for them.

All the groups considered in Chapters II–VIII have a very simple structure. More complicated groups are considered in Chapters IX–XI; they are the groups of motions of the $n$-dimensional sphere, of the $n$-dimensional Lobačevskiĭ space and of the $n$-dimensional euclidean space. Here, however, we do not consider all irreducible representations of these groups, but only the so-called representations of class 1, and we do not consider all matrix elements of representations, but only the matrix elements of "null column". This is found to be sufficient for constructing the theory of Gegenbauer polynomials and harmonic polynomials, and also for obtaining the decomposition of functions on the $n$-dimensional sphere and hyperboloid. In Chapter XI further properties of Bessel functions are derived.

Unfortunately the size of the book did not permit going into certain questions connected with the theory of representations and special functions. Thus the group of motions of the $n$-dimensional pseudo-euclidean space has not been considered. Hardly any light has been thrown upon asymptotic properties of special functions. However, the integral representations of these functions, presented in the book, allow one to obtain their asymptotic expansions by the usual methods (see [6], [62]).

The author expresses his profound gratitude to I. M. Gel'fand, whose advice and instructions were made use of throughout the work on the book, and whose part in creating the book is difficult to overrate. The author also thanks M. A. Naimark, M. I. Graev, I. I. Pjateckiĭ-Šapiro, F. A. Berezin, F. I. Karpelevič, S. G. Gindikin, and D. P. Želobenko, with whom questions of the theory of group representations and the theory of special functions were often discussed. Numerous suggestions of M. I. Graev influenced the treatment of some questions and the final arrangement of the material.

The conversations with Ja. A. Smorodinskiĭ and his students, on the applications of the theory of group representations and integral transformations to physics, were highly useful for the author.

V. V. Cukerman took upon himself the difficult task of verifying the formulas. Great attention was paid to the manuscript, at all stages of its passage, by A. Z. Ryvkin. S. A. Vilenkina went through several variants of the manuscript. To them I express my gratitude for their help, which considerably facilitated preparation of the manuscript for printing.

*N. Ja. Vilenkin*

TABLE OF CONTENTS

# INTRODUCTION

The special functions of mathematical physics appear most often in solving partial differential equations by the method of separation of variables, or in finding eigenfunctions of differential operators in certain curvilinear systems of coordinates. But the differential operators of mathematical physics usually possess definite properties of invariance. Thus the Laplace operator

$$\Delta=\frac{\partial^2}{\partial x^2}+\frac{\partial^2}{\partial y^2}+\frac{\partial^2}{\partial z^2}$$

is invariant for motions of the euclidean space, the wave operator

$$\square=\frac{\partial^2}{\partial x^2}+\frac{\partial^2}{\partial y^2}+\frac{\partial^2}{\partial z^2}-\frac{\partial^2}{\partial t^2}$$

is invariant for transformations of the Lorentz group, and so on. This facilitates the search for eigenfunctions of the operators. The theory of group representations also allows one to consider the invariance of the operators of mathematical physics.

Namely, suppose that linear operator $A$ is invariant for some transformation group $G$. One can show that these transformations take eigenfunctions of the operator into eigenfunctions corresponding to the same eigenvalue. By the same token, with elements of group $G$ one associates a linear transformation $T(g)$ in the space of eigenfunctions; moreover, one has the equality

$$T(g_1)\,T(g_2)=T(g_1g_2). \tag{1}$$

Operator functions on groups, possessing property (1), are called group representations. In this way eigenfunctions of invariant operators are connected with representations of groups, relative to which the operators are invariant. The knowledge of these representations facilitates the search for eigenfunctions and allows one to explain their behavior under transformations of the given group.

Operators of the representation $T(g)$ can be given in matrix form, on choosing some basis in the representation space. In this connection one meets numerical functions on the group, namely the matrix elements of the representation. But elements of groups essential for mathematical physics are usually given by numerical parameters. For example, elements of the group of motions of the euclidean plane are given by coordinates $(a, b)$ of the image of point $O(0, 0)$ and by the angle of rotation $\phi$; elements of the group of rotations of the three-dimensional euclidean space are given by the Euler angles $\phi$, $\theta$, $\psi$, and so on. In this way, by studying group representations, we arrive at numerical functions of several variables.

Clearly it is desirable to express these functions by functions of the least possible number of variables, and preferably by functions of one variable. It has been found that such expression is possible for certain groups (group of rotations of three-dimensional space, group of motions of the euclidean plane, etc.). For these groups one can choose a basis in the representation space in such a way that that elements of some subgroup $H$ are given by diagonal matrices with exponential functions in the principal diagonal. The remaining elements of the group can be represented in the form $h_1 \theta h_2$, where $h_1, h_2 \in H$ and $\theta(t)$ runs through a certain one-parameter manifold.

By utilizing this fact one succeeds in expressing the matrix elements of representations of such groups by means of the exponential function and functions of the one parameter $t$. It has been found that these functions coincide with the classical special functions of mathematical physics. Thus, representations of the group of motions of the euclidean plane were found to be connected with the Bessel functions $J_n(t)$, representations of the group of rotations of the three-dimensional euclidean space with the Legendre and Jacobi functions, and so on. We note that the appearance of the exponential function is not accidental: functions of the form $e^{int}$, where $n$ is an integer, give representations of the group of rotations of the euclidean plane.

On studying more complicated groups (group of rotations of $n$-dimensional euclidean space, group of all motions of this space, the Lorentz group, etc.), it was found that not all matrix elements of representations of these groups are expressed by means of known special functions. One could obtain such expressions only for some of the matrix elements, and for the remainder one needed functions not encountered earlier in mathematical analysis. These new functions possess properties just as varied as classical special functions (see [121]).

In this way there exists a connection between special functions and matrix elements of group representations. It must be noted that this connection depends on the choice of the subgroup $H$ whose elements are expressed, in the given realization, by diagonal (or, in the general case, by block-diagonal) matrices. Therefore different special functions may turn out to be connected with one and the same group, depending on the subgroup $H$ for which operators $T(h)$ have been "diagonalized". For example, if $G$ is the group of real unimodular matrices of the second order, and $H$ is the subgroup of matrices of the form

$$\begin{pmatrix} \cos\theta & -\sin\theta \\ \sin\theta & \cos\theta \end{pmatrix},$$

then conical functions appear. But if the subgroup of diagonal matrices is chosen as $H$, then the hypergeometric function appears. Lastly, if $H$ is the subgroup of triangular matrices of the form

$$\begin{pmatrix} 1 & 0 \\ t & 1 \end{pmatrix},$$

then we get Hankel functions.

The establishment of a connection between matrix elements of representations and special functions indicated a general method of deriving the properties of these functions. For example, it follows from equality (1) that matrix elements of representations satisfy the relation

$$t_{ij}(g_1 g_2) = \sum_k t_{ik}(g_1)\, t_{kj}(g_2). \tag{2}$$

But the matrix elements are expressed by means of special functions. Therefore formula (2) leads to formulas for special functions. In this way one obtains, in particular, composition theorems for the functions of Bessel, Legendre, Gegenbauer etc. If one of the elements in equality (2) is taken to be "infinitely close to the group identity", then (2) turns into a recurrence relation for the corresponding special functions.

The group-theoretic approach leads to a natural treatment of integral representations of special functions. If, in the space of the representation $T(g)$, an orthonormal basis $\{e_k\}$ is chosen, then the matrix elements are given by the formula

$$t_{ij}(g) = (T(g)\mathbf{e}_j, \mathbf{e}_i). \tag{3}$$

But the representation space is most frequently realized in the form of some functional space (for example, as the space of eigenfunctions of an invariant operator), and the scalar product of the space in the form of an integral. Therefore the right hand side of formula (3) is expressed in the form an integral, and the left hand side reduces to special functions. This gives integral representations for special functions.

It is not always possible to choose the basis in the representation space in such a way that elements of a given subgroup are expressed by diagonal (or block-diagonal) matrices. Sometimes one has to choose a realization of the representation, under which elements of $H$ are expressed by operators of multiplication by a function (continual analogue of a diagonal matrix). In these cases operators of the representation $T(g)$ take the form of integral operators whose kernels are expressed by means of special functions. This leads to various integral relations between special functions; in particular it leads to continual analogues of composition theorems.

The group-theoretic approach to special functions is closely associated with the harmonic analysis of functions. As a typical example, we consider the representation

$$T(g_\alpha)f(\varphi)=f(\varphi+\alpha)$$

of the group of rotations of the circle ($f(\phi)$ is a function on the circle, which depends on the angle $\phi$, and $g_\alpha$ is the rotation by an angle $\alpha$). We expand the function $f(\phi)$ in a Fourier series:

$$f(\varphi)=\sum_{n=-\infty}^{\infty} c_n e^{in\varphi}.$$

The one-dimensional spaces of functions of the form $c_n e^{in\phi}$ remains invariant under the transformations $T(g_\alpha)$:

$$T(g_\alpha)e^{in\varphi}=e^{in(\varphi+\alpha)}=e^{in\alpha}e^{in\varphi}.$$

One says that the space of functions $f(\phi)$ on the circle is decomposed into one-dimensional invariant subspaces, and the representation $T(g)$ into representations $T_n(g_\alpha)=e^{in\alpha}$.

An analogous problem appears in several other cases. Suppose that a group $G$ acts on some set $\mathfrak{M}$ and let $\mathfrak{L}$ be the space of functions on this set. One is required to decompose this space into minimum spaces, invariant for the group $G$ (i.e. containing together with each function $f(x)$ all its displacements $f(gx)$). In analogy with the example considered above, this problem is called the problem of harmonic analysis of the space $\mathfrak{L}$.

Depending upon the properties of the group $G$, various cases arise. If $G$ is commutative and compact (as is the case for the group of rotations of the circle), one obtains a decomposition of $\mathfrak{L}$ as the direct sum of one-dimensional invariant subspaces. If $G$ is commutative but not compact, then one obtains a decomposition as the "continuous direct sum" of one-dimensional spaces. A typical example is the expression of functions on the line as a Fourier integral:

$$f(x)=\int_{-\infty}^{\infty} F(\lambda)\, e^{i\lambda x} d\lambda.$$

The functions $e^{i\lambda x}$, on which the expression is based here, give one-dimensional representations of the additive group of real numbers:

$$e^{i\lambda(x+y)} = e^{i\lambda x} e^{i\lambda y},$$

and the parameter $\lambda$, specifying the representation, runs through the whole line.

For noncommutative compact groups (for example the group of unitary matrices) one obtains a decomposition into a denumerable set of finite-dimensional invariant subspaces. The harmonic analysis for noncommutative noncompact groups (for example the Lorentz group) is the most complicated. Here one usually gets decompositions as continuous direct sums of infinite-dimensional subspaces.

Harmonic analysis on groups is connected with the decomposition of functions on groups and homogeneous spaces, in matrix elements of representations. Since these matrix elements are expressed by means of special functions, one obtains decompositions into series and integrals of special functions. This leads to a group-theoretic treatment of certain integral transformations encountered in mathematical physics (the transformations of Mehler-Fock, Hankel, Kontorovič-Lebedev Olevskiĭ and others), and also of expansions in series of special functions. The group-theoretic point of view allows one to carry over, in a natural manner, ordinary harmonic analysis (the theory of positive definite functions, the Plancherel formula, etc.) to such decompositions. For concrete harmonic analysis on groups the horisphere method, developed by I. M. Gel'fand and M. I. Graev ([134], [141], [142]), has proved to be useful.

Other integral transformations connected with the theory of representations arise in writing down the representation operators themselves as integral operators. The kernels of these operators are expressed by means of special functions. In this case the inversion formulas are linked with the fact that the operators $T(g)$ and $T(g^{-1})$ are mutually reciprocal.

The group-theoretic interpretation of special functions allows one to find their natural generalizations. For example, by considering, instead of real second order matrices, matrices with elements from a discrete normed field, I. M. Gel'fand and M. I. Graev [143] obtained special functions on such fields. In this way one also obtains special functions of matrix argument (another approach to these functions depends upon a study of homogeneous cones; see [158]). The special functions appearing in this way possess many of the properties of special functions of a numerical argument. With their help one can construct a harmonic analysis on groups of matrices with $\mathfrak{p}$-adic elements, and in several other cases not covered by classical analysis.

Finally, as we have already said, in the theory of representations there arise new special functions, a study of which leads to new relations for the usual functions too. In particular, orthogonal vector functions with polynomial elements apparently appear here.

CHAPTER I

# GROUP REPRESENTATIONS

This chapter is of an introductory nature. It contains those basic concepts and results of the theory of group representations which will be used in the book. In addition, we show in it how representations of transformation groups are constructed and also examine the connection between eigenfunctions of invariant operators and the theory of group representations. Finally, at the end of the chapter, we study the matrix elements of irreducible unitary representations of compact groups. It is shown that these elements form a complete orthogonal system of functions on the group relative to an invariant measure. This establishes a connection between the theory of group representations and orthogonal decompositions.

## §1. BASIC CONCEPTS OF THE THEORY OF REPRESENTATIONS

1. **Definition.** The concept of a group representation is a far-reaching generalization of the concept of the exponential function. The exponential function $e^{ax}$ can be defined as the continuous solution of the functional equation

$$f(x+y)=f(x)f(y), \tag{1}$$

satisfying the initial condition $f'(0) = a$.

On generalizing this equation to any group $G$, we are led to consider scalar functions on $G$ satisfying the equation

$$f(g_1g_2)=f(g_1)f(g_2). \tag{2}$$

However, for noncommutative groups there are not many such functions, since equality (2) implies that they must satisfy the relation

$$f(g_1g_2)=f(g_1)f(g_2)=f(g_2)f(g_1)=f(g_2g_1).$$

Therefore scalar functions satisfying equation (2) are inadequate for the purpose of expanding an arbitrary function $F(g)$ on the group $G$.

In order to obtain a sufficiently good supply of solutions of equation (2), one has to turn from scalar functions to functions whose values are matrices or linear transformations. Since matrix multiplication is noncommutative there is a sufficiently large supply of solutions of such a form. We are thus led to consider solutions of the functional equation

$$T(g_1 g_2) = T(g_1)\, T(g_2).$$

where $g_1$ and $g_2$ are elements of the given group $G$, and $T(g)$ is a function on this group taking values in the set of linear transformations of some linear space $\mathfrak{R}$. These solutions are called *representations of the group* $G$. In what follows we shall consider representations of continuous groups, i.e. groups in which the concept of convergence of a sequence of elements is, in some way, defined. Here we shall require that the operators $T(g)$ depend continuously on $g$, i.e. $\lim_{n\to\infty} g_n = g$ should imply that $\lim_{n\to\infty} T(g_n) = T(g)$. Clearly the requirement of continuity of the representation depends on how the convergence of the operators $T(g)$ is defined (for example, one can speak of weakly continuous, strongly continuous representations etc.). We shall require, in addition, that all operators $T(g)$ be continuous in $\mathfrak{R}$, i.e. $\lim_{n\to\infty} x_n = x$ should imply that $\lim_{n\to\infty} T(g)x_n = T(g)x$. The precise meaning of this requirement also depends on the choice of convergence in $\mathfrak{R}$.

Further, we shall restrict ourselves to the consideration of nonsingular representations, i.e. we shall require that the operators $T(g)$ have continuous inverses.

Thus, by a *representation of the group* $G$, we shall mean a continuous function $T(g)$ on this group, taking values in the group of nonsingular continuous linear transformations of the linear space $\mathfrak{R}$ and satisfying the functional equation

$$T(g_1 g_2) = T(g_1)\, T(g_2). \tag{3}$$

It easily follows from this equation that $T(g^{-1}) = T^{-1}(g)$ and $T(e) = E$, where $e$ is the unity element of the group $G$, and $E$ is the identity operator in $\mathfrak{R}$.

The equalities $T(g_1 g_2) = T(g_1)T(g_2)$ and $T(g^{-1}) = T^{-1}(g)$ show that $T(g)$ is a homomorphic mapping of $G$ into the group of nonsingular continuous linear transformations of $\mathfrak{R}$.

A representation $T(g)$ is called *faithful* if only for the unity element $e$ of $G$ do we have $T(e) = E$ and *trivial* (or the *identity* representation) if $T(g) = E$ for all elements $g$ of $G$.

The linear space $\mathfrak{R}$, in which the operators $T(g)$ act, is called the *space of the representation* $T(g)$. If this space is finite-dimensional, then the representation $T(g)$ is also called *finite-dimensional*.

In the infinite-dimensional case we shall, as a rule, consider representations by operators in Hilbert spaces. Only occasionally will it be necessary to deal with representations in pre-Hilbert spaces, which turn into Hilbert spaces on completion with respect to a norm of the form $\|x\| = \sqrt{(x, x)}$, where $(x, y)$ is the scalar product in $\mathfrak{L}$.

2. The matrix notation for representations. A representation $T(g)$ of a group $G$ has been defined as an operator function on this group satisfying the functional equation

$$T(g_1 g_2) = T(g_1)\, T(g_2). \tag{1}$$

An analyst naturally prefers to deal with functions which take numerical values. In order to pass from abstract functions to functions taking numerical values, we make use of the matrix notation for operators.

We first consider the case when the space of the representation $T(g)$ is finite-dimensional. In this space we choose a basis $\mathbf{e}_1, \cdots, \mathbf{e}_n$. The operator $T(g)$ transforms the basis element $\mathbf{e}_j$ into $T(g)\mathbf{e}_j$. Decomposing $T(g)\mathbf{e}_j$ into basis elements, we get

$$T(g)\,\mathbf{e}_j = \sum_{i=1}^{n} t_{ij}(g)\,\mathbf{e}_i. \tag{2}$$

Thus with each operator of the representation $T(g)$ we associate the matrix

$$(T(g)) \equiv (t_{ij}(g)), \qquad 1 \leqslant i, j \leqslant n, \tag{3}$$

or, what is the same thing, a collection of $n^2$ numerical functions $t_{ij}(g)$ on the group. From the continuity of $T(g)$ it follows that the functions $t_{ij}(g)$ are continuous.

Since the multiplication of operators means the multiplication of corresponding matrices, from the functional equation (1) we get the system of $n^2$ equalities

$$t_{ij}(g_1 g_2) = \sum_{k=1}^{n} t_{ik}(g_1)\, t_{kj}(g_2), \qquad 1 \leqslant i, j \leqslant n \tag{4}$$

for functions $t_{ij}(g)$.

In this way we can define an $n$-dimensional representation of the group $G$ as the collection of $n^2$ continuous numerical functions $t_{ij}(g)$, $g \in G$, satisfying the system of functional equations (4), and such that $\operatorname{Det}(t_{ij}(g)) \neq 0$.

Of course, the matrix notation depends on the choice of the basis $\{\mathbf{e}_i\}$ in the representation space. If $A$ is a nonsingular operator, mapping the space $\mathfrak{L}$

onto itself, and $f_i = Ae_i$, then in the basis $\{f_i\}$ the matrix of the representation $T(g)$ has the form

$$(A^{-1})(T(g))(A). \tag{5}$$

Here $(A)$ denotes the matrix of the operator $A$ in the basis $\{e_k\}$ i.e. a matrix such that

$$\mathbf{f}_j = \sum_{i=1}^{n} a_{ij}\mathbf{e}_i. \tag{6}$$

Thus on going over to another basis in $\mathfrak{L}$ the matrix $(T(g))$ is replaced by an equivalent matrix.

It is clear that if $(t_{ij}(g))$ is the matrix of the representation $T(g)$, then $\overline{(t_{ij}(g))}$ is the matrix of a certain representation $\overline{T(g)}$. It should be borne in mind that $\overline{T(g)}$ depends not only on $T(g)$, but also on the choice of the basis in $\mathfrak{L}$.

If the space of the representation $T(g)$ is euclidean, then one usually chooses an orthogonal normalized (orthonormal) basis in it:

$$(\mathbf{e}_i, \mathbf{e}_j) = \delta_{ij}, \tag{7}$$

where $\delta_{ij}$ is the Kronecker symbol [1]. In this case the matrix elements are calculated according to the formula

$$t_{ij}(g) = (T(g)\mathbf{e}_j, \mathbf{e}_i). \tag{8}$$

In order to obtain this formula it is sufficient to take the scalar product of both sides of equality (2) with $e_i$.

We now consider an infinite-dimensional representation $T(g)$. As we have already stated, in this case we shall take the representation space to be Hilbertian. In this Hilbert space we choose an orthonormal basis $\{e_i\}$, $i = 1, 2, \cdots, n, \cdots$. In this basis we have

$$T(g)\mathbf{e}_j = \sum_{i=1}^{\infty} t_{ij}(g)\mathbf{e}_i. \tag{9}$$

On taking the scalar product of both sides of equality (9) with $e_i$, we get

$$t_{ij}(g) = (T(g)\mathbf{e}_j, \mathbf{e}_i), \quad 1 \leqslant i, j < \infty. \tag{10}$$

Thus with each operator $T(g)$ we associate the infinite matrix $(T(g))$ with

---

1) $\delta_{ij} = \begin{cases} 0 \text{ for } i \neq j \\ 1 \text{ for } i = j \end{cases}.$

elements $t_{ij}(g)$. It is easy to show that in multiplying operators, the matrices are multiplied according to the usual rule, i.e.

$$t_{ij}(g_1g_2) = \sum_{k=1}^{\infty} t_{ik}(g_1)\, t_{kj}(g_2). \tag{11}$$

In fact, by formula (10) we have

$$t_{ij}(g_1g_2) = (T(g_1g_2)\,\mathbf{e}_j, \mathbf{e}_i) = (T(g_1)\,T(g_2)\,\mathbf{e}_j, \mathbf{e}_i) = (T(g_2)\,\mathbf{e}_j, T(g_1)^*\,\mathbf{e}_i). \tag{12}$$

The scalar product of vectors $T(g_2)\mathbf{e}_j$ and $T(g_1)^*\mathbf{e}_i$ is expressed as follows:

$$(T(g_2)\,\mathbf{e}_j, T(g_1)^*\,\mathbf{e}_i) = \sum_{k=1}^{\infty} (T(g_2)\,\mathbf{e}_j, \mathbf{e}_k)\,\overline{(T(g_1)^*\,\mathbf{e}_i, \mathbf{e}_k)}$$

$$= \sum_{k=1}^{\infty} (T(g_2)\,\mathbf{e}_j, \mathbf{e}_k)\,(T(g_1)\,\mathbf{e}_k, \mathbf{e}_i) = \sum_{k=1}^{\infty} t_{ik}(g_1)\, t_{kj}(g_2). \tag{13}$$

From (12) and (13) we get equality (11).

3. **Equivalent representations.** Let $T(g)$ be a representation of the group $G$ in the space $\mathfrak{L}_1$ and let $A$ be a linear mapping of $\mathfrak{L}_1$ into a space $\mathfrak{L}_2$, with a continuous inverse $A^{-1}$. Then the equality

$$Q(g) = A\,T(g)\,A^{-1} \tag{1}$$

determines a representation of $G$ in the space $\mathfrak{L}_2$. Indeed, we have

$$Q(g_1)\,Q(g_2) = A\,T(g_1)\,A^{-1}A\,T(g_2)\,A^{-1} = A\,T(g_1)\,T(g_2)\,A^{-1}$$
$$= A\,T(g_1g_2)\,A^{-1} = Q(g_1g_2).$$

It is clear that in this way one can, starting with a given representation $T(g)$, construct as many "new" representations of the same group as desired. These representations are regarded as equivalent to $T(g)$.

Thus representations $T(g)$ and $Q(g)$ of the group $G$ in spaces $\mathfrak{L}_1$ and $\mathfrak{L}_2$ are *equivalent* if there exists a linear operator $A$, mapping $\mathfrak{L}_1$ into $\mathfrak{L}_2$, with an inverse linear operator $A^{-1}$, and such that

$$Q(g) = A\,T(g)\,A^{-1}.$$

The concept of equivalence of representations is reflexive, symmetric and transitive: every representation is equivalent to itself; if a representation $T(g)$ is equivalent to a representation $Q(g)$, then $Q(g)$ is equivalent to $T(g)$; if $T(g)$ is equivalent to $Q(g)$ and $Q(g)$ is equivalent to a representation $R(g)$, then $T(g)$ is equivalent to $R(g)$. Therefore *the set of all representations of the group $G$ can be divided into equivalence classes of representations.* Henceforth we shall not distinguish between equivalent representations, i.e. we shall study the properties of equivalence classes of representations.

*If representations $T(g)$ and $Q(g)$ are equivalent, then, for a suitable choice of the bases, they are given by identical matrices.* Namely, if in the space $\mathfrak{L}_1$ of the representation $T(g)$ one chooses a basis $\{\mathbf{e}_k\}$, then in the space $\mathfrak{L}_2$ of $Q(g)$ one must choose the basis $\{A\mathbf{e}_k\}$. We have

$$Q(g)A\mathbf{e}_k = AT(g)\mathbf{e}_k.$$

Therefore, if

$$T(g)\mathbf{e}_k = \sum t_{jk}(g)\mathbf{e}_j,$$

then

$$Q(g)A\mathbf{e}_k = A\sum t_{jk}(g)\mathbf{e}_j = \sum t_{jk}(g)A\mathbf{e}_j.$$

Thus the representation $T(g)$ in the basis $\{\mathbf{e}_k\}$ has the same matrix as the representation $Q(g)$ in the basis $\{A\mathbf{e}_k\}$.

4. Adjoint representations. Let $T(g)$ be a representation of a group $G$ in a space $\mathfrak{L}$. We denote by $\mathfrak{L}'$ the space adjoint to $\mathfrak{L}$, i.e. the space of linear functionals in $\mathfrak{L}$. The equality

$$T'(g)\mathbf{f}(\mathbf{x}) = \mathbf{f}(T(g^{-1})\mathbf{x}), \qquad \mathbf{f}\in\mathfrak{L}',\ \mathbf{x}\in\mathfrak{L} \tag{1}$$

determines a representation of $G$. In fact,

$$\begin{aligned}T(g_1)T(g_2)\mathbf{f}(\mathbf{x}) = T'(g_1)\mathbf{f}(T(g_2^{-1})\mathbf{x}) &= \mathbf{f}(T(g_2^{-1})T(g_1^{-1})\mathbf{x})\\ &= \mathbf{f}(T((g_1g_2)^{-1})\mathbf{x}) = T'(g_1g_2)\mathbf{f}(\mathbf{x}),\end{aligned}$$

and therefore

$$T'(g_1)T'(g_2) = T'(g_1g_2).$$

The representation $T'(g)$ is called *adjoint* to the representation $T(g)$.

In the space $\mathfrak{L}$ of the representation $T(g)$ we choose a basis $\{\mathbf{e}_i\}$ and denote by $\mathbf{f}_k$ a linear functional such that $\mathbf{f}_k(\mathbf{e}_i) = \delta_{ki}$. The linear functionals $\{\mathbf{f}_i\}$ form a basis in the space $\mathfrak{L}'$, called *biorthogonal* to the basis $\{\mathbf{e}_i\}$. We shall now prove the following assertion.

*If the matrix of the representation $T(g)$ in the basis $\{\mathbf{e}_i\}$ is equal to $(t_{ij}(g))$, then the matrix of the adjoint representation in the biorthogonal basis has the form $(t_{ji}(g^{-1}))$.* In other words, it is obtained from $(t_{ij}(g))$ by transposition and the replacement of $g$ by $g^{-1}$.

In fact, owing to the biorthogonality of bases, we have

$$\begin{aligned}T'(g)\mathbf{f}_j(\mathbf{e}_k) &= \mathbf{f}_j(T(g^{-1})\mathbf{e}_k) = \mathbf{f}_j\Big(\sum_i t_{ik}(g^{-1})\mathbf{e}_i\Big)\\ &= \sum_i t_{ik}(g^{-1})\mathbf{f}_j(\mathbf{e}_i) = t_{jk}(g^{-1}) = \sum_i t_{ji}(g^{-1})\mathbf{f}_i(\mathbf{e}_k)\end{aligned}$$

and therefore

$$T'(g)\,\mathfrak{f}_j=\sum_i t_{ji}(g^{-1})\,\mathfrak{f}_i.$$

This proves our assertion.

If in the space $\mathfrak{L}$ of the representation $T(g)$ a nonsingular bilinear form $B(\mathbf{x}, \mathbf{y})$ is given, then to each element $\mathbf{y}$ of $\mathfrak{L}$ there corresponds a linear functional $\mathbf{y}(\mathbf{x})$, defined by the equality

$$\mathbf{y}(\mathbf{x})=B(\mathbf{x}, \mathbf{y}). \tag{2}$$

Since the form $B(\mathbf{x}, \mathbf{y})$ is nonsingular, equality (2) establishes an imbedding of $\mathfrak{L}$ in $\mathfrak{L}'$.

Restricting the representation $T'(g)$ to the image of $\mathfrak{L}$, we obtain a representation of $G$ in $\mathfrak{L}$. It is also called *adjoint* to $T(g)$ (relative to the bilinear from $B(\mathbf{x}, \mathbf{y})$).

5. Hermitian-adjoint representations. Unitary representations. Suppose that in a space $\mathfrak{H}$ we are given a nondegenerate scalar product $(\mathbf{x}, \mathbf{y})$ (linear in $\mathbf{x}$ and antilinear in $\mathbf{y}$). The operator $A^*$ is called *Hermitian-adjoint* to $A$ relative to this scalar product if for any two vectors $\mathbf{x}$ and $\mathbf{y}$ of $\mathfrak{H}$ we have

$$(A\mathbf{x}, \mathbf{y})=(\mathbf{x}, A^*\mathbf{y}).$$

We show that if $T(g)$ is a representation of the group $G$ in the space $\mathfrak{H}$, then $T^*(g^{-1})$ is also a representation of this group. Indeed, since $(AB)^* = B^*A^*$, we have

$$T^*(g_1^{-1})\,T^*(g_2^{-1})=[T(g_2^{-1})\,T(g_1^{-1})]^*=T^*(g_2^{-1}g_1^{-1})=T^*[(g_1g_2)^{-1}].$$

The representation $T^*(g^{-1})$ is called *Hermitian-adjoint* to $T(g)$ and is denoted by $\tilde{T}(g)$.

In $\mathfrak{H}$ we choose a basis $\{\mathbf{e}_i\}$, orthonormalized relative to the scalar product $(\mathbf{x}, \mathbf{y})$, i.e. such that $(\mathbf{e}_i, \mathbf{e}_j) = \delta_{ij}$. We show that in this basis the matrix of the representation $\tilde{T}(g)$ is obtained from the matrix of the representation $T(g)$ by going over to the Hermitian-adjoint matrix (i.e. by transposing and replacing all elements by their complex conjugates) and replacing $g$ by $g^{-1}$. In other words, we show that

$$\tilde{t}_{ij}(g)=\overline{t_{ji}(g^{-1})}.$$

Indeed, from formula (8) of subsection 2 we have

$$\begin{aligned}\tilde{t}_{ij}(g)&=(\tilde{T}(g)\mathbf{e}_j, \mathbf{e}_i)=(T^*(g^{-1})\mathbf{e}_j, \mathbf{e}_i)=(\mathbf{e}_j, T(g^{-1})\mathbf{e}_i)\\&=\overline{(T(g^{-1})\mathbf{e}_i, \mathbf{e}_j)}=\overline{t_{ji}(g^{-1})}.\end{aligned}$$

A representation $T(g)$ of a group $G$ in a space $\mathfrak{H}$ is called unitary relative to the scalar product $(\mathbf{x}, \mathbf{y})$ if the operators $T(g)$ leave the scalar product invariant,i.e. if for all vectors $\mathbf{x}$ and $\mathbf{y}$ of $\mathfrak{H}$ and all elements $g$ of $G$ one has the equality

$$(T(g)\mathbf{x},\ T(g)\mathbf{y}) = (\mathbf{x},\ \mathbf{y}). \tag{1}$$

In this case we have $T^*(g)T(g) = E$, and therefore $T^*(g) = T^{-1}(g) = T(g^{-1})$.

This equality can be written in the form $T(g) = \widetilde{T}(g)$. Thus *a representation $T(g)$ is unitary if it coincides with its Hermitian-adjoint representation.*

Taking into consideration the matrix form of the Hermitian-adjoint representation in an orthonormal basis, we obtain the following assertion.

*In an orthonormal basis the matrix of a unitary representation is unitary, i.e.*

$$(t_{ij}(g))^{-1} = \overline{(t_{ji}(g))}. \tag{2}$$

Hence for such a matrix we have the equalities

$$\sum_j t_{ij}(g)\overline{t_{kj}(g)} = \sum_j t_{ji}(g)\overline{t_{jk}(g)} = \delta_{ik}. \tag{3}$$

It is clear that if the representation $T(g)$ is unitary and the basis $\{\mathbf{e}_k\}$ is orthonormal, then $\overline{T}(g)$ (see subsection 2) is also unitary.

Unitary representations play a very important part in the theory of group representations. For instance, it shall be shown below that any finite-dimensional representation of a compact group is unitary relative to a certain scalar product.

6. **Invariant subspaces. Irreducible representations.** A subspace $\mathfrak{L}_1$ of the space of the representation $T(g)$ is called *invariant* if $\mathbf{x} \in \mathfrak{L}_1$ implies that for all $g \in G$ we have $T(g)\mathbf{x} \in \mathfrak{L}_1$. In other words, all operators of the representation $T(g)$ transform vectors of the subspace $\mathfrak{L}_1$ into vectors of the same subspace. For infinite-dimensional representations we shall consider only closed invariant subspaces.

For every representation $T(g)$ there are at least two invariant subspaces, namely the null subspace and the whole space $\mathfrak{L}$ of this representation. These invariant subspaces are called *trivial*. If a representation $T(g)$ possesses only trivial invariant subspaces it is called *irreducible*. A representation with nontrivial invariant subspaces is *reducible*.

With every reducible representation $T(g)$ of the group $G$, two new representations of this group are associated. The first of these is obtained if one considers operators of the representation $T(g)$ only on the subspace $\mathfrak{L}_1$. We shall call this representation the *restriction* of $T(g)$ to the invariant subspace $\mathfrak{L}_1$. The second representation is constructed in the factor-space $\mathfrak{L}/\mathfrak{L}_1$, i.e. in the linear space having the cosets $\mathbf{x} + \mathfrak{L}_1$ as elements. It follows from the invariance of $\mathfrak{L}_1$ that $T(g)$ transforms the coset $\mathbf{x} + \mathfrak{L}_1$ into the coset $T(g)\mathbf{x} + \mathfrak{L}_1$. This determines a representation of $G$ in the factor-space $\mathfrak{L}/\mathfrak{L}_1$.

We show that if a representation $T(g)$ is reducible, then, in a suitably chosen basis, the matrix of this representation has the form

$$\begin{pmatrix} T_1(g) & A(g) \\ 0 & T_2(g) \end{pmatrix}. \tag{1}$$

Namely, if the invariant subspace $\mathfrak{L}_1$ has finite dimension $k$, then the basis $\{\mathbf{e}_i\}$ in $\mathfrak{L}$ must be chosen in such a way that the vectors $\{\mathbf{e}_i\}$, $1 \leq i \leq k$, give the basis in the subspace $\mathfrak{L}_1$. Then, for $1 \leq j \leq k$, vectors $T(g)\mathbf{e}_j$ belong to $\mathfrak{L}_1$ and consequently can be decomposed in the basis vectors $\{\mathbf{e}_i\}$, $1 \leq i \leq k$, of this subspace:

$$T(g)\mathbf{e}_j = \sum_{i=1}^{k} t_{ij}(g)\mathbf{e}_i.$$

Hence, for $i > k$, $1 \leq j \leq k$, the matrix elements $t_{ij}(g)$ are equal to zero, and consequently the representation matrix has the form (1).

If the invariant subspace is infinite-dimensional, then we number the basis vectors by indices $i$, $-\infty < i < \infty$, and choose a basis $\{\mathbf{e}_i\}$ such that the vectors $\{\mathbf{e}_i\}$, $-\infty < i < 0$, form the basis in $\mathfrak{L}_1$. Then the matrix of the representation $T(g)$ will have the form (1).

7. **Decomposition of a representation as a direct sum.** Generally speaking it is not possible to choose a basis in the space of an irreducible representation such that the matrix $A(g)$ in (1) of subsection 6 becomes zero. But if the invariant subspace $\mathfrak{L}_1$ has an invariant complementary subspace $\mathfrak{L}_2$ then such a choice is possible. A subspace $\mathfrak{L}_2$ is called *complementary* to $\mathfrak{L}_1$ if every vector $\mathbf{x}$ of $\mathfrak{L}$ can be uniquely expressed in the form $\mathbf{x} = \mathbf{x}_1 + \mathbf{x}_2$, where $\mathbf{x}_1 \in \mathfrak{L}_1$, $\mathbf{x}_2 \in \mathfrak{L}_2$ (in other words, if $\mathfrak{L}_1$ and $\mathfrak{L}_2$ make up the whole space $\mathfrak{L}$ and have zero intersection). The space $\mathfrak{L}$ in that case is said to be the *direct sum* of the subspaces $\mathfrak{L}_1$ and $\mathfrak{L}_2$, and one writes $\mathfrak{L} = \mathfrak{L}_1 + \mathfrak{L}_2$.

So let us suppose that the space $\mathfrak{L}$ of the representation $T(g)$ is the direct sum of invariant subspaces $\mathfrak{L}_1$ and $\mathfrak{L}_2$. We denote by $T_1(g)$ the restriction of $T(g)$ to $\mathfrak{L}_1$ and by $T_2(g)$ the restriction to $\mathfrak{L}_2$ (clearly $T_2(g)$ is equivalent to the representation in the factor-space $\mathfrak{L}/\mathfrak{L}_1$). For any vector $\mathbf{x}$ of $\mathfrak{L}$ we have

$$T(g)\,\mathbf{x} = T(g)(\mathbf{x}_1 + \mathbf{x}_2) = T(g)\,\mathbf{x}_1 + T(g)\,\mathbf{x}_2 = T_1(g)\,\mathbf{x}_1 + T_2(g)\,\mathbf{x}_2,$$

where $\mathbf{x}_1 \in \mathfrak{L}_1$, $\mathbf{x}_2 \in \mathfrak{L}_2$. This shows that $T(g)$ is uniquely determined by specifying $T_1(g)$ and $T_2(g)$. $T(g)$ is said to be the *direct sum* of $T_1(g)$ and $T_2(g)$, and one writes

$$T(g) = T_1(g) + T_2(g).$$

Thus, if the space $\mathfrak{L}$ of the representation $T(g)$ is the direct sum of the invariant subspaces $\mathfrak{L}_1$ and $\mathfrak{L}_2$, then $T(g)$ is the direct sum of its restrictions $T_1(g)$ and $T_2(g)$ to these subspaces.

Suppose that the space $\mathfrak{L}$ of the representation $T(g)$ is decomposed as the direct sum of invariant subspaces $\mathfrak{L}_1$ and $\mathfrak{L}_2$. We choose a basis $\{\mathbf{e}_i\}$ in $\mathfrak{L}$ such that the vectors $\mathbf{e}_i$, belonging to the invariant subspaces $\mathfrak{L}_1$ and $\mathfrak{L}_2$, provide a basis in them. For such a choice the matrix of the representation $T(g)$ takes the form

$$\begin{pmatrix} T_1(g) & 0 \\ 0 & T_2(g) \end{pmatrix}.$$

The definition of a direct sum of representations can be naturally extended to the case of several summands. In this case, for a suitably chosen basis, the representation matrix has the form

$$\begin{pmatrix} T_1(g) & 0 & \ldots & 0 \\ 0 & T_2(g) & \ldots & 0 \\ \ldots & \ldots & \ldots & \ldots \\ 0 & 0 & \ldots & T_n(g) \end{pmatrix}.$$

We now introduce the concept of a direct orthogonal sum of representations. Suppose that in the Hilbert space $\mathfrak{H}$ of representation $T(g)$ there are pairwise orthogonal invariant subspaces $\mathfrak{H}_k$, $1 \leq k < \infty$, such that every element $\mathbf{x}$ of $\mathfrak{H}$ can be decomposed into a convergent series $x = \sum_{k=1}^{\infty} x_k$, where $x_k \in \mathfrak{H}_k$. In that case one writes

$$\mathfrak{H} = \sum_{k=1}^{\infty} \oplus\, \mathfrak{H}_k$$

and says that $\mathfrak{H}$ is the *direct orthogonal sum* of the invariant subspaces $\mathfrak{H}_k$.

We denote the restriction of $T(g)$ to the subspace $\mathfrak{H}_k$ by $T\,(g)$. It is clear that if $\mathbf{x} = \sum_{k=1}^{\infty} \mathbf{x}_k$, where $\mathbf{x}_k \in \mathfrak{H}_k$, then

$$T(g)\,\mathbf{x} = \sum_{k=1}^{\infty} T_k(g)\,\mathbf{x}_k,$$

where $T_k(g)\mathbf{x}_k \in \mathfrak{H}_k$. Therefore the representation $T(g)$ is uniquely determined by specifying the representations $T_k(g)$. $T(g)$ is said to be the *orthogonal direct sum* of the representations $T_k(g)$, and we write

$$T(g) = \sum_{k=1}^{\infty} \oplus T_k(g)$$

When there is no risk of ambiguity we write simply

$$\mathfrak{H} = \sum_{k=1}^{\infty} \mathfrak{H}_k$$

and

$$T(g) = \sum_{k=1}^{\infty} T_k(g).$$

8. **Complete reducibility of unitary representations.** A representation $T(g)$ is said to be *completely reducible* if it is the direct sum of irreducible representations. For completely reducible representations one can choose the basis in such a way that the representation matrix is block-diagonal:

$$\begin{pmatrix} T_1(g) & 0 & \ldots & 0 \\ 0 & T_2(g) & \ldots & 0 \\ \ldots & \ldots & \ldots & \ldots \\ 0 & 0 & \ldots & T_n(g) \end{pmatrix},$$

where the principal diagonal consists of matrices of the irreducible representations $T_k(g)$ of the group $G$ (in the infinite-dimensional case the principal diagonal may contain infinitely many matrices $T_k(g)$).

We shall now show that *all finite-dimensional unitary representations are completely reducible*. For this, we need the following lemma.

**Lemma.** *Let the representation* $T(g)$ *be unitary relative to the scalar product* $(\mathbf{x}, \mathbf{y})$ *and let* $\mathfrak{L}_1$ *be an invariant subspace in the space* $\mathfrak{L}$ *of this representation. Then the orthogonal complement*[1] $\mathfrak{L}_2$ *of* $\mathfrak{L}_1$ *is also invariant.*

To prove the lemma we choose any vector $\mathbf{x}$ of $\mathfrak{L}_2$. Then for any vector $\mathbf{y}$ of $\mathfrak{L}_1$, since $T(g)$ is unitary, we have

1) The orthogonal complement of a subspace $\mathfrak{L}_1$ consists of the totality of all vectors $\mathbf{x} \in \mathfrak{L}$ such that $(\mathbf{x}, \mathbf{y}) = 0$ for all vectors $\mathbf{y}$ of $\mathfrak{L}_1$.

$$(T(g)\mathbf{x},\ \mathbf{y}) = (\mathbf{x},\ T(g^{-1})\mathbf{y}).$$

But, because of the invariance of $\mathfrak{L}_1$, the vector $T(g^{-1})\mathbf{y}$ belongs to $\mathfrak{L}_1$, so that $(\mathbf{x}, T(g^{-1})\mathbf{y}) = 0$ and consequently $(T(g)\mathbf{x}, \mathbf{y}) = 0$ too. This proves that $T(g)\mathbf{x} \in \mathfrak{L}_2$, i.e. $\mathfrak{L}_2$ is an invariant subspace. The lemma is proved.

We now note that the space $\mathfrak{L}$ is the direct sum of the subspace $\mathfrak{L}_1$ and of its orthogonal complement $\mathfrak{L}_2$. Therefore if $\mathfrak{L}_1$ is invariant, then $\mathfrak{L}$ is the direct sum of invariant subspaces $\mathfrak{L}_1$ and $\mathfrak{L}_2$.

Moreover, the restrictions of the representation $T(g)$ to the subspaces $\mathfrak{L}_1$ and $\mathfrak{L}_2$ are also unitary. Continuing this process, we arrive, in a finite number of steps, at irreducible invariant subspaces and obtain the decomposition of $T(g)$ as the direct sum of irreducible representations.

For infinite-dimensional representations the situation is more complicated, since in infinite-dimensional spaces there may be descending chains of invariant subspaces. Therefore all infinite-dimensional unitary representations are not completely reducible. However, in all cases encountered by us infinite-dimensional unitary representations can be decomposed as the continuous direct sum of irreducible representations (concerning the concept of a continuous direct sum of representations, see the Supplement to the chapter). If the measure for which the continuous direct sum is constructed is concentrated at a denumerable set of points, one obtains a decomposition of the representation as the orthogonal direct sum of irreducible representations.

9. **The Kronecker product of representations.** We define the Kronecker product of representations. Let $T(g)$ and $Q(g)$ be representations of the group $G$, acting in linear spaces $\mathfrak{L}$ and $\mathfrak{M}$ respectively. With each element $g$ of $G$ we associate an operator $R(g) = T(g) \otimes Q(g)$, the *Kronecker product* of the operators $T(g)$ and $Q(g)$ (for the definition of the Kronecker product of operators and the properties of this product, see the Supplement to this chapter). We show that $R(g)$ is a representation of the group $G$. For this we use the equality

$$(AB) \otimes (CD) = (A \otimes C)(B \otimes D).$$

It follows from this equality that

$$\begin{aligned} R(g_1g_2) = T(g_1g_2) \otimes Q(g_1g_2) &= (T(g_1)\,T(g_2)) \otimes (Q(g_1)\,Q(g_2)) \\ &= (T(g_1) \otimes Q(g_1))\,(T(g_2) \otimes Q(g_2)) = R(g_1)\,R(g_2). \end{aligned}$$

Thus $R(g_1g_2) = R(g_1)R(g_2)$, i.e. $R(g)$ is a representation of the group $G$. It is called the *Kronecker* or *tensor product* of the representations $T(g)$ and $Q(g)$.

The Kronecker product of classes of equivalent representations possesses the properties of commutativity and associativity. Moreover, this product is dis-distributive with respect to compositition of representations:

$$T(g)\otimes(Q_1(g)+Q_2(g))=T(g)\otimes Q_1(g)+T(g)\otimes Q_2(g).$$

Let us clarify the form of the matrix obtained as a result of taking the Kronecker product of two representations. Suppose that the representation $T(g)$ is realized in the space $\mathfrak{L}$, the representation $Q(g)$ in the space $\mathfrak{M}$, and that $\{\mathbf{e}_i\}$, $\{\mathbf{f}_j\}$ are the bases of these spaces. In the Supplement to this chapter it is shown that one can choose $\mathbf{e}_i \otimes \mathbf{f}_j$ as the basis in $\mathfrak{L}\otimes \mathfrak{M}$. The matrix elements of the operator $T(g) \otimes Q(g)$ in this basis are products of matrix elements of the operators $T(g)$ and $Q(g)$. Thus the matrix elements of the operator $R(g) = T(g) \otimes Q(g)$ are enumerated by two pairs of numbers $(ij)$, $(km)$, and have the form

$$r_{ij,km}(g)=t_{ik}(g)\,q_{jm}(g).$$

10. Characters of representations. The matrix notation for representations suffers from the defect that it depends not only on the representation but also on the choice of basis in the representation space. On passing to another basis the representation matrix is replaced by an equivalent matrix. It is therefore natural to study functions of matrix elements which remain invariant on passing from a given matrix to an equivalent one. The simplest of such functions is the *trace* of a matrix, i.e. the sum of diagonal elements of the matrix:

$$\operatorname{Tr} A=\sum_i a_{ii}. \tag{1}$$

The trace of a representation matrix is called the *character* of the corresponding representation, and is denoted by $\chi_T(g)$. Thus if $(T(g)) = (t_{ij}(g))$, then

$$\chi_T(g)=\sum_i t_{ii}(g). \tag{2}$$

From the properties of the trace of a matrix, it follows that the character of a representation does not depend on the choice of basis in the representation space. Since, by a suitable choice of basis, the matrices of equivalent representations can be made identical, *the character of a representation depends only on the equivalence class of representations.*

For fixed $T$ the character $\chi_T(g)$ is a function on the group $G$. Since

$$T(g_1^{-1}gg_1)=T^{-1}(g_1)\,T(g)\,T(g_1),$$

we have

$$\operatorname{Tr} T(g_1^{-1}gg_1)=\operatorname{Tr} T(g)$$

so that

$$\chi_T(g_1^{-1}gg_1)=\chi_T(g). \tag{3}$$

*Thus the character of the representation $T(g)$ is a function on $G$, constant on classes of conjugate elements*

Corresponding to operations on representations we have similar operations on their characters. Namely, *the character of the sum of two representations is equal to the sum of their characters, and the character of the Kronecker product of two representations is equal to the product of their characters*:

$$\chi_{T+Q}(g)=\chi_T(g)+\chi_Q(g), \tag{4}$$

$$\chi_{T\otimes Q}(g)=\chi_T(g)\,\chi_Q(g). \tag{5}$$

Indeed, if the matrices of the representations $T(g)$ and $Q(g)$ are $(t_{ij}(g))$ and $(q_{ij}(g))$ respectively, then

$$\chi_{T+Q}(g)=\sum_i t_{ii}(g)+\sum_j q_{jj}(g)=\chi_T(g)+\chi_Q(g)$$

and

$$\chi_{T\otimes Q}(g)=\sum_{i,j} r_{ij,\,ij}(g)=\sum_{i,j} t_{ii}(g)\,q_{jj}(g)$$

$$=\sum_i t_{ii}(g)\sum_j q_{jj}(g)=\chi_T(g)\,\chi_Q(g).$$

**11. Infinitesimal operators of a representation.** [1]) Suppose that with each real number $t$ we associate an element $g(t)$ of the group $G$. If for all $t$ and $s$ we have the equality

$$g(t)g(s)=g(t+s), \tag{1}$$

then $g(t)$ is called a *one-parameter subgroup* of $G$. It follows from equality (1) that $g(0)=e$ and $g(-t)=g^{-1}(t)$ (in what follows, we shall exclude the case $g(t)\equiv e$).

As examples of one-parameter groups one can cite the subgroup of rotations around a fixed axis in the group of rotations of euclidean space, the subgroup of

1) We only mention briefly the basic facts relating to infinitesimal operators of representations. A more detailed account can be found, for example, in [39].

parallel translations along a fixed direction in the group of motions of euclidean space, and so on.

Let $g(t)$ be a one-parameter subgroup of $G$, and let $T(g)$ be a representation of this group. The limit

$$A=\lim_{t\to 0}\frac{T(g(t))-E}{t}\equiv\frac{dT(g(t))}{dt}\Big|_{t=0}, \tag{2}$$

if it exists, defines an operator $A$, called the *infinitesimal operator* of the representation $T(g)$, corresponding to the one-parameter subgroup $g(t)$.

If $G$ is a matrix group (i.e. if its elements are nonsingular matrices of the $n$th order, and matrix multiplication is the group operation) and if the representation $T(g)$ is finite-dimensional, then infinitesimal operators are defined on the whole representation space. The situation is more complicated for infinite-dimensional representations, since, generally speaking, infinitesimal operators of infinite-dimensional representations are unbounded (for example, the representation operators act in the space of functions with square-integrable modulus on some manifold, and the infinitesimal operators are differential). Utilizing the theory of generalized functions, one can, in the infinite-dimensional case too, extend infinitesimal operators to the whole representation space. But we shall not go into that here.

Let $A$ be the infinitesimal operator of the representation $T(g)$ corresponding to the one-parameter subgroup $g(t)$. We denote by $\mathfrak{L}_1$ the (in general, open) subspace on which the following operators are defined:

$$e^{tA}\equiv 1+tA+\frac{t^2A^2}{2!}+\ldots+\frac{t^nA^n}{n!}+\ldots \tag{3}$$

It follows from equality (3) that $(e^{tA})' = Ae^{tA}$, and therefore

$$\frac{d(e^{tA})}{dt}\Big|_{t=0}=A. \tag{4}$$

Utilizing the uniqueness theorem for solutions of differential equations, we deduce from equalities (2) and (4) that in the finite-dimensional case,

$$T(g(t))=e^{tA}. \tag{5}$$

*Thus in the finite-dimensional case the representation $T(g)$ is determined for elements of the one-parameter subgroup $g(t)$ by specifying the infinitesimal operator $A$ of this subgroup.* The situation is slightly more complicated in the infinite-dimensional case, but, for all the cases which we shall come across in this book, the assertion remains valid.

The groups encountered by us in the sequel will have "sufficiently many" one-parameter subgroups. Namely, one-parameter subgroups pass through every element $g$ of these groups. Therefore representations $T(g)$ for such groups are determined by specifying their infinitesimal operators. The space of infinitesimal operators turns out to be finite-dimensional, and its dimension is equal to the dimension of the group, i.e. the number of parameters giving the elements of the group. For example, for the group of motions of the euclidean plane, the space of infinitesimal operators of any representation is three-dimensional, since any motion of the plane is given by three parameters, i.e. the angle of rotation and coordinates of the point into which the origin is sent.

Thus, on passing to infinitesimal operators, we replace the infinite set of operators of the representation $T(g)$ by a finite set, namely the basis of the space of infinitesimal operators. This leads to a substantial simplification in problems of representation theory. For example, every subspace invariant for the representation $T(g)$ is also invariant for its infinitesimal operators. Therefore to prove the irreducibility of $T(g)$ it is sufficient to show that in the space $\mathfrak{L}$ of this representation there is no nontrivial subspace invariant for basic infinitesimal operators.

Henceforth, as a rule, we shall consider only matrix groups. For them one can speak of infinitesimal operators (more precisely, matrices) of one-parameter subgroups themselves:

$$a = \left.\frac{dg(t)}{dt}\right|_{t=0}.$$

We shall call $a$ the *tangent matrix* to the one-parameter subgroup.

A matrix group is called a *linear Lie group* if there exists a neighborhood of the identity matrix in this group that is homeomorphic to the unit ball. For linear Lie groups the totality of tangent matrices to one-parametric subgroups forms a linear space, such that, together with any two matrices $a$ and $b$, their *commutator* $[a, b] = ab - ba$ belongs to it. Such linear spaces of matrices are called *matrix Lie algebras*. Thus, with each linear Lie group $G$ a matrix Lie algebra $\mathfrak{G}$ is associated. One can show that this Lie algebra determines the group $G$ to within a local isomorphism. In other words, suppose that two Lie groups $G$ and $H$ have one and the same Lie algebra. Then in these groups one can select neighborhoods $U$ and $V$ respectively, and determine a one-to-one mapping $v = \phi(u)$ of $U$ onto $V$ such that

$$\varphi(u_1 u_2) = \varphi(u_1)\varphi(u_2).$$

Let $g(t)$ be a one-parametric subgroup of the Lie group $G$, let $T(g)$ be a representation of this group and let $A$ be the infinitesimal operator of the representation $T(g)$ corresponding to the subgroup $g(t)$. With the tangent matrix $a$ of $g(t)$ we associate the operator $A = dT(g(t))/dt|_{t=0}$. One can show that this correspondence is a linear mapping of the Lie algebra of $G$ onto the space of infinitesimal operators of the representation $T(g)$ under which the matrix commutator becomes the operator commutator: If $a \to A$ and $b \to B$, then $[a, b] \to [A, B]$ (where $[A, B] = AB - BA$). In other words, *the representation $T(g)$ of a group $G$ gives the infinitesimal representation $a \to A$ of its Lie algebra.* In the case of finite-dimensional representations the infinitesimal representation uniquely determines the representation $T(g)$. Namely, suppose that matrices $a_1, \cdots, a_n$ form the basis of the Lie algebra of $G$, and

$$g = \exp[t_1 a_1 + \ldots + t_n a_n],$$

where $t_1, \cdots, t_n$ are real numbers. Let us denote by $A_k$, $1 \le k \le n$, the infinitesimal operators of $T(g)$ corresponding to the one-parameter subgroups $e^{ta_k}$. Then it can be shown that

$$T(g) = \exp[t_1 A_1 + \ldots + t_n A_n].$$

The situation is slightly more complicated in the infinite-dimensional case, since the infinitesimal operators are, generally speaking, unbounded, and so are not defined on the whole representation space.

## §2. TRANSFORMATION GROUPS AND THEIR REPRESENTATIONS

In this section we describe some constructions of representations for transformation groups. In particular, we consider regular representations of groups and representations of groups induced by representations of subgroups.

1. **Transformation groups.** By a *transformation* of a set $\mathfrak{M}$ we mean a one-to-one mapping of this set onto itself. We shall denote the image of an element $x$ of $\mathfrak{M}$ under a mapping $g$ by $gx$. Let $G$ be some group. We say that $G$ is a *transformation group* of the set $\mathfrak{M}$ if, with each element $g$ of this group, one can associate a transformation $x \to gx$ in $\mathfrak{M}$, where the following conditions are satisfied.

1) With the identity element $e$ of $G$ is associated the identity transformation of the set $\mathfrak{M}$: $ex \equiv x$.

2) For any two elements $g_1$ and $g_2$ of $G$, we have

$$(g_1g_2)\,x = g_1\,(g_2x). \tag{1}$$

As examples of transformation groups we have the group of all permutations of $n$ elements, the group of nonsingular linear transformations of $n$-dimensional linear space, the group of orthogonal transformations of euclidean space, the group of parallel translations of the $n$-dimensional euclidean space and so on.

The transformation group $G$ of the set $\mathfrak{M}$ is called *effective* if for every element $g \neq e$ of this group there is a point $x$ of $\mathfrak{M}$ such that $gx \neq x$. Henceforth we shall only consider effective transformation groups.

We note that every group $G$ can be regarded as a transformation group of the set $\mathfrak{M}$ consisting of elements of the same group. Namely, with each element $g_0$ of $G$ we associate a left shift of the group, transforming each element $g$ into the element $g_0g$. Another realization of $G$ as a transformation group is obtained if with each element $g_0$ we associate a right shift, transforming $g$ into $gg_0^{-1}$.

2. **Transitive transformation groups.** Let $G$ be an (effective) transformation group of a set $\mathfrak{M}$. If for any two elements $x$ and $y$ of $\mathfrak{M}$ there is an element $g$ of $G$ such that $gx = y$, then $G$ is called a *transitive* transformation group of $\mathfrak{M}$. The set $\mathfrak{M}$ in which the transitive transformation group $G$ acts, is called a *homogeneous space with transformation group* $G$. For example, the group of rotations of euclidean space is not transitive, since points situated at different distances from the origin cannot be sent into one another by a rotation (around the origin). At the same time, this group is a transitive group of transformations of the unit sphere into itself. Thus the sphere is a homogeneous space relative to the group of rotations of euclidean space, but euclidean space itself is not a homogeneous space relative to this group.

If $G$ is not transitive, then $\mathfrak{M}$ splits into nonoverlapping classes of elements such that two elements of any one class can be sent into one another by a transformation of group $G$ but elements of different classes cannot be sent into another by such transformations. These classes of elements are called *transitivity classes*. For example, spheres of different radii with centers at the origin are transitivity classes for the group of rotations of euclidean space.

Let $G$ be a transitive transformation group of $\mathfrak{M}$, and let $a$ be some fixed point of this set. We consider all elements $h$ of $G$ which leave the point $a$ fixed, i.e. such that $ha = a$. Since $h_1 a = a$ and $h_2 a = a$ imply that $h_1^{-1} a = a$ and $(h_1 h_2)a = a$, the set of such elements forms a subgroup $H$ of $G$. It is called the *stationary subgroup* of the point $a$.

*If a transformation $g$ carries the point $a$ into another point $x$, then all other transformations $g_1$ carrying $a$ into $x$ have the form $g_1 = gh$, where $h \in H$.* In other words, the set of such transformations is a left coset $gH$ with respect to the subgroup $H$. Since $G$ is transitive, this establishes a one-to-one correspondence between the points $x$ of the set $\mathfrak{M}$ and the left cosets with respect to $H$. Moreover, the transformations $g \to gx$ turn into left multiplication by $g$: If to the point $x$ corresponds the coset $g_1 H$, then to the point $gx$ corresponds the coset $gg_1 H$.

We note that a stationary subgroup is not uniquely determined by specifying the homogeneous space $\mathfrak{M}$, but depends also on the choice of $a$. Let us explain how the stationary subgroups of different points are connected with each other. Let $H$ be the stationary subgroup of the point $a$, and let the transformation $g$ carry $a$ into $b$. Then transformations of the form $ghg^{-1}$, $h \in H$, and only such transformations, leave the point $b$ fixed. Thus as the stationary subgroup of $b$ we have the subgroup $gHg^{-1}$, conjugate to the stationary subgroup $H$ of $a$. Since $G$ acts transitively on $\mathfrak{M}$, all stationary subgroups of the points of $\mathfrak{M}$ are conjugate among themselves.

Thus we have proved that *to each homogeneous space $\mathfrak{M}$ for the group $G$ there corresponds a class of conjugate subgroups in $G$, namely the stationary subgroups of the points of $\mathfrak{M}$. Conversely, to each class of conjugate subgroups corresponds some homogeneous space.* In order to construct this space, we take one of the mutually conjugate subgroups $H$ and denote by $\mathfrak{M}$ the space of left cosets with respect to $H$. With each element $g$ of $G$ we associate a transformation in $\mathfrak{M}$, carrying the coset $g_1 H$ into the coset $gg_1 H$. This takes $G$ into the transitive transformation group of $\mathfrak{M}$, and $H$ into the stationary subgroup of the coset $eH = H$. In this case we shall write

$$\mathfrak{M} = G/H.$$

Another realization of $G$ as the group of motions of the homogeneous space $\mathfrak{M}$ is obtained if elements of $\mathfrak{M}$ are identified with right cosets with respect to $H$, and the motions are determined by the rule that $Hg_1 \to Hg_1 g^{-1}$. To pass from one reali zation to the other we make use of the transformation $g \to g^{-1}$.

We introduce the concept of a sphere in a homogeneous space. We take any point $y$ of $\mathfrak{M}$ and apply to it all the motions $h$ of the stationary subgroup of the point $a$. We obtain the set $Hy$ which we shall call *the sphere with center at point a, passing through point y*. It is clear that in the realization of $\mathfrak{M}$ as the space of left cosets with respect to $H$, the sphere is realized as a two-sided coset, i.e. as a set of elements of the form $HgH$, where $g$ is fixed.

We illustrate the concepts introduced above with the example of the group of motions of euclidean space. As the stationary subgroup of any point a we have the rotations of the space around this point. If y is another point, then under the rotations leaving a fixed the point y describes the euclidean sphere with center at a, passing through y.

Let us consider another example. Let $G = SO(n)$ be the group of rotations of the $n$-dimensional euclidean space around the origin. Clearly the unit sphere $S^{n-1}$ with center at the origin is a homogeneous space with the group of motions $SO(n)$. On the sphere we take the point $N(0, \cdots, 0, 1)$ (the north pole of the sphere). The stationary subgroup for this point is the group $SO(n-1)$ of euclidean rotations of the $(n-1)$-dimensional subspace $x_n = 0$. Therefore

$$S^{n-1} = SO(n)/SO(n-1).$$

3. **Invariant measures.** Suppose that, on a set $\mathfrak{M}$ with transformation group $G$, a denumerably additive measure $\mu$ is given. One says that this measure is *invariant* for (relative to) the group $G$ if, for any measurable subset $A$ of $\mathfrak{M}$ and any element $g$ of $G$, the equality

$$\mu(A) = \mu(gA)$$

is satisfied ($gA$ denotes the image of set $A$ under transformation $g$). For example, the Lebesgue measure in $n$-dimensional euclidean space is invariant for the group of motions of this space (and even for the group of nonhomogeneous linear transformations with determinant 1), Lebesgue measure on the sphere is invariant for the group of rotations of the sphere around the center, and so on.

In particular, a measure $\mu$ on the group $G$ is called *left-invariant* if for any measurable subset $A \subset G$ and any element $g$ of $G$ we have

$$\mu(A) = \mu(gA)$$

(here $gA$ is the totality of elements of the form $ga$, $a \in A$). If

$$\nu(A) = \nu(Ag),$$

then the measure $\nu$ on the group $G$ is called *right-invariant*. It is clear that if the measure $\mu(A)$ on a group is left-invariant, then $\nu(A) = \mu(A^{-1})$ [1] is right-invariant.

If the measure $\mu$ on a set $\mathfrak{M}$ is invariant relative to $G$, then $\int f(x)d\mu(x)$ possesses the following invariance property:

$$\int f(x)\,d\mu(x) = \int f(gx)\,d\mu(x).$$

In particular, if $\mu$ is a left-invariant measure on $G$, then the integral with respect to this measure possesses the following invariance property for left shifts:

$$\int f(g)\,d\mu(g) = \int f(g_0 g)\,d\mu(g).$$

Similarly, for a right-invariant measure $\nu$ we have

$$\int f(g)\,d\nu(g) = \int f(gg_0)\,d\nu(g).$$

4. **Group representations by shift operators.** We consider representations of groups by shift operators on homogeneous spaces. Let $G$ be a transformation group of the set $\mathfrak{M}$, and let $\mathfrak{L}$ be a linear space consisting of functions $f(x)$, $x \in \mathfrak{M}$, with values in a linear space $\mathfrak{N}$. We shall call $\mathfrak{L}$ *invariant* for motions of $G$ if together with any function $\mathfrak{f}(x)$ it contains all functions $\mathfrak{f}(gx)$, $g \in G$.

For each invariant subspace $\mathfrak{L}$ we have the following representation of the group $G$ by shift operators:

$$T(g)\,\mathfrak{f}(x) = \mathfrak{f}(g^{-1}x), \qquad g \in G,\ \mathfrak{f} \in \mathfrak{L}. \tag{1}$$

In order to prove that equality (1) determines a representation of $G$, it is sufficient to note that for any two elements $g_1$ and $g_2$ of $G$ we have

$$T(g_1)\,T(g_2)\,\mathfrak{f}(x) = T(g_1)\,\mathfrak{f}(g_2^{-1}x)$$
$$= \mathfrak{f}(g_2^{-1}g_1^{-1}x) = \mathfrak{f}((g_1g_2)^{-1}x) = T(g_1g_2)\,\mathfrak{f}(x).$$

If in $\mathfrak{M}$ we have a measure $\mu$, invariant relative to $G$, then as $\mathfrak{L}$ we can take the space of scalar functions $f(x)$ whose squares are integrable with respect to this measure. In this case the representation $T(g)$ is unitary relative to the scalar product

---

1) $A^{-1}$ is the set of all elements of the form $a^{-1}$, where $a \in A$.

$$(f_1, f_2) = \int_{\mathfrak{M}} f_1(x)\overline{f_2(x)}\, d\mu(x).$$

In fact, we have

$$(T(g)f_1, T(g)f_2) = \int_{\mathfrak{M}} T(g)f_1(x)\overline{T(g)f_2(x)}\, d\mu(x)$$
$$= \int_{\mathfrak{M}} f_1(g^{-1}x)\overline{f_2(g^{-1}x)}\, d\mu(x).$$

In this integral we make the substitution $g^{-1}x = y$ and use the invariance of $\mu$. We get

$$(T(g)f_1, T(g)f_2) = \int_{\mathfrak{M}} f_1(y)\overline{f_2(y)}\, d\mu(y) = (f_1, f_2).$$

This proves that $T(g)$ is unitary.

The above construction is applicable, in particular, to spaces of functions on $G$ itself, regarded as a shift group. If one takes the left shifts $g \to g_0 g$, then the representation operators have the form

$$L(g_0)f(g) = f(g_0^{-1}g). \tag{2}$$

But if one takes right shifts, they have the form

$$R(g_0)f(g) = f(gg_0). \tag{3}$$

In the case when there is a left-invariant measure on $G$, one usually takes as $\mathfrak{L}$ the space of those functions whose squares are integrable with respect to this measure. The representation $L(g)$ is then called the *left-regular representation* of $G$. Similarly, if $\mathfrak{L}$ is the space of those functions whose squares are integrable with respect to a right-invariant measure, then $R(g)$ is called the *right-regular representation* of $G$.

We prove that any irreducible representation $T(g)$ of $G$ is equivalent to the representation by shift operators in some space of scalar functions on this group. For this we choose any vector $\mathbf{a} \neq \mathbf{0}$ in the space $L$ of the representation $T(g)$, and with each vector $\mathbf{f}$ of $L$ we associate a scalar function $f(g)$ on the group $G$:

$$f(g) = (T(g)\mathbf{f}, \mathbf{a}).^{1)} \tag{4}$$

The equality

$$(T(g)T(g_0)\mathbf{f}, \mathbf{a}) = (T(gg_0)\mathbf{f}, \mathbf{a}) = f(gg_0)$$

shows that, on replacing $\mathbf{f}$ by $T(g_0)\mathbf{f}$, the function $f(g)$ becomes

1) $(\mathbf{f}, \mathbf{a})$ is some scalar product in the representation space.

$$Q(g_0)f(g) = f(gg_0).$$

In order to prove that the representation $Q(g)$ is equivalent to $T(g)$, it is necessary to prove that the kernel of the mapping $\mathbf{f} \to f(g)$ is equal to zero. But if $f(g) \equiv 0$, then for all $g_0$ we have $f(gg_0) = 0$. Therefore if the vector $\mathbf{f}$ belongs to the kernel $\mathfrak{L}_1$ of the mapping $\mathbf{f} \to f(g)$ (i.e. turns into zero under this mapping), then the vector $T(g_0)\mathbf{f}$ also belongs to this kernel. But then $\mathfrak{L}_1$ is an invariant subspace in $\mathfrak{L}$. It cannot coincide with the whole space $\mathfrak{L}$, since $a(e) = (\mathbf{a}, \mathbf{a}) > 0$. Consequently, owing to the irreducibility of $T(g)$, the subspace $\mathfrak{L}_1$ is the null subspace. This proves the equivalence of $T(g)$ and $Q(g)$.

If the functions $f(g) \in \mathfrak{L}^2(G)$, i.e. if their squares are integrable with respect to a right-invariant measure, and the mapping $\mathbf{f} \to f(g)$ is bicontinuous relative to convergence in $\mathfrak{L}$ and $\mathfrak{L}^2(G)$, then the representation $T(g)$ is equivalent to the irreducible part of the regular representation of $G$. For many groups (in particular for all compact groups; see §4) all irreducible representations are obtained by decomposing the regular representation into irreducible components.

Let us denote the image of $\mathfrak{L}$ under the mapping $\mathbf{f} \to f(g)$ by $\mathfrak{M}$. As the basis in $\mathfrak{M}$ it is natural to choose the images of basis vectors $\{\mathbf{e}_i\}$ of $\mathfrak{L}$. We shall assume that the basis $\{e_i\}$ is chosen in such a way that the invariant vector $\mathbf{a}$ is one of the basis vectors, say $\mathbf{a} = \mathbf{e}_1$. In that case basis vectors $\mathbf{e}_j$ become the matrix elements of the first row of the representation $T(g)$. In fact,

$$e_j(g) = (T(g)\mathbf{e}_j, \mathbf{a}) = (T(g)\mathbf{e}_j, \mathbf{e}_1) = t_{1j}(g),$$

whence our assertion follows. From

$$Q(g_0)t_{1j}(g) = t_{1j}(gg_0) = \sum_k t_{kj}(g_0)t_{1k}(g)$$

it is evident that in the basis $\{t_{1j}(g)\}$ the representation $Q(g)$ is given by the same matrix as the representation $T(g)$ in the basis $\{\mathbf{e}_j\}$.

**5. Representations of class 1. Spherical functions.** Let $T(g)$ be an irreducible representation of the group $G$ in the space $\mathfrak{L}$, and let $H$ be a subgroup of this group. A vector $\mathbf{a}$ in $\mathfrak{L}$ is called *invariant* relative to $H$ if for all $h \in H$ we have $T(h)\mathbf{a} = \mathbf{a}$. A representation $T(g)$ is called a *representation of class* 1 relative to $H$ if in its space there are nonzero vectors invariant relative to $H$ and the restriction of $T(g)$ to $H$ is unitary:

$$(T(h)\mathbf{x}, T(h)\mathbf{y}) = (\mathbf{x}, \mathbf{y}), \qquad h \in H. \tag{1}$$

If in the space $\mathfrak{L}$ of any representation of class 1 relative to $H$ there is only one normalized invariant vector a, then we shall call $H$ *massive*.

Let $T(g)$ be a representation of the group $G$ of class 1 relative to the massive subgroup $H$, and let a be a normalized invariant vector in the space $\mathfrak{L}$ of this representation. We apply to a the construction of subsection 4, i.e. with each vector f of $\mathfrak{L}$ we associate the function

$$f(g)=(T(g)\,\mathbf{f},\ \mathbf{a}) \tag{2}$$

on $G$. These functions are called *spherical functions of the representation* $T(g)$ *relative to* $H$. As was shown above, the representation $T(g)$ is equivalent to the representation

$$Q(g_0)f(g)=f(gg_0) \tag{3}$$

in the space of spherical functions.

Spherical functions can be regarded as functions on the homogeneous manifold $\mathfrak{M}$ with group of motions $G$ and stationary subgroup $H$. Indeed, from the fact that the restriction of $T(g)$ to $H$ is unitary, we have the equality

$$\begin{aligned} f(hg)&=(T(hg)\,\mathbf{f},\ \mathbf{a})=(T(h)\,T(g)\,\mathbf{f},\ \mathbf{a})\\ &=(T(g)\,\mathbf{f},\ T(h^{-1})\,\mathbf{a})=(T(g)\,\mathbf{f},\ \mathbf{a})=f(g), \qquad h\in H. \end{aligned}$$

Thus *spherical functions of* $T(g)$ *are constant on the right cosets* $H_g$ *with respect to* $H$. But the set of right cosets w.r.t. $H$ forms a homogeneous space with group of motions $G$ and stationary subgroup $H$:

$$Hg\rightarrow Hgg_0^{-1}.$$

Therefore $f(g)$ can be regarded as a function on $\mathfrak{M}=G/H$.

Among spherical functions of the representation $T(g)$ of class 1, the most interesting is the *zonal spherical function*, corresponding to the invariant vector a:

$$a(g)=(T(g)\,\mathbf{a},\ \mathbf{a}). \tag{4}$$

This function is constant on the two-sided cosets $HgH$ w.r.t. $H$.

In fact, by what has already been proved $a(h_1gh_2)=a(gh_2)$, where $h_1, h_2\in H$ But $T(h_2)\mathbf{a}=\mathbf{a}$ and therefore

$$a(gh_2)=(T(gh_2)\,\mathbf{a},\ \mathbf{a})=(T(g)\,T(h_2)\,\mathbf{a},\ \mathbf{a})=(T(g)\,\mathbf{a},\ \mathbf{a})=a(g).$$

Thus

$$a(h_1 g h_2) = a(g). \tag{5}$$

As we remarked in subsection 2, two-sided cosets w.r.t. $H$ correspond to spheres in homogeneous spaces. Consequently *zonal spherical functions are constant on spheres.*

If in space $\mathfrak{L}$ one chooses a basis such that $a = e_1$, then the zonal spherical function is none other than the matrix element $t_{11}(g)$. The remaining matrix elements $t_{1j}(g)$ of the first row are usually called *associated* spherical functions. They, as we have said, are constant on right cosets w.r.t. $H$.

Sometimes the matrix elements of the first column are also called associated spherical functions. They are

$$t_{i1}(g) = (T(g)\mathbf{e}_1, \mathbf{e}_i) \equiv (T(g)\mathbf{a}, \mathbf{e}_i). \tag{6}$$

These functions are constant on left cosets w.r.t. $H$:

$$t_{i1}(gh) = t_{i1}(g). \tag{7}$$

One can show that if $G$ is the group of rotations of the three-dimensional euclidean space, and $H$ is the subgroup of rotations of the palne, then the spherical functions of irreducible unitary representations of $G$ are the classical spherical functions $Y_l^n(\phi, \theta)$. Moreover, the zonal spherical functions are Legendre polynomials $P_l(\cos\theta)$. But if $G$ is the group of rotations of the $n$-dimensional euclidean space and $H$ is the subgroup of rotations of the $(n-1)$-dimensional euclidean space, then spherical functions are expressed by means of Gegenbauer polynomials.[1)]

6. **Induced representations.** The regular representation is a special case of induced representations. The latter are constructed in the following manner. We take some subgroup $H$ of a group $G$ and its representation $Q(h)$ in a Hilbert space $\mathfrak{H}$. We denote by $\mathfrak{L}$ the set of vector functions $\mathbf{f}(g)$ given on $G$, taking values in $\mathfrak{H}$ and possessing the following properties.

1) For any element a of $\mathfrak{H}$, the scalar function $\phi(g) = (\mathbf{f}(g), \mathbf{a})$ on $G$ is measurable relative to the left-invariant measure $dg$ on $G$, as well as $(\mathbf{f}(g), \mathbf{f}(g))$, where

$$\int (\mathbf{f}(g), \mathbf{f}(g))\, dg < +\infty. \tag{1}$$

2) For any element $h$ of $H$ we have

1) See Chapter IX.

$$\mathfrak{f}(gh) = Q(h)\,\mathfrak{f}(g). \tag{2}$$

We show that this space is invariant for left-shift operators

$$T(g_0)\,\mathfrak{f}(g) = \mathfrak{f}(g_0^{-1}g). \tag{3}$$

Condition 1) is obviously satisfied by $\mathfrak{f}(g_0^{-1}g)$, and condition 2) is satisfied because

$$T(g_0)\,\mathfrak{f}(gh) = \mathfrak{f}(g_0^{-1}gh) = Q(h)\,\mathfrak{f}(g_0^{-1}g) = Q(h)\,T(g_0)\,\mathfrak{f}(g).$$

It follows from the invariance of $\mathfrak{L}$ that $T(g)$ is a representation of $G$. It is called *the representation induced by the representation $Q(h)$ of the subgroup $H$.*

The regular representation of $G$, considered above, is induced by the trivial representation of the identity subgroup $\{e\}$. Further, the trivial representation of any subgroup $H$: $Q(h) = E$ induces the representation

$$T(g)\,\mathfrak{f}(x) = \mathfrak{f}(g^{-1}x), \qquad x \in \mathfrak{M}$$

in the space of functions on the homogeneous space $\mathfrak{M} = G/H$.

7. **Representations of groups with operator factors.** We now consider a more general construction of representations for transformation groups, in which shifts are combined with multiplications by an operator function $A(x, g)$, depending on the point $x \in \mathfrak{M}$ and the element $g$ of the transformation group $G$ of the set $\mathfrak{M}$.

We assume that $\mathfrak{L}$ is a linear space of functions given on the set $\mathfrak{M}$ and taking values in the linear space $\mathfrak{N}$. Suppose that, with each element $x$ of $\mathfrak{M}$ and each element $g$ of $G$, a nonsingular operator $A(x, g)$ is associated in the space $\mathfrak{N}$, and for each function $\mathfrak{f}(x)$ of $\mathfrak{L}$ the function

$$T(g)\,\mathfrak{f}(x) = A(x, g)\,\mathfrak{f}(g^{-1}x) \tag{1}$$

belongs to $\mathfrak{L}$ (it is evident that for fixed $g$ the function $T(g)\mathfrak{f}(x)$ is given on $\mathfrak{M}$ and takes values in $\mathfrak{N}$).

We clarify the form $A(x, g)$ must have in order that $T(g)$ be a representation of group $G$, i.e. in order that one may have

$$T(g_1)\,T(g_2) = T(g_1g_2). \tag{2}$$

It is obvious from formula (1) that

$$T(g_1g_2)\,\mathfrak{f}(x) = A(x, g_1g_2)\,\mathfrak{f}((g_1g_2)^{-1}x) = A(x, g_1g_2)\,\mathfrak{f}(g_2^{-1}g_1^{-1}x),$$

and

$$T(g_1)\,T(g_2)\,\mathfrak{f}(x) = T(g_1)\,A(x, g_2)\,\mathfrak{f}(g_2^{-1}x) = A(x, g_1)\,A(g_1^{-1}x, g_2)\,\mathfrak{f}(g_2^{-1}g_1^{-1}x).$$

Therefore, in order that equality (1) may give a representation of $G$, the function

$A(x, g)$ must satisfy the functional equation

$$A(x, g_1g_2) = A(x, g_1)\, A(g_1^{-1}x, g_2). \tag{3}$$

The converse is also true: if the operator function $A(x, g)$ satisfies condition (3), then equality (1) determines a representation of $G$.

We note a few properties of $A(x, g)$, ensuing from (3). Putting $g_1 = e$, $g_2 = g$ we get

$$A(x, g) = A(x, e)\, A(x, g).$$

Therefore $A(x, e)$ is the identity operator. Further, putting $g_1 = g$, $g_2 = g^{-1}$, we get

$$E = A(x, g)\, A(g^{-1}x, g^{-1}),$$

i.e.

$$A^{-1}(x, g) = A(g^{-1}x, g^{-1}). \tag{4}$$

Since the set $\mathfrak{M}$ is homogeneous for the transformation group $G$, $A(x, g)$ can be expressed by means of a function of the form $B(g)$. Let us choose some point $a$ in $\mathfrak{M}$. Owing to the homogeneity of $\mathfrak{M}$, for any other point $x$ there is a $g_x$ such that $g_x^{-1}a = x$. On putting $g_1 = g_x$, $g_2 = g$ in (3), we get

$$A(x, g) \equiv A(g_x^{-1}a, g) = A^{-1}(a, g_x)\, A(a, g_xg).$$

Putting $A(a, g) = B(g)$, we have

$$A(x, g) = B^{-1}(g_x)\, B(g_xg), \tag{5}$$

with $g_x^{-1}a = x$.

It follows from (3) that if $h$ is an element of the stationary subgroup $H$ of element $a$, then the

$$A(a, hg) = A(a, h)\, A(h^{-1}a, g) = A(a, h)\, A(a, g).$$

This means that

$$B(hg) = B(h)\, B(g). \tag{6}$$

In particular, if $h_1$ and $h_2$ belong to $H$,

$$B(h_1h_2) = B(h_1)\, B(h_2). \tag{7}$$

Thus $B(h)$ is a representation of $H$.

We have proved the following assertion.

*Let $\mathfrak{M}$ be a homogeneous space with transformation group $G$. Let $\mathfrak{L}$ be a linear space of vector functions on $\mathfrak{M}$, invariant for operators of the form*

$$T(g)\mathfrak{f}(x) = A(x, g)\mathfrak{f}(g^{-1}x). \tag{8}$$

*For $T(g)$ to be a representation of $G$, it is necessary and sufficient that $A(x, g)$ have the form*

$$A(x, g) = B^{-1}(g_x) B(g_x g), \tag{9}$$

*where $g_x^{-1} a = x$ and $B(g)$ is an operator function on $G$ satisfying the functional equation*

$$B(hg) = B(h) B(g), \qquad h \in H, \quad g \in G \tag{10}$$

*($H$ is the stationary subgroup of the point $a$).*

We note that representations (8), corresponding to different choices of $a$, are equivalent to each other. Further, let $C(x)$ be an operator function on $\mathfrak{M}$ such that $C^{-1}(x)$ exists for all $x$ and $C(a) = E$. Let

$$B_1(g_x) = B(g_x) C(g_x^{-1} a). \tag{11}$$

It is clear that if $h \in H$, i.e. if $h^{-1} a = a$, then

$$B_1(hg) = B(hg) C(g^{-1} h^{-1} a) = B(h) B(g) C(g^{-1} a) = B(h) B_1(g). \tag{12}$$

In particular, $B_1(h) = B(h) B_1(e) = B(h)$. Therefore equality (12) can be rewritten as

$$B_1(hg) = B_1(h) B_1(g). \tag{12'}$$

Hence $B_1(g)$ determines a representation of $G$, and

$$Q(g)\mathfrak{f}(x) = A_1(x, g)\mathfrak{f}(g^{-1}x),$$

where

$$A_1(x, g) = B_1^{-1}(g_x) B_1(g_x g).$$

This representation is equivalent to the representation (8). The equivalence is established by replacing $\mathfrak{f}(x)$ in (8) by $C^{-1}(x)\mathfrak{f}(x)$.

8. **Some examples.** Henceforth we shall come across numerous examples of representations of transformation groups; practically the whole book is devoted to these representations and their matrix elements. We mention a few typical examples (omitting proofs, which will be given in the appropriate chapters).

We begin with one of the simplest transformation groups, the group of rotations of the circle. The regular representation of this group is constructed in the space of functions

$$F(\varphi) \equiv f(\cos\varphi, \ \sin\varphi)$$

on the unit circle and is given by the formula

$$T(g_\alpha) F(\varphi) = F(\varphi + \alpha) \tag{1}$$

($g_\alpha$ is the rotation of the circle by an angle $\alpha$).

We shall now take the $n$-dimensional analogue of this group, namely the group $SO(n)$ of rotations of the $n$-dimensional euclidean space $E_n$. The transformations of this group carry the unit sphere $S^{n-1}$ into itself. Therefore it is natural to construct representations of $SO(n)$ in the space of functions $f(\xi)$ on this sphere. The representation operators are given here by the formula

$$T(g)f(\xi)=f(g^{-1}\xi). \tag{2}$$

It will be shown in Chapter IX that this representation is reducible. In order to obtain irreducible representations it is necessary to take on $S^{n-1}$ spaces of functions $f(\xi)$, such that $r^k f(\mathbf{x}/r)$ ($r^2 = x_1^2 + \cdots + x_n^2$) is a homogeneous harmonic polynomial of degree $k$. The representation $T(g)$ is decomposed as the direct orthogonal sum of representations in these spaces. In this way the theory of representations of the group $SO(n)$ turns out to be connected with the theory of harmonic polynomials of $n$ variables.

Let us now consider the group $SL(n, R)$ of those linear transformations of the $n$-dimensional linear space $E_n$ whose determinant is equal to 1. We denote by $\mathfrak{L}$ the space of all functions $f(\mathbf{x})$ in $E_n$ with square-integrable modulus, and take

$$T(g)f(\mathbf{x})=f(g^{-1}\mathbf{x}). \tag{3}$$

We obtain a representation of the group $SL(n, R)$. This representation is reducible. In order to obtain irreducible representations we consider homogeneous functions. A function $f(\mathbf{x})$ in $E_n$ is called *homogeneous with degree of homogeneity* $\chi = (\omega, \epsilon)$ if for any real number $a$ one has the equality

$$f(a\mathbf{x})=|a|^{\omega}\operatorname{sign}^{\epsilon} a\, f(\mathbf{x}).$$

Here $\omega$ is any complex number, and $\epsilon$ takes the values 0 and 1; $\operatorname{sign} a$ is the sign of the number $a$.

It is easy to see that the space of homogeneous functions is invariant for shifts; if the function $f(\mathbf{x})$ is homogeneous, then $f(g^{-1}\mathbf{x})$ is also a homogeneous function of the same degree. Therefore the formula

$$T(g)f(\mathbf{x})=f(g^{-1}\mathbf{x}) \tag{4}$$

determines a representation of $SL(n, R)$ in the space of homogeneous functions of the given degree. If $\omega$ is a positive integer the representation (4) is reducible. The homogeneous polynomials of degree $\omega$ of $n$ variables form an invariant subspace in this space.

## §3. INVARIANT OPERATORS AND THE THEORY OF REPRESENTATIONS

1. **Operators permutable with representations.** Suppose that $T(g)$ and $Q(g)$ are two representations of a group $G$ in the linear spaces $\mathfrak{L}_1$ and $\mathfrak{L}_2$ respectively A linear operator $A$ mapping $\mathfrak{L}_1$ into $\mathfrak{L}_2$ is said to be *permutable* with these representations if for any element $g$ of $G$ we have

$$AT(g) = Q(g)A. \tag{1}$$

In particular, if $A$ is a linear operator in the space $\mathfrak{L}$ of the representation $T(g)$ such that

$$AT(g) = T(g)A, \tag{2}$$

then $A$ is called *permutable* with the representation $T(g)$.

*Suppose that an operator $A$ in the space $\mathfrak{L}$ is permutable with the representation $T(g)$ of the group $G$ in $\mathfrak{L}$, and let $\mathbf{x}$ be an eigenvector of this operator corresponding to the eigenvalue $\lambda$. Then all operators $T(g)\mathbf{x}$ are eigenvectors for $A$, corresponding to the same eigenvalue $\lambda$.*

In fact, by hypothesis, we have

$$AT(g)\mathbf{x} = T(g)A\mathbf{x}.$$

But $A\mathbf{x} = \lambda\mathbf{x}$ and therefore

$$AT(g)\mathbf{x} = \lambda T(g)\mathbf{x}.$$

This proves our assertion.

It follows that *the eigenfunctions of an operator $A$, permutable with the representation $T(g)$, corresponding to a given eigenvalue $\lambda$, form an invariant subspace in the representation space.*

Hence an operator $A$, permutable with an irreducible representation $T(g)$, can have at the most one eigenvalue. Otherwise the subspace of eigenvectors corresponding to some eigenvalue would be a nontrivial invariant subspace.

We prove a stronger assertion:

*Any bounded operator $A$ in the space $\mathfrak{L}$ of an irreducible representation $T(g)$, permutable with all operators of $T(g)$, is a multiple of the identity operator $E$, $A = \lambda E$.*

Let us first consider the case when $A$ is selfadjoint.[1] In this case one

---

1) We may recall that $\mathfrak{L}$ is either a finite-dimensional or a Hilbert space. Therefore we can assume that in it there is a scalar product $(\mathbf{x}, \mathbf{y})$ (in general, noninvariant relative to $T(g)$).

has the spectral decomposition

$$A = \int\limits_{-\infty} \lambda \, dP(\lambda), \tag{3}$$

where, as we know (see [40]), all the operators $P(\lambda)$ are permutable with operators of $T(g)$. But the $P(\lambda)$ are projection operators and therefore, by virtue of the irreducibility of $T(g)$, the equality

$$P(\lambda)\, T(g) = T(g)\, P(\lambda)$$

can hold either if $P(\lambda) = 0$ or if $P(\lambda) = E$. Since $(P(\lambda)\mathbf{x}, \mathbf{x})$ is a nondecreasing function, there exists a value $\lambda_0$ such that $P(\lambda) = 0$ for $\lambda = \lambda_0$ and $P(\lambda) = E$ for $\lambda \geq \lambda_0$. It follows from equality (2) that in this case $A = \lambda_0 E$.

Thus when $T(g)$ is irreducible, any bounded selfadjoint operator $A$ permutable with the representation operators is a multiple of the identity operator. But then any bounded operator $B$ permutable with the operators $T(g)$ is also a multiple of the identity operator. In order to see this, it is sufficient to write $B$ in the form $B = B_1 + iB_2$, where $B_1$ and $B_2$ are selfadjoint operators.

We introduce the following definition.

A representation $T(g)$ is *operator-irreducible* if any bounded operator permutable with it is a multiple of the identity operator.

So, we have proved that *any irreducible representation in a Hilbert space is operator-irreducible.*

The converse assertion, generally speaking, is false even for finite-dimensional representations. But if a representation $T(g)$ in a Hilbert space $\mathfrak{H}$ is unitary, then its operator-irreducibility implies ordinary irreducibility. Indeed, let the unitary representation $T(g)$ be operator-irreducible and let $\mathfrak{H}_1$ be an invariany subspace of it. We denote by $P$ the operator of orthogonal projection on $\mathfrak{H}_1$. Since the orthogonal complement of $\mathfrak{H}_1$ is also invariant (see §1.8) we have

$$PT(g) = T(g)\, P.$$

This equality, by virtue of the operator-irreducibility of $T(g)$, implies that either $P = 0$ or $P = E$, and so the invariant subspace $\mathfrak{H}_1$ is trivial.

We have thus proved that, *for unitary representations, the concepts of operator-irreducibility and ordinary irreducibility coincide.*

2. **Schur's lemma.** In many problems of representation theory the following assertion concerning operators permutable with irreducible finite-dimensional representations is found to be useful.

**Schur's lemma.** *Suppose that $T(g)$ and $Q(g)$ are finite-dimensional irreducible representations of the group $G$ in the spaces $\mathfrak{L}_1$ and $\mathfrak{L}_2$ respectively, and that the operator $A$ mapping $\mathfrak{L}_1$ into $\mathfrak{L}_2$ is permutable with these representations. Then either $A$ is the null operator, or it has an inverse (and consequently $T(g)$ and $Q(g)$ are equivalent). In the second case $A$ is uniquely determined (to within multiplication by a scalar).*

In order to prove the lemma we consider two linear subspaces connected with the operator $A$: the subspace $\mathfrak{M}_1$ in $\mathfrak{L}_1$, consisting of vectors $\mathbf{x} \in \mathfrak{L}_1$ such that $A\mathbf{x} = 0$, and the subspace $\mathfrak{M}_2$ in $\mathfrak{L}_2$, consisting of vectors $\mathbf{y} \in \mathfrak{L}_2$ having the form $\mathbf{y} = A\mathbf{x}$, $\mathbf{x} \in \mathfrak{L}_1$ (i.e. we consider the kernel $A^{-1}(0)$ of the mapping $A$ and the image $A(\mathfrak{L}_1)$ of $\mathfrak{L}_1$ under the mapping).

The permutability of $A$ with $T(g)$ and $Q(g)$ implies that these subspaces are invariant. Indeed, if $\mathbf{x} \in \mathfrak{M}_1$ then $A\mathbf{x} = 0$. But then

$$AT(g)\,\mathbf{x} = Q(g)\,A\mathbf{x} = Q(g)\,\mathbf{0} = \mathbf{0},$$

and so $T(g)\mathbf{x} \in \mathfrak{M}_1$. Similarly, if $\mathbf{y} \in \mathfrak{M}_2$ then $\mathbf{y}$ has the form $\mathbf{y} = A\mathbf{x}$, $\mathbf{x} \in \mathfrak{L}_1$, and

$$Q(g)\,\mathbf{y} = Q(g)A\mathbf{x} = AT(g)\,\mathbf{x}.$$

Since $T(g)\mathbf{x} \in \mathfrak{L}_1$, we have $Q(g)\mathbf{y} \in \mathfrak{M}_2$. By the same token it is proved that $\mathfrak{M}_1$ is invariant for $T(g)$, and $\mathfrak{M}_2$ for $Q(g)$.

But $T(g)$ and $Q(g)$ are irreducible, and therefore $\mathfrak{M}_1$ and $\mathfrak{M}_2$ are either null, or coincide respectively with $\mathfrak{L}_1$ and $\mathfrak{L}_2$. The following cases are possible here.

a) If $\mathfrak{M}_2 = 0$ the entire space $\mathfrak{L}_1$ is mapped into the null element by $A$, and so $A = 0$.

b) If $\mathfrak{M}_1 = 0$ and $\mathfrak{M}_2 = \mathfrak{L}_2$, then as the image of $\mathfrak{L}_1$ we have the entire space $\mathfrak{L}_2$, where only the null element is mapped into the null element of $\mathfrak{L}_2$. In this case $A$ has an inverse operator $A^{-1}$, and consequently $T(g)$ and $Q(g)$ are equivalent.

c) Lastly, the case $\mathfrak{M}_1 = \mathfrak{L}_1$ and $\mathfrak{M}_2 = \mathfrak{L}_2$ can occur only if both the spaces $\mathfrak{L}_1$ and $\mathfrak{L}_2$ consist only of the null elements. In this case $A = 0$.

We have still to show that in case b) the operator $A$ is uniquely determined (apart from a constant factor). Let us assume that some operator $B$ is also permutable with $T(g)$ and $Q(g)$:

$$BT(g) = Q(g)\,B.$$

Then all operators of the form $A - \lambda B$, where $\lambda$ is a number, are also permutable with $T(g)$ and $Q(g)$. We choose $\lambda$ such that $A - \lambda B$ is singular, i.e. such that

$$\operatorname{Det}(A - \lambda B) = 0.$$

Then the operator $A - \lambda B$, permutable with $T(g)$ and $Q(g)$, does not have an inverse. Therefore, by virtue of what we have proved above, it is the null operator, $A - \lambda B = 0$. But then $A = \lambda B$. This proves the lemma.

3. **Implications of Schur's lemma.** By making use of Schur's lemma, one can prove a number of important assertions concerning group representations. In particular, the operator-irreducibility of finite-dimensional irreducible representations is a direct consequence of Schur's lemma. In fact, suppose that the representation $T(g)$ is irreducible and $AT(g) = T(g)A$. Since by Schur's lemma an operator commuting with an irreducible representation is defined uniquely to within a scalar factor, we get $A = \lambda E$.

As another important corollary of Schur's lemma we have the result that *irreducible finite-dimensional representations $T(g)$ of commutative groups are one-dimensional.* Indeed, if the group $G$ is commutative, then for any two of its elements $g$ and $h$ we have $gh = hg$, so that

$$T(g)\,T(h) = T(h)\,T(g).$$

For fixed $h$ the operator $T(h)$ is permutable with all operators $T(g)$, and by virtue of the irreducibility of $T(g)$ we have $T(h) = \lambda(h)E$. But a representation of such a form can be irreducible only in the case when it is one-dimensional.

The following theorem, which we shall use later on, also follows from Schur's lemma.

**Theorem.** *Suppose that a representation $T(g)$ of a group $G$ in a Hilbert space $\mathfrak{H}$ is the orthogonal direct sum of pairwise nonequivalent irreducible representations $T_i(g)$ in subspaces $\mathfrak{H}_i$, and let $\mathfrak{F}$ be an invariant subspace in $\mathfrak{H}$ such that the restriction of $T(g)$ to $\mathfrak{F}$ is completely reducible. Then $\mathfrak{F}$ is an orthogonal direct sum of some of the subspaces $\mathfrak{H}_i$:*

$$\mathfrak{F} = \sum_n \mathfrak{H}_{i_n}.$$

We first prove the following lemma.

**Lemma.** *Suppose that the representation $T(g)$ of the group $G$ is the orthogonal direct sum of pairwise nonequivalent irreducible representations $T_i(g)$ of this group. Then any minimal invariant subspace $\mathfrak{F}$ of the space $\mathfrak{H}$ of the representation $T(g)$ coincides with one of the spaces $\mathfrak{H}_i$.*

Indeed, let us denote by $P_i$ the operator of orthogonal projection of the subspace $\mathfrak{F}$ on the subspace $\mathfrak{H}_i$ of the representation $T_i(g)$. The pairwise orthogonality of the subspaces $\mathfrak{H}_i$ implies that

$$P_i Q(g) = T_i(g) P_i,$$

where $Q(g)$ is the restriction of $T(g)$ to $\mathfrak{F}$. Since $T_i(g)$ and $Q(g)$ are irreducible, by Schur's lemma we find that if $T_i(g)$ and $Q(g)$ are nonequivalent then $P_i = 0$. Since the $T_i(g)$ are pairwise nonequivalent, only one of the operators $P_i$ can be non zero, say $P_j \neq 0$, and $P_i = 0$ for $i \neq j$. But then $\mathfrak{F}$ must lie in $\mathfrak{H}_j$ and be an invariant subspace in $\mathfrak{H}_j$. Since $T_j(g)$ is irreducible, $\mathfrak{F}$ coincides with $\mathfrak{H}_j$.

To prove the theorem it is sufficient to decompose $\mathfrak{F}$ into minimal invariant subspaces and apply the lemma to each of them.

Finally, utilizing Schur's lemma, it is easy to prove the following assertion.

*Let $T(g)$ be an irreducible unitary representation of a group $G$ in a Hilbert space $\mathfrak{H}$, and let $\{\mathbf{x}, \mathbf{y}\}$ be a bounded Hermitian form in $\mathfrak{H}$, invariant for $T(g)$:*

$$\{T(g)\mathbf{x},\ T(g)\mathbf{y}\} = \{\mathbf{x},\ \mathbf{y}\}. \tag{1}$$

*Then $\{\mathbf{x}, \mathbf{y}\}$ differs from the scalar product $(\mathbf{x}, \mathbf{y})$ in $\mathfrak{H}$ only by a scalar factor:*

$$\{\mathbf{x},\ \mathbf{y}\} = \lambda(\mathbf{x},\ \mathbf{y}).$$

In fact, the form $\{\mathbf{x}, \mathbf{y}\}$ can be written as

$$\{\mathbf{x},\ \mathbf{y}\} = (A\mathbf{x},\ \mathbf{y}),$$

where $A$ is a bounded operator in $\mathfrak{H}$. From equality (1) and the fact that $T(g)$ is unitary, it follows that

$$(AT(g)\mathbf{x},\ T(g)\mathbf{y}) = (A\mathbf{x},\ \mathbf{y}) = (T(g)A\mathbf{x},\ T(g)\mathbf{y}),$$

so that

$$AT(g) = T(g)A.$$

Thus $A$ is permutable with $T(g)$, and therefore, by virtue of the irreducibility of $T(g)$, it is a multiple of the identity operator: $A = \lambda E$. Consequently

$$\{\mathbf{x},\ \mathbf{y}\} = (A\mathbf{x},\ \mathbf{y}) = \lambda(\mathbf{x},\ \mathbf{y}).$$

4. **Invariant operators.** The solution of many problems of mathematical physics is facilitated by the fact that they possess the property of invariance for certain transformations. For example, the equations describing the motion of a solid in a Coulomb field are invariant for rotations around the charged point, the Laplace operator $\Delta$ is invariant for motions of euclidean space, the wave operator $\Box$ for transformations of the Lorentz group, and so on.

Several integral operators of mathematical physics are also invariant. For example, the integral operator with the kernel

$$K(x, y)=\frac{1}{\sqrt{(x_1-y_1)^2+(x_2-y_2)^2+(x_3-y_3)^2}}$$

is invariant for the group of motions of euclidean space: the operator kernel is unaltered if the points x and y are displaced simultaneously.

We introduce the general definition of an operator invariant for transformations of some group. Let $G$ be a transformation group of the set $\mathfrak{M}$, and let $\mathfrak{L}$ be an invariant linear space of vector functions on $\mathfrak{M}$ with values in the linear space $\mathfrak{N}$. An operator $A$ in $\mathfrak{L}$ is called invariant relative to $G$ if it is permutable with the shift operators

$$T(g)\mathbf{f}(x)=\mathbf{f}(g^{-1}x),$$

i.e. if for any element $g$ of $G$ we have

$$AT(g)=T(g)A. \tag{1}$$

For example, if $K(x, y)$ is a kernel, depending on points x and y of $\mathfrak{M}$, and invariant for $G$:

$$K(gx, gy)=K(x, y),$$

where $\mu$ is an invariant measure on $\mathfrak{M}$ and $f(x)$ is a scalar function, then the integral operator

$$Af(x)=\int K(x, y)f(y)\,d\mu(y) \tag{2}$$

is invariant. Indeed, $AT(g)f(x) = \int K(x, y)f(g^{-1}y)d\mu(y)$. Making the substitution $g^{-1}y = z$, by virtue of the invariance of $K(x, y)$ and $\mu$, we get

$$AT(g)f(x)=\int K(g^{-1}x, z)f(z)\,d\mu(z)=(Af)(g^{-1}x)=T(g)Af(x).$$

The results of subsection 1 lead to the following assertion.

*The space of eigenfunctions of an invariant operator $A$, corresponding to the eigenvalue $\lambda$, is invariant for shift operators. To each eigenvalue $\lambda$ of the invariant operator $A$ corresponds a representation $T_\lambda(g)$ of the group $G$. It is constructed in the space $\mathfrak{L}_\lambda$ of eigenfunctions of $A$, corresponding to the eigenvalue $\lambda$, and is given by the formula*

$$T_\lambda(g)f(x)=f(g^{-1}x). \tag{3}$$

In the case when the set $\mathfrak{M}$ on which the group $G$ acts is homogeneous, the connection between eigenfunctions of invariant operators and representations of $G$ becomes particularly close. In this case elements of the space $\mathfrak{L}_\lambda$ are linear combinations of matrix elements of $T_\lambda(g)$. In fact, let us choose a basis

$\phi_1(x), \cdots. \phi_m(x), \cdots$ in the space $\mathfrak{L}_\lambda$, and fix a point $a \in \mathfrak{M}$. Then we have

$$T_\lambda(g)\varphi_m(a) \equiv \varphi_m(g^{-1}a) = \sum_k t_{km}(g)\varphi_k(a).$$

But by the homogeneity of $\mathfrak{M}$ any point belonging to it can be written in the form $x = g_x^{-1}a$, so that

$$\varphi_m(x) = \varphi_m(g_x^{-1}a) = \sum_k \varphi_k(a)\, t_{km}(g_x). \tag{4}$$

Our assertion is thus proved.

If the representation $T_\lambda(g)$ is reducible, then one similarly proves that the eigenfunctions belonging to an invariant subspace $\mathfrak{L}'$ are linear combinations of matrix elements of the restrictions of $T_\lambda(g)$ to $\mathfrak{L}'$. In particular, if $T_\lambda(g)$ is completely reducible, then any element of $\mathfrak{L}_\lambda$ is a linear combination of matrix elements of irreducible representations of $G$.

## §4. REPRESENTATIONS OF COMPACT GROUPS

1. **Matrix groups. Compact and locally compact groups.** In this book we shall only consider matrix groups, i.e. groups whose elements are nonsingular matrices of one and the same order, and for which matrix multiplication serves as the group operation.

For matrices the convergence concept is well-defined: matrices $g_1, \cdots, g_n, \cdots, g_n = (g_{ij}^{(n)})$, *converge* to the matrix $g = (g_{ij})$ if for arbitrary $i$ and $j$ we have $\lim_{n\to\infty} g_{ij}^{(n)} = g_{ij}$. Therefore in matrix groups one can speak of convergence of a sequence of elements. Evidently, if

$$\lim_{n\to\infty} g_n = g \text{ and } \lim_{n\to\infty} h_n = h,$$

then

$$\lim_{n\to\infty} g_n h_n = gh \text{ and } \lim g_n^{-1} = g^{-1}.$$

Thus the group operation in matrix groups is continuous relative to the topology chosen.

Since passage to the limit is defined in a matrix group, one can define, in a natural manner, the concept of limit of a function $f(g)$ given on such a group, its continuity, and so on.

We shall call a matrix group $G$ *compact* if from any sequence $\{g_n\}$ of its elements it is possible to extract a convergent subsequence. In order that a matrix group $G$ be compact, it is necessary and sufficient that

1) the matrix elements of all the matrices appearing in $G$ be bounded as a whole,

2) the set of matrices $\{g\}$ appearing in $G$ be closed in the set of all matrices of the given order.

As an example of a compact group we have the group $SO(n)$ of orthogonal matrices of the $n$th order with real elements. The orthogonality of the matrix $g = (g_{ij})$ implies that

$$\sum_{j=1}^{n} g_{ij}^2 = 1$$

so that all the elements of an orthogonal matrix are not greater than 1 in absolute value. Further, a matrix $g$ belongs to $SO(n)$ if and only if $gg' = e$.[1] Let $\lim_{k\to\infty} g_k = g$ and $g_k \in SO(n)$. Passing to the limit in the equality $g_k g_k' = e$ we get $gg' = e$, so that $g \in SO(n)$.

Consequently $SO(n)$ forms a closed subset in the set of all matrices. The group $SU(n)$ of unitary unimodular matrices of the $n$th order is also compact.

A matrix group $G$ is called *locally compact* if the set of its elements $g = (g_{ij})$, satisfying the condition $|g_{ij} - \delta_{ij}| \leq 1$, where $\delta_{ij}$ is the Kronecker symbol, is compact. As an example of a locally compact group, we have the group of all nonsingular matrices of the $n$th order. This group is not compact, since, for example, from the sequence of matrices $\{me\}$ ($e$ is the identity matrix), one cannot select a convergent subsequence.

One of the most important properties of a locally compact group is the availability on a left-invariantmeasure in it. More precisely, in any locally compact group $G$ there is a left-invariant measure, which takes finite nonzero values on open subsets in $G$ having a compact closure. With the help of this measure we define a left-invariant integral on the locally compact group $G$:

$$\int f(g)\, dg = \int f(g_0 g)\, dg.$$

One can show that, on compact groups, a measure which is left invariant is also right invariant:

$$\int f(g)\, dg = \int f(g g_0)\, dg.$$

Moreover, the measure of the whole compact group $G$ is finite. We shall not prove these assertions, since in concrete cases the invariant measure is constructed in

1) $g'$ is the transpose of the matrix $g$.

an explicit form.

If a subgroup $H$ of the locally compact group $G$ is compact, then, starting from a left-invariant measure $dg$ in $G$, we define the measure $dx$ on the homogeneous space $\mathfrak{M} = G/H$ by putting

$$\int_{\mathfrak{M}} f(x)\, dx = \int_{G} f[\varphi(g)]\, dg.$$

Here $\phi(g)$ is the mapping $g \to gH$ of $G$ onto $\mathfrak{M}$. One can show that the measure $dx$ is invariant and takes finite nonzero values on open sets in $\mathfrak{M}$ having a compact closure.

2. **Complete reducibility of representations of compact groups.** We pass on to the consideration of representations of compact groups by bounded operators in a Hilbert space (in particular by linear transformations in a finite-dimensional space, since one can always introduce a scalar product in such a space). We show that these representations reduce to unitary ones. More precisely, we have

**Theorem 1.** *Let $T(g)$ be a representation of the compact group $G$ by bounded operators in a Hilbert space $\mathfrak{H}$. Then in this space one can introduce a scalar product $(\mathbf{x}, \mathbf{y})_1$, invariant relative to all operators $T(g)$, i.e. such that*

$$(T(g_0)\,\mathbf{x},\ T(g_0)\,\mathbf{y})_1 = (\mathbf{x}, \mathbf{y})_1 \tag{1}$$

*for all vectors $\mathbf{x}$ and $\mathbf{y}$ of $\mathfrak{H}$ and all elements $g_0$ of $G$.*

To prove the theorem we take

$$(\mathbf{x}, \mathbf{y})_1 = \int (T(g)\,\mathbf{x},\ T(g)\,\mathbf{y})\, dg, \tag{2}$$

where $dg$ is the invariant measure on $G$ and $(\mathbf{x}, \mathbf{y})$ is the scalar product in $\mathfrak{H}$. Integral (2) is convergent because of the boundedness of the operators $T(g)$ and the finiteness of the measure $dg$. From the invariance of this measure, it follows that equality (1) holds.

The above theorem can also be formulated as follows.

**Theorem 1 .** *Any representation $T(g)$ of the compact group $G$ by bounded operators in a Hilbert space is unitary for a certain scalar product in the representation space.*

By virtue of the results of §1.8 this theorem implies the complete reducibility of all finite-dimensional representations of compact groups. A corresponding theorem is valid for infinite-dimensional unitary representations of compact groups; the space of such representations is decomposed as a direct sum of the finite-dimensional invariant subspaces in which irreducible representations of $G$ are induced (see [264]).

Let us apply these results to representations of compact commutative groups Since all irreducible representations of commutative groups are one-dimensional, and representations of compact groups are completely reducible, we obtain the following assertion.

*Any representation of a compact commutative group $G$ by bounded operators in a Hilbert space is the direct sum of one-dimensional representations.*

We note that Theorem 1′ is, in general, false for noncompact groups. For example, if $G$ is the additive group of real numbers, then the one-dimensional representation $x \to e^x$ is not unitary relative to any scalar product. The proof given by us for compact groups is not valid for noncompact groups, since the invariant measure of the whole noncompact group is infinite, and the integral

$$\int (T(g)\mathbf{x},\ T(g)\mathbf{y})\, dg$$

is divergent.

3. Fourier series on compact groups. In this book we shall study orthogonal systems of functions arising from irreducible representations of compact groups. These systems of functions are connected with the matrix elements of such representations.

We divide the set of all irreducible unitary representations of the compact group $G$ into classes of equivalent representations, and choose one representation from each class. The resulting system of representations $\{T_\alpha(g)\}$ is called *a complete system of pairwise nonequivalent irreducible unitary representations of the group $G$.* We denote the set of indices by $A$.

The following theorem is fundamental.

Theorem 1. *Let $\{T_\alpha(g)\}$, $\alpha \in A$, be a complete system of pairwise nonequivalent irreducible unitary representations of the compact group $G$, and let the dimension of the representation $T_\alpha(g)$ be $d_\alpha$, and its matrix elements be $t_{ij}^\alpha(g)$, $1 \leq i, j \leq d_\alpha$. Then the functions*

$$\sqrt{d_\alpha}\, t_{ij}^\alpha(g), \quad 1 \leqslant i, j \leqslant d_\alpha, \ \alpha \in A \tag{1}$$

*form a complete orthonormal system on $G$ for a normalized invariant measure $dg$ on this group.*

Thus

$$\int t_{ij}^{\alpha}(g)\overline{t_{mn}^{\beta}(g)}\,dg=0, \tag{2}$$

if $(\alpha, i, j)$ is not equal to $(\beta, m, n)$, and

$$\int |t_{ij}^{\alpha}(g)|^2\,dg=\frac{1}{d_\alpha}. \tag{3}$$

From the completeness of the system $\{t_{ij}^{\alpha}(g)\}$, it follows that *any function* $f(g)$ *on* $G$, *such that*

$$\int |f(g)|^2\,dg<+\infty,$$

*can be decomposed into a mean-convergent Fourier series in the functions* $t_{ij}^{\alpha}(g)$:

$$f(g)=\sum_{\alpha\in A}\sum_{i,j=1}^{d_\alpha} a_{ij}^{\alpha}t_{ij}^{\alpha}(g). \tag{4}$$

*The Fourier coefficients of this series are calculated according to the formula*

$$a_{ij}^{\alpha}=d_\alpha\int f(g)\overline{t_{ij}^{\alpha}(g)}\,dg. \tag{5}$$

*Moreover, we have the Parseval equality*

$$\int |f(g)|^2\,dg=\sum_{\alpha\in A}\frac{1}{d_\alpha}\sum_{i,j=1}^{d_\alpha}|a_{ij}^{\alpha}|^2. \tag{6}$$

In the case when $G$ is the group of rotations of the circle, its irreducible unitary representations have the form $e^{in\phi}$ (see Chapter II, §1.2). For this special case the theorem signifies orthonormalization and completeness of the system of functions $\{e^{in\phi}\}$ on the circle for measure $d\phi/2\pi$.

We pass on to the proof of the theorem. In order to prove the orthogonality of the matrix elements corresponding to nonequivalent representations $T_\alpha(g)$ and $T_\beta(g)$, we consider the matrix

$$A_{ij}=\int T_\alpha(g)B_{ij}T_\beta(g^{-1})\,dg, \tag{7}$$

where all the elements $B_{ij}$ are equal to zero, apart from the element $b_{ij}$ which is equal to 1. From the invariance of the measure $dg$ it easily follows that $A_{ij}$ commutes with $T_\alpha(g)$ and $T_\beta(g)$:

$$T_\alpha(g_0)A_{ij}=A_{ij}T_\beta(g_0),\quad g_0\in G. \tag{8}$$

Hence by Schur's lemma $A_{ij}$ is equal to zero. But the elements of this matrix are integrals of the form $\int t_{ki}^{\alpha}(g)t_{jp}^{\beta}(g^{-1})dg$. Because the representation $T_\beta(g)$ is unitary, we have $t^{\beta}_{jp}(g^{-1}) = \overline{t^{\beta}_{pj}(g)}$, and so $A_{ij}$ consists of integrals of the form

$$\int t^{\alpha}_{ki}(g)\overline{t^{\beta}_{pj}(g)}\,dg.$$

Therefore the vanishing of $A_{ij}$ implies that

$$\int t^{\alpha}_{ki}(g)\overline{t^{\beta}_{pj}(g)}\,dg = 0, \quad \alpha \neq \beta. \tag{9}$$

Similarly one proves that, for $(k, i) \neq (p, j)$,

$$\int t^{\alpha}_{ki}(g)\overline{t^{\alpha}_{pj}(g)}\,dg = 0. \tag{9'}$$

To calculate (3) we consider the matrix

$$A_{ii} = \int T_{\alpha}(g) B_{ii} T_{\alpha}(g^{-1})\,dg, \tag{10}$$

where $B_{ii}$ has the same meaning as above. It is easy to verify that this matrix commutes with the matrices of the representation $T_{\alpha}(g)$ and so, by Schur's lemma, is a scalar matrix. Hence

$$\int |t^{\alpha}_{ki}(g)|^2\,dg = \lambda,$$

where $\lambda$ does not depend on $k$. In order to find $\lambda$, we observe that the trace of the matrix $A_{ii}$ is equal to $d_{\alpha}\lambda$. On the other hand, it is equal to the trace of $B_{ii}$, i.e. 1. Therefore $\lambda = 1/d_{\alpha}$.

We have thus proved the orthogonality of the system of functions $\{t^{\alpha}_{ij}(g)\}$ and the normalization condition (3). We now come to the most difficult part of the theorem, namely the proof of completeness of this system (the completeness of the system of matrix elements of irreducible unitary representations of a compact group has been proved by F. Peter and H. Weyl [284]).

So, we have to prove that *any function $f(g)$ on the compact group $G$ with an integrable square and orthogonal to all functions of the system $\{t^{\alpha}_{ij}(g)\}$ is equal to zero.*

Let us consider first the case when $f(g)$ is continuous and possesses Hermitian symmetry: $f(g^{-1}) = \overline{f(g)}$. With this function we associate the kernel

$$K(g, h) = f(gh^{-1}), \quad h, g \in G. \tag{11}$$

It possesses Hermitian symmetry:

$$K(g, h) = f(gh^{-1}) = \overline{f(hg^{-1})} = \overline{K(h, g)}.$$

Moreover, by virtue of the continuity of the function $f(g)$ and the finiteness of invariant measure on the compact group $G$, the integral operator with kernel $K(g, h)$ is completely continuous.

Therefore if $f(g)$ is not identically equal to zero there exists a nonzero eigenvalue $\lambda$ of the integral operator

$$A\varphi(g)=\int f(gh^{-1})\varphi(h)\,dh \tag{12}$$

with finite-dimensional characteristic subspace $\mathfrak{L}_\lambda$.[1] We show that the existence of such a subspace is incompatible with the assumption of orthogonality of $f(g)$ to all functions $\{t_{ij}^{\alpha}(g)\}$.

For this we first prove that *the equalities*

$$\int f(g)\overline{t_{ij}^{\alpha}(g)}\,dg=0,\qquad \alpha\in A,\quad 1\leqslant i,j\leqslant d_\alpha \tag{13}$$

*imply the orthogonality of all functions $\phi(g)$ of $\mathfrak{L}_\lambda$ to functions of the system* $\{t_{ij}^{\alpha}(g)\}$.

Indeed, as $\lambda\neq 0$ and

$$t_{ij}^{\alpha}(gh)=\sum_k t_{ik}^{\alpha}(g)\,t_{kj}^{\alpha}(h), \tag{14}$$

so, for $A\phi(g)=\lambda\phi(g)$, we have

$$\begin{aligned}\int \varphi(g)\overline{t_{ij}^{\alpha}(g)}\,dg&=\frac{1}{\lambda}\int A\varphi(g)\overline{t_{ij}^{\alpha}(g)}\,dg\\&=\frac{1}{\lambda}\int\int f(gh^{-1})\varphi(h)\overline{t_{ij}^{\alpha}(g)}\,dg\,dh\\&=\frac{1}{\lambda}\int\int f(g)\varphi(h)\overline{t_{ij}^{\alpha}(gh)}\,dg\,dh\\&=\frac{1}{\lambda}\sum_k\int f(g)\overline{t_{ik}^{\alpha}(g)}\,dg\int\varphi(h)\overline{t_{kj}^{\alpha}(h)}\,dh.\end{aligned}$$

It follows from (13) that all the terms of the last sum are equal to zero. Hence

$$\int\varphi(g)\overline{t_{ij}^{\alpha}(g)}\,dg=0. \tag{15}$$

We now note that the kernel $K(g,h)=f(gh^{-1})$ of the operator $A$ is invariant for right shifts:

$$K(gg_0,hg_0)=f(gg_0(hg_0)^{-1})=f(gh^{-1})=K(g,h).$$

Consequently $A$ is itself invariant. But then, owing to the results of § 3.4, the subspace $\mathfrak{L}_\lambda$ is invariant for shifts, and the formula $T_\lambda(g_0)f(g)=f(gg_0)$. gives a representation of $G$ in $\mathfrak{L}_\lambda$.

The representation $T_\lambda(g)$ is completely reducible, being a finite-dimensional

1) See [34] or [1], where a corresponding theorem is proved for any completely continuous selfadjoint operator.

representation of the compact group $G$. Consequently, as was shown in §3.4, any function $\phi(g)$ of $\mathfrak{L}_\lambda$ is a linear combination of matrix elements of the irreducible representations of $G$.

We choose in $\mathfrak{L}_\lambda$ a nonzero function $\phi(g)$. We have shown that it is a linear combination of matrix elements of the irreducible representations of $G$. But every irreducible representation of the group $G$ is equivalent to one of the representations $T_\alpha(g)$, and therefore its matrix elements are linear combinations of functions $\{t_{ij}^\alpha(g)\}$, $1 \le i, j \le d_\alpha$. Consequently $\phi(g)$ is also a linear combination of the functions $\{t_{ij}^\alpha(g)\}$. On the other hand, we have shown that $\phi(g)$ is orthogonal to all the functions of $\{t_{ij}^\alpha(g)\}$. By virtue of the orthogonality of this system it follows that $\phi(g) = 0$, contrary to our assumption.

This contradiction shows that there does not exist a nonzero continuous Hermitian-symmetric function orthogonal to all the matrix elements of irreducible unitary representations of this group.

The general case is easily reduced to that considered above. Namely, if $F(g)$ is a nonzero square-integrable function, orthogonal to all functions $\{t_{ij}^\alpha(g)\}$, then let us first construct the function

$$\psi(g) = \int F(gh)\overline{F(h)}\,dh.$$

It is also orthogonal to all functions $\{t_{ij}^\alpha(g)\}$. In fact,

$$\begin{aligned}\int \psi(g)\overline{t_{ij}^\alpha(g)}\,dg &= \int\int F(gh)\overline{F(h)}\,\overline{t_{ij}^\alpha(g)}\,dg\,dh \\ &= \int\int F(g)\overline{F(h)}\,\overline{t_{ij}^\alpha(gh^{-1})}\,dg\,dh \\ &= \sum_k \int F(g)\overline{t_{ik}^\alpha(g)}\,dg \int \overline{F(h)\,t_{kj}^\alpha(h^{-1})}\,dh = 0,\end{aligned}$$

as

$$\int F(g)\overline{t_{ik}^\alpha(g)}\,dg = 0.$$

A simple verification shows that $\psi(g)$ is continuous. Moreover, we have

$$\psi(e) = \int |F(h)|^2\,dh > 0.$$

Further, we construct the Hermitian symmetric function

$$f(g) = \psi(g) + \overline{\psi(g^{-1})}.$$

Since $f(e) = 2\psi(e) > 0$ this function is nonzero, and it is easily seen to be orthogonal to all functions $\{t_{ij}^\alpha(g)\}$. But we have proved above that such a function does not exist. Consequently every square-integrable function orthogonal to all the functions of the system $\{t_{ij}^\alpha(g)\}$ is equal to zero. The completeness of the system $\{t_{ij}^{\alpha}(g)\}$ is thus proved.

4. **Harmonic analysis of functions on compact groups.** The results of the previous subsection allow one to solve the problem of decomposition into irreducible components of the regular representation of a compact group $G$. We recall that the right regular representation is constructed in the space $\mathfrak{L}^2(G)$ of functions $f(g)$ on the group $G$, having integrable squares with respect to any invariant measure, and is given by the formula

$$R(g_0)f(g)=f(gg_0). \tag{1}$$

If $G$ is the group of rotations of the circle, then its regular representation has the form

$$T(g_\alpha)f(\varphi)=f(\varphi+\alpha)$$

(see §2.8), and the decomposition of the regular representation into irreducible ones reduces to the Fourier series expansion

$$f(\varphi)=\sum c_n e^{in\varphi}$$

in matrix elements $e^{in\phi}$ of irreducible unitary representations of this group (in this special case such representations are one-dimensional).

The decomposition of the regular representation of any compact group is closely connected with the expansion of the functions $f(g)$ in a Fourier series. In analogy with the decomposition of functions on the circle, it is called *the harmonic analysis of functions on* $G$.

Let us denote by $\mathfrak{H}_i^{\alpha}$ the subspace of functions $f(g)$ which are linear combinations of the functions

$$t_{ij}^{\alpha}(g), \qquad 1\leqslant j\leqslant d_\alpha. \tag{2}$$

Since for arbitrary $j$ we have

$$R(g_0)t_{ij}^{\alpha}(g)=t_{ij}^{\alpha}(gg_0)=\sum_k t_{kj}^{\alpha}(g_0)t_{ik}^{\alpha}(g)\in\mathfrak{H}_i^{\alpha}, \tag{3}$$

this space is invariant for $R(g)$.

As a basis in $\mathfrak{H}_i^{\alpha}$, we choose the functions $t_{ij}^{\alpha}(g)$. It immediately follows from formula (3) that the restriction of the representation $R(g)$ to $\mathfrak{H}_i^{\alpha}$ is then given by the matrix $(t_{ij}^{\alpha}(g))$. Consequently it is equivalent to the representation $T_\alpha(g)$ and is therefore irreducible.

We have thus constructed subspaces $\mathfrak{H}_i^{\alpha}$ in $\mathfrak{L}^2(G)$, invariant for right-shift operators, such that the restriction $R_i^{\alpha}(g)$ of the representation $R(g)$ to $\mathfrak{H}_i^{\alpha}$ is equivalent to the irreducible unitary representation $T_\alpha(g)$.

It remains to show that $\mathfrak{L}^2(G)$ is the direct sum of the subspaces $\mathfrak{H}_i^\alpha$:

$$\mathfrak{L}^2(G) = \sum_\alpha \sum_{i=1}^{d_\alpha} \mathfrak{H}_i^\alpha,$$

i.e. that any function $f(g)$ of $\mathfrak{L}^2(G)$ has a uniquely determined decomposition into functions $f_i^\alpha(g)$ of the subspaces $\mathfrak{H}_i^\alpha$. For this we observe that, by formula (4) of subsection 3,

$$f(g) = \sum_{\alpha \in A} \sum_{i,j=1}^{d_\alpha} c_{ij}^\alpha t_{ij}^\alpha(g). \tag{4}$$

We group together the terms of this series belonging to the subspace $\mathfrak{H}_i^\alpha$. We obtain

$$f(g) = \sum_{\alpha \in A} \sum_{i=1}^{d_\alpha} f_i^\alpha(g), \tag{5}$$

where

$$f_i^\alpha(g) \equiv \sum_{j=1}^{d_\alpha} c_{ij}^\alpha t_{ij}^\alpha(g) \in \mathfrak{H}_i^\alpha. \tag{6}$$

We have thus constructed the decomposition of functions $f(g)$ into functions $f_i^\alpha(g)$ of $\mathfrak{H}_i^\alpha$. This decomposition is unique because of the mutual orthogonality of the $\mathfrak{H}_i^\alpha$. In conclusion we note that the representations $R_i^\alpha(g)$, $1 \le i \le d_\alpha$, are equivalent to $T_\alpha(g)$ and consequently to each other. Thus the representation $T_\alpha(g)$ appears in the decomposition of the representation $R(g)$ with multiplicity $d_\alpha$, equal to the dimension of this representation:

$$R(g) = \sum_\alpha d_\alpha T_\alpha(g), \tag{7}$$

Thus we have proved

**Theorem 1.** *The regular representation of a compact group $G$ is decomposed as the direct sum of irreducible unitary representations of this group. Each irreducible unitary representation of $G$ appears in the decomposition with a multiplicity equal to the dimension of this representation. The invariant subspaces in which the irreducible summands of the right regular representation are realized consist of functions of the form*

$$f(g) = \sum_{j=1}^{d_\alpha} c_{ij}^\alpha t_{ij}^\alpha(g). \tag{8}$$

Similar assertions hold for the left regular representation, where in this case the minimal invariant subspaces consist of functions of the form

$$f(g)=\sum_{i=1}^{d_\alpha} c_{ij}^\alpha t_{ij}^\alpha(g). \tag{9}$$

We also note that *every irreducible representation $T(g)$ of the compact group $G$ appears in the decomposition of the regular representation of this group.*

To prove this theorem we construct, according to §2.4, the mapping

$$\mathfrak{f} \to f(g) \equiv (T(g)\mathfrak{f},\ \mathbf{a}) \tag{10}$$

of the space $\mathfrak{L}$ of the representation $T(g)$ into the space of functions on $G$($\mathbf{a} \neq 0$ is an arbitrary vector of $\mathfrak{L}$). By virtue of the continuity of $T(g)$ the functions $f(g)$ are continuous on $G$. Therefore, owing to the finiteness of invariant measure, the functions $f(g)$ have integrable squares, $f(g) \in \mathfrak{L}^2(G)$.

We have constructed the mapping (10) of $\mathfrak{L}$ into $\mathfrak{L}^2(G)$. As shown in §2.4, under this mapping only the zero element of $\mathfrak{L}$ is transformed into zero, while to operators of $T(g)$ correspond right shifts.

Thus $T(g)$ is equivalent to the restriction of the regular representation to the invariant subspace $\mathfrak{L}_1$, the image of $\mathfrak{L}$ under mapping (10). By virtue of the complete reducibility of a regular representation of the compact group $G$, the representation $T(g)$ appears in the decomposition of a regular representation into irreducible ones.

5. **Decomposition of functions on homogeneous spaces.** In applications it is necessary to decompose not only functions given on a compact group, but also functions on homogeneous spaces with a compact group of motions (for example, not functions on the group of rotations of the sphere, but functions on the sphere). We have seen in §2.2 that a homogeneous space $\mathfrak{M}$ with the group of motions $G$ can be realized as a space of left cosets with respect to the stationary subgroup $H$ of some point $a \in \mathfrak{M}$. Moreover, the motions are given by the formula $g_0H \to gg_0H$. It is clear that the decomposition of functions on the homogeneous space $\mathfrak{M}$ reduces to the decomposition of functions on the group, constant on left cosets with respect to the subgroup $H$, i.e. such that $f(g) = f(gh)$.

We restrict ourselves to the consideration of the most important case when $H$ is massive (see §2.5). From the system of pairwise nonequivalent irreducible unitary representations $\{T_\alpha(g)\}$ of $G$ we choose representations of class 1 relative to $H$ and denote the set of these representations by $A_0$. In the spaces $\mathfrak{L}_\alpha$

of the representations $T_\alpha(g)$, $\alpha \in A_0$, we choose an orthonormal basis $\{e_i^\alpha\}$, such that the vectors $e_1^\alpha$ are invariant for the operators $T_\alpha(h)$, $h \in H$. It is clear that for such a choice of the basis we have $t_{11}^\alpha(h) = 1$.

We consider matrix elements $t_{i1}^\alpha(g)$ of the representations $T_\alpha(g)$, $\alpha \in A_0$. They are constant on left cosets w.r.t. $H$. Indeed,

$$t_{i1}^\alpha(gh) = (T_\alpha(gh)e_1^\alpha, e_i^\alpha) = (T_\alpha(g)T_\alpha(h)e_1^\alpha, e_i^\alpha) = (T_\alpha(g)e_1^\alpha, e_i^\alpha) = t_{i1}^\alpha(g). \tag{1}$$

One naturally expects to decompose, in these matrix elements, those functions on $G$ which are constant on left cosets w.r.t. $H$. We shall now prove that this is a valid premise.

So, let $f(g)$ be a function on the compact group $G$, constant on left cosets w.r.t. the massive subgroup $H$, and having an integrable square.

From the results of subsection 3, we have

$$f(g) = \sum_{\alpha \in A} \sum_{i,j=1}^{d_\alpha} c_{ij}^\alpha t_{ij}^\alpha(g), \tag{2}$$

where

$$c_{ij}^\alpha = d_\alpha \int f(g) \overline{t_{ij}^\alpha(g)}\, dg. \tag{3}$$

One has to show that decomposition (2) only contains terms for which $\alpha \in A_0$ and $j = 1$.

Since by hypothesis for any $h \in H$ we have $f(g) = f(gh^{-1})$, it follows that

$$c_{ij}^\alpha = d_\alpha \int f(gh^{-1}) \overline{t_{ij}^\alpha(g)}\, dg.$$

We integrate this over $H$:

$$c_{ij}^\alpha = d_\alpha \int\int f(gh^{-1}) \overline{t_{ij}^\alpha(g)}\, dg\, dh = d_\alpha \int\int f(g) \overline{t_{ij}^\alpha(gh)}\, dg\, dh$$

$$= d_\alpha \sum_{k=1}^{d_\alpha} \int_G f(g) \overline{t_{ik}^\alpha(g)}\, dg \int_H \overline{t_{kj}^\alpha(h)}\, dh.$$

Let us consider the integral $\int \overline{t_{kj}^\alpha(h)}dh$. If $\alpha \bar{\in} A_0$, then in the space of the representation $T_\alpha(g)$ there is no vector invariant for the operators $T_\alpha(h)$, $h \in H$. Therefore the decomposition of the representation $T_\alpha(h)$ of $H$ into irreducible ones does not contain the identity representation. But then all matrix elements $t_{kj}^\alpha(h)$ of this representation are linear combinations of matrix elements of irreducible, different from the identity representation, and are therefore orthogonal to the function $t(h) = 1$. This shows that if $\alpha \bar{\in} A_0$, then $\int t_{kj}^\alpha(h)dh = 0$.

Now let $\alpha \in A_0$. Because of the choice of basis we have $t_{11}^\alpha(h) = 1$ and so $t_{1j}^\alpha(h) = t_{ji}^\alpha(h) = 0$ for $j \neq 1$, $h \in H$. Since in $\mathfrak{L}_\alpha$ there is only one normalized vector invariant for $T_\alpha(h)$, $h \in H$, all the other matrix elements $t_{kj}^\alpha(h)$, $k > 1$ $j > 1$, are linear combinations of matrix elements of irreducible unitary representations, different from the identity representation. Hence if $\alpha \in A_0$ but $k \neq 1$ or $j \neq 1$, then

$$\int \overline{t_{kj}^\alpha(h)}\,dh = 0.$$

It follows that if $\alpha \bar{\in} A_0$ or $j \neq 1$ we have $c_{ij}^\alpha = 0$. Therefore

$$f(g) = \sum_{\alpha \in A_0} \sum_{i=1}^{d_\alpha} c_{i1}^\alpha t_{i1}^\alpha(g).$$

Here

$$c_{i1}^\alpha = d_\alpha \int f(g) \overline{t_{i1}^\alpha(g)}\, dg.$$

Of course, since the function $f(g)$ and the matrix element $t_{i1}^\alpha(g)$ are constant on left cosets w.r.t. $H$, the integral (3) can be rewritten as an integral over the homogeneous space $\mathfrak{M} = G/H$.

We have proved our assertion: *functions on compact groups, constant on left cosets w.r.t. massive subgroups, can be decomposed in series of the form* (2), *where $A_0$ is the set of pairwise nonequivalent irreducible unitary representations of class* 1, *and the $t_{i1}^\alpha(g)$ are matrix elements of columns, corresponding to the basis vectors $\mathbf{e}_1^\alpha$, such that $T_\alpha(h)\mathbf{e}_1^\alpha = \mathbf{e}_1^\alpha$, $h \in H$.*

Let us consider the representation

$$T(g)f(x) = f(g^{-1}x) \tag{4}$$

in the space of functions on the homogeneous space $\mathfrak{M} = G/H$ with the compact group of motions $G$ and the massive stationary subgroup $H$. It easily follows from (2) that this representation can be decomposed as the direct sum of representations of class 1 relative to the subgroup $H$, where each representation of class 1 appears only once in the decomposition (since in the decomposition (2) the matrix elements of only one row of the matrix $\{t_{ij}^\alpha(g)\}$, $\alpha \in A_0$, are involved).

In just the same way one constructs the decomposition of functions constant on right cosets w.r.t. the massive subgroup $H$.

Combining the results obtained above, we arrive at the following assertion.

*Let $H$ be a massive subgroup of the compact group $G$. Then any function $f(g)$, constant on two-sided cosets $HgH$ w.r.t. the subgroup $H$, can be decomposed in zonal spherical functions $t_{11}^{\alpha}(g)$, $\alpha \in A_0$, of unitary irreducible representations of class* 1:

$$f(g) = \sum_{\alpha \in A_0} c_\alpha t_{11}^{\alpha}(g).$$

*The coefficients of the decomposition are given by the formulas*

$$c_\alpha = d_\alpha \int f(g) \overline{t_{11}^{\alpha}(g)}\, dg. \tag{5}$$

*The integration in formulas* (5) *reduces to integration over the set of spheres in the homogeneous space* $\mathfrak{M} = G/H$.

6. **Convolution of functions on a group.** By the *convolution* of two functions $f_1(g)$ and $f_2(g)$, given on a compact group $G$, we mean the function

$$f_1 * f_2(g) = \int f_1(gg_1^{-1}) f_2(g_1)\, dg_1. \tag{1}$$

The operation of convolution is associative but not commutative, i.e. one has the equality

$$(f_1 * f_2) * f_3(g) = f_1 * (f_2 * f_3(g)), \tag{2}$$

but, generally speaking, $f_1 * f_2(g) \neq f_2 * f_1(g)$.

*If $f_1(g)$ and $f_2(g)$ are continuous, then their convolution is a continuous function too.*

Let us find out how the Fourier coefficients of the convolution $f(g)$ of two functions $f_1(g)$ and $f_2(g)$ are expressed by means of the Fourier coefficients $a_{ij}^{\alpha}$ and $b_{ij}^{\alpha}$ of the convoluted functions. In the equality

$$c_{ij}^{\alpha} = \int f(g) \overline{t_{ij}^{\alpha}(g)}\, dg \tag{3}$$

we replace $f(g)$ by its expression (1). Changing the order of integration, we get

$$c_{ij}^{\alpha} = \int\int f_1(gg_1^{-1}) f_2(g_1) \overline{t_{ij}^{\alpha}(g)}\, dg\, dg_1 = \int\int f_1(g) f_2(g_1) \overline{t_{ij}^{\alpha}(gg_1)}\, dg\, dg_1.$$

But

$$t_{ij}^{\alpha}(gg_1) = \sum_k t_{ik}^{\alpha}(g) t_{kj}^{\alpha}(g_1). \tag{4}$$

Hence

$$c_{ij}^{\alpha} = \sum_k \int f_1(g) \overline{t_{ik}^{\alpha}(g)}\, dg \int f_2(g) \overline{t_{kj}^{\alpha}(g_1)}\, dg_1 = \sum_k a_{ik}^{\alpha} b_{kj}^{\alpha}.$$

In other words, we have proved

**Theorem 1.** *Suppose that $f(g)$ is the convolution of the functions $f_1(g)$ and $f_2(g)$, given on the group $G$, and that the Fourier coefficients of $f_1(g)$ are equal to $a_{ij}^\alpha$, and those of $f_2(g)$ equal to $b_{ij}^\alpha$. Then the Fourier coefficients of $f(g)$ are given by*

$$c_{ij}^\alpha = \sum_k a_{ik}^\alpha b_{kj}^\alpha. \tag{5}$$

This theorem can be given in a more descriptive form. We arrange the Fourier coefficients of $f(g)$ corresponding to one and the same irreducible representation $T^\alpha(g)$ in the form of a matrix $(C^\alpha)$. Then to $f(g)$ corresponds an infinite aggregate of matrices $(C^\alpha)$.

Theorem 1 signifies that if to $f_1(g)$ corresponds the aggregate of matrices $(A^\alpha)$, and to $f_2(g)$ the aggregate of matrices $(B^\alpha)$, then to their convolution $f(g) = f_1(g) * f_2(g)$ corresponds the aggregate of matrices

$$(C^\alpha) = (A^\alpha)(B^\alpha).$$

In other words, *to the convolution function corresponds the product of matrices composed of the Fourier coefficients of the convoluted functions.*

The aggregate of continuous functions given on $G$ forms a ring if the addition of functions (and multiplication of a function by a scalar) has the usual meaning, and for multiplication one takes the convolution of functions. This ring is called the *group ring* of $G$. On associating with each function $f(g)$ of the group ring some matrix $(C^\alpha)$ of its Fourier coefficients, we obtain a homomorphic mapping of the group ring into the matrix ring. These mappings are in one-to-one correspondence with classes of irreducible unitary representations of $G$.

**7. Decomposition of central functions.** A function $f(g)$ on a group $G$ is called *central* if for any two elements $g$ and $g_1$ of this group one has the equality

$$f(gg_1) = f(g_1g), \tag{1}$$

that is to say, one has the equality

$$f(g_1^{-1}gg_1) = f(g). \tag{1'}$$

Thus *central functions are constant on classes of conjugate elements of $G$.* As examples of central functions we have the characters $\chi(g)$ of representations $T(g)$ of $G$, since they satisfy the functional equation

$$\chi(g_1^{-1}gg_1) = \chi(g) \tag{2}$$

(see §1.10).

If $f(g)$ is a central function on $G$, then for any continuous finite function $\phi(g)$ on this group one has the equality

$$f * \varphi(g) = \varphi * f(g). \tag{3}$$

In fact, since $f(g)$ is a central function, $f(gg_1^{-1}) = f(g_1^{-1}g)$, and so

$$f * \varphi(g) = \int f(gg_1^{-1})\varphi(g_1)\,dg_1 = \int f(g_1^{-1}g)\varphi(g_1)\,dg_1$$
$$= \int \varphi(gg_1^{-1}) f(g_1)\,dg_1 = \varphi * f(g).$$

Equality (3) shows that central functions belong to the center of the group ring of $G$, i.e. they are permutable with all elements of the ring.

Now let us look a little more closely at the form taken by the decomposition of a central function $f(g)$ in matrix elements of irreducible unitary representations. Let

$$f(g) = \sum_{\alpha \in A} \sum_{i,j=1}^{d_\alpha} c_{ij}^\alpha t_{ij}^\alpha(g). \tag{4}$$

We denote by $(C_\alpha)$ the matrix consisting of the coefficients $c_{ij}^\alpha$ of the function $f(g)$. Since $f * t_{ij}^\alpha(g) = t_{ij}^\alpha * f(g)$, we have the equality

$$(C_\alpha)(A)_{ij}^\alpha = (A)_{ij}^\alpha(C_\alpha), \tag{5}$$

where $(A)_{ij}^\alpha$ is the matrix of Fourier coefficients of the function $t_{ij}^\alpha(g)$ (see subsection 6). But it is evident that all elements of this matrix are equal to zero, apart from the element $a_{ij}^\alpha$ which is equal to 1. Since equality (5) is valid for all $i$ and $j$, $C_\alpha$ is a scalar matrix, i.e. $c_{ij}^\alpha = 0$ for $i \neq j$ and $c_{11}^\alpha = \cdots = c_{d_\alpha d_\alpha}^\alpha = c_\alpha$.

From these values for the coefficients $c_{ij}^\alpha$, it follows that the decomposition of $f(g)$ in a Fourier series has the form

$$f(g) = \sum_{\alpha \in A} c_\alpha \sum_{i=1}^{d_\alpha} t_{ii}^\alpha(g) = \sum_{\alpha \in A} c_\alpha \chi_\alpha(g).$$

We have thus proved that every central function $f(g)$ with integrable square can be decomposed in a series of characters $\chi_\alpha(g)$ of the irreducible unitary representations of the group $G$:

$$f(g) = \sum_{\alpha \in A} c_\alpha \chi_\alpha(g). \tag{6}$$

The coefficients $c_\alpha$ are expressed by the formula

$$c_\alpha \equiv c_{ii}^\alpha = d_\alpha \int f(g)\, \overline{t_{ii}^\alpha(g)}\, dg, \quad 1 \leqslant i \leqslant d_\alpha.$$

Summing these equalities for $i$ from 1 to $d_\alpha$, and dividing by $d_\alpha$, we obtain

$$c_\alpha = \int f(g)\, \overline{\chi_\alpha(g)}\, dg. \tag{7}$$

We note that *the characters $\chi_\alpha(g)$ of a complete system of pairwise nonequivalent irreducible unitary representations of the compact group $G$ form an orthogonal normalized system of functions.*

In fact, from the orthogonality relations for matrix elements of irreducible unitary representations we have for $\alpha \neq \beta$

$$\int \chi_\alpha(g)\, \overline{\chi_\beta(g)}\, dg = \sum_{i=1}^{d_\alpha} \sum_{j=1}^{d_\beta} \int t_{ii}^\alpha(g)\, \overline{t_{jj}^\beta(g)}\, dg = 0. \tag{8}$$

But if $\alpha = \beta$, then

$$\int |\chi_\alpha(g)|^2\, dg = \sum_{i=1}^{d_\alpha} \sum_{j=1}^{d_\alpha} \int t_{ii}^\alpha(g)\, \overline{t_{jj}^\alpha(g)}\, dg.$$

The terms for which $i \neq j$ are equal to zero, and the terms for which $i = j$ are equal to $1/d_\alpha$. Therefore

$$\int |\chi_\alpha(g)|^2\, dg = 1. \tag{9}$$

Formulas (6)–(9) show that every central function with integrable square is decomposed in a mean-convergent Fourier series according to the orthogonal system $\{\chi_\alpha(g)\}$. In other words, *the system of functions $\{\chi_\alpha(g)\}$ forms an orthogonal normalized basis in the Hilbert space of central functions with square-integrable moduli.* Hence, for central functions, the following Parseval equality holds:

$$\int |f(g)|^2\, dg = \sum_{\alpha \in A} |c_\alpha|^2. \tag{10}$$

The decomposition according to the system of functions $\{\chi_\alpha(g)\}$ allows one to determine the multiplicity with which a given irreducible unitary representation $T_\alpha(g)$ of the compact group $G$ appears in the decomposition of the reducible representation $T(g)$. If the character of $T(g)$ is equal to $\chi(g)$, then, by the results of §1.10,

$$\chi(g) = \sum_{\alpha \in A} a_\alpha \chi_\alpha(g). \tag{11}$$

where $a_\alpha$ is the multiplicity with which $T_\alpha(g)$ appears in the decomposition of $T(g)$. By formula (7) the Fourier coefficient $a_\alpha$ is expressed by the equality

$$a_\alpha = \int \chi(g)\overline{\chi_\alpha(g)}\, dg. \tag{12}$$

We note a special case. Let $T_\alpha(g)$ and $T_\beta(g)$ be irreducible unitary representations of the group $G$. Then the character of their tensor product $T_\alpha(g) \otimes T_\beta(g)$ is equal to $\chi_\alpha(g)\chi_\beta(g)$. Hence $T_\gamma(g)$ appears in the decomposition of the tensor product $T_\alpha(g) \otimes T_\beta(g)$ with the multiplicity

$$a_{\alpha\beta\gamma} = \int \chi_\alpha(g)\chi_\beta(g)\overline{\chi_\gamma(g)}\, dg. \tag{13}$$

We denote by $\overline{T}(g)$ that representation of $G$ which is complex conjugate to $T(g)$ (i.e. which has matrix elements $(\overline{t_{ij}(g)})$). Evidently,

$$\chi_{\overline{T}}(g) = \overline{\chi_T(g)}.$$

From equality (13) we get the following symmetry relation.

*Let $T_\alpha(g)$, $T_\beta(g)$, $T_\gamma(g)$ be irreducible unitary representations of the compact group $G$. Then the representation $T_\gamma(g)$ appears in the decomposition of the representation $T_\alpha(g) \otimes T_\beta(g)$ with the same multiplicity with which the representation $\overline{T}_\alpha(g)$ appears in the decomposition of $T_\beta(g) \otimes \overline{T}_\gamma(g)$, or the representation $\overline{T}_\beta(g)$ in the decomposition $T_\alpha(g) \otimes \overline{T}_\gamma(g)$.*

## SUPPLEMENT TO CHAPTER 1

### Some facts concerning linear spaces

We assume that the reader possesses a knowledge of the theory of linear spaces to the extent of I. M. Gel'fand's book "Lectures on linear algebra" and elements of the theory of Hilbert spaces.

1. **Kronecker or tensor product of linear spaces and operators.** Let $\mathfrak{L}_1$ and $\mathfrak{L}_2$ be finite-dimensional linear spaces. We denote by $\mathfrak{L}_1'$ and $\mathfrak{L}_2'$ the adjoints of these spaces (i.e. the spaces whose elements are linear functionals in $\mathfrak{L}_1$ and $\mathfrak{L}_2$).

By the *Kronecker* or *tensor product* of the spaces $\mathfrak{L}_1$ and $\mathfrak{L}_2$ is meant the linear space $\mathcal{L}(\mathfrak{L}_2', \mathfrak{L}_1)$ whose elements are the linear mappings of $\mathfrak{L}_2'$ into $\mathfrak{L}_1$.

We shall denote the tensor product of $\mathfrak{L}_1$ and $\mathfrak{L}_2$ by $\mathfrak{L}_1 \otimes \mathfrak{L}_2$.

To each linear mapping $A$ of $\mathfrak{L}_2'$ into $\mathfrak{L}_1$ corresponds the adjoint mapping $A'$ of $\mathfrak{L}_1$ into $\mathfrak{L}_2$. Because of the finiteness of $\mathfrak{L}_1$ and $\mathfrak{L}_2$, the correspondence $A \to A'$ determines an isomorphic mapping of $\mathfrak{L}(\mathfrak{L}_2', \mathfrak{L}_1)$ onto $\mathcal{L}(\mathfrak{L}_1', \mathfrak{L}_2)$. Thus, if one considers linear spaces to within an isomorphism, a tensor product possesses the property of commutativity:

$$\mathfrak{L}_1 \otimes \mathfrak{L}_2 = \mathfrak{L}_2 \otimes \mathfrak{L}_1. \tag{1}$$

Besides, this product is associative:

$$(\mathfrak{L}_1 \otimes \mathfrak{L}_2) \otimes \mathfrak{L}_3 = \mathfrak{L}_1 \otimes (\mathfrak{L}_2 \otimes \mathfrak{L}_3). \tag{2}$$

With each pair $(\mathbf{x}, \mathbf{y})$, $\mathbf{x} \in \mathfrak{L}_1$, $\mathbf{y} \in \mathfrak{L}_2$, we associate an operator $\mathbf{x} \otimes \mathbf{y}$ defined by the equality

$$(\mathbf{x} \otimes \mathbf{y})\,\mathbf{f} = \mathbf{f}(\mathbf{y})\,\mathbf{x}, \tag{3}$$

where $\mathbf{f} \in \mathfrak{L}_2'$. Since $\mathbf{f}(\mathbf{y})$ is a number, this operator maps $\mathfrak{L}_2'$ into $\mathfrak{L}_1$. Thus

$$\mathbf{x} \otimes \mathbf{y} \in \mathcal{L}(\mathfrak{L}_2', \mathfrak{L}_1) = \mathfrak{L}_1 \otimes \mathfrak{L}_2.$$

One has the obvious relations

$$(\alpha \mathbf{x}_1 + \beta \mathbf{x}_2) \otimes \mathbf{y} = \alpha(\mathbf{x}_1 \otimes \mathbf{y}) + \beta(\mathbf{x}_2 \otimes \mathbf{y}) \tag{4}$$

and

$$\mathbf{x} \quad (\alpha \mathbf{y}_1 + \beta \mathbf{y}_2) = \alpha(\mathbf{x} \otimes \mathbf{y}_1) + \beta(\mathbf{x} \otimes \mathbf{y}_2). \tag{5}$$

It is easy to show that any linear operator $X$, mapping $\mathfrak{L}_2'$ into $\mathfrak{L}_1$, can be written as a linear combination of operators of the form $\mathbf{x} \otimes \mathbf{y}$. Therefore, if $\{\mathbf{e}_i\}$, $1 \le i \le n$, is the basis in $\mathfrak{L}_1$, and $\{\mathbf{g}_j\}$, $1 \le j \le m$, is the basis in $\mathfrak{L}_2$, then the elements $\mathbf{h}_{ij} = \mathbf{e}_i \otimes \mathbf{g}_j$, $1 \le i \le n$, $1 \le j \le m$, form a basis in $\mathfrak{L}_1 \otimes \mathfrak{L}_2$. Moreover, if $\mathbf{x} = \sum_{i=1}^n \alpha_i \mathbf{e}_i$ and $\mathbf{y} = \sum_{j=1}^m \beta_j \mathbf{g}_j$, then

$$\mathbf{x} \otimes \mathbf{y} = \sum_{i=1}^{n} \sum_{j=1}^{m} \alpha_i \beta_j (\mathbf{e}_i \otimes \mathbf{g}_j). \tag{6}$$

Let us now define the Kronecker (or tensor) product of operators $A$ and $B$, acting in the spaces $\mathfrak{L}_1$ and $\mathfrak{L}_2$ respectively. Let $X$ be a linear operator which maps $\mathfrak{L}_2'$ into $\mathfrak{L}_1$. Evidently $AXB^*$ maps $\mathfrak{L}_2'$ into $\mathfrak{L}_1$, i.e. also belongs to $\mathfrak{L}(\mathfrak{L}_2', \mathfrak{L}_1)$. The mapping $X \to AXB'$ is a linear operator in $\mathfrak{L}(\mathfrak{L}_2', \mathfrak{L}_1)$. We shall call this operator the *Kronecker product of operators* $A$ and $B$ and denote it by $A \otimes B$. Thus,

$$(A \otimes B)X = AXB', \tag{7}$$

where $X \in \mathfrak{L}(\mathfrak{L}_2', \mathfrak{L}_1)$:

From equality (7) we get the following properties of the Kronecker product of operators:

$$A \otimes (\alpha B_1 + \beta B_2) = \alpha A \otimes B_1 + \beta A \otimes B_2, \tag{8}$$

$$(\alpha A_1 + \beta A_2) \otimes B = \alpha A_1 \otimes B + \beta A_2 \otimes B \tag{9}$$

and

$$A_1 A_2 \otimes B_1 B_2 = (A_1 \otimes B_1)(A_2 \otimes B_2). \tag{10}$$

In order to prove, for example, equality (10), it is sufficient to note that

$$(A_1 A_2 \otimes B_1 B_2) X = A_1 A_2 X B_2' B_1' = A_1 (A_2 X B_2') B_1' = (A_1 \otimes B_1)(A_2 \otimes B_2) X.$$

If the operator $X$ of $\mathfrak{L}(\mathfrak{L}_2', \mathfrak{L}_1)$ has the form $X = \mathbf{x} \otimes \mathbf{y}$, $\mathbf{x} \in \mathfrak{L}_1$, $\mathbf{y} \in \mathfrak{L}_2$ (i.e. if $X\mathbf{f} = \mathbf{f}(\mathbf{y})\mathbf{x}$, $\mathbf{f} \in \mathfrak{L}_2'$), then it is easy to verify that

$$(A \otimes B) X = (A \otimes B)(\mathbf{x} \otimes \mathbf{y}) = A\mathbf{x} \otimes B\mathbf{y}. \tag{11}$$

In particular,

$$(A \otimes B)(\mathbf{e}_i \otimes \mathbf{g}_j) = A\mathbf{e}_i \otimes B\mathbf{g}_j.$$

Hence, if the matrix of $A$ in the basis $\{\mathbf{e}_i\}$ is equal to $(a_{ij})$ and the matrix of $B$ in the basis $\{\mathbf{g}_r\}$ is equal to $(b_{rs})$, then the matrix of $A \otimes B$ in the basis $\mathbf{h}_{ir} = \mathbf{e}_i \otimes \mathbf{g}_r$ is given by

$$(c_{ir,js}) = (a_{ij} b_{rs}). \tag{12}$$

Thus the elements of this matrix are possible pairwise products of the elements of matrices $(a_{ij})$ and $(b_{rs})$.

We also note that if $\mathfrak{L}_1$ and $\mathfrak{L}_2$ are unitary linear spaces (i.e. complex linear spaces in which scalar products with the usual properties are defined), then their Kronecker product $\mathfrak{L}_1 \otimes \mathfrak{L}_2$ is also a unitary space. The scalar product in $\mathfrak{L}_1 \otimes \mathfrak{L}_2$ is defined by the equality

$$(\mathbf{x}_1 \otimes \mathbf{y}_1, \mathbf{x}_2 \otimes \mathbf{y}_2) = (\mathbf{x}_1, \mathbf{x}_2)_1 (\mathbf{y}_1, \mathbf{y}_2)_2, \tag{13}$$

where $(\mathbf{x}_1, \mathbf{x}_2)_1$ and $(\mathbf{y}_1, \mathbf{y}_2)_2$ are the scalar products in $\mathfrak{L}_1$ and $\mathfrak{H}_2$. Moreover, an orthogonal normalized basis in $\mathfrak{L}_1 \otimes \mathfrak{L}_2$ is given by $\mathbf{h}_{ij} = \mathbf{e}_i \otimes \mathbf{g}_j$, where $\{\mathbf{e}_i\}$ is the orthogonal normalized basis in $\mathfrak{L}_1$ and $\{\mathbf{g}_j\}$ is the orthogonal normalized basis in $\mathfrak{L}_2$.

2. **Operators of Hilbert-Schmidt type.** We now pass on to the consideration of infinite-dimensional spaces. An operator $A$, mapping a Hilbert space $\mathfrak{H}_1$ into a Hilbert space $\mathfrak{H}_2$, is called an *operator of Hilbert-Schmidt type* if for any orthonormal system of vectors $\{\mathbf{e}_n\}$ in $\mathfrak{H}_1$ the following series is convergent:

$$\sum_{n=1}^{\infty} \| A\mathbf{e}_n \|^2. \tag{1}$$

One can easily show that for $A$ to be an operator of Hilbert-Schmidt type, it is sufficient that condition (1) be satisfied for one orthonormal basis in $\mathfrak{H}_1$.

Let $\mathfrak{L}^2_\sigma$ be the space of functions $f(x)$ on a set $X$ having square-integrable moduli relative to the measure $\sigma$, and $\mathfrak{L}^2_\rho$ the space of functions $F(y)$ on a set $Y$, having square-integrable moduli relative to the measure $\rho$. If the function $K(x, y)$ is such that [1]

$$\iint |K(x, y)|^2 \, d\sigma(x) \, d\rho(y) < +\infty,$$

then the integral operator

$$\int K(x, y) f(x) \, d\sigma(x)$$

is an operator of Hilbert-Schmidt type.

If $A$ and $B$ are operators of Hilbert-Schmidt type, then $\alpha A + \beta B$, where $\alpha$ and $\beta$ are complex numbers, is also an operator of Hilbert-Schmidt type. Thus operators of Hilbert-Schmidt type form a linear space. We shall denote it by $\mathcal{H}(\mathfrak{H}_1, \mathfrak{H}_2)$.

We introduce a scalar product in $\mathcal{H}(\mathfrak{H}_1, \mathfrak{H}_2)$. Namely, if $A$ and $B$ are operators of Hilbert-Schmidt type, mapping $\mathfrak{H}_1$ into $\mathfrak{H}_2$, then we take

$$(A, B) = \sum_{n=1}^{\infty} (Ae_n, Be_n) \equiv \operatorname{Tr}(B^*A), \tag{2}$$

where $\{e_n\}$ is some orthonormal basis in $\mathfrak{H}_1$. The series (2) converges, since

$$\sum_{n=1}^{\infty} |(Ae_n, Be_n)| \leq \frac{1}{2}\left(\sum_{n=1}^{\infty} \|Ae_n\|^2 + \sum_{n=1}^{\infty} \|Be_n\|^2\right) < +\infty.$$

One can easily verify that $(A, B)$ does not depend on the choice of the basis $\{e_n\}$, possesses all the properties of a scalar product, and that $\mathcal{H}(\mathfrak{H}_1, \mathfrak{H}_2)$, with this scalar product, is a Hilbert space.

We also note the following properties of operators of Hilbert-Schmidt type.

1) *The operator $A^*$, adjoint to an operator of Hilbert-Schmidt type, is an operator of the same type.*

2) *If one of the operators $A$ and $B$ is an operator of Hilbert-Schmidt type, and the other is bounded, then their product $AB$ is an operator of Hilbert-Schmidt type.*

3. **The tensor product of Hilbert spaces.** Let $\mathfrak{H}_1$ and $\mathfrak{H}_2$ be Hilbert spaces. By their *tensor product* we mean the Hilbert space $\mathcal{H}(\hat{\mathfrak{H}}_2, \mathfrak{H}_1)$ consisting of

---

1) If integration is extended over the entire range of variables we shall usually omit the reference to the domain.

antilinear operators [1)] of Hilbert-Schmidt type which map $\mathfrak{H}_2$ into $\mathfrak{H}_1$. We shall denote $\hat{\mathcal{H}}(\mathfrak{H}_2, \mathfrak{H}_1)$ also by $\mathfrak{H}_1 \otimes \mathfrak{H}_2$. We have already seen that $\mathfrak{H}_1 \otimes \mathfrak{H}_2$ is a Hilbert space.

As the example considered in subsection 2 shows, the tensor product of $\mathfrak{L}^2_\sigma$ and $\mathfrak{L}^2_\rho$ is the space $\mathfrak{L}^2_{\sigma\otimes\rho}$ consisting of functions $K(x, y)$ for which

$$\int\int |K(x, y)|^2 \, d\sigma(x) \, d\rho(y) < +\infty. \tag{1}$$

To each pair of vectors $(\mathbf{x}, \mathbf{y})$, $\mathbf{x} \in \mathfrak{H}_1$, $\mathbf{y} \in \mathfrak{H}_2$, corresponds an element $\mathbf{x} \otimes \mathbf{y}$ of $\mathfrak{H}_1 \otimes \mathfrak{H}_2$, mapping the vector $\mathbf{z} \in \mathfrak{H}_2$ into the vector

$$(\mathbf{x} \otimes \mathbf{y})\mathbf{z} = (\mathbf{y}, \mathbf{z})\mathbf{x} \tag{2}$$

of $\mathfrak{H}_1$. The scalar product of two elements $\mathbf{x}_1 \otimes \mathbf{y}_1$ and $\mathbf{x}_2 \otimes \mathbf{y}_2$ is given by the formula

$$(\mathbf{x}_1 \otimes \mathbf{y}_1, \mathbf{x}_2 \otimes \mathbf{y}_2) = (\mathbf{x}_1, \mathbf{x}_2)_1 (\mathbf{y}_1, \mathbf{y}_2)_2. \tag{3}$$

In fact, if $\{\mathbf{f}_n\}$ is an orthonormal basis in $\mathfrak{H}_2$, then

$$\begin{aligned}(\mathbf{x}_1 \otimes \mathbf{y}_1, \mathbf{x}_2 \otimes \mathbf{y}_2) &= \sum_{n=1}^{\infty} ((\mathbf{x}_1 \otimes \mathbf{y}_1)\mathbf{f}_n, (\mathbf{x}_2 \otimes \mathbf{y}_2)\mathbf{f}_n)_1 \\ &= \sum_{n=1}^{\infty} ((\mathbf{y}_1, \mathbf{f}_n)_2 \mathbf{x}_1, (\mathbf{y}_2, \mathbf{f}_n)_2 \mathbf{x}_2)_1 \\ &= (\mathbf{x}_1, \mathbf{x}_2)_1 \sum_{n=1}^{\infty} (\mathbf{y}_1, \mathbf{f}_n)_2 \overline{(\mathbf{y}_2, \mathbf{f}_n)_2} = (\mathbf{x}_1, \mathbf{x}_2)_1 (\mathbf{y}_1, \mathbf{y}_2)_2.\end{aligned}$$

It follows from (3) that if $\{\mathbf{e}_n\}$ is an orthonormal basis in $\mathfrak{H}_1$ and $\{\mathbf{f}_n\}$ is an orthonormal basis in $\mathfrak{H}_2$, then the elements $\mathbf{e}_i \otimes \mathbf{f}_j$ form an orthonormal basis in $\mathfrak{H}_1 \otimes \mathfrak{H}_2$.

We now define the Kronecker product of bounded operators. Let $A$ and $B$ be bounded operators in $\mathfrak{H}_1$ and $\mathfrak{H}_2$ respectively, and let $X$ be an element of $\mathfrak{H}_1 \otimes \mathfrak{H}_2$, i.e. an antilinear operator of Hilbert-Schmidt type mapping $\mathfrak{H}_2$ into $\mathfrak{H}_1$. Then the operator $AXB^*$ also maps $\mathfrak{H}_2$ into $\mathfrak{H}_1$, and is an operator of Hilbert-Schmidt type. Consequently the mapping $X \to AXB^*$ is a (linear) operator in $\mathfrak{H}_1 \otimes \mathfrak{H}_2$. We shall denote it by $A \otimes B$ and call it the *Kronecker product* of the operators $A$ and $B$.

If $\mathbf{x} \in \mathfrak{H}_1$, $\mathbf{y} \in \mathfrak{H}_2$, then

$$(A \otimes B)(\mathbf{x} \otimes \mathbf{y}) = A\mathbf{x} \otimes B\mathbf{y}. \tag{4}$$

---

1) An operator $A$, mapping $\mathfrak{H}_2$ into $\mathfrak{H}_1$, is called *antilinear* if for any two vectors $\mathbf{x}, \mathbf{y}$ of $\mathfrak{H}_2$ and any complex numbers $\alpha$, $\beta$ we have

$$A(\alpha\mathbf{x} + \beta\mathbf{y}) = \bar{\alpha} A\mathbf{x} + \bar{\beta} A\mathbf{y}.$$

Indeed, for any element z of $\mathfrak{H}_2$, because of equality (2), we have

$$(A\otimes B)(\mathbf{x}\otimes\mathbf{y})\mathbf{z}=A(\mathbf{x}\otimes\mathbf{y})B^*\mathbf{z}=A(\mathbf{y},B^*\mathbf{z})\mathbf{x}=(B\mathbf{y},\mathbf{z})A\mathbf{x}=(A\mathbf{x}\otimes B\mathbf{y})\mathbf{z}.$$

Equality (4) is thus proved.

The tensor product of operators in Hilbert spaces possesses properties analo gous to the properties of the tensor product of operators in finite-dimensional spaces. Namely,

$$(\alpha A_1+\beta A_2)\otimes B=\alpha(A_1\otimes B)+\beta(A_2\otimes B), \tag{5}$$

$$A\otimes(\alpha B_1+\beta B_2)=\alpha(A\otimes B_1)+\beta(A\otimes B_2), \tag{6}$$

$$(A_1\otimes B_1)(A_2\otimes B_2)=A_1A_2\otimes B_1B_2. \tag{7}$$

These properties are proved in exactly the same way as in the finite-dimensional case.

Let us now find the matrix elements of the operator $A\otimes B$. Let $\{\mathbf{e}_i\}$ be an orthonormal basis in $\mathfrak{H}_1$, $\{\mathbf{f}_m\}$ an orthonormal basis in $\mathfrak{H}_2$. Then, as noted above, $\{\mathbf{e}_i\otimes\mathbf{f}_m\}$ is an orthonormal basis in $\mathfrak{H}_1\otimes\mathfrak{H}_2$. The matrix elements of $C=A\otimes B$ are expressed, in this basis, by the formula

$$c_{im,\,jn}=((A\otimes B)(\mathbf{e}_j\otimes\mathbf{f}_n),\ (\mathbf{e}_i\otimes\mathbf{f}_m)).$$

But by (3) and (4) we have

$$\begin{aligned}((A\otimes B)(\mathbf{e}_j\otimes\mathbf{f}_n),\ (\mathbf{e}_i\otimes\mathbf{f}_m))&=((A\mathbf{e}_j\otimes B\mathbf{f}_n),\ (\mathbf{e}_i\otimes\mathbf{f}_m))\\&=(A\mathbf{e}_j,\mathbf{e}_i)_1(B\mathbf{f}_n,\mathbf{f}_m)_2=a_{ij}b_{mn}.\end{aligned}$$

Thus

$$c_{im,\,jn}=a_{ij}b_{mn}. \tag{8}$$

We have shown that the matrix elements of $A\otimes B$ in the basis $\{\mathbf{e}_i\otimes\mathbf{f}_m\}$ are products of the matrix elements of $A$ and $B$ in the bases $\{\mathbf{e}_i\}$ and $\{\mathbf{f}_m\}$ respectively.

Finally, we show that the Kronecker product of two unitary operators is a unitary operator. Since the operators $\mathbf{e}_i\otimes\mathbf{f}_n$ form an orthonormal basis in $\mathfrak{H}_1\otimes\mathfrak{H}_2$, it is sufficient to show that if the operators $U$ and $V$, in the spaces $\mathfrak{H}_1$ and $\mathfrak{H}_2$ respectively, are unitary, then

$$((U\otimes V)(\mathbf{e}_i\otimes\mathbf{f}_m),\ (U\otimes V)(\mathbf{e}_j\otimes\mathbf{f}_n))=(\mathbf{e}_i\otimes\mathbf{f}_m,\ \mathbf{e}_j\otimes\mathbf{f}_n). \tag{9}$$

But by (3) and (4) we have

$$\begin{aligned}((U\otimes V)(\mathbf{e}_i\otimes\mathbf{f}_m),\ (U\otimes V)(\mathbf{e}_j\otimes\mathbf{f}_n))&=(U\mathbf{e}_i\otimes V\mathbf{f}_m,\ U\mathbf{e}_j\otimes V\mathbf{f}_n)\\&=(U\mathbf{e}_i,\ U\mathbf{e}_j)_1(V\mathbf{f}_m,\ V\mathbf{f}_n)_2.\end{aligned}$$

Because $U$ and $V$ are unitary, we have

$$(U\mathbf{e}_i,\ U\mathbf{e}_j)_1(V\mathbf{f}_m,\ V\mathbf{f}_n)_2=(\mathbf{e}_i,\mathbf{e}_j)_1(\mathbf{f}_m,\mathbf{f}_n)_2.$$

On the other hand.

$$(\mathbf{e}_i \otimes \mathbf{f}_m, \mathbf{e}_j \otimes \mathbf{f}_n) = (\mathbf{e}_i, \mathbf{e}_j)_1 (\mathbf{f}_m, \mathbf{f}_n)_2.$$

This proves equality (9) and hence the fact that $U \otimes V$ is unitary.

**4. Denumerably-Hilbert spaces. Kernel spaces.** Suppose that in a Hilbert space $\Phi$ we are given a denumerable system of scalar products $(\phi, \psi)_k$, $1 \leq k < \infty$, such that for any element $\phi$ of $\Phi$ we have

$$(\varphi, \varphi)_1 \leqslant (\varphi, \varphi)_2 \leqslant \cdots \leqslant (\varphi, \varphi)_k \leqslant \cdots \tag{1}$$

We assume that $\|\phi\|_k^2 = (\phi, \phi)_k$ for arbitrary $k$. We shall call a sequence of elements $\phi_1, \cdots, \phi_n, \cdots$ of $\Phi$ *fundamental* if for arbitrary $k$ we have $\lim_{m,n\to\infty} \|\phi_m - \phi_n\|_k = 0$. The space $\Phi$ is called complete if for any fundamental sequence $\phi_1, \cdots, \phi_n, \cdots$ there is an element $\phi \in \Phi$, such that $\lim_{n\to\infty} \|\phi - \phi_n\|_k = 0$ for all $k$. A linear space $\Phi$ in which a scalar product with the above-mentioned properties is given is called *denumerably-Hilbert* if it is complete.

Completing the denumerably-Hilbert space $\Phi$ with respect to the norm $\|\phi\|_k$, we obtain a Hilbert space $\Phi_k$. Moreover, it follows from the equality $\|\phi\|_m \leq \|\phi\|_n$, $m < n$, that for $n > m$ there exists a continuous linear mapping $T_m^n$ of $\Phi_n$ into $\Phi_m$ which leaves elements of $\Phi$ fixed.

The denumerably-Hilbert space $\Phi$ is called a *kernel space* if for any $m$ there is an $n$ such that the mapping $T_m^n$ is an operator of Hilbert-Schmidt type. As an example of a denumerably-Hilbert kernel space, we have the space $K(a, b)$ of all infinitely differentiable functions $\phi(x)$ on the segment $[a, b]$. In this space the scalar products $(\phi, \psi)_n$ are defined by the formulas

$$(\varphi, \psi)_n = \sum_{k=0}^{n} \int_a^b \varphi^{(k)}(x) \overline{\psi^{(k)}(x)}\, dx. \tag{2}$$

Later on we shall need one of the properties of kernel spaces connected with the realization of these spaces as spaces of functions. It is well known that a Hilbert space can be realized in the form of a space of functions. These realizations are constructed in the following manner. We choose a positive measure $\sigma$ on some set $X$ (for example, on the real line), and denote by $\mathfrak{L}_\sigma^2$ the space of all functions $\phi(x)$ for which the following integral is convergent:

$$\int |\varphi(x)|^2\, d\sigma(x).$$

Introducing a scalar product in the space $\mathfrak{L}_\sigma^2$ by the formula

$$(\varphi, \psi) = \int \varphi(x) \overline{\psi(x)}\, d\sigma(x),$$

we obtain a Hilbert space. More precisely, as elements of $\mathfrak{L}_\sigma^2$ we do not have the individual functions $\phi(x)$, but classes of functions which differ from each other only by a set of zero $\sigma$-measure.

This realization suffers from the defect that on associating with a function $\phi(x)$ of $\mathfrak{L}^2_\sigma$ its value at some point $x_0$, we do not, generally speaking, get a continuous linear functional in this space (the points at which the nonzero measure is concentrated constitute the exception). Moreover, since the function $\phi(x)$ of the space $\mathfrak{L}^2_\sigma$ is only defined to within its values at a set of zero $\sigma$-measure, we cannot speak of its values at a fixed point $x_0$. But in many problems connected with the decomposition of group representations it seems desirable to regard the value of a function at a point as a linear functional. The example, considered above, of the kernel space of infinitely differentiable functions on a segment suggests how this can be done; for the space of infinitely differentiable functions the value at a point is a continuous linear functional.

In the general case, let us consider the kernel space $\Phi$. Suppose that in this space another scalar product $(\phi, \psi)$ is given, and that for a certain $n$ we have $(\phi, \phi) \le (\phi, \phi)_n$ for all $\phi \in \Phi$. Let us complete $\Phi$ with respect to the norm $\|\phi\|^2 = (\phi, \phi)$. We get a Hilbert space $\mathfrak{H}$. Moreover, there exists a continuous mapping of $\Phi$ into $\mathfrak{H}$. If $\mathfrak{H}$ is realized in the form of a space of functions, then this also gives a realization of $\Phi$ in the form of a space of functions. Thus we have (see [14])

**Theorem 1.** *Let $\phi \to \phi(x)$ be a realization of the kernel space $\Phi$ in the form of a space of functions, induced by the corresponding realization of the Hilbert space $\mathfrak{H}$, $\mathfrak{H} \supset \Phi$, in the form of a space $\mathfrak{L}^2_\sigma$. Then, with each value of $x$ one can associate a linear functional $\lambda_x$ in $\Phi$ such that for any element of this space the equality $\lambda_x\phi = \phi(x)$ is satisfied for almost all values of $x$ (relative to the measure $\sigma$).*

5. **Orthogonal direct sum of Hilbert spaces.** Suppose one is given Hilbert spaces $\mathfrak{H}_1, \cdots, \mathfrak{H}_n, \cdots$. By the *orthogonal direct sum* of these spaces is meant the Hilbert space

$$\mathfrak{H} = \sum_{n=1}^{\infty} \oplus \mathfrak{H}_n,$$

whose elements are the sequences

$$\xi = (\mathbf{h}_1, \ldots, \mathbf{h}_n, \ldots), \ \mathbf{h}_n \in \mathfrak{H}_n.$$

Only those sequences $\xi$ here are taken for which $\sum_{n=1}^{\infty} \|\mathbf{h}_n\|_n^2 < +\infty$, where $\|\mathbf{h}\|_n$ is the norm in $\mathfrak{H}_n$.

Linear operations in the space are defined by coordinates; if $\xi = (\mathbf{h}_1, \cdots, \mathbf{h}_n, \cdots)$, $\eta = (\mathbf{g}_1, \cdots, \mathbf{g}_n, \cdots)$, then $\xi + \eta = (\mathbf{h}_1 + \mathbf{g}_1, \cdots, \mathbf{h}_n + \mathbf{g}_n, \cdots)$ and $\alpha\xi = (\alpha\mathbf{h}_1, \cdots, \alpha\mathbf{h}_k, \cdots)$. The scalar product in $\mathfrak{H}$ is defined by the equality

$$(\xi, \eta) = \sum_{n=1}^{\infty} (\mathbf{h}_n, \mathbf{g}_n)_n,$$

where $(\mathbf{h}, \mathbf{g})_n$ is the scalar product in $\mathfrak{H}_n$.

If $A_n$, $n = 1, 2, \cdots$, are operators in the spaces $\mathfrak{H}_n$, where the norms of these operators are bounded as a whole, then to them corresponds an operator $A$ in $\mathfrak{H}$, defined by the formula $A\xi = (A_1\mathbf{h}_1, \cdots, A_n\mathbf{h}_n, \cdots)$.

Let $\mathfrak{H}$ be a Hilbert space and $\mathfrak{H}_1, \cdots, \mathfrak{H}_n, \cdots$ its subspaces, such that

1) the subspaces $\mathfrak{H}_n$ are pairwise orthogonal, and

2) the smallest closed subspace in $\mathfrak{H}$ containing all the subspaces $\mathfrak{H}_n$ coincides with $\mathfrak{H}$.

Then one says that $\mathfrak{H}$ is the *orthogonal direct sum* of its subspaces $\mathfrak{H}_n$, and one writes

$$\mathfrak{H} = \sum_{n=1}^{\infty} \oplus \mathfrak{H}_n.$$

The connection between the two concepts of an orthogonal direct sum introduced here is evident.

6. **The continuous direct sum of Hilbert spaces.** The concept of an orthogonal direct sum of Hilbert spaces can be slightly generalized by considering the direct sums of Hilbert spaces $\mathfrak{H}_1, \cdots, \mathfrak{H}_n, \cdots$, taken with positive weights $\mu_1, \cdots, \mu_n, \cdots$. In this case the scalar product is defined by the formula

$$(\xi, \eta) = \sum_{n=1}^{\infty} \mu_n (\mathbf{h}_n, \mathbf{g}_n)_n. \tag{1}$$

This equality can be written in the form

$$(\xi, \eta) = \int_X (\mathbf{h}(x), \mathbf{g}(x))_x \, d\mu(x), \tag{2}$$

where $X$ denotes the set consisting of the points $1, 2, \cdots, n, \cdots$, and $\mu(x)$ is the measure on this set equal to $\mu_n$ at the point $n$. In accordance with this the orthogonal direct sum of Hilbert spaces $\mathfrak{H}_1, \cdots, \mathfrak{H}_n, \cdots$, taken with weights $\mu_1, \cdots, \mu_n, \cdots$, can be written in the form

$$\mathfrak{H} = \int_X \oplus \mathfrak{H}(x) \, d\mu(x). \tag{3}$$

We now generalize the concept of an orthogonal direct sum of Hilbert spaces, taken with given weights, without the requirement that the number of terms be denumerable. In other words, we consider some set $X$ on which a positive measure $\mu$ is given. Let a Hilbert space $\mathfrak{H}(x)$ be associated with each point of this set. We shall regard all these spaces as having one and the same dimension. In that case all spaces $\mathfrak{H}(x)$ can be identified with one and the same Hilbert space $\mathfrak{H}$ of the same dimension.

We denote by $\mathcal{H}$ the space whose elements are the vector functions $\xi = \mathbf{h}(x)$ on the set $X$, taking values in the space $\mathfrak{H}$, such that

1) for any element $\mathbf{h}$ of $\mathfrak{H}$ the numerical function $(\mathbf{h}(x), \mathbf{h})$ is measurable with respect to the measure $\mu$, and

2) the numerical function $\|\mathbf{h}(x)\|$ has square-integrable modulus with respect to $\mu$:

$$\int \| \mathbf{h}(x) \|^2 \, d\mu(x) < +\infty. \tag{4}$$

We define linear operators in the space $\mathcal{H}$ by taking, for vector functions $\xi = \mathbf{h}(x)$, $\eta = \mathbf{g}(x)$,

$$\left.\begin{aligned} \xi + \eta &= \mathbf{h}(x) + \mathbf{g}(x), \\ \alpha\xi &= \alpha\mathbf{h}(x) \end{aligned}\right\} \tag{5}$$

and introduce a scalar product by taking

$$(\xi, \eta) = \int \mathbf{h}(x) \overline{\mathbf{g}(x)} \, d\mu(x). \tag{6}$$

It easily follows from conditions 1) and 2) that this product is defined for any two elements $\xi$ and $\eta$ of $\mathcal{H}$. Further, it is easy to verify that $\mathcal{H}$ satisfies all the axioms of a Hilbert space. In particular, it is complete. We shall call $\mathcal{H}$ the *continuous direct sum of the spaces $\mathfrak{H}(x)$ relative to the measure $\mu$*, and write

$$\mathcal{H} = \int \oplus \, \mathfrak{H}(x) \, d\mu(x). \tag{7}$$

The restriction that the spaces $\mathfrak{H}(x)$ all have one and the same dimension is inessential. It is sufficient to assume that the dimension $n(x)$ of $\mathfrak{H}(x)$ is a measurable function. Then the set $X$ is broken up into measurable sets $X_\infty$, $X_1, \ldots, X_n, \ldots$, such that on $X_n$ we have $n(x) = n$ ($n$ can also take the value $\infty$). Taking

$$\int_X \oplus \, \mathfrak{H}(x) \, d\mu(x) = \sum_{n=1}^{\infty} \int_{X_n} \oplus \, \mathfrak{H}(x) \, d\mu(x), \tag{8}$$

we obtain the definition of continuous direct sum of Hilbert spaces in this case too.

Let $\Phi$ be a denumerably-Hilbert kernel space and $(\phi, \psi)$ the scalar product in it, such that, for all elements $\phi$ of $\Phi$, $(\phi, \phi) < (\phi, \phi)_n$ for a certain $n$. We denote by $\mathfrak{H}$ the completion of $\Phi$ with respect to the norm $\|\phi\|^2 = (\phi, \phi)$. Then we have the following theorem, generalizing Theorem 1 of subsection 4.

**Theorem 1.** *Let* $\Phi \subset \mathfrak{H}$ *be a kernel space and*

$$\mathbf{h} \to \xi = \mathbf{h}(x)$$

*an isometric imbedding of* $\mathfrak{H}$ *in the space*

$$\mathcal{H} = \int\limits_X \oplus \mathfrak{H}(x)\, d\mu(x).$$

*Then for any* $x$ *there exists a kernel operator* $T_X$, *mapping* $\Phi$ *into* $\mathcal{H}$, *such that for* $\phi \in \Phi$ *the functions* $\phi(x)$ *and* $T_x(\phi)$ *differ only on a set of measure zero.*

The reader can find a proof of this theorem too in [14] (Russian pp. 151–153).

We consider a few examples of continuous direct sums of Hilbert spaces. If the set $X$ is discrete the continuous direct sum turns into the orthogonal direct sum of Hilbert spaces with given weights. But if all the spaces $\mathfrak{H}(x)$ are one-dimensional, then $\int \oplus\ \mathfrak{H}(x) d\mu(x)$ is none other than the space $\mathfrak{L}^2_\mu$ of functions on the set $X$ having square-integrable moduli relative to the measure $\mu$.

7. The decomposition of operators as the continuous direct sum of operators. Let the Hilbert space $\mathcal{H}$ be the continuous direct sum of spaces $\mathfrak{H}(x)$,

$$\mathcal{H} = \int\limits_X \oplus \mathfrak{H}(x)\, d\mu(x) \tag{1}$$

and let $\phi \to \phi(x)$ be the imbedding of the kernel space $\Phi$ into $\mathcal{H}$. By Theorem 1 of subsection 6, for all $x \in X$ there exists an operator $T_x$ such that for any $\phi \in \Phi$ we have $\phi(x) = T_x(\phi)$ for almost all $x$.

Let $A$ be an operator in $\mathcal{H}$ which leaves $\Phi$ invariant. We shall say that the spaces $\mathfrak{H}(x)$ are invariant for this operator if for almost all $x$ we have

$$AT_x = T_x A. \tag{2}$$

Let us denote $AT_x$ by $A_x$. Then for any element $\phi \in \Phi$ we have

$$(A_x \varphi)(x) = A\varphi(x). \tag{3}$$

We shall say in this case that $A$ is decomposed as the continuous direct sum of the operators $A(x)$.

If all the operators of a representation $A(g)$ are decomposed as the continuous direct sum of operators $A_x(g)$ in accordance with the decomposition (1), then we say that *the representation* $A(g)$ *is decomposed as the continuous direct sum of the representations* $A_x(g)$.

CHAPTER II

# THE ADDITIVE GROUP OF REAL NUMBERS AND THE EXPONENTIAL FUNCTION. FOURIER SERIES AND INTEGRALS

We begin the study of the connection between the theory of representations and special functions with an analysis of two model examples: the additive group $R$ of real numbers and the group $SO(2)$ of rotations of the plane. Representations of the first of these are connected with the exponential function, and of the second, with the trigonometrical functions.

## § 1. THE EXPONENTIAL AND TRIGONOMETRICAL FUNCTIONS

1. Irreducible unitary representations of the group $R$. We denote by $R$ the group whose elements are real numbers and for which addition of numbers is the group operation. Let us find the irreducible unitary representations of this group. Naturally, we restrict ourselves here to continuous representations.

In §3.3 of Chapter I it was shown that all irreducible unitary representations of a compact group are one-dimensional. Therefore, one has to find continuous functions of a real variable, satisfying the functional equation

$$f(x)f(y)=f(x+y) \tag{1}$$

and the condition for being unitary:

$$|f(x)|=1. \tag{2}$$

It is well known that all continuous solutions of equation (1) have the form

$$f(x)=e^{ax}.$$

For completeness, we give the proof of this assertion. We first prove that any continuous solution of the functional equation (1) is differentiable. For this we take an infinitely differentiable finite function $\phi(x)$, such that $\int_{-\infty}^{\infty} f(x)\phi(x)dx \neq 0$ (such a function exists if $f(x)\not\equiv 0$). We multiply both sides of equality (1) by $\phi(y)$, and integrate with respect to $y$ from $-\infty$ to $\infty$:

$$f(x) \int_{-\infty}^{\infty} f(y)\,\varphi(y)\,dy = \int_{-\infty}^{\infty} f(x+y)\,\varphi(y)\,dy = \int_{-\infty}^{\infty} f(y)\,\varphi(y-x)\,dy$$

(the integral converges because $\phi(y)$ is finite). Owing to the infinite differentiability of $\phi(x)$, the right hand side of this equality is infinitely differentiable with respect to $x$. Consequently, the left hand of this equality, i.e. $f(x)$, is also infinitely differentiable.

We differentiate both sides of equality (1) with respect to $y$ and put $y = 0$. We get the differential equation

$$f'(0)\,f(x) = f'(x). \tag{3}$$

Further, from equality (1) we have $f(0) = 1$. But the solutions of equation (3) for which $f(0) = 1$ have the form

$$f(x) = e^{ax},$$

where $a = f'(0)$. Our assertion is thus proved.

So, all one-dimensional continuous representations of the group $R$ have the form $f(x) = e^{ax}$, where $a$ is some number, generally speaking complex. Let us determine the values of $a$ for which these representations are unitary, i.e. when $|e^{ax}| = 1$. We differentiate the equality $e^{ax}\,\overline{e^{ax}} = 1$ with respect to $x$, and put $x = 0$. We get $a + \bar{a} = 0$, i.e. $a$ is a purely imaginary number. Thus, the representation $e^{ax}$ of the group $R$ is unitary if and only if $a$ is a purely imaginary number $a = bi$.

We note that the power function

$$y = t^a, \quad t > 0 \tag{4}$$

is connected with representations of the group $R_+$ of positive numbers, where multiplication of numbers is the group operation. The mapping $t = e^x$ establishes an isomorphism between the groups $R$ and $R_+$. Under this mapping, to the representation $e^{ax}$ of $R$ corresponds the representation $t^a$ of $R_+$. For purely imaginary $a$ these representations are unitary.

2. The group of rotations of the plane and the trigonometrical functions. We pass on to the consideration of irreducible unitary representations of the group $SO(2)$ of rotations of the euclidean plane around the origin, i.e. the group of one-dimensional linear transformations of two variables, preserving the quadratic form $x^2 + y^2$ and having a positive determinant.

It is well known that a rotation of the plane is given by a real number $\phi$, namely, the angle of rotation, and that the matrix of the corresponding linear transformation has the form

$$g(\varphi)=\begin{pmatrix}\cos\varphi & -\sin\varphi\\ \sin\varphi & \cos\varphi\end{pmatrix}. \tag{1}$$

The functions $\sin\phi$ and $\cos\phi$, the trigonometrical functions, are connected with the exponential function by Euler's formulas:

$$e^{\pm i\varphi}=\cos\varphi\pm i\sin\varphi, \tag{2}$$

$$\cos\varphi=\frac{e^{i\varphi}+e^{-i\varphi}}{2}, \tag{3}$$

$$\sin\varphi=\frac{e^{i\varphi}-e^{-i\varphi}}{2i}. \tag{4}$$

In taking the product of rotations, angles of rotation are added, and so

$$g(\varphi)g(\psi)=g(\varphi+\psi) \tag{5}$$

or, in matrix form,

$$\begin{pmatrix}\cos\varphi & -\sin\varphi\\ \sin\varphi & \cos\varphi\end{pmatrix}\begin{pmatrix}\cos\psi & -\sin\psi\\ \sin\psi & \cos\psi\end{pmatrix}=\begin{pmatrix}\cos(\varphi+\psi) & -\sin(\varphi+\psi)\\ \sin(\varphi+\psi) & \cos(\varphi+\psi)\end{pmatrix}.$$

Multiplying out the matrices on the left, and then comparing the matrix elements on the left and on the right, we obtain the addition formulas for trigonometrical functions:

$$\cos(\varphi+\psi)=\cos\varphi\cos\psi-\sin\varphi\sin\psi, \tag{6}$$

$$\sin(\varphi+\psi)=\sin\varphi\cos\psi+\cos\varphi\sin\psi. \tag{7}$$

Later, in the same way, we shall derive composition formulas for nonelementary functions.

Let us now find the irreducible unitary representations of the group $SO(2)$. For this we note that, because of equality (5), the mapping $\phi\to g(\phi)$ is a homomorphic mapping of the additive group $R$ of real numbers onto $SO(2)$. The kernel of this mapping consists of numbers of the form $\phi=2\pi k$, where $k$ is an integer; to these and only these numbers correspond the identity rotations of the plane. Thus, the group $SO(2)$ is a factor-group of the group $R$ with respect to the subgroup $Z_{2\pi}$ of numbers of the form $2\pi k$:

$$SO(2)=R/Z_{2\pi}.$$

This observation allows one to reduce the search for irreducible unitary representations of the group $SO(2)$ to the same problem for the group $R$. Namely, if

$f(\phi)$ is an irreducible unitary representation of $SO(2)$, then the equality

$$F(\varphi+2\pi k)=f(\varphi), \qquad 0\leqslant\varphi<2\pi \tag{8}$$

determines an irreducible unitary representation of $R$. Here, however, one does not obtain arbitrary representations of $R$, but only those for which $F(2\pi)=f(0)=1$.

Thus, owing to the results of subsection 1, all irreducible unitary representations of $SO(2)$ must have the form $f(\phi)=e^{ai\phi}$. It remains for us to clarify the values of $a$ for which one would have

$$f(2\pi)=e^{2ai\pi}=1.$$

By virtue of Euler's formulas we have

$$e^{2ai\pi}=\cos 2a\pi+i\sin 2a\pi.$$

Therefore we must have

$$\cos 2a\pi=1,$$
$$\sin 2a\pi=0.$$

Hence $a$ is an integer.

We have thus proved that the irreducible unitary representations of $SO(2)$ have the form $e^{in\phi}$, where $n$ is an integer.

We note that $SO(2)$ is compact, and so, by virtue of §4.2 of Chapter I, any finite-dimensional representation of it is unitary. Therefore the assertion proved above can be formulated as follows.

*Any finite-dimensional irreducible representation of the group $SO(2)$ is unitary and one-dimensional, and has the form*

$$f(\varphi)=e^{in\varphi}, \tag{9}$$

*where $n$ is an integer.*

We have considered above the two-dimensional representation

$$g(\varphi)=\begin{pmatrix}\cos\varphi & -\sin\varphi\\ \sin\varphi & \cos\varphi\end{pmatrix}$$

of $SO(2)$. In order to decompose it into irreducible ones, we replace $g(\phi)$ by the equivalent representation $a^{-1}g(\phi)a$, where

$$a=\frac{1}{\sqrt{2}}\begin{pmatrix}1 & -i\\ -i & 1\end{pmatrix}.$$

A simple calculation shows that

$$a^{-1} g(\varphi) a = \begin{pmatrix} \cos\varphi + i\sin\varphi & 0 \\ 0 & \cos\varphi - i\sin\varphi \end{pmatrix}. \tag{10}$$

Applying the Euler formula, equality (10) can be rewritten as follows:

$$a^{-1} g(\varphi) a = \begin{pmatrix} e^{i\varphi} & 0 \\ 0 & e^{-i\varphi} \end{pmatrix}. \tag{11}$$

Hence $g(\phi)$ is the direct sum of the one-dimensional representations $e^{i\phi}$ and $e^{-i\phi}$.

3. The group of hyperbolic rotations of the plane and the hyperbolic functions. By a hyperbolic rotation of the plane one means a homogeneous linear transformation of the plane, which keeps the form $x^2 - y^2$ invariant and transforms the quadrants $-x \leq y \leq x$ and $-y \leq x \leq y$ into themselves. The aggregate of all hyperbolic rotations forms a group, which we shall denote by $SH(2)$.

A rotation of the coordinate axes by $-\pi/4$ transforms the form $x^2 - y^2$ into $-2xy$, and we get linear transformations preserving the form $xy$ and transforming the quadrants $x > 0$, $y > 0$ and $x > 0$, $y < 0$ into themselves. Clearly, such transformations have the form

$$\left.\begin{aligned} x' &= \frac{x}{a}, \\ y' &= ay, \end{aligned}\right\} \tag{1}$$

where $a > 0$.

Let $a = e^t$. As $a$ varies from 0 to $+\infty$, the parameter $t$ varies from $-\infty$ to $\infty$. The matrix of the transformation

$$\begin{aligned} x' &= e^{-t} x, \\ y' &= e^t y \end{aligned}$$

has the form

$$g(t) = \begin{pmatrix} e^{-t} & 0 \\ 0 & e^t \end{pmatrix}. \tag{2}$$

Since

$$g(t_1 + t_2) = g(t_1)\, g(t_2),$$

the group $SH(2)$ is isomorphic to the additive group $R$ of real numbers.

In the initial coordinate system, a hyperbolic rotation is given by a matrix of the form $h(t) = s^{-1} g(t) s$, where

$$s=\begin{pmatrix}\cos\frac{\pi}{4} & -\sin\frac{\pi}{4}\\ \sin\frac{\pi}{4} & \cos\frac{\pi}{4}\end{pmatrix}=\begin{pmatrix}\frac{\sqrt{2}}{2} & -\frac{\sqrt{2}}{2}\\ \frac{\sqrt{2}}{2} & \frac{\sqrt{2}}{2}\end{pmatrix}.$$

We have

$$h(t)=s^{-1}g(t)\,s=\begin{pmatrix}\frac{e^t+e^{-t}}{2} & \frac{e^t-e^{-t}}{2}\\ \frac{e^t-e^{-t}}{2} & \frac{e^t+e^{-t}}{2}\end{pmatrix}. \tag{3}$$

We introduce the notation

$$\frac{e^t+e^{-t}}{2}=\operatorname{ch} t,\qquad \frac{e^t-e^{-t}}{2}=\operatorname{sh} t \tag{4}$$

and call $\operatorname{ch} t$ the hyperbolic cosine of $t$, and $\operatorname{sh} t$ the hyperbolic sine of $t$. Formula (3) has the form

$$h(t)=\begin{pmatrix}\operatorname{ch} t & \operatorname{sh} t\\ \operatorname{sh} t & \operatorname{ch} t\end{pmatrix}. \tag{5}$$

It is evident from equality (3) that $h(t)$ is a two-dimensional representation of the group $SH(2)$, equivalent to $g(t)$. Therefore

$$h(t_1)\,h(t_2)=h(t_1+t_2). \tag{6}$$

In equality (6) we replace $h(t_1)$, $h(t_2)$ and $h(t_1+t_2)$ by their expressions (5), multiply out the matrices in the left-hand side, and compare the corresponding matrix elements on the left and on the right. We obtain the addition formulas for hyperbolic functions:

$$\operatorname{ch}(t_1+t_2)=\operatorname{ch} t_1\operatorname{ch} t_2+\operatorname{sh} t_1\operatorname{sh} t_2,$$
$$\operatorname{sh}(t_1+t_2)=\operatorname{sh} t_1\operatorname{ch} t_2+\operatorname{ch} t_1\operatorname{sh} t_2.$$

However, these formulas can also be obtained from the addition formulas for the trigonometrical functions if one observes that, by virtue of Euler's formulas, we have

$$\operatorname{ch} t=\cos it,$$
$$\operatorname{sh} t=\frac{1}{i}\sin it.$$

4. The complex form of the group $SO(2)$. The only distinction between the groups $SO(2)$ and $SH(2)$ is that the first of these consists of transformations preserving the form $x^2+y^2$, and the second of transformations preserving the form $x^2-y^2$. It is clear that on passing from a real linear space to a complex one this distinction

disappears. One says that the groups $SO(2)$ and $SH(2)$ are two real forms of one and the same complex group.

This complex group $SO(2, C)$ consists of matrices of the form

$$g(z) = \begin{pmatrix} \cos z & -\sin z \\ \sin z & \cos z \end{pmatrix}, \tag{1}$$

where $z$ runs through the complex plane. In other words, it is obtained from $SO(2)$ by replacing the real parameter $\phi$ by the complex number $z$.

The group $SO(2)$ is a subgroup of $SO(2, C)$, corresponding to real values of $z$. Therefore it is also called the real form of the group $SO(2, C)$, and $SO(2, C)$ is called the complex form of the group $SO(2)$ or, in other words, the complexification of $SO(2)$.

The group $SH(2)$ is another real form of $SO(2, C)$. It is obtained for purely imaginary values of $z$ (i.e. for real values of the parameter $w = iz$). Namely, if $z = ti$, then

$$g(ti) = \begin{pmatrix} \operatorname{ch} t & -i \operatorname{sh} t \\ i \operatorname{sh} t & \operatorname{ch} t \end{pmatrix}. \tag{2}$$

We replace each matrix (2) by a matrix $a^{-1}g(ti)a$, where $a = \begin{pmatrix} 1 & 0 \\ 0 & i \end{pmatrix}$. Then we get matrices of the form

$$\begin{pmatrix} \operatorname{ch} t & \operatorname{sh} t \\ \operatorname{sh} t & \operatorname{ch} t \end{pmatrix}, \tag{3}$$

i.e. the matrices of the group $SH(2)$.

Let us look into the connection between the representations of the groups $SO(2)$ and $SO(2, C)$. The irreducible unitary representations of $SO(2)$ have the form $e^{in\phi}$, as we have seen above. But $e^{in\phi}$ is an analytic function of $\phi$, and so we can also consider it for complex values of $\phi$. In this way we obtain representations $e^{inz}$ of the group $SO(2, C)$, which are "analytic continuations (extensions)" of representations $e^{in\phi}$ from the subgroup $SO(2)$ to the whole group $SO(2, C)$. Since these representations are one-dimensional, they are irreducible. But they are not unitary, because, in general, $|e^{inz}| \neq 1$.

## §2. FOURIER SERIES

1. Invariant integration on $SO(2)$. We shall study functions on the group $SO(2)$. Each element $g$ of this group is given by a real number $\phi$ (the angle of rotation), defined to within a multiple of $2\pi$. Therefore functions on $SO(2)$ can be identified with functions on the real axis having the period $2\pi$ (or, what is the same thing, with functions on the circle).

We define integration on $SO(2)$ by the formula

$$\int f(g)\,dg = \frac{1}{2\pi}\int_0^{2\pi} f(\varphi)\,d\varphi. \tag{1}$$

The factor $1/2\pi$ has been introduced here to ensure that the measure of the whole group $SO(2)$ is normalized, i.e. equals unity. The integral (1) possesses the following invariance property:

$$\int f(gg_0)\,dg = \int f(g)\,dg.$$

Indeed, the element $gg_0$ of the group $SO(2)$ is given by the number $\phi + \phi_0$. Therefore

$$\int f(gg_0)\,dg = \frac{1}{2\pi}\int_0^{2\pi} f(\varphi+\varphi_0)\,d\varphi = \frac{1}{2\pi}\int_{\varphi_0}^{\varphi_0+2\pi} f(\varphi)\,d\varphi.$$

Hence, by virtue of the periodicity of $f(\phi)$, we have

$$\int f(gg_0)\,dg = \frac{1}{2\pi}\int_0^{2\pi} f(\varphi)\,d\varphi = \int f(g)\,dg.$$

2. Trigonometrical system of functions. Fourier series. We have proved above that the functions

$$\{e^{in\varphi}\}, \tag{1}$$

where $0 \le \phi < 2\pi$, and $n$ runs through all the integers, form a complete system of pairwise nonequivalent unitary representations of $SO(2)$. Therefore, owing to the results of §4.3 of Chapter I these functions form a complete orthononormal system of functions on $SO(2)$ (relative to the invariant normalized measure $(1/2\pi)d\phi$). In other words,

$$\frac{1}{2\pi}\int_0^{2\pi} e^{im\varphi}\overline{e^{in\varphi}}\,d\varphi = \delta_{mn}, \tag{2}$$

where $\delta_{mn}$ is the Kronecker symbol.

Besides, any function $f(\phi)$ with integrable square on $SO(2)$ can be expanded in a mean-convergent series

$$f(\varphi)=\sum_{n=-\infty}^{\infty} c_n e^{in\varphi}, \tag{3}$$

whose coefficients are expressed through $f(\phi)$ by the formulas

$$c_n=\frac{1}{2\pi}\int_0^{2\pi} f(\varphi)e^{-in\varphi}\,d\varphi \tag{4}$$

(see formulas (4) and (5) of Chapter I, §3.4). The series of (3) is called the Fourier series of $f(\phi)$, and the numbers $c_n$ are called the Fourier coefficients of this function.

We denote by $\mathfrak{H}$ the space of all functions $f(\phi)$ on $SO(2)$ having integrable squares.

It follows from the completeness of the system $\{e^{in\phi}\}$ that, for every function $f(\phi)$ of $\mathfrak{H}$, one has the Parseval equality

$$\frac{1}{2\pi}\int_0^{2\pi}|f(\varphi)|^2\,d\varphi=\sum_{n=-\infty}^{\infty}|c_n|^2. \tag{5}$$

Thus the space $\mathfrak{H}$ is the direct sum of the one-dimensional spaces $\mathfrak{H}_n$, consisting of functions of the form $\{a_n e^{in\phi}\}$.

3. Decomposition of the regular representations of $SO(2)$. Let us now consider the regular representation of the group $SO(2)$. According to §2.4 of Chapter I, this representation is constructed in the space $\mathfrak{H}$ of functions $f(\phi)$, $0\leq\phi\leq 2\pi$, with square-integrable moduli, and is given by the formula

$$R(g(\alpha))f(\varphi)=f(\varphi+\alpha). \tag{1}$$

As we have shown above, $\mathfrak{H}$ is the direct sum of one-dimensional spaces $\mathfrak{H}_n$, consisting of functions of the form $\{a_n e^{in\phi}\}$. It is easy to see that these spaces are invariant for operators $R(g)$. In fact,

$$R(g(\alpha))e^{in\varphi}=e^{in(\varphi+\alpha)}=e^{in\alpha}e^{in\varphi}. \tag{2}$$

Moreover, it follows from (2) that $R(g)$ induces, in each of the spaces $\mathfrak{H}_n$, the operator $e^{in\alpha}$ of a unitary irreducible representation of $SO(2)$. By the same token we have proved

**Theorem 1.** *The regular representation $R(g)$ of the group $SO(2)$ is the direct sum of irreducible unitary representations $T_n(g) = e^{in\alpha}$ of this group, where each of these representations appears once in the decomposition of the regular representations:*

$$R(g) = \sum_{n=-\infty}^{\infty} T_n(g). \tag{3}$$

4. **Decomposition of infinitely differentiable functions.** In several cases it is not sufficient to know that the Fourier series of the function $f(\phi)$ converges to it in the mean, and one requires a stronger assertion (for example, the absolute or uniform convergence of this series). Such assertions are valid, for instance, if $f(\phi)$ is infinitely differentiable. We call a function $f(\phi)$ on $SO(2)$ infinitely differentiable if the corresponding periodic function on the real axis is infinitely differentiable. In particular, for such functions, for any $k$ one must have

$$f^{(k)}(0) = f^{(k)}(2\pi). \tag{1}$$

**Lemma 1.** *If the function $f(\phi)$ on the group $SO(2)$ is infinitely differentiable then its Fourier coefficients decrease rapidly, i.e. for all $k$ one has the relation*

$$\lim_{|n|\to\infty} n^k c_n = 0. \tag{2}$$

Indeed, integrating $k+1$ times by parts the integral in the formula

$$c_n = \frac{1}{2\pi}\int_0^{2\pi} f(\varphi)\, e^{-in\varphi} d\varphi,$$

we get

$$c_n = \frac{i^{k+1}}{2\pi n^{k+1}}\int_0^{2\pi} f^{(k+1)}(\varphi)\, e^{-in\varphi} d\varphi$$

(the integrated terms vanish because of the fact that $f^{(j)}(0) = f^{(j)}(2\pi)$). The infinite differentiability of $f(\phi)$ implies the existence of the integral on the right hand side of this equality. Hence

$$\lim_{|n|\to\infty} n^k c_n = 0.$$

This lemma leads to

**Theorem 1.** *If the function $f(\phi)$ on the group $SO(2)$ is infinitely differentiable, then the Fourier series of this function converges absolutely and uniformly to $f(\phi)$.*

For the proof it is sufficient to observe that for any $k$ we have $c_n = o(1/n^k)$.

Therefore the Fourier series of $f(\phi)$ is uniformly and absolutely convergent. Since it converges in the mean to $f(\phi)$, its sum is equal to $f(\phi)$.

There exist many highly subtle theorems on the convergence of Fourier series. However, we only need the theorem on convergence in the mean and the theorem on convergence of the expansion of an infinitely differentiable function.

Clearly, if $f(\phi)$ is infinitely differentiable, then the Fourier series of all its derivatives are also absolutely and uniformly convergent. Moreover, the Fourier coefficients of $f^{(k)}(\phi)$ are equal to $(in)^k c_n$, where $c_n$ are the Fourier coefficients of $f(\phi)$.

We have proved that the Fourier coefficients of infinitely differentiable functions decrease rapidly. The converse is also true.

If the Fourier coefficients $c_n$ of the function $f(\phi)$ decrease rapidly, then $f(\phi)$ is infinitely differentiable.

Indeed, the rapid decrease of the Fourier coefficients implies that the series

$$\sum_{n=-\infty}^{\infty} (in)^k c_n e^{in\varphi}$$

is absolutely and uniformly convergent for all $k$. Therefore the Fourier expansion $f(\phi) = \sum_{n=-\infty}^{\infty} c_n e^{in\phi}$ can be differentiated term-by-term any number of times.

## §3. THE FOURIER INTEGRAL

1. The regular representation of the group $R$. We return to the study of representations of the group $R$, the additive group of real numbers. In §1 we considered one-dimensional representations of this group. These representations were given by a complex parameter $a$ and had the form $T_a(g) = T_a(x) = e^{ax}$. For purely imaginary $a$, $a = i\lambda$, the representation $T_a(x)$ is unitary. Here we shall consider the regular representation of $R$, and give the decomposition of this representation into irreducible unitary representations $T_{i\lambda}(x)$.

We may recall that, according to §2.4 of Chapter I, the regular representation of $R$ is constructed in the space $\mathfrak{H}$ of functions $f(x)$ given on $R$ (i.e. on the real axis $-\infty < x < \infty$), and such that

$$\int_{-\infty}^{\infty} |f(x)|^2 dx < +\infty. \tag{1}$$

To each element of the group $R$, i.e. to each real number $x_0$, corresponds the operator

$R(x_0)$, transforming $f(x)$ into

$$R(x_0)f(x)=f(x+x_0). \tag{2}$$

As stated in Chapter I, $R(x)$ is the regular representation of $R$.

Because of the fact that $R$ is noncompact, the decomposition of the representation $R(x)$ into irreducible ones is more complicated than in the case of the group $SO(2)$. Namely, we do not get a decomposition of $\mathfrak{H}$ as the orthogonal direct sum of invariant subspaces, but only a decomposition of $\mathfrak{H}$ as a continuous direct sum. In other words, we shall show that every function $f(x)$ of the space $\mathfrak{H}$ can be represented in the form of a "continuous linear combination" of the functions $e^{i\lambda x}$:

$$f(x)=\int_{-\infty}^{\infty} F(\lambda)e^{i\lambda x}\,d\lambda, \quad -\infty<x<\infty. \tag{3}$$

Comparing this decomposition with the Fourier series expansion

$$f(x)=\sum_{n=-\infty}^{\infty} c_n e^{inx}, \qquad 0\leqslant x<2\pi \tag{4}$$

we see that $F(\lambda)$ is analogous to the Fourier coefficient $c_n$.

It is evident from (3) that the decomposition of the function $R(x_0)f(x)=f(x+x_0)$ has the form

$$R(x_0)f(x)=\int e^{i\lambda(x+x_0)}\,d\lambda=\int_{-\infty}^{\infty} e^{i\lambda x_0}F(\lambda)\,e^{i\lambda x}d\lambda. \tag{5}$$

Thus, under the shift $f(x)\to f(x+x_0)$ every "Fourier coefficient" $F(\lambda)$ is multiplied by $e^{i\lambda x_0}$. In other words, for fixed $\lambda$, to the operator $R(x_0)$ corresponds the operator of multiplication by $e^{i\lambda x_0}$, i.e. the operator-irreducible unitary representation of $R$. The decomposition (3) thus gives the decomposition of the regular representation as the continuous direct sum of irreducible unitary representations

$$R(x)=\int_{-\infty}^{\infty} T_{i\lambda}(x)\,d\lambda. \tag{6}$$

2. The Fourier transformation and its properties. In this and the following subsections we shall prove the feasibility of expanding functions $f(x)$ of the space $\mathfrak{H}$ in the form given by formula (3) above. Let $F(\lambda)$ be a continuous absolutely integrable function. By its Fourier transform we mean the function

$$f(x)=\int_{-\infty}^{\infty} F(\lambda)\,e^{i\lambda x}d\lambda. \tag{1}$$

The integral (1) is absolutely convergent because of the absolute integrability of $F(\lambda)$.

Generally speaking, the Fourier transform of a continuous absolutely integrable function may not be an integrable function. In order to obtain a symmetric theory, it is necessary to restrict the class of functions to be studied.

We shall say that the function $F(\lambda)$ decreases rapidly for $|\lambda| \to \infty$ if, for arbitrary $n$, we have

$$\lim_{|\lambda|\to\infty} |\lambda|^n F(\lambda) = 0. \tag{2}$$

We denote by $\mathfrak{S}$ the space of all infinitely differentiable functions on the line whose derivatives of all orders decrease rapidly for $|\lambda| \to \infty$. As an example of a function of the space $\mathfrak{S}$ we have $e^{-\lambda^2}$.

We shall now prove that the Fourier transform of a function of $\mathfrak{S}$ is a function of the same space. For this we prove the following two lemmas.

**Lemma 1.** *If the function $F(\lambda)$ decreases rapidly, then its Fourier transform is infinitely differentiable.*

Indeed, let

$$f(x) = \int_{-\infty}^{\infty} F(\lambda) e^{i\lambda x} d\lambda. \tag{3}$$

Since $F(\lambda)$ decreases rapidly, for arbitrary $k$ the integral

$$\int_{-\infty}^{\infty} \lambda^k F(\lambda) d\lambda$$

is absolutely convergent. Therefore, in the right hand side of equality (3) one can differentiate with respect to $x$ under the integral sign:

$$f^{(k)}(x) = \int_{-\infty}^{\infty} (i\lambda)^k F(\lambda) e^{i\lambda x} d\lambda. \tag{4}$$

This proves the infinite differentiability of $f(x)$.

**Lemma 2.** *If the function $F(\lambda)$ is infinitely differentiable, and all its derivatives are absolutely integrable and tend to zero for $|\lambda| \to \infty$, then the Fourier transform $f(x)$ of $F(\lambda)$ decreases rapidly for $|x| \to \infty$.*

In fact, integrating (3) by parts $(n+1)$ times, we get

$$f(x) = \frac{i^{n+1}}{x^{n+1}} \int_{-\infty}^{\infty} F^{(n+1)}(\lambda) e^{i\lambda x} d\lambda \tag{5}$$

(the integrated terms vanish because $\lim_{|\lambda|\to\infty} F^{(k)}(\lambda) = 0$). According to the premise of the theorem, $\int_{-\infty}^{\infty} F^{(n+1)}(\lambda) e^{i\lambda x}\, d\lambda$ is absolutely convergent; consequently

$$\lim_{|x|\to\infty} |x|^n f(x) = 0.$$

Hence $f(x)$ decreases rapidly for $|x| \to \infty$.

Lemmas 1 and 2 immediately lead to

**Theorem 1.** *The Fourier transform $f(x)$ of the function $F(\lambda)$ of the space $\mathfrak{S}$ is a function of the same space.*

3. **Inversion formula.** Our aim now is the derivation of a formula which allows the regeneration of the function $F(\lambda)$ by its Fourier transform $f(x)$. We reduce this problem to the analogous problem for Fourier series.

We shall call a function $\phi(\lambda)$, given on the real axis, conditionally periodic (with parameter $t$) if

$$\varphi(\lambda + 1) = e^{-it} \varphi(\lambda). \tag{1}$$

For $t = 0$, conditionally periodic functions become ordinary periodic functions ($\phi(\lambda + 1) = \phi(\lambda)$); for $t = \pi$, these functions may be called antiperiodic ($\phi(\lambda + 1) = -\phi(\lambda)$).

Suppose that $F(\lambda)$ is a function of the space $\mathfrak{S}$ (i.e. suppose that function $F(\lambda)$ decreases rapidly along with all its derivatives). Let

$$F_t(\lambda) = \sum_{n=-\infty}^{\infty} F(\lambda + n) e^{int}. \tag{2}$$

The rapid decrease of $F(\lambda)$ implies that the series (2) is absolutely and uniformly convergent on the segment $[0, 1]$. Moreover,

$$\begin{aligned} F_t(\lambda + 1) &= \sum_{n=-\infty}^{\infty} F(\lambda + n + 1) e^{int} \\ &= e^{-it} \sum_{n=-\infty}^{\infty} F(\lambda + n) e^{int} = e^{-it} F_t(\lambda). \end{aligned} \tag{3}$$

Therefore $F_t(\lambda)$ is conditionally periodic (with the parameter $t$). Further, it is evident that the series obtained from series (2) by termwise differentiation with respect to $t$ and $\lambda$, also converges uniformly and absolutely:

$$F_t^{(k)}(\lambda) = \sum_{n=-\infty}^{\infty} F^{(k)}(\lambda + n) e^{int}, \tag{4}$$

$$\frac{\partial^k F_t(\lambda)}{\partial t^k} = \sum_{n=-\infty}^{\infty} (in)^k F(\lambda + n) e^{int}. \tag{5}$$

The principal role in the derivation of the inversion formula is played by the fact that every function $F(\lambda)$ of the space $\mathfrak{S}$ can be decomposed in conditionally periodic functions, i.e. can be represented in the form

$$F(\lambda) = \frac{1}{2\pi} \int_0^{2\pi} F_t(\lambda)\, dt, \tag{6}$$

where $F_t(\lambda)$ are functions infinitely differentiable with respect to $\lambda$ and $t$ and conditionally periodic with period 1:

$$F_t(\lambda + 1) = e^{-it} F_t(\lambda). \tag{7}$$

In fact, let us define functions $F_t(\lambda)$ in relation (6) by formula (2). Then

$$\frac{1}{2\pi} \int_0^{2\pi} F_t(\lambda)\, dt = \frac{1}{2\pi} \sum_{n=-\infty}^{\infty} \left[ F(\lambda + n) \int_0^{2\pi} e^{int} dt \right] = F(\lambda).$$

Thus formula (2) can be regarded as the "inversion formula" for the decomposition, given by formula (6), of the function $F(\lambda)$ of the space $\mathfrak{S}$ in conditionally periodic functions.

We now use the fact that the functions $F_t(\lambda) e^{it\lambda}$ are infinitely differentiable with respect to $\lambda$ and are periodic functions (with period 1). Owing to Theorem 1 of §2.4, they can be expanded in a Fourier series

$$e^{it\lambda} F_t(\lambda) = \sum_{n=-\infty}^{\infty} a_n(t) e^{-2\pi in\lambda}, \tag{8}$$

and, moreover, this series is uniformly and absolutely convergent to $e^{it\lambda} F_t(\lambda)$. Hence

$$F_t(\lambda) = \sum_{n=-\infty}^{\infty} a_n(t) e^{-i\lambda(t+2\pi n)}. \tag{9}$$

The coefficients $a_n(t)$ of series (9) are calculated by Fourier's formula:

$$a_n(t) = \int_0^1 F_t(\lambda) e^{i\lambda(t+2\pi n)} d\lambda. \tag{10}$$

We show that $a_n(t) = f(t + 2\pi n)$, where $f(x)$ is the Fourier transform of $F(\lambda)$. For this we substitute, in equality (10), in place of $F_t(\lambda)$ the series (2) and change the order of integration and summation (which is possible because of the rapid decrease of $F(\lambda)$). We get

$$a_n(t) = \sum_{m=-\infty}^{\infty} \int_0^1 F(\lambda + m) e^{i(\lambda+m)(t+2\pi n)} d\lambda \tag{11}$$

(for the sake of symmetry, the factor $e^{2\pi i n m}$, which is equal to 1, has been inserted under the integral sign). If we put $\lambda + m = \mu$ in equality (11), it takes the form

$$a_n(t) = \sum_{m=-\infty}^{\infty} \int_m^{m+1} F(\mu) e^{i\mu(t+2\pi n)} d\mu$$

$$= \int_{-\infty}^{\infty} F(\mu) e^{i\mu(t+2\pi n)} d\mu = f(t + 2\pi n). \tag{12}$$

We can now prove the inversion formula for the Fourier transform, i.e. obtain an expression of the function $F(\lambda)$ in terms of its Fourier transform $f(x)$. Let us substitute expression (9) for the conditionally periodic functions in integral (6). Taking formula (12) into account, we obtain

$$F(\lambda) = \frac{1}{2\pi} \int_0^{2\pi} \left[ \sum_{n=-\infty}^{\infty} a_n(t) e^{-i\lambda(t+2\pi n)} \right] dt$$

$$= \frac{1}{2\pi} \int_0^{2\pi} \left[ \sum_{n=-\infty}^{\infty} f(t + 2\pi n) e^{-i\lambda(t+2\pi n)} \right] dt$$

$$= \frac{1}{2\pi} \sum_{n=-\infty}^{\infty} \int_0^{2\pi} f(t + 2\pi n) e^{-i\lambda(t+2\pi n)} dt$$

$$= \frac{1}{2\pi} \sum_{n=-\infty}^{\infty} \int_{2\pi n}^{2\pi(n+1)} f(x) e^{-i\lambda x} dx = \frac{1}{2\pi} \int_{-\infty}^{\infty} f(x) e^{-i\lambda x} dx$$

(one can change the order of integration and summation because of the absolute convergence of $\int_{-\infty}^{\infty} f(x) dx$). This completes the derivation of the inversion formula for the Fourier transform.

We may note that the inversion formula

$$F(\lambda) = \frac{1}{2\pi} \int_{-\infty}^{\infty} f(x) e^{-i\lambda x} dx \tag{13}$$

is not essentially different from the formula of the Fourier transform itself:

$$f(x)=\int_{-\infty}^{\infty} F(\lambda)\,e^{i\lambda x}d\lambda. \tag{14}$$

It follows at once that the image of the space $\mathfrak{S}$ under the mapping $F(\lambda)\to f(x)$ is the entire space $\mathfrak{S}$. Namely, if $f(x)$ is a function of $\mathfrak{S}$, it is the Fourier transform of the function $2\pi\phi(-\lambda)$, where $\phi(\lambda)$ is the Fourier transform of the function $f(x)$.

4. The Plancherel formula. We prove that the Fourier transform is (apart from a constant factor $2\pi$) an isometric mapping of the space $\mathfrak{S}$ onto itself relative to the norm

$$\|F\|^2=\int_{-\infty}^{\infty} |F(\lambda)|^2\,d\lambda.$$

In other words, we show that if $f(x)$ is the Fourier transform of function $F(\lambda)$, then

$$\int_{-\infty}^{\infty} |F(\lambda)|^2 d\lambda=\frac{1}{2\pi}\int_{-\infty}^{\infty} |f(x)|^2 dx. \tag{1}$$

For this we note that, because of the Parseval equality, one has the relation

$$\int_0^1 |F_t(\lambda)|^2 d\lambda=\sum_{n=-\infty}^{\infty} |a_n(t)|^2. \tag{2}$$

Let us integrate this equality with respect to $t$ from $0$ to $2\pi$ (which is possible since a series consisting of positive functions, which converges to a continuous function, does so uniformly). We obtain

$$\begin{aligned}\int_0^{2\pi}\int_0^1 |F_t(t)|^2 d\lambda\,dt &= \sum_{n=-\infty}^{\infty}\int_0^{2\pi} |a_n(t)|^2 dt=\sum_{n=-\infty}^{\infty}\int_0^{2\pi} |f(t+2\pi n)|^2\,dt\\ &=\sum_{n=-\infty}^{\infty}\int_{2\pi n}^{2\pi(n+1)} |f(x)|^2 dx=\int_{-\infty}^{\infty} |f(x)|^2\,dx. \quad (3)\end{aligned}$$

It remains for us to show that the integral in the left-hand side of this equality coincides (apart from a factor $2\pi$) with $\int_{-\infty}^{\infty}|F(\lambda)|^2 d\lambda$, i.e. for the decomposition (6) of subsection 3 one has the "Plancherel formula":

$$\int_{-\infty}^{\infty} |F(\lambda)|^2\,d\lambda=\frac{1}{2\pi}\int_0^{2\pi}\int_0^1 |F_t(\lambda)|^2 dt\,d\lambda. \tag{4}$$

In fact, by formula (2) of subsection 3, the functions $F(\lambda+n)$ are the Fourier coefficients for $F_t(\lambda)$ (as functions of $t$), and consequently, by virtue of the Parseval equality,

$$\frac{1}{2\pi}\int_0^{2\pi} |F_t(\lambda)|^2 dt = \sum_{n=-\infty}^{\infty} |F(\lambda+n)|^2.$$

We integrate this equality with respect to $\lambda$ from 0 to 1. We get

$$\frac{1}{2\pi}\int_0^1\int_0^{2\pi} |F_t(\lambda)|^2 dt\, d\lambda = \sum_{n=-\infty}^{\infty}\int_0^1 |F(\lambda+n)|^2 d\lambda$$

$$= \sum_{n=-\infty}^{\infty}\int_n^{n+1} |F(\lambda)|^2 d\lambda = \int_{-\infty}^{\infty} |F(\lambda)|^2 d\lambda.$$

Equality (4) is thus proved.

It follows from equalities (3) and (4) that

$$\int_{-\infty}^{\infty} |F(x)|^2 dx = \frac{1}{2\pi}\int_{-\infty}^{\infty} |f(y)|^2 dy, \tag{5}$$

i.e. our assertion is proved.

5. The transformation of square-integrable functions. We can now determine the Fourier transformation for functions $F(\xi)$ with square-integrable moduli. For this, let us first observe that functions of the space $\mathfrak{S}$ are everywhere dense in the space $\mathfrak{H}$ of square-integrable functions. In fact, let $\phi(\lambda)$ be an infinitely differentiable nonnegative finite function such that $\phi(0) = 1$ and $\int_{-\infty}^{\infty} \phi(\lambda)d\lambda = 1$. We put

$$F_n(\lambda) = n\varphi\left(\frac{\lambda}{n}\right)\int_{-\infty}^{\infty} F(\lambda-\tau)\,\varphi(n\tau)\,d\tau. \tag{1}$$

It is easy to show that all the functions $F_n(\lambda)$ are infinitely differentiable and finite, and that the sequence of functions $F_n(\lambda)$ converges in the mean to the function $F(\lambda)$.

We have thus proved that the functions of the space $\mathfrak{S}$ are everywhere dense in $\mathfrak{H}$. But the Fourier transformation is an isometric (apart from a factor $1/2\pi$) mapping of space $\mathfrak{S}$ onto itself. Hence this transformation can be extended by continuity to the whole space $\mathfrak{H}$.

In this way, to each function $F(\lambda)$ of the space $\mathfrak{H}$ corresponds a function

$f(x)$ of the same space—the Fourier transform of $F(\lambda)$. This function can be determined in the following manner: let $F_n(\lambda)$ be a sequence of functions of $\mathfrak{S}$, which is convergent in the mean to $F(\lambda)$,

$$F(\lambda) = \underset{n \to \infty}{\text{l. i. m.}}\ F_n(\lambda).$$

Let us denote by $f_n(x)$ the Fourier transform of $F_n(\lambda)$. Then the sequence of functions $f_n(x)$ converges in the mean. Its limit is the function $f(x)$.

It is easy to prove that

$$f(x) = \underset{n \to \infty}{\text{l. i. m.}} \int_{n}^{n} F(\lambda)\, e^{i\lambda x} d\lambda \tag{2}$$

and

$$F(\lambda) = \underset{n \to \infty}{\text{l. i. m.}} \int_{-n}^{n} f(x)\, e^{-i\lambda x} dx. \tag{3}$$

In addition, for any function $F(\lambda)$ of $\mathfrak{H}$, one has the Plancherel equality

$$\int_{-\infty}^{\infty} |F(\lambda)|^2 d\lambda = \frac{1}{2\pi} \int_{-\infty}^{\infty} |f(x)|^2 dx, \tag{4}$$

where both integrals are understood in the sense of convergence in the mean.

6. The Fourier integral for functions of several variables. Let us consider the $n$-dimensional linear space $R^n$. This space is the direct sum of $n$ copies of the group $R$. Its irreducible representations are the Kronecker products of the irreducible representations $e^{\lambda x}$ of $R$. In other words, any irreducible representation of $R^n$ is given by $n$ complex numbers $\lambda = (\lambda_1, \cdots, \lambda_n)$, and has the form

$$T_\lambda(\mathbf{x}) = e^{\lambda_1 x_1} \ldots e^{\lambda_n x_n} = e^{(\lambda, \mathbf{x})}, \tag{1}$$

where

$$(\lambda, \mathbf{x}) = \lambda_1 x_1 + \ldots + \lambda_n x_n. \tag{2}$$

If all the numbers $\lambda_k$ are purely imaginary, the representation $T_\lambda$ is unitary.

The regular representation of $R^n$ is decomposed into irreducible ones in exactly the same way as in the one-dimensional case. This decomposition is connected with the Fourier integral for functions of $n$ variables.

If $F(\lambda)$ is a function of the space $\mathfrak{S}$ (i.e. an infinitely differentiable function on $R^n$, all of whose derivatives decrease rapidly), then the integral

$$f(\mathbf{x}) = \int F(\lambda)\, e^{i(\lambda, \mathbf{x})} d\lambda \tag{3}$$

($\lambda_1, \cdots, \lambda_n$ being real) converges for all x. The function $f(\mathbf{x})$ is called the Fourier transform of $F(\lambda)$ and also belongs to $\mathfrak{S}$.

The inversion formula for (3) has the form

$$F(\lambda)=\frac{1}{(2\pi)^n}\int f(\mathbf{x})\, e^{-i(\lambda,\, \mathbf{x})}\, d\mathbf{x}. \tag{4}$$

Moreover, one has the Plancherel equality:

$$\int |F(\lambda)|^2\, d\lambda=\frac{1}{(2\pi)^n}\int |f(\mathbf{x})|^2\, d\mathbf{x}. \tag{5}$$

With the help of this equality it is easy to extend the Fourier transformation to all functions $F(\lambda)$ having square-integrable moduli.

## §4. FOURIER TRANSFORMATION IN THE COMPLEX DOMAIN

1. **Definition.** If the function $F(\lambda)$ belongs to the space $\mathfrak{S}$, its Fourier transform

$$f(z)=\int_{-\infty}^{\infty} F(\lambda)\, e^{i\lambda z} d\lambda \tag{1}$$

is defined, in general, only for real values of $z$. But if $F(\lambda)$ decreases exponentially at infinity, then (1) has a meaning for certain complex values of $z$ too.

In fact, suppose that for $\lambda \to +\infty$ we have $F(\lambda) = O(e^{-a\lambda})$, where $a$ is some fixed number. Then the integral (1) converges in the strip $-a < \operatorname{Im} z \leq 0$. Namely, if $z = x + iy$, where $-a < y \leq 0$, then there is an $\epsilon > 0$ such that $|e^{i\lambda z}| < e^{(a-\epsilon)\lambda}$, and so for $-a < y \leq 0$, $\lambda > 0$, we have

$$|F(\lambda)\, e^{i\lambda z}| < Ce^{-\epsilon\lambda}$$

Consequently, (1) converges on the positive semiaxis. Convergence on the negative semiaxis follows from the fact that, for $F(\lambda) \in \mathfrak{S}$, $\lambda < 0$, $\operatorname{Im} z \leq 0$, the function $F(\lambda)e^{i\lambda z}$ decreases rapidly.

In exactly the same way, if for $\lambda \to -\infty$ we have $F(\lambda) = O(e^{b\lambda})$, then (1) converges for values of $z$ belonging to the strip $0 \leq \operatorname{Im} z < b$. Thus we have proved that if $F(\lambda) \in \mathfrak{S}$ and

$$F(\lambda)=\begin{cases} O(e^{-a\lambda}), & \lambda \to +\infty, \quad a>0, \\ O(e^{b\lambda}), & \lambda \to -\infty, \quad b>0, \end{cases} \tag{2}$$

then the integral (1) is defined in the strip $-a < \operatorname{Im} z < b$. In this case one says that the Fourier transform $f(z)$ of the function $F(\lambda)$ is defined in the strip mentioned. Clearly, in this strip it is an analytic function of $z$.

If the infinitely differentiable function $F(\lambda)$ is finite, i.e. equal to zero for

$|\lambda| > a$, the integral (1) converges in the whole complex plane, and so $f(z)$ is an entire (integral) analytic function. For any $b > a$ one has, moreover, the inequality

$$|f(x+iy)| < Ce^{b|y|}. \tag{3}$$

Indeed, since for $|\lambda| > a$ we have $F(\lambda) = 0$,

$$|f(x+iy)| \leqslant \int_{-a}^{a} |F(\lambda)| e^{-\lambda y} d\lambda = e^{b|y|} \int_{-a}^{a} |F(\lambda)| e^{-b|y|-\lambda y} d\lambda.$$

Since $b > a \geq |\lambda|$, we have $e^{-b|y|-\lambda y} \leq 1$, and therefore

$$|f(x+iy)| < Ce^{b|y|}.$$

Entire analytic functions satisfying an inequality of the form (3) for some $b > 0$ are called entire analytic functions of exponential type.

*Thus the Fourier transform of an infinitely differentiable finite function is an entire analytic function of exponential type. Moreover, since $F(\lambda) \in \mathfrak{S}$, the function $f(z)$, considered on the real axis, belongs to the space* $\mathfrak{S}$ (see §3.3).

One can prove the converse assertion.

*If $f(z)$ is an entire analytic function of exponential type, and $f(x)$, $-\infty < x < \infty$, belongs to the space $\mathfrak{S}$, then $f(z)$ is the Fourier transform of an infinitely differentiable finite function* (see [18]).

We now show that if $F(\lambda)$ satisfies condition (2), then the contour of integration in the Fourier inversion formula

$$F(\lambda) = \frac{1}{2\pi} \int_{-\infty}^{\infty} f(z) e^{-i\lambda z} dz$$

can be replaced by any contour parallel to it and lying in the strip $-a < \operatorname{Im} z < b$, i.e. one has the formula

$$F(\lambda) = \frac{1}{2\pi} \int_{-\infty+ci}^{\infty+ci} f(z) e^{i\lambda z} dz, \tag{4}$$

where $-a < c < b$.

In fact, it follows from condition (2) that, for $-a < c < b$, the function $e^{-\lambda c} F(\lambda)$ belongs to $\mathfrak{S}$. The Fourier transform of this function has the form

$$f_c(z) = \int_{-\infty}^{\infty} F(\lambda) e^{-\lambda c + i\lambda z} d\lambda = f(z+ic).$$

By virtue of the Fourier inversion formula we have

$$e^{-\lambda c} F(\lambda) = \frac{1}{2\pi} \int_{-\infty}^{\infty} f(z + ic) e^{-i\lambda z} dz.$$

On making the substitution $w = z + ic$ in this integral, we get the required equality (4).

For finite functions $\Phi(\lambda)$ we shall often write the Fourier transformation in the form

$$\varphi(z) = \int_{-\infty}^{+\infty} \Phi(\lambda) e^{\lambda z} d\lambda. \tag{5}$$

In this case the inversion formula is written as follows:

$$\Phi(\lambda) = \frac{1}{2\pi i} \int_{a-i\infty}^{a+i\infty} \varphi(z) e^{-\lambda z} dz. \tag{6}$$

2. Transformation of square-integrable functions. As in the case of real variables, the Fourier transformation in the complex domain can be extended to the class of square-integrable functions. Namely, the following assertion is valid.

**Theorem 1.** *Let $F(\lambda)$ be a measurable function, square-integrable on every finite segment, and such that*

$$F(\lambda) = \begin{cases} O(e^{-a\lambda}), & \lambda \to +\infty, \\ O(e^{b\lambda}), & \lambda \to -\infty. \end{cases} \tag{1}$$

*Then the integral*

$$f(z) = \int_{-\infty}^{\infty} F(\lambda) e^{i\lambda z} d\lambda \tag{2}$$

*is absolutely and uniformly convergent in every strip $-a + \epsilon < y < b - \epsilon$, $z = x + iy$, $\epsilon > 0$. The function $f(z)$ analytically depends on $z$ in the strip $-a < y < b$, where in every strip $-a + \epsilon < y < b - \epsilon$ the integral*

$$\int_{-\infty}^{\infty} |f(x + iy)|^2 dx \tag{3}$$

*is bounded.*

The absolute and uniform convergence of (2), and consequently the analyticity of $f(z)$, follows immediately from the estimates (1). In order to prove the boundedness of the integral (3) in the strip $-a + \epsilon < y < b - \epsilon$, we apply the Plancherel formula to the function $F(\lambda)e^{-\lambda y}$. The Fourier transform $f(x, y)$ of this function has the form

$$f(x, y)=\int_{-\infty}^{\infty} F(\lambda) e^{-\lambda y} e^{i\lambda x} d\lambda = \int_{-\infty}^{\infty} F(\lambda) e^{i\lambda(x+iy)} d\lambda = f(x+iy).$$

Thus, for $-a+\epsilon<y<b-\epsilon$, because of estimates (1), we have

$$\int_{-\infty}^{\infty} |f(x+iy)|^2 dx \equiv \int_{-\infty}^{\infty} |f(x, y)|^2 dx$$

$$=2\pi \int_{-\infty}^{\infty} |F(\lambda) e^{-\lambda y}|^2 d\lambda \leqslant C_1 \int_0^{\infty} e^{-2\epsilon\lambda} d\lambda + C_2 \int_{-\infty}^{0} e^{2\epsilon\lambda} d\lambda < C.$$

The inversion theorem in this case is formulated as follows.

**Theorem 2.** *Let $f(z)$ be analytic in the strip $-a \le y \le b$ and on its boundary, where $a>0$, $b>0$, and let*

$$\int_{-\infty}^{\infty} |f(x+iy)|^2 dx < C, \qquad -a \leqslant y \leqslant b. \tag{4}$$

*Then $f(z)$ is bounded in every internal strip $-a+\epsilon \le y \le b-\epsilon$, and there exists a function $F(\lambda)$ such that*

$$\int_{-\infty}^{\infty} |F(\lambda) e^{a\lambda}|^2 d\lambda < +\infty, \quad \int_{-\infty}^{\infty} |F(\lambda) e^{-b\lambda}|^2 d\lambda < +\infty.$$

*Moreover, on the segment $-a \le y \le b$ we have*

$$f(x+iy)=\int_{-\infty}^{\infty} F(\lambda) e^{i\lambda(x+iy)} d\lambda, \tag{5}$$

*where the integral is understood in the sense of convergence in the mean. The function $F(\lambda)$ is expressed through $f(z)$ by the formula*

$$F(\lambda)=\frac{1}{2\pi} \int_{-\infty}^{\infty} f(x) e^{-i\lambda x} dx, \tag{6}$$

*where the integral is again understood in the sense of convergence in the mean.*

The proof of this theorem is only slightly more complicated than the proof of Theorem 1.

We shall now investigate the Fourier transforms of functions, given on the half line $0 \le \lambda < \infty$. It follows from the equality

$$\int_{-\infty}^{\infty} |f(x+iy)|^2 dx = 2\pi \int_{-\infty}^{\infty} |F(\lambda) e^{-\lambda y}|^2 d\lambda \tag{7}$$

that the function $f(x+iy)$ is analytic in the upper half plane and has a square-integrable modulus for all $y \ge 0$ if and only if, for all $y \ge 0$, the following integral converges:

$$\int_{-\infty}^{\infty} |F(\lambda)e^{-\lambda y}|^2 d\lambda.$$

Moreover, $\int_{-\infty}^{\infty} |f(x+iy)|^2 dx$ is bounded with respect to $y$ if and only if the integral

$$\int_{-\infty}^{\infty} |F(\lambda)e^{-\lambda y}|^2 d\lambda \tag{8}$$

is bounded on the half line $0 \le y < +\infty$.

But the integral (7) is bounded on the half line $0 \le y < +\infty$ if and only if $F(\lambda) = 0$ for $\lambda < 0$ and $\int_0^\infty |F(\lambda)|^2 d\lambda < +\infty$. In fact, if these conditions are satisfied, then

$$\int_{-\infty}^{\infty} |F(\lambda)e^{-\lambda y}|^2 d\lambda \leqslant \int_0^{\infty} |F(\lambda)|^2 d\lambda.$$

On the other hand, if $F(\lambda)$ is not equal to zero on a set of positive measure, lying on the half line $-\infty < \lambda \le 0$, then one can find a segment $[-a, -b]$, $b > 0$, such that

$$\int_{-a}^{-b} |F(\lambda)|^2 d\lambda = I > 0.$$

In that case

$$\int_{-\infty}^{\infty} |F(\lambda)e^{-\lambda y}|^2 d\lambda > \int_{-a}^{-b} |F(\lambda)e^{-\lambda y}|^2 d\lambda >$$
$$> e^{2by} \int_{-a}^{-b} |F(\lambda)|^2 d\lambda = e^{2by} I \to \infty \qquad \text{for } b \to \infty,$$

which contradicts the boundedness of (8). In order to prove the convergence of $\int_0^\infty |F(\lambda)|^2 d\lambda$, it is sufficient to put $y = 0$ in (7).

Thus we have proved the following theorem.

Theorem 3. *The class of functions $f(x+iy)$, analytic for $y > 0$ and such that*

$$\int_{-\infty}^{\infty} |f(x+iy)|^2 dx < C, \qquad y > 0, \tag{9}$$

*coincides with the class of functions representable in the form*

$$f(z) = \int_0^{\infty} F(\lambda) e^{i\lambda z} d\lambda, \tag{10}$$

*where the integral converges in the mean, and*

$$\int_0^\infty |F(\lambda)|^2 d\lambda < \infty. \tag{11}$$

*Moreover,*

$$\underset{y\to+0}{\text{l. i. m.}} f(x+iy) = \int_0^\infty F(\lambda) e^{i\lambda x} d\lambda, \tag{12}$$

*where the integral is understood in the sense of convergence in the mean.*

Under additional conditions of smoothness and rapid decrease, imposed on $F(\lambda)$, the limit in the mean in equalities (10) and (12) can be replaced by the ordinary limit.

3. The Mellin transform. Let us associate, with each function $F(\lambda)$ on the axis $-\infty < \lambda < \infty$, the function $\Phi(t) = F(\ln t)$, given on the strip $0 < t < \infty$. We elucidate upon the existence of the Fourier transformation under such a mapping. For this, in the formula

$$f(z) = \int_{-\infty}^{\infty} F(\lambda) e^{i\lambda z} d\lambda \tag{1}$$

we make the substitution $\lambda = \ln t$. We obtain

$$f(z) = \int_{-\infty}^{\infty} \Phi(t) t^{iz-1} dt. \tag{2}$$

Here the inversion formula takes the form

$$\Phi(t) = \frac{1}{2\pi} \int_0^\infty f(x) t^{-ix} dx. \tag{3}$$

If the function $F(\lambda)$ is such that its Fourier transform can be continued in the complex domain, then the transform (2) can also be continued in the complex domain, and the inversion formula takes the form

$$\Phi(t) = \frac{1}{2\pi} \int_{-\infty+ci}^{\infty+ci} f(z) t^{iz} dz. \tag{4}$$

Let us now assume that $\Phi(t)$ is infinitely differentiable on the strip $0 < t < \infty$ and that the functions $t^{c_1 - 1}\Phi(t)$ and $t^{-c_2-1}\Phi(t)$, $c_1 > 0$, $c_2 > 0$, are integrable. Then the function $f(z)$, $z = x + iy$, is defined in the strip $-c_1 < y < c_2$. We denote $iz$ by $w$, $w = u + iv$, and put $f(-iw) = \mathfrak{F}(w)$. Then (2) takes the form

$$\mathfrak{F}(w)=\int_0^\infty \Phi(t)\,t^{w-1}\,dt. \tag{5}$$

The inversion formula now becomes

$$\Phi(t)=\frac{1}{2\pi i}\int_{-i\infty}^{i\infty}\mathfrak{F}(w)\,t^{-w}\,dw. \tag{6}$$

Here, instead of the imaginary axis, one can integrate over any straight line parallel to this axis and lying in the strip $-c_2<u<c_1$. In other words,

$$\Phi(t)=\frac{1}{2\pi i}\int_{c-i\infty}^{c+i\infty}\mathfrak{F}(w)\,t^{-w}\,dw, \quad -c_2<c<c_1. \tag{7}$$

Formula (5) is called the Mellin transformation, and (6) is the inversion formula for this transformation. This transformation is connected with representations of the group $R_+$ of positive numbers, in which multiplication of numbers is the group operation.

The Plancherel theorem for the Fourier transformation easily yields the following equality for the Mellin transformation:

$$\int_0^\infty |\Phi(t)|^2\frac{dt}{t}=\frac{1}{2\pi}\int_{-\infty}^{\infty}|\mathfrak{F}(iy)|^2\,dy. \tag{8}$$

We shall call it the analogue of the Plancherel formula for this transformation.

Formula (8) allows us to extend the Mellin transformation to all functions $\Phi(t)$ for which the following integral converges:

$$\|\Phi\|^2=\int_0^\infty|\Phi(t)|^2\frac{dt}{t}. \tag{9}$$

Here the integrals in (5) and (6) have to be understood in the sense of convergence in the mean.

The following integral transformation is related to the Mellin transformation. Let us assume that, in the Mellin formulas, $\Phi(t)$ is analytic at the point $t=0$ and in a domain containing the positive real semiaxis. We consider the integral

$$\int_\Gamma \Phi(z)(-z)^{w-1}\,dz, \tag{10}$$

where $(-z)^{w-1}$ is defined as $e^{(w-1)\ln(-z)}$, $\ln(-z)$ takes real values on the negative real semiaxis, and $\Gamma$ is a contour which runs from infinity parallel to the positive real axis, goes round the point $z=0$ in the positive direction and then again returns to infinity parallel to the positive real semiaxis. We shall contract

$\Gamma$ to the real axis. The part of the contour $\Gamma$ lying above the real axis gives

$$\int_{-\infty}^{\infty} \Phi(t)\, e^{(w-1)(\ln t - i\pi)}\, dt = e^{-iw\pi} \int_{0}^{\infty} \Phi(t)\, t^{w-1}\, dt,$$

and the part lying below the real axis gives

$$\int_{0}^{\infty} \Phi(t)\, e^{(w-1)(\ln t + i\pi)}\, dt = -e^{iw\pi} \int_{0}^{\infty} \Phi(t)\, t^{w-1}\, dt.$$

Therefore

$$\int_{\Gamma} \Phi(z)\,(-z)^{w-1}\, dz = -2i \sin w\pi\, \mathfrak{F}(w). \tag{11}$$

Putting

$$\pi \chi(w) = \mathfrak{F}(w) \sin \pi w,$$

we obtain the dual formulas:

$$\chi(w) = -\frac{1}{2\pi i} \int_{\Gamma} \Phi(z)\,(-z)^{w-1}\, dz, \tag{12}$$

$$\Phi(z) = \frac{1}{2i} \int_{c-i\infty}^{c+i\infty} \frac{\chi(w)\, z^{-w}}{\sin \pi w}\, dw. \tag{13}$$

CHAPTER III

# THE GROUP OF SECOND ORDER UNITARY MATRICES AND THE POLYNOMIALS OF LEGENDRE AND JACOBI

In this chapter we shall consider representations of the group $SU(2)$ of unitary unimodular matrices of the second order, closely related to the group of rotations of three-dimensional euclidean space. We shall see that, for a suitable choice of basis in the representation space, the matrix elements of representations of this group are expressed by means of Jacobi polynomials. Moreover, zonal spherical functions (see Chapter 1, §2.5) can be expressed by means of Legendre polynomials, and associated spherical functions by means of associated Legendre polynomials. Starting from the connection between the matrix elements of representations of $SU(2)$ and the orthogonal polynomials of Jacobi and Legendre, we shall derive a number of properties of these polynomials.

## §1. THE GROUP $SU(2)$

**1. Parametrization.** We denote by $SU(2)$ the set of unitary unimodular matrices of the second order. In other words, $SU(2)$ consists of matrices $u = \begin{pmatrix} \alpha & \beta \\ \gamma & \delta \end{pmatrix}$ such that $u^* = u^{-1}$ and $\operatorname{Det} u = 1$ ($u^*$ denotes the Hermitian adjoint of $u$). If $u_1 \in SU(2)$ and $u_2 \in SU(2)$ then

$$(u_1u_2)^* = u_2^*u_1^* = u_2^{-1}u_1^{-1} = (u_1u_2)^{-1}$$

and

$$\operatorname{Det}(u_1u_2) = 1.$$

Therefore $u_1u_2 \in SU(2)$. It is easy to show that $u_1^{-1} \in SU(2)$. Hence the set of matrices $SU(2)$ is a group.

Suppose that the matrix $u = \begin{pmatrix} \alpha & \beta \\ \gamma & \delta \end{pmatrix}$ belongs to the group $SU(2)$. Since $u^* = \begin{pmatrix} \bar\alpha & \bar\gamma \\ \bar\beta & \bar\delta \end{pmatrix}$ and $u^{-1} = \begin{pmatrix} \delta & -\beta \\ -\gamma & \alpha \end{pmatrix}$ the equality $u^* = u^{-1}$ is equivalent to the relations $\delta = \bar{\alpha}$ and $\gamma = -\bar{\beta}$. Thus any matrix $u$ of $SU(2)$ has the form

$$u = \begin{pmatrix} \alpha & \beta \\ -\bar{\beta} & \bar{\alpha} \end{pmatrix}. \tag{1}$$

Since $\operatorname{Det} u = 1$, we have $|\alpha|^2 + |\beta|^2 = 1$. Conversely, if $u$ is a matrix of the form (1), where $|\alpha|^2 + |\beta|^2 = 1$, then $u$ belongs to $SU(2)$.

Thus each element $u$ of $SU(2)$ is uniquely determined by a pair of complex numbers $(\alpha, \beta)$ such that $|\alpha|^2 + |\beta|^2 = 1$.

If we put $\alpha = \alpha_1 + i\alpha_2$ and $\beta = \beta_1 + i\beta_2$, then the equality $|\alpha|^2 + |\beta|^2 = 1$ implies that $|\alpha_1|^2 + |\alpha_2|^2 + |\beta_1|^2 + |\beta_2|^2 = 1$. It follows that $SU(2)$, as a topological space, is homeomorphic to the sphere in four-dimensional real space.

Complex numbers $(\alpha, \beta)$, such that $|\alpha|^2 + |\beta|^2 = 1$, are given by three parameters. As these parameters one can take, for example $|\alpha|$, $\arg \alpha$ and $\arg \beta$. If $\alpha\beta \neq 0$, then, instead of these parameters, it is more convenient to choose others, called Euler angles. The parameters $\phi, \theta, \psi$ are connected with $|\alpha|$, $\arg \alpha$, $\arg \beta$ by the following relations:

$$|\alpha| = \cos\frac{\theta}{2}, \qquad \operatorname{Arg} \alpha = \frac{\varphi + \psi}{2}, \qquad \operatorname{Arg} \beta = \frac{\varphi - \psi + \pi}{2}. \tag{2}$$

The values of $\phi, \theta, \psi$ are not determined uniquely by (2). We shall require, in addition, that these parameters belong to the domain

$$0 \leqslant \varphi < 2\pi, \quad 0 < \theta < \pi, \quad -2\pi \leqslant \psi < 2\pi. \tag{3}$$

One can easily verify that, to each pair of complex numbers $(\alpha, \beta)$ such that $\alpha\beta \neq 0$ and $|\alpha|^2 + |\beta|^2 = 1$, there corresponds one and only one point $(\phi, \theta, \psi)$ of this domain for which equalities (2) are satisfied.

The Euler angles $\phi, \theta, \psi$ are analogous to geographical coordinates on the sphere in three-dimensional euclidean space. Just as, on this sphere, geographical coordinates are not uniquely determined at the north and south poles, the parameters $\phi, \theta, \psi$ are not uniquely determined if $\alpha = 0$ or $\beta = 0$. But the set of matrices for which the parametrization introduced above is not unique has a smaller dimension than the whole group $SU(2)$.

It follows from (2) that $|\beta| = \sin\theta/2$, and that the matrix $u = u(\phi, \theta, \psi)$ with the given Euler angles has the form

$$u=\begin{pmatrix} \cos\frac{\theta}{2}e^{\frac{i(\varphi+\psi)}{2}} & i\sin\frac{\theta}{2}e^{\frac{i(\varphi-\psi)}{2}} \\ i\sin\frac{\theta}{2}e^{\frac{i(\psi-\varphi)}{2}} & \cos\frac{\theta}{2}e^{-\frac{i(\varphi+\psi)}{2}} \end{pmatrix}. \tag{4}$$

Relations (1) and (4) immediately lead to

$$\cos\theta = 2|\alpha|^2 - 1, \tag{5}$$

$$e^{i\varphi} = -\frac{\alpha\beta i}{|\alpha||\beta|}, \tag{5'}$$

$$e^{\frac{i\psi}{2}} = \frac{\alpha e^{-\frac{i\varphi}{2}}}{|\alpha|}. \tag{5''}$$

An important part is played in the sequel by the following factorization of unitary matrices, directly derivable from (4):

$$u(\varphi,\ \theta,\ \psi)=\begin{pmatrix} e^{\frac{i\varphi}{2}} & 0 \\ 0 & e^{-\frac{i\varphi}{2}} \end{pmatrix}\begin{pmatrix} \cos\frac{\theta}{2} & i\sin\frac{\theta}{2} \\ i\sin\frac{\theta}{2} & \cos\frac{\theta}{2} \end{pmatrix}\begin{pmatrix} e^{\frac{i\psi}{2}} & 0 \\ 0 & e^{-\frac{i\psi}{2}} \end{pmatrix}$$

$$\equiv u(\varphi,\ 0,\ 0)\,u(0,\ \theta,\ 0)\,u(0,\ 0,\ \psi)^{1)}. \tag{6}$$

The diagonal matrices

$$\begin{pmatrix} e^{\frac{i\varphi}{2}} & 0 \\ 0 & e^{-\frac{i\varphi}{2}} \end{pmatrix}$$

form a one-parametric subgroup in $SU(2)$. Thus every matrix $u$ of $SU(2)$ lies in the two-sided coset with respect to this subgroup, containing a matrix of the form

$$\begin{pmatrix} \cos\frac{\theta}{2} & i\sin\frac{\theta}{2} \\ i\sin\frac{\theta}{2} & \cos\frac{\theta}{2} \end{pmatrix}. \tag{7}$$

We note that matrices of the form (7) also constitute a one-parameter subgroup in $SU(2)$.

**2. The Euler angles of the product of two matrices.** Let $u = u_1u_2$ be the product of two matrices $u_1$ and $u_2$ of $SU(2)$. We denote the Euler angles of $u$ by $\phi$, $\theta$, $\psi$, of $u_1$ by $\phi_1$, $\theta_1$, $\psi_1$, and of $u_2$ by $\phi_2$, $\theta_2$, $\psi_2$. We shall express the Euler angles of $u$ in terms of the Euler angles of its factors.

1) We may note that $u(\psi,0,0) = u(0,0,\psi)$.

Let us first consider the case when $\phi_1 = \psi_1 = \psi_2 = 0$. Then

$$u = \begin{pmatrix} \cos\frac{\theta_1}{2} & i\sin\frac{\theta_1}{2} \\ i\sin\frac{\theta_1}{2} & \cos\frac{\theta_1}{2} \end{pmatrix} \begin{pmatrix} \cos\frac{\theta_2}{2} e^{\frac{i\varphi_2}{2}} & i\sin\frac{\theta_2}{2} e^{\frac{i\varphi_2}{2}} \\ i\sin\frac{\theta_2}{2} e^{-\frac{i\varphi_2}{2}} & \cos\frac{\theta_2}{2} e^{-\frac{i\varphi_2}{2}} \end{pmatrix}. \tag{1}$$

On multiplying out the matrices in the right-hand side of the equality and applying formulas (5) and (5′) above, we obtain

$$\cos\theta = \cos\theta_1\cos\theta_2 - \sin\theta_1\sin\theta_2\cos\varphi_2, \tag{2}$$

$$e^{i\varphi} = \frac{\sin\theta_1\cos\theta_2 + \cos\theta_1\sin\theta_2\cos\varphi_2 + i\sin\theta_2\sin\varphi_2}{\sin\theta}, \tag{2′}$$

$$e^{\frac{i(\varphi+\psi)}{2}} = \frac{\cos\frac{\theta_1}{2}\cos\frac{\theta_2}{2}e^{\frac{i\varphi_2}{2}} - \sin\frac{\theta_1}{2}\sin\frac{\theta_2}{2}e^{-\frac{i\varphi_2}{2}}}{\cos\frac{\theta}{2}}. \tag{2″}$$

It easily follows from (2′) and (2″) that

$$\operatorname{tg}\varphi = \frac{\sin\theta_2\sin\varphi_2}{\cos\theta_1\sin\theta_2\cos\varphi_2 + \sin\theta_1\cos\theta_2}, \tag{3′}$$

$$\operatorname{tg}\psi = \frac{\sin\theta_1\sin\varphi_2}{\sin\theta_1\cos\theta_2\cos\varphi_2 + \cos\theta_1\sin\theta_2}. \tag{3″}$$

But (3′) and (3″) do not determine $\phi$ and $\psi$ uniquely in the domain $0 \le \phi < 2\pi$, $-2\pi \le \psi < 2\pi$.

The problem is thus solved in the case when $\phi_1 = \psi_1 = \psi_2 = 0$. We can pass to the general case without any difficulty. In fact, owing to equality (6) of subsection 1, we have

$$\begin{aligned} &u(\varphi_1, \theta_1, \psi_1)\, u(\varphi_2, \theta_2, \psi_2) \\ &= u(\varphi_1, 0, 0)\, u(0, \theta_1, 0)\, u(0, 0, \psi_1)\, u(\varphi_2, 0, 0)\, u(0, \theta_2, 0)\, u(0, 0, \psi_2). \end{aligned} \tag{4}$$

But it is evident that

$$u(0, 0, \psi_1)\, u(\varphi_2, 0, 0) = u(\varphi_2 + \psi_1, 0, 0). \tag{5}$$

Moreover, a left-multiplication of the matrix $u(\phi, \theta, \psi)$ by the matrix $u(\phi_1, 0, 0)$ increases the Euler angle $\phi$ by $\phi_1$, and leaves the other Euler angles unchanged Similarly, a right-multiplication of $u(\phi, \theta, \psi)$ by $u(0, 0, \psi_2)$ increases the Euler angle $\psi$ by $\psi_2$.

Hence, in the general case, the formulas for Euler angles have the same form as (2)–(2″) with the only difference that $\phi_2$ must be replaced by $\phi_2 + \psi_1$, and $\phi$ and $\psi$ by $\phi - \phi_1$ and $\psi - \psi_2$. We leave the writing of the final formulas to the reader.

**3. Lie algebra.** Since each element of $SU(2)$ is given by three parameters $\phi$, $\theta$, $\psi$, this group is a three-dimensional manifold in the linear space of all second order matrices $\begin{Vmatrix}\alpha & \beta\\ \gamma & \delta\end{Vmatrix}$. Let us construct the tangent space to $SU(2)$ at the point $e = \begin{Vmatrix}1 & 0\\ 0 & 1\end{Vmatrix}$. For this we draw three curves through $e$, and find the tangent vectors to these curves. If the curves are chosen such that the tangent vectors to them are linearly independent, then these tangent vectors shall give the basis of the tangent space to $SU(2)$ at the point $e$.

It is best to choose, as the curves drawn through the point $e$, one-parametric subgroups, i.e. curves $g(t)$ such that

$$g(t+s)=g(t)\,g(s), \qquad -\infty<t,s<\infty. \tag{1}$$

We choose the following three one-parametric subgroups: the set $\Omega_1$ of all matrices of the form

$$\omega_1(t)=\begin{pmatrix}\cos\frac{t}{2} & i\sin\frac{t}{2}\\ i\sin\frac{t}{2} & \cos\frac{t}{2}\end{pmatrix}, \tag{1'}$$

the set $\Omega_2$ of matrices of the form

$$\omega_2(t)=\begin{pmatrix}\cos\frac{t}{2} & -\sin\frac{t}{2}\\ \sin\frac{t}{2} & \cos\frac{t}{2}\end{pmatrix} \tag{2}$$

and the set $\Omega_3$ of diagonal matrices

$$\omega_3(t)=\begin{pmatrix}e^{\frac{it}{2}} & 0\\ 0 & e^{-\frac{it}{2}}\end{pmatrix}. \tag{3}$$

A simple calculation shows that the tangent matrices $a_1$, $a_2$, $a_3$ to these subgroups have the form

$$a_1=\left.\frac{d\omega_1(t)}{dt}\right|_{t=0}=\frac{i}{2}\begin{pmatrix}0 & 1\\ 1 & 0\end{pmatrix}, \tag{4}$$

$$a_2=\left.\frac{d\omega_2(t)}{dt}\right|_{t=0}=\frac{1}{2}\begin{pmatrix}0 & -1\\ 1 & 0\end{pmatrix}, \tag{5}$$

$$a_3=\left.\frac{d\omega_3(t)}{dt}\right|_{t=0}=\frac{i}{2}\begin{pmatrix}1 & 0\\ 0 & -1\end{pmatrix}. \tag{6}$$

The matrices $a_1, a_2, a_3$ are linearly independent and consequently form a basis of the Lie algebra of $SU(2)$. The Lie algebra of this group thus consists of the matrices $x_1a_1 + x_2a_2 + x_3a_3$, where $x_1, x_2, x_3$ are real numbers.

The commutation relations for matrices $a_1, a_2, a_3$ have the following form:

$$[a_1, a_2] = a_3, \tag{7}$$

$$[a_2, a_3] = a_1, \tag{7'}$$

$$[a_3, a_1] = a_2. \tag{7''}$$

Relations (7), (7′) and (7″) admit a simple geometrical interpretation: they coincide with the formulas for vector products of unit coordinate vectors in three-dimensional euclidean space. Hence, if

$$a = x_1a_1 + x_2a_2 + x_3a_3 \text{ and } b = y_1a_1 + y_2a_2 + y_3a_3$$

are two matrices of the Lie algebra of $SU(2)$, then

$$[a, b] = z_1a_1 + z_2a_2 + z_3a_3,$$

where the vector $(z_1, z_2, z_3)$ is the vector product of vectors $(x_1, x_2, x_3)$ and $(y_1, y_2, y_3)$.

**4. Complexification.** Let us now consider the complexification of the Lie algebra of $SU(2)$, i.e. the complex linear space spanned by the matrices $a_1, a_2, a_3$. Since the traces of $a_1, a_2, a_3$ are equal to zero, all matrices of this linear space also have zero traces. On the other hand, it is evident that any second order matrix with zero trace is a linear combination (with complex coefficients) of $a_1$, $a_2, a_3$. Thus the complexification of the Lie algebra of $SU(2)$ is the space of second order complex matrices whose trace is equal to zero. But the trace of the product of two matrices is unchanged on permuting the factors, and so the trace of the commutator of any two matrices is equal to zero. Hence complex matrices with trace zero form a real Lie algebra. The matrices $a_1, a_2, a_3, ia_1, ia_2, ia_3$ are the basis of this algebra.

We show that the algebra constructed above is the tangent space at the point $e = \begin{pmatrix} 1 & 0 \\ 0 & 1 \end{pmatrix}$ to the group $SL(2, C)$ of all unimodular complex matrices of the second order. In fact, let

$$g(t) = \begin{pmatrix} \alpha(t) & \beta(t) \\ \gamma(t) & \delta(t) \end{pmatrix}$$

be a one-parametric subgroup of the group $SL(2, C)$. Then for all $t$ we have

$$\alpha(t)\delta(t) - \beta(t)\gamma(t) = 1.$$

We differentiate this identity with respect to $t$ and put $t = 0$. As $g(0) = e$ so that $\alpha(0) = \delta(0) = 1$, $\beta(0) = \gamma(0) = 0$, we have $\alpha'(0) + \delta'(0) = 0$. Thus, tangent matrices to one-parametric subgroups of $SL(2, C)$ have zero trace.

In order to complete the proof of our assertion, we have to show now that the dimension of the Lie algebra of $SL(2, C)$ coincides with the dimension of the space of matrices with zero trace. But the matrices with zero trace are linear combinations of $a_1, a_2, a_3$, and so the complex dimension of the space of such matrices is equal to 3. On the other hand, because of the relation $\alpha\delta - \beta\gamma = 1$, each element of $SL(2, C)$ is determined by three complex parameters (for example, for $\gamma \neq 0$, by the parameters $\alpha, \gamma, \delta$), and so the complex dimension of the Lie algebra of this group is also equal to 3. Our assertion is thus completely proved: The Lie algebra of $SL(2, C)$ is the space of all complex second order matrices with trace zero.

The group $SL(2, C)$ is called the complexification of $SU(2)$, and $SU(2)$ is one of the real forms of $SL(2, C)$. We indicate how one can directly pass from $SU(2)$ to $SL(2, C)$. As shown above, each element of $SU(2)$ is determined by three real parameters $\phi, \theta, \psi$ —the Euler angles. Moreover, the matrix with Euler angles $\phi, \theta, \psi$ has the form (4) of subsection 1. Let us give complex values to the Euler angles $\phi, \theta, \psi$. It is easy to show that matrices of the form (4) of subsection 1 will then be unimodular complex matrices of the second order. Moreover, any matrix of $SL(2, C)$ can be represented in such a form. This representation is unique for almost all elements of $SL(2, C)$, if one restricts oneself to the values of $\phi, \theta$, $\psi$ belonging to the domain

$$\left.\begin{aligned} 0 \leqslant \operatorname{Re} \theta \leqslant \pi, \\ 0 \leqslant \operatorname{Re} \varphi < 2\pi, \\ 2\pi \leqslant \operatorname{Re} \psi < 2\pi. \end{aligned}\right\} \tag{1}$$

In other words, when $\phi, \theta, \psi$ vary in this domain, the matrix $g(\phi, \theta, \psi)$ runs through the whole group $SL(2, C)$, and almost all (i.e. all, apart from a set of lesser dimension) matrices $g$ of $SL(2, C)$ are encountered only once in this process.

We shall call $\phi, \theta, \psi$ the complex Euler angles of $g$. Clearly, the formulas of subsection 2 for calculating the Euler angles of the product of two matrices remain valid even after the complexification of $SU(2)$.

**5. Connection with the rotation group.** By a rotation $g$ of three-dimensional euclidean space one means a linear transformation of this space, which preserves the distance between points of this space and the origin, and which does not alter the orientation of this space. The set $SO(3)$ of all the rotations is a group. Evidently, $SO(3)$ is isomorphic to the group of real orthogonal matrices of the third order whose determinant is equal to 1.

We establish a connection between the groups $SU(2)$ and $SO(3)$. For this we associate with each vector $\mathbf{x}(x_1, x_2, x_3)$ of three-dimensional euclidean space a complex second order matrix of the form

$$h_{\mathbf{x}} = \begin{pmatrix} x_3 & x_1 + ix_2 \\ x_1 - ix_2 & -x_3 \end{pmatrix}. \tag{1}$$

The space of matrices of the form (1) consists of all Hermitian matrices $g$ whose trace is equal to zero.

Further, with each matrix $u$ of $SU(2)$ we associate a transformation $T(u)$, which transforms a matrix $h_{\mathbf{x}}$ of the form (1) into the matrix

$$T(u)\, h_{\mathbf{x}} = uh_{\mathbf{x}}u^*. \tag{2}$$

Since $u^* = u^{-1}$ for unitary matrices, the traces of $h_{\mathbf{x}}$ and $T(u)h_{\mathbf{x}} = uh_{\mathbf{x}}u^{-1}$ coincide, and therefore the trace of $T(u)h_{\mathbf{x}}$ is equal to zero. In addition,

$$(T(u)\, h_{\mathbf{x}})^* = (uh_{\mathbf{x}}\, u^*)^* = uh_{\mathbf{x}}^* u^* = uh_{\mathbf{x}}u^* = T(u)\, h_{\mathbf{x}}, \tag{3}$$

and so $T(u)h_{\mathbf{x}}$ is Hermitian. Consequently this matrix has the form

$$T(u)\, h_{\mathbf{x}} = \begin{pmatrix} y_3 & y_1 + iy_2 \\ y_1 - iy_2 & -y_3 \end{pmatrix} = h_{\mathbf{y}}, \tag{4}$$

where $\mathbf{y}(y_1, y_2, y_3)$ is a vector of three-dimensional euclidean space.

It is evident from (2) that the coordinates of $\mathbf{y}$ are linear combinations of the coordinates of $\mathbf{x}$, and therefore $T(u)$ can be regarded as a linear transformation in three-dimensional euclidean space. We show that this transformation is a rotation of the euclidean space. For this we observe that $\operatorname{Det} h_{\mathbf{x}} = -x_1^2 - x_2^2 - x_3^2$, and so $-\operatorname{Det} h_{\mathbf{x}}$ is equal to the quadrate of distance of the point $\mathbf{x}(x_1, x_2, x_3)$ from the origin. But $u^* = u^{-1}$, and therefore

$$\operatorname{Det} T(u)\, h_{\mathbf{x}} = \operatorname{Det}(uh_{\mathbf{x}}\, u^*) = \operatorname{Det} h_{\mathbf{x}}. \tag{5}$$

Hence the transformation $T(u)$ leaves $\operatorname{Det} h_{\mathbf{x}}$ unaltered, and consequently it does not alter the distance from the origin of points of the euclidean space. Finally, it is easy to verify that the determinant of $T(u)$ in the euclidean space is equal to 1, and therefore this transformation does not alter the orientation.

So, with each unitary unimodular matrix $u$ we have associated the rotation

$T(u)$ in three-dimensional euclidean space. One can easily show that $T(u)$ is a homomorphic mapping of $SU(2)$ onto the whole group $SO(3)$, where the homomorphism kernel consists of the two matrices $e$ and $-e$.

As $T(u)$ is one-to-one on a sufficiently small neighborhood of elements of $SU(2)$, one says that $SU(2)$ and $SO(3)$ are locally isomorphic.

The connection between $SU(2)$ and $SO(3)$ admits a simple geometrical interpretation. Namely, let us consider, in three-dimensional euclidean space, the sphere

$$x_1^2+x_2^2+x_3^2=\frac{1}{4}. \tag{6}$$

With each point of this sphere we associate its image $M(x_1, x_2, -\frac{1}{2})$ under stereographic projection of the sphere on the plane $x_3=-\frac{1}{2}$ from the point $N(0, 0, \frac{1}{2})$ or, what is the same thing, the complex number $w=x_1+ix_2$. It can then be shown that to the linear-fractional transformation

$$\hat{w}=\frac{\alpha w-\bar{\beta}}{\beta w+\bar{\alpha}}, \qquad |\alpha|^2+|\beta|^2=1 \tag{7}$$

of the plane $z=-\frac{1}{2}$ there corresponds a rotation $g$ of the sphere, where

$$g=T(u), \quad u=\begin{pmatrix} \alpha & \beta \\ -\bar{\beta} & \bar{\alpha} \end{pmatrix}.$$

**6. The Euler angles of rotations.** From the local isomorphism of $SU(2)$ and $SO(3)$ it follows that rotations of three-dimensional euclidean space can be given by the three Euler angles $\phi$, $\theta$, $\psi$. Moreover, unlike $SU(2)$, the Euler angle $\psi$ varies between the limits $0$ and $2\pi$ (and not between $-2\pi$ and $2\pi$). This is linked with the fact that one and the same rotation of three-dimensional space corresponds to the matrices $u$ and $-u$.

Let us explain the geometrical meaning of Euler angles for rotations of three-dimensional space. It is easy to verify that to matrices of the form

$$\omega_1(t)=\begin{pmatrix} \cos\frac{t}{2} & i\sin\frac{t}{2} \\ i\sin\frac{t}{2} & \cos\frac{t}{2} \end{pmatrix} \tag{1}$$

correspond rotations of three-dimensional euclidean space by the angle $t$ around the axis $Ox_1$, to the matrices

$$\omega_2(t)=\begin{pmatrix} \cos\frac{t}{2} & -\sin\frac{t}{2} \\ \sin\frac{t}{2} & \cos\frac{t}{2} \end{pmatrix} \tag{2}$$

rotations by the angle $t$ around the axis $Ox_2$, and, lastly, to the matrices

$$\omega_3(t)=\begin{pmatrix} e^{\frac{it}{2}} & 0 \\ 0 & e^{-\frac{it}{2}} \end{pmatrix} \tag{3}$$

rotations by the angle $t$ around the axis $Ox_3$.

But it was shown in subsection 1 that the matrix $u$ of $SU(2)$, corresponding to the Euler angles $\phi$, $\theta$, $\psi$, has the form

$$u(\varphi, \theta, \psi)=u(\varphi, 0, 0)\,u(0, \theta, 0)\,u(\psi, 0, 0),$$

where $u(\phi, 0, 0)$ and $u(\psi, 0, 0)$ are matrices of the form (3), and $u(0, \theta, 0)$ is a matrix of the form (1). Hence a rotation $g$ of the three-dimensional euclidean space $E_3$, given by Euler angles $\phi$, $\theta$, $\psi$, is the product of a rotation by the angle $\psi$ around the axis $Ox_3$, a rotation by the angle $\theta$ around the axis $Ox_1$, and a rotation by the angle $\phi$ around the axis $Ox_3$. It follows that the rotation $g(\phi, \theta, \psi)$ has the matrix form

$$g(\varphi, \theta, \psi)=g(\varphi, 0, 0)\,g(0, \theta, 0)\,g(\psi, 0, 0)$$
$$=\begin{pmatrix} \cos\varphi & -\sin\varphi & 0 \\ \sin\varphi & \cos\varphi & 0 \\ 0 & 0 & 1 \end{pmatrix}\begin{pmatrix} 1 & 0 & 0 \\ 0 & \cos\theta & -\sin\theta \\ 0 & \sin\theta & \cos\theta \end{pmatrix}\begin{pmatrix} \cos\psi & -\sin\psi & 0 \\ \sin\psi & \cos\psi & 0 \\ 0 & 0 & 1 \end{pmatrix}$$
$$=\begin{pmatrix} \cos\varphi\cos\psi-\sin\varphi\sin\psi\cos\theta & -\cos\varphi\sin\psi-\sin\varphi\cos\psi\cos\theta & \sin\varphi\sin\theta \\ \sin\varphi\cos\psi+\cos\varphi\sin\psi\cos\theta & -\sin\varphi\sin\psi+\cos\varphi\cos\psi\cos\theta & -\cos\varphi\sin\theta \\ \sin\psi\sin\theta & \cos\psi\sin\theta & \cos\theta \end{pmatrix}.$$

Let us also dwell on the geometrical interpretation of tangent matrices for one-parametric subgroups of $SO(3)$. It is geometrically evident that any one-parametric subgroup $g(t)$ of $SO(3)$ is a subgroup of rotations around the fixed axis 1 in $E_3$, where $t$ is proportional to the angle of rotation. Therefore it is natural to specify the one-parametric subgroup $g(t)$ by the vectors $\mathbf{x} = x_1\mathbf{e}_1 + x_2\mathbf{e}_2 + x_3\mathbf{e}_3$, directed along the axis of rotations, whose length is equal to the angular velocity of rotation. Under such a correspondence, to the one-parametric subgroups $\omega_1(t)$, $\omega_2(t)$, $\omega_3(t)$ (see subsection 3) correspond the coordinate unit vectors $\mathbf{e}_1$, $\mathbf{e}_2$, $\mathbf{e}_3$.

It is not difficult to show that if to the one-parametric subgroup $g(t)$ corresponds the vector $\mathbf{x} = x_1\mathbf{e}_1 + x_2\mathbf{e}_2 + x_3\mathbf{e}_3$, then the tangent matrix $a$ of this subgroup has the form $a = x_1a_1 + x_2a_2 + x_3a_3$, where $a_1$, $a_2$, $a_3$ are tangent matrices to the subgroups $\omega_1(t)$, $\omega_2(t)$, $\omega_3(t)$. We omit the proof.

Utilizing the geometrical interpretation of the tangent matrices to one-parametric subgroups, one can easily express the tangent matrix of the subgroup

$\tilde{g}(t) = g_0 g(t) g_0^{-1}$ in terms of the tangent matrix to the subgroup $g(t)$. To do this, we observe that if the rotation $g(t)$ leaves the axis $\mathbf{1}$ fixed, then the rotation $g_0 g(t) g_0^{-1}$ leaves the axis $g_0\mathbf{1}$ (obtained from $\mathbf{1}$ by a rotation $g_0$) fixed. Indeed, it follows from $g(t)\mathbf{1} = \mathbf{1}$ that

$$g_0 g(t) g_0^{-1}(g_0\mathbf{1}) = g_0\mathbf{1}.$$

The velocity of $g(t)$ and $g_0 g(t) g_0^{-1}$ is the same.

We thus arrive at the following conclusion: if the one-parametric subgroup $g(t)$ has the tangent matrix

$$a = x_1 a_1 + x_2 a_2 + x_3 a_3,$$

then the tangent matrix of $g_0 g(t) g_0^{-1}$ has the form

$$b = y_1 a_1 + y_2 a_2 + y_3 a_3,$$

where

$$\mathbf{y} = g_0 \mathbf{x}, \qquad \mathbf{x} = (x_1, x_2, x_3), \qquad \mathbf{y} = (y_1, y_2, y_3).$$

7. **The sphere as a homogeneous space.** The group $SO(3)$ of rotations of three-dimensional euclidean space is a transitive group of transformations of the unit sphere $S^2$ in this space. Moreover, the stationary subgroup of a point of the sphere is the subgroup of rotations around the axis, passing through this point. The stationary subgroup of the point $M(0, 0, 1)$ is the subgroup of rotations around the axis $Oz$. We have seen in the preceding subsection that to these rotations correspond diagonal matrices of $SU(2)$, i.e. matrices of the form

$$\begin{pmatrix} e^{\frac{it}{2}} & 0 \\ 0 & e^{-\frac{it}{2}} \end{pmatrix}.$$

Hence the unit sphere $S^2$ is a homogeneous space with the group of motions $SU(2)$ and the stationary subgroup $\Omega$, consisting of matrices of the form (1). By virtue of Chapter I, §1.3, this means that $S^2$ can be identified with the space of left cosets of $SU(2)$ with respect to the subgroup $\Omega$. In this way, functions $f(\xi)$ given on $S^2$ can be regarded as functions $f(u)$ on $SU(2)$, constant on left cosets with respect to $\Omega$, i.e. such that

$$f(uh) = f(u), \qquad h \in \Omega.$$

One can easily verify that if the Euler angles of $u$ are equal to $\phi, \theta, \psi$, then there corresponds to it a point $M$ of the sphere, with the Cartesian coordinates

$$M(\sin\theta\sin\varphi, \ -\sin\theta\cos\varphi, \ \cos\theta).$$

Evidently, the spherical coordinates of $M$ are equal to $-\pi/2+\phi$ and $\theta$.

## §2. IRREDUCIBLE UNITARY REPRESENTATIONS $T_l(u)$

In this section we shall describe irreducible unitary representations of the group $SU(2)$. Since $SU(2)$ is compact, these representations are finite-dimensional. In order to construct them, let us consider first one class of representations of $SL(2, C)$ of unimodular complex matrices of the second order. Restricting these representations to $SU(2)$, we shall get its irreducible unitary representations.

**1. Representations in spaces of homogeneous polynomials.** To each unitary complex matrix $g=\begin{Vmatrix}\alpha & \beta\\ \gamma & \delta\end{Vmatrix}$ corresponds the linear transformation

$$\left.\begin{aligned} w_1 &= \alpha z_1+\gamma z_2,\\ w_2 &= \beta z_1+\delta z_2\end{aligned}\right\} \tag{1}$$

of two-dimensional complex linear space. Associated with this transformation is the operator

$$T(g)f(z_1, z_2)=f(\alpha z_1+\gamma z_2, \ \beta z_1+\delta z_2) \tag{2}$$

in the space of functions of two complex variables.

Evidently, $T(g_1g_2)=T(g_1)T(g_2)$, and therefore $T(g)$ is a representation of $SL(2, C)$. This representation is reducible, since the space of functions of two complex variables contains a subspace invariant for the transformations $T(g)$. For example, any homogeneous polynomial of two variables is transformed by $T(g)$ into a homogeneous polynomial of the same degree.

Let $l$ be an integer or a semi-integer. We denote by $\mathfrak{H}_l$ the space of all homogeneous polynomials

$$f(z_1, z_2)=\sum_{n=-l}^{l} a_n z_1^{l-n} z_2^{l+n} \tag{3}$$

of degree $2l$, and by $T_l(g)$ the restriction of $T(g)$ to the space $\mathfrak{H}_l$. We shall see below that $T_l(g)$ is an irreducible representation of $SL(2, C)$. Moreover, this representation remains irreducible if one restricts it to the subgroup $SU(2)$ of $SL(2, C)$ (see subsection 3).

It will be convenient to make use of other realizations of the representations

$T_l(g)$ in the sequel. In order to obtain these realizations, we observe that the homogeneous polynomial $f(z_1, z_2)$ is uniquely determined by its values on any "contour" $\Gamma$ intersecting the arbitrary complex line $a_1 z_1 + a_2 z_2 = 0$ at one point. Therefore $\mathfrak{H}_l$ can be characterized not as the space of homogeneous polynomials of two complex variables, but as the space of functions on the contour $\Gamma$.

Let us consider the complex line $z_2 = 1$ in two-dimensional complex space. This line intersects every line passing through the origin (except the line $z_2=0$) at one and only one point. Therefore every polynomial $f(z_1, z_2)$ is uniquely determined by its values on the line $z_2 = 1$. With each polynomial $f(z_1, z_2)$ of $\mathfrak{H}_l$ we associate the following polynomial of degree $2l$ in one variable:

$$\varphi(z_1) = f(z_1, 1) = \sum_{n=-l}^{l} a_n z_1^{l-n}. \tag{4}$$

Evidently $f(z_1, z_2)$ is determined by $\phi(z_1)$ as follows:

$$f(z_1, z_2) = z_2^{2l} \varphi\left(\frac{z_1}{z_2}\right). \tag{5}$$

We shall denote the space of polynomials of degree $2l$ in one variable by the same letter $\mathfrak{H}_l$. Let us find the operators in this space, corresponding to the operators of the representation $T_l(g)$. To do this we observe that to the polynomial $\phi(z)$ corresponds the homogeneous polynomial $f(z_1, z_2)$ of two variables, given by formula (5). The operator $T_l(g)$ transforms this polynomial into the polynomial

$$f_g(z_1, z_2) = f(\alpha z_1 + \gamma z_2, \beta z_1 + \delta z_2). \tag{6}$$

Because of the homogeneity of $f(z_1, z_2)$, we have

$$f_g(z_1, z_2) = (\beta z_1 + \delta z_2)^{2l} f\left(\frac{\alpha z_1 + \gamma z_2}{\beta z_1 + \delta z_2}, 1\right)$$
$$= (\beta z_1 + \delta z_2)^{2l} \varphi\left(\frac{\alpha z_1 + \gamma z_2}{\beta z_1 + \delta z_2}\right). \tag{7}$$

Therefore the polynomial $\phi_g(z) = f_g(z, 1)$, corresponding to $f_g(z_1, z_2)$, is expressed through $\phi(z)$ by the formula

$$\varphi_g(z) = (\beta z + \delta)^{2l} \varphi\left(\frac{\alpha z + \gamma}{\beta z + \delta}\right).$$

Thus, if $\mathfrak{H}_l$ is realized as the space of polynomials of degree $2l$ in one variable, the representation operator $T_l(g)$ is given by the formula

$$T_l(g)\varphi(z) = (\beta z + \delta)^{2l} \varphi\left(\frac{\alpha z + \gamma}{\beta z + \delta}\right). \tag{8}$$

We may also mention the realization of $T_l(g)$ in the space of trigonometrical polynomials of order $l$. For this, with each polynomial $F(z)$ of degree $2l$ we associate the trigonometrical polynomial

$$\Phi(e^{i\varphi}) = e^{-il\varphi} F(e^{i\varphi}). \tag{9}$$

In other words, we put

$$\Phi(e^{i\varphi}) = \sum_{n=-l}^{l} a_n e^{-in\varphi}. \tag{9'}$$

Following the same arguments as above, we find that, under the realization of $\mathfrak{H}_l$ as the space of trigonometric polynomials of order $l$, the operators $T_l(g)$ are realized as follows:

$$T_l(g)\Phi(e^{i\varphi}) = e^{-il\varphi}(\alpha e^{i\varphi}+\gamma)^l(\beta e^{i\varphi}+\delta)^l\,\Phi\left(\frac{\alpha e^{i\varphi}+\gamma}{\beta e^{i\varphi}+\delta}\right). \tag{10}$$

So, we have constructed a representation $T_l(g)$ of $SL(2, C)$, and indicated various realizations of this representation. We now restrict this representation to $SU(2) \subset SL(2, C)$. This means that in formulas (8) and (10) one must put $\delta = \bar{\alpha}$, $\gamma = -\bar{\beta}$. This gives a representation of $SU(2)$. We shall use the same notation for it as for $SL(2, C)$, i.e. we shall write $T_l(u)$, $u \in SU(2)$:

$$T_l(u)\varphi(z) = (\beta z+\bar{\alpha})^{2l}\,\varphi\left(\frac{\alpha z-\bar{\beta}}{\beta z+\bar{\alpha}}\right), \quad u = \begin{pmatrix} \alpha & \beta \\ -\bar{\beta} & \bar{\alpha} \end{pmatrix}.$$

**2. Infinitesimal operators of the representation $T_l(u)$.** In the following subsection it shall be shown that the representation $T_l(u)$ of the group $SU(2)$ is irreducible. For this we need an expression for the infinitesimal operators of $T_l(u)$. We utilize the realization of this space in the space of polynomials of degree $2l$ in one variable.

To the matrix

$$\omega(t) = \begin{pmatrix} \cos\frac{t}{2} & i\sin\frac{t}{2} \\ i\sin\frac{t}{2} & \cos\frac{t}{2} \end{pmatrix} \tag{1}$$

of the subgroup $\Omega_1$ (see §1.3) corresponds the operator $T_l[\omega(t)]$, transforming the polynomials $\phi(x)$ into

$$T_l[\omega(t)]\varphi(x) = \left(ix\sin\frac{t}{2}+\cos\frac{t}{2}\right)^{2l}\varphi\left(\frac{x\cos\frac{t}{2}+i\sin\frac{t}{2}}{ix\sin\frac{t}{2}+\cos\frac{t}{2}}\right).$$

We differentiate this equality with respect to $t$, and put $t = 0$. We find that to the subgroup $\Omega_1$ corresponds the infinitesimal operator

$$A_1 = ilx + \frac{i}{2}(1 - x^2)\frac{d}{dx}. \tag{2}$$

One similarly proves that to the one-parametric subgroups $\Omega_2$ and $\Omega_3$ correspond the infinitesimal operators

$$A_2 = -lx + \frac{1}{2}(1 + x^2)\frac{d}{dx} \tag{3}$$

and

$$A_3 = i\left(x\frac{d}{dx} - l\right). \tag{4}$$

Let us apply the infinitesimal operators $A_1$, $A_2$, $A_3$ to the basis monomials $x^{l-n}$, $-l \le n \le l$:

$$A_1 x^{l-n} = \frac{i}{2}(l - n)\, x^{l-n-1} + \frac{i}{2}(l + n)\, x^{l-n+1}, \tag{5}$$

$$A_2 x^{l-n} = \frac{1}{2}(l - n)\, x^{l-n-1} - \frac{1}{2}(l + n)\, x^{l-n+1}, \tag{6}$$

$$A_3 x^{l-n} = -inx^{l-n}. \tag{7}$$

Thus, infinitesimal operators $A_1$ and $A_2$ transform the basis monomials $x^{l-n}$ into linear combinations of "adjacent" monomials $x^{l-n-1}$ and $x^{l-n+1}$. It is not difficult to write down operators $H_+$ and $H_-$, transforming $x^{l-n}$ into polynomials proportional to $x^{l-n-1}$ and $x^{l-n+1}$. Namely, we take $H_+ = iA_1 - A_2$ and $H_- = iA_1 + A_2$. In other words, we take

$$H_+ = -\frac{d}{dx}, \tag{8}$$

$$H_- = -2lx + x^2\frac{d}{dx}. \tag{9}$$

In addition, we introduce an operator $H_3$ by taking

$$H_3 = iA_3 = l - x\frac{d}{dx}. \tag{10}$$

Formulas (5)–(7) immediately yield

$$H_+ x^{l-n} = (n - l)\, x^{l-n-1}, \tag{11}$$

$$H_- x^{l-n} = -(n + l)\, x^{l-n+1}, \tag{12}$$

$$H_3 x^{l-n} = nx^{l-n}. \tag{13}$$

It is obvious from these formulas that $x^{l-m}$ is an eigenfunction of $H_3$, corresponding to the eigenvalue $n$. The operator $H_+$ transforms this function into an eigenfunction of $H_3$, corresponding to the eigenvalue $n + 1$. But $H_-$ transforms $x^{l-n}$ into the eigenfunction corresponding to the eigenvalue $n - 1$. The operator

$H_+$ transforms into zero the function 1, corresponding to the largest eigenvalue $l$, and operator $H_-$—the function $x^{2l}$, corresponding to the smallest eigenvalue $-l$.

3. Irreducibility. We prove that the representation $T_l(u)$ of the group $SU(2)$ is irreducible. For this it is sufficient to show that the space $\mathfrak{H}_l$ does not contain a nontrivial subspace invariant for the operators $A_1$, $A_2$, $A_3$ (see Chapter 1, §1.10). Since $A_1$, $A_2$, $A_3$ are linear combinations of $H_+$, $H_-$, $H_3$, it is sufficient to demonstrate the absence of a nontrivial subspace invariant for $H_+$, $H_-$, $H_3$.

Let us denote by $\mathfrak{H}_{ln}$ the subspace in $\mathfrak{H}_l$, consisting of the monomials $a_n x^{l-n}$. It follows from formula (13), above, that this subspace is invariant for $H_3$, and that to different values of $n$ correspond different eigenvalues of $H_3$ in $\mathfrak{H}_{ln}$.

We assume that a subspace $\mathfrak{T}$ in $\mathfrak{H}_l$ is invariant for operators $H_+$, $H_-$, $H_3$. By virtue of Chapter 1, §3.3, the invariance of the subspace $\mathfrak{T}$ for $H_3$ implies that it is a direct sum of some of the subspaces $\mathfrak{H}_{ln}$, i.e.

$$\mathfrak{T} = \mathfrak{H}_{ln_1} + \cdots + \mathfrak{H}_{ln_s}. \tag{1}$$

Therefore, if $\mathfrak{T} \neq 0$, it contains one of the monomials $x^{l-n}$.

Let us now use the invariance of $\mathfrak{T}$ for the operators $H_+$ and $H_-$. It follows that $\mathfrak{T}$ contains all functions of the form $H_+^s x^{l-n}$ and $H_-^s x^{l-n}$. If $0 \leq s \leq l-n$, then, owing to formula (11) of subsection 2, we have

$$H_+^s x^{l-n} = \alpha x^{l-n-s}, \tag{2}$$

where $\alpha \neq 0$. Similarly, if $0 \leq s \leq l+n$, then we have

$$H_-^s x^{l-n} = \beta x^{l-n+s}, \tag{3}$$

where $\beta \neq 0$. Hence $\mathfrak{T}$ contains all the monomials $x^{l-m}$, $-l \leq m \leq l$, and so coincides with $\mathfrak{H}_l$. The irreducibility of $T_l(u)$ is thus proved. The representation $T_l(g)$ of the group $SL(2, C)$ is, a fortiori, reducible.

We may note that the irreducible representations $T_l(u)$ of the group $SU(2)$ constructed by us, are pairwise nonequivalent. Indeed, representations with different values of $l$ have different dimensions.

4. An invariant scalar product. Since the group $SU(2)$ is compact, in the space $\mathfrak{H}$ of its finite-dimensional representation $T_l(u)$ there exists an invariant scalar product (see Chapter 1, §4.2). In order to obtain an explicit expression

for it, it is sufficient to calculate the scalar products $(x^{l-k}, x^{l-m})$, where $-l \leq k \leq l$, $-l \leq m \leq l$.

We shall first show that for $k \neq m$ we have $(x^{l-k}, x^{l-m}) = 0$. For this we use the invariance of the scalar product for operators $T_l(h)$, where

$$h = \begin{pmatrix} e^{\frac{it}{2}} & 0 \\ 0 & e^{-\frac{it}{2}} \end{pmatrix}.$$

One can easily see that $T_l(h)$ transforms $x^{l-k}$ into $e^{-ikt}x^{l-k}$. Therefore

$$(x^{l-k}, x^{l-m}) = (T_l(h)\,x^{l-k}, T_l(h)\,x^{l-m}) = e^{-i(k-m)t}(x^{l-k}, x^{l-m}).$$

It follows from this relation that if $k \neq m$, then $(x^{l-k}, x^{l-m}) = 0$.

Thus we have only to calculate $(x^{l-k}, x^{l-k})$, $-l \leq k \leq l$. To do so, we use the equality

$$(x^{l-k}, x^{l-k+1}) = (T_l(u)\,x^{l-k}, T_l(u)\,x^{l-k+1}),$$

where

$$u = \begin{pmatrix} \cos\frac{t}{2} & -\sin\frac{t}{2} \\ \sin\frac{t}{2} & \cos\frac{t}{2} \end{pmatrix}. \tag{1}$$

We differentiate both sides with respect to $t$ and put $t = 0$ (i.e. we pass to the infinitesimal operators). We obtain

$$0 = (A_2x^{l-k}, x^{l-k+1}) + (x^{l-k}, A_2x^{l-k+1}).$$

Hence, owing to formula (6) of subsection 2, we have

$$0 = -(l+k)(x^{l-k+1}, x^{l-k+1}) + (l-k+1)(x^{l-k}, x^{l-k}). \tag{2}$$

The invariant scalar product is thus determined to within a constant factor. We choose this factor such that the equality $(1, 1) = (2l)!$ holds. Then the recurrence relation (2) gives

$$(x^{l-k}, x^{l-k}) = (l-k)!\,(l+k)!, \quad -l \leqslant k \leqslant l. \tag{3}$$

This completes the determination of the invariant scalar product. It follows from formula (3) that the system of functions

$$\psi_k(x) = \frac{x^{l-k}}{\sqrt{(l-k)!\,(l+k)!}}, \quad -l \leqslant k \leqslant l \tag{4}$$

forms an orthogonal normalized basis in the space $\mathfrak{H}_l$. This basis consists of the eigenfunctions of the operators $T_l(h)$, where

$$h=\begin{pmatrix} e^{\frac{it}{2}} & 0 \\ 0 & e^{-\frac{it}{2}} \end{pmatrix} \tag{5}$$

Moreover, we have

$$(\varphi, x^{l-k}) = (l-k)!(l+k)!a_k \tag{6}$$

where

$$\varphi(x)=\sum_{k=-l}^{l} a_k x^{l-k}.$$

In conclusion we note that the formulas of subsection 2 yield the equalities

$$\left.\begin{aligned} H_+\psi_n(x) &= -\sqrt{(l-n)(l+n+1)}\,\psi_{n+1}(x), \\ H_-\psi_n(x) &= -\sqrt{(l+n)(l-n+1)}\,\psi_{n-1}(x), \\ H_3\psi_n(x) &= n\psi_n(x). \end{aligned}\right\} \tag{6}$$

**5. Completeness of the system of representations $T_l(u)$.** We have constructed the system $T_l(u)$ of pairwise nonequivalent unitary irreducible representations of $SU(2)$. It turns out that these representations exhaust all (to within equivalence) irreducible unitary representations of $SU(2)$. In other words, we have

**Theorem 1.** *Every irreducible unitary representation $T(u)$ of the group $SU(2)$ is equivalent to one of the representations $T_l(u)$, $l = 0, \frac{1}{2}, 1, \cdots$.*

We shall omit the proof of this theorem, referring the reader to [16], where in Chapter 1, §2.3 this theorem is proved for the group $SO(3)$. The proof is carried over without any changes to $SU(2)$. The number $l$ is called the weight of the representation $T(u)$.

Let $T(u)$ be an irreducible unitary representation with weight $l$, of the group $SU(2)$ in the space $\mathfrak{H}$. It follows from Theorem 1 that in $\mathfrak{H}$ there exists an orthonormal basis $f_{-l}, \cdots, f_l$ such that the operators $T(u)$ are given in this basis by the same matrices as the operators $T_l(u)$ in the basis $\{\psi_k(x)\}$. This basis in $\mathfrak{H}$ will be called canonical. It is easy to see that the canonical basis is uniquely determined apart from a common scalar factor $\lambda$, such that $|\lambda| = 1$. Namely, the canonical basis consists of the normalized eigenvectors of the operators $T(h)$, where

$$h=\begin{pmatrix} e^{\frac{it}{2}} & 0 \\ 0 & e^{-\frac{it}{2}} \end{pmatrix}.$$

## §3. MATRIX ELEMENTS OF THE REPRESENTATIONS $T_l(g)$. THE POLYNOMIALS OF LEGENDRE AND JACOBI

In this section we shall calculate the matrix elements of irreducible unitary representations $T_l(u)$ of the group $SU(2)$. To begin with, we shall calculate the matrix elements for representations $T_l(g)$ of the group $SL(2, C)$, obtained on the complexification of $SU(2)$, and then, passing to real values of the parameters, we shall obtain the matrix elements of the representations $T_l(u)$. It will be shown that these matrix elements are expressed by means of the exponential function and the polynomials $P^l_{mn}(z)$, closely related to the classical polynomials of Jacobi and Legendre.

**1. Calculation of the matrix elements.** In §2 we constructed the representations $T_l(g)$ of the group $SL(2, C)$. They are given by the formula

$$T_l(g)\varphi(x)=(\beta x+\delta)^{2l}\varphi\left(\frac{\alpha x+\gamma}{\beta x+\delta}\right), \tag{1}$$

where $\phi(x)$ is a polynomial of degree $2l$ in $x$, and $g=\begin{pmatrix}\alpha & \beta \\ \gamma & \delta\end{pmatrix}$. In the space $\mathfrak{H}_l$ of such polynomials we choose a basis, consisting of the monomials

$$\psi_n(x)=\frac{x^{l-n}}{\sqrt{(l-n)!\,(l+n)!}}, \qquad -l\leqslant n\leqslant l. \tag{2}$$

It has been shown in §2.4 that this basis is orthonormal relative to the scalar product in $\mathfrak{H}_l$, and invariant for the operators $T_l(u)$, $u \in SU(2)$. It follows that to operators $T_l(u)$ of irreducible unitary representations of $SU(2)$ there correspond unitary matrices in this basis.

For calculating the matrix elements, we use the formula

$$a_{ij}=(A\mathbf{e}_j, \mathbf{e}_i), \tag{3}$$

where $\{\mathbf{e}_i\}$ is an orthonormal basis. In our case this formula takes the form

$$t^l_{mn}(g)=(T_l(g)\psi_n, \psi_m)=\frac{(T_l(g)x^{l-n}, x^{l-m})}{\sqrt{(l-m)!\,(l+m)!\,(l-n)!\,(l+n)!}}. \tag{4}$$

But

$$T_l(g)x^{l-n}=(\alpha x+\gamma)^{l-n}(\beta x+\delta)^{l+n}. \tag{5}$$

It follows from formula (4) that

$$t^l_{mn}(g)=\frac{((\alpha x+\gamma)^{l-n}(\beta x+\delta)^{l+n},\ x^{l-m})}{\sqrt{(l-m)!\,(l+m)!\,(l-n)!\,(l+n)!}}, \tag{6}$$

We simplify this expression, remembering that $(x^{l-k},\ x^{l-m})=0$ for $k\neq m$ and $(x^{l-m},\ x^{l-m})=(l-m)!\,(l+m)!$. We get

$$\begin{aligned} t^l_{mn}(g)&=\sqrt{\frac{(l-m)!\,(l+m)!}{(l-n)!\,(l+n)!}}\sum_{j=M}^{N}C^{l-m-j}_{l-n}C^{j}_{l+n}\alpha^{l-m-j}\beta^j\gamma^{m+j-n}\delta^{l+n-j}\\ &=\sqrt{(l-m)!\,(l+m)!\,(l-n)!\,(l+n)!}\ \alpha^{l-m}\gamma^{m-n}\delta^{l+n}\\ &\times\sum_{j=M}^{N}\frac{1}{j!\,(l-m-j)!\,(l+n-j)!\,(m-n+j)!}\left(\frac{\beta\gamma}{\alpha\delta}\right)^j, \end{aligned}\tag{7}$$

where $M=\max(0,\ n-m)$, $N=\min(l-m,\ l+n)$.

Thus we have found an expression for the matrix elements of $T_l(g)$ in terms of the elements of the matrix $g$. We may note that this expression is actually independent of $\beta$, since, as $g$ is unimodular, we have $\beta\gamma=\alpha\delta-1$.

**2. Various expressions for the matrix elements.** Let us derive some other expression for the matrix elements $t^l_{mn}(g)$ of the representations $T_l(g)$. The first of these is connected with the factorization of matrices as the product of upper and lower triangular matrices. Namely, if $g=\begin{pmatrix}\alpha&\beta\\ \gamma&\delta\end{pmatrix}$ and $\delta\neq 0$, then $g=kz$, where $k=\begin{pmatrix}\delta^{-1}&\beta\\ 0&\delta\end{pmatrix}$ and $z=\begin{pmatrix}1&0\\ \gamma/\delta&1\end{pmatrix}$.

It follows from the equality $g=kz$ that $T_l(g)=T_l(k)\,T_l(z)$. Since the multiplication of operators means the multiplication of their matrices, we have

$$t^l_{mn}(g)=\sum_{j=-l}^{l}t^l_{mj}(k)\,t^l_{jn}(z). \tag{1}$$

This equality reduces the problem of calculating the matrix elements $t^l_{mn}(g)$ to the calculation of these elements for triangular matrices $k$ and $z$. But for $k$ we have $\gamma=0$. Hence it follows from equality (7) of the preceding subsection that $t^l_{mn}(k)$ vanishes for $m>n$, and for $m\leq n$ we have only the term for which $j=n-m$:

$$t^l_{mn}(k)=\sqrt{\frac{(l-m)!\,(l+n)!}{(l+m)!\,(l-n)!}}\ \frac{\beta^{n-m}\delta^{m+n}}{(n-m)!}. \tag{2}$$

Similarly, for $m<n$ we have $t^l_{mn}(z)=0$, and for $m\geq n$

$$t^l_{mn}(z)=\sqrt{\frac{(l+m)!\,(l-n)!}{(l-m)!\,(l+n)!}}\ \frac{\gamma^{m-n}}{(m-n)!\,\delta^{m-n}}. \tag{3}$$

On substituting in formula (1) the expressions for $t^l_{mj}(k)$ and $t^l_{jn}(z)$ obtained from (2) and (3), we get

$$t^l_{mn}(g)=\sqrt{\frac{(l-m)!\,(l-n)!}{(l+m)!\,(l+n)!}\frac{\delta^{m+n}}{\beta^m\gamma^n}}\sum_{j=\max(m,n)}^{l}\frac{(l+j)!}{(l-j)!\,(j-m)!\,(j-n)!}(\beta\gamma)^j.$$

This formula is suitable for the matrix elements $t^l_{mn}(g)$, $g=\begin{Vmatrix}\alpha & \beta\\ \gamma & \delta\end{Vmatrix}$ such that $\delta\neq 0$. For $\delta=0$, i.e. $g=\begin{Vmatrix}\alpha & \beta\\ -1/\beta & 0\end{Vmatrix}$, the matrix elements $t^l_{mn}(g)$ are directly calculated from formula (7) of subsection 1. We get

$$t^l_{mn}(g)=\begin{cases}0 & \text{for } m+n>0,\\ \sqrt{\frac{(l-m)!\,(l-n)!}{(l+m)!\,(l+n)!}}\frac{(-1)^{l+m}\beta^{n-m}}{(-m-n)!\,\alpha^{m+n}} & \text{for } m+n\leqslant 0.\end{cases} \tag{5}$$

If $\alpha=\delta=0$, then the same formula (7) gives $t^l_{mn}(g)=0$ for $m+n\neq 0$, and

$$t^l_{m,-m}(g)=(-1)^{l-m}\gamma^{2m}. \tag{6}$$

In particular, if $g=\begin{Vmatrix}0 & i\\ i & 0\end{Vmatrix}$, then $t^l_{mn}(g)=0$ for $m+n\neq 0$, and $t^l_{m,-m}(g)=i^{2l}$.

Another expression for the matrix elements of operator $T_l(g)$ is obtained in the following manner. It follows from the definition of the matrix elements that

$$T_l(g)\psi_n(x)=\sum_{m=-l}^{l}t^l_{mn}(g)\psi_m(x), \tag{1}$$

where

$$\psi_n(x)=\frac{x^{l-n}}{\sqrt{(l-n)!\,(l+n)!}}. \tag{8}$$

As

$$T_l(g)\psi_n(x)=\frac{(\alpha x+\gamma)^{l-n}(\beta x+\delta)^{l+n}}{\sqrt{(l-n)!\,(l+n)!}}, \tag{9}$$

$t^l_{mn}(g)$ is equal to the coefficient of $x^{l-m}$ in the expansion of (9) in powers of $x$, multiplied by $\sqrt{(l-m)!\,(l+m)!}$. By Taylor's theorem we have

$$t^l_{mn}(g)=\sqrt{\frac{(l+m)!}{(l-n)!\,(l+n)!\,(l-m)!}}\frac{d^{l-m}}{dx^{l-m}}[(\alpha x+\gamma)^{l-n}(\beta x+\delta)^{l+n}]_{x=0}. \tag{10}$$

To simplify this expression, we make the substitution $y=\alpha(\beta x+\delta)$ in it. It follows from the equality $\alpha\delta-\beta\gamma=1$ that $\alpha x+\gamma=(y-1)/\beta$, and so

$$t^l_{mn}(g)=\sqrt{\frac{(l+m)!}{(l-n)!\,(l+n)!\,(l-m)!}}\frac{\beta^{n-m}}{\alpha^{m+n}}\frac{d^{l-m}}{dy^{l-m}}[y^{l+n}(y-1)^{l-n}]_{y=\alpha\delta}.$$

Replacing $y$ by $z+1$ here and remembering that $\alpha\delta - 1 = \beta\gamma$, we get

$$t^l_{mn}(g) = \sqrt{\frac{(l+m)!}{(l-n)!\,(l+n)!\,(l-m)!}}\,\frac{\beta^{n-m}}{\alpha^{n+m}}\,\frac{d^{l-m}}{dz^{l-m}}\left[z^{l-n}(z+1)^{l+n}\right]_{z=\beta\gamma}. \tag{11}$$

Finally, we derive an integral representation for the matrix elements $t^l_{mn}(g)$. For this it is convenient to use the realization in the space of trigonometrical polynomials of order $2l$. Owing to formula (10) of §2.1, we have

$$T_l(g)\,e^{-in\varphi} = (\alpha e^{i\varphi} + \gamma)^{l-n}(\beta e^{i\varphi} + \delta)^{l+n}e^{-il\varphi}. \tag{12}$$

Therefore

$$t^l_{mn}(g) = \frac{(T_l(g)\,e^{-in\varphi},\ e^{-im\varphi})}{\sqrt{(l-m)!\,(l+m)!\,(l-n)!\,(l+n)!}}$$
$$= \frac{((\alpha e^{i\varphi}+\gamma)^{l-n}(\beta e^{i\varphi}+\delta)^{l+n}e^{-il\varphi},\ e^{-im\varphi})}{\sqrt{(l-m)!\,(l+m)!\,(l-n)!\,(l+n)!}}. \tag{13}$$

In order to represent this expression in the form of an integral, we observe that for $k \neq m$,

$$(e^{-ik\varphi},\ e^{\,im\varphi}) = \frac{1}{2\pi}\int_0^{2\pi} e^{\,ik\varphi}e^{im\varphi}\,d\varphi = 0,$$

and for $k = -m$,

$$(e^{-im\varphi},\ e^{-im\varphi}) = (l-m)!\,(l+m)! = \frac{(l-m)!\,(l+m)!}{2\pi}\int_0^{2\pi} e^{-im\varphi}e^{im\varphi}\,d\varphi.$$

Therefore any trigonometrical polynomial $\Phi(e^{i\phi})$ of $\mathfrak{H}_l$ satisfies the equality

$$(\Phi(e^{i\varphi}),\ e^{\,im\varphi}) = \frac{(l-m)!\,(l+m)!}{2\pi}\int_0^{2\pi}\Phi(e^{i\varphi})\,e^{im\varphi}\,d\varphi.$$

On applying this formula to (13), we get

$$t^l_{mn}(g) = \frac{1}{2\pi}\sqrt{\frac{(l-m)!\,(l+m)!}{(l-n)!\,(l+n)!}}\int_0^{2\pi}(\alpha e^{i\varphi}+\gamma)^{l-n}(\beta e^{i\varphi}+\delta)^{l+n}e^{i(m-l)\varphi}\,d\varphi. \tag{14}$$

With the substitution $e^{i\phi} = z$, integral (14) can be written in the form of a contour integral:

$$t^l_{mn}(g) = \frac{1}{2\pi i}\sqrt{\frac{(l-m)!\,(l+m)!}{(l-n)!\,(l+n)!}}\oint_\Gamma(\alpha z+\gamma)^{l-n}(\beta z+\delta)^{l+n}z^{m-l-1}\,dz, \tag{15}$$

where $\Gamma$ is the unit circle, described counterclockwise. We may mention that by calculating (15) with the help of residues we arrive anew at formula (10).

3. Expression in terms of the Euler angles. In this subsection we shall obtain an expression for the matrix elements $t^l_{mn}(g)$ in terms of the Euler angles $\phi, \theta, \psi$ of the matrix $g$. By formula (6) of §1.1,

$$g(\varphi, \theta, \psi)=g(\varphi, 0, 0)\,g(0, \theta, 0)\,g(0, 0, \psi). \tag{1}$$

Therefore

$$T_l\,[g(\varphi, \theta, \psi)]=T_l\,[g(\varphi, 0, 0)]\,T_l\,[g(0, \theta, 0)]\,T_l\,[g(0, 0, \psi)]. \tag{2}$$

Thus the search for the matrix of the operator $T_l(g)$ in the general case is reduced to the search for the matrix form of operators $T_l\,[g(\phi, 0, 0)]$, $T_l\,[g(0, \theta, 0)]$ and $T_l\,[g(0, 0, \psi)]$. But $g(\phi, 0, 0)$ is a diagonal matrix:

$$g(\varphi, 0, 0)=\begin{pmatrix} e^{\frac{i\varphi}{2}} & 0 \\ 0 & e^{-\frac{i\varphi}{2}} \end{pmatrix}.$$

For such matrices, by formula (1) of subsection 1, we have

$$T_l\,[g(\varphi, 0, 0)]\,x^{l-n}=e^{-in\varphi}x^{l-n}. \tag{2'}$$

Therefore the matrix of $T_l\,[g(\phi, 0, 0)]$ is the diagonal matrix with the expressions $e^{-in\phi}$, $-l \le n \le l$, on the principal diagonal. The matrix of $T_l\,[g(0, 0, \psi)]$ has a similar form.

Let us denote the matrix elements of $T_l\,[g(0, \theta, 0)]$ by $t^l_{mn}(\theta)$. Then, by virtue of the diagonality of the matrices of $T_l\,[g(\phi, 0, 0)]$ and $T_l\,[g(0, 0, \psi)]$, we have the formula

$$t^l_{mn}(g)=t^l_{mm}[g(\varphi, 0, 0)]\,t^l_{mn}(\theta)\,t^l_{nn}\,[g(0, 0, \psi)]=e^{-i(m\varphi+n\psi)}t^l_{mn}(\theta). \tag{3}$$

It remains for us to obtain an explicit expression for $t^l_{mn}(\theta)$. The matrix $g(0, \theta, 0)$ has the form

$$g(0, \theta, 0)=\begin{pmatrix} \cos\frac{\theta}{2} & i\sin\frac{\theta}{2} \\ i\sin\frac{\theta}{2} & \cos\frac{\theta}{2} \end{pmatrix}, \qquad 0 \leqslant \operatorname{Re}\theta < \pi.$$

Hence, for calculating $t^l_{mn}(\theta)$, it is necessary to put $\alpha = \delta = \cos\theta/2$, $\beta = \gamma = i\sin\theta/2$ in formula (4) of subsection 2. We obtain

$$t^l_{mn}(\theta)=i^{-m-n}\sqrt{\frac{(l-m)!\,(l-n)!}{(l+m)!\,(l+n)!}}\operatorname{ctg}^{m+n}\frac{\theta}{2}$$
$$\times\sum_{j=\max(m,n)}^{l}\frac{(l+j)!\,i^{2j}}{(l-j)!\,(j-m)!\,(j-n)!}\sin^{2j}\frac{\theta}{2}. \tag{4}$$

As remarked in §1.4, the parameter $\theta$ varies in the domain $0\le \operatorname{Re}\theta<\pi$. But, in this domain, to different values of $\theta$ correspond different values of $z=\cos\theta$. Therefore $t^l_{mn}(\theta)$ can be regarded as a function of $\cos\theta$. In accordance with this, we put

$$t^l_{mn}(\theta)=P^l_{mn}(\cos\theta),\qquad 0\le\operatorname{Re}\theta<\pi \tag{5}$$

and write formula (3) as follows:

$$t^l_{mn}(g)=e^{i(m\varphi+n\psi)}P^l_{mn}(\cos\theta). \tag{6}$$

4. **Various expressions for the functions $P^l_{mn}(z)$.** We have just introduced certain functions $P^l_{mn}(z)$ which express the matrix elements of the irreducible unitary representation $T_l(g)$ of the group $SL(2, C)$. We shall now study the properties of these functions. Let us first obtain an explicit expansion for the function $P^l_{mn}(z)$. Formulas (4) and (5) above give

$$P^l_{mn}(z)=i^{-m-n}\sqrt{\frac{(l-m)!\,(l-n)!}{(l+m)!\,(l+n)!}}\left(\frac{1+z}{1-z}\right)^{\frac{m+n}{2}}$$
$$\times\sum_{j=\max(m,n)}^{l}\frac{(l+j)!\,i^{2j}}{(l-j)!\,(j-m)!\,(j-n)!}\left(\frac{1-z}{2}\right)^{j}. \tag{1}$$

The expression $((1+z)/(1-z))^{(m+n)/2}$ in this formula is two-valued (it will be recalled that $m$ and $n$ are either both integral or both semi-integral). In order to determine $P^l_{mn}(z)$ uniquely, let us recollect that the parameter $\theta$ varies in the domain $0\le\operatorname{Re}\theta<\pi$. The mapping $z=\cos\theta$ carries this domain into the $z$-plane with the real axis cut along the rays $(-\infty,-1)$ and $(1,\infty)$. But in such a cut plane the function $((1+z)/(1-z))^{(m+n)/2}$ is single valued.

We choose that branch of this function which takes positive values on the segment $(-1, 1)$ of the real axis.

Utilizing the formulas proved in subsections 1 and 2 for $t^l_{mn}(g)$, it is easy to obtain other expressions for $P^l_{mn}(z)$. On applying, for example, formula (7) of subsection 1 to the matrix

$$g(\theta)=\begin{pmatrix}\cos\frac{\theta}{2} & i\sin\frac{\theta}{2}\\ i\sin\frac{\theta}{2} & \cos\frac{\theta}{2}\end{pmatrix},$$

we obtain

$$P^l_{mn}(z)=i^{m-n}\sqrt{(l-m)!\,(l+m)!\,(l-n)!\,(l+n)!}\left(\frac{1-z}{1+z}\right)^{\frac{m-n}{2}}\left(\frac{1+z}{2}\right)^l$$
$$\times\sum_{j=M}^{N}\frac{(-1)^j}{j!\,(l-m-j)!\,(l+n-j)!\,(m-n+j)!}\left(\frac{1-z}{1+z}\right)^j, \tag{2}$$

where $M=\max(0,n-m)$, $N=\min(l+n,l-m)$.

Next, let us apply formula (11) of subsection 2 to the matrix $g(\theta)$. Replacing $z$ by $(z-1)/2$, we easily obtain

$$P^l_{mn}(z)=\frac{(-1)^{l-n}i^{n-m}}{2^l}\sqrt{\frac{(l+m)!}{(l-n)!\,(l+n)!\,(l-m)!}}$$
$$\times(1+z)^{\frac{-m-n}{2}}(1-z)^{\frac{n-m}{2}}\frac{d^{l-m}}{dz^{l-m}}[(1-z)^{l-n}(1+z)^{l+n}]. \tag{3}$$

Lastly, from the integral representations for $t^l_{mn}(g)$, given by formulas (14) and (15) of subsection 2, we find:

$$P^l_{mn}(z)=\frac{1}{2\pi}\sqrt{\frac{(l-m)!\,(l+m)!}{(l-n)!\,(l+n)!}}$$
$$\times\int_0^{2\pi}\left(\cos\frac{\theta}{2}e^{\frac{i\varphi}{2}}+i\sin\frac{\theta}{2}e^{-\frac{i\varphi}{2}}\right)^{l-n}\left(i\sin\frac{\theta}{2}e^{\frac{i\varphi}{2}}+\cos\frac{\theta}{2}e^{-\frac{i\varphi}{2}}\right)^{l+n}e^{im\varphi}d\varphi$$
$$=\frac{1}{2\pi i}\sqrt{\frac{(l-m)!\,(l+m)!}{(l-n)!\,(l+n)!}}\oint_\Gamma\left(t\cos\frac{\theta}{2}+i\sin\frac{\theta}{2}\right)^{l-n} \tag{4}$$
$$\times\left(it\sin\frac{\theta}{2}+\cos\frac{\theta}{2}\right)^{l+n}t^{m-l-1}dt,$$

where $z=\cos\theta$, $|t|=1$.

5. Special values of $P^l_{mn}(z)$. Let us calculate the value of $P^l_{mn}(z)$ for $z=1$ (i.e. for $\theta=0$). If $\theta=0$, then $g(\theta)$ is the identity matrix. Therefore the matrix of $T_l(g(\theta))$ for $\theta=0$ is also an identity matrix. Since $t^l_{mn}(\theta)\equiv P^l_{mn}(\cos\theta)$, we have

$$P^l_{mn}(1)=\delta_{mn}, \tag{1}$$

where $\delta_{mn}$ is the Kronecker symbol.

Further, let $z=-1$, i.e. $\theta=\pi$. Then $g(\theta)$ takes the form

$$g(\pi)=\begin{pmatrix}0 & i\\ i & 0\end{pmatrix}.$$

As shown in subsection 2, for this matrix $t^l_{mn}(g) = 0$ for $m+n\neq 0$ and $t^l_{m,-m}(g) = i^{2l}$. Thus, $P^l_{mn}(-1) = 0$ for $m+n\neq 0$ and $P^l_{m,-m}(-1) = i^{2l}$.

We now look at the function $P^l_{mn}(z)$ for $m = l$. In this case the sum in formula (1) of subsection 4 contains only the term corresponding to the value $j = l$, and there fore

$$P^l_{ln}(z)=\frac{i^{l-n}}{2^l}\sqrt{\frac{(2l)!}{(l-n)!\,(l+n)!}}(1-z)^{\frac{l-n}{2}}(1+z)^{\frac{l+n}{2}}. \tag{2}$$

In particular,

$$P^l_{ll}(z)=\left(\frac{1+z}{2}\right)^l, \tag{3}$$

$$P^l_{l,-l}(z)=i^{2l}\left(\frac{1-z}{2}\right)^l, \tag{4}$$

$$P^l_{l0}(z)=\frac{i^l\sqrt{(2l)!}}{2^l l!}(1-z^2)^{\frac{l}{2}}. \tag{5}$$

Formulas (2)–(5) can be written as follows:

$$P^l_{ln}(\cos\theta)=i^{l-n}\sqrt{\frac{(2l)!}{(l-n)!\,(l+n)!}}\sin^{l-n}\frac{\theta}{2}\cos^{l+n}\frac{\theta}{2}, \tag{6}$$

$$P^l_{ll}(\cos\theta)=\cos^{2l}\frac{\theta}{2}, \tag{7}$$

$$P^l_{l,-l}(\cos\theta)=i^{2l}\sin^{2l}\frac{\theta}{2}, \tag{8}$$

$$P^l_{l0}(\cos\theta)=\frac{i^l\sqrt{(2l)!}}{2^l l!}\sin^l\theta. \tag{9}$$

We shall not calculate $P^l_{ml}(z)$ etc., since it will be shown below that $P^l_{mn}(z)$ is symmetric relative to the indices $m$ and $n$.

6. The symmetry relations. The functions $P^l_{mn}(z)$ depend not only on the argument $z$, but also on the indices $l$, $m$, $n$. We examine the properties of symmetry of $P^l_{mn}(z)$ for the indices $m$ and $n$. We show first that $P^l_{mn}(z)$ is unaltered if one changes the signs of both the indices $m$ and $n$. For this we use the equality

$$g(\pi)\,g(\theta)=g(\theta)\,g(\pi),$$

where, for brevity,

$$g(\theta)=\begin{pmatrix}\cos\frac{\theta}{2} & i\sin\frac{\theta}{2}\\ i\sin\frac{\theta}{2} & \cos\frac{\theta}{2}\end{pmatrix}.$$

It follows from this equality that

$$T_l(\pi)\, T_l(\theta) = T_l(\theta)\, T_l(\pi). \tag{1}$$

But the matrix elements of the operator $T_l(\theta)$ are equal to $P^l_{mn}(\cos\theta)$. The matrix elements for $T_l(\pi)$ were found by us in the preceding subsection. It was shown there that $t^l_{mn}(\pi) = 0$ for $m + n \neq 0$ and $t^l_{m,-m}(\pi) = i^{2l}$.

On replacing $T_l(\pi)$ and $T_l(\theta)$ in (1) by the corresponding matrices, multiplying the latter out and comparing corresponding elements, we obtain

$$P^l_{m,\,-n}(\cos\theta) = P^l_{-m,\,n}(\cos\theta).$$

This immediately leads to the relation

$$P^l_{mn}(z) = P^l_{-m,\,-n}(z). \tag{2}$$

We now prove that $P^l_{mn}(z)$ are unchanged if one permutes $m$ and $n$. For this we use formula (1) of subsection 4, which gives an explicit expression for $P^l_{mn}(z)$. Since this formula is symmetric in $m$ and $n$, we have

$$P^l_{mn}(z) = P^l_{nm}(z). \tag{3}$$

Equalities (2) and (3) show that the functions $P^l_{mn}(z)$ do not really depend on $m$ and $n$, but on $|m+n|$ and $|m-n|$.

Relation (3) can also be obtained from formula (2) of subsection 4, by making use of the following considerations. The matrix

$$g(\theta) = \begin{pmatrix} \cos\frac{\theta}{2} & i\sin\frac{\theta}{2} \\ i\sin\frac{\theta}{2} & \cos\frac{\theta}{2} \end{pmatrix}$$

has the inverse

$$g(-\theta) = \begin{pmatrix} \cos\frac{\theta}{2} & -i\sin\frac{\theta}{2} \\ -i\sin\frac{\theta}{2} & \cos\frac{\theta}{2} \end{pmatrix},$$

the Euler angles for which are $\pi$, $\theta$, $-\pi$. Hence the matrix elements of the operator $T^l_{mn}(-\theta)$ have the form

$$t^l_{mn}(-\theta) = (-1)^{n-m} P^l_{mn}(\cos\theta). \tag{4}$$

But for real values of $\theta$ the matrix $g(\theta)$ is unitary. Since $T_l(u)$ is a unitary representation of $SU(2)$, we have $T_l(-\theta) = T_l^*(\theta)$. Hence $t^l_{mn}(-\theta) = \overline{t^l_{nm}(\theta)}$ and, by (4),

$$(-1)^{n-m} P^l_{mn}(\cos\theta) = \overline{P^l_{nm}(\cos\theta)}.$$

But $\overline{P^l_{nm}(\cos\theta)} = (-1)^{n-m} P^l_{nm}(\cos\theta)$, as $P^l_{nm}(\cos\theta)$ contains the factor $i^{n-m}$. Thus the equality

$$P^l_{mn}(\cos\theta) = P^l_{nm}(\cos\theta) \tag{5}$$

is valid for real values of $\theta$, i.e. $P^l_{mn}(x) = P^l_{nm}(x)$ is valid for $-1 < x < 1$. Since $P^l_{mn}(z)$ depends analytically on $z$ in a plane cut along the rays $-\infty < x < -1$ and $1 < x < \infty$, equality (5) is valid in the whole of this plane.

Finally, we note the relation

$$P^l_{mn}(-z) = i^{2l-2m-2n} P^l_{m,-n}(z), \tag{6}$$

directly derivable from formula (3) of subsection 4.

7. **The matrices $T_l(\theta)$.** Utilizing formula (4) of subsection 4, we find that the matrices $T_l(\theta)$ for $l = 0, \frac{1}{2}, 1, 3/2$ have the following form:

$$T_0(\theta) = 1,$$

$$T_{\frac{1}{2}}(\theta) = \begin{pmatrix} \cos\frac{\theta}{2} & i\sin\frac{\theta}{2} \\ i\sin\frac{\theta}{2} & \cos\frac{\theta}{2} \end{pmatrix}, \tag{1}$$

$$T_1(\theta) = \begin{pmatrix} \cos^2\frac{\theta}{2} & \frac{i\sin\theta}{\sqrt{2}} & -\sin^2\frac{\theta}{2} \\ \frac{i\sin\theta}{\sqrt{2}} & \cos\theta & \frac{i\sin\theta}{\sqrt{2}} \\ -\sin^2\frac{\theta}{2} & \frac{i\sin\theta}{\sqrt{2}} & \cos^2\frac{\theta}{2} \end{pmatrix}. \tag{2}$$

$$T_{\frac{3}{2}}(\theta) = \begin{pmatrix} \cos^3\frac{\theta}{2} & i\sqrt{3}\sin\frac{\theta}{2}\cos^2\frac{\theta}{2} & -\sqrt{3}\sin^2\frac{\theta}{2}\cos\frac{\theta}{2} & -i\sin^3\frac{\theta}{2} \\ i\sqrt{3}\sin\frac{\theta}{2}\cos^2\frac{\theta}{2} & \cos^3\frac{\theta}{2}-2\cos\frac{\theta}{2}\sin^2\frac{\theta}{2} & 2i\cos^2\frac{\theta}{2}\sin\frac{\theta}{2}-i\sin^3\frac{\theta}{2} & -\sqrt{3}\sin^2\frac{\theta}{2}\cos\frac{\theta}{2} \\ -\sqrt{3}\sin^2\frac{\theta}{2}\cos\frac{\theta}{2} & 2i\cos^2\frac{\theta}{2}\sin\frac{\theta}{2}-i\sin^3\frac{\theta}{2} & \cos^3\frac{\theta}{2}-2\cos\frac{\theta}{2}\sin^2\frac{\theta}{2} & i\sqrt{3}\sin\frac{\theta}{2}\cos^2\frac{\theta}{2} \\ -i\sin^3\frac{\theta}{2} & -\sqrt{3}\sin^2\frac{\theta}{2}\cos\frac{\theta}{2} & i\sqrt{3}\sin\frac{\theta}{2}\cos^2\frac{\theta}{2} & \cos^3\frac{\theta}{2} \end{pmatrix}. \tag{3}$$

8. **The circuit relations.** The functions $P^l_{mn}(z)$ are defined in the complex plane, cut along the rays $(-\infty, -1)$ and $(1, \infty)$. On the two sides of the cuts these functions take different values. It follows from formula (3) of subsection 4 that if $x > 1$, then

$$P^l_{mn}(x + i0) = (-1)^{m-n} P^l_{mn}(x - i0). \tag{1}$$

Similarly, if $x < -1$, then

$$P^l_{mn}(x+i0) = (-1)^{m+n} P^l_{mn}(x-i0). \tag{3}$$

9. **Connection with the classical orthogonal polynomials.** We have introduced the functions $P^l_{mn}(z)$, and obtained various representations for them. We shall now ascertain the connection of the functions $P^l_{mn}(z)$ with certain classical ortho gonal polynomials, namely the Jacobi polynomials, the associated Legendre polynomials and the Legendre polynomials. This connection allows us to derive, from each of the properties of the functions $P^l_{mn}(z)$, the corresponding properties of the Jacobi and Legendre polynomials.

The Jacobi polynomials are defined by the formula

$$P_k^{(\alpha,\beta)}(z) = \frac{(-1)^k}{2^k k!}(1-z)^{-\alpha}(1+z)^{-\beta}\frac{d^k}{dz^k}[(1-z)^{\alpha+k}(1+z)^{\beta+k}]. \tag{1}$$

On comparing this formula with formula (3) of subsection 4, we obtain

$$P_k^{(\alpha,\beta)}(z) = 2^m i^{n-m}\sqrt{\frac{(l-n)!\,(l+n)!}{(l-m)!\,(l+m)!}}(1-z)^{\frac{n-m}{2}}(1+z)^{\frac{-n-m}{2}}P^l_{mn}(z), \tag{2}$$

where

$$l = k+\frac{\alpha+\beta}{2},\quad m = \frac{\alpha+\beta}{2},\quad n = \frac{\beta-\alpha}{2}. \tag{3}$$

It is obvious from (3) that $\alpha = m - n$ and $\beta = m + n$ are integers. Thus the functions $P^l_{mn}(z)$ reduce to that special case of Jacobi polynomials for which $\alpha$ and $\beta$ are integers.

We shall now dwell on two important special cases of Jacobi polynomials, i.e. the Legendre polynomials and the associated Legendre functions. The Legendre polynomials are defined by the equality

$$P_l(z) = \frac{(-1)^l}{2^l l!}\frac{d^l}{dz^l}(1-z^2)^l. \tag{4}$$

In other words, $P_l(z) = P_l^{(0,0)}(z)$. Comparing formula (4) with formula (3) of subsection 4, we get

$$P_l(z) = P^l_{00}(z). \tag{5}$$

The associated Legendre functions $P_l^m(z)$, $m \geq 0$, are defined by the formula ($l$, $m$ are integers)

$$P_l^m(z) = \frac{(-1)^{m+l}(1-z^2)^{\frac{m}{2}}}{2^l l!}\frac{d^{m+l}}{dz^{m+l}}(1-z^2)^l, \tag{6}$$

i.e.

$$P_l^m(z)=\frac{2^m(l+m)!}{l!}(1-z^2)^{-\frac{m}{2}}P_{l+m}^{(-m,-m)}(z). \tag{7}$$

A comparison with formula (3) of subsection 4 leads to the equality

$$P_l^m(z)=i^m\sqrt{\frac{(l+m)!}{(l-m)!}}P^l_{-m,0}(z). \tag{8}$$

We have seen in subsection 6 that $P^l_{-m,0}(z)=P^l_{m,0}(z)$. Therefore equality (8) can be rewritten as follows:

$$P_l^m(z)=i^m\sqrt{\frac{(l+m)!}{(l-m)!}}P^l_{m0}(z),\qquad m\geqslant 0. \tag{9}$$

For $m<0$ we define $P_l^m(z)$ by the same formula (9). Then

$$P_l^{-m}(z)=(-1)^m\frac{(l-m)!}{(l+m)!}P_l^m(z). \tag{10}$$

Formulas (1)–(4) of subsection 4 yield the following representations for the functions $P_l^m(z)$:

$$P_l^m(z)=\left(\frac{1+z}{1-z}\right)^{\frac{m}{2}}\sum_{j=\max(m,0)}^{l}\frac{(-1)^j(l+j)!}{(l-j)!(j-m)!j!}\left(\frac{1-z}{2}\right)^j; \tag{11}$$

$$P_l^m(z)=(-1)^m l!(l+m)!\left(\frac{1-z}{1+z}\right)^{\frac{m}{2}}\left(\frac{1+z}{2}\right)^l \times\sum_{M}^{N}\frac{(-1)^j}{j!(l-m-j)!(l-j)!(m+j)!}\left(\frac{1-z}{1+z}\right)^j, \tag{12}$$

where $M=\max(0,-m)$, $N=\min(l,l-m)$;

$$\begin{aligned}P_l^m(\cos\theta)&=\frac{i^m(l+m)!}{2\pi l!}\int_0^{2\pi}(\cos\theta+i\sin\theta\cos\varphi)^l e^{im\varphi}\,d\varphi\\&=\frac{i^{m+l}(l+m)!\sin^l\theta}{2^{l+1}\pi i l!}\oint_\Gamma(z^2-2iz\operatorname{ctg}\theta+1)^l z^{m-l-1}\,dz;\end{aligned} \tag{13}$$

$$\begin{aligned}P_l^m(z)&=\frac{(-1)^{l+m}}{2^l l!}(1-z^2)^{\frac{m}{2}}\frac{d^{l+m}}{dz^{l+m}}(1-z^2)^l\\&=\frac{(-1)^l(l+m)!}{2^l l!(l-m)!}(1-z^2)^{-\frac{m}{2}}\frac{d^{l-m}}{dz^{l-m}}(1-z^2)^l.\end{aligned} \tag{14}$$

Putting $m=0$, we get the following formulas for Legendre polynomials:

$$P_l(z)=\sum_{j=0}^{l}\frac{(-1)^j(l+j)!}{(l-j)!\,(j!)^2}\left(\frac{1-z}{2}\right)^j$$
$$=(l!)^2\left(\frac{1+z}{2}\right)^l\sum_{j=0}^{l}\frac{(-1)^j}{(j!)^2\,((l-j)!)^2}\left(\frac{1-z}{1+z}\right)^j; \tag{15}$$

$$P_l(\cos\theta)=\frac{1}{2\pi}\int_0^{2\pi}(\cos\theta+i\sin\theta\cos\varphi)^l\,d\varphi$$
$$=\frac{i^l\sin^l\theta}{2^{l+1}\pi i}\oint_\Gamma(z^2-2iz\operatorname{ctg}\theta+1)^l z^{-l-1}\,dz. \tag{16}$$

Comparing formulas (4) and (6), we find that for $m\geq 0$,

$$P_l^m(z)=(-1)^m(1-z^2)^{\frac{m}{2}}\frac{d^mP_l(z)}{dz^m}. \tag{17}$$

We mention a few other representations of functions $P_l^m(z)$ and $P_l(z)$. In formula (13) let us expand $(\cos\theta+i\sin\theta\cos\phi)^l$ by the binomial theorem. Since, for integral $k$ and $m$, $k\geq 0$, $m\geq 0$,

$$\frac{1}{2\pi}\int_0^{2\pi}\cos^k\varphi e^{im\varphi}\,d\varphi=\frac{1}{2^{k+1}\pi}\int_0^{2\pi}(e^{i\varphi}+e^{-i\varphi})^k e^{im\varphi}\,d\varphi$$
$$=\begin{cases}0, \text{ if } k \text{ and } m \text{ have different parity, or } k<m,\\ \dfrac{(m+2r)!}{2^{m+2r}r!\,(m+r)!}, \quad \text{if} \quad k=m+2r, \quad r\geqslant 0,\end{cases} \tag{18}$$

and hence for $m\geq 0$,

$$P_l^m(z)=(-1)^m(l+m)!\left(\frac{\operatorname{tg}\theta}{2}\right)^m\cos^l\theta$$
$$\times\sum_{r=0}^{\left[\frac{l-m}{2}\right]}\frac{(-1)^r}{r!\,(m+r)!\,(l-m-2r)!}\left(\frac{\operatorname{tg}\theta}{2}\right)^{2r}. \tag{19}$$

In particular,

$$P_l(z)=l!\cos^l\theta\sum_{r=0}^{\left[\frac{l}{2}\right]}\frac{(-1)^r}{(r!)^2\,(l-2r)!}\left(\frac{\operatorname{tg}\theta}{2}\right)^{2r}. \tag{20}$$

Further, we note that

$$P_l^m(z)=\frac{(-1)^{l+m}}{2^l}(1-z^2)^{\frac{m}{2}}\sum_k\frac{(-1)^k(2k)!\,z^{2k-l-m}}{k!\,(l-k)!\,(2k-l-m)!}$$

$$=\frac{(-1)^l(l+m)!}{2^l(l-m)!}(1-z^2)^{-\frac{m}{2}}\sum_k\frac{(-1)^k(2k)!\,z^{2k-l+m}}{k!\,(l-k)!\,(2k-l+m)!},\qquad(21)$$

where the summation is carried out over integral values of $k$ such that $0\le k\le l$ and $2k\ge l+m$ ($2k\ge l-m$ in the second sum). For proving this equality it is sufficient to expand $(1-z^2)^l$ in formula (14) by the binomial theorem, and differentiate the resulting expression term by term.

In particular, it follows from formula (21) that

$$P_l(z)=\frac{(-1)^l}{2^l}\sum_{2k\ge l}^{k=l}\frac{(-1)^k(2k)!\,z^{2k-l}}{k!\,(l-k)!\,(2k-l)!}.\qquad(22)$$

In conclusion we obtain an expansion of the Legendre polynomial $P_l(\cos\theta)$ in a Fourier series. For this let us rewrite formula (16) in the form

$$P_l(\cos\theta)=\frac{1}{2\pi}\int_0^{2\pi}\left(e^{i\theta}\cos^2\frac{\varphi}{2}+e^{-i\theta}\sin^2\frac{\varphi}{2}\right)^l d\varphi.\qquad(16')$$

We expand $[e^{i\theta}\cos^2(\phi/2)+e^{-i\theta}\sin^2(\phi/2)]^l$ by the binomial theorem and take into account the fact that, for $0\le k\le l$

$$\frac{1}{2\pi}\int_0^{2\pi}\cos^{2k}\frac{\varphi}{2}\sin^{2l-2k}\frac{\varphi}{2}\,d\varphi=\frac{\Gamma\left(k+\frac{1}{2}\right)\Gamma\left(l-k+\frac{1}{2}\right)}{l!\,\pi}\qquad(23)$$

(see Chapter V, §1.7). We thus obtain the equality

$$P_l(\cos\theta)=\frac{e^{-il\theta}}{\pi}\sum_{k=0}^{l}\frac{\Gamma\left(k+\frac{1}{2}\right)\Gamma\left(l-k+\frac{1}{2}\right)}{k!\,(l-k)!}e^{2ik\theta}.\qquad(24)$$

Since

$$e^{i\theta}=\cos\theta+i\sin\theta=z+\sqrt{z^2-1},$$

equality (24) gives the expansion of $P_l(z)$ in powers of $z+\sqrt{z^2-1}$.

**10. The Legendre polynomials as zonal spherical functions.** We have defined the Legendre polynomials with the help of the equalities

$$P_l(\cos\theta)=P_{00}^l(\cos\theta)=t_{00}^l(g),\qquad g=g(0,\,\theta,\,0),\qquad(1)$$

where $l$ is an integer. But the basis function

$$\psi_0(x)=\frac{x^l}{l!}$$

is invariant for all operators of the form $T_l(h)$, where

$$h=\begin{pmatrix} e^{\frac{it}{2}} & 0 \\ 0 & e^{-\frac{it}{2}} \end{pmatrix}. \qquad (2)$$

According to Chapter I, §2.5, the matrix element $t^l_{00}(g)$ in this case is called the zonal spherical function of the representation $T_l(g)$ relative to the subgroup of matrices of the form (2). Thus $P_l(\cos\theta)$ is the zonal spherical function of the representation $T_l(g)$. It follows from the properties of zonal spherical functions that, for $g = g(\phi, \theta, \psi)$,

$$t^l_{00}(g) = P_l(\cos\theta).$$

Let us now consider the associated spherical functions $t^l_{k0}(g)$ (see Chapter I, §2.5). It follows from formula (6) of subsection 3 that

$$t^l_{k0}(g) = e^{-ik\varphi} P^l_{k0}(\cos\theta)$$

or, by virtue of equality (9) of subsection 9,

$$t^l_{k0}(g) = i^{-k}\sqrt{\frac{(l-k)!}{(l+k)!}}\, e^{-ik\varphi} P^k_l(\cos\theta), \qquad -l\leqslant k\leqslant l.$$

It was observed in Chapter I, §2.5, that the associated spherical functions can be regarded as functions on the homogeneous space $\mathfrak{M} = G/H$. In our case this homogeneous space can be identified with the sphere $S^2$ (see §1.7). Therefore $t^l_{k0}(g)$ are functions on $S^2$.

It is usual to denote the functions $t^l_{k0}(g)$, regarded as functions on the sphere by

$$Y_{lk}(\varphi, \theta), \quad 0\leqslant\varphi<2\pi, \quad 0\leqslant\theta<\pi.$$

Here $\phi$ and $\theta$ are none other than the geographical coordinates on the spheres, where the latitude is reckoned not from the equator, but from a pole.

It will be shown below that the associated spherical functions $Y_{lk}(\phi, \theta)$, $l = 0, 1, \cdots, -l \le k \le l$, form a complete orthogonal system of functions on $S^2$.

## §4. FUNCTIONAL RELATIONS FOR THE FUNCTIONS $P^l_{mn}(z)$

1. **The composition theorem.** Several important properties of the functions $P^l_{mn}(z)$ are connected with the composition theorem for these functions. In order to derive this composition theorem, we use the relation

$$T_l(g_1g_2) = T_l(g_1)\, T_l(g_2). \tag{1}$$

It follows from this relation that

$$t^l_{mn}(g_1g_2) = \sum_{k=-l}^{l} t^l_{mk}(g_1)\, t^l_{kn}(g_2). \tag{2}$$

We apply equality (2) to the matrices $g_1$ and $g_2$, the Euler angles of which are equal to $0, \theta_1, 0$ and $\phi_2, \theta_2, 0$ respectively. For these matrices

$$t^l_{mk}(g_1) = P^l_{mk}(\cos\theta_1) \tag{3}$$

and

$$t^l_{kn}(g_2) = e^{-ik\varphi_2} P^l_{kn}(\cos\theta_2) \tag{4}$$

(see formula (6) of §3.3).

The matrix element $t^l_{mn}(g_1g_2)$ has the form

$$t^l_{mn}(g_1g_2) = e^{-i(m\varphi+n\psi)} P^l_{mn}(\cos\theta), \tag{5}$$

where $\phi, \theta, \psi$ are the Euler angles of $g_1g_2$. According to §1.2, these Euler angles are expressed in terms of the Euler angles $\theta_1, \phi_2, \theta_2$ by the formulas

$$\cos\theta = \cos\theta_1\cos\theta_2 - \sin\theta_1\sin\theta_2\cos\varphi_2, \tag{6}$$

$$e^{i\varphi} = \frac{\sin\theta_1\cos\theta_2 + \cos\theta_1\sin\theta_2\cos\varphi_2 + i\sin\theta_2\sin\varphi_2}{\sin\theta}, \tag{6'}$$

$$e^{\frac{i(\varphi+\psi)}{2}} = \frac{\cos\frac{\theta_1}{2}\cos\frac{\theta_2}{2}e^{\frac{i\varphi_2}{2}} - \sin\frac{\theta_1}{2}\sin\frac{\theta_2}{2}e^{-\frac{i\varphi_2}{2}}}{\cos\frac{\theta}{2}}, \tag{6''}$$

where

$$0 \leqslant \operatorname{Re}\theta < \pi,$$
$$0 \leqslant \operatorname{Re}\varphi < 2\pi,$$
$$-2\pi \leqslant \operatorname{Re}\psi < 2\pi.$$

Let us substitute the values of $t^l_{mn}(g_1g_2)$, $t^l_{mk}(g_1)$ and $t^l_{kn}(g_2)$ in (2). We thus obtain the following composition theorem for functions $P^l_{mn}(z)$:

$$e^{-i(m\varphi+n\psi)} P^l_{mn}(\cos\theta) = \sum_{k=-l}^{l} e^{-ik\varphi_2} P^l_{mk}(\cos\theta_1)\, P^l_{kn}(\cos\theta_2), \tag{7}$$

where the angles $\phi, \psi, \theta, \theta_1, \phi_2, \theta_2$ are connected by the relations (6)–(6″).

We look into some special cases of the composition theorem. Let $\phi_2 = 0$. Then, if $\operatorname{Re}(\theta_1 + \theta_2) < \pi$, we have $\theta = \theta_1 + \theta_2$, $\phi = \psi = 0$, and so formula (7) takes the form

$$P^l_{mn}[\cos(\theta_1+\theta_2)] = \sum_{k=-l}^{l} P^l_{mk}(\cos\theta_1) P^l_{kn}(\cos\theta_2). \tag{8}$$

But if $\operatorname{Re}(\theta_1+\theta_2) > \pi$, then $\theta = 2\pi - \theta_1 - \theta_2$, $\phi = \pi$ and $\psi = \pi$. Therefore (7) takes the form

$$P^l_{mn}[\cos(\theta_1+\theta_2)] = (-1)^{m+n} \sum_{k=-l}^{l} P^l_{mk}(\cos\theta_1) P^l_{kn}(\cos\theta_2). \tag{9}$$

Now let $\phi_2 = \pi$. If $\operatorname{Re}\theta_1 \geq \operatorname{Re}\theta_2$, then $\theta = \theta_1 - \theta_2$, $\phi = 0$, $\psi = \pi$, and so we get

$$P^l_{mn}[\cos(\theta_1-\theta_2)] = \sum_{k=-l}^{l} (-1)^{n-k} P^l_{mk}(\cos\theta_1) P^l_{kn}(\cos\theta_2). \tag{10}$$

In particular, for $\theta_1 = \theta_2 = \theta$, we have

$$\sum_{k=-l}^{l} (-1)^{n-k} P^l_{mk}(\cos\theta) P^l_{kn}(\cos\theta) = \delta_{mn}. \tag{11}$$

For real values of $\theta$ one has the equality

$$\overline{P^l_{mn}(\cos\theta)} = (-1)^{n-m} P^l_{mn}(\cos\theta)$$

(see §3.6). Thus, for real $\theta_1$ and $\theta_2$, formula (10) can be written as follows:

$$P^l_{mn}[\cos(\theta_1-\theta_2)] = \sum_{k=-l}^{l} P^l_{mk}(\cos\theta_1) \overline{P^l_{nk}(\cos\theta_2)}, \tag{10'}$$

where $\theta_1 \geq \theta_2$. For $\theta_1 = \theta_2$ we get

$$\sum_{k=-l}^{l} P^l_{mk}(\cos\theta_1) \overline{P^l_{nk}(\cos\theta_1)} = P^l_{mn}(1) = \delta_{mn}. \tag{11'}$$

Equality (11′) has a simple group-theoretic meaning. For real $\theta$ the matrix $g(0, \theta, 0)$ belongs to the subgroup $SU(2)$. But $T_l(u)$ is a unitary representation of this subgroup, and consequently the matrix with elements $P^l_{mn}(\cos\theta)$ is unitary.

In the sequel we shall also need the case when $\phi_2 = \pi/2$. In this case formulas (6)–(6″) take the following form:

$$\cos\theta = \cos\theta_1 \cos\theta_2, \tag{12}$$

$$e^{i\varphi} = \frac{\sin\theta_1\cos\theta_2 + i\sin\theta_2}{\sin\theta}, \tag{12'}$$

$$e^{\frac{i(\varphi+\psi)}{2}} = \frac{\sqrt{2}\left[\cos\frac{\theta_1+\theta_2}{2} + i\cos\frac{\theta_1-\theta_2}{2}\right]}{2\cos\frac{\theta}{2}}. \tag{12''}$$

It is convenient to replace (12′)–(12″) by

$$\operatorname{tg}\varphi=\frac{\sin\theta_2}{\sin\theta_1\cos\theta_2}, \tag{13'}$$

$$\operatorname{tg}\psi=\frac{\sin\theta_1}{\cos\theta_1\sin\theta_2}, \tag{13''}$$

directly derivable from (3′), (3″) of §1.2. Thus,

$$e^{-i(m\varphi+n\psi)}P^l_{mn}(\cos\theta)=\sum_{k=-l}^{l} i^{-k}P^l_{mk}(\cos\theta_1)P^l_{kn}(\cos\theta_2), \tag{14}$$

where $\phi$, $\psi$, $\theta$, $\theta_1$, $\theta_2$ are connected by the relations (12)–(12″).

2. **The composition theorem for Legendre polynomials.** The composition theorem for the functions $P^l_{mn}(z)$ yields, as special cases, composition theorems for Legendre polynomials and for associated Legendre functions. We have defined these polynomials by the formulas

$$P_l(z)=P^l_{00}(z) \tag{1}$$

and

$$P^m_l(z)=i^m\sqrt{\frac{(l+m)!}{(l-m)!}}P^l_{m0}(z) \tag{2}$$

(see §3.9).

On putting $n=0$ in formula (7) of subsection 1 and using relations (1) and (2), we get

$$e^{-im\varphi}P^m_l(\cos\theta)$$
$$=i^m\sqrt{\frac{(l+m)!}{(l-m)!}}\sum_{k=-l}^{l} i^{-k}\sqrt{\frac{(l-k)!}{(l+k)!}}e^{-ik\varphi_2}P^l_{mk}(\cos\theta_1)P^k_l(\cos\theta_2), \tag{3}$$

where the numbers $\phi$, $\theta$, $\phi_2$, $\theta_1$, $\theta_2$ are connected by the relations (6)–(6″) of subsection 1.

However, if we put $m=0$, $n=0$, then we have

$$P_l(\cos\theta_1\cos\theta_2-\sin\theta_1\sin\theta_2\cos\varphi_2)$$
$$=\sum_{k=-l}^{l}(-1)^k\frac{(l-k)!}{(l+k)!}e^{-ik\varphi_2}P^k_l(\cos\theta_1)P^k_l(\cos\theta_2). \tag{4}$$

This equality can be simplified as follows. Owing to the symmetry relation $P^l_{m0}(z)=P^l_{-m,0}(z)$ and formula (2), we have

$$P^{-m}_l(z)=(-1)^m\frac{(l-m)!}{(l+m)!}P^m_l(z).$$

Hence it follows from equality (4) that the Legendre polynomials $P_l(z)$ satisfy the following composition theorem:

$$P_l(\cos\theta_1\cos\theta_2-\sin\theta_1\sin\theta_2\cos\varphi_2)$$
$$=\sum_{k=-l}^{l} e^{-ik\varphi_2}P_l^k(\cos\theta_1)P_l^{-k}(\cos\theta_2). \tag{5}$$

3. **The multiplication formula.** Suppose that the Euler angle $\phi_2$ in the composition formula (7) of subsection 1 is real. Then (7) can be regarded as an expansion in Fourier series of the function $e^{-i(m\phi+n\psi)}P^l_{mn}(\cos\theta)$ (where $\phi$, $\psi$ and $\theta$ depend on $\phi_2$ according to (6)–(6″)). Therefore

$$P^l_{mk}(\cos\theta_1)P^l_{kn}(\cos\theta_2)=\frac{1}{2\pi}\int_{-\pi}^{\pi} e^{i(k\varphi_2-m\varphi-n\psi)}P^l_{mn}(\cos\theta)d\varphi_2. \tag{1}$$

We shall call this equality the multiplication formula for the functions $P^l_{mn}(z)$.

We note a special case of formula (1). Let us put $m=n=0$ in it. Because of relation (2) of subsection 2, we have

$$\frac{1}{2\pi}\int_{-\pi}^{\pi} e^{ik\varphi_2}P_l(\cos\theta_1\cos\theta_2-\sin\theta_1\sin\theta_2\cos\varphi_2)\,d\varphi_2$$
$$=(-1)^k\frac{(l-k)!}{(l+k)!}P_l^k(\cos\theta_1)P_l^k(\cos\theta_2)=P_l^k(\cos\theta_1)P_l^{-k}(\cos\theta_2). \tag{2}$$

Since $P_l(\cos\theta_1\cos\theta_2-\sin\theta_1\sin\theta_2\cos\phi_2)$ is an even function of $\phi_2$, this equality can be rewritten as follows:

$$\frac{1}{\pi}\int_0^{\pi} P_l(\cos\theta_1\cos\theta_2-\sin\theta_1\sin\theta_2\cos\varphi_2)\cos k\varphi_2\,d\varphi_2$$
$$=P_l^k(\cos\theta_1)P_l^{-k}(\cos\theta_2). \tag{2'}$$

If we put $k=0$, we get

$$\frac{1}{2\pi}\int_{-\pi}^{\pi} P_l(\cos\theta_1\cos\theta_2-\sin\theta_1\sin\theta_2\cos\varphi_2)\,d\varphi_2$$
$$=P_l(\cos\theta_1)P_l(\cos\theta_2). \tag{3}$$

The geometrical significance of formula (2) consists of the following. We choose a point $A$ on the unit sphere, such that $\cup AN=\theta_1$ ($N$ is the north pole of the sphere), and describe on the sphere a circle with center at point $A$ and spherical radius $\theta_2$. Let us denote by $\theta$ the polar distance of a point $B$ of this circle, such that the arc $AB$ of the great circle forms an angle $\phi_2$ with the meridian $AN$. Formula (2) signifies that $P_l^k(\cos\theta_1)P_l^{-k}(\cos\theta_2)$ is the mean value of $P_l(\cos\theta)\,e^{ik\phi_2}$ on this circle. In particular, $P_l(\cos\theta_1)P_l(\cos\theta_2)$ is the mean value of $P_l(\cos\theta)$.

We transform formula (2′). We assume that $\theta_1$, $\theta_2$, $\phi_2$ are real numbers such that

$0 \leq \theta_1 < \pi$, $0 \leq \theta_2 < \pi$, $0 \leq \theta_1 + \theta_2 < \pi$[1], and make the change of variable

$$\cos\theta = \cos\theta_1\cos\theta_2 - \sin\theta_1\sin\theta_2\cos\varphi_2. \tag{4}$$

We shall denote by $T_n(x)$ the function $\cos(n \arccos x)$. It is called the Čebyšev polynomial of the first kind. It follows from equality (4) that

$$\cos k\varphi_2 = T_k\left(\frac{\cos\theta_1\cos\theta_2 - \cos\theta}{\sin\theta_1\sin\theta_2}\right). \tag{5}$$

In addition, equality (4) gives

$$d\varphi_2 = \frac{-\sin\theta\, d\theta}{\sqrt{[\cos\theta - \cos(\theta_1+\theta_2)][\cos(\theta_1-\theta_2) - \cos\theta]}}. \tag{6}$$

Since the variable $\theta$ varies from $\theta_1 + \theta_2$ to $|\theta_1 - \theta_2|$ as $\phi_2$ varies from 0 to $\pi$, the change of variable made above transforms integral (2′) to the form

$$\frac{1}{\pi}\int_{|\theta_1-\theta_2|}^{\theta_1+\theta_2} P_l(\cos\theta)\, T_k\left(\frac{\cos\theta_1\cos\theta_2 - \cos\theta}{\sin\theta_1\sin\theta_2}\right) \times \frac{\sin\theta\, d\theta}{\sqrt{[\cos\theta - \cos(\theta_1+\theta_2)][\cos(\theta_1-\theta_2) - \cos\theta]}} = P_l^k(\cos\theta_1)\, P_l^{-k}(\cos\theta_2). \tag{7}$$

The expression in the denominator of the left hand side has a simple geometrical meaning: it is equal to the area of the spherical triangle with sides $\theta_1$, $\theta_2$, $\theta$, divided by $4\pi^2$.

4. **Recurrence formulas.** We shall now derive formulas connecting functions $P^l_{mn}(z)$ whose lower indices differ from each other by 1. These formulas, called the recurrence formulas for functions $P^l_{mn}(z)$, can be regarded as the infinitesimal form of the composition theorem. They are obtained from the composition theorem for infinitely small $\theta_2$. In order to obtain them it is necessary to differentiate both sides of formula (7) of subsection 1 with respect to $\theta_2$, and put $\theta_2 = 0$. It will be more convenient to utilize, instead of the general formula (7), its special cases, corresponding to the values $\phi_2 = 0$ and $\phi_2 = \pi/2$ (see formulas (8) and (14) of subsection 1).

As a preliminary, let us find the value of $d[P^l_{mn}(\cos\theta)]/d\theta$ for $\theta = 0$. To do this, we differentiate both sides of formula (4) of §3.4 with respect to $\theta$, and put

[1] If the last condition is not satisfied, then $\theta_1$ and $\theta_2$ must be replaced by $\pi - \theta_1$ and $\pi - \theta_2$.

$\theta = 0$. We obtain

$$\frac{d}{d\theta}[P^l_{mn}(\cos\theta)]_{\theta=0} = \frac{i}{4\pi}\sqrt{\frac{(l-m)!\,(l+m)!}{(l-n)!\,(l+n)!}} \times \int_0^{2\pi} [(l-n)e^{-i(n+1)\varphi} + (l+n)e^{-i(n-1)\varphi}]\, e^{im\varphi}\, d\varphi. \qquad (1)$$

Evidently, the right hand side of this equality is equal to zero if $m$ is different from $n+1$ or $n-1$. But if $m = n+1$ then it follows from formula (1) that

$$\frac{d}{d\theta}[P^l_{n+1,\,n}(\cos\theta)]_{\theta=0} = \frac{i}{2}\sqrt{(l-n)(l+n+1)}. \qquad (2)$$

Similarly, we get

$$\frac{d}{d\theta}[P^l_{n-1,\,n}(\cos\theta)]_{\theta=0} = \frac{i}{2}\sqrt{(l+n)(l-n+1)}. \qquad (2')$$

Let us now pass on to the derivation of the recurrence relations. We differentiate both sides of equality (8) of subsection 1 with respect to $\theta_2$, put $\theta_2 = 0$ and utilize relations (2) and (2′). Replacing $\cos\theta_1$ by $z$, we obtain the required recurrence relation:

$$\sqrt{1-z^2}\,\frac{dP^l_{mn}(z)}{dz} = -\frac{i}{2}\left[\sqrt{(l-n)(l+n+1)}\,P^l_{m,\,n+1}(z) + \sqrt{(l+n)(l-n+1)}\,P^l_{m,\,n-1}(z)\right]. \qquad (3)$$

In order to derive a second recurrence relation, we utilize the special case of the composition theorem corresponding to $\phi_2 = \pi/2$. We differentiate both sides of formula (14) of subsection 1 with respect to $\theta_2$, put $\theta_2 = 0$ and utilize relations (12)–(12″) of subsection 1. We easily obtain

$$i\left[m\frac{d\varphi}{d\theta_2} + n\frac{d\psi}{d\theta_2}\right]_{\theta_2=0} P^l_{mn}(\cos\theta_1) - \frac{dP^l_{mn}(\cos\theta_1)}{d\theta_1}\,\frac{d\theta}{d\theta_2}\bigg|_{\theta_2=0} = \frac{1}{2}\left[\sqrt{(l+n)(l-n+1)}\,P^l_{m,\,n-1}(\cos\theta_1) - \sqrt{(l-n)(l+n+1)}\,P^l_{m,\,n+1}(\cos\theta_1)\right], \qquad (4)$$

where $\theta, \phi, \psi$ are expressed in terms of $\theta_1$ and $\theta_2$ according to formulas (12)–(12″) of subsection 1.

It remains for us to find the values of the derivatives $d\phi/d\theta_2$, $d\psi/d\theta_2$, $d\theta/d\theta_2$ for $\theta_2 = 0$. For this we differentiate equality (12) of subsection 1 with respect to $\theta_2$, and put $\theta_2 = 0$. Since, for $\theta_2 = 0$, we have $\theta = \theta_1$, $\phi = 0$, $\psi = \phi_2 = \pi/2$, we get $d\theta/d\theta_2|_{\theta_2=0} = 0$. In exactly the same way, formulas (12′) and (12″) of subsection 1 yield

$$\left.\frac{d\varphi}{d\theta_2}\right|_{\theta_2=0}=\frac{1}{\sin\theta_1} \text{ and } \left.\frac{d\psi}{d\theta_2}\right|_{\theta_2=0}=-\operatorname{ctg}\theta_1.$$

Substituting these values of the derivatives in formula (4), and replacing $\cos\theta_1$ by $z$, we get

$$i\left[\frac{m-nz}{\sqrt{1-z^2}}\right]P^l_{mn}(z) =\frac{1}{2}\left[\sqrt{(l+n)(l-n+1)}\,P^l_{m,\,n-1}(z) - \sqrt{(l-n)(l+n+1)}\,P^l_{m,\,n+1}(z)\right]. \tag{5}$$

Recurrence formulas (3) and (5) easily lead to the recurrence formulas

$$\sqrt{1-z^2}\,\frac{dP^l_{mn}(z)}{dz}+\frac{nz-m}{\sqrt{1-z^2}}\,P^l_{mn}(z) =-i\sqrt{(l-n)(l+n+1)}\,P^l_{m,\,n+1}(z) \tag{6}$$

and

$$\sqrt{1-z^2}\,\frac{dP^l_{mn}(z)}{dz}-\frac{nz-m}{\sqrt{1-z^2}}\,P^l_{mn}(z) =-i\sqrt{(l+n)(l-n+1)}\,P^l_{m,\,n-1}(z). \tag{7}$$

Owing to the symmetry relations, it follows from (6) and (7) that

$$\sqrt{1-z^2}\,\frac{dP^l_{mn}(z)}{dz}+\frac{mz-n}{\sqrt{1-z^2}}\,P^l_{mn}(z) =-i\sqrt{(l-m)(l+m+1)}\,P^l_{m+1,\,n}(z) \tag{8}$$

and

$$\sqrt{1-z^2}\,\frac{dP^l_{mn}(z)}{dz}-\frac{mz-n}{\sqrt{1-z^2}}\,P^l_{mn}(z) =-i\sqrt{(l+m)(l-m+1)}\,P^l_{m-1,\,n}(z). \tag{9}$$

Subtracting (7) from (6), we get a recurrence relation connecting three functions $P^l_{mn}(z)$ with contiguous indices:

$$2\,\frac{nz-m}{\sqrt{1-z^2}}\,P^l_{mn}(z) =i\left[\sqrt{(l+n)(l-n+1)}\,P^l_{m,\,n-1}(z) - \sqrt{(l-n)(l+n+1)}\,P^l_{m,\,n+1}(z)\right]. \tag{10}$$

However, if we add (6) and (7), we get

$$\sqrt{1-z^2}\,\frac{dP^l_{mn}(z)}{dz}$$
$$=-\frac{i}{2}\left[\sqrt{(l-n)(l+n+1)}\,P^l_{m,n+1}(z)\right.$$
$$\left.+\sqrt{(l+n)(l-n+1)}\,P^l_{m,n-1}(z)\right]. \tag{11}$$

Let us put $m=0$ in (6) and (7), and use the equality

$$P^l_{0n}(z)=i^{\,n}\sqrt{\frac{(l-n)!}{(l+n)!}}\,P^n_l(z). \tag{12}$$

We thus obtain the following recurrence relations for the associated Legendre functions $P^m_l(z)$:

$$\sqrt{1-z^2}\,\frac{dP^n_l(z)}{dz}+\frac{nz}{\sqrt{1-z^2}}P^n_l(z)=-P^{n+1}_l(z) \tag{12}$$

and

$$\sqrt{1-z^2}\,\frac{dP^n_l(z)}{dz}-\frac{nz}{\sqrt{1-z^2}}P^n_l(z)=(l+n)(l+n+1)P^{n-1}_l(z). \tag{13}$$

5. The differential equation. It follows from formula (6) above that the differential operator

$$\sqrt{1-z^2}\,\frac{d}{dz}+\frac{nz-m}{\sqrt{1-z^2}} \tag{1}$$

transforms $P^l_{mn}(z)$ into $-i\sqrt{(l-n)(l+n+1)}\;P^l_{m,n+1}(z)$. However, from formula (7) it follows that the operator

$$\sqrt{1-z^2}\,\frac{d}{dz}-\frac{(n+1)z-m}{\sqrt{1-z^2}} \tag{2}$$

transforms $P^l_{m,n+1}(z)$ into $-i\sqrt{(l+n+1)(l-n)}\;P^l_{mn}(z)$. Thus the differential operator

$$\left(\sqrt{1-z^2}\,\frac{d}{dz}-\frac{(n+1)z-m}{\sqrt{1-z^2}}\right)\left(\sqrt{1-z^2}\,\frac{d}{dz}+\frac{nz-m}{\sqrt{1-z^2}}\right) \tag{3}$$

transforms $P^l_{mn}(z)$ into the function $-(l-n)(l+n+1)P^l_{mn}(z)$.

In other words, we have proved that the function $P^l_{mn}(z)$ satisfies the differential equation

$$\left(\sqrt{1-z^2}\,\frac{d}{dz}-\frac{(n+1)z-m}{\sqrt{1-z^2}}\right)\left(\sqrt{1-z^2}\,\frac{d}{dz}+\frac{nz-m}{\sqrt{1-z^2}}\right)P^l_{mn}(z)$$
$$=-(l-n)(l+n+1)\,P^l_{mn}(z). \tag{4}$$

Let us simplify this equation. This leads us to the equation

$$(1-z^2)\frac{d^2P^l_{mn}(z)}{dz^2}-2z\frac{dP^l_{mn}(z)}{dz}-\frac{m^2+n^2-2mnz}{1-z^2}P^l_{mn}(z)$$
$$=-l(l+1)P^l_{mn}(z). \qquad (5)$$

Hence the functions $P^l_{mn}(z)$ are eigenfunctions of the second order differential operator

$$(1-z^2)\frac{d^2}{dz^2}-2z\frac{d}{dz}-\frac{m^2+n^2-2mnz}{1-z^2}, \qquad (6)$$

corresponding to the eigenvalue $-l(l+1)$. The functions assume finite values at the points $z=\pm 1$.

It can be shown that these conditions determine the functions $P^l_{mn}(z)$ uniquely apart from a constant factor. In other words, any eigenfunction of the operator (6) which corresponds to the eigenvalue $-l(l+1)$ and takes finite values at the points $z=\pm 1$, is proportional to $P^l_{mn}(z)$.

It can be proved that if $m$ and $n$ are simultaneously integers or semi-integers, then the eigenfunctions of the operator (6), finite for $z=\pm 1$, exist only for eigenvalues of the form $\lambda=-l(l+1)$.. Here $l$ is an integer or a semi-integer simultaneously with $m$ and $n$.

For $n=0$, as a special case, the differential equation (5) gives the following differential equation for associated Legendre functions:

$$(1-z^2)\frac{d^2P^m_l(z)}{dz^2}-2z\frac{dP^m_l(z)}{dz}+\left[l(l+1)-\frac{m^2}{1-z^2}\right]P^m_l(z)=0. \qquad (7)$$

But if one takes not only $n=0$ but also $m=0$, one obtains the differential equation for Legendre polynomials:

$$(1-z^2)\frac{d^2P_l(z)}{dz^2}-2z\frac{dP_l(z)}{dz}+l(l+1)P_l(z)=0. \qquad (8)$$

6. **Infinitesimal operators of the regular representation.** Recurrence relations for the functions $P^l_{mn}(z)$ are connected with the infinitesimal operators of the regular representation.

Let $\omega(t)$ be a one-parametric subgroup of the group $SU(2)$. Operators of the right-regular representation, corresponding to elements of this subgroup, transform the functions $f(u)$ into $R(\omega(t))f(u)=f(u\omega(t))$. Therefore the infinitesimal operator of the representation $R(u)$, corresponding to the one-parametric subgroup $\omega(t)$, transforms $f(u)$ into the value of the function $df(u\omega(t))/dt$ for $t=0$. This operator is defined, at all events, in the space of infinitely differentiable functions on the group $SU(2)$.

We denote by $\phi(t)$, $\theta(t)$, $\psi(t)$ the Euler angles of the element $u\omega(t)$. Then we have

$$\left.\frac{df(u\omega(t))}{dt}\right|_{t=0}=\frac{\partial f}{\partial\varphi}\varphi'(0)+\frac{\partial f}{\partial\theta}\theta'(0)+\frac{\partial f}{\partial\psi}\psi'(0). \tag{1}$$

Hence the infinitesimal operator $\hat{A}_\omega$ corresponding to the subgroup $\omega(t)$ has the form

$$\hat{A}_\omega=\varphi'(0)\frac{\partial}{\partial\varphi}+\theta'(0)\frac{\partial}{\partial\theta}+\psi'(0)\frac{\partial}{\partial\psi}. \tag{2}$$

The calculation of $\hat{A}_\omega$ thus reduces to a calculation of the derivatives $\phi'(t)$, $\theta'(t)$, $\psi'(t)$ for $t=0$.

Let us calculate the infinitesimal operators $\hat{A}_1$, $\hat{A}_2$, $\hat{A}_3$, corresponding to the one-parametric subgroups $\Omega_1$, $\Omega_2$, $\Omega_3$ (see §1.3). It will be recalled that the subgroup $\Omega_3$ consists of matrices of the form

$$\omega_3(t)=\begin{pmatrix} e^{\frac{it}{2}} & 0 \\ 0 & e^{-\frac{it}{2}} \end{pmatrix}.$$

Let $u=u(\phi,\theta,\psi)$ be a matrix with the Euler angles $\phi,\theta,\psi$. Then the Euler angles of $u\omega_3(t)$ are equal to $\phi,\theta,\psi+t$. Hence

$$\varphi'(0)=0,\quad \theta'(0)=0,\quad \psi'(0)=1.$$

Thus we have proved that the infinitesimal operator $\hat{A}_3$ corresponding to the subgroup $\Omega_3$ has the form

$$\hat{A}_3=\frac{\partial}{\partial\psi}. \tag{3}$$

We pass on to the calculation of the infinitesimal operator $\hat{A}_1$, corresponding to the subgroup $\Omega_1$. This subgroup consists of matrices of the form

$$\omega_1(t)=\begin{pmatrix} \cos\frac{t}{2} & i\sin\frac{t}{2} \\ i\sin\frac{t}{2} & \cos\frac{t}{2} \end{pmatrix},$$

the Euler angles of which are equal to $0, t, 0$. Employing formulas (2)–(2″) of §1.2, we find that the Euler angles $\phi(t)$, $\theta(t)$, $\psi(t)$ of $u\omega(t)$ are connected with the Euler angles $\phi,\theta,\psi$ of $u$ by the relations

$$\cos\theta(t)=\cos\theta\cos t-\sin\theta\sin t\cos\psi, \tag{4}$$

$$e^{i\varphi(t)} = e^{i\varphi}\frac{\sin\theta\cos t + \cos\theta\sin t\cos\psi + i\sin t\sin\psi}{\sin\theta(t)}, \tag{4'}$$

$$e^{\frac{i[\varphi(t)+\psi(t)]}{2}} = e^{\frac{i\varphi}{2}}\frac{\cos\frac{\theta}{2}\cos\frac{t}{2}e^{\frac{i\psi}{2}} - \sin\frac{\theta}{2}\sin\frac{t}{2}e^{-\frac{i\psi}{2}}}{\cos\frac{\theta(t)}{2}}. \tag{4''}$$

For calculating the derivatives $\phi'(t)$, $\theta'(t)$, $\psi'(t)$ for $t = 0$, we differentiate both sides of each of the equalities (4)–(4″) with respect to $t$ and put $t = 0$. Since

$$\varphi(0) = \varphi, \quad \theta(0) = \theta, \quad \psi(0) = \psi,$$

we have

$$\theta'(0) = \cos\psi, \quad \varphi'(0) = \frac{\sin\psi}{\sin\theta}, \quad \psi'(0) = -\operatorname{ctg}\theta\sin\psi.$$

Hence the infinitesimal operator $\hat{A}_1$ corresponding to $\Omega_1$ has the form

$$\hat{A}_1 = \cos\psi\frac{\partial}{\partial\theta} + \frac{\sin\psi}{\sin\theta}\frac{\partial}{\partial\varphi} - \operatorname{ctg}\theta\sin\psi\frac{\partial}{\partial\psi}. \tag{5}$$

The operator $\hat{A}_2$, corresponding to the subgroup $\Omega_2$ of matrices

$$\omega_2(t) = \begin{pmatrix} \cos\frac{t}{2} & -\sin\frac{t}{2} \\ \sin\frac{t}{2} & \cos\frac{t}{2} \end{pmatrix},$$

is calculated in exactly the same way. As a result, we get

$$\hat{A}_2 = -\sin\psi\frac{\partial}{\partial\theta} + \frac{\cos\psi}{\sin\theta}\frac{\partial}{\partial\varphi} - \operatorname{ctg}\theta\cos\psi\frac{\partial}{\partial\psi}. \tag{6}$$

Instead of the operators $\hat{A}_1$, $\hat{A}_2$, $\hat{A}_3$, it is convenient to take their linear combinations

$$\hat{H}_+ = i\hat{A}_1 - \hat{A}_2, \quad \hat{H}_- = i\hat{A}_1 + \hat{A}_2, \quad \hat{H}_3 = i\hat{A}_3.$$

From (3), (5) ,and (6) it follows that these operators are given by the equalities

$$\hat{H}_+ = e^{-i\psi}\left[i\frac{\partial}{\partial\theta} - \frac{1}{\sin\theta}\frac{\partial}{\partial\varphi} + \operatorname{ctg}\theta\frac{\partial}{\partial\psi}\right], \tag{7}$$

$$\hat{H}_- = e^{i\psi}\left[i\frac{\partial}{\partial\theta} + \frac{1}{\sin\theta}\frac{\partial}{\partial\varphi} - \operatorname{ctg}\theta\frac{\partial}{\partial\psi}\right], \tag{8}$$

$$\hat{H}_3 = i\frac{\partial}{\partial\psi}. \tag{9}$$

7. **Infinitesimal operators and recurrence formulas.** We shall now give another derivation of the recurrence relations for the functions $P^l_{mn}(z)$, based on the realization of the representations $T_l(u)$ in the space of functions on the group

$SU(2)$. It was shown in Chapter I, §2.4, that if $T(g)$ is a representation of the group $G$ in the space $\mathfrak{H}$, and $\mathbf{a}$ is a vector of $\mathfrak{H}$, then the equality

$$R(g_0)f(g)=f(gg_0),$$

where

$$f(g)=(T(g)\,\mathbf{f},\ \mathbf{a}),\qquad \mathbf{f}\in\mathfrak{H},\quad g\in G,$$

gives a representation of $G$, equivalent to $T(g)$. Moreover, the matrix of $T(g)$ in the basis $\{\mathbf{e}_n\}$ coincides with the matrix of $R(g)$ in the basis

$$\varphi_n(g)=(T(g)\,\mathbf{e}_n,\ \mathbf{a}).$$

Let us apply these results to the group $SU(2)$ and its irreducible representations $T_l(u)$. As the basis in $\mathfrak{H}_l$ we choose the functions

$$\psi_n(x)=\frac{x^{l-n}}{\sqrt{(l-n)!\,(l+n)!}},\qquad -l\leqslant n\leqslant l,$$

and as the vector $\mathbf{a}$ we take the basis function $\psi_m(x)$. The mapping $\mathbf{f}\longrightarrow f(u)$ takes this basis into a basis consisting of the functions $(T_l(u)\,\psi_n,\ \psi_m)$, i.e. of the elements $t^l_{mn}(u)$ of the $m$th row of the matrix of the representation $T_l(u)$ in the basis $\{\psi_n(x)\}$. We shall denote the representation of $SU(2)$ in the space $\mathfrak{H}_{lm}$ of functions $f(u)$ by $R_{lm}(u)$. Because of what we have said above, the matrix of $R_{lm}(u)$ in the basis $\{t^l_{mn}(u)\}$, $-l\leq n\leq l$, coincides with the matrix of $T_l(u)$ in the basis $\{\psi_n(x)\}$, $-l\leq n\leq l$.

But if the representation matrices coincide, then the matrices of infinitesimal operators also coincide, since they are uniquely determined by the representation operators. Let us denote the infinitesimal operators of $R_{lm}(u)$ by $A^{lm}_k$, $1\leq k\leq 3$. Taking into consideration formulas (6) of §4.2 and the remark made concerning the coinciding of matrices, we get

$$H^{lm}_{+}t^l_{mn}(u)=-\sqrt{(l-n)(l+n+1)}\,t^l_{m,\,n+1}(u),\tag{1}$$
$$H^{lm}_{-}t^l_{mn}(u)=-\sqrt{(l+n)(l-n+1)}\,t^l_{m,\,n-1}(u),\tag{2}$$
$$H^{lm}_{3}t^l_{mn}(u)=nt^l_{mn}(u),\tag{3}$$

where

$$H^{lm}_{+}=iA^{lm}_1-A^{lm}_2,\quad H^{lm}_{-}=iA^{lm}_1+A^{lm}_2,\quad H^{lm}_3=iA^{lm}_3.$$

In order to obtain the required recurrence relations, we have to calculate the infinitesimal operators $A^{lm}_k$ and substitute their expressions, as well as the expressions for $t^l_{mn}(u)$, in formulas (1)–(3). For calculating the infinitesimal operators, we observe that $R_{lm}(u)$ is the restriction to the subspace $\mathfrak{H}_{lm}$ of the right-regular representation

$$R(u_0)f(u)=f(uu_0).$$

Therefore the operators $H_+^{lm}$, $H_-^{lm}$ and $H_3^{lm}$ are given by the same formulas as the operators $\hat{H}_+$, $\hat{H}_-$, $\hat{H}_3$ (see formulas (7)–(9) of subsection 6). We substitute the expressions for $H_+^{lm}$ and $H_-^{lm}$ and the expression (6) of §3.3 for $t^l_{mn}(u)$ in formulas (1)–(3). After simple manipulations we obtain the following recurrence formulas for $P^l_{mn}(z)$:

$$\sqrt{1-z^2}\,\frac{dP^l_{mn}(z)}{dz}-\frac{m-nz}{\sqrt{1-z^2}}P^l_{mn}(z)$$
$$=-i\sqrt{(l-n)(l+n+1)}\,P^l_{m,n+1}(z)$$

and

$$\sqrt{1-z^2}\,\frac{dP^l_{mn}(z)}{dz}+\frac{m-nz}{\sqrt{1-z^2}}P^l_{mn}(z)$$
$$=-i\sqrt{(l+n)(l-n+1)}\,P^l_{m,n-1}(z).$$

8. The Laplace operator. The expressions found for the operators $A_k^{lm}$, $k=1, 2, 3$, allow one to give a new derivation of the differential equation for functions $P^l_{mn}(z)$. To that end, we consider the operator

$$\Delta_{lm}=(A_1^{lm})^2+(A_2^{lm})^2+(A_3^{lm})^2.$$

Since the operators $A_k^{lm}$ are given by the same matrices as the operators $A_k$, $k=1, 2, 3$, the matrix of the operator $\Delta_{lm}$ coincides with the matrix of the operator $A_1^2+A_2^2+A_3^2$. But we easily find from formulas (5)–(7) of §2.2 that

$$A_1^2+A_2^2+A_3^2=-l(l+1)E.$$

Consequently,

$$\Delta_{lm}=-l(l+1)E \text{ (in the space } \mathfrak{H}_{lm}). \tag{1}$$

In order to obtain an explicit expression for the operator $\Delta_{lm}$ in terms of the Euler angles, we use expressions (3), (5) and (6) of subsection 6. We obtain

$$\Delta_{lm}=\frac{\partial^2}{\partial\theta^2}+\operatorname{ctg}\theta\,\frac{\partial}{\partial\theta}+\frac{1}{\sin^2\theta}\left(\frac{\partial^2}{\partial\varphi^2}-2\cos\theta\,\frac{\partial^2}{\partial\varphi\,\partial\psi}+\frac{\partial^2}{\partial\psi^2}\right). \tag{2}$$

Since the matrix elements $t^l_{mn}(u)$, $-l\le n\le l$, belong to the space $\mathfrak{H}_{lm}$, they satisfy the equality

$$\Delta_{lm}t^l_{mn}(u)=-l(l+1)\,t^l_{mn}(u).$$

Substituting the expression (2) for $\Delta_{lm}$, and the expression (6) of §3.3 for $t^l_{mn}(u)$, we easily obtain the following differential equation for functions $P^l_{mn}(z)$:

$$(1-z^2)\frac{d^2P^l_{mn}(z)}{dz^2}-2z\frac{dP^l_{mn}(z)}{dz}-\frac{m^2+n^2-2mnz}{1-z^2}P^l_{mn}(z)=-l(l+1)P^l_{mn}(z).$$

We note that the operators $A^{mn}_k$ are the restriction to $\mathfrak{H}_{lm}$ of $\hat{A}_k$, $k=1, 2, 3$, of the right-regular representation $R(g)$. Therefore $\Delta_{lm}$ is the restriction to $\mathfrak{H}_{lm}$ of the operator

$$\Delta=\hat{A}_1^2+\hat{A}_2^2+\hat{A}_3^2.$$

This operator is called the Laplace operator on the group $SU(2)$, and is expressed in terms of the Euler angles by the same formula as $\Delta_{lm}$ (see (2)):

$$\Delta=\frac{\partial^2}{\partial\theta^2}+\operatorname{ctg}\theta\frac{\partial}{\partial\theta}+\frac{1}{\sin^2\theta}\left(\frac{\partial^2}{\partial\varphi^2}-2\cos\theta\frac{\partial^2}{\partial\varphi\,\partial\psi}+\frac{\partial^2}{\partial\psi^2}\right).$$

A remarkable property of the Laplace operator $\Delta$ is its permutability with right-shift operators:

$$\Delta R(u)=R(u)\Delta,$$

where

$$R(u_0)f(u)=f(uu_0).$$

This assertion can be proved, for example, by decomposing the right-regular representation $R(g)$ into irreducible ones, and by observing that the restriction of $\Delta$ to $\mathfrak{H}_{lm}$ is a multiple of the identity operator (see formula (1)). Consequently $\Delta$ is permutable with all irreducible components $R_{lm}(g)$, and so it is permutable with the operators $R(g)$ as well.

Another method of proof, which does not utilize formula (1), consists of the following. We first show that $\Delta$ permutes with the infinitesimal operators $\hat{A}_1$, $\hat{A}_2$, $\hat{A}_3$. As remarked in Chapter I, §1.11, the infinitesimal operators of a representation satisfy the same commutation relations as the corresponding elements of the Lie algebra of the given group. But it was shown in §1.3 that the tangent matrices $a_1$, $a_2$, $a_3$ to the one-parametric subgroups $\Omega_1$, $\Omega_2$, $\Omega_3$ satisfy the commutation relations

$$[a_1, a_2]=a_3,$$
$$[a_2, a_3]=a_1,$$
$$[a_3, a_1]=a_2.$$

Therefore one has the equalities

$$[\hat{A}_1, \hat{A}_2] = \hat{A}_3,$$
$$[\hat{A}_2, \hat{A}_3] = \hat{A}_1,$$
$$[\hat{A}_3, \hat{A}_1] = \hat{A}_2.$$

Utilizing these equalities, we find

$$[\hat{A}_2^2, \hat{A}_1] = \hat{A}_2\hat{A}_2\hat{A}_1 - \hat{A}_1\hat{A}_2\hat{A}_2 = \hat{A}_2\hat{A}_2\hat{A}_1 - \hat{A}_2\hat{A}_1\hat{A}_2 - \hat{A}_3\hat{A}_2$$
$$= \hat{A}_2[\hat{A}_2, \hat{A}_1] - \hat{A}_3\hat{A}_2 = -\hat{A}_2\hat{A}_3 - \hat{A}_3\hat{A}_2.$$

Similarly, it can be proved that

$$[\hat{A}_3^2, A_1] = \hat{A}_2\hat{A}_3 + \hat{A}_3\hat{A}_2.$$

Lastly, $[\hat{A}_1^2, \hat{A}_1] = 0$. It follows from these equalities that

$$[\Delta, \hat{A}_1] = [\hat{A}_1^2 + \hat{A}_2^2 + \hat{A}_3^2, \hat{A}_1] = 0.$$

In exactly the same way one proves that $[\Delta, \hat{A}_2] = 0$ and $[\Delta, \hat{A}_3] = 0$.

According to §1.10 of Chapter I, the operators of representation $R(u)$ are expressed in terms of the infinitesimal operators $\hat{A}_k$; $k = 1, 2, 3$, by the formulas

$$R(u) = \exp(t_1\hat{A}_1 + t_2\hat{A}_2 + t_3\hat{A}_3),$$

where

$$u = \exp(t_1a_1 + t_2a_2 + t_3a_3).$$

Hence it follows from the equality $[\Delta, \hat{A}_k] = 0$ that the operators $R(u)$ are permutable with the Laplace operator $\Delta$.

One can show that the Laplace operator $\Delta$ on the group $SU(2)$ permutes not only with right shifts, but also with left shifts:

$$\Delta L(u) = L(u)\Delta,$$

where

$$L(u_0)f(u) = f(u_0^{-1}u).$$

Let us restrict $\Delta$ to the subspace $\mathfrak{H}_0$ of functions $f(u)$ constant on right cosets with respect to the subgroup $\Omega$, consisting of the matrices

$$\begin{pmatrix} e^{\frac{it}{2}} & 0 \\ 0 & e^{-\frac{it}{2}} \end{pmatrix},$$

or, what is the same thing, on the space of functions given on the unit sphere $S^2$. Clearly, functions $f(u)$ of $\mathfrak{H}_0$ do not depend on the Euler angle $\varphi$, and therefore the Laplace operator on the unit sphere $S^2$ has the form

$$\Delta_0 = \frac{\partial^2}{\partial\theta^2} + \operatorname{ctg}\theta\frac{\partial}{\partial\theta} + \frac{1}{\sin^2\theta}\frac{\partial^2}{\partial\psi^2}.$$

Here $\psi$, $\theta$ denote geographical coordinates on the sphere. The operator $\Delta_0$ is the angular part of the Laplace operator in three-dimensional euclidean space.

9. Further recurrence relations. Recurrence relations of another type can be obtained if one uses the properties, proved in §1.6, of tangent matrices to one-parametric subgroups. Namely, we have proved that if $a = x_1a_1 + x_2a_2 + x_3a_3$ is a tangent matrix to the subgroup $g(t)$, then the tangent matrix to the subgroup $g_0g(t)g_0^{-1}$, $g_0 \in SO(3)$, has the form $b = y_1a_1 + y_2a_2 + y_3a_3$, where $\mathbf{y} = g_0\mathbf{x}$. Hence for any representation $T(g)$ one has the equality

$$T(g_0)A_{\mathbf{x}}T(g_0^{-1}) = A_{\mathbf{y}},$$

where $A_{\mathbf{x}}$ is the infinitesimal operator corresponding to the one-parametric subgroup $g(t)$, and $A_{\mathbf{y}}$, $\mathbf{y} = g_0\mathbf{x}$, is the infinitesimal operator corresponding to the one-parametric subgroup $g_0g(t)g_0^{-1}$. Thus

$$T(g_0)A_{\mathbf{x}} = A_{g_0\mathbf{x}}T(g_0). \tag{1}$$

Let us put $\mathbf{x} = (1, 0, 0)$ in (1). The components of the vector $g_0\mathbf{x}$ will be elements of the first column of the matrix of the rotation $g_0$ with Euler angles $\phi$, $\theta$, $\psi$ (see §1.6). Thus we find from equality (1), for $\mathbf{x} = (1, 0, 0)$, that

$$\begin{aligned} T(g_0)A_1 &= (\cos\varphi\cos\psi - \sin\varphi\sin\psi\cos\theta)A_1T(g_0) \\ &+ (\sin\varphi\cos\psi + \cos\varphi\sin\psi\cos\theta)A_2T(g_0) + \sin\psi\sin\theta A_3T(g_0). \end{aligned}$$

We pass from the matrices $A_1$, $A_2$, $A_3$ to $H_+$, $H_-$, $H_3$ by the formulas

$$\begin{aligned} H_+ &= iA_1 - A_2, \\ H_- &= iA_1 + A_2, \\ H_3 &= iA_3. \end{aligned}$$

On doing so and collecting the coefficients of $H_+$ and $H_-$ in the right-hand side, we obtain the relation

$$\begin{aligned} T(g_0)(H_+ + H_-) &= e^{-i\varphi}(\cos\psi - i\sin\psi\cos\theta)H_+T(g_0) \\ &+ e^{i\varphi}(\cos\psi + i\sin\psi\cos\theta)H_-T(g_0) + 2\sin\psi\sin\theta H_3T(g_0), \end{aligned}$$

where $g_0 = g(\phi, \theta, \psi)$.

Let us apply the transformations which stand on the two sides of this equality to the vector $\mathbf{f}_n$ of the canonical basis, and compare the coefficients of $\mathbf{f}_m$ in the expressions obtained. We have

$$T(g_0)\mathfrak{f}_n=\sum_m t_{mn}(g_0)\mathfrak{f}_m,$$
$$H_+\mathfrak{f}_n=-\sqrt{(l-n)(l+n+1)}\,\mathfrak{f}_{n+1},$$
$$H_-\mathfrak{f}_n=-\sqrt{(l+n)(l-n+1)}\,\mathfrak{f}_{n-1},$$
$$H_3\mathfrak{f}_n=n\mathfrak{f}_n.$$

Taking into account these relations and also the expression for the matrix elements $t^l_{mn}(g_0)$ in terms of the Euler angles $\phi$, $\theta$, $\psi$, we get

$$\begin{aligned}\sqrt{(l-n)(l+n+1)}\,e^{-i\psi}P^l_{m,n+1}(\cos\theta)&\\ +\sqrt{(l+n)(l-n+1)}\,e^{i\psi}P^l_{m,n-1}(\cos\theta)&\\ =\sqrt{(l-m+1)(l+m)}(\cos\psi-i\sin\psi\cos\theta)P^l_{m-1,n}(\cos\theta)&\\ +\sqrt{(l+m+1)(l-m)}(\cos\psi+i\sin\psi\cos\theta)P^l_{m+1,n}(\cos\theta)&\\ -2m\sin\psi\sin\theta P^l_{mn}(\cos\theta).&\end{aligned}$$

This equality holds for all values of $\psi$. Let us put $\cos\psi=(e^{i\psi}+e^{-i\psi})/2$ and $\sin\psi=(e^{i\psi}-e^{-i\psi})/2i$ in it, and equate the coefficients of $e^{i\psi}$ and $e^{-i\psi}$ in the left and right hand sides. We obtain the required recurrence formulas

$$\begin{aligned}\sqrt{(l+n)(l-n+1)}\,P^l_{m,n-1}(\cos\theta)&=\tfrac{1}{2}\sqrt{(l-m+1)(l+m)}\\ &\times(1-\cos\theta)P^l_{m-1,n}(\cos\theta)+\tfrac{1}{2}\sqrt{(l+m+1)(l-m)}\\ &\times(1+\cos\theta)P^l_{m+1,n}(\cos\theta)+mi\sin\theta P^l_{mn}(\cos\theta)\end{aligned}\tag{2}$$

and

$$\begin{aligned}\sqrt{(l-n)(l+n+1)}\,P^l_{m,n+1}(\cos\theta)&=\tfrac{1}{2}\sqrt{(l-m+1)(l+m)}\\ &\times(1+\cos\theta)P^l_{m-1,n}(\cos\theta)+\tfrac{1}{2}\sqrt{(l+m+1)(l-m)}\\ &\times(1-\cos\theta)P^l_{m+1,n}(\cos\theta)-mi\sin\theta P^l_{mn}(\cos\theta).\end{aligned}\tag{3}$$

These formulas connect three adjacent elements of the $n$th column with elements of the $(n-1)$th or $(n+1)$th column. On using the symmetry relations for functions $P^l_{mn}(z)$, we get analogous formulas which connect three adjacent elements of one row with elements of the row above or below it. We shall leave it to the reader to write these formulas.

If we take $\mathbf{x}=(0,1,0)$ in (1), we obtain the same recurrence formulas. For $\mathbf{x}=(0,0,1)$ one obtains the formulas of subsection 4.

## §5. GENERATING FUNCTIONS FOR $P^l_{mn}(z)$

In this section we shall indicate a new method of deriving recurrence formulas, based on the use of generating functions. $f(z, h)$ is called a generating function for the functions $\phi_k(z)$ if its expansion in powers of $h$ has the form

$$f(z, h) = \sum_{k=0}^{\infty} \alpha_k \varphi_k(z) h^k.$$

We shall consider two kinds of generating functions for $P^l_{mn}(z)$. One of these gives functions $P^l_{mn}(z)$ with fixed values of $l$ and $n$, and the second gives functions with fixed values of $m$ and $n$.

1. The case of fixed $l$ and $n$. In order to obtain the generating function for $P^l_{mn}(z)$ for fixed values of $l$ and $n$, we rewrite equality (4) of §3.4 in the form

$$\frac{P^l_{mn}(\cos\theta)}{\sqrt{(l-m)!\,(l+m)!}} = \frac{1}{2\pi\sqrt{(l-n)!\,(l+n)!}} \int_0^{2\pi} \left(\cos\frac{\theta}{2} e^{\frac{i\varphi}{2}} + i\sin\frac{\theta}{2} e^{-\frac{i\varphi}{2}}\right)^{l-n} \times \left(i\sin\frac{\theta}{2} e^{\frac{i\varphi}{2}} + \cos\frac{\theta}{2} e^{-\frac{i\varphi}{2}}\right)^{l+n} e^{im\varphi}\, d\varphi. \qquad (1)$$

This formula signifies that the functions $P^l_{mn}(\cos\theta)/\sqrt{(l-m)!\,(l+m)!}$ are Fourier coefficients for

$$\frac{1}{\sqrt{(l-n)!\,(l+n)!}} \left(\cos\frac{\theta}{2} e^{\frac{i\varphi}{2}} + i\sin\frac{\theta}{2} e^{-\frac{i\varphi}{2}}\right)^{l-n} \times \left(i\sin\frac{\theta}{2} e^{\frac{i\varphi}{2}} + \cos\frac{\theta}{2} e^{-\frac{i\varphi}{2}}\right)^{l+n}.$$

Thus

$$\frac{1}{\sqrt{(l-n)!\,(l+n)!}} \left(\cos\frac{\theta}{2} e^{\frac{i\varphi}{2}} + i\sin\frac{\theta}{2} e^{-\frac{i\varphi}{2}}\right)^{l-n} \times \left(i\sin\frac{\theta}{2} e^{\frac{i\varphi}{2}} + \cos\frac{\theta}{2} e^{-\frac{i\varphi}{2}}\right)^{l+n} = \sum_{m=-l}^{l} \frac{P^l_{mn}(\cos\theta)}{\sqrt{(l-m)!\,(l+m)!}} e^{-im\varphi}. \qquad (2)$$

On multiplying both sides of this equality by $e^{il\phi}$ and replacing $e^{i\phi}$ by $w$, we obtain

$$F(w)\equiv\frac{1}{\sqrt{(l-n)!\,(l+n)!}}\left(w\cos\frac{\theta}{2}+i\sin\frac{\theta}{2}\right)^{l-n}\left(iw\sin\frac{\theta}{2}+\cos\frac{\theta}{2}\right)^{l+n}$$
$$=\sum_{m=-l}^{l}\frac{P^l_{mn}(\cos\theta)}{\sqrt{(l-m)!\,(l+m)!}}w^{l-m}. \tag{3}$$

Hence $F(w)$ is a generating function for $P^l_{mn}(\cos\theta)$.

From equality (3) one can derive certain curious identities for the functions $P^l_{mn}(\cos\theta)$. For example, putting $w=1$ we get

$$\sum_{m=-l}^{l}\frac{P^l_{mn}(\cos\theta)}{\sqrt{(l-m)!\,(l+m)!}}=\frac{e^{il\theta}}{\sqrt{(l-n)!\,(l+n)!}}. \tag{4}$$

But if we put $w=-1$, we get

$$\sum_{m=-l}^{l}\frac{(-1)^{n-m}P^l_{mn}(\cos\theta)}{\sqrt{(l-m)!\,(l+m)!}}=\frac{e^{-il\theta}}{\sqrt{(l-n)!\,(l+n)!}}. \tag{4'}$$

Lastly, putting $n=0$ and $w=\pm i$, we have

$$\sum_{m=-l}^{l}\frac{i^{\pm m}P^l_{m0}(\cos\theta)}{\sqrt{(l-m)!\,(l+m)!}}=\frac{\cos^l\theta}{l!}. \tag{5}$$

Taking into consideration equality (9) of §3.9, one can rewrite formula (5) as follows:

$$\sum_{m=-l}^{l}\frac{(\pm 1)^m P^m_l(\cos\theta)}{(l+m)!}=\frac{\cos^l\theta}{l!}. \tag{6}$$

We shall now show how one can derive recurrence formulas for functions $P^l_{mn}(\cos\theta)$ from the expansion (3). On differentiating both sides of this equality with respect to $\theta$, we obtain

$$\frac{i\sqrt{l-n}}{2\sqrt{(l-n-1)!\,(l+n)!}}\left(w\cos\frac{\theta}{2}+i\sin\frac{\theta}{2}\right)^{l-n-1}$$
$$\times\left(iw\sin\frac{\theta}{2}+\cos\frac{\theta}{2}\right)^{l+n+1}+\frac{i\sqrt{l+n}}{2\sqrt{(l-n)!\,(l+n-1)!}}$$
$$\times\left(w\cos\frac{\theta}{2}+i\sin\frac{\theta}{2}\right)^{l-n+1}\left(iw\sin\frac{\theta}{2}+\cos\frac{\theta}{2}\right)^{l+n-1}$$
$$=\sum_{m=-l}^{l}\frac{\frac{d}{d\theta}[P^l_{mn}(\cos\theta)]}{\sqrt{(l-m)!\,(l+m)!}}w^{l-m}.$$

The left-hand side of this equality is a linear combination of expressions of the same type as the left-hand side of equality (3), with the only difference that

$n$ is replaced in one term by $n+1$, and in the other by $n-1$. Employing the expansion (3) for these expressions, and comparing the coefficients of $w^{l-m}$, we get

$$\frac{d}{d\theta}[P^l_{mn}(\cos\theta)] = \frac{i}{2}\left[\sqrt{(l-n)(l+n+1)}\,P^l_{m,n+1}(\cos\theta) + \sqrt{(l+n)(l-n+1)}\,P^l_{m,n-1}(\cos\theta)\right]. \qquad (7)$$

In other words,

$$\sqrt{1-z^2}\,\frac{dP^l_{mn}(z)}{dz} = -\frac{i}{2}\left[\sqrt{(l-n)(l+n+1)}\,P^l_{m,n-1}(z) + \sqrt{(l+n)(l-n+1)}\,P^l_{m,n-1}(z)\right]. \qquad (7')$$

We have thus derived anew formula (11) of §4.4.

In order to derive formula (10) of §4.4, we multiply both sides of (3) by the corresponding sides of the identity

$$\cos\frac{\theta}{2}\left(w\cos\frac{\theta}{2}+i\sin\frac{\theta}{2}\right) - i\sin\frac{\theta}{2}\left(iw\sin\frac{\theta}{2}+\cos\frac{\theta}{2}\right) = w$$

and differentiate with respect to $w$. We obtain

$$\frac{1}{\sqrt{(l-n)!\,(l+n)!}}\left[(l+1-n\cos\theta)\left(w\cos\frac{\theta}{2}+i\sin\frac{\theta}{2}\right)^{l-n}\right.$$
$$\times\left(iw\sin\frac{\theta}{2}+\cos\frac{\theta}{2}\right)^{l+n} + \frac{i(l+n)\sin\theta}{2}\left(w\cos\frac{\theta}{2}+i\sin\frac{\theta}{2}\right)^{l-n+1}$$
$$\times\left(iw\sin\frac{\theta}{2}+\cos\frac{\theta}{2}\right)^{l+n-1} - \frac{i(l-n)\sin\theta}{2}\left(w\cos\frac{\theta}{2}+i\sin\frac{\theta}{2}\right)^{l-n-1}$$
$$\left.\times\left(iw\sin\frac{\theta}{2}+\cos\frac{\theta}{2}\right)^{l+n+1}\right] = \sum_{m=-l}^{l}\frac{(l-m+1)\,P^l_{mn}(\cos\theta)}{\sqrt{(l-m)!\,(l+m)!}}\,w^{l-m}.$$

The expression in square brackets is a linear combination of expressions, analogous to the left-hand side of (3), with the only difference that in some terms $n$ is replaced by $n-1$ or $n+1$. Expanding these expressions according to (3) and comparing the coefficients of $w^{l-m}$, we arrive at the equality

$$i(m-n\cos\theta)\,P^l_{mn}(\cos\theta) = \frac{\sin\theta}{2}\left[\sqrt{(l+n)(l-n+1)}\,P^l_{m,n+1}(\cos\theta) - \sqrt{(l-n)(l+n+1)}\,P^l_{m,n+1}(\cos\theta)\right]. \qquad (8)$$

Replacing $\cos\theta$ by $z$ here, we get formula (10) of §4.4.

2. Recurrence formulas for different values of $l$. With the help of the generating function derived in the preceding subsection, one can easily obtain recurrence formulas connecting functions $P^l_{mn}(z)$ with different values of $l$. For

example, on differentiating both sides of expansion (3) of subsection 1 with respect to $w$, we obtain

$$\frac{1}{\sqrt{(l-n)!\,(l+n)!}}\left[(l-n)\cos\frac{\theta}{2}\left(w\cos\frac{\theta}{2}+i\sin\frac{\theta}{2}\right)^{l-n-1}\right.$$
$$\times\left(iw\sin\frac{\theta}{2}+\cos\frac{\theta}{2}\right)^{l+n}+i(l+n)\sin\frac{\theta}{2}\left(w\cos\frac{\theta}{2}+i\sin\frac{\theta}{2}\right)^{l-n}$$
$$\left.\times\left(iw\sin\frac{\theta}{2}+\cos\frac{\theta}{2}\right)^{l+n-1}\right]=\sum_{m=-l}^{l}\frac{(l-m)\,P^l_{mn}(\cos\theta)}{\sqrt{(l-m)!(l+m)!}}\,w^{l-m-1}. \tag{1}$$

For both the terms in the left-hand side of this equality let us employ expansion (3) (with $l$ replaced by $l-\frac{1}{2}$), and compare the coefficients of like powers of $w$:

$$\sqrt{l-n}\cos\frac{\theta}{2}P^{l-1/2}_{m+1/2,\,n+1/2}(\cos\theta)$$
$$+i\sqrt{l+n}\sin\frac{\theta}{2}P^{l-1/2}_{m+1/2,\,n-1/2}(\cos\theta)=\sqrt{l-m}\,P^l_{mn}(\cos\theta). \tag{2}$$

Then, on multiplying both sides of the same equality (3) by the expression $w\cos\theta/2+i\sin\theta/2$, we get

$$\frac{1}{\sqrt{(l-n)!\,(l+n)!}}\left(w\cos\frac{\theta}{2}+i\sin\frac{\theta}{2}\right)^{l-n+1}\left(iw\sin\frac{\theta}{2}+\cos\frac{\theta}{2}\right)^{l+n}$$
$$=\sum_{m=-l}^{l}\frac{\cos\frac{\theta}{2}P^l_{mn}(\cos\theta)\,w^{l-m+1}+i\sin\frac{\theta}{2}P^l_{mn}(\cos\theta)\,w^{l-m}}{\sqrt{(l-m)!\,(l+m)!}}.$$

The left-hand side of this equality has the same form as the left-hand side of equality (3) with $l$ replaced by $l+\frac{1}{2}$ and $n$ by $n-\frac{1}{2}$. Comparing the coefficients of $w^{l-m+1}$, we find

$$\sqrt{l-n+1}\,P^{l+1/2}_{m-1/2,\,n-1/2}(\cos\theta)$$
$$=\sqrt{l-m+1}\cos\frac{\theta}{2}P^l_{mn}(\cos\theta)+i\sqrt{l+m}\sin\frac{\theta}{2}P^l_{m-1,\,n}(\cos\theta). \tag{3}$$

Similarly, multiplying both sides of equality (3) of subsection 1 by the expression $iw\sin\theta/2+\cos\theta/2$, we get

$$\sqrt{l+n+1}\,P^{l+1/2}_{m-1/2,\,n+1/2}(\cos\theta)$$
$$=i\sqrt{l-m+1}\sin\frac{\theta}{2}P^l_{mn}(\cos\theta)+\sqrt{l+m}\cos\frac{\theta}{2}P^l_{m-1,\,n}(\cos\theta). \tag{4}$$

It follows immediately from (3) and (4) that

$$-i\sqrt{l-n+1}\sin\frac{\theta}{2}P^{l+1/2}_{m+1/2,\,n-1/2}(\cos\theta)$$
$$+\sqrt{l+n+1}\cos\frac{\theta}{2}P^{l+1/2}_{m+1/2,\,n+1/2}(\cos\theta)=\sqrt{l+m+1}\,P^l_{mn}(\cos\theta) \tag{5}$$

and

$$\sqrt{l-n+1}\cos\frac{\theta}{2}P^{l+1/2}_{m-1/2,\,n-1/2}(\cos\theta)$$
$$-i\sqrt{l+n+1}\sin\frac{\theta}{2}P^{l+1/2}_{m-1/2,\,n+1/2}(\cos\theta)$$
$$=\sqrt{l-m+1}\,P^l_{mn}(\cos\theta). \tag{6}$$

A relation between the functions $P^l_{mn}(\cos\theta)$, the upper indices of which differ by 1, is derived in exactly the same way. For example, let us multiply both sides of equality (3) of subsection 1 by the corresponding sides of the identity

$$\left(w\cos\frac{\theta}{2}+i\sin\frac{\theta}{2}\right)\left(iw\sin\frac{\theta}{2}+\cos\frac{\theta}{2}\right)=\frac{i(w^2+1)}{2}\sin\theta+w\cos\theta.$$

Then we easily find that

$$\sqrt{(l-n+1)(l+n+1)}\,P^{l+1}_{mn}(\cos\theta)$$
$$=\sqrt{(l-m+1)(l+m+1)}\cos\theta\,P^l_{mn}(\cos\theta)$$
$$+\frac{i\sin\theta}{2}\Big[\sqrt{(l-m)(l-m+1)}\,P^l_{m+1,\,n}(\cos\theta)$$
$$+\sqrt{(l+m)(l+m+1)}\,P^l_{m-1,\,n}(\cos\theta)\Big], \tag{7}$$

Other relations are obtained by a double application of relation (3) or (4):

$$\sqrt{(l-n)(l-n+1)}\,P^{l+1}_{mn}(\cos\theta)$$
$$=\sqrt{(l-m)(l-m+1)}\cos^2\frac{\theta}{2}P^l_{m+1,\,n+1}(\cos\theta)$$
$$+i\sqrt{(l-m+1)(l+m+1)}\sin\theta\,P^l_{m,\,n+1}(\cos\theta)$$
$$-\sqrt{(l+m)(l+m+1)}\sin^2\frac{\theta}{2}P^l_{m-1,\,n+1}(\cos\theta) \tag{8}$$

and

$$\sqrt{(l+n)(l+n+1)}\,P^{l+1}_{mn}(\cos\theta)$$
$$=-\sqrt{(l-m)(l-m+1)}\sin^2\frac{\theta}{2}P^l_{m+1,\,n-1}(\cos\theta)$$
$$+i\sqrt{(l-m+1)(l+m+1)}\sin\theta\,P^l_{m,\,n-1}(\cos\theta)$$
$$+\sqrt{(l+m)(l+m+1)}\cos^2\frac{\theta}{2}P^l_{m-1,\,n-1}(\cos\theta). \tag{9}$$

We obtain further relations, connecting functions $P^l_{mn}(z)$ with different values of $l$, by considering the following two expansions:

$$\frac{1}{\sqrt{(l-n)!\,(l+n)!}}\left(w\cos\frac{\theta}{2}+i\sin\frac{\theta}{2}\right)^{l-n}$$
$$\times\left(iw\sin\frac{\theta}{2}+\cos\frac{\theta}{2}\right)^{l+n}=\sum_{m=-l}^{l}\frac{P^l_{mn}(\cos\theta)}{\sqrt{(l-m)!\,(l+m)!}}w^{l-m}$$

and

$$\frac{1}{\sqrt{(s-r)!\,(s+r)!}}\left(w\cos\frac{\theta}{2}+i\sin\frac{\theta}{2}\right)^{s-r}$$
$$\times\left(iw\sin\frac{\theta}{2}+\cos\frac{\theta}{2}\right)^{s+r}=\sum_{q=-s}^{s}\frac{P^s_{qr}(\cos\theta)}{\sqrt{(s-q)!\,(s+q)!}}w^{s-q}.$$

Multiplying these equalities together side by side, we get

$$\frac{1}{\sqrt{(l-n)!\,(l+n)!\,(s-r)!\,(s+r)!}}\left(w\cos\frac{\theta}{2}+i\sin\frac{\theta}{2}\right)^{l+s-n-r}$$
$$\times\left(iw\sin\frac{\theta}{2}+\cos\frac{\theta}{2}\right)^{l+s+n+r}$$
$$=\sum_{q=-l-s}^{l+s}\left[\sum_{m=L}^{M}\frac{P^l_{mn}(\cos\theta)\,P^s_{q-m,\,r}(\cos\theta)}{\sqrt{(l-m)!\,(l+m)!\,(s-q+m)!\,(s+q-m)!}}\right]w^{l+s-q},$$

where $L=\max(-l,\ q-s)$, $M=\min(l,\ s+q)$. Employing expansion (3) of subsection 1 for the left hand side of this equality, and comparing the coefficients of like powers of $w$, we find that

$$P^{l+s}_{q,\,n+r}(\cos\theta)=A^{ls}_{nr}\sum_{m=L}^{M}B^{ls}_{mqr}P^l_{mn}(\cos\theta)\,P^s_{q-m,\,r}(\cos\theta),\tag{10}$$

and

$$A^{ls}_{nr}=\sqrt{\frac{(l-n)!\,(l+n)!}{(l+s-n-r)!\,(l+s+n+r)!}},$$
$$B^{ls}_{mqr}=\sqrt{\frac{(s-r)!\,(s+r)!\,(l+s-q)!\,(l+s+q)!}{(l-m)!\,(l+m)!\,(s-q+m)!\,(s+q-m)!}}.$$

To obtain further formulas we use the following equality, proved in §4.1:

$$\sum_{n=-l}^{l}(-1)^{k-n}P^l_{mn}(\cos\theta)\,P^l_{nk}(\cos\theta)=\delta_{mk}.$$

We multiply both sides of equality (10) by $(-1)^{k-n}P^l_{nk}/A^{ls}_{nr}$ and sum the relations thus obtained for $n$ running from $-l$ to $l$. We find that, for $L\le k\le M$,

$$P^s_{q-k,\,r}(\cos\theta)=\frac{1}{B^{ls}_{kqr}}\sum_{n=-l}^{l}\frac{(-1)^{k-n}P^l_{nk}(\cos\theta)\,P^{l+s}_{q,\,n+r}(\cos\theta)}{A^{ls}_{nr}}\tag{11}$$

and for all other $k$ this sum is equal to zero.

In conclusion, we mention another expression for $P^l_{m0}(\cos\theta)$, where $l$ is an integer. We have

$$\left(w \cos \frac{\theta}{2} + i \sin \frac{\theta}{2}\right)^l \left(iw \sin \frac{\theta}{2} + \cos \frac{\theta}{2}\right)^l = w^l \left(\cos \theta + i \frac{w^2+1}{2w} \sin \theta\right)^l .$$

On applying relation (3) of subsection 1 to both sides of this identity, we obtain

$$\sum_{m=-l}^{l} \frac{l!\, P^l_{m0}(\cos \theta)}{\sqrt{(l-m)!\,(l+m)!}} w^{l-m}$$
$$= \sum_{s=-\frac{l}{2}}^{\frac{l}{2}} \sqrt{\frac{l!}{\left(\frac{l}{2}-s\right)!\,\left(\frac{l}{2}+s\right)!}}\, P^{\frac{l}{2}}_{s,\,-\frac{l}{2}}(\cos 2\theta)\, w^{\frac{l}{2}+s} \left(\frac{w^2+1}{2}\right)^{\frac{l}{2}-s}.$$

Further, we remove the brackets in $((w^2+1)/2)^{l/2-s}$, and take into consideration expression (6) of §3.5 for $P^{l/2}_{s,-l/2}(\cos 2\theta)$. Comparing the coefficients of like powers of $w$, we obtain

$$P^l_{m0}(\cos \theta) = \sum_{s=-\frac{l}{2}-m}^{\frac{l}{2}} \frac{i^{\frac{l}{2}+s} \sqrt{(l-m)!\,(l+m)!} \cos^{\frac{l}{2}-s} \theta \sin^{\frac{l}{2}+s} \theta}{2^{\frac{l}{2}-s} \left(\frac{l}{2}+s\right)!\; k!\left(\frac{l}{2}-s-k\right)!} \tag{12}$$

Hence, owing to formula (9) of §3.9, we have

$$P^m_l(\cos \theta) = \sum_{s+2k=\frac{l}{2}-m}^{\frac{l}{2}} \frac{i^{\frac{l}{2}+s+m} (l+m)! \cos^{\frac{l}{2}-s} \theta \sin^{\frac{l}{2}+s} \theta}{2^{\frac{l}{2}-s} \left(\frac{l}{2}+s\right)!\; k!\left(\frac{l}{2}-s-k\right)!}. \tag{13}$$

3. **The case of fixed $m$ and $n$.** In order to derive a generating function for $P^l_{mn}(\cos \theta)$ for fixed values of $m$ and $n$, we need a new integral representation for these functions. It is obtained from the integral representation (4) of §3.4 by changing the variable of integration. Namely, let us write this integral representation in the form

$$P^l_{mn}(\cos \theta) = \frac{1}{2\pi i} \sqrt{\frac{(l-m)!\,(l+m)!}{(l-n)!\,(l+n)!}}$$
$$\times \oint_{|t|=1} \left[\cos \theta + \frac{i}{2} \sin \theta \,(t+t^{-1})\right]^{l-n} \left(it \sin \frac{\theta}{2} + \cos \frac{\theta}{2}\right)^{2n} t^{m-n-1} dt$$

and assume that $m \geq n \geq 0$.[1] Because of Cauchy's theorem, we can replace the

[1] This restriction is inessential, since, by virtue of the symmetry relations (§3.6),

$$P^l_{mn}(\cos \theta) = P^l_{nm}(\cos \theta) = P^l_{-m,\,-n}(\cos \theta) = P^l_{-n,\,-m}(\cos \theta)$$

and

$$P^l_{mn}(-z) = i^{2l-2m-2n} P^l_{m,\,-n}(z).$$

path of integration by the contour $\Gamma'$: $|t| = a$, where $a > 1$. We now make the change of variable

$$w = \cos\theta + \frac{i}{2}\sin\theta\,(t + t^{-1}).$$

When $t$ traverses the contour $\Gamma'$, the variable $w$ traverses a contour $\Gamma''$, containing the segment $[e^{-i\theta}, e^{i\theta}]$, counterclockwise. The variable $t$ is uniquely determined by the variable $w$ if the $w$-plane is cut along the segment $[e^{-i\theta}, e^{i\theta}]$.

Namely, we have

$$t = \frac{w - \cos\theta + \sqrt{w^2 - 2w\cos\theta + 1}}{i\sin\theta}, \tag{1}$$

where the sign of the radical is determined by the condition $|t| > 1$.

Without losing generality, the contour $\Gamma''$ can be taken to be the circle $|w| = b$, where $b > 1$.

After the change of variable, we get

$$P^l_{mn}(\cos\theta) = \frac{1}{2\pi i}\sqrt{\frac{(l-m)!\,(l+m)!}{(l-n)!\,(l+n)!}} \times \oint_{\Gamma''} w^{l-n} t^{m-n}\left(\cos\frac{\theta}{2} + it\sin\frac{\theta}{2}\right)^{2n} \frac{dw}{\sqrt{w^2 - 2w\cos\theta + 1}}, \tag{2}$$

where $t$ is given by formula (1), and the radical sign is chosen such that $|t| > 1$.

In particular, if $m = n = 0$, we get the integral representation of Legendre polynomials:

$$P_l(\cos\theta) = \frac{1}{2\pi i}\oint_{\Gamma''} \frac{w^l\,dw}{\sqrt{w^2 - 2w\cos\theta + 1}}. \tag{2'}$$

In order to obtain a generating function for $P^l_{mn}(\cos\theta)$, we make the substitution $w = 1/h$ in integral (2), and make use of Cauchy's formula for the coefficients of a Taylor series (see [35]):

$$\sum_{l=n}^{\infty} \sqrt{\frac{(l-n)!\,(l+n)!}{(l-m)!\,(l+m)!}}\, P^l_{mn}(\cos\theta)\, h^{l-n} = \frac{t^{m-n}\left(it\sin\frac{\theta}{2} + \cos\frac{\theta}{2}\right)^{2n}}{\sqrt{1 - 2h\cos\theta + h^2}}, \tag{3}$$

where

$$t = \frac{1 - h\cos\theta + \sqrt{1 - 2h\cos\theta + h^2}}{ih\sin\theta}, \tag{4}$$

and the radical sign is chosen such that $|t| > 1$.

If $n$ is a negative integer, $m \geq |n|$, we get a similar formula:

$$\sum_{l=-n}^{\infty} \sqrt{\frac{(l-n)!\,(l+n)!}{(l-m)!\,(l+m)!}}\, P^l_{mn}(\cos\theta)\, h^{l+n} = \frac{t^{m+n}\left(t\cos\frac{\theta}{2}+i\sin\frac{\theta}{2}\right)^{-2n}}{\sqrt{1-2h\cos\theta+h^2}}, \tag{5}$$

where $t$ has the same value.

Formulas (3) and (5) remain valid also in the case when $l$ is a semi-integer, with the only difference that the summation has to be carried out not over integral, but over semi-integral values of $l$.

Formula (3) takes a particularly simple form for $m = n = 0$. This gives the generating function for Legendre polynomials:

$$\sum_{l=0}^{\infty} P_l(\cos\theta)\, h^l = \frac{1}{\sqrt{1-2h\cos\theta+h^2}}. \tag{6}$$

This formula is often taken as defining the Legendre polynomials. The latter are therefore also called the Legendre coefficients.

For $n = 0$, by utilizing the equality

$$P_l^m(z) = i^m \sqrt{\frac{(l+m)!}{(l-m)!}}\, P^l_{m0}(z),$$

we get the generating function for associated Legendre functions:

$$\sum_{l=0}^{\infty} \frac{l!}{(l+m)!}\, P_l^m(\cos\theta) h^l = \frac{(it)^m}{\sqrt{1-2h\cos\theta+h^2}}, \tag{7}$$

where $t$ is given by formula (4).

4. The Dirichlet-Murphy integral representations. It was proved in the preceding subsection that

$$P_l(\cos\theta) = \frac{1}{2\pi i} \oint_{\Gamma''} \frac{z^l\,dz}{\sqrt{1-2z\cos\theta+z^2}}, \tag{1}$$

where $\Gamma''$ is a circle of radius $a > 1$ with center at the origin, and

$$\left|\cos\theta - z + \sqrt{1-2z\cos\theta+z^2}\right| > \sin\theta. \tag{2}$$

The function $\sqrt{1-2z\cos\theta+z^2}$ has two branch-points $\alpha = e^{-i\theta}$ and $\beta = e^{i\theta}$, lying within the circle $\Gamma''$ (it will be recalled that we are now considering values of $\theta$ lying on the segment $[0, \pi]$). We contract the circle to the branch-line, i.e. the segment which joins the points $\alpha$ and $\beta$. The integrals over the infinitely small circles surrounding points $\alpha$ and $\beta$ give zero contribution, and the contributions from the segments $[\alpha+0, \beta+0]$ and $[\beta-0, \alpha-0]$ are equal, since $\sqrt{1-2z\cos\theta+z^2}$ has opposite signs at corresponding points of these segments. Hence

$$P_l(\cos\theta)=\frac{1}{\pi i}\int_{\alpha+0}^{\beta+0}\frac{z^l dz}{\sqrt{1-2z\cos\theta+z^2}}=\frac{1}{\pi i}\int_{\beta-0}^{\alpha-0}\frac{z^l dz}{\sqrt{1-2z\cos\theta+z^2}}. \tag{3}$$

It follows from formula (2) that the real part of the radical is positive in the first integral, and negative in the second.

Let us now deform the segments $[\alpha+0,\ \beta+0]$ and $[\alpha-0,\ \beta-0]$ into arcs of the unit circle, joining points $\alpha$ and $\beta$. It follows from equality (3) that

$$P_l(\cos\theta)=\frac{1}{\pi}\int_{-\theta}^{\theta}\frac{e^{i(l+1)\varphi}\,d\varphi}{\sqrt{1-2e^{i\varphi}\cos\theta+e^{2i\varphi}}}=\frac{2}{\pi}\int_0^{\theta}\frac{\cos\left(l+\frac{1}{2}\right)\varphi\,d\varphi}{\sqrt{2(\cos\varphi-\cos\theta)}} \tag{4}$$

and

$$P_l(\cos\theta)=\frac{1}{\pi}\int_{\theta}^{2\pi-\theta}\frac{e^{i(l+1)\varphi}\,d\varphi}{\sqrt{1-2e^{i\varphi}\cos\theta+e^{2i\varphi}}}=\frac{2}{\pi}\int_{\theta}^{\pi}\frac{\sin\left(l+\frac{1}{2}\right)\varphi\,d\varphi}{\sqrt{2(\cos\theta-\cos\varphi)}}. \tag{5}$$

Indeed, owing to inequality (2), for $-\theta<\phi<\theta$ we have

$$\sqrt{1-2e^{i\varphi}\cos\theta+e^{2i\varphi}}=e^{\frac{i\varphi}{2}}\sqrt{2(\cos\varphi-\cos\theta)},$$

and for $\theta<\phi<2\pi-\theta$ we have

$$\sqrt{1-2e^{i\varphi}\cos\theta+e^{2i\varphi}}=ie^{\frac{i\varphi}{2}}\sqrt{2(\cos\theta-\cos\varphi)},$$

where the radicals in the right-hand sides are understood in the sense of arithmetical value.

Thus, we have proved that

$$P_l(\cos\theta)=\frac{2}{\pi}\int_0^{\theta}\frac{\cos\left(l+\frac{1}{2}\right)\varphi\,d\varphi}{\sqrt{2(\cos\varphi-\cos\theta)}}=\frac{2}{\pi}\int_{\theta}^{\pi}\frac{\sin\left(l+\frac{1}{2}\right)\varphi\,d\varphi}{\sqrt{2(\cos\theta-\cos\varphi)}},$$

where the radicals are understood in the arithmetical sense. These formulas were established by Dirichlet and Murphy.

5. Recurrence formulas for the Legendre polynomials. We now derive recurrence formulas connecting the Legendre polynomials $P_l(z)$ with different indices $l$. We differentiate both sides of the equality

$$(1-2hz+h^2)^{-\frac{1}{2}}=\sum_{l=0}^{\infty}P_l(z)h^l \tag{1}$$

with respect to $h$, and multiply the resulting equality by $1-2hz+h^2$. We obtain

$$(z-h)(1-2hz+h^2)^{-\frac{1}{2}}=(1-2hz+h^2)\sum_{l=1}^{\infty}lP_l(z)h^{l-1}, \tag{2}$$

i.e.

$$(z-h)\sum_{l=0}^{\infty}P_l(z)h^l=(1-2hz+h^2)\sum_{l=1}^{\infty}lP_l(z)h^{l-1}. \tag{3}$$

Let us compare the coefficients of $h^{l-1}$. We easily obtain

$$lP_l(z)-(2l-1)zP_{l-1}(z)+(l-1)P_{l-2}(z)=0. \tag{4}$$

Using equality (4), it is easy to calculate the polynomials $P_l(z)$, starting from $P_0(z)=1$ and $P_1(z)=z$.

Equality (4) shows that the Legendre polynomials possess the same properties as the Sturm polynomials. No two consecutive Legendre polynomials can simultaneously vanish. If one of the Legendre polynomials is equal to zero for some $z$, then for this value of $z$ the polynomials adjacent to it have different signs. With the help of the same arguments as are used in Sturm's theorem, we find that between two adjacent roots of the polynomials $P_{l+1}(z)$ lies one and only one root of the polynomial $P_l(z)$. Hence all the roots of the polynomial $P_l(z)$ are real. It is not difficult to show that they lie between $-1$ and $1$. For this it is sufficient to use the expansion (4) of §3.9 for Legendre polynomials and make use of Pólya's theorem.

We now differentiate both sides of equality (1) with respect to $z$, and compare the resulting equality

$$(1-2hz+h^2)^{-\frac{3}{2}}=\sum_{l=1}^{\infty}h^{l-1}\frac{dP_l(z)}{dz} \tag{5}$$

with the equality

$$(z-h)(1-2hz+h^2)^{-\frac{3}{2}}=\sum_{l=1}^{\infty}lh^{l-1}P_l(z)$$

(see formula (2)). We obtain

$$(z-h)\sum_{l=1}^{\infty}h^{l-1}\frac{dP_l(z)}{dz}=\sum_{l=1}^{\infty}lh^{l-1}P_l(z). \tag{6}$$

Comparing the coefficients of $h^{l-1}$, we get the recurrence formula

$$lP_l(z) = z\frac{dP_l(z)}{dz} - \frac{dP_{l-1}(z)}{dz}. \tag{7}$$

Further, let us use the identity

$$(1-2hz+h^2)^{-\frac{1}{2}} = (1-2hz+h^2)^{-\frac{3}{2}}(1-hz) - h(z-h)(1-2hz+h^2)^{-\frac{3}{2}}.$$

Employing relations (1), (2) and (5), we get

$$\sum_{l=0}^{\infty} P_l(z)h^l = (1-hz)\sum_{l=1}^{\infty} h^{l-1}\frac{dP_l(z)}{dz} - \sum_{l=1}^{\infty} lh^l P_l(z). \tag{8}$$

Comparing the coefficients of $h^l$, we have

$$(l+1)P_l(z) = \frac{dP_{l+1}(z)}{dz} - z\frac{dP_l(z)}{dz}. \tag{9}$$

Adding (7) and (9), we get

$$(2l+1)P_l(z) = \frac{d}{dz}[P_{l+1}(z) - P_{l-1}(z)]. \tag{10}$$

But if we eliminate $P_l(z)$ from (7) and (9), we find

$$(2l+1)z\frac{dP_l(z)}{dz} = l\frac{dP_{l+1}(z)}{dz} + (l+1)\frac{dP_{l-1}(z)}{dz}. \tag{11}$$

We may also note the equality

$$(1-z^2)\frac{dP_l(z)}{dz} = l[P_{l-1}(z) - zP_l(z)] = (l+1)[zP_l(z) - P_{l+1}(z)], \tag{12}$$

ensuing from equalities (7) and (9), and the Legendre equation

$$\frac{d}{dz}\left[\left(1-z^2\right)\frac{dP_l(z)}{dz}\right] + l(l+1)P_l(z) = 0$$

(see §4.5, formula (8)).

## §6. EXPANSION OF FUNCTIONS ON THE GROUP $SU(2)$

In this section we shall study expansions of functions $f(u)$ on the group $SU(2)$ in series of matrix elements $t^l_{mn}(u)$. As a special case of these results, we shall obtain decompositions of functions and vector fields on the sphere in three-dimensional euclidean space.

1. An invariant measure. Since $SU(2)$ is compact, an invariant measure $du$ exists on it. In other words, there exists a measure $du$ such that

$$\int f(u)\,du=\int f(u_0u)\,du=\int f(uu_0)\,du=\int f(u^{-1})\,du \tag{1}$$

for all continuous functions $f(u)$ and all elements $u_0$ of $SU(2)$. Let us find an expression for this measure in terms of the parameters of the group $SU(2)$.

We consider first a group $G$, consisting of nonsingular matrices of the form $\begin{pmatrix}\alpha & \beta\\ -\bar\beta & \bar\alpha\end{pmatrix}$. Its elements are given by arbitrary pairs $(\alpha, \beta)$ of complex numbers. On right multiplication of an element of $G$ by the element $u_0 = \begin{pmatrix}\alpha_0 & \beta_0\\ -\bar\beta_0 & \bar\alpha_0\end{pmatrix}$ of $SU(2)$, the parameters $\alpha$, $\beta$ undergo the linear transformation

$$\left.\begin{aligned}\alpha^1&=\alpha_0\alpha-\bar\beta_0\beta,\\ \beta^1&=\beta_0\beta+\bar\alpha_0\beta,\end{aligned}\right\} \tag{2}$$

the determinant of which is equal to unity. Hence the measure $dg = d\alpha\, d\bar\alpha\, d\beta\, d\bar\beta$ [1] on $G$ is invariant for right multiplication by elements of $SU(2)$.

In place of the parameters $\alpha = \alpha_1 + i\alpha_2$ and $\beta = \beta_1 + i\beta_2$, we introduce parameters $r$, $\phi$, $\theta$, $\psi$ connected with $\alpha$ and $\beta$ by the equalities

$$\left.\begin{aligned}\alpha&=r\cos\frac{\theta}{2}e^{\frac{i(\varphi+\psi)}{2}},\\ \beta&=ir\sin\frac{\theta}{2}e^{\frac{i(\varphi-\psi)}{2}}.\end{aligned}\right\} \tag{3}$$

It follows from equality (4) of §1.1 that for $r = 1$ the matrix $\begin{pmatrix}\alpha & \beta\\ -\bar\beta & \bar\alpha\end{pmatrix}$ belongs to $SU(2)$, where $\phi$, $\theta$, $\psi$, are the Euler angles of this matrix. A simple calculation shows that

$$d\alpha\, d\bar\alpha\, d\beta\, d\bar\beta=\frac{1}{2}r^3\sin\theta\,dr\,d\theta\,d\varphi\,d\psi. \tag{4}$$

But elements of $SU(2)$ are characterized by the condition $r = 1$. Therefore an invariant integral on $SU(2)$ is given in Euler angles by the formula

$$\int\limits_{SU(2)} f(u)\,du=\frac{1}{16\pi^2}\int\limits_{-2\pi}^{2\pi}\int\limits_0^{2\pi}\int\limits_0^{\pi} f(\varphi,\ \theta,\ \psi)\sin\theta\,d\theta\,d\varphi\,d\psi \tag{5}$$

(the factor $1/16\pi^2$ is chosen in order that the measure of the whole group $SU(2)$ will be equal to 1). We note that, because of the compactness of the group $SU(2)$,

[1] For $\alpha = \alpha_1 + i\alpha_2$, we take $d\alpha\, d\bar\alpha = -2i\,d\alpha_1\, d\alpha_2$

integral (5) is not only right-invariant, but also left-invariant:

$$\int f(u)\,du = \int f(uu_0)\,du = \int f(u_0u)\,du = \int f(u^{-1})\,du. \tag{6}$$

The expression

$$\frac{1}{16\pi^2}\sin\theta\,d\theta\,d\varphi\,d\psi \tag{7}$$

for the invariant measure on $SU(2)$ has a simple geometrical significance. It differs only in the factor $\frac{1}{2}$ from the product of the normalized euclidean measure $(1/4\pi)\ \sin\theta\,d\theta\,d\phi$ on the two-dimensional sphere and the normalized euclidean measure $(1/2\pi)\,d\psi$ on the circle.

2. Orthogonality relations for the functions $P^l_{mn}(z)$. Now we shall apply to $SU(2)$ the theorems of orthogonality and completeness of the system of matrix elements of pairwise nonequivalent irreducible unitary representations of a compact group (see §4.3 of Chapter I). Since the dimension of the representation $T_l(u)$ of the group $SU(2)$ is equal to $2l+1$, it follows from these theorems that the functions $\sqrt{2l+1}\ t^l_{mn}(u)$ form a complete orthonormal system of functions relative to the invariant measure $du$ on this group. Here the index $l$ runs through all possible integral and semi-integral nonnegative values, and the indices $m$ and $n$ run through the values $-l, -l+1, \cdots, l-1, l$.

In other words, the functions $t^l_{mn}(u)$ satisfy the relations

$$\int\limits_{SU(2)} t^l_{mn}(u)\,\overline{t^s_{pq}(u)}\,du = \frac{1}{2l+1}\,\delta_{ls}\,\delta_{mp}\delta_{nq}. \tag{1}$$

We substitute, in formula (1), the expression

$$t^l_{mn}(\varphi,\ \theta,\ \psi) = e^{-i(m\varphi+n\psi)}P^l_{mn}(\cos\theta) \tag{2}$$

for the matrix elements, and recall that the invariant measure $du$ on $SU(2)$ is given by the formula

$$du = \frac{1}{16\pi^2}\sin\theta\,d\varphi\,d\theta\,d\psi. \tag{3}$$

We find that if $l \neq s$, or $m \neq p$, or $n \neq q$, then

$$\int\limits_{-2\pi}^{2\pi}\int\limits_0^{2\pi}\int\limits_0^{\pi} P^l_{mn}(\cos\theta)\,\overline{P^s_{pq}(\cos\theta)}\sin\theta\,e^{i(p-m)\varphi}e^{i(q-n)\psi}\,d\theta\,d\varphi\,d\psi = 0. \tag{4}$$

In the cases when $p \neq m$ or $q \neq n$, one cannot draw any conclusion about the functions $P^l_{mn}$ from relation (4), since the left hand side vanishes owing to the exponential functions. Suppose now that $p = m$ and $q = n$. Then it follows from equality (4) that, for $l \neq s$,

$$\int_0^\pi P^l_{mn}(\cos\theta)\overline{P^s_{mn}(\cos\theta)}\sin\theta\, d\theta = 0. \tag{5}$$

Similarly, it follows from (1) that

$$\int_0^\pi |P^l_{mn}(\cos\theta)|^2 \sin\theta\, d\theta = \frac{2}{2l+1}. \tag{6}$$

Putting $\cos\theta = x$ in these equalities, we get the orthogonality relations for functions $P^l_{mn}(x)$:

$$\int_{-1}^{1} P^l_{mn}(x)\overline{P^s_{mn}(x)}\, dx = \frac{2}{2l+1}\delta_{ls}. \tag{7}$$

The formulas derived above lead to the following results: for fixed values of $l$ and $n$, $l \geq n \geq 0$, the sequence of functions

$$\sqrt{\frac{2l+1}{2}}P^l_{ln}(x),\ \sqrt{\frac{2l+3}{2}}P^{l+1}_{ln}(x), \ldots, \sqrt{\frac{2l+2k+1}{2}}P^{l+k}_{ln}(x), \ldots \tag{8}$$

is orthogonal and normalized on the segment $[-1, 1]$ relative to the measure $dx$.

Let us consider some special cases. For $n = 0$, we find that the sequence of functions

$$\sqrt{\frac{2l+1}{2}}P^l_{l0}(x), \ldots, \sqrt{\frac{2l+2k+1}{2}}P^{l+k}_{l0}(x), \ldots \tag{9}$$

is orthogonal and normalized. But

$$P^l_{l+k}(x) = i^l\sqrt{\frac{(2l+k)!}{k!}}P^{l+k}_{l0}(x), \tag{10}$$

where $P^m_l(x)$ is the associated Legendre function. Hence the sequence of functions

$$\sqrt{\frac{(2l+2k+1)\,k!}{2\,(2l+k)!}}P^l_{l+k}(x), \qquad k = 0, 1, \ldots \tag{11}$$

is orthogonal and normalized on the segment $[-1, 1]$.

In particular, for $l = 0$ the sequence of Legendre polynomials

$$\sqrt{\frac{2k+1}{2}}P_k(x), \qquad k = 0, 1, \ldots \tag{12}$$

is orthogonal and normalized on the segment $[-1, 1]$.

In conclusion, let us consider the question of orthogonality of the Jacobi polynomials $P_k^{(\alpha,\beta)}(x)$. As shown in §3.9, these polynomials are connected with the functions $P^l_{mn}(x)$ by the relations

$$P_k^{(\alpha,\beta)}(x) = 2^{\frac{\alpha+\beta}{2}} i^{-\alpha} \sqrt{\frac{(k+\alpha)!\,(k+\beta)!}{k!\,(k+\alpha+\beta)!}} (1-x)^{-\frac{\alpha}{2}} (1+x)^{-\frac{\beta}{2}} P^{k+\frac{\alpha+\beta}{2}}_{\frac{\alpha+\beta}{2},\frac{\beta-\alpha}{2}}(x), \tag{13}$$

where $\alpha$, $\beta$, $k$ are integers.

But, as we have shown above, the functions

$$\sqrt{\frac{\alpha+\beta+2k+1}{2}}\, P^{k+\frac{\alpha+\beta}{2}}_{\frac{\alpha+\beta}{2},\frac{\beta-\alpha}{2}}(x), \qquad k=0,\ 1,\ \ldots$$

form an orthonormal system on the segment $[-1, 1]$. Hence, the polynomials

$$2^{-\frac{\alpha+\beta+1}{2}} i^{\alpha} \sqrt{\frac{k!\,(k+\alpha+\beta)!\,(\alpha+\beta+2k+1)}{(k+\alpha)!\,(k+\beta)!}}\, P_k^{(\alpha,\beta)}(x), \qquad k=0,\ 1,\ 2,\ \ldots \tag{14}$$

form an orthonormal system on the segment $[-1, 1]$ relative to the weight $(1-x)^{\alpha}(1+x)^{\beta}$.[1]

Let us apply the results obtained above for calculating, in a finite form, a sum containing associated Legendre functions. From the orthogonality of Legendre polynomials and formula (7) of §4.3, it follows that $((2l+1)/2)P_l^k(\cos\theta_1)\times P_l^{-k}(\cos\theta_2)$ is the coefficient in the expansion of

$$f(\theta;\ \theta_1,\ \theta_2) = \begin{cases} \frac{1}{\pi} T_k\left(\frac{\cos\theta_1\cos\theta_2-\cos\theta}{\sin\theta_1\sin\theta_2}\right) \\ \times \{[\cos\theta-\cos(\theta_1+\theta_2)]\,[\cos(\theta_1-\theta_2)-\cos\theta]\}^{-\frac{1}{2}}, \\ \qquad\qquad \text{if } \ |\theta_1-\theta_2|\leqslant\theta\leqslant\theta_1+\theta_2, \\ 0, \text{ if } \theta \text{ is a point of the segment } [0,\pi], \text{ not belonging to the segment} \\ \quad [|\theta_1-\theta_2|,\ \theta_1+\theta_2]. \end{cases}$$

in Legendre polynomials $P_l(\cos\theta)$. Hence

[1] The system of functions $f_1(x),\cdots,f_n(x),\cdots$ is called orthonormal relative to a weight $p(x)$ on the segment $[a, b]$ if

$$\int_a^b f_m(x)\overline{f_n(x)}\,p(x)\,dx=\delta_{mn}.$$

$$\sum_{l=0}^{\infty} \frac{2l+1}{2} P_l^k(\cos\theta_1) P_l^{-k}(\cos\theta_2) P_l(\cos\theta) = f(\theta;\ \theta_1,\ \theta_2). \tag{15}$$

In particular

$$\sum_{l=0}^{\infty} \frac{2l+1}{2} P_l(\cos\theta_1) P_l(\cos\theta_2) P_l(\cos\theta)$$

$$= \begin{cases} \frac{1}{\pi}\left\{[\cos\theta - \cos(\theta_1+\theta_2)][\cos(\theta_1-\theta_2) - \cos\theta]\right\}^{-\frac{1}{2}}, \\ \qquad\qquad \text{if } |\theta_1-\theta_2| \leqslant \theta \leqslant \theta_1+\theta_2, \\ 0, \text{ if } \theta \text{ is a point of the segment } [0, \pi], \text{ not belonging to the segment} \\ \quad [|\theta_1-\theta_2|, \theta_1+\theta_2]. \end{cases}$$

3. **Expansions in series of functions $P^l_{mn}(x)$.** Let $f(u)$ be an arbitrary function on $SU(2)$ for which $\int |f(u)|^2\, du$ is convergent. We have proved above that the functions $\sqrt{2l+1}\, t^l_{mn}(u)$ form a complete orthonormal system relative to the invariant measure $du$. It follows that $f(u)$ can be expanded in a Fourier series of functions $t^l_{mn}(u)$, and also that this series converges in the mean.

The Fourier series of $f(u)$ has the form

$$f(u) = \sum_l \sum_{m=-l}^{l} \sum_{n=-l}^{l} \alpha^l_{mn} t^l_{mn}(u), \tag{1}$$

where $l = 0, ½, 1, \cdots$ and the Fourier coefficients $\alpha^l_{mn}$ are given by the formulas

$$\alpha^l_{mn} = (2l+1)\int f(u)\, \overline{t^l_{mn}(u)}\, du. \tag{2}$$

Let us go over to Euler parameters. Utilizing relations (2) and (3) of subsection 2, we obtain the following result:

*Any function $f(\phi, \theta, \psi)$, $0 \le \phi < 2\pi$, $0 \le \theta < \pi$, $-2\pi < \psi < 2\pi$, belonging to the space $\mathfrak{L}^2$, i.e. such that*

$$\int_{-2\pi}^{2\pi}\int_0^{2\pi}\int_0^{\pi} |f(\varphi, \theta, \psi)|^2 \sin\theta\, d\theta\, d\varphi\, d\psi < +\infty, \tag{3}$$

*can be expanded in a mean-convergent series*

$$f(\varphi, \theta, \psi) = \sum_l \sum_{m=-l}^{l} \sum_{n=-l}^{l} \alpha^l_{mn} e^{-i(m\varphi+n\psi)} P^l_{mn}(\cos\theta), \tag{4}$$

*where*[1)]

$$\alpha_{mn}^{l}=\frac{(-1)^{m-n}(2l+1)}{16\pi^{3}} \times \int_{-2\pi}^{2\pi}\int_{0}^{2\pi}\int_{0}^{\pi} f(\varphi, \theta, \psi)\, e^{i(m\varphi+n\psi)} P_{mn}^{l}(\cos\theta)\sin\theta\, d\theta\, d\varphi\, d\psi. \tag{5}$$

Moreover, it follows from the Parseval equality that

$$\sum_{l}\frac{1}{2l+1}\sum_{m=-l}^{l}\sum_{n=-l}^{l}|\alpha_{mn}^{l}|^{2}=\frac{1}{16\pi^{2}}\int_{-2\pi}^{2\pi}\int_{0}^{2\pi}\int_{0}^{\pi}|f(\varphi, \theta, \psi)|^{2}\sin\theta\, d\theta\, d\varphi\, d\psi. \tag{6}$$

4. **Certain subspaces of functions.** We shall denote by $\mathfrak{L}_n^2$ the subspace in $\mathfrak{L}^2$ consisting of functions $f(u)$ such that, for all diagonal matrices

$$h=\begin{pmatrix} e^{\frac{it}{2}} & 0 \\ 0 & e^{-\frac{it}{2}} \end{pmatrix} \tag{1}$$

of $SU(2)$, one has the equality

$$f(uh)=e^{int}f(u). \tag{2}$$

If the Euler angles of the matrix $g$ are equal to $\phi, \theta, \psi$, then

$$g(\varphi, \theta, \psi)=g(\varphi, \theta, 0)g(0, 0, \psi), \text{where } g(0, 0, \psi)=\begin{pmatrix} e^{\frac{i\psi}{2}} & 0 \\ 0 & e^{-\frac{i\psi}{2}} \end{pmatrix}$$

is a diagonal matrix. Thus, for functions $f(g)$ of $\mathfrak{L}_n^2$, we have

$$f(g)\equiv f(\varphi, \theta, \psi)=e^{in\psi}f(\varphi, \theta, 0). \tag{3}$$

Conversely, if a function $f(g)$ on $SU(2)$ has the form (3), it belongs to $\mathfrak{L}_n^2$.

To $\mathfrak{L}_n^2$ belong, in particular, all the matrix elements $t_{mn}^l(g)$ for a given $n$:

$$t_{mn}^{l}(g)=e^{-in\psi}e^{-im\varphi}P_{mn}^{l}(\cos\theta),$$
$$l=|n|, |n|+1, \ldots, |n|+k, \ldots, \quad -l\leq m\leq l. \tag{4}$$

We shall show that these functions form an orthogonal basis in the space $\mathfrak{L}_n^2$. In other words, it will be shown that any function $f(u)$ of $\mathfrak{L}_n^2$ can be expanded in a Fourier series of the form

1) It will be recalled that $\overline{P_{mn}^{l}(\cos\theta)}=(-1)^{m-n}P_{mn}^{l}(\cos\theta)$.

$$f(u)=\sum_{l=|n|}^{\infty}\sum_{m=-l}^{l}\alpha_m^l t_{mn}^l(u)=e^{-in\psi}\sum_{l=|n|}^{\infty}\sum_{m=-l}^{l}\alpha_m^l e^{-im\varphi}P_{mn}^l(\cos\theta), \tag{5}$$

where the Fourier coefficients are given by

$$\alpha_m^l=\frac{(-1)^{m-n}(2l+1)}{4\pi}\int_0^{2\pi}\int_0^{\pi} f(\varphi,\theta,0)\,e^{im\varphi}P_{mn}^l(\cos\theta)\sin\theta\,d\theta\,d\varphi. \tag{6}$$

For the proof, it is sufficient to show that, for $k\neq n$, we have

$$\int f(u)\overline{t_{mk}^l(u)}\,du=0.$$

In this integral we make the substitution $u=vh$, where

$$h=\begin{pmatrix} e^{\frac{it}{2}} & 0\\ 0 & e^{-\frac{it}{2}}\end{pmatrix}.$$

By the invariance of the measure $du$, we obtain

$$\alpha_m^l=\int f(vh)\overline{t_{mk}^l(vh)}\,dv.$$

But, because $f(u)$ belongs to $\mathfrak{L}_n^2$,

$$f(vh)=e^{-int}f(v),$$

and, by virtue of formula (4),

$$t_{mk}^l(vh)=e^{-ikt}t_{mk}^l(v).$$

Therefore

$$\alpha_m^l=e^{i(k-n)t}\alpha_m^l.$$

Since for $k\neq n$ we have $e^{i(k-n)t}\not\equiv 1$, it follows that $\alpha_m^l=0$. Our assertion is thus proved.

In particular, functions $f(u)$ on $SU(2)$, which do not depend on the Euler angle $\psi$, i.e. such that

$$f(uh)=f(u)\ \text{ for }\ h=\begin{pmatrix} e^{it/2} & 0\\ 0 & e^{-it/2}\end{pmatrix},$$

are expanded in the Fourier series

$$f(u)=\sum_{l=0}^{\infty}\sum_{m=-l}^{l}\alpha_m^l e^{-im\varphi}P_{m0}^l(\cos\theta), \tag{7}$$

where

$$\alpha^l_m = \frac{(-1)^m(2l+1)}{4\pi}\int_0^{2\pi}\int_0^{\pi} f(\varphi,\ \theta,\ 0)\, e^{im\varphi}P^l_{m0}(\cos\theta)\sin\theta\, d\theta\, d\varphi.$$

Taking into account equality (9) of §3.9, we see that such functions are expanded in a series of the form

$$f(u) = \sum_{l=0}^{\infty}\sum_{m=-l}^{l} \beta^m_l e^{-im\varphi}P^m_l(\cos\theta),$$

where $P^m_l(\cos\theta)$ are associated Legendre functions. The Fourier coefficients are equal to

$$\beta^m_l = \frac{2l+1}{4\pi}\frac{(l-m)!}{(l+m)!}\int_0^{2\pi}\int_0^{\pi} f(\varphi,\ \theta,\ 0)\, e^{im\varphi}P^m_l(\cos\theta)\sin\theta\, d\theta\, d\varphi.$$

In just the same way, one solves the problem of expanding functions which satisfy for

$$h = \begin{pmatrix} e^{\frac{it}{2}} & 0 \\ 0 & e^{-\frac{it}{2}} \end{pmatrix}$$

the functional equation

$$f(hu) = e^{-imt}f(u).$$

We denote the space of such functions by ${}_m\mathfrak{L}^2$. As above, we can prove that any function $f(u)$ of ${}_m\mathfrak{L}^2$ is expanded in a series of the form

$$f(u) = e^{-im\varphi}\sum_{l=|m|}^{\infty}\sum_{n=-l}^{l} \alpha^l_n e^{-in\psi}P^l_{mn}(\cos\theta),$$

where

$$\alpha^l_n = \frac{(-1)^{m-n}(2l+1)}{8\pi}\int_{-2\pi}^{2\pi}\int_0^{\pi} f(0,\ \theta,\ \psi)\, e^{in\psi}P^l_{mn}(\cos\theta)\sin\theta\, d\theta\, d\varphi.$$

In particular, functions $f(u)$ which do not depend on the Euler angle $\phi$ are expanded in a series of the form

$$f(u) = \sum_{l=0}^{\infty}\sum_{n=-l}^{l} \beta^n_l e^{-in\psi}P^n_l(\cos\theta),$$

where

$$\beta^n_l = \frac{2l+1}{8\pi}\frac{(l-n)!}{(l+n)!}\int_{-2\pi}^{2\pi}\int_0^{\pi} f(0,\ \theta,\ \psi)\, e^{in\psi}P^n_l(\cos\theta)\sin\theta\, d\theta\, d\psi.$$

Let us denote by ${}_m\mathfrak{L}^2_n$ the intersection of the subspaces $\mathfrak{L}^2_n$ and ${}_m\mathfrak{L}^2$. This subspace consists of functions $f(u)$ such that

$$f(h_1 u h_2) = e^{-i(mt_1 + nt_2)} f(u) \tag{16}$$

for any two matrices

$$h_1 = \begin{pmatrix} e^{\frac{it_1}{2}} & 0 \\ 0 & e^{-\frac{it_1}{2}} \end{pmatrix} \text{ and } h_2 = \begin{pmatrix} e^{\frac{it_2}{2}} & 0 \\ 0 & e^{-\frac{it_2}{2}} \end{pmatrix}.$$

By virtue of the results obtained above, we have the following assertion

*Any function $f(u)$ of ${}_m\mathfrak{L}^2_n$ can be expanded in a Fourier series of the form*

$$f(u) = e^{-i(m\varphi + n\psi)} \sum_{l=\max(|m|,|n|)}^{\infty} \alpha_l P^l_{mn}(\cos\theta), \tag{17}$$

*where the Fourier coefficients are given by*

$$\alpha_l = \frac{(-1)^{m-n}(2l+1)}{2} \int_0^{\pi} f(0, \theta, 0) P^l_{mn}(\cos\theta)\sin\theta\, d\theta. \tag{18}$$

In particular, it follows that any function $f(u)$ on $SU(2)$ which does not depend on the Euler angles $\phi$ and $\psi$ is expanded in a series of the form

$$f(u) = \sum_{l=0}^{\infty} \alpha_l P^l_{00}(\cos\theta), \tag{19}$$

where

$$\alpha_l = \frac{2l+1}{2} \int_0^{\pi} f(0, \theta, 0) P^l_{00}(\cos\theta)\sin\theta\, d\theta. \tag{20}$$

Since $P^l_{00}(\cos\theta) \equiv P_l(\cos\theta)$, this series can be rewritten as follows:

$$f(u) = \sum_{l=0}^{\infty} \alpha_l P_l(\cos\theta), \tag{21}$$

where $P_l(x)$ are Legendre polynomials, and

$$\alpha_l = \frac{2l+1}{2} \int_0^{\pi} f(\theta) P_l(\cos\theta)\sin\theta\, d\theta. \tag{22}$$

5. **Expansions of functions on the sphere.** In the preceding subsection we constructed an expansion of functions $f(u)$ on the group $SU(2)$, constant on left cosets with respect to the subgroup $\Omega$ of diagonal matrices. But it was shown in §1.7 that the space of such left cosets can be identified with the unit sphere $S^2$ in three-dimensional euclidean space. Moreover, to the coset $u\Omega$, where $u = u(\phi, \theta, \psi)$, corresponds the point $\xi$ of the sphere with spherical coordinates $-\pi/2 + \phi$ and $\theta$.

Let us apply the expansion of subsection 4 to functions on the sphere. We obtain the following results.

Suppose that $f(\xi)$ is a function on the unit sphere $S^2$, such that

$$\int |f(\xi)|^2 d\xi < \infty \tag{1}$$

(here $d\xi$ denotes the normalized euclidean measure on $S^2$, $d\xi = (1/4\pi)\sin\theta\, d\theta\, d\phi$) Then[1]

$$f(\xi) \equiv f(\varphi, \theta) = \sum_{l=0}^{\infty} \sum_{m=-l}^{l} \beta_l^m e^{-im\varphi} P_l^m(\cos\theta), \tag{2}$$

where

$$\beta_l^m = \frac{2l+1}{4\pi} \frac{(l-m)!}{(l+m)!} \int_0^{2\pi} \int_0^{\pi} f(\varphi, \theta) e^{im\varphi} P_l^m(\cos\theta) \sin\theta\, d\theta\, d\varphi. \tag{3}$$

Moreover,

$$(2l+1) \int |f(\xi)|^2 d\xi = \sum_{l=0}^{\infty} \frac{1}{2l+1} \sum_{m=-l}^{l} \frac{(l+m)!}{(l-m)!} |\beta_l^m|^2. \tag{4}$$

This expansion shows that the functions

$$\sqrt{\frac{(2l+1)(l-k)!}{(l+k)!}}\, P_l^k(\cos\theta) e^{-ik\varphi} \equiv i^k \sqrt{2l+1}\, Y_{kl}(\varphi, \theta), \tag{5}$$
$$l = 0, 1, \dots; \quad -l \leq k \leq l$$

form an orthogonal normalized basis on the sphere (relative to the normalized euclidean measure).

The expansion of functions on the sphere according to the basis (5) is convenient, since associated spherical functions transform very simply under rotations of the sphere (it will be recalled that they are not essentially different from the matrix elements of irreducible unitary representations of this group).

6. Decomposition of fields of quantities on the sphere. In applied problems it becomes necessary to decompose not only scalar functions on the sphere, but also such entities as vector, tensor and other fields. In this connection it is, of course, required that the components into which these fields are decomposed should transform simply under rotations of the sphere. We now show how such decompositions are carried out.

---

[1] Here $\phi$ and $\theta$ are spherical coordinates of the point $\xi(\xi_1, \xi_2, \xi_3)$:

$$\xi_1 = \sin\varphi \sin\theta, \quad \xi_2 = \cos\varphi \sin\theta, \quad \xi_3 = \cos\theta.$$

We first introduce the concept of a quantity field on the sphere. Let $T_l(u)$ be an irreducible unitary representation of the group $SU(2)$, and let $\mathfrak{F}$ be the space of vector functions $\mathbf{f}(\xi)$ on the sphere $S^2$, taking values in the space $\mathfrak{H}_l$ of the representation $T_l(u)$. With each element $u$ of $SU(2)$, we associate the operator

$$V(u)\,\mathbf{f}(\xi)=T_l(u)\,\mathbf{f}(u^{-1}\xi) \tag{1}$$

(it will be recalled that $S^2$ is a homogeneous space with the group of motions $SU(2)$ and the stationary subgroup $\Omega$, consisting of the diagonal matrices $h=\begin{pmatrix} e^{it/2} & 0 \\ 0 & e^{-it/2}\end{pmatrix}$).

It is easy to verify that $V(u)$ is a group representation. We shall call the space $\mathfrak{F}$ with the representation $V(u)$ of the group $SU(2)$, given in it, a field of quantities on the sphere, transforming according to the irreducible unitary representation $T_l(u)$ (for example, for $l=1$ we get a vector field on the sphere).

In order to decompose the representation $V(u)$ into irreducible ones, we pass from functions $\mathbf{f}(\xi)$ on the sphere to functions $\phi(u)$ on $SU(2)$ by the formulas

$$\varphi(u)=V(u^{-1})\,\mathbf{f}(\xi_0) \tag{2}$$

(here $\xi_0$ denotes the point $\xi_0(0,0,1)$ of the sphere).

It is easy to show that the functions $\phi(u)$ satisfy the equation

$$\varphi(uh)=T_l(h^{-1})\varphi(u), \qquad h\in\Omega. \tag{3}$$

Moreover, to the operators $V(u_0)$ correspond the operators

$$\hat{V}(u_0)\,\varphi(u)=\varphi(u_0^{-1}u). \tag{4}$$

We shall denote the space of functions $\phi(u)$ also by $\mathfrak{F}$.

We decompose the space $\mathfrak{F}$ as the direct sum of one-dimensional subspaces $\mathfrak{F}_k$, $-l\le k\le l$, invariant for the operators $\phi(u)\to\phi(uh)$, $h\in\Omega$. For this, with the function $\phi(u)$ of $\mathfrak{F}$ we associate the functions $\phi_k(u)$, $-l\le k\le l$, defined by the formula

$$\varphi_k(u)=\frac{1}{2\pi}\int_0^{2\pi}\varphi\,[uh(t)]\,e^{-ikt}dt, \tag{5}$$

where

$$h(t)=\begin{pmatrix} e^{\frac{it}{2}} & 0 \\ 0 & e^{-\frac{it}{2}}\end{pmatrix}. \tag{6}$$

Clearly,

$$\varphi_k[uh(a)] = e^{ika}\varphi_k(u) \tag{1}$$

and therefore the space $\mathfrak{F}_k$ of functions of the form (5) is invariant for the transformations $\phi(u) \to \phi(uh)$, $h \in \Omega$. Moreover, one has

$$\varphi(u) = \sum_{k=-l}^{l} \varphi_k(u) \tag{8}$$

and

$$[\hat{V}(u_0)\varphi]_k(u) = \varphi_k(u_0^{-1}u). \tag{9}$$

These equalities show that the representation $\hat{V}(u)$ is the direct sum of representations $\hat{V}_k(u)$ of $SU(2)$, induced by the representations $h(t) \to e^{ikt}$ of the subgroup $\Omega$. We already know how to decompose representations of such a form into irreducible ones (see subsection 4). The problem of decomposing the representations $V(u)$ is thus solved.

## §7. THE CHARACTERS OF THE REPRESENTATIONS $T_l(u)$

1. Calculation of characters. We pass on to the consideration of the characters of representations $T_l(u)$. According to §1.10 of Chapter I, the trace of the matrix of a representation $T_l(u)$ of the group $SU(2)$ is called the character of this representation. Thus, the character $\chi_l(u)$ of $T_l(u)$ is given by the formula

$$\chi_l(u) = \sum_{m=-l}^{l} t^l_{mm}(u) \tag{1}$$

or, in Euler angles, by the formula

$$\chi_l(\varphi, \theta, \psi) = \sum_{m=-l}^{l} e^{-im(\varphi+\psi)} P^l_{mm}(\cos\theta). \tag{2}$$

This formula is not convenient, since it expresses the character as a function of three variables: $\phi$, $\psi$, $\theta$. But actually, as we shall show, the character $\chi_l(u)$ is a function of only one variable. Indeed, according to §1.10 of Chapter I, the character of a representation is a function on the group, constant on classes of conjugate elements. In our case this means that for any two elements $u_1$ and $u$ of $SU(2)$ one has the equality

$$\chi_l(u_1 u u_1^{-1}) = \chi_l(u). \tag{3}$$

In order to show that $\chi_l(u)$ is a function of one variable, it is sufficient to prove that classes of conjugate elements in $SU(2)$ are given by one parameter.

But it is well known from linear algebra that any unitary unimodular matrix $u$ can be written in the form $u = u_1 \delta u_1^{-1}$, where $u_1 \in SU(2)$ and $\delta$ is a diagonal matrix of the form

$$\delta = \begin{pmatrix} e^{\frac{it}{2}} & 0 \\ 0 & e^{-\frac{it}{2}} \end{pmatrix}. \tag{4}$$

The numbers $\lambda = e^{it/2}$ and $1/\lambda = e^{-it/2}$ are the eigenvalues of $u$. Moreover, among the matrices equivalent to $u$ there is only one other diagonal matrix, viz. the matrix $\delta'$ obtained from $\delta$ by interchanging the diagonal elements.

Hence each conjugacy class of elements in $SU(2)$ is given by one parameter $t$, varying in the limits $-2\pi \leq t \leq 2\pi$, where the parameters $t$ and $-t$ give one and the same class. Therefore we can regard the characters $\chi_l(u)$ as functions of one variable $t$, which ranges from 0 to $2\pi$.

The parameter $t$ has a simple geometrical significance—it is equal to the angle of rotation which corresponds to the matrix $u$. Thus a conjugacy class of elements in $SU(2)$ consists of matrices to which correspond rotations by one and the same angle in three-dimensional euclidean space.

Let us now derive an explicit expression for the $\chi_l(u)$ as functions of $t$. To do this we observe that, under the representation $T_l(u)$, to the diagonal matrix $\delta$ corresponds a diagonal matrix $T_l(\delta)$ of order $2l+1$, having the numbers $e^{-ikt}$, $-l \leq k \leq l$, on its principal diagonal.

Let $u = u_1 \delta u_1^{-1}$. Since characters are constant on conjugacy classes of elements,

$$\chi_l(u) = \chi_l(\delta) = \mathrm{Tr}\,[T_l(\delta)] = \sum_{k=-l}^{l} e^{-ikt}. \tag{5}$$

Summing the geometrical progression, we get

$$\chi_l(u) = \frac{e^{i(l+1)t} - e^{-ilt}}{e^{it} - 1} = \frac{\sin\left(l + \frac{1}{2}\right)t}{\sin\frac{t}{2}}, \tag{5'}$$

where $e^{it/2}$ and $e^{-it/2}$ are the eigenvalues of $u$.

As is well known, the eigenvalues of a matrix are the roots of its characteristic equation. For the matrix

$$u = \begin{pmatrix} \alpha & \beta \\ -\bar{\beta} & \bar{\alpha} \end{pmatrix}, \quad |\alpha|^2 + |\beta|^2 = 1$$

the characteristic equation has the form

$$\begin{vmatrix} \alpha-\lambda & \beta \\ -\bar{\beta} & \bar{\alpha}-\lambda \end{vmatrix}=0,$$

i.e.

$$\lambda^2-2\lambda\operatorname{Re}\alpha+1=0.$$

The roots of this equation are given by the formula

$$\lambda_{1,2}\equiv e^{\pm\frac{it}{2}}=\operatorname{Re}\alpha\pm i\sqrt{1-(\operatorname{Re}\alpha)^2}.$$

Hence

$$\cos\frac{t}{2}=\operatorname{Re}\alpha. \tag{6}$$

When the Euler angles of $u$ are equal to $\phi$, $\theta$, $\psi$, we have

$$\alpha=\cos\frac{\theta}{2}e^{\frac{i(\varphi+\psi)}{2}}$$

and so

$$\operatorname{Re}\alpha=\cos\frac{t}{2}=\cos\frac{\theta}{2}\cos\frac{\varphi+\psi}{2}. \tag{7}$$

Therefore

$$\chi_l(u)=\frac{\sin\left(l+\frac{1}{2}\right)t}{\sin\frac{t}{2}}, \tag{8}$$

where $\cos t/2$ is expressed in terms of Euler angles by formula (7). We have obtained above another expression for $\chi_l(u)$ in terms of Euler angles (see formula (2)). Comparing these two expressions, we obtain the equality

$$\sum_{m=-l}^{l} e^{-im(\varphi+\psi)}P^l_{mm}(\cos\theta)=\frac{\sin\left(l+\frac{1}{2}\right)t}{\sin\frac{t}{2}}, \tag{9}$$

where $\cos t/2=(\cos\theta/2)\cos((\phi+\psi)/2)$. In particular, for $\phi=\psi=0$, we have

$$\sum_{m=-l}^{l} P^l_{mm}(\cos\theta)=\frac{\sin\left(l+\frac{1}{2}\right)\theta}{\sin\frac{\theta}{2}}. \tag{10}$$

2. Orthogonality of characters. As shown in §4.7 of Chapter I, the characters of irreducible unitary representations of a compact group form an orthonormal system of functions. Thus for the group $SU(2)$ we have

$$\int \chi_m(u)\overline{\chi_n(u)}\,du = \delta_{mn}. \tag{1}$$

It was shown above that

$$\chi_m(u) = \frac{\sin\left(m+\frac{1}{2}\right)t}{\sin\frac{t}{2}}, \tag{2}$$

where $e^{\pm it/2}$ are the eigenvalues of the matrix $u$. In order to obtain, from formula (1), an orthogonality relation for the functions $\sin(m+\frac{1}{2})t$, it is necessary to express the invariant measure $du$ in new parameters, one of which is $t$.

We first write down an integral in the parameters $\alpha$, $\beta$, where $u = \begin{pmatrix} \alpha & \beta \\ -\bar{\beta} & \bar{\alpha} \end{pmatrix}$. It easily follows from formula (5) of §6.1 that

$$\int f(u)\,du = \frac{1}{4\pi^2}\int_0^1\int_0^{2\pi}\int_0^{2\pi} f(r,\,\tau,\,\sigma)\,d\sigma\,d\tau\,dr, \tag{3}$$

where $r = |\alpha|^2$, $\tau = \arg\alpha$, $\sigma = \arg\beta$. In this integral, let us pass from the variables $r$, $\tau$ to the variables $\alpha_1$, $\alpha_2$, where $\alpha = \alpha_1 + i\alpha_2$. We obtain

$$\int f(u)\,du = \frac{1}{2\pi^2}\int_{-1}^{1} d\alpha_1 \int_{-\sqrt{1-\alpha_1^2}}^{\sqrt{1-\alpha_1^2}} d\alpha_2 \int_0^{2\pi} f(u)\,d\sigma. \tag{4}$$

Since $\alpha_1 = \cos t/2$, this formula can be written as follows:

$$\int f(u)\,du = \frac{1}{4\pi^2}\int_0^{2\pi} \sin\frac{t}{2}\,dt \int_{-\sin\frac{t}{2}}^{\sin\frac{t}{2}} d\alpha_2 \int_0^{2\pi} f(u)\,d\sigma. \tag{5}$$

We have thus expressed the invariant integral over $SU(2)$ in the parameters $t$, $\alpha_2$, $\sigma$.

If the function $f(u)$ is constant on conjugacy classes of elements, i.e. depends only on the parameter $t$, $f(u) = F(t)$, then it follows from formula (5) that

$$\int f(u)\,du = \frac{1}{\pi}\int_0^{2\pi} F(t)\sin^2\frac{t}{2}\,dt. \tag{6}$$

3. **Decomposition of central functions.** According to §4.7 of Chapter I, a function $\phi(u)$ on $SU(2)$ is called central if it is constant on conjugacy classes of elements in $SU(2)$. We have shown above that conjugacy classes of elements in $SU(2)$ are determined by one parameter $t$. Therefore central functions are

functions of one variable $t$:

$$\varphi(u)=F(t). \tag{1}$$

It was shown in §4.7 of Chapter I that the characters of irreducible unitary representations of a compact group form a complete orthonormal system in the space of central functions on this group.

Applying this assertion to $SU(2)$ and to the function $f(t) = F(t)\sin t/2$, we get the following result.

*Let $f(t)$ be a function on the segment $[0, 2\pi]$, such that*

$$\int_0^{2\pi} |f(t)|^2\, dt < +\infty. \tag{2}$$

*Then this function can be expanded in the mean-convergent series*

$$f(t)=\sum a_n \sin\left(n+\frac{1}{2}\right)t. \tag{3}$$

*where $n$ runs through the values* $0, \frac{1}{2}, 1, \cdots$, *and*

$$a_n=\frac{1}{\pi}\int_0^{2\pi} f(t)\sin\left(n+\frac{1}{2}\right)t\, dt. \tag{4}$$

Hence it easily follows that the system of functions $\{\sin kx\}$, $k = 1, 2, \cdots$, is complete on the segment $[0, \pi]$.

## §8. THE CLEBSCH-GORDAN COEFFICIENTS

1. **The Kronecker product of the representations $T_l(u)$.** Let $T_{l_1}(u)$ be an irreducible unitary representation of weight $l_1$ of the group $SU(2)$, and let $T_{l_2}(u)$ be an irreducible unitary representation of weight $l_2$ of the same group. Let us consider the Kronecker product

$$T(u)= T_{l_1}(u)\otimes T_{l_2}(u). \tag{1}$$

Our aim is to decompose this product into irreducible ones.

In the first place, we clarify which representations appear in the decomposition of $T(u)$ and with what multiplicity they appear. By virtue of §4.7 of Chapter I, for this we have to decompose the product of the characters of $T_{l_1}(u)$ and $T_{l_2}(u)$ as the sum of the characters of irreducible representations.

Let $e^{it/2}$ and $e^{-it/2}$ be the eigenvalues of the matrix $u$. As shown in §7.1, we have

$$\chi_{l_1}(u)=\sum_{k=-l_1}^{l_1} e^{-ikt} \text{ and } \chi_{l_2}(u)=\sum_{m=-l_2}^{l_2} e^{-imt}.$$

Hence, for $l_1 \geq l_2$,

$$\chi_{l_1}(u)\chi_{l_2}(u)=\sum_{m=-l_2}^{l_2} \varepsilon^m \frac{\varepsilon^{l_1+1}-\varepsilon^{-l_1}}{\varepsilon-1}, \tag{2}$$

where for brevity we have put $e^{-it}=\epsilon$. Thus,

$$\chi_{l_1}(u)\chi_{l_2}(u)=\sum_{m=-l_2}^{l_2} \frac{\varepsilon^{l_1+m+1}-\varepsilon^{m-l_1}}{\varepsilon-1}$$

$$=\frac{1}{\varepsilon-1}[\varepsilon^{l_1+l_2+1}+\ldots+\varepsilon^{l_1-l_2+1}-\varepsilon^{l_2-l_1}-\ldots-\varepsilon^{-l_1-l_2}]. \tag{3}$$

On combining pairwise positive and negative terms, the sum of whose indices is equal to unity, we obtain

$$\chi_{l_1}(u)\chi_{l_2}(u)=\sum_{l=l_1-l_2}^{l_1+l_2} \frac{\varepsilon^{l+1}-\varepsilon^{-l}}{\varepsilon-1}=\sum_{l=l_1-l_2}^{l_1+l_2}\chi_l(u). \tag{4}$$

For $l_2 \geq l_1$ the summation starts from $l = l_2 - l_1$. Thus

$$\chi_{l_1}(u)\chi_{l_2}(u)=\sum_{l=|l_1-l_2|}^{l_1+l_2}\chi_l(u). \tag{5}$$

This equality signifies that, in the decomposition of the representation $T_{l_1}(u) \otimes T_{l_2}(u)$, all those representations $T_l(u)$ appear for which $|l_1 - l_2| \leq l \leq l_1 + l_2$ and $l$ is an integer or a semi-integer simultaneously with $l_1 + l_2$; moreover, each of these representations appears only once.

In other words,

$$T_{l_1}(u)\otimes T_{l_2}(u)=\sum_{l=|l_1-l_2|}^{l_1+l_2} T_l(u). \tag{6}$$

In the sequel we shall denote by $\mathfrak{H}_1$ and $\mathfrak{H}_2$ the spaces of the representations $T_{l_1}(u)$ and $T_{l_2}(u)$, and by $\mathfrak{F}_l$ invariant subspaces in $\mathfrak{H}_1 \otimes \mathfrak{H}_2$ in which the representations $T_l(u)$, $|l_1 - l_2| \leq l \leq l_1 + l_2$, are realized. Clearly,

$$\mathfrak{H}_1\otimes\mathfrak{H}_2=\sum_{l=|l_1-l_2|}^{l_1+l_2}\mathfrak{F}_l.$$

Formula (13) of §4.7, Chapter I, gives an expansion for the multiplicity with which $T_l(u)$ appears in $T_{l_1}(u) \otimes T_{l_2}(u)$, in terms of the characters of these representations. If in this formula we substitute the expression (5′) of §7.1 for $\chi_l(u)$ and the expression (3) of §7.2 for $du$, we arrive at the following conclusion: the integral

$$\frac{1}{\pi}\int_0^{2\pi} \sin\left(l_1+\frac{1}{2}\right)t \sin\left(l_2+\frac{1}{2}\right)t \frac{\sin\left(l+\frac{1}{2}\right)t}{\sin\frac{t}{2}}\,dt,$$

where $l_1$, $l_2$, $l$ are integers or semi-integers, is equal to unity in the case when $2(l_1+l_2+l)$ is even and a triangle with sides $2l_1$, $2l_2$, $2l$ exists; otherwise it is equal to zero.

2. **Bases in the space $\mathfrak{H}_1 \otimes \mathfrak{H}_2$.** In the space $\mathfrak{H}_1 \otimes \mathfrak{H}_2$ of the representation $T_{l_1}(u) \otimes T_{l_2}(u)$ there are two bases of interest to us. The first of these is constituted by the pairwise products

$$\mathbf{f}_j \otimes \mathbf{h}_k, \quad -l_1 \leqslant j \leqslant l_1, \quad -l_2 \leqslant k \leqslant l_2 \tag{1}$$

of vectors of the canonical bases of $\mathfrak{H}_1$ and $\mathfrak{H}_2$ [$\mathbf{f}_j$, $-l_1 \le j \le l_1$, being the canonical basis in $\mathfrak{H}_1$, and $\mathbf{h}_k$, $-l_2 \le k \le l_2$, the canonical basis in $\mathfrak{H}_2$]. According to the Supplement to Chapter I, this basis is orthonormal in $\mathfrak{H}_1 \otimes \mathfrak{H}_2$.

The basis $\mathbf{f}_j \otimes \mathbf{h}_k$ in $\mathfrak{H}_1 \otimes \mathfrak{H}_2$ is convenient inasmuch as the matrix elements of $T_{l_1}(u) \otimes T_{l_2}(u)$ in this basis are products of the matrix elements of $T_{l_1}(u)$ and $T_{l_2}(u)$ in the bases $\mathbf{f}_j$ and $\mathbf{h}_k$ respectively. We shall denote these matrix elements by $\alpha_{(jk),(j'k')}(u)$. Thus,

$$\alpha_{(jk),(j'k')}(u) = t^{l_1}_{jj'}(u)\, t^{l_2}_{kk'}(u). \tag{2}$$

The second orthonormal basis in $\mathfrak{H}_1 \otimes \mathfrak{H}_2$ consists of the vectors of canonical bases of the irreducible subspaces $\mathfrak{F}_l$ into which $\mathfrak{H}_1 \otimes \mathfrak{H}_2$ is decomposed (see formula (6) of subsection 1). We shall denote these vectors by

$$\mathbf{a}^l_m, \quad |l_1 - l_2| \leqslant l \leqslant l_1 + l_2, \quad -l \leqslant m \leqslant l. \tag{3}$$

This basis is convenient because in it the matrix of $T_{l_1}(u) \otimes T_{l_2}(u)$ is block-diagonal, where the principal diagonal is composed of the matrix of the irreducible unitary representations $T_l(u)$, $|l_1 - l_2| \le l \le l_1 + l_2$, of the group $SU(2)$. We denote the elements of the matrix of $T_{l_1}(u) \otimes T_{l_2}(u)$ in the basis $\mathbf{a}^l_m$ by $\beta_{(lm),(l'm')}(u)$. It follows from the block-diagonality of this matrix that

$B_{(lm),(l'm')}=0$ if $l\neq l'$. But if $l=l'$ then $\beta_{(lm),(l'm')}=t^l_{mm'}(u)$. Thus,

$$\beta_{(lm),(l'm')}(u)=\delta_{ll'}t^l_{mm'}(u). \tag{4}$$

Since the bases $\mathbf{f}_j$, $\mathbf{h}_k$, $\mathbf{a}^l_m$ are canonical, the matrix elements $t^{l_1}_{jj'}(u)$, $t^{l_2}_{kk'}(u)$, $t^l_{mm'}(u)$ in formulas (2) and (4) are expressed in terms of Euler angles by formula (6) of §3.3:

$$t^{l_1}_{jj'}(\varphi,\ \theta,\ \psi)=e^{-i(j\varphi+j'\psi)}P^{l_1}_{jj'}(\cos\theta), \tag{5}$$

and so on. Since $\mathbf{f}_j\otimes\mathbf{h}_k$ and $\mathbf{a}^l_m$ are orthonormal bases in one and the same space $\mathfrak{H}_1\otimes\mathfrak{H}_2$, there exists a unitary matrix $C$ which transforms $\mathbf{a}^l_m$ into $\mathbf{f}_j\otimes\mathbf{h}_k$. The rows of this matrix are indicated by the pair of indices $(l, m)$, and the columns by the pair of indices $(j, k)$ (these indices vary between different limits $-l_1\leq j\leq l_1$, $-l_2\leq k\leq l_2$ and $|l_1-l_2|\leq l\leq|l_1+l_2|$, $-l\leq m\leq l$). Thus,

$$\mathbf{f}_j\otimes\mathbf{h}_k=\sum_{l=|l_1-l_2|}^{|l_1+l_2|}\ \sum_{m=-l}^{l}C_{(lm),(jk)}\mathbf{a}^l_m. \tag{6}$$

Since $C$ is unitary, it follows that

$$\mathbf{a}^l_m=\sum_{j=-l_1}^{l_1}\ \sum_{k=-l_2}^{l_2}\overline{C_{(lm),(jk)}}\mathbf{f}_j\otimes\mathbf{h}_k. \tag{7}$$

In many cases it is useful to mention explicitly the dependence of $C$ on the parameters $l_1$ and $l_2$. Therefore, instead of $C_{(lm),(jk)}$ one writes $C(l_1, l_2, l; j, k, m)$ or, more concisely, $C(\mathbf{1}, \mathbf{j})$, where $\mathbf{1}=(l_1, l_2, l)$, $\mathbf{j}=(j, k, m)$,

$$C(\mathbf{1},\ \mathbf{j})\equiv C(l_1,\ l_2,\ l;\ j,\ k,\ m)\equiv C_{(lm),(jk)}. \tag{8}$$

The numbers $C(l_1, l_2, l; j, k, m)$ are called the Clebsch-Gordan coefficients (in works on physics they are often incorrectly called the Clebsch-Jordan coefficients).

The Clebsch-Gordan coefficients are not uniquely defined. This has to do with the fact that the vectors of a canonical basis are only determined to within a common factor with modulus unity (after multiplication by such a factor the basis remains orthonormal and canonical). Therefore the coefficients $C(l_1, l_2, l; j, k, m)$ of (6) are defined to within a factor $\alpha_l$ such that $|\alpha_l|=1$. On using this indeterminacy, one can ensure that

$$C(l_1,\ l_2,\ l;\ l_1,\ -l_2,\ l_1-l_2)\geqslant 0. \tag{9}$$

It will be shown below that this relation implies that all the coefficients $C(\mathbf{1}, \mathbf{j})$.

are real.

3. **Calculation of the Clebsch–Gordan coefficents.** From the fact that the bases $\mathbf{f}_j \otimes \mathbf{h}_k$ and $\mathbf{a}_m^l$ are connected by relations (6) and (7) of subsection 2, it follows that the matrices of the operator $T_{l_1}(u) \otimes T_{l_2}(u)$ in these bases are connected by the relation

$$(\alpha_{(jk),\,(j'k')}(u)) = C^* \left(\beta_{(lm),\,(l'm')}(u)\right) C. \tag{1}$$

Hence

$$\alpha_{(jk),\,(j'k')}(u) = \sum_{l,\, l',\, m,\, m'} \overline{C_{(lm),\,(jk)}}\, \beta_{(lm),\,(l'm')}(u)\, C_{(l'm'),\,(j'k')}. \tag{2}$$

The summation in this formula extends over values of $l,\ l',\ m,\ m'$ such that $|l_1 - l_2| \leq l,\ l' \leq l_1 + l_2;\ -l \leq m \leq l;\ -l' \leq m' \leq l'$.

Taking into account formulas (2) and (4) of subsection 2, we rewrite this equality as follows:

$$t^{l_1}_{jj'}(u)\, t^{l_2}_{kk'}(u) = \sum_{l,\, m,\, m'} C(\mathbf{l},\ \mathbf{j}')\, \overline{C(\mathbf{l},\ \mathbf{j})}\, t^l_{mm'}(u), \tag{3}$$

where $\mathbf{l} = (l_1,\ l_2,\ l)$, $\mathbf{j}' = (j',\ k',\ m')$. The summation in (3) extends over values of $l,\ m,\ m'$ such that $|l_1 - l_2| \leq l \leq l_1 + l_2,\ -l \leq m,\ m' \leq l$.

Let us multiply both sides of (3) by $\overline{t^l_{mm'}(u)}$ and integrate over the whole group $SU(2)$. By virtue of the orthonormality of the system of functions $\sqrt{2l+1}\, t^l_{mm'}(u)$ (see §6.2), we get

$$C(\mathbf{l},\ \mathbf{j}')\, \overline{C(\mathbf{l},\ \mathbf{j})} = (2l+1) \int t^{l_1}_{jj'}(u)\, t^{l_2}_{kk'}(u)\, \overline{t^l_{mm'}(u)}\, du. \tag{4}$$

Taking into account expression (5) of subsection 2 for the matrix elements $t^{l_1}_{jj'}(u)$ etc, and also the expression

$$du = \frac{1}{16\pi^2} \sin\theta\, d\theta\, d\varphi\, d\psi \tag{5}$$

for the invariant measure, we find that integral (4) is nonzero only if $j + k = m$, $j' + k' = m'$ (otherwise integration with respect to $\phi$ or $\psi$ gives zero). Therefore the Clebsch-Gordan coefficients $C(l_1,\ l_2,\ l;\ j,\ k,\ m)$ are nonzero only if $j+k=m$.[1)]

[1)] This result can also be obtained on applying, to both sides of equality (6) of subsection 2, the operator $T_{l_1}(u) \otimes T_{l_2}(u)$, where $u = u(\phi,\ 0,\ 0)$. Then the left-hand side of the equality is multiplied by $e^{-i(j+k)\phi}$, and the terms containing $\mathbf{a}^l_m$, by $e^{-im\phi}$. The linear independence of the bases yields the results that, in equality (6) of subsection 2, only those terms appear for which $m = j + k$.

Henceforth we shall assume that in (4) we have $j+k=m$, $j'+k'=m'$. Substituting explicit expressions for the matrix elements and measure $du$ in this equality, making the substitution $\cos\theta = x$ and carrying out the integration with respect to $\phi$ and $\psi$, we obtain

$$C(\mathbf{l},\ \mathbf{j}')\overline{C(\mathbf{l},\ \mathbf{j})}=\frac{2l+1}{2}\int_{-1}^{1}P^{l_1}_{jj'}(x)P^{l_2}_{kk'}(x)\overline{P^{l}_{j+k,\ j'+k'}(x)}\,dx, \tag{6}$$

where

$$\mathbf{l}=(l_1,\ l_2,\ l),\ \mathbf{j}=(j,\ k,\ j+k),\ \mathbf{j}'=(j',\ k',\ j'+k').$$

Let us put $j'=l_1$ and $k'=-l_2$ in this equality. By formulas (6) of §3.5 and (2), (3) §3.6, we have

$$P^{l_1}_{jl_1}(x)=\frac{i^{l_1-j}}{2^{l_1}}\sqrt{\frac{(2l_1)!}{(l_1-j)!\,(l_1+j)!}}(1-x)^{\frac{l_1-j}{2}}(1+x)^{\frac{l_1+j}{2}} \tag{7}$$

and

$$P^{l_2}_{k,\,-l_2}(x)=\frac{i^{l_2+k}}{2^{l_2}}\sqrt{\frac{(2l_2)!}{(l_2+k)!\,(l_2-k)!}}(1-x)^{\frac{l_2+k}{2}}(1+x)^{\frac{l_2-k}{2}}. \tag{8}$$

In addition, by formula (3) of §3.4,

$$\begin{aligned}P^{l}_{j+k,\,l_1-l_2}(x)=&\frac{i^{-j-k-l_1+l_2+2l}}{2^{l}}\sqrt{\frac{(l+j+k)!}{(l-j-k)!\,(l+l_1-l_2)!\,(l-l_1+l_2)!}}\\&\times(1-x)^{-\frac{j+k+l_2-l_1}{2}}(1+x)^{\frac{l_2-l_1-j-k}{2}}\\&\times\frac{d^{l-j-k}}{dx^{l-j-k}}[(1-x)^{l-l_1+l_2}(1+x)^{l+l_1-l_2}].\end{aligned} \tag{9}$$

On substituting these expansions in (6), we obtain

$$\begin{aligned}&C(l_1,\ l_2,\ l;\ l_1,\ -l_2,\ l_1-l_2)\overline{C(l_1,\ l_2,\ l;\ j,\ k,\ j+k)}\\&=\frac{(-1)^{-l+l_1+k}(2l+1)}{2^{l+l_1+l_2+1}}\sqrt{\frac{(2l_1)!\,(2l_2)!\,(l+j+k)!}{(l_1-j)!\,(l_1+j)!\,(l_2-k)!\,(l_2+k)!}}\\&\times\sqrt{\frac{1}{(l-j-k)!\,(l+l_1-l_2)!\,(l-l_1+l_2)!}}\\&\times\int_{-1}^{1}(1-x)^{l_1-j}(1+x)^{l_2-k}\frac{d^{l-j-k}}{dx^{l-j-k}}[(1-x)^{l-l_1+l_2}(1+x)^{l+l_1-l_2}]\,dx.\end{aligned} \tag{10}$$

In order to find $C(l_1,\ l_2,\ l;\ l_1,\ -l_2,\ l_1-l_2)$ from this equality, we put $j=l_1$, $k=-l_2$ in it:

$$|C(l_1, l_2, l; l_1, -l_2, l_1-l_2)|^2 = \frac{(2l+1)(-1)^{-l+l_1-l_2}}{2^{l+l_1+l_2+1}(l-l_1+l_2)!}$$
$$\times \int_{-1}^{1} (1+x)^{2l_2} \frac{d^{l-l_1+l_2}}{dx^{l-l_1+l_2}} [(1-x)^{l-l_1+l_2}(1+x)^{l+l_1-l_2}]\, dx.$$

Let us integrate by parts $l - l_1 + l_2$ times. Since the integrated terms vanish on substituting the limits of integration, we get

$$|C(l_1, l_2, l; l_1, -l_2, l_1-l_2)|^2$$
$$= \frac{2l+1}{2^{l+l_1+l_2+1}} \frac{(2l_2)!}{(l-l_1+l_2)!\,(l_1+l_2-l)!} \int_{-1}^{1} (1+x)^{2l_1}(1-x)^{l-l_1+l_2}\, dx$$
$$= \frac{(2l+1)(2l_1)!\,(2l_2)!}{(l_1+l_2-l)!\,(l_1+l_2+l+1)!} \tag{11}$$

(see formulas (1) and (3) of §1.7, Chapter V, below).

As the Clebsch-Gordan coefficients are normalized in such a way that the inequality $C(l_1, l_2, l; l_1, -l_2, l_1 - l_2) \geq 0$ is satisfied, it follows from formula (11) that

$$C(l_1, l_2, l; l_1, -l_2, l_1-l_2) = \sqrt{\frac{(2l+1)(2l_1)!\,(2l_2)!}{(l_1+l_2-l)!\,(l_1+l_2+l+1)!}}\,. \tag{12}$$

On substituting this in (10), we obtain

$$C(l_1, l_2, l; j, k, j+k) = \frac{(-1)^{-l+l_1+k}}{2^{l+l_1+l_2+1}}$$
$$\times \sqrt{\frac{(2l+1)(l+j+k)!\,(l_1+l_2-l)!\,(l_1+l_2+l+1)!}{(l_1-j)!\,(l_1+j)!\,(l_2-k)!\,(l_2+k)!\,(l-j-k)!\,(l+l_1-l_2)!\,(l-l_1+l_2)!}}$$
$$\times \int_{-1}^{1} (1-x)^{l_1-j}(1+x)^{l_2-k} \frac{d^{l-j-k}}{dx^{l-j-k}} [(1-x)^{l-l_1+l_2}(1+x)^{l+l_1-l_2}]\, dx. \tag{13}$$

It is obvious from this expression that, under the normalization chosen by us, all Clebsch-Gordan coefficients are real.

Another expression for $C(l_1, l_2, l; j, k, j+k)$ can be obtained on observing that $P^l_{j+k, l_1-l_2}(x) = P^l_{l_1-l_2, j+k}(x)$ and substituting the corresponding expression in formula (6). As a result, we get

$$C(l_1, l_2, l; j, k, j+k) = \frac{(-1)^{-l+l_1+k}}{2^{l+l_1+l_2+1}}$$
$$\times \sqrt{\frac{(2l+1)(l+l_1-l_2)!\,(l_1+l_2-l)!\,(l_1+l_2+l+1)!}{(l_1-j)!\,(l_1+j)!\,(l_2-k)!\,(l_2+k)!\,(l+j+k)!\,(l-j-k)!\,(l-l_1+l_2)!}}$$
$$\times \int_{-1}^{1}(1-x)^{l_2+k}(1+x)^{l_2-k}\frac{d^{l-l_1+l_2}}{dx^{l-l_1+l_2}}[(1-x)^{l-j-k}(1+x)^{l+j+k}]\,dx. \quad (13')$$

From these expressions for the Clebsch-Gordan coefficients, it is easy to express these coefficients in the form of sums. For this it is sufficient to apply Leibnitz's formula to the integrands, and integrate the resulting expressions term by term. Thus we easily get

$$C(l_1, l_2, l; j, k, j+k) = (-1)^{-l+l_1+k}$$
$$\times \sqrt{\frac{(2l+1)(l+j+k)!\,(l_1+l_2-l)!\,(l-j-k)!\,(l-l_1+l_2)!\,(l+l_1-l_2)!}{(l_1+l_2+l+1)!\,(l_1-j)!\,(l_1+j)!\,(l_2-k)!\,(l_2+k)!}}$$
$$\times \sum_{s=M}^{N}\frac{(-1)^s(l+l_2-j-s)!\,(l_1+j+s)!}{s!\,(l-j-k-s)!\,(l-l_1+l_2-s)!\,(l_1-l_2+j+k+s)!}, \quad (14)$$

where

$$M = \max(0,\ l_2-l_1-j-k),\quad N = \min(l-j-k,\ l-l_1+l_2).$$

Another representation of $C(1, j)$ in the form of a sum is obtained if, in formula (13), one integrates by parts $l-j-k$ times, and observes that all the integrated terms vanish. Applying Leibnitz's formula and integrating term by term, we find

$$C(l_1, l_2, l; j, k, j+k) = (-1)^{l_1-j}$$
$$\times \sqrt{\frac{(2l+1)(l+j+k)!\,(l-j-k)!\,(l_1-j)!\,(l_2-k)!\,(l_1+l_2-l)!}{(l_1+j)!\,(l_2+k)!\,(l+l_1-l_2)!\,(l-l_1+l_2)!\,(l_1+l_2+l+1)!}}$$
$$\times \sum_{s=M_1}^{N_1}\frac{(-1)^s(l_1+j+s)!\,(l+l_2-j-s)!}{s!\,(l-j-k-s)!\,(l_1-j-s)!\,(l_2-l+j+s)!}, \quad (14')$$

where

$$M_1 = \max(0,\ l-l_2-j),\quad N = \min(l-j-k,\ l_1-j).$$

Yet another representation of $C(1, j)$ in the form of a sum is obtained if one puts $j' = l_1$, $k' = -l_2$ in formula (6), substitutes the expression for $P^l_{j+k,l_1-l_2}(z)$ given by (1) of §3.4 and carries out term-by-term integration:

$$C(l_1, l_2, l; j, k, j+k)$$
$$=\sqrt{\frac{(2l+1)(l_1+j)!(l-j-k)!(l-l_1+l_2)!(l_1+l_2-l)!(l_1+l_2+l+1)!}{(l_1-j)!(l_2+k)!(l_2-k)!(l+j+k)!(l+l_1-l_2)!}}$$
$$\times \sum_{s=\max(j+k,\, l_1-l_2)}^{l} \frac{(-1)^{l_1+k-s}(l+s)!(l_2+s-j)!}{(l-s)!(s-j-k)!(s-l_1+l_2)!(l_1+l_2+s+1)!}. \qquad (14'')$$

If $l_1 - l_2$ is an integer, then the summation in this formula is extended over integral values of $s$, and if $l_1 - l_2$ is a semi-integer, the summation is extended over semi-integral values of $s$.

4. The symmetry relations. Like the functions $P_{mn}(x)$, the Clebsch-Gordan coefficients possess a number of symmetry relations. We show first that

$$C(l_1, l_2, l; j, k, j+k) =(-1)^{l-l_1-l_2} C(l_1, l_2, l; -j, -k, -j-k). \qquad (1)$$

For this it is sufficient, in formula (13′) of subsection 3, to replace $j$ by $-j$ and $k$ by $-k$, make the substitution $x = -y$ and note that $k + l_2$ is an integer.

One similarly proves that

$$C(l_1, l_2, l; j, k, j+k)=(-1)^{l-l_1-l_2} C(l_2, l_1, l; k, j, j+k). \qquad (2)$$

For this it is necessary to use formula (13) of subsection 3, interchange $(l_1, j)$ and $(l_2, k)$ in it, and make the substitution $x = -y$.

In order to obtain the symmetry relations which follow, we make a change of parameters in formula (13) of subsection 3:

$$\left.\begin{aligned} l_1&=\frac{l_1'+l_2'+j'+k'}{2}, & l_2&=\frac{l_1'+l_2'-j'-k'}{2}, \\ j&=\frac{l_1'-l_2'+j'-k'}{2}, & k&=\frac{l_1'-l_2'-j'+k'}{2}. \end{aligned}\right\} \qquad (3)$$

Comparing the result with formula (13′) of subsection 3, we obtain

$$C\left(\frac{l_1+l_2+j+k}{2}, \frac{l_1+l_2-j-k}{2}, l; \frac{l_1-l_2+j-k}{2}, \frac{l_1-l_2-j+k}{2}, l_1-l_2\right) = C(l_1, l_2, l; j, k, j+k). \qquad (4)$$

Further, in formula (13) let us integrate by parts $l - j - k$ times and put

$$\left.\begin{aligned} l_1&=\frac{l'+l_1'-k'}{2}, & l_2&=\frac{l'+l_1'+k'}{2}, \quad l=l_2', \\ j&=\frac{l_1'-l'+k'}{2}+j', & k&=\frac{l_1'-l'-k'}{2}-j'. \end{aligned}\right\} \qquad (5)$$

Comparing the result with formula (13′), we obtain the following symmetry relation

$$C\left(\frac{l+l_1-k}{2}, \frac{l+l_1+k}{2}, l_2; \frac{l_1-l+k}{2}+j, \frac{l_1-l-k}{2}-j, l_1-l\right)$$
$$= (-1)^{l_1-j}\sqrt{\frac{2l_2+1}{2l+1}}\, C(l_1, l_2, l; j, k, j+k). \qquad (6)$$

Combining relations (4) and (6), we get

$$C(l_1, l_2, l; j, k, j+k) = (-1)^{l_1-j}\sqrt{\frac{2l+1}{2l_2+1}}\, C(l_1, l, l_2; j, -j-k, -k). \qquad (7)$$

Further, combining this with (1) and (2), we get

$$C(l_1, l_2, l; j, k, j+k)$$
$$= (-1)^{l-l_1-k}\left(\frac{2l+1}{2l_1+1}\right)^{\frac{1}{2}} C(l_2, l, l_1; k, -j-k, -j)$$
$$= (-1)^{l-l_2+j}\left(\frac{2l+1}{2l_2+1}\right)^{\frac{1}{2}} C(l, l_1, l_2; -j-k, j, -k)$$
$$= (-1)^{l_2+k}\left(\frac{2l+1}{2l_1+1}\right)^{\frac{1}{2}} C(l, l_2, l_1; -j-k, k, -j). \qquad (7')$$

It is convenient to write these symmetry relations in terms of Wigner's symbol $\begin{Bmatrix} l_1 & l_2 & l_3 \\ m_1 & m_2 & m_3 \end{Bmatrix}$, which we introduce by writing

$$C(l_1, l_2, l; j, k, j+k) = (-1)^{l_1-l_2+j+k}\sqrt{2l+1}\begin{pmatrix} l_1 & l_2 & l \\ j & k & -j-k \end{pmatrix}. \qquad (8)$$

The symmetry relations for Wigner's symbol are formulated in the following way: with this symbol we associate a third order matrix

$$\begin{pmatrix} j_1 & j_2 & j_3 \\ m_1 & m_2 & m_3 \end{pmatrix} \to \begin{pmatrix} -j_1+j_2+j_3 & j_1-j_2+j_3 & j_1+j_2-j_3 \\ j_1-m_1 & j_2-m_2 & j_3-m_3 \\ j_1+m_1 & j_2+m_2 & j_3+m_3 \end{pmatrix}. \qquad (9)$$

For an even permutation of the rows or columns of this matrix, or under its transposition, Wigner's symbol is unchanged, and for an odd permutation of the rows or columns it must be multiplied by $(-1)^{j_1+j_2+j_3}$.

5. Some special values. Let us state a few cases when the Clebsch-Gordan coefficients consist of a single term. Let $l_1 = j$. Then only the term for $s = 0$ remains in formula (14′) of subsection 3, and we get

$$C(l_1, l_2, l; l_1, k, l_1+k)$$
$$=\sqrt{\frac{(2l+1)(l+l_1+k)!(l_2-k)!(2l_1)!(l+l_2-l_1)!}{(l-l_1-k)!(l_2-l+l_1)!(l_2+k)!(l-l_2+l_1)!(l+l_2+l_1+1)!}}. \quad (1)$$

Utilizing the symmetry relations just derived, we get one-term expressions in the cases:

$$\text{a) } j=\pm l_1, \quad \text{b) } k=\pm l_2, \quad \text{c) } j+k=\pm l.$$

For example,

$$C(l_1, l_2, l; j, l-j, l)=(-1)^{l_1-j}$$
$$\times\sqrt{\frac{(2l+1)!(l_1+l_2-l)!(l_1+j)!(l_2+l-j)!}{(l-l_1+l_2)!(l_1+l_2+l+1)!(l_1-j)!(l_2-l+j)!(l+l_1-l_2)!}}. \quad (2)$$

We next consider the case when $l=l_1+l_2$. In this case only the term for $s=l_1-j$ remains in formula (14′) of subsection 3. We get

$$C(l_1, l_2, l_1+l_2; j, k, j+k)$$
$$=\sqrt{\frac{(l_1+l_2+j+k)!(l_1+l_2-j-k)!(2l_1)!(2l_2)!}{(l_1-j)!(l_2-k)!(l_1+j)!(l_2+k)!(2l_1+2l_2)!}}. \quad (3)$$

With the help of the symmetry relations, this leads to one-term expressions in the cases $l=l_2-l_1$ and $l=l_1-l_2$. For example, for $l_1\geq l_2$ we have

$$C(l_1, l_2, l_1-l_2; j, k, j+k)=(-1)^{l_2+k}$$
$$\times\sqrt{\frac{(2l_1-2l_2+1)!(2l_2)!(l_1-j)!(l_1+j)!}{(l_1-l_2+j+k)!(l_1-l_2-j-k)!(2l_1+1)!(l_2-k)!(l_2+k)!}}. \quad (4)$$

A one-term expression for the Clebsch-Gordan coefficients is also obtained in the case when $j=k=0$. It follows from formula (13′) that

$$C(l_1, l_2, l; 0, 0, 0)=\frac{(-1)^{l_1-l}}{2^{l+l_1+l_2+1}\, l_1!\, l_2!\, l!}$$
$$\times\sqrt{\frac{(2l+1)(l+l_1-l_2)!(l_1+l_2-l)!(l_1+l_2+l+1)!}{(l-l_1+l_2)!}}$$
$$\times\int_{-1}^{1}(1-x^2)^{l_2}\frac{d^{l-l_1+l_2}}{dx^{l-l_1+l_2}}(1-x^2)^l\,dx.$$

Clearly, this integral is equal to zero if $l+l_1+l_2$ is an odd number (in that case the integrand is an odd function). Now let $l+l_1+l_2=2g$, where $g$ is an integer One can easily show that

$$J(p, s, q) = \int_{-1}^{1} (1-x^2)^p \frac{d^{2s}}{dx^{2s}} (1-x^2)^q \, dx$$

$$= \frac{(-1)^s 2^{2p+2q-2s+1} p!\, q!\, (2s)!\, (p+q-2s)!\, (p+q-s)!}{s!\, (p-s)!\, (q-s)!\, (2p+2q-2s+1)!}. \quad (5)$$

Indeed, for $s = 0$ the value of the integral (5) is given by the equality

$$J(p, 0, q) = \int_{-1}^{1} (1-x^2)^{p+q} dx = \frac{2^{2p+2q+1} [(p+q)!]^2}{(2p+2q+1)!} \quad (6)$$

in accordance with formula (5). Since

$$\frac{d^{2s}}{dx^{2s}} (1-x^2)^q = \frac{d^{2s-2}}{dx^{2s-2}} [2q(2q-2)(1-x^2)^{q-2} - 2q(2q-1)(1-x^2)^{q-1}],$$

we have

$$J(p, s, q) = 2q(2q-2) J(p, s-1, q-2) - 2q(2q-1) J(p, s-1, q-1). \quad (7)$$

Evidently, the recurrence relation (7) with the initial condition (6) has a unique solution. Substitution show that this solution is given by (5).

It follows from (5) that

$$C(l_1, l_2, l; 0, 0, 0)$$
$$= (-1)^{g-l} \sqrt{\frac{(2l+1)(l+l_1-l_2)!\,(l_1+l_2-l)!\,(l-l_1+l_2)!}{(l_1+l_2+l+1)!}}$$
$$\times \frac{g!}{(g-l_1)!\,(g-l_2)!\,(g-l)!}, \quad (8)$$

where $2g = l + l_1 + l_2$.

Hence, owing to the symmetry relations (4) of subsection 4, we find that if $2g = l + 2l_1$ is even, then

$$C(l_1, l_1, l; j, j, 2j)$$
$$= (-1)^{g-l} \sqrt{\frac{(2l+1)(2j+l)!\,(2l_1-l)!\,(l-2j)!}{(2l_1+l+1)!}}$$
$$\times \frac{g!}{(g-l_1-j)!\,(g-l_1+j)!\,(g-l)!}. \quad (9)$$

But if $2g$ is odd, then $C(l_1, l_1, l; k, j, 2j) = 0$.

6. **Expansion of products of functions $P^l_{mn}(z)$.** It was shown in subsection 3 that

$$t^{l_1}_{jj'}(u)\, t^{l_2}_{kk'}(u)$$
$$= \sum_{l=|l_2-l_1|}^{l_1+l_2} C(l_1, l_2, l; j; k; j+k)\, C(l_1, l_2, l; j'; k'; j'+k')\, t^l_{j+k, j'+k'}(u) \quad (1)$$

(it will be recalled that the Clebsch-Gordan coefficients are real in the normalization chosen by us).

In this equality we substitute the expressions for functions $t^{l_1}_{jj'}(u)$, $t^{l_2}_{kk'}(u)$, $t^{l}_{j+k,j'+k'}(u)$ in terms of Euler angles (see formula (8) of §3.3). After cancellation by $e^{-i[(j+k)\phi+(j'+k')\psi]}$ we get the equality

$$\begin{aligned} &P^{l_1}_{jj'}(z)\,P^{l_2}_{kk'}(z)\\ &=\sum_{l=|l_1-l_2|}^{l_1+l_2} C(l_1,\ l_2,\ l;\ j,\ k,\ j+k)\,C(l_1,\ l_2,\ l;\ j',\ k',\ j'+k')\\ &\qquad\qquad\times P^{l}_{j+k,\ j'+k'}(z). \end{aligned} \tag{2}$$

This formula allows us to expand the product of the functions $P^{l_1}_{jj'}(z)$ and $P^{l_2}_{kk'}(z)$ in the functions $P^{l}_{j+k,j'+k'}(z)$. Only those functions $P^{l}_{j+k,j'+k'}(z)$ for which $|l_1-l_2|\le l\le l_1+l_2$ and $2l$ has the same parity as $2l_1+2l_2$ appear in the expansion. Expansion (2) is called the Clebsch-Gordan series.

In formula (2) let us put $j=k=j'=k'=0$. Since $P^{l}_{00}(z)=P_l(z)$, and the value of $C(l_1,\ l_2,\ l;\ 0,\ 0,\ 0)$ is given by formula (6) of subsection 5, we get

$$\begin{aligned} &P_{l_1}(z)\,P_{l_2}(z)\\ &=\sum_{l=|l_1-l_2|}^{l_1+l_2}\frac{(2l+1)(l+l_1-l_2)!\,(l_1+l_2-l)!\,(l-l_1+l_2)!\,(g!)^2}{(l_1+l_2+l+1)!\,[(g-l_1)!\,(g-l_2)!\,(g-l)!]^2}\,P_l(z), \end{aligned} \tag{3}$$

where $g=(l_1+l_2+l)/2$, and the summation extends over values of $l$ having the same parity as $l_1+l_2$.

Formula (2) allows one to derive an expansion of the functions $P^{l_1}_{jj'}(z)$ in associated Legendre functions. Let $j\ge|j'|$. On putting $l_2=j,\ k=-j,\ k'=j$ in formula (2), we obtain

$$\begin{aligned} &P^{l_1}_{jj'}(z)\,P^{j}_{-j,\,j}(z)\\ &=\sum_{l=|l_1-j|}^{l_1+j} C(l_1,\ j,\ l;\ j,\ -j,\ 0)\,C(l_1,\ j,\ l;\ j',\ j,\ j'+j)\,P^{l}_{0,\ j'+j}(z). \end{aligned} \tag{4}$$

Since

$$P^{j}_{-j,\,j}(z)=i^{2j}\left(\frac{1-z}{2}\right)^{j},$$

$$P^{l}_{0,\ j'+j}(z)=i^{-j-j'}\sqrt{\frac{(l-j-j')!}{(l+j+j')!}}\,P^{j+j'}_{l}(z),$$

$$\begin{aligned} &C(l_1,\ j,\ l;\ j,\ -j,\ 0)\\ &=\sqrt{\frac{(2l+1)(l+l_1-j)!\,(2j)!\,(l_1+j)!}{(l_1-j)!\,(l-l_1+j)!\,(l_1-l+j)!\,(l_1+l+j+1)!}} \end{aligned}$$

and

$$C(l_1, j, l; j', j, j'+j) = (-1)^{-l+l_1+j} \times \sqrt{\frac{(2l+1)(l+j+j')!\,(l_1-j')!\,(l+l_1-j)(2j)!}{(l+l_1+j+1)!\,(l-l_1+j)!\,(l_1-l+j)!\,(l-j-j')!\,(l_1+j')!}},$$

we get

$$\left(\frac{1-z}{2}\right)^j P^{l_1}_{jj'}(z) = i^{2l_1-j-j'}(2j)! \sqrt{\frac{(l_1+j)!\,(l_1-j')!}{(l_1-j)!\,(l_1+j')!}} \times \sum_{l=|l_1-j|}^{l_1+j} (-1)^l \frac{(2l+1)(l+l_1-j)!}{(l+l_1+j+1)!\,(l-l_1+j)!\,(l_1-l+j)!} P_l^{j+j'}(z). \quad (5)$$

If we put $l_2 = j$, $k = -j$, $k' = -j'$ in (2), then we similarly get the expansion of $P^{l_1}_{jj'}(z)$ in $P_l(z)$:

$$P^{l_1}_{jj'}(z) = 2^j i^{j'-j} \sqrt{\frac{(j-j')!\,(j+j')!}{(2j)!}}(1-z)^{\frac{j-j'}{2}}(1+z)^{\frac{-j-j'}{2}} \times \sum_{l=|l_1-j|}^{l_1+j} C(l_1, j, l; j, -j, 0)\, C(l_1, j, l; j', -j', 0)\, P_l(z). \quad (6)$$

However, in this case the second Clebsch-Gordan coefficient does not reduce to a single term.

Formula (2) also leads to numerous recurrence relations for the functions $P^l_{mn}(z)$. In order to obtain these, it is necessary to give $l_2$ in formula (2) the values $l_2 = \frac{1}{2}$ or 1.

Let us take $l_2 = \frac{1}{2}$, $k = k' = \frac{1}{2}$ and observe that $P^{\frac{1}{2}}_{\frac{1}{2},\frac{1}{2}}(\cos\theta) = \cos\frac{\theta}{2}$, and that, by formula (14′) of subsection 3,

$$C\left(l_1, \frac{1}{2}, l_1-\frac{1}{2}; j, \frac{1}{2}, j+\frac{1}{2}\right) = -\sqrt{\frac{l_1-j}{2l_1+1}},$$

$$C\left(l_1, \frac{1}{2}, l_1+\frac{1}{2}; j, \frac{1}{2}, j+\frac{1}{2}\right) = \sqrt{\frac{l_1+j+1}{2l_1+1}}.$$

Then we get the recurrence relation

$$\cos\frac{\theta}{2} P^{l_1}_{jj'}(\cos\theta) = \frac{\sqrt{(l_1-j)(l_1-j')}}{2l_1+1} P^{l_1-1/2}_{j+1/2,\, j'+1/2}(\cos\theta) + \frac{\sqrt{(l_1+j+1)(l_1+j'+1)}}{2l_1+1} P^{l_1+1/2}_{j+1/2,\, j'+1/2}(\cos\theta).$$

In just the same way, putting $l_2 = k = \frac{1}{2}$, $k' = -\frac{1}{2}$, we get

$$i \sin\frac{\theta}{2} P^{l_1}_{jj'}(\cos\theta) = -\frac{\sqrt{(l_1-j)(l_1+j')}}{2l_1+1} P^{l_1-1/2}_{j+1/2,\, j'-1/2}(\cos\theta)$$
$$+\frac{\sqrt{(l_1+j+1)(l_1-j'+1)}}{2l_1+1} P^{l_1+1/2}_{j+1/2,\, j'-1/2}(\cos\theta).$$

For $l_2 = 1$ we get nine recurrence relations. We shall write down only one of these, obtained for $k = k' = 0$:

$$\frac{\sqrt{(l_1+k+1)(l_1-k+1)(l_1+j+1)(l_1-j+1)}}{(2l_1+1)(l_1+1)} P^{l_1+1}_{jk}(\cos\theta)$$
$$+\frac{kj}{l_1(l_1+1)} P^{l}_{jk}(\cos\theta) + \frac{\sqrt{(l_1+k)(l_1-k)(l_1+j)(l_1-j)}}{l_1(2l_1+1)} P^{l_1-1}_{jk}(\cos\theta)$$
$$= \cos\theta\, P^{l_1}_{jk}(\cos\theta).$$

7. **The connection with Jacobi polynomials.** In this subsection it will be shown that the Clebsch-Gordan coefficients are the finite-difference analogue of Jacobi polynomials. For this we need the finite-difference analogue of a power, the so-called symbolic power. Let $n$ be an integer and $a$ an arbitrary complex number. By the symbolic $n$th power of the number $a$, denoted by $a^{(n)}$, we shall mean the expression

$$a^{(n)} = \frac{\Gamma(a+1)}{\Gamma(a-n+1)}. \tag{1}$$

Further, we introduce the finite-difference analogue of a derivative, namely the operator

$$\Delta f(a) = f(a+1) - f(a). \tag{2}$$

It is easy to verify that

$$\Delta a^{(n)} = n a^{(n-1)}, \tag{3}$$

$$\Delta^k f(a) = \sum_{s=0}^{k} (-1)^{k-s} \frac{k!}{s!\,(k-s)!} f(a+s). \tag{4}$$

With the help of formula (4) and equality (14′) of subsection 3 we get

$$C(l_1, l_2, l; j, k, j+k)$$
$$= (-1)^{l_1-l+k} \sqrt{\frac{(2l+1)(l+j+k)!\,(l_1+l_2-l)!}{(l-j-k)!\,(l+l_1-l_2)!\,(l-l_1+l_2)!\,(l+l_1+l_2+1)!}}$$
$$\times \sqrt{\frac{(l_1-j)^{(l_1-l_2-j-k)}}{(l_1+j)^{(l_1-l_2+j+k)}}}\, \Delta^{l-j-k}\left[\frac{(l_1+j)^{(l+l_1-l_2)}}{(l_1-j)^{(l_1-l-l_2)}}\right].$$

This formula is the finite-difference analogue of the formula (3) of §3.4 for the functions $P^l_{mn}(z)$.

We have thus proved that the Clebsch-Gordan coefficients are the finite-difference analogue of the functions $P^l_{mn}(z)$, i.e. the matrix elements of irreducible unitary representations of the group $SU(2)$.

From this analogy it is natural to presume that the Clebsch-Gordan coefficients possess properties analogous to the corresponding properties of the functions $P^l_{mn}(z)$. A number of such properties shall be given below.

8. Recurrence formulas. Let us establish recurrence formulas for the Clebsch-Gordan coefficients. To do so, we use

$$\mathbf{f}_j \otimes \mathbf{h}_k = \sum_{l=|l_1-l_2|}^{l_1+l_2} C(l_1, l_2, l; j, k, j+k)\, \mathbf{a}^l_{j+k}, \tag{1}$$

where $\{\mathbf{f}_j\}$ and $\{\mathbf{h}_k\}$ are the canonical bases in the spaces $\mathfrak{H}_1$ and $\mathfrak{H}_2$, and $\{\mathbf{a}^l_m\}$ is the canonical basis in the spaces $F_l$ into which $\mathfrak{H}_1 \otimes \mathfrak{H}_2$ is decomposed. We apply the infinitesimal operator $H_+$ to both sides of this equality. Since

$$H_+(\mathbf{f}_j \otimes \mathbf{h}_k) = H_+\mathbf{f}_j \otimes \mathbf{h}_k + \mathbf{f}_j \otimes H_+\mathbf{h}_k, \tag{2}$$

and $\{\mathbf{f}_j\}$ and $\{\mathbf{h}_k\}$ are canonical bases, we get

$$\begin{aligned} H_+(\mathbf{f}_j \otimes \mathbf{h}_k) = &- \sqrt{(l_1-j)(l_1+j+1)}\, \mathbf{f}_{j+1} \otimes \mathbf{h}_k \\ &- \sqrt{(l_2-k)(l_2+k+1)}\, \mathbf{f}_j \otimes \mathbf{h}_{k+1} \end{aligned} \tag{3}$$

(see formula (6) of §2.4). Similarly, we have

$$H_+\mathbf{a}^l_{j+k} = -\sqrt{(l-j-k)(l+j+k+1)}\, \mathbf{a}^l_{j+k+1}.$$

We substitute these results in (2) and decompose $\mathbf{f}_{j+1} \otimes \mathbf{h}_k$, $\mathbf{f}_j \otimes \mathbf{h}_{k+1}$ according to (1). Comparing the coefficients of $\mathbf{a}^l_{j+k+1}$, we obtain

$$\begin{aligned} &\sqrt{(l_1-j)(l_1+j+1)}\, C(l_1, l_2, l; j+1, k, j+k+1) \\ &\quad + \sqrt{(l_2-k)(l_2+k+1)}\, C(l_1, l_2, l; j, k+1, j+k+1) \\ &\quad = \sqrt{(l-j-k)(l+j+k+1)}\, C(l_1, l_2, l; j, k, j+k). \end{aligned} \tag{4}$$

In exactly the same way, applying the operator $H_-$ to both sides of the decomposition, we find that

$$\begin{aligned} &\sqrt{(l_1+j)(l_1-j+1)}\, C(l_1, l_2, l; j-1, k, j+k-1) + \\ &\quad + \sqrt{(l_2+k)(l_2-k+1)}\, C(l_1, l_2, l; j, k-1, j+k-1) \\ &\quad = \sqrt{(l+j+k)(l-j-k+1)}\, C(l_1, l_2, l; j, k, j+k). \end{aligned} \tag{5}$$

The application of $H_-H_+$ leads to the following difference equation for the Clebsch-Gordan coefficients:

$$\begin{aligned}[&(l_1-j)(l_1+j+1)+(l_2-k)(l_2+k+1)\\ &\quad -(l-j-k)(l+j+k+1)]\,C(l_1,l_2,l;j;k;j+k)\\ &\quad +\sqrt{(l_1-j)(l_1+j+1)(l_2+k)(l_2-k+1)}\times\\ &\qquad\times C(l_1,l_2,l;j+1,k-1;j+k)\\ &\quad +\sqrt{(l_1+j)(l_1-j+1)(l_2-k)(l_2+k+1)}\times\\ &\qquad\times C(l_1,l_2,l;j+1,k-1,j+k)=0. \qquad (6)\end{aligned}$$

This equation is the analogue of the second order differential equation for the functions $P^l_{mn}(z)$ (see §4.5).

We shall now derive recurrence relations for the Clebsch-Gordan coefficients, connecting coefficients with different values of $l_1$, $l_2$ and $l$. For this we use the identity

$$\begin{aligned}&\frac{d^s}{dx^s}[(1-x)^p(1+x)^q]\\ &\quad =q\frac{d^{s-1}}{dx^{s-1}}[(1-x)^p(1+x)^{q-1}]-p\frac{d^{s-1}}{dx^{s-1}}[(1-x)^{p-1}(1+x)^q].\end{aligned}$$

Let us apply this identity to equality (13) of subsection 3. After simple manipulations, we get

$$\begin{aligned}C(l_1,l_2,l;j,k,j+k)&=\sqrt{\frac{2l+1}{2l(l-j-k)(l_1+l_2-l+1)}}\\ &\times\Big[\sqrt{(l-l_1+l_2)(l_1+j+1)}\,C\Big(l_1+\frac{1}{2},l_2,l-\frac{1}{2};j+\frac{1}{2},k,j+k+\frac{1}{2}\Big)\\ &-\sqrt{(l+l_1-l_2)(l_2+k+1)}\,C\Big(l_1,l_2+\frac{1}{2},l-\frac{1}{2};j,k+\frac{1}{2},j+k+\frac{1}{2}\Big)\Big]. \qquad (7)\end{aligned}$$

In just the same way, from (13′) we get

$$\begin{aligned}C(l_1,l_2,l;j,k,j+k)&=\sqrt{\frac{2l+1}{2l(l_1+l_2-l+1)(l-l_1+l_2)}}\\ &\times\Big[\sqrt{(l-j-k)(l_1+j+1)}\,C\Big(l_1+\frac{1}{2},l_2,l-\frac{1}{2};j+\frac{1}{2},k,j+k+\frac{1}{2}\Big)\\ &-\sqrt{(l+j+k)(l_1-j+1)}\,C\Big(l_1+\frac{1}{2},l_2,l-\frac{1}{2};j-\frac{1}{2},k,j+k-\frac{1}{2}\Big)\Big]. \qquad (8)\end{aligned}$$

9. A generating function. A number of new properties of the Clebsch-Gordan coefficients follow from the identity

$$A \equiv (u_1v_2 - u_2v_1)^{l_1+l_2-l}(u_1x_1+u_2x_2)^{l_1-l_2+l}(v_1x_1+v_2x_2)^{-l_1+l_2+l}$$
$$=(-1)^{l_1+l_2-l}\sqrt{\frac{(l_1+l_2+l+1)!\,(l_1+l_2-l)!\,(l_1-l_2+l)!\,(l+l_2-l_1)!}{2l+1}}$$
$$\times \sum_{j=-l_1}^{l_1}\sum_{k=-l_2}^{l_2} C(l_1, l_2, l; j, k, j+k)\, U_j^{l_1} V_k^{l_2} X_{j+k}^l,$$

where

$$\left.\begin{aligned} U_j^{l_1} &= \frac{u_1^{l_1-j}u_2^{l_1+j}}{\sqrt{(l_1+j)!(l_1-j)!}}; \qquad V_k^{l_2} = \frac{v_1^{l_2-k}v_2^{l_2+k}}{\sqrt{(l_2+k)!(l_2-k)!}};\\ X_m^l &= \frac{x_1^{l-m}x_2^{l+m}}{\sqrt{(l+m)!\,(l-m)!}}. \end{aligned}\right\} \tag{1}$$

This identity is proved in the following way.

We realize the representation $T_{l_1}(g)$ in the space $\mathfrak{H}_{l_1}$ of homogeneous polynomials of degree $2l_1$ in two variables.[1] The monomials $U_j^{l_1}$ form a canonical basis in this space. Similarly, the monomials $V_k^{l_2}$ form a canonical basis in the space $\mathfrak{H}_{l_2}$ of the representation $T_{l_2}(g)$. Therefore $U_j^{l_1}V_k^{l_2}$, $-l_1 \le j \le l_1$, $-l_2 \le k \le l_2$, form an orthonormal basis in the space $\mathfrak{H}_{l_1} \otimes \mathfrak{H}_{l_2}$ of the representation $T_{l_1}(g) \otimes T_{l_2}(g)$. The representation $T_{l_1}(g) \otimes T_{l_2}(g)$ of the group $SU(2)$ in the space $\mathfrak{H}_{l_1} \otimes \mathfrak{H}_{l_2}$ is connected with the following transformations of variables:

$$\left.\begin{aligned} u_1 &\to \alpha u_1 - \bar{\beta} u_2,\\ u_2 &\to \beta u_1 + \bar{\alpha} u_2; \end{aligned}\right\} \tag{2'}$$

$$\left.\begin{aligned} v_1 &\to \alpha v_1 - \bar{\beta} v_2,\\ v_2 &\to \beta v_1 + \bar{\alpha} v_2. \end{aligned}\right\} \tag{2''}$$

Let us introduce variables $x_1$ and $x_2$, which transform according to the formulas

$$\left.\begin{aligned} x_1 &\to \bar{\alpha} x_1 - \beta x_2,\\ x_2 &\to \bar{\beta} x_1 + \alpha x_2 \end{aligned}\right\} \tag{3}$$

(i.e. contragradient to the variables $u_1$ and $u_2$), and consider the identity

$$(u_1x_1+u_2x_2)^{2l} = (2l)!\sum_{m=-l}^{l} U_m^l X_m^l, \tag{4}$$

[1] See formulas (4) of §2.4 and (5) of §2.1.

where

$$U_s^l = \frac{u_1^{l-s} u_2^{l+s}}{\sqrt{(l+s)!\,(l-s)!}}.$$

The left-hand side of this expression is invariant for the transformations (2′) and (3). Therefore, under the transformations indicated, the quantities $U_m^l$ transform contragradiently to the quantities $X_m^l$.

On the other hand, the expression $A$ is also invariant for the transformations (2′), (2″) and (3). Therefore if one writes $A$ in the form

$$A = \sum_{m=-l}^{l} W_m^l X_m^l, \tag{5}$$

the coefficients $W_m^l$ (which depend on $u_1, u_2, v_1, v_2$) transform contragradiently to $X_m^l$ under transformations (2′), (2″). But then they are transformed just as $U_m^l$, i.e. according to the representation $T_l(g)$.

This means that the coefficients $W_m^l$ in (5) transform according to the representation $T_l(g)$. Therefore the polynomials $W_m^l$ can differ from the elements of the canonical basis in the irreducible invariant subspace $\mathfrak{H}_l$ of $\mathfrak{H}_{l_1} \otimes \mathfrak{H}_{l_2}$ in a common scalar factor only.

Taking into account equality (7) of subsection 2, we conclude that[1])

$$W_m^l = \rho(l_1, l_2, l) \sum_{j=-l_1}^{l_1} \sum_{k=-l_2}^{l_2} C(l_1, l_2, l; j, k, m)\, U_j^{l_1} V_k^{l_2},$$

and therefore

$$\begin{aligned} A &\equiv (u_1 v_2 - u_2 v_1)^{l_1 + l_2 - l} (u_1 x_1 + u_2 x_2)^{l_1 - l_2 + l} (v_1 x_1 + v_2 x_2)^{-l_1 + l_2 + l} \\ &= \rho(l_1, l_2, l) \sum_{j=-l_1}^{l_1} \sum_{k=-l_2}^{l_2} C(l_1, l_2, l; j, k, j+k)\, U_j^{l_1} V_k^{l_2} X_{j+k}^l. \end{aligned} \tag{6}$$

Removing the brackets in the left-hand side of this equality, we easily obtain

$$\begin{aligned} &\rho(l_1, l_2, l)\, C(l_1, l_2, l; j, k, j+k) \\ &= (-1)^{l_1 + l_2 - l} (l_1 + l_2 - l)!\,(l_1 - l_2 + l)!\,(l + l_2 - l_1)! \\ &\times \sum_{s=T}^{Q} \frac{(-1)^s \sqrt{(l_1+j)!\,(l_1-j)!\,(l_2+k)!\,(l_2-k)!\,(l+j+k)!\,(l-j-k)!}}{(l_1-j-s)!\,(l-l_2+j+s)!\,(l_2+k-s)!\,(l-l_1-k+s)!\,s!\,(l_1+l_2-l-s)!}, \end{aligned} \tag{7}$$

where

[1]) It will be recalled that in the normalization chosen by us the Clebsch-Gordan coefficients are real, and also that only the coefficients for which $m = j + k$ are nonzero.

$$T=\max(0,\ k+l_1-l,\ l_2-l-j),\quad Q=\min(l_1-j,\ l_2+k,\ l_1+l_2-l).$$

In order to find the normalizing factor $\rho(l_1, l_2, l)$, let us put $j = l_1$, $k = -l_2$ in (7) and make use of formula (12) of subsection 3. For these values of $j$ and $k$, only the term for $s = 0$ remains in the sum in (7), and we get the equality

$$\rho(l_1, l_2, l)\sqrt{\frac{(2l+1)(2l_1)!(2l_2)!}{(l_1+l_2-l)!(l_1+l_2+l+1)!}} = (-1)^{l_1+l_2-l}\sqrt{(2l_1)!(2l_2)!(l+l_1-l_2)!(l-l_1+l_2)!}.$$

Hence

$$\rho(l_1, l_2, l) = (-1)^{l_1+l_2-l}\sqrt{\frac{(l_1+l_2-l)!(l+l_1-l_2)!(l-l_1+l_2)!(l_1+l_2+l+1)!}{2l+1}}. \tag{8}$$

Consequently, from formula (6) we get

$$\begin{aligned}(u_1v_1-u_2v_2)^{l_1+l_2-l}&(u_1x_2+u_2x_1)^{l_1-l_2+l}(v_1x_1+v_2x_2)^{-l_1+l_2+l}\\ &=\sqrt{\frac{(l_1+l_2-l)!(l+l_1-l_2)!(l-l_1+l_2)!(l_1+l_2+l+1)!}{2l+1}}\\ &\times x(-1)^{l_1+l_2-l}\sum_{j=-l_1}^{l_1}\sum_{k=-l_2}^{l_2} C(l_1, l_2, l; j, k, j+k)\,U_j^{l_1}V_k^{l_2}X_{j+k}^{l},\end{aligned} \tag{9}$$

where $U_j^{l_1}$, $V_k^{l_2}$ and $X_{j+k}^{l}$ are given by (1).

Our assertion is thus proved. Incidentally, we have obtained a new expansion for the Clebsch-Gordan coefficients:

$$\begin{aligned}C(l_1, l_2, l; j, k, j+k) &=\sqrt{\frac{(2l+1)(l_1+l_2-l)!(l_1-l_2+l)!(l+l_2-l_1)!}{(l_1+l_2+l+1)!}}\\ &\times\sum_s\frac{(-1)^s\sqrt{(l_1+j)!(l_1-j)!(l_2+k)!(l_2-k)!(l+j+k)!(l-j-k)!}}{(l_1-j-s)!(l-l_2+j+s)!(l_2+k-s)!(l-l_1-k+s)!\,s!\,(l_1+l_2-l-s)!}.\end{aligned} \tag{10}$$

Identity (9) gives a generating function for the Clebsch-Gordan coefficients. Like the generating function for $P^l_{mn}(x)$, it can be utilized for obtaining a number of new properties of these coefficients.

Let $l_1 = l_1' + l_1''$, $l_2 = l_2' + l_2''$, $l = l' + l''$. Then the following identity holds:

$$\begin{aligned}(u_1v_2-u_2v_1)^{l_1+l_2-l}&(u_1x_1+u_2x_2)^{l_1-l_2+l}(v_1x_1+v_2x_2)^{-l_1+l_2+l}\\ &=[(u_1v_2-u_2v_1)^{l_1'+l_2'-l'}(u_1x_1+u_2x_2)^{l_1'-l_2'+l'}(v_1x_1+v_2x_2)^{-l_1'+l_2'+l'}]\\ &\times[(u_1v_2-u_2v_1)^{l_1''+l_2''-l''}(u_1x_1+u_2x_2)^{l_1''-l_2''+l''}(v_1x_1+v_2x_2)^{-l_1''+l_2''+l''}]\end{aligned} \tag{11}$$

We apply (9) to both sides of this equality, substitute the expressions for $U_j^{l_1}$ etc. from (1), and then compare coefficients on the left and on the right. We obtain

$$\alpha C(l_1, l_2, l; j, k, j+k) = \sum_{\substack{j'+j''=j \\ k'+k''=k}} \alpha'\alpha'' C(l_1', l_2', l'; j', k', j'+k')\, C(l_1'', l_2'', l''; j'', k'', j''+k''). \quad (12)$$

Here for brevity we have put

$$\alpha = \sqrt{\frac{(l_1+l_2-l)!\,(l+l_1-l_2)!\,(l-l_1+l_2)!\,(l_1+l_2+l+1)!}{(2l+1)\,(l_1+j)!\,(l_1-j)!\,(l_2+k)!\,(l_2-k)!\,(l+j+k)!\,(l-j-k)!}} \quad (13)$$

and $\alpha'$ and $\alpha''$ have similar meanings ($l_1$, $l_2$, $l$, $j$, $k$ being replaced by $l_1'$, $l_2'$, $l'$, $j'$, $k'$ or $l_1''$, $l_2''$, $l''$, $j''$, $k''$).

Let us put $l_1'' = ½$, $l_2'' = ½$, $l'' = 0$ in (12). Then $j''$ and $k''$ can only assume the values $j'' = ½$, $k'' = ½$ or $j'' = -½$, $k'' = ½$. It follows from formulas (1) of subsection 5 and (1) of subsection 4 that

$$C\left(\frac{1}{2}, \frac{1}{2}, 0; \frac{1}{2}, -\frac{1}{2}, 0\right) = -C\left(\frac{1}{2}, \frac{1}{2}, 0; -\frac{1}{2}, \frac{1}{2}, 0\right) = \frac{1}{\sqrt{2}},$$

and so we get

$$\begin{aligned} C(l_1, l_2, l; j, k, j+k) &= \sqrt{\frac{1}{(l_1+l_2-l)(l_1+l_2+l+1)}} \\ &\times \Big[\sqrt{(l_1+j)(l_2-k)}\, C\left(l_1-\frac{1}{2}, l_2-\frac{1}{2}, l; j-\frac{1}{2}, k+\frac{1}{2}, j+k\right) \\ &- \sqrt{(l_1-j)(l_2+k)}\, C\left(l_1-\frac{1}{2}, l_2-\frac{1}{2}, l; j+\frac{1}{2}, k-\frac{1}{2}, j+k\right)\Big]. \quad (13) \end{aligned}$$

Similarly, putting $l_1'' = ½$, $l_2'' = 0$, $l'' = ½$, we get

$$\begin{aligned} C(l_1, l_2, l; j, k, j+k) &= \sqrt{\frac{2l+1}{2l(l+l_1-l_2)(l_1+l_2+l+1)}} \\ &\times \Big[\sqrt{(l_1+j)(l+k+j)}\, C\left(l_1-\frac{1}{2}, l_2, l-\frac{1}{2}; j-\frac{1}{2}, k, k+j-\frac{1}{2}\right) \\ &+ \sqrt{(l_1-j)(l-k-j)}\, C\left(l_1-\frac{1}{2}, l_2, l-\frac{1}{2}; j+\frac{1}{2}, k, k+j+\frac{1}{2}\right)\Big]. \quad (14) \end{aligned}$$

CHAPTER IV

# REPRESENTATIONS OF THE GROUP OF MOTIONS OF THE PLANE AND BESSEL FUNCTIONS

In this chapter we shall study Bessel functions. Just as the study of Legendre and Jacobi polynomials was found to be connected with representations of the group of rotations of the three-dimensional euclidean space, the study of Bessel functions is connected with representations of the group $M(2)$ of motions of euclidean plane. However, the situation is more complicated here, because the group of motions of the plane is noncompact.

## §1. THE GROUP $M(2)$

1. **Definition.** By motions of the euclidean plane one means transformations of this group into itself, which preserve the distance between points and do not change the orientation of the plane. As examples of motions we have parallel translations of the plane and rotations of the plane around some point. The set of all motions of the plane forms a group. We shall denote it by $M(2)$.

Let us take a system of cartesian coordinates on the plane. It is well known from analytical geometry that every motion of the plane is given, in this coordinate system, in the following way: a motion $g$ carries the point $P(x, y)$ into the point $P'(x', y')$, where

$$\left.\begin{aligned} x' &= x\cos\alpha - y\sin\alpha + a, \\ y' &= x\sin\alpha + y\cos\alpha + b. \end{aligned}\right\} \tag{1}$$

Obviously, the numbers $a$, $b$, $\alpha$ uniquely determine the motion $g$ and are uniquely determined by this motion (the parameter $\alpha$ is determined to within a summand, which is a multiple of $2\pi$). Therefore each element $g$ of $M(2)$ is given by three parameters $a$, $b$, $\alpha$, where

$$\left.\begin{array}{l} -\infty < a < \infty, \\ -\infty < b < \infty, \\ 0 \leqslant \alpha < 2\pi. \end{array}\right\} \tag{2}$$

We shall denote the element $g$ of the group $M(2)$, given by parameters $a, b, \alpha$, by $g(a, b, \alpha)$.

We indicate some other realizations of $M(2)$. With each motion $g(a, b, \alpha)$ we associate the matrix

$$T(g) = \begin{pmatrix} \cos\alpha & -\sin\alpha & a \\ \sin\alpha & \cos\alpha & b \\ 0 & 0 & 1 \end{pmatrix}. \tag{3}$$

A simple calculation shows that

$$T(g_1)\, T(g_2) = T(g_1 g_2).$$

Hence $T(g)$ is a representation of $M(2)$. This representation is faithful, i.e. $T(g_1) \neq T(g_2)$ if $g_1 \neq g_2$. In this way $M(2)$ is realized as the group of real third order matrices having the form (3).

$M(2)$ can also be realized with the help of complex second order matrices. Namely, with the motion $g(a, b, \alpha)$ we associate the matrix

$$Q(g) = \begin{pmatrix} e^{i\alpha} & z \\ 0 & 1 \end{pmatrix}, \tag{4}$$

where $z = a + bi$. It is easy to verify that $Q(g_1)Q(g_2) = Q(g_1 g_2)$ and $Q(g_1) \neq Q(g_2)$ if $g_1 \neq g_2$.

2. Parametrizations. One of the parametrizations of $M(2)$ has already been indicated. Namely, it was shown that each element $g$ of this group is given by three numbers $a, b, \alpha$. Let us find out how parameters of the product of two elements of the group are expressed in terms of the parameters of its factors. Let $g_1 = g(a_1, b_1, \alpha_1)$ and $g_2 = g(a_2, b_2, \alpha_2)$. On multiplying the matrices of $T(g_1)$ and $T(g_2)$, we get

$$T(g_1 g_2) = T(g_1)\, T(g_2)$$
$$= \begin{pmatrix} \cos(\alpha_1 + \alpha_2) & -\sin(\alpha_1 + \alpha_2) & a_1 + a_2\cos\alpha_1 - b_2\sin\alpha_1 \\ \sin(\alpha_1 + \alpha_2) & \cos(\alpha_1 + \alpha_2) & b_1 + a_2\sin\alpha_1 + b_2\cos\alpha_1 \\ 0 & 0 & 1 \end{pmatrix}. \tag{1}$$

Hence the parameters $a, b, \alpha$ of the element $g = g_1 g_2$ are expressed by the formulas

$$a = a_1 + a_2 \cos \alpha_1 - b_2 \sin \alpha_1, \tag{2}$$

$$b = b_1 + a_2 \sin \alpha_1 + b_2 \cos \alpha_1, \tag{2'}$$

$$\alpha = \alpha_1 + \alpha_2. \tag{2''}$$

These formulas can be written in the following form. We denote the vectors $(a_1, b_1)$ by $\mathbf{x}$ and $(a_1, b_2)$ by $\mathbf{y}$. Further, let us denote by $\mathbf{y}_\alpha$ the vector obtained from $\mathbf{y}$ through a rotation by an angle $\alpha$. Then equalities (2)–(2$''$) take the form

$$g(\mathbf{x}, \alpha) g(\mathbf{y}, \beta) = g(\mathbf{x} + \mathbf{y}_\alpha, \alpha + \beta). \tag{3}$$

It follows from formula (3) that if $g = g(\mathbf{x}, \alpha)$, then

$$g^{-1} = g(-\mathbf{x}_{-\alpha}, 2\pi - \alpha). \tag{4}$$

$M(2)$ is an example of an extensive class of groups, the cross products of commutative groups and their automorphism groups. Namely, let $B$ be some commutative group and let $A$ be the group whose elements are automorphisms of $B$. We consider pairs of the form $(b, a)$, where $b \in B$, $a \in A$, and define multiplication of these pairs by the formula

$$(b_1, a_1)(b_2, a_2) = (b_1 + a_1 b_2, a_1 a_2)$$

($a_1 b_2$ is the element into which the element $b_2$ of $B$ is carried under the automorphism $a_1$). One can easily verify that the set $G$ of pairs $(b, a)$, $a \in A$, $b \in B$, forms a group relative to this multiplication. Elements of the form $(b, e)$, where $e$ is the identity automorphism, constitute a normal divisor in $G$, isomorphic to the group $B$, and elements of the form $(0, a)$ form a subgroup isomorphic to $A$.

The group $g$ is called the cross (or semidirect) product of $B$ and $A$, and denoted by $G = B * A$. The group $M(2)$ of motions of the euclidean plane is the cross product of two-dimensional real linear space and the group of euclidean rotations of this space:

$$M(2) = E_2 * SO(2). \tag{5}$$

Along with the parameters $a$, $b$, $\alpha$, it is convenient to take other parameters in the group $M(2)$, analogous to the Euler angles for the group of rotations of three-dimensional euclidean space. These parameters are defined in the following way. We give polar coordinates to the vector $\mathbf{x} = (a, b)$, i.e. we put $a = r \cos \phi$, $b = r \sin \phi$. As the parameters determining a motion $g(a, b, \alpha)$, we take the numbers $r$, $\phi$, $\alpha$. We shall write $g(a, b, \alpha) = g(r, \phi, \alpha)$, and hope that this does not lead to misunderstandings. Evidently the parameters $r, \phi, \alpha$ vary between the following limits:

$$\left.\begin{array}{l} 0 \leqslant r < \infty, \\ 0 \leqslant \varphi < 2\pi, \\ 0 \leqslant \alpha < 2\pi. \end{array}\right\} \tag{6}$$

The analogue of the factorization (6) of §1.1 of Chapter III for $M(2)$ is

$$g(r, \varphi, \alpha) = g(0, 0, \varphi) g(r, 0, 0) g(0, 0, \alpha - \varphi). \tag{7}$$

The motions $g(0, 0, \phi)$ and $g(0, 0, \alpha - \phi)$ are rotations around the origin, and $g(r, 0, 0)$ is a (parallel) translation by $r > 0$ along the axis $Ox$.

If $g_1 = g(r_1, 0, \alpha_1)$ and $g_2 = g(r_2, 0, 0)$, then it easily follows from (2)–(2″) that $g_1 g_2 = g(r, \phi, \alpha)$, where

$$r = \sqrt{r_1^2 + r_2^2 + 2r_1 r_2 \cos \alpha_1}, \tag{8}$$

$$e^{i\varphi} = \frac{r_1 + r_2 e^{i\alpha_1}}{r}, \tag{8′}$$

$$\alpha = \alpha_1. \tag{8″}$$

The geometrical significance of these formulas is clear from Figure 1.

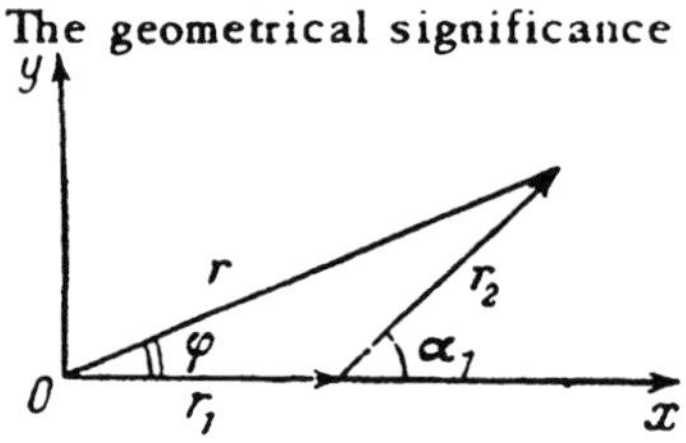

Figure 1.

The general case is easily reduced to one considered above. Namely, in order to find parameters of the product $g_1 g_2$ of motions $g_1 = g(r_1, \phi_1, \alpha_1)$ and $g_2(r_2, \phi_2, \alpha_2)$, it is necessary to replace $\alpha_1$ by $\alpha_1 + \phi_2 - \phi_1$, $\phi$ by $\phi - \phi_1$ and $\alpha$ by $\alpha - \alpha_2$ in formulas (8)–(8″). Indeed, from (7) and the equality

$$g(0, 0, \alpha_1 - \varphi_1) g(0, 0, \varphi_2) = g(0, 0, \alpha_1 + \varphi_2 - \varphi_1) \tag{9}$$

it follows that

$$\begin{aligned} g_1 g_2 = g(0, 0, \varphi_1) g(r_1, 0, 0) & \\ \times g(0, 0, \alpha_1 + \varphi_2 - \varphi_1) & \\ \times g(r_2, 0, 0) g(0, 0, \alpha_2 - \varphi_2). & \end{aligned} \tag{10}$$

Parameters of the motion $g' = g(r, 0, 0) g(0, 0, \alpha_1 + \phi_2 - \phi_1) g(r_2, 0, 0)$ can be calculated by (8)–(8″) (with $\alpha_1$ replaced by $\alpha_1 + \phi_2 - \phi_1$). But the multiplication of the motion $g'$ by $g(0, 0, \phi_1)$ on the left and $g(0, 0, \alpha_2 - \phi_2)$ on the right increases $\phi$ by $\phi_1$, and $\alpha$ by $\alpha_2 + \phi_1 - \phi_2$. This easily leads to our assertion.

3. **The Lie algebra.** Let us now calculate the Lie algebra of $M(2)$. We consider three one-parametric subgroups in $M(2)$: the subgroup $\Omega_1$, consisting of

matrices of the form

$$\omega_1(t)=\begin{pmatrix}1 & 0 & t\\ 0 & 1 & 0\\ 0 & 0 & 1\end{pmatrix}, \tag{1}$$

the subgroup $\Omega_2$, consisting of matrices of the form

$$\omega_2(t)=\begin{pmatrix}1 & 0 & 0\\ 0 & 1 & t\\ 0 & 0 & 1\end{pmatrix} \tag{1'}$$

and the subgroup $\Omega_3$, consisting of matrices of the torm

$$\omega_3(t)=\begin{pmatrix}\cos t & -\sin t & 0\\ \sin t & \cos t & 0\\ 0 & 0 & 1\end{pmatrix}. \tag{1''}$$

Evidently, the tangent matrix $a_1 = d\omega_1(t)/dt|_{t=0}$ for $\Omega_1$ has the form

$$a_1=\begin{pmatrix}0 & 0 & 1\\ 0 & 0 & 0\\ 0 & 0 & 0\end{pmatrix}. \tag{2}$$

Similarly, the tangent matrix for $\Omega_2$ has the form

$$a_2=\begin{pmatrix}0 & 0 & 0\\ 0 & 0 & 1\\ 0 & 0 & 0\end{pmatrix}, \tag{2'}$$

and for $\Omega_3$, the form

$$a_3=\begin{pmatrix}0 & -1 & 0\\ 1 & 0 & 0\\ 0 & 0 & 0\end{pmatrix}. \tag{2''}$$

Since $M(2)$ is given by three parameters and the matrices $a_1, a_2, a_3$ are linearly independent, they form a basis in the Lie algebra of $M(2)$.

The commutation relations for $a_1, a_2, a_3$ have the following form.

$$[a_1, a_2]=0, \tag{3}$$
$$[a_2, a_3]=a_1, \tag{3'}$$
$$[a_3, a_1]=a_2, \tag{3''}$$

where $[a, b] = ab - ba$.

We see that these relations resemble the commutation relations, given in

§1.3 of Chapter III, for the Lie algebra of the group $SU(2)$ (or, what is the same thing, for the Lie algebra of the group $SO(3)$, locally isomorphic to $SU(2)$). The only difference is that in the first of these relations $a_3$ is replaced by zero. The Lie algebra of $M(2)$ can be obtained by a limiting process from the Lie algebra of $SO(3)$. For this it is necessary to replace $a_1$ by $Ra_1$ by $a_2$ by $Ra_2$ in the Lie algebra of $SO(3)$, and then make $R$ tend to infinity. It will be shown in §7.1 that $M(2)$ itself is obtained by a similar limiting process from $SO(3)$.

4. Complexification. We construct the complex Lie group, one of whose real forms is $M(2)$. To do this we shall assume that in $M(2)$ the parameters $a$, $b$, $\alpha$ are complex numbers. Obviously, $\alpha$ is defined here to within a multiple of $2\pi$. Therefore these parameters vary in the following domains: $a$ and $b$ run through the whole complex plane, and $\alpha$ runs through the strip $0 \le \operatorname{Re} \alpha < 2\pi$.

It is evident that the parametric equations (2)–(2″) of subsection 2 retain their form on passing to complex values of the parameters. We denote the group thus constructed by $M(2, C)$. The group $M(2)$ is a subgroup of $M(2, C)$, corresponding to real values of the parameters $a$, $b$, $\alpha$.

Along with $a$, $b$, $\alpha$, in the group $M(2, C)$ one can utilize the parameters $r$, $\phi$, $\alpha$ connected with $a$, $b$, $\alpha$ by the equalities $a = r\cos\phi$, $b = r\sin\phi$. These parameters vary in the domains

$$\operatorname{Re} r > 0, \quad 0 \leqslant \operatorname{Re} \varphi < 2\pi, \quad 0 \leqslant \operatorname{Re} \alpha < 2\pi;$$

the parametric equations (8)–(8″) of subsection 2 remain valid for them.

We may also note that the Lie algebra of $M(2, C)$ consists of linear combinations of matrices $a_1$, $a_2$, $a_3$ (see subsection 3) with complex coefficients.

## §2. IRREDUCIBLE UNITARY REPRESENTATIONS OF THE GROUP $M(2)$

1. The description of representations. We shall now describe the irreducible representations of $M(2)$. We denote by $\mathfrak{D}$ the space of infinitely differentiable functions $f(\mathbf{x})$ on the circle $x_1^2 + x_2^2 = 1$. Let us fix a complex number $R$ and associate with each element $g(\mathbf{a}, \alpha)$ of $M(2)$ an operator $T_R(g)$ which transforms $f(\mathbf{x})$ into the function

$$T_R(g) f(\mathbf{x}) = e^{R(\mathbf{a},\, \mathbf{x})} f(\mathbf{x}_{-\alpha}). \tag{1}$$

Here $\mathbf{x}_{-\alpha}$ is the vector into which $\mathbf{x}$ is transformed under a rotation by an angle $-\alpha$, and $(\mathbf{a}, \mathbf{x}) = a_1 x_1 + a_2 x_2$.

We show that $T_R(g)$ is a representation of $M(2)$. Let $g_1 = g(a, \alpha)$ and $g_2 = g(b, \beta)$. Then

$$T_R(g_1)\,T_R(g_2)f(\mathbf{x}) = T_R(g_1)\,e^{R(\mathbf{b},\,\mathbf{x})}f(\mathbf{x}_{-\beta}) = e^{R(\mathbf{a},\,\mathbf{x})}e^{R(\mathbf{b},\,\mathbf{x}_{-\alpha})}f(\mathbf{x}_{-\alpha-\beta}).$$

Since $(\mathbf{b}, \mathbf{x}_{-\alpha}) = (\mathbf{b}_\alpha, \mathbf{x})$, therefore

$$T_R(g_1)\,T_R(g_2)f(\mathbf{x}) = e^{R(\mathbf{a}+\mathbf{b}_\alpha,\,\mathbf{x})}f(\mathbf{x}_{-\alpha-\beta}).$$

On the other hand, by formula (3) of §1.2, we have

$$g_1g_2 = g(\mathbf{a}, \alpha)\,g(\mathbf{b}, \beta) = g(\mathbf{a}+\mathbf{b}_\alpha, \alpha+\beta)$$

and so

$$T_R(g_1g_2)f(\mathbf{x}) = e^{R(\mathbf{a}+\mathbf{b}_\alpha,\,\mathbf{x})}f(\mathbf{x}_{-\alpha-\beta}).$$

Hence $T_R(g_1g_2) = T_R(g_1)\,T_R(g_2)$. By the same token it is proved that $T_R(g)$ is a representation of $M(2)$.

The parametric equation of the circle $x_1^2 + x_2^2 = 1$ has the form

$$\left.\begin{aligned} x_1 &= \cos\psi, \\ x_2 &= \sin\psi, \end{aligned}\quad 0 \leqslant \psi < 2\pi.\right\} \tag{2}$$

Therefore we can regard functions $f(\mathbf{x})$ of the space $\mathfrak{D}$ as functions of $\psi$:

$$f(\mathbf{x}) \equiv f(\psi).$$

The operators $T_R(g)$ are then written down as follows:

$$T_R(g)f(\psi) = e^{Rr\cos(\psi-\varphi)}f(\psi-\alpha), \tag{3}$$

where $\mathbf{a} = (r\cos\phi, r\sin\phi)$, $g = g(\mathbf{a}, \alpha)$.

We introduce a scalar product in $\mathfrak{D}$ by putting

$$(f_1, f_2) = \frac{1}{2\pi}\int_0^{2\pi} f_1(\psi)\overline{f_2(\psi)}\,d\psi. \tag{4}$$

On completing $\mathfrak{D}$ with respect to this scalar product, we obtain a Hilbert space $\mathfrak{H}$. Clearly, $T_R(g)$ is unitary relative to the scalar product (4) if and only if $R = i\rho$ is a purely imaginary number.

2. Infinitesimal operators. Let us calculate the infinitesimal operators of the representations $T_R(g)$. The operator $T_R(\omega_1(t))$, where

$$\omega_1(t) = \begin{pmatrix} 1 & 0 & t \\ 0 & 1 & 0 \\ 0 & 0 & 1 \end{pmatrix}, \tag{1}$$

transforms the function $f(\psi)$ into

$$T_R(\omega_1(t))f(\psi)=e^{Rt\cos\psi}f(\psi).$$

Hence

$$A_1=\left.\frac{dT_R(\omega_1(t))}{dt}\right|_{t=0}=R\cos\psi, \tag{2}$$

i.e. $A_1$ is the operator of multiplication by $R\cos\psi$.

In just the same way, one proves that the infinitesimal operator $A_2$, corresponding to the subgroup $\Omega_2$ of matrices of the form

$$\omega_2(t)=\begin{pmatrix}1&0&0\\0&1&t\\0&0&1\end{pmatrix}, \tag{3}$$

is given by the formula

$$A_2=R\sin\psi. \tag{4}$$

The operator $A_3$, corresponding to the subgroup $\Omega_3$ of matrices of the form

$$\omega_3(t)=\begin{pmatrix}\cos t&-\sin t&0\\\sin t&\cos t&0\\0&0&1\end{pmatrix}, \tag{5}$$

is given by the formula

$$A_3=-\frac{d}{d\psi}. \tag{6}$$

3. Irreducibility of representations. We shall prove that for $R\neq 0$ the representations $T_R(g)$ are irreducible. Let us first observe that a restriction of $T_R(g)$ gives the representation

$$T_R(\omega_3(\alpha))f(\psi)=f(\psi-\alpha) \tag{1}$$

of the subgroup $\Omega_3$ of rotations of the plane. But this representation is none other than the regular representation of the group of rotations of the plane. It was shown in §2.3 of Chapter II that the one-dimensional subspaces $\mathfrak{H}_n$, consisting of functions of the form $a_n e^{in\psi}$, are invariant subspaces in $\mathfrak{H}$ for the representation (1). The restrictions of $T_R(\omega_3(\alpha))$ to the various subspaces $\mathfrak{H}_n$ here are not equivalent among themselves.

Let $\mathfrak{L}$ be a nonnull subspace in $\mathfrak{H}$, invariant for all operators $T_R(g)$. Then, in particular, $\mathfrak{T}$ is invariant for all operators $T_R(\omega_3(\alpha))$ too. Consequently, by virtue of §3.3 of Chapter I, it must be an orthogonal direct sum of some of the spaces $\mathfrak{H}_n$:

$$\mathfrak{T}=\sum_k \mathfrak{H}_{n_k}. \tag{2}$$

Therefore $\mathfrak{T}$ contains at least one of the functions $e^{in\psi}$.

It remains for us to show that if the invariant subspace $\mathfrak{T}$ contains one of the functions $e^{in\psi}$, then it also contains the remaining functions $e^{ik\psi}$, $-\infty < k < \infty$. But if $\mathfrak{T}$ is invariant for all operators $T_R(g)$, then it is invariant for the infinitesimal operators $A_1, A_2, A_3$ as well. Let us explain how these operators act on the basis functions $e^{ik\psi}$. It follows from formulas (2), (4) and (6) of subsection 2 that

$$A_1 e^{ik\psi}=R\cos\psi e^{ik\psi}, \tag{3}$$

$$A_2 e^{ik\psi}=R\sin\psi e^{ik\psi}, \tag{3'}$$

$$A_3 e^{ik\psi}=-ike^{ik\psi}. \tag{3''}$$

We introduce the linear combinations $H_+=A_1+iA_2$, $H_-=A_1-iA_2$ of $A_1$ and $A_2$. The subspace $\mathfrak{T}$ is invariant for these operators. But $H_+$ transforms $e^{ik\psi}$ into the function

$$H_+ e^{ik\psi}=R\,e^{i(k+1)\psi}. \tag{4}$$

The operator $H_-$ transforms $e^{ik\psi}$ into

$$H_- e^{ik\psi}=R\,e^{i(k-1)\psi}. \tag{4'}$$

Hence, if the invariant subspace $\mathfrak{T}$ contains one of the functions $e^{in\psi}$, it contains all functions $e^{ik\psi}$, $-\infty < k < \infty$, i.e. it coincides with $\mathfrak{H}$. Thus we have proved that any nonnull invariant subspace in $\mathfrak{H}$ coincides with $\mathfrak{H}$, i.e. we have proved the irreducibility of $T_R(g)$ for $R\neq 0$.

If $R=0$, the representation $T_R(g)$ takes the form

$$T_0(g)f(\psi)=f(\psi-\alpha),\quad g=(x,\alpha). \tag{5}$$

This representation, as we remarked above, is reducible. It is decomposed as the direct sum of the one-dimensional representations

$$T_{0n}(g) = e^{in\alpha}. \tag{6}$$

It can be proved that the representations $T_R(g)$, $R \neq 0$, and $T_{0n}(g)$, $-\infty < n < \infty$ exhaust all the irreducible representations of the group $M(2)$ (see [297]).

4. Representations of cross products. The construction of the representations $T_R(g)$ of the group $M(2)$, described above, is a special case of the general construction of representations of cross products. Let $G = B * A$ be the cross product of a commutative group $B$ and its automorphism group $A$. Let us denote by $X$ the group whose elements are one-dimensional representations of $B$, i.e. scalar functions $\chi(b)$, $b \in B$, such that

$$\chi(b_1 + b_2) = \chi(b_1)\chi(b_2). \tag{1}$$

To each automorphism $a$ of $B$ corresponds an automorphism $\hat{a}$ of $X$, defined by the formula

$$\hat{a}\chi(b) = \chi(a^{-1}b), \tag{2}$$

where $\hat{a}_1\hat{a}_2 = (a_1a_2)\hat{}$. Therefore $A$ is isomorphic to the group of automorphisms $\hat{A}$ for $X$.

We decompose $X$ into transitivity classes for transformations $\hat{A}$. Let $\Phi$ be one of these classes, and let $f(\phi)$ be a function given on $\Phi$. With each element $g = (b, a)$ of $G$, we associate an operator $T(g)$, defined by the formula

$$T(g)f(\varphi) = \varphi(b)f(\hat{a}^{-1}\varphi)^{1)}. \tag{3}$$

Since, for $g_1 = (b_1, a_1)$, $g_2 = (b_2, a_2)$, we have

$$T(g_1)T(g_2)f(\varphi) = T(g_1)\varphi(b_2)f(\hat{a}_2^{-1}\varphi) = \varphi(b_1)\hat{a}_1^{-1}\varphi(b_2)f_2(\hat{a}_2^{-1}\hat{a}_1^{-1}\varphi)$$
$$= \varphi(b_1 + a_1b_2)f[(\hat{a}_1\hat{a}_2)^{-1}\varphi]$$

and

$$(b_1, a_1)(b_2, a_2) = (b_1 + a_1b_2, a_1a_2),$$

it follows that $T(g)$ is a representation of $G$ in the space of functions on $\Phi$.

As $B$ we choose the two-dimensional real linear space, and as $A$, the group of euclidean rotations. One-dimensional representations of $B$ have the form

$$\chi(b) = \exp[\chi_1b_1 + \chi_2b_2],$$

where $\chi_1$ and $\chi_2$ are arbitrary complex numbers and $b = (b_1, b_2)$. Hence $\chi$ is the two-dimensional complex linear space. Transitivity classes for $A$ consist of

1) It will be recalled that $\phi \in \Phi \subset X$, and is therefore a function on $B$.

complex number pairs of the form $(R\cos\alpha, R\sin\alpha)$ ($R$ complex). It is easy to verify that formula (3) leads in this case to the representations $T_R(g)$ of $M(2)$, considered above.

## §3. MATRIX ELEMENTS OF REPRESENTATIONS $T_R(g)$ AND BESSEL FUNCTIONS

1. Calculation of matrix elements. In this section we shall calculate the matrix elements $t^R_{mn}(g)$ of irreducible representations $T_R(g)$ of the group $M(2)$ of motions of the euclidean plane. According to §2, these representations are constructed in the space $\mathfrak{H}$ of functions $f(\psi)$ given on the circle and such that

$$(f, f)=\frac{1}{2\pi}\int_0^{2\pi}|f(\psi)|^2\,d\psi<\infty. \tag{1}$$

The representations $T_R(g)$ are given by the formula

$$T_R(g)f(\psi)=e^{Rr\cos(\psi-\varphi)}f(\psi-\alpha), \tag{2}$$

where $g=g(r, \phi, \alpha)$ and $R\neq 0$.

Evidently, the operators $T_R(g)$ are unitary in $\mathfrak{H}$ if and only if $R$ is a purely imaginary number, $R=i\rho$.

In the space $\mathfrak{H}$ we choose an orthonormal basis, consisting of the functions $\{e^{in\psi}\}$, i.e. the eigenfunctions of the operators $T_R(\omega)$, $\omega\in\Omega_3$, where $\Omega_3$ is the subgroup of rotations of the plane. In this basis, the matrix elements $t^R_{mn}(g)$ are given by

$$t^R_{mn}(g)=(T_R(g)e^{in\psi},\ e^{im\psi}). \tag{3}$$

Taking into account expression (4) of §2.1 for the scalar product in $\mathfrak{H}$ and equality (2), we get

$$t^R_{mn}(g)=\frac{e^{-in\alpha}}{2\pi}\int_0^{2\pi}e^{Rr\cos(\psi-\varphi)}e^{i(n-m)\psi}\,d\psi. \tag{4}$$

Let $r=\phi=0$, i.e. let the motion $g$ be a rotation by an angle $\alpha$ around the origin. Then, because of the orthogonality of the functions $e^{in\psi}$, we have

$$t^R_{mn}(g)\equiv t^R_{mn}(\alpha)=e^{-in\alpha}\delta_{mn}. \tag{5}$$

Thus, to the rotation by an angle $\alpha$ around the origin corresponds the diagonal matrix $T_R(\alpha)$, having the functions $e^{-in\alpha}$, $-\infty < n < \infty$, on its principal diagonal.

Now let $\phi = \alpha = 0$. Then $g$ is a shift by $r > 0$ along the axis $Ox$, and formula (4) takes the form

$$t_{mn}^R(g) \equiv t_{mn}^R(r) = \frac{1}{2\pi} \int_0^{2\pi} e^{Rr\cos\psi + i(n-m)\psi}\, d\psi. \tag{6}$$

The substitution $\psi = \pi/2 - \theta$ transforms (6) to the form

$$t_{mn}^R(r) = \frac{i^{n-m}}{2\pi} \int_0^{2\pi} e^{Rr\sin\theta - i(n-m)\theta}\, d\theta. \tag{7}$$

We introduce the notation

$$J_n(x) = \frac{1}{2\pi} \int_0^{2\pi} e^{ix\sin\theta - in\theta}\, d\theta, \tag{8}$$

where $n$ is an integer, and call $J_n(x)$ the $n$th Bessel function of $x$. Utilizing this notation, formula (7) can be rewritten as follows:

$$t_{mn}^R(r) = i^{n-m} J_{n-m}(-iRr). \tag{9}$$

Thus the element $t_{mn}^R(r)$ of the matrix $T_R(r)$, corresponding to a shift by $r > 0$ along the axis $Ox$, depends on $Rr$ and the difference $n - m$ of the indices, and, the factor $i^{n-m}$ apart, is essentially the same as the $(n-m)$th Bessel function of argument $-iRr$.

In order to obtain the expression for $t_{mn}^R(g)$ in the general case, $g = g(r, \phi, \alpha)$, it is sufficient to make the substitution $\psi - \phi = \pi/2 - \theta$ in (4). We obtain

$$t_{mn}^R(g) = i^{n-m} e^{-i[n\alpha + (m-n)\varphi]} J_{n-m}(-iRr). \tag{10}$$

However, formula (10) can also be directly derived from the factorization (7) of §1.2. Because of this factorization, we have

$$T_R(g) = T_R(\varphi)\, T_R(r)\, T_R(\alpha - \varphi). \tag{11}$$

Since the matrices $T_R(\phi)$ and $T_R(\alpha - \phi)$ are diagonal, with functions $e^{-in\phi}$ and $e^{-in(\alpha-\phi)}$ on their principal diagonals, and $t_{mn}^R(r) = i^{n-m} J_{n-m}(-irR)$, we arrive at formula (10).

As remarked above, for purely imaginary values of $R$, $R = i\rho$, the representations $T_R(g)$ are unitary. The matrix elements of these representations are expressed by means of Bessel functions of real argument:

$$t_{mn}^{i\rho}(g) = i^{n-m} e^{-i[n\alpha + (m-n)\varphi]} J_{n-m}(\rho r).$$

If $g$ is the identity transformation, $g = g(0, 0, 0)$, then $T_R(g)$ is the identity operator and to it corresponds the identity matrix. Hence $J_{n-m}(0) = \delta_{mn}$, $J_0(0) = 1$ and $J_n(0) = 0$ for $n \neq 0$.

2. The relation between Bessel functions with equal and opposite indices. We shall now prove the formula

$$J_n(z) = (-1)^n J_{-n}(z). \tag{1}$$

For this let us introduce in the space $\mathfrak{H}$ of functions $f(\psi)$ on the circle an operator $Q$[1], defined by the formula

$$Qf(\psi) = f(-\psi). \tag{2}$$

This operator commutes with the operator $T_R(g) \equiv T_R(r)$, where $g = g(r, 0, 0)$ is a shift along the axis $Ox$. In fact,

$$QT_R(r) f(\psi) = Qe^{Rr\cos\psi} f(\psi) = e^{Rr\cos\psi} f(-\psi)$$

and

$$T_R(r)\, Qf(\psi) = T_R(r) f(-\psi) = e^{Rr\cos\psi} f(-\psi),$$

and therefore

$$QT_R(r) = T_R(r)\, Q. \tag{3}$$

The operator $Q$ transforms the basis element $e^{in\psi}$ into $e^{-in\psi}$, and so the matrix of $Q$ has the form $(q_{mn})$, where $q_{m,-m} = 1$ and $q_{mn} = 0$ if $m + n \neq 0$. But then equality (3) gives

$$t^R_{-m,n}(r) = t^R_{m,-n}(r). \tag{4}$$

Hence, taking into account formula (9) of subsection 1, we get

$$i^{n+m} J_{n+m}(-iRr) = i^{-n-m} J_{-n-m}(-iRr).$$

Let us put $m = 0$, $iRr = z$ in this equality. Then we have

$$J_n(z) = (-1)^n J_{-n}(z).$$

Equality (1) is thus proved.

---

1) The group $M(2)$ is a subgroup in the group of motions and mappings of the plane. $Q$ is an operator in the representation of this wider group, corresponding to a mapping into the axis $Ox$.

3. **Expansion of Bessel functions in power series.** We now derive the expansion of Bessel functions in a power series in $x$. To do this we use the integral representation of these functions (see formula (8) above). If in this representation we expand $e^{ix\sin\psi}$ in powers of $ix\sin\psi$ and integrate term by term, we get

$$J_n(x)=\sum_{k=0}^{\infty} a_k x^k,$$

where

$$a_k=\frac{1}{2\pi k!}\int_0^{2\pi} e^{\,in\psi}(i\sin\psi)^k\,d\psi.$$

But by Euler's formula we have

$$(i\sin\psi)^k=\frac{(e^{\psi i}-e^{-\psi i})^k}{2^k}=\frac{\sum\limits_{m=0}^{k}(-1)^m C_k^m e^{(k-2m)\psi i}}{2^k}.$$

We substitute this expression in (1). We find that $a_k$ is nonzero only if $k-n$ is an even nonnegative number, $k-n=2m$, $m\geq 0$. If $k=n+2m$,

$$a_k=\frac{(-1)^m}{2^k m!\,(k-m)!}=\frac{(-1)^m}{2^{n+2m}m!(n+m)!}.$$

Therefore

$$J_n(x)=\left(\frac{x}{2}\right)^n\sum_{m=0}^{\infty}\frac{(-1)^m x^{2m}}{2^{2m}m!\,(n+m)!}.$$

This expansion is evidently convergent for all complex values of $x$. Bessel functions are integral functions of $x$. It is obvious from formula (3) that for real values of $x$ the function $J_n(x)$ assumes real values.

## §4. FUNCTIONAL RELATIONS FOR BESSEL FUNCTIONS

1. **The composition theorem.** The composition theorem for Bessel functions is derived just as it was done in §4.1 of Chapter III for the functions $P^l_{mn}(\cos\theta)$. We shall start from the equality $T_R(g_1g_2)=T_R(g_1)T_R(g_2)$. This equality can be written down as follows in terms of matrix elements:

$$t^R_{mn}(g_1g_2)=\sum_{k=-\infty}^{\infty} t^R_{mk}(g_1)\,t^R_{kn}(g_2).$$

Let us put $g_1=g(r_1,0,0)$ and $g_2=g(r_2,\phi_2,0)$ in this relation. Then the parameters $r,\phi,\alpha$, determining the motion $g=g_1g_2$, are expressed in terms of the parameters $r_1,r_2,\phi_2$ by the formulas

$$r=\sqrt{r_1^2+r_2^2+2r_1r_2\cos\varphi_2}, \tag{2}$$

$$e^{i\varphi}=\frac{r_1+r_2e^{i\varphi_2}}{r}, \tag{2'}$$

$$\alpha=0 \tag{2''}$$

(see §1.2).

In (1) we substitute the expression for matrix elements given by formula (10) of §3.1, and put $m = 0$, $R = i$. We easily obtain

$$e^{in\varphi}J_n(r)=\sum_{k=-\infty}^{\infty} e^{ik\varphi_2}J_{n-k}(r_1)\,J_k(r_2), \tag{3}$$

where $r, r_1, r_2, \phi, \phi_2$ are connected by relations (2)–(2″). The diagrammatic connection of the quantities $r, r_1, r_2, \phi, \phi_2$ is given by Figure 2. Formula (3) is called the composition formula for Bessel functions. In particular, for $n = 0$, it gives

$$J_0(r)=\sum_{k=-\infty}^{\infty}(-1)^k e^{ik\varphi_2}J_k(r_1)\,J_k(r_2). \tag{4}$$

We also note the following special cases of formula (3). For $\phi_2 = 0$ we have $r = r_1 + r_2$, $\phi = 0$, and therefore

$$J_n(r_1+r_2)=\sum_{k=-\infty}^{\infty} J_{n-k}(r_1)\,J_k(r_2). \tag{5}$$

Further, for $\phi_2 = \pi$, $r_1 \geq r_2$ we have $r = r_1 - r_2$, $\phi = 0$. Hence, for $r_1 \geq r_2$

$$J_n(r_1-r_2)=\sum_{k=-\infty}^{\infty}(-1)^k J_{n-k}(r_1)\,J_k(r_2). \tag{6}$$

Finally, if $\phi_2 = \pi/2$, then formula (3) gives

$$\left(\frac{r_1+ir_2}{r_1-ir_2}\right)^{\frac{n}{2}} J_n\left(\sqrt{r_1^2+r_2^2}\right)=\sum_{k=-\infty}^{\infty} i^k J_{n-k}(r_1)\,J_k(r_2). \tag{7}$$

Let us put $r_1 = r_2 = r$ in (6). Taking into account formula (1) of §3.2, we obtain

$$\sum_{k=-\infty}^{\infty} J_{n+k}(r)\,J_k(r)=J_n(0)=\begin{cases}1 & \text{for } n=0,\\ 0 & \text{for } n\neq 0,\end{cases}$$

(Jansen's formula). Jansen's formula can also be deduced from the fact that the matrix $(t^i_{mn}(r))$ is unitary, and therefore $(t^i_{mn}(r))(t^i_{mn}(r))^* = e$, where $e$ is the identity matrix.

2. The multiplication formula. We multiply both sides of the composition formula (3) of subsection 1 by $e^{-im\phi_2}/2\pi$, and integrate with respect to $\phi_2$ from 0 to $2\pi$. By the orthogonality of the functions $e^{in\phi_2}$, all the terms reduce to zero except for the term corresponding to $k = m$. Hence

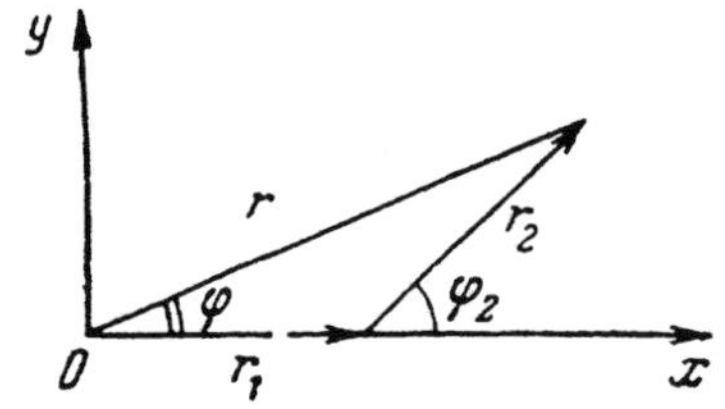

Figure 2.

$$\frac{1}{2\pi}\int_0^{2\pi} e^{i(n\varphi - m\varphi_2)} J_n(r)\, d\varphi_2 = J_{n-m}(r_1)\, J_m(r_2), \tag{1}$$

where $\phi, r_1, r_2, r$ are connected by relations (2)–(2″) of subsection 1. We shall call equality (1) the multiplication formula for Bessel functions.

We may note the geometrical significance of formula (1). For fixed $r_1$ and $r_2$ the point with polar coordinates $(r, \phi)$ describes, as $\phi_2$ varies from 0 to $2\pi$, a circle with center at the point $A(r_1, 0)$ and radius $r_2$. Thus the expression on the right hand side of formula (1) is nothing but the mean value of the function $e^{i(n\phi - m\phi_2)} J_n(r)$ on this circle.

We mention a special case of formula (1). For $r_1 = r_2 = R$ we have $r = 2R\cos\phi_2/2$, $\phi = \phi_2/2$, and therefore

$$J_{n-m}(R)\, J_m(R) = \frac{1}{\pi}\int_0^{\pi} e^{i(n-2m)\varphi} J_n(2R\cos\varphi)\, d\varphi. \tag{2}$$

Let us make a change of variable in formula (1) by taking $r$ as the variable of integration. As $\phi_2$ varies from 0 to $\pi$, $r$ varies from $r_1 + r_2$ to $|r_1 - r_2|$, and as $\phi_2$ varies from $\pi$ to $2\pi$, it varies from $|r_1 - r_2|$ to $r_1 + r_2$. Moreover,

$$\frac{dr}{d\varphi_2} = \mp \frac{\sqrt{4r_1^2 r_2^2 - (r^2 - r_1^2 - r_2^2)^2}}{2r},$$

where for $0 \leq \phi_2 \leq \pi$ we choose the $-$ sign, and for $\pi \leq \phi_2 \leq 2\pi$ the $+$ sign. Therefore we have

$$J_{n-m}(r_1)\, J_m(r_2) = \frac{2}{\pi}\int_{|r_1 - r_2|}^{r_1 + r_2} e^{i(n\varphi - m\varphi_2)} \frac{rJ_n(r)\, dr}{\sqrt{4r_1^2 r_2^2 - (r^2 - r_1^2 - r_2^2)^2}}, \tag{3}$$

where $\phi$ and $\phi_2$ are connected with $r$ by formulas (2)–(2″) of subsection 1.

For $m = n = 0$ equality (3) takes a particularly simple form. Namely, we have

$$J_0(r_1)\,J_0(r_2)=\frac{2}{\pi}\int\limits_{|r_1-r_2|}^{r_1+r_2}\frac{rJ_0(r)\,dr}{\sqrt{4r_1^2r_2^2-(r^2-r_1^2-r_2^2)^2}}. \qquad (4)$$

The expression in the denominator of the integrand is equal to four times the area of the triangle with sides $r_1$, $r_2$, $r$.

3. Recurrence formulas. As in the case of functions $P^l_{mn}(\cos\theta)$, recurrence formulas for Bessel functions follow from the composition formula. Namely, if $r_2$ is taken to be an infinitely small quantity, then the composition formula leads to recurrence relations.

We shall first find the values of the derivatives of Bessel functions for $x = 0$. On differentiating formula (8) of §3.1 with respect to $x$ and putting $x = 0$, we obtain

$$J'_n(0)=\frac{i}{2\pi}\int\limits_0^{2\pi}e^{-in\theta}\sin\theta\,d\theta=\frac{1}{4\pi}\int\limits_0^{2\pi}[e^{-(n-1)i\theta}-e^{-(n+1)i\theta}]\,d\theta. \qquad (1)$$

This integral is nonzero only for $n = \pm 1$. Moreover,

$$J'_1(0)=-J'_{-1}(0)=\frac{1}{2}.$$

Let us now differentiate both sides of equality (5) of subsection 1 with respec to $r_2$ and put $r_2 = 0$. Utilizing the values of $J'_n(0)$, found above, and replacing $r_1$ by $x$, we get the recurrence formula

$$2J'_n(x)=J_{n-1}(x)-J_{n+1}(x). \qquad (2)$$

In just the same way, differentiating both sides of formula (7) of subsection 1 with respect to $r_2$ and putting $r_2 = 0$, $r_1 = x$, we get

$$\frac{2n}{x}J_n(x)=J_{n-1}(x)+J_{n+1}(x). \qquad (3)$$

Adding and subtracting formulas (2) and (3), we find

$$J_{n-1}(x)=\frac{n}{x}J_n(x)+J'_n(x), \qquad (4)$$

$$J_{n+1}(x)=\frac{n}{x}J_n(x)-J'_n(x). \qquad (5)$$

These formulas can also be written in the form

$$\frac{d}{dx}\{x^n J_n(x)\} = x^n J_{n-1}(x), \tag{6}$$

$$\frac{d}{dx}\{x^{-n} J_n(x)\} = -x^{-n} J_{n+1}(x), \tag{7}$$

or in the form

$$J_{n-1}(x) = \left(\frac{n}{x} + \frac{d}{dx}\right) J_n(x), \tag{8}$$

$$J_{n+1}(x) = \left(\frac{n}{x} - \frac{d}{dx}\right) J_n(x). \tag{9}$$

4. The differential equation. Recurrence formulas (8) and (9) of subsection 3 immediately lead to the differential equation for Bessel functions. We apply the differential operator $(n-1)/x - d/dx$ to both sides of equality (8). Then, because of equality (9), we obtain

$$J_n(x) = \left(\frac{n-1}{x} - \frac{d}{dx}\right)\left(\frac{n}{x} + \frac{d}{dx}\right) J_n(x) \tag{1}$$

or

$$x^2 J_n''(x) + x J_n'(x) + (x^2 - n^2) J_n(x) = 0. \tag{1$'$}$$

Equation (1′), satisfied by the function $J_n(x)$, is called the Bessel differential equation. We shall give another derivation of this equation, and of the recurrence formulas, below.

Thus we have proved that $J_n(x)$ is an eigenfunction of the differential operator

$$x^2 \frac{d^2}{dx^2} + x\frac{d}{dx} + x^2, \tag{2}$$

corresponding to the eigenvalue $n^2$ under the boundary conditions: $J_n(x)$ is finite for $x = 0$. It can be proved that these are the only eigenfunctions (apart from constant multiples) of the operator (2), finite for $x = 0$.

5. The generating function. The other method of deriving recurrence relations for Bessel functions is based on an application of the generating function. In order to obtain the latter, we use the integral representation

$$J_n(x) = \frac{1}{2\pi}\int_0^{2\pi} e^{ix\sin\theta - in\theta}\, d\theta \tag{1}$$

of Bessel functions. Formula (1) signifies that $J_n(x)$ is the $n$th Fourier coefficient of the function $f(\theta) = e^{ix\sin\theta}$. Therefore we have the equality

$$e^{ix\sin\theta} = \sum_{n=-\infty}^{\infty} J_n(x)\, e^{in\theta}. \tag{2}$$

Thus $e^{ix \sin\theta}$ is a generating function for $J_n(x)$.

Equality (2) yields a number of relations for Bessel functions. On putting $\theta = \pi/2$ in this equality, we obtain

$$\sum_{n=-\infty}^{\infty} i^n J_n(x) = e^{ix}. \tag{3}$$

Similarly, putting $\theta = -\pi/2$, we find

$$\sum_{n=-\infty}^{\infty} i^{-n} J_n(x) = e^{-ix}. \tag{4}$$

Lastly, putting $\theta = 0$, we get

$$\sum_{n=-\infty}^{\infty} J_n(x) = 1. \tag{5}$$

6. The recurrence relations. We shall use the generating function found above to obtain a new derivation of the recurrence relations for Bessel functions. Let us differentiate equality (2) of subsection 5 with respect to $x$. Because of Euler's formula, we find

$$\frac{1}{2}(e^{i\theta} - e^{-i\theta}) e^{ix \sin\theta} = \sum_{n=-\infty}^{\infty} J_n'(x) e^{in\theta}. \tag{1}$$

We employ expansion (2) of subsection 5 for the expression in the left hand side, and compare coefficients of $e^{in\theta}$ on the left and on the right:

$$J_n'(x) = \frac{1}{2}(J_{n-1}(x) - J_{n+1}(x)). \tag{2}$$

Further, let us differentiate both sides of expansion (2) of subsection 5 with respect to $\theta$, and employ Euler's formula:

$$\frac{1}{2}(e^{i\theta} + e^{-i\theta}) x e^{ix \sin\theta} = \sum_{n=-\infty}^{\infty} n J_n(x) e^{in\theta}. \tag{3}$$

On expanding the expression in the left hand side according to the same formula (2) and comparing coefficients of $e^{in\theta}$, we get the recurrence formula

$$J_{n-1}(x) + J_{n+1}(x) = \frac{2n}{x} J_n(x). \tag{4}$$

In this way, we have obtained a new derivation of formulas (2) and (3) of subsection 3.

## § 5. DECOMPOSITIONS OF THE REPRESENTATIONS OF THE GROUP $M(2)$ AND THE FOURIER-BESSEL TRANSFORMATION

Just as functions on the sphere are expanded in series of spherical functions, one can decompose functions on the euclidean plane in series and integrals of Bessel functions. This is related to the decomposition of certain representations of the group $M(2)$ into irreducible ones and the decomposition of spaces of irreducible representations into subspaces invariant for operators $T(\omega)$, $\omega \in \Omega_3$ ($\Omega_3$ is the subgroup of rotations of the plane).

1. The quasi-regular representation. Since the elements of $M(2)$ are motions of the plane, it is natural to construct the representations of this group in the space of functions $f(x, y) \equiv f(\mathbf{x})$, given on the plane. We shall denote by $\mathfrak{L}^2$ the space of functions of two variables $f(x, y)$, $-\infty < x < \infty$, $-\infty < y < \infty$, such that

$$\int_{-\infty}^{\infty}\int_{-\infty}^{\infty} |f(x, y)|^2\, dx\, dy < +\infty. \tag{1}$$

With each element $g$ of $M(2)$ we associate an operator $L(g)$, transforming $f(\mathbf{x})$ into the function

$$L(g) f(\mathbf{x}) = f(g^{-1}\mathbf{x}). \tag{2}$$

Here, as usual, $g\mathbf{x}$ denotes the point into which a point $x$ of the plane is carried by the motion $g$.

Equality (2) can also be written in the form

$$L(g) f(\mathbf{x}) = f[(\mathbf{x} - \mathbf{a})_{-\alpha}], \tag{2'}$$

where $g = g(\mathbf{a}, \alpha)$. Since

$$L(g_1) L(g_2) f(\mathbf{x}) = L(g_1) f(g_2^{-1}\mathbf{x})$$
$$= f(g_2^{-1} g_1^{-1}\mathbf{x}) = f[(g_1 g_2)^{-1}\mathbf{x}] = L(g_1 g_2) f(\mathbf{x}),$$

$L(g)$ is a representation of $M(2)$. This representation is unitary because of the invariance of euclidean measure $d\mathbf{x} \equiv dx\,dy$ for shifts of the plane

$$\int |f(g^{-1}\mathbf{x})|^2\, d\mathbf{x} = \int |f(\mathbf{x})|^2\, d(g\mathbf{x}) = \int |f(\mathbf{x})|^2\, d\mathbf{x}.$$

We shall call $L(g)$ the quasi-regular representation of $M(2)$.

We need the decomposition of this representation into irreducible ones. We use the Fourier transformation for this. With each function $f(\mathbf{x})$ of $\mathfrak{L}^2$ we associate its Fourier transform

$$F(\mathbf{y}) = \int f(\mathbf{x}) e^{i(\mathbf{x}, \mathbf{y})} d\mathbf{x},$$

where $(\mathbf{x}, \mathbf{y}) = x_1 y_1 + x_2 y_2$.

Let us find out how the operators $L(g)$ of the quasi-regular representation are transformed on passing from $f(\mathbf{x})$ to $F(\mathbf{y})$. The Fourier transform of the function $L(g) f(\mathbf{x}) = f(g^{-1}\mathbf{x})$ is

$$F_g(\mathbf{y}) = \int f(g^{-1}\mathbf{x}) e^{i(\mathbf{x}, \mathbf{y})} d\mathbf{x}. \tag{3}$$

The invariance of $d\mathbf{x}$ for shifts implies that

$$F_g(\mathbf{y}) = \int f(\mathbf{x}) e^{i(g\mathbf{x}, \mathbf{y})} d\mathbf{x}. \tag{3'}$$

If $g = g(\mathbf{b}, \alpha)$ (see §1.1), then $g$ transforms $\mathbf{x}$ into the vector $\mathbf{x}_\alpha + \mathbf{b}$, where $\mathbf{x}_\alpha$ is the vector into which $\mathbf{x}$ is carried by a rotation by angle $\alpha$. Therefore

$$\begin{aligned} F_g(\mathbf{y}) &= \int f(\mathbf{x}) e^{i(\mathbf{x}_\alpha + \mathbf{b}, \mathbf{y})} d\mathbf{x} \\ &= e^{i(\mathbf{b}, \mathbf{y})} \int f(\mathbf{x}) e^{i(\mathbf{x}, \mathbf{y}_{-\alpha})} d\mathbf{x} = e^{i(\mathbf{b}, \mathbf{y})} F(\mathbf{y}_{-\alpha}). \end{aligned} \tag{4}$$

Thus we have proved that, under the Fourier transformation, the operator $L(g)$ is carried into the operator $\hat{L}(g)$:

$$\hat{L}(g) F(\mathbf{y}) = e^{i(\mathbf{b}, \mathbf{y})} F(\mathbf{y}_{-\alpha}), \tag{5}$$

where $g = g(\mathbf{b}, \alpha)$.

The representation $\hat{L}(g)$ of $M(2)$ is equivalent to the representation $L(g)$, as $L(g) = Q \hat{L}(g) Q^{-1}$, where $Q$ is the Fourier transform operator. Therefore the problem of decomposing the quasi-regular representation is equivalent to the problem of decomposing $\hat{L}(g)$. We pass on to a solution of this problem.

Functions $F(\mathbf{y})$, which are Fourier transforms of the functions $f(\mathbf{x})$ of $\mathfrak{L}^2$, form a Hilbert space $\hat{\mathfrak{L}}^2$, consisting of functions, for which

$$\|F\|^2 = \int |F(\mathbf{y})|^2 dy < \infty. \tag{6}$$

We shall now decompose this Hilbert space as the continuous direct sum of Hilbert spaces $\hat{\mathfrak{H}}_R$, such that to the operators $\hat{L}(g)$ correspond operators of the irreducible unitary representations of $M(2)$ in $\hat{\mathfrak{H}}_R$.

We denote by $\hat{\mathfrak{H}}_R$ the Hilbert space of functions $\Phi(\psi)$ given on the circle $|\mathbf{y}| = R$, and such that

$$\|\Phi\|_R^2 \equiv \frac{1}{2\pi}\int_0^{2\pi} |\Phi(\psi)|^2 d\psi < +\infty. \tag{7}$$

Let $F_R(\psi) = F(R\cos\psi, R\sin\psi)$. The equality

$$\int |F(y)|^2 dy = \int_0^\infty R\,dR \int_0^{2\pi} |F(R\cos\psi, R\sin\psi)|^2 d\psi = 2\pi \int_0^\infty \|F_R\|_R^2 R\,dR \tag{8}$$

shows that $\hat{\mathfrak{L}}_2$ is the continuous direct sum of the spaces $\hat{\mathfrak{H}}_R$:

$$\hat{\mathfrak{L}}^2 = 2\pi \int_0^\infty \hat{\mathfrak{H}}_R R\,dR. \tag{9}$$

Since the functions $F(y)$ of $\hat{\mathfrak{L}}^2$ may, generally speaking, not have a meaning at isolated points, we must make our assertion more precise. We first consider the space $\hat{\mathfrak{S}}$ of infinitely differentiable functions $F(y)$ which decrease rapidly together with all their derivatives for $|y| \to \infty$. For every such function the value of $F_R(\psi) = F(R\cos\psi, R\sin\psi)$ is defined. This determines a decomposition of the pre-Hilbert space $\hat{\mathfrak{S}}$ as the continuous direct sum of spaces $\hat{\mathfrak{S}}_R$, consisting of infinitely differentiable functions on circles. Completing these spaces afterwards with respect to the appropriate norms, we obtain the decomposition (9).

Let us clarify how the operators of the representation $\hat{L}(g)$ are decomposed. To do this it is sufficient to note that the operators $\hat{L}(g)$ leads to rotations around the origin and multiplication by the function $e^{iRr\cos(\psi-\phi)}$. Therefore the spaces $\hat{\mathfrak{H}}_R$ of functions given on circles of radius $R$ are invariant for them. Hence it immediately follows that to $\hat{L}(g)$ there correspond the operators $\hat{L}_R(g)$ in the spaces $\hat{\mathfrak{H}}_R$, having the form

$$\hat{L}_R(g) F_R(\psi) = e^{iRr\cos(\psi-\varphi)} F_R(\psi - \alpha), \tag{10}$$

where

$$g = g(\mathbf{a}, \alpha), \quad \mathbf{a} = (r\cos\varphi, r\sin\varphi).$$

Moreover, it follows from (5) that

$$[\hat{L}(g) F(\mathbf{y})]_R = \hat{L}_R(g) F_R(\mathbf{y}). \tag{11}$$

Thus $\hat{L}(g)$ is the continuous direct sum of the representations $\hat{L}_R(g)$:

$$\hat{L}(g) = 2\pi \int_0^\infty \hat{L}_R(g) R\,dR. \tag{12}$$

The representations $\hat{L}_R(g)$, defined by formula (10), coincide with the irreducible unitary representations $T_{iR}(g)$ of the group $M(2)$, constructed in §2.1. Hence the equality (12) gives the decomposition of $\hat{L}(g)$ as the continuous direct sum of irreducible unitary representations. Since $\hat{L}(g)$ and $L(g)$ are equivalent, this equality also gives the decomposition of the quasi-regular representation. We shall make this assertion more precise below.

2. The Fourier-Bessel transformation. In the space $\mathfrak{L}^2$ of square-integrable functions $f(x, y)$ we consider the subspaces $\mathfrak{L}^2_n$, consisting of functions of the form

$$f(r\cos\alpha,\ r\sin\alpha)=\varphi(r)e^{in\alpha}, \tag{1}$$

where

$$\int_0^\infty |\varphi(r)|^2 r\,dr<+\infty.$$

Let us ascertain what form the Fourier transform takes for functions of the space $\mathfrak{L}^2_n$.

We replace $f(\mathbf{x})$ by the expression $\phi(r)e^{in\alpha}$ in formula (2) of subsection 1, and put $F(\mathbf{y}) = F(R\cos\psi,\ R\sin\psi)$. Going over to polar coordinates, we get

$$F(R\cos\psi,\quad R\sin\psi)=\int_0^{2\pi}\int_0^\infty \varphi(r)e^{i[Rr\cos(\alpha-\psi)+n\alpha]}r\,dr\,d\alpha$$

$$=e^{in\psi}\int_0^\infty \varphi(r)r\left[\int_0^{2\pi} e^{i[Rr\cos\alpha+n\alpha]}d\alpha\right]dr.$$

Since by formula (8) of §3.1 we have

$$\int_0^{2\pi} e^{i[Rr\cos\alpha+n\alpha]}d\alpha=2\pi i^n J_n(Rr),$$

it follows that this equality can be written in the form

$$F(R\cos\psi,\ R\sin\psi)=2\pi i^n e^{in\psi}\int_0^\infty \varphi(r)J_n(Rr)r\,dr. \tag{2}$$

Hence the Fourier transform of the function

$$f(\mathbf{x})=\varphi(r)e^{in\alpha},\quad \mathbf{x}=(r\cos\alpha,\ r\sin\alpha)$$

is the function

$$F(\mathbf{y})=2\pi i^n\Phi(R)e^{in\psi},\quad \mathbf{y}=(R\cos\psi,\quad R\sin\psi),$$

where

$$\Phi(R)=\int_0^\infty \varphi(r)\,J_n(Rr)\,r\,dr. \tag{3}$$

This transformation is called the Fourier-Bessel transformation. We have thus proved that, under Fourier transformation on the plane, the "radial part" $\phi(r)$ of functions of the subspace $\mathfrak{L}_n^2$ undergoes the Fourier-Bessel transformation.

Employing the inverse Fourier transform on the plane, we get the inversion formula for the Fourier-Bessel transformation:

$$\varphi(r)=\int_0^\infty \Phi(R)\,J_n(Rr)\,R\,dR. \tag{4}$$

Utilizing the Plancherel formula for the Fourier transformation, it is not difficult to show that functions $\phi(r)$ and $\Phi(R)$ are connected by the equality

$$\int_0^\infty |\varphi(r)|^2\,r\,dr=\int_0^\infty |\Phi(R)|^2\,R\,dR. \tag{5}$$

This is called the analogue of the Plancherel formula for the Fourier-Bessel transformation.

Equality (5) shows that the Fourier-Bessel transformation is an isometric mapping of functions of $\mathfrak{L}_n^2$ onto functions of $\hat{\mathfrak{L}}_n^2$, i.e. onto functions of the form $e^{in\psi}\Phi(R)$, where

$$\int_0^\infty |\Phi(R)|^2 R\,dR<+\infty.$$

3. The decomposition of the quasi-regular representation. We obtained in subsection 1 a decomposition of the quasi-regular representation of $M(2)$ into irreducible ones. However, this decomposition is too complicated, since we utilized the Fourier transform twice. We now determine a straightforward decomposition of the quasi-regular representation.

The space $\mathfrak{L}^2$ is the orthogonal direct sum of subspaces $\mathfrak{L}_n^2$:

$$\mathfrak{L}^2=\sum_{n=-\infty}^{\infty}\mathfrak{L}_n^2.$$

Indeed, any function $f(r\cos\alpha, r\sin\alpha)$ of $\mathfrak{L}^2$ can be written as a Fourier series in $\alpha$:

$$f(r\cos\alpha,\ r\sin\alpha)=\sum_{n=-\infty}^{\infty}\varphi_n(r)\,e^{in\alpha}, \tag{1}$$

where

$$\varphi_n(r)=\frac{1}{2\pi}\int_0^{2\pi} f(r\cos\alpha,\ r\sin\alpha)\,e^{-in\alpha}d\alpha. \tag{2}$$

Let us apply formula (4) of subsection 2—the Fourier-Bessel transformation—to the terms of expansion (1). We obtain

$$f(r\cos\alpha,\ r\sin\alpha)=\sum_{n=-\infty}^{\infty}{}' e^{in\alpha}\int_0^{\infty}\Phi_n(R)\,J_n(Rr)\,R\,dR, . \tag{3}$$

where, by virtue of formulas (3) of subsection 2 and (2),

$$\Phi_n(R)=\int_0^{\infty}\varphi_n(\rho)\,J_n(R\rho)\,\rho\,d\rho$$

$$=\frac{1}{2\pi}\int_0^{\infty}\int_0^{2\pi} f(\rho\cos\theta,\ \rho\sin\theta)\,e^{-in\theta}J_n(R\rho)\,\rho\,d\theta\,d\rho. \tag{4}$$

We substitute the expression (4) for $\Phi_n(R)$ in formula (3), and change the order of summation and integration. Taking into account the composition theorem for Bessel functions $J_0(R)$ (see formula (4) of §4.1), we get

$$f(r\cos\alpha,\ r\sin\alpha)=\frac{1}{2\pi}\int_0^{\infty} R\,dR$$

$$\times\int_0^{2\pi} f(\rho\cos\theta,\ \rho\sin\theta)\,J_0\left(R\sqrt{r^2+\rho^2-2r\rho\cos(\alpha-\theta)}\right)\rho\,d\theta\,d\rho. \tag{5}$$

This equality can be rewritten as

$$f(\mathbf{x})=\int_0^{\infty} f_R(\mathbf{x})\,R\,dR, \tag{6}$$

where

$$f_R(r\cos\alpha,\ r\sin\alpha)$$

$$=\frac{1}{2\pi}\int_0^{2\pi} f(\rho\cos\theta,\ \rho\sin\theta)\,J_0\left(R\sqrt{r^2+\rho^2-2r\rho\cos(\alpha-\theta)}\right)\rho\,d\theta\,d\rho. \tag{7}$$

It is not difficult to show that the space $\mathfrak{H}_R$ of functions $f_R(x)$, expressed by formula (7), is invariant for shifts of the plane. These shifts give the representation $L_R(g)$ of $M(2)$:

$$L_R(g)f_R(\mathbf{x})=f_R(g^{-1}\mathbf{x}). \tag{8}$$

In addition,

$$[L(g)f(\mathbf{x})]_R=L_R(g)f_R(\mathbf{x})=f_R(g^{-1}x). \tag{9}$$

Therefore equality (6) determines a decomposition of the quasi-regular representation as the continuous direct sum of the irreducible unitary representations $L_R(g)$:

$$L(g)=\int_0^\infty L_R(g)\,R\,dR. \tag{10}$$

This decomposition gives an explicit form for the Fourier transformation of decomposition (12) of subsection 1.

We give an example where the Fourier-Bessel transformation is used. We apply formulas (3) and (4) of subsection 2 to equality (3) of §4.2, having written it in the form

$$J_{n-m}(Rr_1)\,J_m(Rr_2)=\frac{2}{\pi}\int_{|r_1-r_2|}^{r_1+r_2}\frac{e^{i(n\varphi-m\varphi_2)}\,J_n(Rr)\,r\,dr}{\sqrt{4r_1^2r_2^2-(r^2-r_1^2-r_2^2)^2}},$$

where

$$r^2=r_1^2+r_2^2+2r_1r_2\cos\varphi_2 \tag{11}$$

and

$$e^{i\varphi}=\frac{r_1+r_2e^{i\varphi_2}}{r}. \tag{11'}$$

Hence, because of formulas (3) and (4) of subsection 2, we get

$$\int_0^\infty J_{n-m}(Rr_1)\,J_m(Rr_2)\,J_n(Rr)\,R\,dR$$

$$=\begin{cases}\dfrac{2}{\pi}\dfrac{e^{i(n\varphi-m\varphi_2)}}{\sqrt{4r_1^2r_2^2-(r^2-r_1^2-r_2^2)^2}}, \\ \text{if}\quad |r_1-r_2|<r<r_1+r_2, \\ 0 \quad \text{otherwise}.\end{cases} \tag{12}$$

Here $\phi_2$ and $\phi$ are connected with $r_1, r_2, r$ by (11),

4. **Infinitesimal operators.** Let us now calculate infinitesimal operators of the quasi-regular representation $L(g)$ of the group $M(2)$. We begin by calculating the infinitesimal operator corresponding to the one-parametric subgroup $\Omega_1$ of elements of the form

$$g=\omega_1(t)=\begin{pmatrix}1 & 0 & t\\ 0 & 1 & 0\\ 0 & 0 & 1\end{pmatrix}. \tag{1}$$

The motion $g^{-1}$ transforms the vector $\mathbf{x}(x, y)$ into the vector $g^{-1}\mathbf{x}=(x-t, y)$.

Therefore $L(\omega_1(t))$ transforms $f(x, y)$ into

$$L(\omega_1(t))f(x, y)=f(x-t, y). \tag{2}$$

Hence

$$A_1=\left.\frac{dL(\omega_1(t))}{dt}\right|_{t=0}=-\frac{\partial}{\partial x}. \tag{3}$$

One similarly proves that

$$A_2=-\frac{\partial}{\partial y}. \tag{4}$$

Finally, $L(\omega_3(t))$, where

$$\omega_3(t)=\begin{pmatrix}\cos t & -\sin t & 0\\ \sin t & \cos t & 0\\ 0 & 0 & 1\end{pmatrix},$$

transforms $f(x, y)$ into

$$L(\omega_3(t)f(x, y))=f(x\cos t+y\sin t, \quad -x\sin t+y\cos t).$$

Therefore

$$A_3=y\frac{\partial}{\partial x}-x\frac{\partial}{\partial y}. \tag{5}$$

In polar coordinates the operators $A_1$, $A_2$, $A_3$ are written down as follows:

$$\left.\begin{aligned}A_1&=-\cos\alpha\frac{\partial}{\partial r}+\frac{\sin\alpha}{r}\frac{\partial}{\partial\alpha},\\ A_2&=-\sin\alpha\frac{\partial}{\partial r}-\frac{\cos\alpha}{r}\frac{\partial}{\partial\alpha},\\ A_3&=-\frac{\partial}{\partial\alpha}.\end{aligned}\right\} \tag{6}$$

Clearly, infinitesimal operators for the representation $L_R(g)$ in $\mathfrak{H}_R$ also have the same form, since it is obtained by restricting the quasi-regular representation. It follows from (6) that

$$\left.\begin{aligned}H_+&=A_1+iA_2=-e^{i\alpha}\left(\frac{\partial}{\partial r}+\frac{i}{r}\frac{\partial}{\partial\alpha}\right),\\ H_-&=A_1-iA_2=-e^{-i\alpha}\left(\frac{\partial}{\partial r}-\frac{i}{r}\frac{\partial}{\partial\alpha}\right).\end{aligned}\right\} \tag{7}$$

With the help of (7) we can give a new derivation of the recurrence relations for Bessel functions. For this we observe that the canonical basis in $\mathfrak{H}_R$ consists of the functions

$$(-i)^n J_n(Rr)\,e^{in\alpha}.$$

Indeed, in $\hat{\mathfrak{H}}_R$ the canonical basis consists of the functions $e^{in\psi}$. But, under the inverse Fourier transformation the functions $e^{in\psi}\delta(|y| - R)$ become

$$\frac{1}{(2\pi)^2}\int e^{in\psi}\delta(|\mathbf{y}| - R)\, e^{-i(\mathbf{x},\,\mathbf{y})}\, d\mathbf{y}$$

$$= \frac{1}{(2\pi)^2}\int_0^{2\pi} e^{-iRr\cos(\psi-\alpha)}\, e^{in\psi}\, d\psi = \frac{1}{2\pi}(-i)^n J_n(Rr)\, e^{in\alpha}$$

(see §3.1). Hence our assertion.

But, as shown in formulas (4) and (4′) of §2.3, the vectors $\mathbf{f}_n \equiv e^{in\psi}$ of the canonical basis satisfy the relations

$$H_+\mathbf{f}_n = iR\mathbf{f}_{n+1}, \tag{8′}$$

$$H_-\mathbf{f}_n = iR\mathbf{f}_{n-1} \tag{8″}$$

(we have replaced $R$ by $iR$ in formulas (4) and (4′) of §2.3, since $\hat{L}_R(g) \sim T_{iR}(g)$; see the end of subsection 1). The same relations must hold for the vectors $(-i)^n e^{in\alpha} J_n(Rr)$ of the canonical basis in $\mathfrak{H}_R$. Thus, from (8′) we get

$$-e^{i\alpha}\left(\frac{\partial}{\partial r} + \frac{i}{r}\frac{\partial}{\partial \alpha}\right)\left[(-i)^n e^{in\alpha} J_n(Rr)\right] = R(-i)^n e^{i(n+1)\alpha} J_{n+1}(Rr).$$

Simplifying this and putting $R = 1$, we get

$$rJ_{n+1}(r) = nJ_n(r) - rJ_n'(r). \tag{9}$$

Similarly, from (8″) we get

$$rJ_{n-1}(r) = nJ_n(r) + rJ_n'(r).$$

It follows from (8′)–(8″) that vectors of the canonical basis satisfy the equality

$$H_+H_-\mathbf{f}_n = -R^2\mathbf{f}_n. \tag{10}$$

We replace $H_+$ and $H_-$ here by the expressions (7), and $\mathbf{f}_n$ by the functions $(-i)^n e^{in\alpha} J_n(Rr)$. We find that Bessel functions satisfy the differential equation

$$e^{i\alpha}\left(\frac{\partial}{\partial r} + \frac{i}{r}\frac{\partial}{\partial \alpha}\right) e^{-i\alpha}\left(\frac{\partial}{\partial r} - \frac{i}{r}\frac{\partial}{\partial \alpha}\right) e^{in\alpha} J_n(Rr) = -R^2 e^{in\alpha} J_n(Rr).$$

Simplifying this and putting $R = 1$, we obtain the Bessel equation

$$J_n''(r) + \frac{1}{r}J_n'(r) + \left(1 - \frac{n^2}{r^2}\right)J_n(r) = 0$$

for the functions $J_n(r)$.

We note in conclusion that the operator $H_+H_-$ coincides with the Laplace operator $\Delta = \partial^2/\partial x^2 + \partial^2/\partial y^2$, written in polar coordinates. Therefore it follows from equality (10) that the functions $e^{in\alpha}J_n(Rr)$ are eigenfunctions of the Laplace operator, corresponding to the eigenvalue $-R^2$.

5. Decomposition of the regular representation. We turn, finally, to the decomposition of the regular representation

$$T(g_0)f(g)=f(g_0^{-1}g) \tag{1}$$

of the group $M(2)$. Here $f(g)$ is a function on $M(2)$, such that

$$\int |f(g)|^2\, dg < +\infty. \tag{2}$$

Let us denote the space of such functions by $\mathfrak{H}$. We shall show that the regular representation is a direct sum of a denumerable set of representations, equivalent to the quasi-regular representation.

According to §1.2, elements $g$ of $M(2)$ can be specified by a vector $\mathbf{x}$ of (parallel) translation and by an angle $\phi$ of rotation around the origin, $g = g(\mathbf{x}, \phi)$ We expand functions $f(\mathbf{x}, \phi)$ of $\mathfrak{H}$ in a Fourier series in $\phi$:

$$f(\mathbf{x}, \varphi)=\sum_{n=-\infty}^{\infty} f_n(\mathbf{x})\, e^{in\varphi}. \tag{3}$$

It follows from this expansion that $\mathfrak{H}$ is the direct orthogonal sum of spaces $\mathfrak{H}_n$, consisting of functions of the form $f_n(\mathbf{x})e^{in\phi}$, where

$$\|f_n\|=\int |f_n(\mathbf{x})|^2\, d\mathbf{x} < +\infty. \tag{4}$$

In addition

$$f_n(\mathbf{x})=\frac{1}{2\pi}\int_0^{2\pi} f(\mathbf{x}, \varphi)\, e^{-in\varphi}\, d\varphi. \tag{5}$$

We show that the $\mathfrak{H}_n$ are invariant for the operators $T(g)$. Indeed, the operator $T(g_0)$, $g_0 = g(\mathbf{b}, \alpha)$, transforms $f(\mathbf{x}, \phi)$ into $f[(\mathbf{x} - \mathbf{b})_{-\alpha}, \phi - \alpha]$ (see formula (3) of §1.2). The Fourier coefficients of the latter have the form

$$f_n^{(g_0)}(\mathbf{x})=\frac{1}{2\pi}\int_0^{2\pi} f[(\mathbf{x}-\mathbf{b})_{-\alpha}, \varphi-\alpha]\, e^{-in\varphi}\, d\varphi$$

$$=\frac{e^{-in\alpha}}{2\pi}\int_0^{2\pi} f[(\mathbf{x}-\mathbf{b})_{-\alpha}, \varphi]\, e^{-in\varphi}\, d\varphi = e^{-in\alpha} f_n\, [(\mathbf{x}-\mathbf{b})_{-\alpha}].$$

Consequently $T(g_0)$ transforms $\mathfrak{H}_n$ into itself and gives, in $\mathfrak{H}_n$, the representation

$$T_n(g_0)f_n(\mathbf{x})=e^{-in\alpha}f_n[(\mathbf{x}-\mathbf{b})_{-\alpha}]. \tag{6}$$

Thus we have

$$\mathfrak{H}=\sum_{n=-\infty}^{\infty}\mathfrak{H}_n, \tag{7}$$

and, moreover,

$$T(g)=\sum_{n=-\infty}^{\infty}T_n(g). \tag{8}$$

We now show that $T_n(g)$ is equivalent to the quasi-regular representation. If we take the Fourier transform of $f_n(x)$, then the representation $T_n(g_0)$ turns into

$$\hat{L}_n(g_0)F_n(\mathbf{y})=e^{-in\alpha}e^{i(\mathbf{b},\mathbf{y})}F_n(\mathbf{y}_{-\alpha}).$$

Let us define an operator $Q$ by the formula

$$\Phi(y)=QF_n(\mathbf{y})=e^{in\varphi}F_n(\mathbf{y}),$$

where $\phi$ is the polar angle of the point $y$. A simple calculation shows that

$$Q\hat{L}_n(g_0)Q^{-1}\Phi(\mathbf{y})=\hat{L}(g_0)\Phi(\mathbf{y}),$$

where $\hat{L}(g)$ is the image of the quasi-regular representation under the Fourier transformation. This proves the equivalence of $\hat{L}_n(g)$ to the quasi-regular representation.

Thus equality (8) gives a decomposition of the regular representation of $M(2)$ into representations $T_n(g)$, equivalent to the quasi-regular representation of the same group. Since we already know how to decompose the quasi-regular representation into irreducible ones, the problem of decomposing the regular representation is solved.

## § 6. PRODUCT OF REPRESENTATIONS

1. The Kronecker product of representations $T_R(g)$. Let

$$T_{iR_1}(g)f(\psi)=e^{iR_1r\cos(\psi-\varphi)}f(\psi-\alpha)$$

and

$$T_{iR_2}(g)f(\psi)=e^{iR_2r\cos(\psi-\varphi)}f(\psi-\alpha)$$

be infinitesimal unitary representations of the group $M(2)$. Let us construct the Kronecker product $T(g)$ of these representations. According to the Supplement to Chapter I, such a product is constructed in the space $\mathfrak{H}$ of functions of two variables $f(\psi_1, \psi_2)$, $0 \le \psi_1 < 2\pi$, $0 \le \psi_2 < 2\psi$, such that

$$\|f\|^2 = \frac{1}{4\pi^2} \int_0^{2\pi} \int_0^{2\pi} |f(\psi_1, \psi_2)|^2 \, d\psi_1 \, d\psi_2 < +\infty.$$

The representation $T(g)$ is given by the formula

$$T(g) f(\psi_1, \psi_2) = e^{ir[R_1 \cos(\psi_1 - \varphi) + R_2 \cos(\psi_2 - \varphi)]} f(\psi_1 - \alpha, \psi_2 - \alpha). \tag{3}$$

We decompose this representation as the continuous direct sum of irreducible representations. We first observe that (3) can be rewritten as

$$T(g) f(\psi_1, \psi_2) = e^{irR\cos(\psi_1 - \varphi + \beta)} f(\psi_1 - \alpha, \varphi_2 - \alpha), \tag{3'}$$

where

$$R^2 = R_1^2 + R_2^2 + 2R_1R_2 \cos(\psi_2 - \psi_1) \tag{4}$$

and

$$e^{i\beta} = \frac{R_1 + R_2 e^{i(\psi_2 - \psi_1)}}{R}. \tag{4'}$$

Here $R$ as well as $\beta$ depends on $R_1$, $R_2$ and $\psi_2 - \psi_1$.

Next we decompose the Hilbert space $\mathfrak{H}$ as a continuous direct sum of Hilbert spaces invariant for the operators $T(g)$. To do this we consider in $\mathfrak{H}$ the subspace $\mathfrak{S}$, consisting of infinitely differentiable functions $f(\psi_1, \psi_2)$. With each function $f(\psi_1, \psi_2)$ of $\mathfrak{S}$ and each number $\mu$, $0 \le \mu < 2\pi$, we associate the function

$$f_\mu(\psi) = f(\psi, \psi + \mu) \tag{5}$$

and denote the set of all functions $f_\mu(\psi)$, corresponding for fixed $\mu$ to functions of $\mathfrak{S}$, by $\mathfrak{S}_\mu$.

The spaces $\mathfrak{S}_\mu$ are pre-Hilbert spaces relative to the norm

$$\|f_\mu\|_\mu^2 = \frac{1}{2\pi} \int_0^{2\pi} |f_\mu(\psi)|^2 \, d\mu.$$

For any function $f(\psi_1, \psi_2)$ of $\mathfrak{H}$ we have

$$\|f\|^2 = \frac{1}{4\pi^2} \int_0^{2\pi} \int_0^{2\pi} |f(\psi_1, \psi_2)|^2 \, d\psi_1 \, d\psi_2$$

$$= \frac{1}{4\pi^2} \int_0^{2\pi} \int_0^{2\pi} |f(\psi, \psi + \mu)|^2 \, d\psi \, d\mu =$$

$$=\frac{1}{2\pi}\int_0^{2\pi}\int_0^{2\pi}|f_\mu(\psi)|^2\,d\psi\,d\mu=\frac{1}{2\pi}\int_0^{2\pi}\|f_\mu\|_\mu^2\,d\mu.$$

Therefore $\mathfrak{H}$ is the continuous direct sum of the spaces $\mathfrak{H}_\mu$, i.e. of the completions of $\mathfrak{S}_\mu$:

$$\mathfrak{H}=\frac{1}{2\pi}\int_0^{2\pi}\mathfrak{H}_\mu\,d\mu. \tag{6}$$

Further,

$$[T(g)f(\psi_1,\ \psi_2)]_\mu=[e^{iRr\cos(\psi_1-\varphi+\beta)}f(\psi_1-\alpha,\ \psi_2-\alpha)]_\mu$$
$$=e^{iRr\cos(\psi_1-\varphi+\beta)}f(\psi_1-\alpha,\ \psi_2+\mu-\alpha),$$

where $R$ and $\beta$ are defined by (4) and (4′) for $\psi_2-\psi_1=\mu$. This equality can be rewritten as follows:

$$[T(g)f(\psi_1,\ \psi_2)]_\mu=T_\mu(g)f_\mu(\psi_1), \tag{7}$$

where for $g=g(r,\ \phi,\ \alpha)$ we have

$$T_\mu(g)f_\mu(\psi)=e^{iRr\cos(\psi-\varphi+\beta)}f_\mu(\psi-\alpha) \tag{8}$$

($R$ and $\beta$ depend on $\mu$ according to (4) and (4′) for $\psi_2-\psi_1=\mu$). Hence $T(g)$ is the continuous direct sum of the representations $T_\mu(g)$:

$$T(g)=\frac{1}{2\pi}\int_0^{2\pi}T_\mu(g)\,d\mu. \tag{9}$$

Formula (8) shows that $T_\mu(g)$ is equivalent to the representation $T_{iR}(g)$ of $M(2)$, and is therefore unitary and irreducible. Thus we have decomposed $T(g)$ as the continuous direct sum of irreducible unitary representations.

**2. Kronecker product and the multiplication formula.** We show that the decomposition of a Kronecker product yields a new derivation of the multiplication formula for Bessel functions. In fact, it follows from formulas (7) and (9) of subsection 1 that

$$(T(g)f_1,\ f_2)=\frac{1}{2\pi}\int_0^{2\pi}([T(g)f_1]_\mu,\ f_{2\mu})_\mu\,d\mu$$
$$=\frac{1}{2\pi}\int_0^{2\pi}(T_\mu(g)f_{1\mu},\ f_{2\mu})_\mu\,d\mu$$
$$=\frac{1}{4\pi^2}\int_0^{2\pi}\int_0^{2\pi}T_\mu(g)f_{1\mu}(\psi)\overline{f_{2\mu}(\psi)}\,d\mu\,d\psi. \tag{1}$$

Let us apply this formula to the functions

$$f_1(\psi_1, \psi_2) = 1, \; f_2(\psi_1, \psi_2) = e^{in\psi_1} e^{im\psi_2}.$$

For these functions we have

$$f_{1\mu}(\psi) = 1,$$
$$T_\mu(g) f_{1\mu}(\psi) = e^{iRr\cos(\psi - \varphi + \beta)}$$

and

$$f_{2\mu}(\psi) = e^{im\mu} e^{i(n+m)\psi}.$$

Therefore, by formula (1), we get

$$\begin{aligned}(T(g) f_1, f_2) &= \frac{1}{4\pi^2} \int_0^{2\pi}\int_0^{2\pi} e^{i[Rr\cos(\psi - \varphi + \beta) - m\mu - (n+m)\psi]}\, d\psi\, d\mu \\ &= \frac{e^{-i(n+m)\varphi}}{4\pi^2} \int_0^{2\pi} e^{-i[m\mu - (n+m)\beta]}\, d\mu \int_0^{2\pi} e^{i[Rr\cos\psi - (n+m)\psi]}\, d\psi \\ &= \frac{e^{-i(n+m)\varphi}\, i^{n+m}}{2\pi} \int_0^{2\pi} e^{-i[m\mu - (n+m)\beta]} J_{n+m}(Rr)\, d\mu, \end{aligned} \tag{2}$$

where

$$R^2 = R_1^2 + R_2^2 + 2R_1R_2\cos\mu,$$
$$e^{i\beta} = \frac{R_1 + R_2 e^{i\mu}}{R}.$$

But $(T(g) f_1, f_2)$ is the matrix element $t_{nm,00}(g)$ of the representation $T(g)$. The matrix elements of a Kronecker product are equal to the products of the matrix elements of its factors, and so

$$t_{nm,00}(g) = t_{n0}^{iR_1}(g)\, t_{m0}^{iR_2}(g).$$

Substituting the values of the matrix elements $t_{n0}^{iR_1}(g)$ and $t_{m0}^{iR_2}(g)$ (see formula (10) of §3.1), we easily obtain

$$J_n(R_1) J_m(R_2) = \frac{1}{2\pi} \int_0^{2\pi} e^{-i[m\mu - (n+m)\beta]} J_{n+m}(R)\, d\mu,$$

where $\beta$ and $R$ are expressed in terms of $R_1$, $R_2$, $\psi$ by formulas (4)–(4″) of subsection 1. This equality is only notationally different from the multiplication formula for the Bessel functions (see §4.2).

## §7. BESSEL FUNCTIONS AND THE FUNCTIONS $P^l_{mn}(x)$

1. The group of motions of the plane and the group of rotations of the sphere. The euclidean plane can, in a sense, be regarded as a sphere of infinite radius; if the radius of the sphere is increased, and, correspondingly, its center is removed to infinity, the sphere in the limit becomes the plane. In accordance with this, the group $M(2)$ of motions of the plane can also be regarded as the limit of the group $SO(3)$ of rotations of the sphere.

More precisely, this means the following. The multiplication of elements in $SO(3)$ is determined by formula (2) of §1.2 of Chapter III. In these formulas we replace the angles $\phi, \theta, \psi, \theta_1, \theta_2, \phi_2$ by $\phi, r/R, \alpha, r_1/R, r_2/R, \alpha_1$ respectively, and take the principal terms for $R \to \infty$. A simple calculation shows that this gives formula (8) of §1.2, determining the multiplication of elements in the group $M(2)$.

Thus $M(2)$ can be regarded as the degenerate form of $SO(3)$ for $R$ tending to $\infty$. This can be seen by examining the Lie algebras of these groups. For $SO(3)$ the commutation relations in the Lie algebra have the form

$$[a_1, a_2] = a_3,$$
$$[a_2, a_3] = a_1,$$
$$[a_3, a_1] = a_2.$$

If one replaces the parameter $t$ in the one-parametric subgroups $\Omega_1$ and $\Omega_2$ by $t/R$, and makes $R$ tend to infinity, then in the limit one obtains the commutation relations

$$[a_1, a_2] = 0,$$
$$[a_2, a_3] = a_1,$$
$$[a_3, a_1] = a_2,$$

determining the Lie algebra of $M(2)$.

2. Bessel functions and Jacobi polynomials. The connection between $SO(3)$ and $M(2)$ naturally indicates a connection between the matrix elements of irreducible unitary representations of these groups. Thus, Bessel functions, i.e. the matrix elements of the representations $T_{i\rho}(g)$ of the group $M(2)$, must arise, through a limiting process, from the functions $P^l_{mn}(\cos\theta)$, i.e. the matrix elements of the representations $T_l(g)$ of $SO(3)$. It turns out that, in this limiting process, not only $R$ but also $l$ must be made to tend to infinity.

In order to obtain the desired formulas, we use the following integral representation for the functions $P^l_{mn}(\cos\theta)$:

$$P^l_{mn}(\cos\theta)=\frac{1}{2\pi}\sqrt{\frac{(l-m)!\,(l+m)!}{(l-n)!\,(l+n)!}}$$
$$\times\int_0^{2\pi}\left(\cos\frac{\theta}{2}e^{i\varphi/2}+i\sin\frac{\theta}{2}e^{-i\varphi/2}\right)^{l-n}$$
$$\times\left(i\sin\frac{\theta}{2}e^{i\varphi/2}+\cos\frac{\theta}{2}e^{-i\varphi/2}\right)^{l+n}e^{im\varphi}\,d\varphi.$$

Let us put $\theta = r/l$ here and take the limit for $l \to \infty$. We obtain

$$\lim_{l\to\infty}P^l_{mn}\left(\cos\frac{r}{l}\right)$$
$$=\frac{1}{2\pi}\lim_{l\to\infty}\int_0^{2\pi}\left(1+\frac{ir}{2l}e^{-i\varphi}\right)^{l-n}\left(1+\frac{ir}{2l}e^{i\varphi}\right)^{l+n}e^{i(m-n)\varphi}\,d\varphi$$
$$=\frac{1}{2\pi}\int_0^{2\pi}e^{ir\cos\varphi}e^{i(m-n)\varphi}\,d\varphi.$$

But, owing to the integral representation of Bessel functions (see formula (8) of §3.1), this equality can be rewritten as follows:

$$\lim_{l\to\infty}P^l_{mn}\left(\cos\frac{r}{l}\right)=i^{m-n}J_{m-n}(r). \tag{1}$$

We note a special case of this formula. For $m = n = 0$ we have

$$\lim_{l\to\infty}P_l\left(\cos\frac{r}{l}\right)=J_0(r).$$

Thus $J_0(r)$ is obtained through a limiting process from the Legendre polynomials.

We may also mention that the Laplace operator on $M(2)$ is obtained from the Laplace operator on $SU(2)$ through a similar limiting process. We omit the details of this calculation.

3. An asymptotic formula for the Clebsch-Gordan coefficients. The connection between the functions $P^l_{mn}(z)$ and Bessel functions allows us to establish an asymptotic formula for the Clebsch-Gordan coefficients in the case of large values of $l_1$, $l_2$, $l$. We use formula (6) of §8.3, Chapter III, for this. Putting $j' = j$, $k' = k$ in that formula and remembering that the Clebsch-Gordan coefficients are real, we find that, for all positive integers $N$, we have

$$[C(Nl_1,\ Nl_2,\ Nl;\ j;\ k;\ j+k)]^2$$
$$= \frac{2Nl+1}{2} \int_{-1}^{1} P_{jj}^{Nl_1}(x)\, P_{kk}^{Nl_2}(x)\, \overline{P_{j+k,\,j+k}^{Nl}(x)}\, dx.$$

Let us make the substitution $x = \cos y/N$ and take the limit for $N \to \infty$. Because of formulas (1) of subsection 2 and (12) of §5.3, we obtain

$$\lim_{N\to\infty} N[C(Nl_1,\ Nl_2,\ Nl;\ j,\ k,\ j+k)]^2$$
$$= l\int_0^\infty J_0(yl_1)\, J_0(yl_2)\, J_0(yl)\, y\, dy$$
$$= \begin{cases} \dfrac{2l}{\pi\sqrt{4l_1^2l_2^2-(l^2-l_1^2-l_2^2)^2}}, \\ \text{if} \quad |l_1-l_2|<l<l_1+l_2, \\ 0 \text{ otherwise.} \end{cases}$$

This expression does not depend on $j$ and $k$. Clearly, the sign of $C(Nl_1, Nl_2, Nl;\ j, k, j+k)$ coincides, for large values of $N$, with the sign of $C(Nl_1, Nl_2, Nl;\ 0, 0, 0)$, i.e. with the sign of $(-1)^{N(g-l)}$, where $l+l_1+l_2=2g$. Hence

$$C(Nl_1,\ Nl_2,\ Nl;\ j,\ k,\ j+k)$$
$$\sim \begin{cases} (-1)^{N(g-l)}\left[\dfrac{2l}{\pi N\sqrt{4l_1^2l_2^2-(l^2-l_1^2-l_2^2)^2}}\right]^{1/2}, \\ \text{if} \quad |l_1-l_2|<l<l_1+l_2, \\ 0 \text{ otherwise.} \end{cases} \tag{1}$$

# CHAPTER V

# REPRESENTATIONS OF THE GROUP OF MOTIONS OF THE PSEUDO-EUCLIDEAN PLANE AND THE FUNCTIONS OF BESSEL AND MACDONALD

It was shown in the preceding chapter that many properties of Bessel functions with integral index are connected with representations of the group of motions of the euclidean plane. In order to pass to Bessel functions with arbitrary index, we shall have to replace the euclidean plane by the pseudo-euclidean one, for which the subgroup of rotations is not compact. We shall show that representations of the group of motions of the pseudo-euclidean plane are given by integral operators whose kernels are expressed by means of the Hankel and Macdonald functions, closely related to Bessel functions with arbitrary index. This will lead to a number of properties of these functions. At the beginning of the chapter we shall consider representations of the group of linear transformations of the straight line—the one-dimensional analogue of the group of motions of the pseudo-euclidean plane.

## §1. REPRESENTATIONS OF THE GROUP OF LINEAR TRANSFORMATIONS OF THE STRAIGHT LINE AND Γ-FUNCTIONS

1. **The group of linear transformations of the straight line.** We study representations of the group $G$ of linear transformations of the straight line which preserve orientation, i.e. transformations of the form $y = ax + b$, where $a > 0$. Each element $g$ of $G$ is determined by two real numbers, $a > 0$ and $b$. We shall denote elements of $G$ by $g(b, a)$.

If $y = a_2x + b_2$ and $z = a_1y + b_1$, then

$$z = a_1a_2x + a_1b_2 + b_1.$$

Therefore the group operation in $G$ is given by the formula

$$g(b_1, a_1)\, g(b_2, a_2) = g(b_1 + a_1b_2,\, a_1a_2). \tag{1}$$

It is obvious from (1) that elements of the form $g(b, 1)$ in $G$ constitute a subgroup, isomorphic to the additive group $R$ of real numbers:

$$g(b_1, 1)\,g(b_2, 1)=g(b_1+b_2, 1).$$

However, elements of the form $g(0, a)$ constitute a subgroup $A$ in $G$, isomorphic to the multiplicative group $R_+$ of positive numbers:

$$g(0, a_1)\,g(0, a_2)=g(0, a_1a_2).$$

The group $G$ is isomorphic to the group of motions of the form $\begin{bmatrix} a & b \\ 0 & 1 \end{bmatrix}$. This immediately follows from the equality

$$\begin{pmatrix} a_1 & b_1 \\ 0 & 1 \end{pmatrix}\begin{pmatrix} a_2 & b_2 \\ 0 & 1 \end{pmatrix}=\begin{pmatrix} a_1a_2 & a_1b_2+b_1 \\ 0 & 1 \end{pmatrix}.$$

$G$ is the cross product of the additive group $R$ of real numbers and the group $R_+$ of its automorphism $x \to ax$, $a < 0$.

2. **Irreducible representations of** $G$. To construct irreducible representations of $G$, we consider representations of $G$ induced by the one-dimensional representations $e^{\lambda b}$ of the subgroup $B$ of elements of the form $g(b, 1)$. According to § 2.6 of Chapter I, the induced representations are constructed in the space $\mathfrak{D}_\lambda$ of functions $f(g)$ on $G$, such that

$$f(g(b_0, 1)\,g(b, a))=e^{\lambda b_0}f(g(b, a)). \tag{1}$$

If one puts $f(g) \equiv f(b, a)$, equality (1) can be written as

$$f(b+b_0, a)=e^{\lambda b_0}f(b, a). \tag{2}$$

Therefore the functions $f(b, a)$ of the space $\mathfrak{D}_\lambda$ have the form

$$f(b, a)=e^{\lambda b}f(0, a)=e^{\lambda b}\varphi(a), \tag{3}$$

where $\phi(a) \equiv f(0, a)$ is a function on the subgroup $A$ (or, what is the same thing, on the ray $0 < a < \infty$).

Operators of the induced representations are expressed by the formula

$$R_\lambda(g_0)f(g)=f(gg_0), \qquad f(g) = \mathfrak{D}_\lambda. \tag{4}$$

We clarify how these operators are written down on passing from functions $f(g)$ to functions $\phi(a)$. Since

$$g(0, a)\,g(b_0, a_0)=g(ab_0, a_0a),$$

we have

$$R_\lambda(g_0)\,\varphi(a)=f(ab_0, a_0a)=e^{\lambda ab_0}\varphi(a_0a).$$

Thus the representation $R_\lambda(g)$ of the group $G$, induced by the representation $e^{\lambda b}$ of the subgroup $B$, is constructed in the space of functions on the semiaxis $0 < x < \infty$ and is given by the formula

$$R_\lambda(g)\varphi(x) = e^{\lambda bx}\varphi(ax) \tag{5}$$

(we have replaced $a$ by $x$ and $a_0$, $b_0$ by $a$, $b$).

Evidently, the operators $R_\lambda(g)$ leave invariant the space $\mathfrak{D}$ of infinitely differentiable finite functions on the semiaxis, which vanish in some neighborhood of the point $x = 0$. In the sequel we shall consider operators $R_\lambda(g)$ in this space.

Let us find the infinitesimal operators of $R_\lambda(g)$, corresponding to one-parametric subgroups $A$ and $B$. To the elements $g(0, e^t)$ of $A$ correspond the operators

$$R_\lambda(g(0, e^t))\varphi(x) = \varphi(e^t x). \tag{6}$$

We differentiate both sides of this equality with respect to $t$ and put $t = 0$. We find that the infinitesimal operator $a_\lambda$ has the form

$$a_\lambda\varphi(x) = x\varphi'(x). \tag{7}$$

In other words,

$$a_\lambda = x\frac{d}{dx}. \tag{7'}$$

In just the same way, it follows from the equality

$$R_\lambda(g(t, 1))\varphi(x) = e^{\lambda tx}\varphi(x)$$

that the infinitesimal operator of the representation $R_\lambda(g)$, corresponding to $B$, has the form

$$b_\lambda\varphi(x) = \lambda x\varphi(x). \tag{8}$$

Therefore

$$b_\lambda = \lambda x. \tag{8'}$$

The operators $a_\lambda$ and $b_\lambda$ are defined in the whole space $\mathfrak{D}$. The commutation relation for $a_\lambda$ and $b_\lambda$ has the form

$$[a_\lambda, b_\lambda] \equiv a_\lambda b_\lambda - b_\lambda a_\lambda = b_\lambda. \tag{9}$$

We show that for $\lambda \neq 0$ the representations $R_\lambda(g)$ of $G$ are operator-irreducible, i.e. any operator $Q$, permutable with the $R_\lambda(g)$, is a multiple of the identity operator. The permutability of $Q$ with the operators of $R_\lambda$ implies its permutability with the infinitesimal operators $a_\lambda$ and $b_\lambda$. But $b_\lambda$ differs from the operator of multiplication by $x$ in a constant factor only. It follows that $Q$, being permutable with all operators $R_\lambda(g)$, must be permutable with all operators of multiplication by polynomials, and hence with all the operators of multiplication by a function (we may recall that the operators act in the space of finite functions).

It is known that the only operators in the space of infinitely differentiable finite functions, permutable with all the operators of multiplication by a function, are operators of multiplication by a function. Therefore $Q$ has the form

$$Q\varphi(x)=q(x)\varphi(x). \tag{10}$$

To calculate the function $q(x)$ we utilize the permutability of $Q$ with the operators

$$R_\lambda(g(0, a))\varphi(x)=\varphi(ax). \tag{11}$$

The equality $QR_\lambda(g(0, a)) = R_\lambda(g(0, a))Q$ means that

$$q(x)\varphi(ax)=q(ax)\varphi(ax)$$

and so $q(x) = q(ax)$. Consequently $q(x)$ is a constant. This proves that any operator $Q$, permutable with all operators $R_\lambda(g)$, is a multiple of the identity operator, and hence the representation $R_\lambda(g)$ is operator-irreducible.

For $\lambda = 0$ the representation $R_\lambda(g)$ takes the form

$$R_0(g)\varphi(x)=\varphi(ax). \tag{12}$$

This representation is none other than the regular representation of $A$, and is therefore reducible. The irreducible components of $R_0(g)$ are given by the unitary one-dimensional representations

$$g(b, a)\to a^{i\rho} \tag{13}$$

of $G$.

We have thus constructed the series $R_\lambda(g)$ of irreducible representations of $G$. We now show that if $\lambda = t\mu$, where $t > 0$ is a positive number, then $R_\lambda(g)$ and $R_\mu(g)$ are equivalent. Indeed, let us denote by $S$ the operator

$$S\varphi(x)=\varphi(tx) \tag{14}$$

in $\mathfrak{D}$. It is easy to verify that

$$S^{-1}R_\lambda(g)S=R_\mu(g) \tag{15}$$

and consequently $R_\lambda(g)$ and $R_\mu(g)$ are equivalent.

So far we have considered the representations $R_\lambda(g)$ in the space $\mathfrak{D}$ of infinitely differentiable finite functions on the semiaxis $0 < x < \infty$, vanishing in some neighborhood of the point $x = 0$. We now introduce a scalar product in this space by putting

$$(\varphi, \psi) = \int_0^\infty \varphi(x)\overline{\psi(x)}\frac{dx}{x} \tag{16}$$

(the measure $dx/x$ is invariant for the similarity transformations $x \to ax$ on the straight line: $d(ax)/ax = dx/x$). Let us determine the values of $\lambda$ for which $R_\lambda(g)$ is unitary relative to this scalar product. We have

$$(R_\lambda(g)\varphi, R_\lambda(g)\psi)$$
$$= \int_0^\infty e^{(\lambda+\bar{\lambda})bx}\varphi(ax)\overline{\psi(ax)}\frac{dx}{x} = \int_0^\infty e^{\frac{(\lambda+\bar{\lambda})bx}{a}}\varphi(x)\overline{\psi(x)}\frac{dx}{x}.$$

Clearly, in order that the equality

$$(R_\lambda(g)\varphi, \quad R_\lambda(g)\psi) = (\varphi, \psi) \tag{17}$$

be satisfied for all $\phi, \psi \in \mathfrak{D}$ and $g \in G$, it is necessary and sufficient that $\lambda + \bar{\lambda} = 0$, i.e. $\lambda$ must be a purely imaginary number. Thus for purely imaginary values of $\lambda$, the representations $R_\lambda(g)$ can be extended to unitary representations in the Hilbert space $\mathfrak{H}$. It is easy to see that, for the remaining values of $\lambda$, the operators $R_\lambda(g)$ are unbounded in $\mathfrak{H}$, and so $R_\lambda(g)$ cannot be extended to a representation in $\mathfrak{H}$. But if $\mathrm{Re}\,\lambda < 0$, then the representation (5) can be extended to a representation in $\mathfrak{H}$ of the semigroup $g(a, b)$, $b > 0$. For $\mathrm{Re}\,\lambda > 0$, the representation of $g(a, b)$, $b < 0$ can be extended.

Since for $t > 0$ the representations $R_\lambda(g)$ and $R_{t\lambda}(g)$ are equivalent, it follows that the irreducible unitary representations $R_{\rho i}(g)$ of the group $G$ are equivalent to one of the following two representations:

$$R_-(g)\varphi(x) = e^{ibx}\varphi(ax)$$

and

$$R_+(g)\varphi(x) = e^{-ibx}\varphi(ax).$$

It can be shown that these representations together with the one-dimensional representations $g(a, b) \to a^{i\rho}$ exhaust, to within equivalence, all irreducible unitary representations of $G$ (see [145]).

3. **Reduction of the operators $R_\lambda(g(0, a))$ to a diagonal form.** In the realization of the representations $R_\lambda(g)$ constructed above (see formula (5) of subsection 2) to elements $g(b, 1)$ of the subgroup $B$ correspond operators of multiplication by $e^{\lambda bx}$. If the values of a function at a point are regarded as the "coordinates' of this function, then the operator of multiplication by a function is the analogue of a diagonal matrix (each "coordinate" is multiplied by a certain number). Therefore one can say that, under the chosen realization of $R_\lambda(g)$, the operators

$R_\lambda(g(b, 1))$ are reduced to a diagonal form.

We now construct another realization of the representations $R_\lambda(g)$, under which operators corresponding to elements $g(0, a)$ of $A$ have diagonal form. To do this we pass from the functions $\phi(x)$ to their Mellin transforms

$$\mathfrak{F}(w) = \int_0^\infty \varphi(x) x^{w-1} dx. \tag{1}$$

It is not difficult to show (cf. §4.3 of Chapter II) that, under this transformation, to the functions $\phi(x)$ of $\mathfrak{D}$ 1) correspond integral analytic functions of $w = u + iv$ which decrease rapidly on the imaginary axis:

$$\lim_{v\to\infty} |v|^n \mathfrak{F}(iv) = 0, \tag{2}$$

and are such that for some $c > 0$ we have the inequality

$$\max_v |\mathfrak{F}(u+iv)| \leqslant |u|^c \max_v |\mathfrak{F}(iv)|. \tag{3}$$

Let us ascertain how the operators $R_\lambda(g)$ are transformed on passing from $\phi(x)$ to $\mathfrak{F}(w)$. We denote by $\mathfrak{F}_g(w)$ the Mellin transform of $R_\lambda(g)\phi(x)$:

$$\mathfrak{F}_g(w) = \int_0^\infty e^{\lambda bx} \varphi(ax) x^{w-1} dx = a^{-w} \int_0^\infty e^{\frac{\lambda bx}{a}} \varphi(x) x^{w-1} dx. \tag{4}$$

For $b = 0$ this formula takes the form

$$\mathfrak{F}_g(w) = a^{-w} \int_0^\infty \varphi(x) x^{w-1} dx = a^{-w}\mathfrak{F}(w). \tag{5}$$

Thus to the elements $g(0, a)$ of $A$ corresponds the operator of multiplication by $a^{-w}$:

$$R_\lambda(g(0, a))\mathfrak{F}(w) = a^{-w}\mathfrak{F}(w). \tag{6}$$

We now clarify the form taken by the operators $R_\lambda(g)$, corresponding to $g(a, b)$ for $b \neq 0$. By the inversion formula for the Mellin transform, we have

$$\varphi(x) = \frac{1}{2\pi i} \int_{c-i\infty}^{c+i\infty} x^{-z}\mathfrak{F}(z)\, dz, \tag{7}$$

where $c$ is any real number. Therefore

$$\mathfrak{F}_g(w) = \frac{a^{-w}}{2\pi i} \int_0^\infty \int_{c-i\infty}^{c+i\infty} e^{\frac{\lambda bx}{a}} x^{w-z-1}\mathfrak{F}(z)\, dz\, dx. \tag{8}$$

Let $b > 0$. Then, for $\operatorname{Re}\lambda < 0$ and $\operatorname{Re} w > c$, the integral (8) is absolutely

1) I.e. to infinitely differentiable finite functions, equal to zero in some neighborhood of the point $x = 0$.

convergent, and the order of integration can be changed. We obtain

$$\mathfrak{F}_g(w) \equiv R_\lambda(g)\mathfrak{F}(w) = \int_{c-i\infty}^{c+i\infty} K(w, z; g)\mathfrak{F}(z)\,dz, \tag{9}$$

where

$$K(w, z; g) = \frac{a^{-w}}{2\pi i}\int_0^\infty e^{\frac{\lambda bx}{a}} x^{w-z-1}\,dx. \tag{10}$$

Utilizing (6) and (10), it is easy to find the infinitesimal operators of $R_\lambda(g)$. Let $g = g(0, e^t)$. Then (6) gives

$$R_\lambda(g)\mathfrak{F}(w) = e^{-tw}\mathfrak{F}(w).$$

Differentiating with respect to $t$ and putting $t = 0$, we get

$$\hat{a}_\lambda\mathfrak{F}(w) \equiv \frac{dR_\lambda(g(0, e^t))}{dt}\bigg|_{t=0}\mathfrak{F}(w) = -w\mathfrak{F}(w). \tag{11}$$

Similarly, the equality

$$R_\lambda(g(t, 1))\mathfrak{F}(w) = \int_0^\infty e^{\lambda tx}\varphi(x)x^{w-1}\,dx$$

gives

$$\hat{b}_\lambda = \frac{dR_\lambda(g(t, 1))}{dt}\bigg|_{t=0}\mathfrak{F}(w) = \lambda\int_0^\infty \varphi(x)x^w\,dx = \lambda\mathfrak{F}(w+1).$$

Therefore the infinitesimal operator corresponding to $B$ has the form

$$\hat{b}_\lambda\mathfrak{F}(w) = \lambda\mathfrak{F}(w+1). \tag{12}$$

4. **Expression of the kernel $K(w, z; g)$ by means of the $\Gamma$-function.** We have obtained an expression of the operators of the representation $R_\lambda(g)$ in the form of integral operators with kernel $K(w, z; g)$. This kernel can be expressed by means of the power function and a special function $\Gamma(z)$, called the gamma function.

We define $\Gamma(z)$ at first only for $\operatorname{Re} z > 0$ by the formula

$$\Gamma(z) = \int_0^\infty e^{-x}x^{z-1}\,dx. \tag{1}$$

Let $\lambda < 0$ and $b > 0$ in the integral (10) of subsection 3. We make the substitution $\lambda bx/a = -t$ in this integral:

$$K(w, z; g) = \frac{a^{-w}\left(-\frac{\lambda b}{a}\right)^{z-w}}{2\pi i}\int_0^\infty e^{-t}t^{w-z-1}\,dt = \frac{\Gamma(w-z)a^{-w}}{2\pi i}\left(-\frac{\lambda b}{a}\right)^{z-w}.$$

Thus, for $\lambda < 0$, $b > 0$, $\mathrm{Re}\, w > \mathrm{Re}\, z$, the kernel $K(w, z; g)$ is expressed by the formula

$$K(w, z; g) = \frac{\Gamma(w - z)\, a^{-w}}{2\pi i} \left(-\frac{\lambda b}{a}\right)^{z-w}. \tag{2}$$

We now extend formula (2) to complex values of $\lambda$ and $b$. For this we observe that, by formula (1), the kernel $K(w, z; g)$ is an analytic function of $-\lambda b$ in the complex plane, cut along the negative semiaxis. On the other hand, it is evident from formula (4) of subsection 3 that $\mathfrak{F}_g(w)$ is also an analytic function of $-\lambda b$. Since we have proved that the equality

$$R_\lambda(g)\mathfrak{F}(w) = \frac{a^{-w}}{2\pi i} \int_{c-i\infty}^{c+i\infty} \Gamma(w - z) \left(-\frac{\lambda b}{a}\right)^{z-w} \mathfrak{F}(z)\, dz \tag{3}$$

holds for $0 < -\lambda b < \infty$, therefore it is valid for all values of $\lambda$ and $b$ such that $-\lambda b$ does not lie on the negative semiaxis.

In particular, one has the formulas

$$R_+(g)\mathfrak{F}(w) = \frac{a^{-w}}{2\pi i} \int_{c-i\infty}^{c+i\infty} \Gamma(w - z) \left(-\frac{bi}{a}\right)^{z-w} \mathfrak{F}(z)\, dz \tag{4}$$

and

$$R_-(g)\mathfrak{F}(w) = \frac{a^{-w}}{2\pi i} \int_{c-i\infty}^{c+i\infty} \Gamma(w - z) \left(\frac{bi}{a}\right)^{z-w} \mathfrak{F}(z)\, dz. \tag{4'}$$

5. Properties of the $\Gamma$-function. The function $\Gamma(z)$ has been defined only for $\mathrm{Re}\, z > 0$. Evidently, in this domain it analytically depends on $z$. We shall define it for $\mathrm{Re}\, z < 0$ with the help of analytic continuation. We shall show that this continuation is uniquely determined. For this we establish the functional equation satisfied by the $\Gamma$-function.

It follows from formulas (12) of subsection 3 and (2) of subsection 4 that the operator $\hat{b}_\lambda R_\lambda(t, 1)$ is an integral operator with the kernel

$$\frac{\lambda\Gamma(w - z + 1)}{2\pi i} (-\lambda t)^{z-w-1}. \tag{1}$$

On the other hand, it follows from the equality

$$\hat{b}_\lambda R_\lambda(t, 1) = \frac{dR_\lambda(s, 1)}{ds} R_\lambda(t, 1)\Big|_{s=0} = \frac{dR_\lambda(s + t, 1)}{ds}\Big|_{s=0}$$

that the kernel of this operator is equal to

$$\left.\frac{dK(w, z; t+s, 1)}{ds}\right|_{s=0}=\frac{\lambda(w-z)\Gamma(w-z)}{2\pi i}(-\lambda t)^{z-w-1}. \tag{2}$$

Comparing (1) and (2), we get

$$\Gamma(w-z+1)=(w-z)\Gamma(w-z).$$

Replacing $w - z$ by $z$ in this equality and noting that $\operatorname{Re}(w - z) > 0$, we get the following result.

*In the halfplane* $\operatorname{Re} z > 0$, *the function* $\Gamma(z)$ *satisfies the functional equation*

$$\Gamma(z+1)=z\Gamma(z). \tag{3}$$

Since

$$\Gamma(1)=\int_0^\infty e^{-t}\,dt=1,$$

it follows from (3) that, for natural values of $n$, we have

$$\Gamma(n+1)=n(n-1)\ldots 1=n!.$$

In other words, $\Gamma(z+1)$ is a generalization of the factorial to nonnatural values of $z$. Clearly, if we put

$$\Gamma(z)=\frac{\Gamma(z+1)}{z}, \tag{4}$$

we get the analytic continuation of $\Gamma(z)$ in the halfplane $\operatorname{Re} z > -1$ (from which the point $z = 0$ is excluded). Similarly, the equality

$$\Gamma(z)=\frac{\Gamma(z+n+1)}{z(z+1)\ldots(z+n)} \tag{5}$$

gives the analytic continuation of $\Gamma(z)$ in the halfplane $\operatorname{Re}(z) > -n-1$ with the exclusion of the points $z = 0, -1, \cdots, -n$.

We show that, at the point $z = -n$, the function $\Gamma(z)$ has a simple pole with the residue

$$\operatorname*{Res}_{z=-n}\Gamma(z)=\frac{(-1)^n}{n!}. \tag{6}$$

Indeed, by formula (5)

$$\lim_{z\to -n}(z+n)\Gamma(z)=\lim_{z\to -n}\frac{\Gamma(z+n+1)}{z(z+1)\ldots(z+n-1)}=\frac{(-1)^n\Gamma(1)}{n!}. \tag{7}$$

But $\Gamma(1) = 1$, and therefore

$$\operatorname*{Res}_{z=-n} \Gamma(z) = \lim_{z \to -n} (z+n)\Gamma(z) = \frac{(-1)^n}{n!}.$$

Let us mention another method of analytically continuing $\Gamma(z)$. We break up integral (1) of subsection 4 into integrals over the segments $[0, 1]$ and $[1, \infty]$:

$$\Gamma(z) = \int_0^1 e^{-x} x^{z-1} dx + \int_1^\infty e^{-x} x^{z-1} dx.$$

Hence

$$\Gamma(z) = \int_1^\infty e^{-x} x^{z-1} dx + \int_0^1 x^{z-1} \left[ e^{-x} - \sum_{k=0}^n \frac{(-1)^k x^k}{k!} \right] dx$$

$$+ \sum_{k=0}^n \frac{(-1)^k}{k!} \int_0^1 x^{k+z-1} dx = \int_1^\infty e^{-x} x^{z-1} dx$$

$$+ \int_0^1 x^{z-1} \left[ e^{-x} - \sum_{k=0}^n \frac{(-1)^k x^k}{k!} \right] dx + \sum_{k=0}^n \frac{(-1)^k}{k!\,(z+k)}. \tag{8}$$

It is evident that the integrals in the right-hand side are convergent for $\operatorname{Re} z > -n-1$, and therefore (8) gives the analytical continuation of $\Gamma(z)$ in the half-plane $\operatorname{Re} z > -n-1$. The analytic contination of $\Gamma(z)$ to the whole plane is given by the formula

$$\Gamma(z) = \int_1^\infty e^{-x} x^{z-1} dx + \sum_{k=0}^\infty \frac{(-1)^k}{k!\,(z+k)}. \tag{9}$$

Thus we have proved that $\Gamma(z)$ is an analytic function in the whole complex plane, with the exception of the points $z = 0, -1, \cdots, -n, \cdots$, where it has simple poles with the residues

$$\operatorname*{Res}_{z=-n} \Gamma(z) = \frac{(-1)^n}{n!}.$$

We shall now show that if $c > 0$, then the function $\Gamma(c+it)$ rapidly decreases for $|t| \to \infty$, i.e. for any $n > 0$

$$\lim_{|t| \to \infty} |t|^n \Gamma(c+it) = 0. \tag{10}$$

For this we put $z = c + it$ and $x = e^s$ in the integral (1) of subsection 4. We obtain

$$\Gamma(c+it)=\int_{-\infty}^{\infty} e^{-e^s+cs}e^{its}ds. \tag{11}$$

Thus $\Gamma(c+it)$ is the Fourier transform of the function $e^{-e^s+cs}$. But this function is infinitely differentiable and for $|s|\to\infty$ it decreases rapidly together with all its derivatives. In other words, $e^{-e^s+cs}$ is a function of the space $\mathfrak{S}$ (see §3.2 of Chapter II). But then, according to §4.1 of Chapter II, the Fourier transform $\Gamma(c+it)$ of this function also belongs to $\mathfrak{S}$ and hence decreases rapidly for $|t|\to\infty$.

6. **The composition theorem for the $\Gamma$-function and its corollaries.** We shall now derive the composition theorem for the $\Gamma$-function. For this we utilize the equality

$$g(b,\ 1)g(1,\ 1)=g(b+1,\ 1). \tag{1}$$

It follows from (1) that

$$R_\lambda(b,\ 1)R_\lambda(1,\ 1)=R_\lambda(b+1,\ 1) \tag{2}$$

and so, for any function $\mathfrak{F}(z)$ of $\mathfrak{S}$ we have (for $\lambda=-1$)

$$\frac{1}{2\pi i}\int_{c_1-i\infty}^{c_1+i\infty}\Gamma(w-u)b^{u-w}du\int_{c-i\infty}^{c+i\infty}\Gamma(u-z)\mathfrak{F}(z)dz = \int_{c-i\infty}^{c+i\infty}\Gamma(w-z)(b+1)^{z-w}\mathfrak{F}(z)dz, \tag{3}$$

where $\operatorname{Re} w>c_1>c$. Because of the rapid decrease of $\Gamma(c-z+it)$ and $\Gamma(w-c-it)$ for $|t|\to\infty$, proved above, one can change the order of integration and obtain

$$\int_{c-i\infty}^{c+i\infty}\Gamma(w-z)(b+1)^{z-w}\mathfrak{F}(z)dz = \frac{1}{2\pi i}\int_{c-i\infty}^{c+i\infty}\mathfrak{F}(z)dz\int_{c_1-i\infty}^{c_1+i\infty}\Gamma(w-u)\Gamma(u-z)b^{u-w}du. \tag{4}$$

Since (4) holds for all functions $\mathfrak{F}(z)$, it implies that

$$\Gamma(w-z)\left(\frac{b+1}{b}\right)^{z-w}=\frac{1}{2\pi i}\int_{c_1-i\infty}^{c_1+i\infty}\Gamma(w-u)\Gamma(u-z)b^{u-z}du, \tag{5}$$

where $\operatorname{Re} w>c_1>\operatorname{Re} z$. On replacing $b$ by $t$, $u-z$ by $u$ and $w-z$ by $w$, we obtain

$$\Gamma(w)\left(\frac{t}{t+1}\right)^w=\frac{1}{2\pi i}\int_{c_1-i\infty}^{c_1+i\infty}\Gamma(w-u)\Gamma(u)t^u du, \tag{5'}$$

where $\operatorname{Re} w > c_1 > 0$. We shall call this the composition formula for the $\Gamma$-function. For $t = 1$ we get the following special case of the composition formula:

$$2^{-w}\Gamma(w)=\frac{1}{2\pi i}\int_{c_1-i\infty}^{c_1+i\infty}\Gamma(w-u)\Gamma(u)\,du. \tag{6}$$

On replacing $u$ by $-u$ in formula (5′), we can see that the function $\Gamma(w)\,(t/(t+1)^w)$ is the inverse Mellin transform for the function $\Gamma(w+u)\Gamma(-u)$. Therefore, owing to the dual Mellin transformation, we have

$$\frac{\Gamma(w+u)\Gamma(-u)}{\Gamma(w)}=\int_0^\infty t^{w+u-1}(1+t)^{-w}\,dt, \tag{7}$$

where $\operatorname{Re} u < 0$, $\operatorname{Re}(w+u) > 0$.

Let us put $w = 1$ in formula (7). As $\Gamma(1) = 1$, we get

$$\Gamma(1+u)\Gamma(-u)=\int_0^\infty t^u(1+t)^{-1}\,dt, \tag{8}$$

where $-1 < \operatorname{Re} u < 0$. We calculate the integral on the right-hand side of (8). To do this we make the substitution $t = x^2$ in it; we obtain

$$\Gamma(1+u)\Gamma(-u)=2\int_0^\infty \frac{x^{2u+1}}{1+x^2}\,dx.$$

We replace integration over $0 < x < \infty$ by integration over the whole real axis $-\infty < x < \infty$, and we go round the point $x = 0$ in the upper half plane. Then, on the negative semiaxis we have $x^{2u+1} = |x|^{2u+1}e^{(2u+1)\pi i}$, and so

$$\int_{-\infty}^\infty \frac{x^{2u+1}dx}{1+x^2}=(1-e^{2u\pi i})\int_0^\infty \frac{x^{2u+1}}{1+x^2}\,dx.$$

Consequently,

$$\Gamma(1+u)\Gamma(-u)=\frac{2}{1-e^{2u\pi i}}\int_{-\infty}^\infty \frac{x^{2u+1}dx}{1+x^2}, \tag{9}$$

where $-1 < \operatorname{Re} u < 0$, and the point $x = 0$ is by-passed in the upper half plane.

Let us add to the real axis a semicircle of infinitely large radius in the upper half plane. The integral over this semicircle gives zero contribution, since $\operatorname{Re} u < 0$. The contour consisting of this semicircle and the real axis encloses the singular point $x = e^{\pi i/2}$ of the integrand. Hence, employing the Cauchy and Euler formulas, we get

$$\Gamma(1+u)\Gamma(-u)=\frac{2\pi e^{u\pi i}}{1-e^{2u\pi i}}=-\frac{\pi}{\sin \pi u}.$$

In other words,

$$\Gamma(x)\Gamma(1-x)=\frac{\pi}{\sin \pi x}. \tag{10}$$

Formula (10) is called the complementation formula for the gamma function. Putting $x = \frac{1}{2}$ in formula (10), we find that $\Gamma^2(\frac{1}{2}) = \pi$ and therefore

$$\Gamma\left(\frac{1}{2}\right)=\sqrt{\pi}. \tag{11}$$

Further, putting $x = \frac{1}{2} + it$ in (10), we get

$$\left|\Gamma\left(\frac{1}{2}+it\right)\right|^2=\frac{\pi}{\sin\left(\frac{\pi}{2}+it\pi\right)}=\frac{\pi}{\operatorname{ch} t\pi}. \tag{12}$$

7. **Beta function and the duplication formula for $\Gamma(x)$.** Closely related to $\Gamma(x)$ is a function $B(x, y)$, called the *beta function*. It is defined by the equality

$$B(x, y)=\int_0^1 t^{x-1}(1-t)^{y-1}dt, \qquad \begin{cases} \operatorname{Re} x>0, \\ \operatorname{Re} y>0. \end{cases} \tag{1}$$

In order to establish the connection between $\Gamma(x)$ and $B(x, y)$, we make the substitution $t/(1 + t) = s$ in the integral (7) of subsection 6. We get

$$\frac{\Gamma(w+u)\,\Gamma(-u)}{\Gamma(w)}=\int_0^1 s^{u+w-1}(1-s)^{-u-1}\,ds. \tag{2}$$

Comparing (1) and (2), we see that

$$B(u+w, -u)=\frac{\Gamma(u+w)\,\Gamma(-u)}{\Gamma(w)}$$

or, in other words,

$$B(x, y) = \frac{\Gamma(x)\Gamma(y)}{\Gamma(x+y)}, \tag{3}$$

To derive the duplication formula for $\Gamma(x)$ we put $x = y$ in formula (3). We obtain

$$\frac{\Gamma^2(x)}{\Gamma(2x)} = B(x, x) = \int_0^1 t^{x-1}(1-t)^{x-1}dt = \int_0^1 \left[\frac{1}{4} - \left(\frac{1}{2} - t\right)^2\right]^{x-1} dt$$

$$= \int_{-\frac{1}{2}}^{\frac{1}{2}} \left(\frac{1}{4} - u^2\right)^{x-1} du = 2\int_0^{\frac{1}{2}} \left(\frac{1}{4} - u^2\right)^{x-1} du.$$

We make the substitution $4u^2 = t$. This gives

$$\frac{\Gamma^2(x)}{\Gamma(2x)} = 2^{-2x+1}\int_0^1 (1-t)^{x-1}t^{-\frac{1}{2}}\,dt$$

$$= 2^{-2x+1} B\left(x, \frac{1}{2}\right) = \frac{2^{-2x+1}\Gamma(x)\,\Gamma\left(\frac{1}{2}\right)}{\Gamma\left(x + \frac{1}{2}\right)}.$$

As $\Gamma(\frac{1}{2}) = \sqrt{\pi}$, it follows from this equality that

$$\Gamma(2x) = \frac{2^{2x-1}\Gamma(x)\,\Gamma\left(x + \frac{1}{2}\right)}{\sqrt{\pi}}. \tag{4}$$

8. **The Fourier transform of the functions $x_+^u$ and $x_-^u$.** We have defined $\Gamma(u)$ for $\operatorname{Re} u > 0$ by the formula

$$\Gamma(u) = \int_0^\infty e^{-x}x^{u-1}\,dx, \tag{1}$$

whence we immediately find that, for $t > 0$,

$$\int_0^\infty e^{-tx}x^{u-1}dx = \Gamma(u)\,t^{-u}. \tag{2}$$

We shall now prove that this formula remains valid for complex values of $t$ such that $\operatorname{Re} t > 0$. To do this we make the substitution $tx = z$ in (2). If $\arg t = \alpha$, we get

$$\int_0^\infty e^{-tx}x^{u-1}\,dx = t^{-u}\int_0^{\infty e^{-i\alpha}} e^{-z}z^{u-1}dz.$$

The integrand has no singularity in the sector bounded by the axis $Ox$ and the ray $(0, \infty e^{-i\alpha})$. Therefore

$$\int_L e^{-z}z^{u-1}dz = 0,$$

where $L$ denotes the contour depicted in Figure 3. It is easy to verify that if $-\pi/2 < \alpha < \pi/2$, then as $\epsilon \to 0$ the integral over the arc $AB$ tends to zero, and as $R \to \infty$ the integral over the arc $CD$ tends to zero. It follows that, for $-\pi/2 < \alpha < \pi/2$,

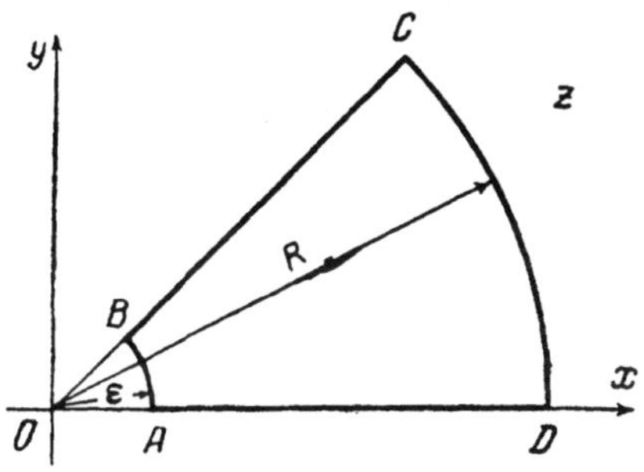

Figure 3

$$\int_0^{\infty e^{-i\alpha}} e^{-z}z^{u-1}dz = \int_0^\infty e^{-x}x^{u-1}dx = \Gamma(u),$$

and hence, for $-\pi/2 < \arg t < \pi/2$,

$$\int_0^\infty e^{-tx}x^{u-1}dx = \Gamma(u)\,t^{-u}.$$

The validity of formula (2) is thereby proved under the condition $-\pi/2 < \arg t < \pi/2$.

Formula (2) remains valid for $\arg t = \pm\pi/2$ also; if $0 < \operatorname{Re} u < 1$, then

$$\int_0^\infty e^{-itx}x^{u-1}dx = \Gamma(u)\,t^{-u}e^{-\frac{\pi u i}{2}} \tag{3}$$

and

$$\int_0^\infty e^{itx}x^{u-1}dx = \Gamma(u)\,t^{-u}e^{\frac{\pi u i}{2}}. \tag{3'}$$

Formulas (3) and (3′) can be regarded as the Fourier transform formulas for the functions $x_-^{u-1}$ and $x_+^{u-1}$, defined by the equalities

$$x_-^{u-1} = \begin{cases} 0 & \text{for } x > 0, \\ |x|^{u-1} & \text{for } x < 0 \end{cases}$$

and

$$x_+^{u-1} = \begin{cases} x^{u-1} & \text{for } x > 0, \\ 0 & \text{for } x < 0. \end{cases}$$

9. **Representations of the group of linear transformations of the straight line, induced by one-dimensional representations of the subgroup $A$.** In subsection 2 we considered irreducible representations of the group $G$, induced by one-dimensional representations of the subgroup $B$ of elements of the form $g(b, 1)$. Let us now consider irreducible representations of this group induced by one-dimensional representations of the subgroup $A$ of elements of the form $g(0, a)$. These one-dimensional representations have the form $g(0, a) \to a^{\sigma}$. Therefore the induced representations are realized in the space $\mathfrak{H}^{\sigma}$ of functions $f(g)$ on $G$ such that

$$f(g(0, a_0)g(b, a)) = a_0^{\sigma} f(g(b, a)). \tag{1}$$

The representation operator is given by the formula

$$R^{\sigma}(g_0) f(g) = f(gg_0). \tag{2}$$

Since $g(0, a_0)g(b, a) = g(a_0, b, a_0, a)$, it follows from (1) that the functions $f(g)$ of $\mathfrak{H}^{\sigma}$ have the form

$$f(g(b, a)) = f\left(g(0, a)\, g\left(\frac{b}{a}, 1\right)\right) = a^{\sigma}\varphi\left(\frac{b}{a}\right), \tag{3}$$

where $\phi(x) \equiv f(g(x, 1))$ is a function on the real line. We clarify how the functions $\phi(x)$ transform under the shifts $g \to gg_0$. Let $\phi(x) = f(g(x, 1))$. We denote by $\phi_{g_0}(x)$ the function corresponding to $f(gg_0)$. Since $g(x, 1)g(b_0, a_0) = g(x + b_0, a_0)$, we have

$$\varphi_{g_0}(x) = a_0^{\sigma} f\left(g\left(\frac{x+b_0}{a_0}, 1\right)\right) = a_0^{\sigma}\varphi\left(\frac{x+b_0}{a_0}\right).$$

Thus the operators $R^{\sigma}(g)$ are realized in the space $\mathfrak{H}$ of functions $\phi(x)$ in the form

$$R^{\sigma}(g)\varphi(x) = a^{\sigma}\varphi\left(\frac{x+b}{a}\right), \quad g = g(b, a). \tag{4}$$

We decompose the representation $R^{\sigma}(g)$ into irreducible components. For this we use the Fourier transformation

$$F(\lambda) = \int_{-\infty}^{\infty} \varphi(x) e^{i\lambda x} dx. \tag{5}$$

Let $F_g(\lambda)$ denote the Fourier transform of the function $R^{\sigma}(g)\phi(x)$. It follows from (4) that

$$F_g(\lambda) = a^{\sigma} \int_{-\infty}^{\infty} \varphi\left(\frac{x+b}{a}\right) e^{i\lambda x} dx = a^{\sigma+1} e^{-i\lambda b} \int_{-\infty}^{\infty} \varphi(x) e^{i\lambda a x} dx = a^{\sigma+1} e^{-i\lambda b} F(a\lambda).$$

Thus under Fourier transformation the operators $R^{\sigma}(g)$ turn into the operators

$$R^{\sigma}(g) F(\lambda) = a^{\sigma+1} e^{-i\lambda b} F(a\lambda). \tag{6}$$

Putting $F(\lambda) = \lambda^{\sigma+1}\Phi(\lambda)$, we can write (6) as follows:

$$R^{\sigma}(g) \Phi(\lambda) = e^{-i\lambda b} \Phi(a\lambda). \tag{7}$$

The representation (7) is reducible. Namely, operators (7) leave invariant the subspaces $\mathfrak{H}_+$ and $\mathfrak{H}_-$, consisting of those functions $\Phi(\lambda)$ which vanish for $\lambda > 0$ and $\lambda < 0$ respectively. On $\mathfrak{H}_+$ the representation (7) coincides with the irreducible unitary representation $R_+(g)$ constructed in subsection 2, and on $\mathfrak{H}_-$ it coincides with the irreducible unitary representation $R_-(g)$.

We have thus proved that representations of $G$, induced by one-dimensional representations of $A$, decompose into the direct sum of irreducible unitary representations (and consequently all these representations are equivalent to each other). Utilizing these results, it is easy to decompose the regular representations of $G$. Namely, let $f(b, a) \equiv f(g)$ be a function on $G$. We denote by $f^{\sigma}(g)$ the function

$$f^{\sigma}(g) = \int_{0}^{\infty} f(tb, ta) t^{-\sigma-1} dt. \tag{8}$$

We show that this function belongs to $\mathfrak{H}^{\sigma}$, i.e. we show that

$$f^{\sigma}(g(0, a_0) g(b, a)) = a_0^{\sigma} f^{\sigma}(g). \tag{9}$$

Indeed, $g(0, a_0)g(b, a) = g(a_0 b, a_0 a)$, and therefore

$$f^{\sigma}(g(0, a_0) g(b, a)) = \int_{0}^{\infty} f(ta_0 b, ta_0 a) t^{-\sigma-1} dt = a_0^{\sigma} \int_{0}^{\infty} f(tb, ta) t^{-\sigma-1} dt = a_0^{\sigma} f^{\sigma}(g).$$

To the operator

$$R(g_0) f(g) = f(gg_0) \tag{10}$$

of regular representation corresponds the operator

$$R^\sigma(g_0) f^\sigma(g) = f^\sigma(gg_0) \tag{11}$$

in the space of functions $f^\sigma(g)$.

We have seen above that $R^\sigma(g)$ decomposes into the direct sum of two irreducible unitary representations, equivalent to the representations $R_+(g)$ and $R_-(g)$. In order to complete the decomposition of the regular representation we have only to express $f(g)$ in terms of its components $f^\sigma(g)$. This is done with the help of the inversion formula for the Mellin transformation. Namely, it follows from (8) that

$$f(tb, ta) = \frac{1}{2\pi i} \int_{c-i\infty}^{c+i\infty} f^\sigma(b, a) t^\sigma d\sigma.$$

Putting $t = 1$ here, we find that

$$f(b, a) = \frac{1}{2\pi i} \int_{c-i\infty}^{c+i\infty} f^\sigma(b, a) d\sigma. \tag{12}$$

Thus we have proved that the regular representation of the group $G$ of linear transformations of the straight line decomposes into the continuous direct sum of representations $R^\sigma(g)$, each of which is the direct sum of two irreducible unitary representations of $G$.

## §2. THE GROUP $MH(2)$ OF MOTIONS OF THE PSEUDO-EUCLIDEAN PLANE

1. **The pseudo-euclidean plane.** By the pseudo-euclidean plane we mean the two-dimensional real linear space in which the following indefinite bilinear form is given:

$$[\mathbf{x}, \mathbf{y}] = x_1 y_1 - x_2 y_2, \quad \mathbf{x} = (x_1, x_2), \quad \mathbf{y} = (y_1, y_2). \tag{1}$$

The distance $r$ between the points $M(x_1, x_2)$ and $N(y_1, y_2)$ of the pseudo-euclidean plane is defined by the formula

$$r^2 = (x_1 - y_1)^2 - (x_2 - y_2)^2 \equiv [\mathbf{x} - \mathbf{y}, \mathbf{x} - \mathbf{y}]. \tag{2}$$

This formula determines the distance, apart from the sign.

In contrast to the case of the euclidean plane, the distance between two points on the pseudo-euclidean plane may be not only real, but also purely imaginary. In particular, the distance between two different points $M(x_1, x_2)$ and $N(y_1, y_2)$ of the pseudo-euclidean plane can be equal to zero. This will be the case if

$$x_1 - y_1 = \pm (x_2 - y_2).$$

We introduce an analogue of the polar coordinate system on the pseudo-euclidean plane. The lines $x_1 = x_2$ and $x_1 = -x_2$ whose points are at zero distance from the origin, divide the plane into four quadrants. In the first of these we have $-x_1 < x_2 < x_1$, and so $r^2 = x_1^2 - x_2^2 > 0$, $x_1 > 0$. We shall assume that in this quad rant $r > 0$. Since $r^2 = x_1^2 - x_2^2$, we can take

$$\left.\begin{aligned} x_1 &= r \operatorname{ch} \theta, \\ x_2 &= r \operatorname{sh} \theta. \end{aligned}\right\} \tag{3}$$

when $r$ varies from 0 to $\infty$, and $\theta$ from $-\infty$ to $\infty$, the point $(x_1, x_2)$ traverses the whole quadrant. Similarly, in the second quadrant $-x_2 < x_1 < x_2$ and $r^2 = x_1^2 - x_2^2 < 0$. Therefore in it we can take

$$\left.\begin{aligned} x_1 &= -ir \operatorname{sh} \theta, \\ x_2 &= -ir \operatorname{ch} \theta, \end{aligned}\right\} \tag{3'}$$

where $r \equiv i|r|$ is a purely imaginary number. Similarly, in the third quadrant $x_1 < x_2 < -x_1$. We have $r^2 = x_1^2 - x_2^2 > 0$, $x_1 < 0$, and we take

$$\left.\begin{aligned} x_1 &= -r \operatorname{ch} \theta, \\ x_2 &= -r \operatorname{sh} \theta, \end{aligned}\right\} \tag{3''}$$

and in the fourth, $x_2 < x_1 < -x_2$; we have $r^2 = x_1^2 - x_2^2 < 0$, $x_2 < 0$:

$$\left.\begin{aligned} x_1 &= -ir \operatorname{sh} \theta, \\ x_2 &= ir \operatorname{ch} \theta. \end{aligned}\right\} \tag{3'''}$$

It is clear that the specifying of a real or purely imaginary number $r$ and a real number $\theta$ uniquely determines a point of the pseudo-euclidean plane.

2. The group $MH(2)$. By a motion of the pseudo-euclidean plane we mean a nonhomogeneous linear transformation of two variables, which does not change orientations, preserves the distance between points of this plane, and transforms points of one of the quadrants into points of the same quadrant. Fvidently, motions of the pseudo-euclidean plane form a group. We shall denote it by $MH(2)$. As in the case of the euclidean plane, one establishes that such transformations are given by the formulas

$$\left.\begin{aligned} x_1' &= x_1 \operatorname{ch} \varphi + x_2 \operatorname{sh} \varphi + a_1, \\ x_2' &= x_1 \operatorname{sh} \varphi + x_2 \operatorname{ch} \varphi + a_2, \end{aligned}\right\} \tag{1}$$

where

$$\left.\begin{aligned} -\infty &< a_1 < \infty, \\ -\infty &< a_2 < \infty, \\ -\infty &< \varphi < \infty. \end{aligned}\right\} \tag{2}$$

Thus each motion $g$ of the pseudo-euclidean plane is given by three real numbers

$a_1$, $a_2$, $\phi$. We shall denote such a motion by $g = g(a_1, a_2, \phi)$.

It follows at once from formula (1) that

$$g(a_1, a_2, \varphi)g(b_1, b_2, \psi) = g(c_1, c_2, \theta), \tag{3}$$

where

$$c_1 = b_1 \operatorname{ch} \varphi + b_2 \operatorname{sh} \varphi + a_1, \tag{4}$$
$$c_2 = b_1 \operatorname{sh} \varphi + b_2 \operatorname{ch} \varphi + a_2, \tag{4'}$$
$$\theta = \varphi + \psi. \tag{4''}$$

These formulas can be interpreted as follows: we denote by **a** the vector $(a_1, a_2)$ and by $\mathbf{a}_\phi$ the vector into which **a** is transformed through a hyperbolic rotation by the "angle" $\phi$:

$$\mathbf{a}_\varphi = (a_1 \operatorname{ch} \varphi + a_2 \operatorname{sh} \varphi,\ a_1 \operatorname{sh} \varphi + a_2 \operatorname{ch} \varphi).$$

Then

$$g(\mathbf{a}, \varphi)g(\mathbf{b}, \psi) = g(\mathbf{a} + \mathbf{b}_\varphi, \varphi + \psi). \tag{5}$$

It is easy to see that the parameters $a_1 = a_2 = \phi = 0$ give the identity motion and that the motion inverse to $g(\mathbf{a}, \phi)$ is $g(-\mathbf{a}_{-\phi}, -\phi)$.

The group $MH(2)$ can be realized as a group of third order matrices. For this, with each motion of the form (1) we associate the matrix

$$A(g) = \begin{pmatrix} \operatorname{ch} \varphi & \operatorname{sh} \varphi & a_1 \\ \operatorname{sh} \varphi & \operatorname{ch} \varphi & a_2 \\ 0 & 0 & 1 \end{pmatrix}. \tag{6}$$

A direct verification shows that this correspondence is a representation of $MH(2)$, i.e. we have

$$A(g_1 g_2) = A(g_1) A(g_2). \tag{7}$$

The representation $A(g)$ is reducible but not completely reducible.

We mention two subgroups of $MH(2)$. The first of these consists of motions of the pseudo-euclidean plane which leave the point $(0, 0)$ fixed. These motions are called *hyperbolic rotations*. They are given by the formulas

$$x_1' = x_1 \operatorname{ch} \varphi + x_2 \operatorname{sh} \varphi,$$
$$x_2' = x_1 \operatorname{sh} \varphi + x_2 \operatorname{ch} \varphi. \tag{8}$$

In other words, for hyperbolic rotations we have $g = g(0, 0, \phi)$. Under hyperbolic rotations, every point of the pseudo-euclidean plane is moved along the line $r^2 = \text{const}$, i.e. along the hyperbola $x_1^2 - x_2^2 = r^2$. If $r^2 > 0$, then points of the hyperbola have the coordinates $x_1 = r \operatorname{ch} \theta$, $x_2 = r \operatorname{sh} \theta$. It is evident from (8) that a hyperbolic rotation transforms the point $M(x_1, x_2)$ into the point $N(x_1', x_2')$, where

$$\left.\begin{aligned} x_1' &= r \operatorname{ch}(\varphi + \theta), \\ x_2' &= r \operatorname{sh}(\varphi + \theta). \end{aligned}\right\} \tag{9}$$

Hence the measure $d\theta$ on the hyperbola is invariant for hyperbolic rotations. We shall denote the subgroup of hyperbolic rotations by $\Omega$.

The second subgroup of interest to us consists of parallel translations:

$$\left.\begin{aligned} x_1' &= x_1 + a_1, \\ x_2' &= x_2 + a_2. \end{aligned}\right\} \tag{10}$$

For parallel translations we have $g = g(a_1, a_2, 0)$.

The elements of the subgroup $A$ of parallel translations are in one-to-one correspondence with points of the pseudo-euclidean plane. Therefore elements of the subgroup $\Omega$ of hyperbolic rotations can be regarded as automorphisms of $A$. The equality

$$g(\mathbf{a}, \varphi)\, g(\mathbf{b}, \psi) = g(\mathbf{a} + \mathbf{b}_\varphi, \varphi + \psi)$$

shows that $MH(2)$ is the cross product of $B$ and the group $\Omega$ of its automorphisms.

3. **Parametrization of $MH(2)$.** Every element $g$ of $MH(2)$ can be specified by three real numbers $\phi$, $a_1$, $a_2$ which vary from $-\infty$ to $\infty$. In many cases it is more convenient to use another parametrization of this group, based on the following proposition.

*Every element $g = g(a_1, a_2, \phi)$ of $MH(2)$ can be represented in the form*

$$g = g(0, 0, \psi)\, h g(0, 0, \varphi - \psi), \tag{1}$$

*where $h$ is a (parallel) translation of one of the following forms:*

$$\left.\begin{aligned} g &= g(\pm r, 0, 0), \quad r > 0, \\ g &= g(0, \pm r, 0), \quad r > 0, \end{aligned}\right\} \tag{2}$$

*or*

$$g = (\pm 1, \pm 1, 0) \tag{3}$$

*(in the last case, all four sign combinations are possible).*

In fact, let us suppose, for example, that $-a_1 < a_2 < a_1$. We put

$$a_1 = r \operatorname{ch} \psi,$$
$$a_2 = r \operatorname{sh} \psi,$$

where $r^2 = a_1^2 - a_2^2$, $r > 0$. Then

$$g(0, 0, \psi) g(r, 0, 0) = g(r \operatorname{ch} \psi, r \operatorname{sh} \psi, \psi) \equiv g(\mathbf{a}, \psi),$$

and therefore

$$g(0, 0, \psi) g(r, 0, 0) g(0, 0, \varphi - \psi)$$
$$= g(\mathbf{a}, \psi) g(\mathbf{0}, \varphi - \psi) = g(\mathbf{a}, \varphi) \equiv g(a_1, a_2, \varphi).$$

For the remaining cases, the proof is similar. In particular, for $r > 0$,

$$g(\pm r, \pm r, \varphi) = g(0, 0, \psi) g(\pm 1, \pm 1, 0) g(0, 0, \varphi - \psi), \tag{4}$$

where $e^{\psi} = r$.

This parametrization of $MH(2)$ is analogous to the parametrization of $M(2)$ considered in §1.2 of Chapter IV.

Motions of the form

$$g(0, 0, \psi) g(r, 0, 0) g(0, 0, \varphi - \psi), \qquad r > 0, \tag{5}$$

transform points of the quadrant $-x_1 < x_2 < x_1$ into points of the same quadrant. Consequently, these motions form a semigroup in the form $MH(2)$.[1)] A similar assertion holds for motions of the form

$$g(0, 0, \psi) g(-r, 0, 0) g(0, 0, \varphi - \psi) \tag{6}$$

etc.

Each of these semigroups consists of motions which transform points of some quadrant into points of the same quadrant.

4. The Lie algebra of $MH(2)$. Let us construct the Lie algebra of $MH(2)$. As the basis of this algebra we choose tangent matrices to the one-parametric subgroups $\Omega_1$, $\Omega_2$, $\Omega_3$, consisting of matrices of the form

$$\omega_1(t) = \begin{pmatrix} 1 & 0 & t \\ 0 & 1 & 0 \\ 0 & 0 & 1 \end{pmatrix}, \tag{1}$$

$$\omega_2(t) = \begin{pmatrix} 1 & 0 & 0 \\ 0 & 1 & t \\ 0 & 0 & 1 \end{pmatrix} \tag{2}$$

---

1) A set of elements of a group forms a semigroup if, together with any two elements, their product also belongs to it.

and

$$\omega_3(t)=\begin{pmatrix} \operatorname{ch} t & \operatorname{sh} t & 0 \\ \operatorname{sh} t & \operatorname{ch} t & 0 \\ 0 & 0 & 1 \end{pmatrix}. \tag{3}$$

We have

$$a_1=\frac{d\omega_1(t)}{dt}\bigg|_{t=0}=\begin{pmatrix} 0 & 0 & 1 \\ 0 & 0 & 0 \\ 0 & 0 & 0 \end{pmatrix}. \tag{4}$$

In just the same way, we find

$$a_2=\begin{pmatrix} 0 & 0 & 0 \\ 0 & 0 & 1 \\ 0 & 0 & 0 \end{pmatrix} \tag{5}$$

and

$$a_3=\begin{pmatrix} 0 & 1 & 0 \\ 1 & 0 & 0 \\ 0 & 0 & 0 \end{pmatrix}. \tag{6}$$

Commutation relations for $a_1$, $a_2$, $a_3$ have the form

$$[a_1, a_2]=0, \tag{7}$$

$$[a_2, a_3]=-a_1, \tag{7'}$$

$$[a_3, a_1]=a_2. \tag{7''}$$

In many cases it is more convenient to use, instead of $a_1$ and $a_2$, their linear combinations:

$$h_+=a_1-a_2=\begin{pmatrix} 0 & 0 & 1 \\ 0 & 0 & -1 \\ 0 & 0 & 0 \end{pmatrix} \tag{8}$$

and

$$h_-=a_1+a_2=\begin{pmatrix} 0 & 0 & 1 \\ 0 & 0 & 1 \\ 0 & 0 & 0 \end{pmatrix}. \tag{8'}$$

They are tangent matrices to one-parametric subgroups $\Xi_+$ and $\Xi_-$, which consist, respectively, of the matrices

$$\xi_+(t)=\begin{pmatrix} 1 & 0 & t \\ 0 & 1 & -t \\ 0 & 0 & 1 \end{pmatrix} \tag{9}$$

and

$$\xi_{-}(t)=\begin{pmatrix}1 & 0 & t\\ 0 & 1 & t\\ 0 & 0 & 1\end{pmatrix}. \tag{9'}$$

The commutation relations for matrices $h_+$, $h_-$, $a_3$ have the form

$$[h_+, h_-]=0, \tag{10}$$
$$[h_+, a_3]=h_+, \tag{10'}$$
$$[h_-, a_3]=h_-. \tag{10''}$$

## §3. REPRESENTATIONS OF $MH(2)$

1. **Irreducible representations.** The irreducible representations of $MH(2)$ are analogous to representations of the group $M(2)$ of motions of the euclidean plane, which are described in §2.1 of Chapter IV. Each such representation is given by a nonzero complex number $R$. A representation $T_R(g)$ of $MH(2)$ is constructed in the space of infinitely differentiable finite numbers $f(\mathbf{x})$ on the hyperbola $x_1^2 - x_2^2 = 1$. Let $g = g(a_1, a_2, \phi)$ be an element of $MH(2)$. We denote by $h(-\phi)$ the hyperbolic rotation

$$\left.\begin{aligned} x_1' &= x_1 \operatorname{ch}\varphi - x_2 \operatorname{sh}\varphi,\\ x_2' &= -x_1 \operatorname{sh}\varphi + x_2 \operatorname{ch}\varphi \end{aligned}\right\} \tag{1}$$

and by a the vector $(a_1, a_2)$. The operator of the representation $T_R(g)$ is then given by the formula

$$T_R(g) f(\mathbf{x}) = e^{-R[\mathbf{a},\, \mathbf{x}]} f(h(-\varphi)\mathbf{x}) \tag{2}$$

(we may recall that the hyperbolic rotation $h(-\phi)$ carries points of the hyperbola into points of the same hyperbola).

It is easily verified that

$$T_R(g_1)\, T_R(g_2) = T_R(g_1 g_2), \tag{3}$$

and so $T_R(g)$ is a representation of $MH(2)$. Points of the hyperbola $x_1^2 - x_2^2 - 1$ can be written in the form $x_1 = \operatorname{ch}\theta$, $x_2 = \operatorname{sh}\theta$. With each function $f(\mathbf{x})$ on the hyperbola we associate a function

$$\hat{f}(\theta) = f(\operatorname{ch}\theta, \operatorname{sh}\theta) \tag{4}$$

of the real variable $\theta$. In the space of these functions $T_R(g)$ is given as follows:

$$T_R(g)\hat{f}(\theta) = e^{R(-a_1 \operatorname{ch}\theta + a_2 \operatorname{sh}\theta)} \hat{f}(\theta - \varphi). \tag{5}$$

We introduce a scalar product in the space of functions $\hat{f}(\theta)$ by putting

$$(\hat{f}_1, \hat{f}_2) = \int_{-\infty}^{\infty} \hat{f}_1(\theta)\overline{\hat{f}_2(\theta)}\, d\theta. \tag{6}$$

Evidently, $T_R(g)$ is unitary relative to this scalar product if and only if $R$ is a purely imaginary number.

2. Another realization of the representations $T_R(g)$ of $MH(2)$. We describe another realization of the representations $T_R(g)$ of $MH(2)$. Under this realization, corresponding to hyperbolic rotations we shall have operators of multiplication by a function. It is constructed in the following way.

With each function

$$\hat{f}(\theta) = f(\operatorname{ch}\theta,\ \operatorname{sh}\theta) \tag{1}$$

of the representation space we associate its Fourier transform $F(\lambda)$:

$$F(\lambda) = \int_{-\infty}^{\infty} \hat{f}(\theta)\, e^{\lambda\theta}\, d\theta. \tag{2}$$

This integral converges for all complex values of $\lambda$, since the functions $\hat{f}(\theta)$ have been taken to be finite and infinitely differentiable. Moreover, the function $F(\lambda)$ is analytic in the whole complex domain, satisfies an inequality of the form

$$|F(\lambda_1 + i\lambda_2)| < Ce^{a|\lambda_1|} \tag{3}$$

and decreases rapidly on every straight line parallel to the imaginary axis (see §4.1 of Chapter II). As we showed there, the inversion formula has the form

$$\hat{f}(\theta) = \frac{1}{2\pi i}\int_{a-i\infty}^{a+i\infty} F(\mu)\, e^{-\mu\theta}\, d\mu. \tag{4}$$

We shall denote by $Q_R(g)$ the operator in the space of functions $F(\lambda)$ corresponding to the operator $T_R(g)$ in the space of functions $\hat{f}(\theta)$. The operators $Q_R(g)$ give another realization of the representation $T_R(g)$ of $MH(2)$.

Let us determine the form of $Q_R(g)$ for elements of the form $g(0, 0, \phi)$. By formula (5) of subsection 1, the operator $T_R(g(0, 0, \phi))$ transforms $\hat{f}(\theta)$ into $\hat{f}(\theta - \phi)$. Thus, for $g = g(0, 0, \phi)$,

$$Q_R(g)\, F(\lambda) = \int_{-\infty}^{\infty} \hat{f}(\theta - \varphi)\, e^{\lambda\theta}\, d\theta = e^{\lambda\varphi}\int_{-\infty}^{\infty} \hat{f}(\theta)\, e^{\lambda\theta}\, d\theta = e^{\lambda\varphi}\, F(\lambda). \tag{5}$$

Hence to hyperbolic rotations there correspond operators of multiplication by $e^{\lambda\phi}$.

By formula (1) of §2.3, any element of $MH(2)$ can be written as a product of hyperbolic rotations and a parallel translation by vectors of the form $(\pm r, 0), (0, \pm r)$ or $(\pm 1, \pm 1)$. Since we have already found the operators corresponding to hyperbolic rotations, we have only to find the operators corresponding to parallel translations of the form stated.

Let $g = g(r, 0, 0)$, $r > 0$. Then, by formula (5) of subsection 1,

$$T_R(g)\hat{f}(\theta) = e^{-Rr\operatorname{ch}\theta}\hat{f}(\theta), \tag{6}$$

and therefore

$$Q_R(g)F(\lambda) = \int_{-\infty}^{\infty} e^{-Rr\operatorname{ch}\theta + \lambda\theta}\hat{f}(\theta)\,d\theta. \tag{7}$$

From (4) we have

$$\hat{f}(\theta) = \frac{1}{2\pi i}\int_{a-i\infty}^{a+i\infty} F(\mu)\,e^{-\mu\theta}\,d\mu$$

and hence

$$Q_R(g)F(\lambda) = \frac{1}{2\pi i}\int_{-\infty}^{\infty} e^{-Rr\operatorname{ch}\theta}\,d\theta \int_{a-i\infty}^{a+i\infty} F(\mu)\,e^{(\lambda-\mu)\theta}\,d\mu. \tag{8}$$

If $\operatorname{Re} R > 0$, then (8) is absolutely convergent for all $\lambda$. We can therefore change the order of integration. Consequently,

$$Q_R(g)F(\lambda) = \int_{a-i\infty}^{a+i\infty} K(\lambda, \mu; R; r)\,F(\mu)\,d\mu, \tag{9}$$

where

$$K(\lambda, \mu; R; r) = \frac{1}{2\pi i}\int_{-\infty}^{\infty} e^{-Rr\operatorname{ch}\theta + (\lambda-\mu)\theta}\,d\theta. \tag{10}$$

We have thus proved that, for $g = g(r, 0, 0)$, $r > 0$, $\operatorname{Re} R > 0$, the operator $Q_R(g)$ is an integral transformation with the kernel $K(\lambda, \mu; R; r)$. This kernel will be expressed below by means of the so-called Macdonald functions, closely related to Bessel functions.

$Q_R(g)F(\lambda)$ has a similar form in the case when $\operatorname{Re} R < 0$ and $g = g(-r, 0, 0)$, $r > 0$. Then $r$ must be replaced by $-r$ in (10).

Utilizing these results, one can easily obtain an expression for $Q_R(g)$ for any element $g = g(a_1, a_2, \phi)$ such that $-a_1 < a_2 < a_1$ (as remarked in §1.3, such elements form a subgroup in $MH(2)$. If $-a_1 < a_2 < a_1$, then $g = g(a_1, a_2, \phi)$ can be presented in the form

$$g = g(0, 0, \psi) g(r, 0, 0) g(0, 0, \varphi - \psi), \tag{11}$$

where $r > 0$ (see §1.3). To the element $g(r, 0, 0)$ corresponds an integral operator with the kernel $K(\lambda, \mu; R; r)$, and to elements $g(0, 0, \phi - \psi)$ and $g(0, 0, \psi)$ correspond the operators of multiplication by $e^{\mu(\phi - \psi)}$ and $e^{\lambda\psi}$ respectively. Therefore

$$Q_R(g) F(\lambda) = \int_{a - i\infty}^{a + i\infty} K(\lambda, \mu; R; g) F(\mu) \, d\mu, \tag{12}$$

where

$$K(\lambda, \mu; R; g) = e^{(\lambda - \mu)\psi + \mu\varphi} K(\lambda, \mu; R; r). \tag{13}$$

For $\operatorname{Re} R < 0$, one can similarly write down the operators $Q_R(g(a_1, a_2, \phi))$ in the case $a_1 < a_2 < -a_1$.

However, for elements of the form $g(0, \pm r, 0)$ one does not obtain an expression in the form of an integral operator for $\operatorname{Re} R \neq 0$, since in this case we arrive at the integral

$$\int_{-\infty}^{\infty} e^{\pm Rr \operatorname{sh} \theta + (\lambda - \mu)\theta} \, d\theta, \tag{14}$$

which diverges for $\operatorname{Re} R \neq 0$ for all values of $\lambda$ and $\mu$. The case $\operatorname{Re} R = 0$ will be studied below.

The kernels of the operators $Q_R(g)$ take a particularly simple form if $g$ is a (parallel) translation in the direction of the straight lines $x_1 = \pm x_2$.

If $\operatorname{Re} R > 0$, the integrals for the elements $g = g(r, -r, 0)$ and $g = g(r, r, 0)$, $r > 0$, converge. Namely, for $g = g(r, -r, 0)$ we have

$$K(\lambda, \mu; R; g) = \frac{1}{2\pi i} \int_{-\infty}^{\infty} e^{-Rre^{\theta} + (\lambda - \mu)\theta} \, d\theta.$$

The substitution $e^{\theta} = t$ transforms this integral into the form

$$K(\lambda, \mu; R; g) = \frac{1}{2\pi i} \int_0^{\infty} e^{-Rrt} t^{\lambda - \mu - 1} \, dt. \tag{15}$$

As shown in §1.4, this integral converges for $\operatorname{Re}(\lambda - \mu) > 0$ and is equal to $(\Gamma(\lambda - \mu)/2\pi i)(Rr)^{\mu - \lambda}$. Hence for $g = g(r, -r, \phi)$, $r > 0$,

$$\left.\begin{aligned} K(\lambda,\ \mu;\ R;\ g) &= \frac{\Gamma(\lambda-\mu)}{2\pi i}(Rr)^{\mu-\lambda}, \\ \operatorname{Re} R > 0, &\quad \operatorname{Re}(\lambda-\mu) > 0. \end{aligned}\right\} \tag{16}$$

In just the same way one proves that if $g = g(r, r, 0)$, then

$$\left.\begin{aligned} K(\lambda,\ \mu;\ R;\ g) &= \frac{\Gamma(\mu-\lambda)}{2\pi i}(Rr)^{\lambda-\mu}, \\ \operatorname{Re} R > 0, &\quad \operatorname{Re}(\mu-\lambda) > 0, \end{aligned}\right\} \tag{17}$$

if $g = g(-r, r, 0)$, then

$$\left.\begin{aligned} K(\lambda,\ \mu;\ R;\ g) &= \frac{\Gamma(\lambda-\mu)}{2\pi i}(-Rr)^{\mu-\lambda}, \\ \operatorname{Re} R < 0, &\quad \operatorname{Re}(\lambda-\mu) > 0, \end{aligned}\right\} \tag{18}$$

and if $g = g(-r, -r, 0)$, then

$$\left.\begin{aligned} K(\lambda,\ \mu;\ R;\ g) &= \frac{\Gamma(\mu-\lambda)}{2\pi i}(-Rr)^{\lambda-\mu}, \\ \operatorname{Re} R < 0, &\quad \operatorname{Re}(\mu-\lambda) > 0. \end{aligned}\right\} \tag{19}$$

3. **The unitary case.** It remains for us to consider the case when $R$ is a purely imaginary number, i.e. when $Q_R(g)$ is unitary (see subsection 1). Let $R = \rho i$, $\rho > 0$. Then to the element $g = g(r, 0, 0)$ corresponds the operator

$$Q_{\rho i}(g)F(\lambda) = \frac{1}{2\pi i}\int_{-\infty}^{\infty} e^{-r\rho i \operatorname{ch}\theta} \int_{a-i\infty}^{a+i\infty} F(\mu)\, e^{(\lambda-\mu)\theta}\, d\mu\, d\theta. \tag{1}$$

The integral (1) is not absolutely convergent, but we can prove that for $-1 < \operatorname{Re}(\lambda - \mu) < 1$ a change in the order of integration is justified, and therefore

$$Q_{\rho i}(g)F(\lambda) = \int_{a-i\infty}^{a+i\infty} K(\lambda,\ \mu;\ \rho i;\ g)\, F(\mu)\, d\mu, \tag{2}$$

where

$$\begin{gathered} K(\lambda,\ \mu;\ \rho i;\ g) = \frac{1}{2\pi i}\int_{-\infty}^{\infty} e^{-r\rho i \operatorname{ch}\theta + (\lambda-\mu)\theta}\, d\theta, \\ -1 < \operatorname{Re}(\lambda-\mu) < 1. \end{gathered} \tag{3}$$

The operator $Q_{\rho i}(g)$ has a similar form for $g = g(0, \pm r, 0)$. In this case the kernel is given by the integral

$$K(\lambda, \mu; \rho l; g)=\frac{1}{2\pi i}\int_{-\infty}^{\infty} e^{\pm r\rho i \operatorname{sh}\theta+(\lambda-\mu)\theta} d\theta. \tag{4}$$

**4. The functions of Macdonald and Hankel.** We introduce a new special function $K_\nu(z)$ by putting, for arbitrary complex $\nu$ and $\operatorname{Re} z > 0$,

$$K_\nu(z)=\int_0^{\infty} e^{-z \operatorname{ch} t} \operatorname{ch} \nu t \, dt=\frac{1}{2}\int_{-\infty}^{\infty} e^{-z \operatorname{ch} t \pm \nu t} dt. \tag{1}$$

Evidently, $K_\nu(z)$ is an analytic function of $\nu$ and $z$, if $z$ varies in the given half plane. For $\operatorname{Re} z \le 0$ we define $K_\nu(z)$ with the help of analytic continuation in $z$. $K_\nu(z)$ is thus uniquely defined in the complex plane, cut along the ray $-\infty < z < 0$; it is call the Macdonald function of $z$ with index $\nu$. It follows from (1) that

$$K_\nu(z)=K_{-\nu}(z). \tag{2}$$

The Macdonald functions are closely related to the so-called Hankel functions of the first and second kind. For $x > 0$, $-1 < \operatorname{Re}\nu < 1$ these functions are defined by

$$H_\nu^{(1)}(x)=\frac{e^{-\frac{\nu\pi i}{2}}}{\pi i}\int_{-\infty}^{\infty} e^{ix \operatorname{ch} t-\nu t} dt \tag{3}$$

and

$$H_\nu^{(2)}(x)=-\frac{e^{\frac{\nu\pi i}{2}}}{\pi i}\int_{-\infty}^{\infty} e^{-ix \operatorname{ch} t-\nu t} dt \tag{3$'$}$$

(it is not difficult to show that, under the conditions stated, these integrals are convergent). Comparing (3) and (3$'$) with (1), we see that

$$H_\nu^{(1)}(x)=\frac{2e^{-\frac{\nu\pi i}{2}}}{\pi i} K_\nu\left(e^{-\frac{\pi i}{2}} x\right) \tag{4}$$

and

$$H_\nu^{(2)}(x)=-\frac{2e^{\frac{\nu\pi i}{2}}}{\pi i} K_\nu\left(e^{\frac{\pi i}{2}} x\right). \tag{4$'$}$$

In other words, the Hankel functions $H_\nu^{(1)}(x)$ and $H_\nu^{(2)}(x)$ coincide, to within a numerical factor, with the values of the Macdonald function on the negative and positive imaginary semi-axes.

Utilizing formulas (4) and (4′), one can continue the Hankel functions to the whole complex plane. In this connection, for the function $H_{(\nu)}^{(1)}(z)$ the plane has to be cut along the ray $-\infty < \operatorname{Im} z < 0$, and for $H_\nu^{(2)}(z)$ along the ray $0 < \operatorname{Im} z < \infty$.

In the sequel we shall also need the integral

$$\int_{-\infty}^{\infty} e^{ix\,\mathrm{sh}\,t - \nu t}\,dt, \quad \text{where } x > 0, \quad -1 < \operatorname{Re}\nu < 1. \tag{5}$$

To calculate this integral we make the change of variable $t = u + \pi i/2$. Since $\mathrm{sh}\,(u + \pi i/2) = i\,\mathrm{ch}\,u$, we get

$$\int_{-\infty}^{\infty} e^{ix\,\mathrm{sh}\,t - \nu t}\,dt = e^{-\frac{\pi\nu i}{2}} \int_{-\infty - \frac{\pi}{2} i}^{\infty - \frac{\pi}{2} i} e^{-x\,\mathrm{ch}\,u - \nu u}\,du.$$

We shift the contour of integration by $\pi i/2$ and use formula (1). We find that, for $x > 0$,

$$\int_{-\infty}^{\infty} e^{ix\,\mathrm{sh}\,t - \nu t}\,dt = 2e^{-\frac{\pi\nu i}{2}} K_\nu(x) \equiv \pi i\, H_\nu^{(1)}(ix). \tag{6}$$

In just the same way one proves that, for $x \leq 0$,

$$\int_{-\infty}^{\infty} e^{-ix\,\mathrm{sh}\,t - \nu t}\,dt = 2e^{\frac{\pi\nu i}{2}} K_\nu(x) \equiv -\pi i\, H_\nu^{(2)}(-ix) = \pi i\, H_{-\nu}^{(1)}(ix). \tag{7}$$

It follows from (4) and (4′) that

$$H_\nu^{(2)}(x) = -e^{\nu\pi i} H_\nu^{(1)}(e^{\pi i} x). \tag{8}$$

Since Hankel functions are analytic, this formula is valid in the whole plane, cut along the ray $0 < \operatorname{Im} z < \infty$,

The equality $K_\nu(z) - K_{-\nu}(z)$ and formulas (4) and (4′) yield

$$H_\nu^{(2)}(z) = -H_{-\nu}^{(1)}(e^{\pi i} z) = e^{\nu\pi i} H_{-\nu}^{(2)}(z). \tag{9}$$

Similarly,

$$H_\nu^{(1)}(z) = -H_{-\nu}^{(2)}(e^{-\pi i} z) = e^{-\nu\pi i} H_{-\nu}^{(1)}(z). \tag{9′}$$

5. **Expression of the kernels of the representation $Q_R(g)$ by means of the Macdonald function.** It follows from formulas (9), (10) of subsection 2 and (1) of

subsection 4 that, for $g = g(r, 0, 0)$, $\operatorname{Re} R > 0$, the operator $Q_R(g)$ is an integral operator whose kernel is expressed as follows by means of the Macdonald function:

$$K(\lambda, \mu; R; r) = \frac{1}{\pi i} K_{\lambda-\mu}(Rr). \tag{1}$$

It was shown in subsection 3 that for $\operatorname{Re} R = 0$ all the operators $Q_R(g)$ are integral operators. In this case it is convenient to express the operator kernels by means of Hankel functions. Let $R = i\rho$, $\rho > 0$. Then, by formula (7) of subsection 4, for the element $g = g(r, 0, 0)$ we have

$$K(\lambda, \mu; i\rho; g) = -\frac{e^{\frac{(\lambda-\mu)\pi i}{2}}}{2} H^{(2)}_{\mu-\lambda}(r\rho), \tag{2}$$

where $-1 < \operatorname{Re}(\lambda - \mu) < 1$.

Similarly, for the element $g = g(-r, 0, 0)$ we get

$$K(\lambda, \mu; i\rho; g) = \frac{e^{\frac{(\mu-\lambda)\pi i}{2}}}{2} H^{(1)}_{\mu-\lambda}(r\rho), \tag{2'}$$

for the element $g = g(0, r, 0)$

$$K(\lambda, \mu; i\rho; g) = \frac{1}{2} H^{(1)}_{\mu-\lambda}(ir\rho) = \frac{1}{\pi i} e^{\frac{(\lambda-\mu)\pi i}{2}} K_{\mu-\lambda}(r\rho) \tag{2''}$$

and for the element $g = g(0, -r, 0)$

$$K(\lambda, \mu; i\rho; g) = \frac{1}{2} H^{(1)}_{\lambda-\mu}(ir\rho) = \frac{1}{\pi i} e^{\frac{(\mu-\lambda)\pi i}{2}} K_{\lambda-\mu}(r\rho). \tag{2'''}$$

For elements of the form $g = g(\pm r, \pm r, 0)$, formulas (16)–(19) of subsection 2 are valid with $R$ replaced by $i\rho$.

6. **Infinitesimal operators of the representations $T_R(g)$ and $Q_R(g)$.** Let us calculate the infinitesimal operators of $T_R(g)$ and $Q_R(g)$. We first find the operator corresponding to the one-parametric subgroup $\Xi_+$:

$$\xi_+(t) = \begin{pmatrix} 1 & 0 & t \\ 0 & 1 & -t \\ 0 & 0 & 1 \end{pmatrix}. \tag{1}$$

The operator $T_R(\xi_+(t))$ transforms the function $\hat{f}(\theta)$ into the function

$$T_R(\xi_+(t))\hat{f}(\theta) = e^{-Rte^\theta} \hat{f}(\theta). \tag{2}$$

On differentiating with respect to $t$ and putting $t = 0$, we obtain

$$A_+\hat{f}(\theta) \equiv \frac{dT_R(\xi_+(t))\hat{f}(\theta)}{dt}\Bigg|_{t=0} = -Re^{\theta}\hat{f}(\theta). \tag{3}$$

In just the same way one proves that to the subgroup $\Xi_-$:

$$\xi_-(t) = \begin{pmatrix} 1 & 0 & t \\ 0 & 1 & t \\ 0 & 0 & 1 \end{pmatrix} \tag{4}$$

there corresponds, for $T_R(g)$, the infinitesimal operator

$$A_-\hat{f}(\theta) = -Re^{-\theta}\hat{f}(\theta), \tag{5}$$

and to the subgroup $\Omega_3$ of hyperbolic rotations, the operator

$$A_3\hat{f}(\theta) = -\frac{d\hat{f}(\theta)}{d\theta}. \tag{6}$$

We clarify what becomes of these operators when one replaces $\hat{f}(\theta)$ by $F(\lambda)$. For this one has to find the Fourier transform of the functions $A_+\hat{f}(\theta)$, $A_-\hat{f}(\theta)$, $A_3\hat{f}(\theta)$. We have

$$H_+F(\lambda) \equiv \int_{-\infty}^{\infty} A_+\hat{f}(\theta)\, e^{\lambda\theta}\, d\theta = -R\int_{-\infty}^{\infty} \hat{f}(\theta)\, e^{(\lambda+1)\theta}\, d\theta = -RF(\lambda+1). \tag{7}$$

Similarly, one proves that

$$H_-F(\lambda) = -RF(\lambda-1). \tag{8}$$

Lastly, integrating by parts, we get

$$H_3F(\lambda) = \int_{-\infty}^{\infty} A_3\hat{f}(\theta)\, e^{\lambda\theta}\, d\theta$$

$$= -\int_{-\infty}^{\infty} \frac{d\hat{f}(\theta)}{d\theta} e^{\lambda\theta}\, d\theta = \lambda\int_{-\infty}^{\infty} \hat{f}(\theta)\, e^{\lambda\theta}\, d\theta = \lambda F(\lambda). \tag{9}$$

7. Irreducibility of $T_R(g)$. Utilizing the expressions obtained above for the infinitesimal operators of the representations $T_R(g)$, it is easy to prove that for $R \neq 0$ these representations are irreducible. We confine ourselves to proving the operator-irreducibility of these representations. So, let $S$ be an operator permutable with all the operators of $T_R(g)$. Then it must also be permutable with the operators of $A_+$, $A_-$, $A_3$.

The operator $A_+$ is the operator of multiplication by $-Re^{\theta}$ in the space $K$ of infinitely differentiable finite functions on the line. Suppose that operator $S$ is permutable with $A_+$. Then it is permutable with all the operators $A_+^n = -R^n e^{n\theta}$. But on any finite segment one can uniformly approximate any continuous function by linear combinations of the functions $e^{n\theta}$. Therefore $S$ is permutable with all operators of multiplication by continuous functions, and consequently is the operator of multiplication by an infinitely differentiable function

$$S\hat{f}(\theta) = s(\theta)\hat{f}(\theta). \tag{1}$$

From the permutability of $S$ with $A_3 = -d/d\theta$ we get $s'(\theta) = 0$ and hence $s(\theta) =$ const.

Thus we have proved that any operator permutable with all operators $T_R(g)$ is the operator of multiplication by a constant, and so $T_R(g)$ is operator-irreducible. It can be proved that these representations are also spatially irreducible.

In case $R = 0$ the representation $T_R(g)$ has the form

$$T_0(g)\hat{f}(\theta) = \hat{f}(\theta - \varphi), \tag{2}$$

i.e. it reduces to the regular representation of the additive group of real numbers. Fourier transformation with respect to $\theta$ gives the decomposition of this representation as a continuous direct sum of one-dimensional representations.

## §4. RECURRENCE FORMULAS AND THE DIFFERENTIAL EQUATION FOR THE MACDONALD AND HANKEL FUNCTIONS

**1. Relations between infinitesimal operators and representation operators.** We assume, for definiteness, that $R = 1$. Let us consider the element

$$g = g(r, 0, 0)g(t, t, 0) = g(r + t, t, 0), \tag{1}$$

$r > 0$, $t > 0$, of the group $MH(2)$. As shown in §2.3 this element can be presented in the form

$$g = g(0, 0, \psi)g(\rho, 0, 0)g(0, 0, -\psi), \tag{2}$$

where

$$\left.\begin{aligned} \rho \operatorname{ch} \psi &= r + t, \\ \rho \operatorname{sh} \psi &= t \end{aligned}\right\} \tag{3}$$

and therefore

$$\left.\begin{aligned} \rho^2 &= r^2 + 2rt, \\ \operatorname{th}\psi &= \frac{t}{r+t}. \end{aligned}\right\} \tag{4}$$

It follows that 1)

$$Q(r,\ 0,\ 0)\,Q(t,\ t,\ 0) = Q(0,\ 0,\ \psi)\,Q(\rho,\ 0,\ 0)\,Q(0,\ 0,\ -\psi). \tag{5}$$

We differentiate both sides with respect to $t$ and put $t = 0$. From (4), for $t = 0$, we have $\rho = r$, $\psi = 0$ and

$$\left.\frac{d\rho}{dt}\right|_{t=0} = 1, \quad \left.\frac{d\psi}{dt}\right|_{t=0} = \frac{1}{r}. \tag{6}$$

Further, $Q(0, 0, 0)$ is the identity operator,

$$\left.\frac{dQ(t,\ t,\ 0)}{dt}\right|_{t=0} = H_-, \tag{7}$$

and

$$\left.\frac{dQ(0,\ 0,\ \psi)}{d\psi}\right|_{\psi=0} = H_3. \tag{8}$$

Therefore, as a result of differentiating (5), we obtain

$$Q(r,\ 0,\ 0)\,H_- = \frac{dQ(r,\ 0,\ 0)}{dr} + \frac{1}{r}\left(H_3Q(r,\ 0,\ 0) - Q(r,\ 0,\ 0)\,H_3\right). \tag{9}$$

One similarly proves the equality

$$Q(r,\ 0,\ 0)\,H_+ = \frac{dQ(r,\ 0,\ 0)}{dr} - \frac{1}{r}\left(H_3Q(r,\ 0,\ 0) - Q(r,\ 0,\ 0)\,H_3\right). \tag{10}$$

2. **Recurrence formulas.** In order to obtain recurrence formulas for Macdonald functions, one has to compare the kernels of operators on the right and left hand sides in (9) and (10) of subsection 1.

For $R = 1$ the operator $H_-$ transforms $F(\mu)$ into $-F(\mu - 1)$. As for the operator $Q(r, 0, 0)$, $r > 0$, for $R = 1$ it is an integral operator with the kernel $(1/\pi i)K_{\lambda-\mu}(r)$. Thus $Q(r, 0, 0)H_-$ transforms $F(\mu)$ into

$$-\frac{1}{\pi i}\int_{a-i\infty}^{a+i\infty} K_{\lambda-\mu}(r)\,F(\mu - 1)\,d\mu.$$

Replacing $\mu - 1$ by $\mu$, we see that $Q(r, 0, 0)H_-$ is an integral operator with the kernel

1) For brevity we write $Q(a, b, \phi)$ instead of $Q_R(g(a, b, \phi))$.

$$-\frac{1}{\pi i} K_{\lambda-\mu-1}(r). \tag{1}$$

In just the same way one proves that the operator kernel in the right hand side of formula (9) of subsection 1 has the form

$$\frac{1}{\pi i}\left[\frac{dK_{\lambda-\mu}(r)}{dr}+\frac{\lambda-\mu}{r}K_{\lambda-\mu}(r)\right]. \tag{2}$$

This leads to the equality

$$-K_{\lambda-\mu-1}(r)=\frac{dK_{\lambda-\mu}(r)}{dr}+\frac{\lambda-\mu}{r}K_{\lambda-\mu}(r). \tag{3}$$

Replacing $\lambda-\mu$ by $\nu$, we obtain

$$K_{\nu-1}(r)=-\frac{dK_\nu(r)}{dr}-\frac{\nu}{r}K_\nu(r). \tag{4}$$

The analyticity of $K_\nu(z)$ in the cut plane implies that this relation holds for all $z$ not belonging to the negative semiaxis:

$$K_{\nu-1}(z)=-\frac{dK_\nu(z)}{dz}-\frac{\nu}{z}K_\nu(z). \tag{5}$$

Similarly, from formula (10) we get

$$K_{\nu+1}(z)=-\frac{dK_\nu(z)}{dz}+\frac{\nu}{z}K_\nu(z). \tag{6}$$

It follows from (5) and (6) that

$$\frac{dK_\nu(z)}{dz}=-\frac{1}{2}[K_{\nu+1}(z)+K_{\nu-1}(z)] \tag{7}$$

and

$$\frac{\nu}{z}K_\nu(z)=\frac{1}{2}[K_{\nu+1}(z)-K_{\nu-1}(z)]. \tag{8}$$

Utilizing formulas (4) and (4′) of §3.4, we get the following recurrence relations for Hankel functions:

$$\frac{dH_\nu^{(1,2)}(z)}{dz}=\frac{1}{2}[H_{\nu-1}^{(1,2)}(z)-H_{\nu+1}^{(1,2)}(z)] \tag{9}$$

and

$$\frac{\nu}{z}H_\nu^{(1,2)}(z)=\frac{1}{2}[H_{\nu-1}^{(1,2)}(z)+H_{\nu+1}^{(1,2)}(z)]. \tag{9'}$$

Clearly, these formulas are valid for any linear combinations of functions $H_\nu^{(1)}(z)$ and $H_\nu^{(2)}(z)$. Henceforth we shall denote such linear combinations by $Z_\nu(z)$ and call them cylindrical functions of $z$ with index $\nu$. Thus,

$$\frac{dZ_\nu(z)}{dz}=\frac{1}{2}[Z_{\nu-1}(z)-Z_{\nu+1}(z)] \tag{10}$$

and

$$\frac{\nu}{z}Z_\nu(z)=\frac{1}{2}[Z_{\nu-1}(z)+Z_{\nu+1}(z)]. \tag{10'}$$

3. **Differential equations for the Macdonald and Hankel functions.** It was shown above that

$$\left(-\frac{d}{dz}+\frac{\nu}{z}\right)K_\nu(z)=K_{\nu+1}(z) \tag{1}$$

and

$$\left(-\frac{d}{dz}-\frac{\nu}{z}\right)K_\nu(z)=K_{\nu-1}(z) \tag{2}$$

It follows that

$$\left(\frac{d}{dz}+\frac{\nu+1}{z}\right)\left(\frac{d}{dz}-\frac{\nu}{z}\right)K_\nu(z)=K_\nu(z). \tag{3}$$

Simplifying this, we obtain the following second order differential equation for the Macdonald functions:

$$\frac{d^2K_\nu(z)}{dz^2}+\frac{1}{z}\,\frac{dK_\nu(z)}{dz}-\frac{z^2+\nu^2}{z^2}K_\nu(z)=0. \tag{4}$$

Since (4) is unaltered on replacing $z$ by $-z$, the function $K_\nu(-z)$ is also a solution of it.

Relations (10) and (10′) of subsection 2 similarly lead to the second order differential equation for cylindrical functions

$$\frac{d^2Z_\nu(z)}{dz^2}+\frac{1}{z}\,\frac{dZ_\nu(z)}{dz}+\frac{z^2-\nu^2}{z^2}Z_\nu(z)=0. \tag{5}$$

Particular solutions of this equation are the Hankel functions $H_\nu^{(1)}(z)$ and $H_\nu^{(2)}(z)$, and also the functions $H_\nu^{(1)}(-z)$ and $H_\nu^{(2)}(-z)$.

4. **Connection between Hankel functions and Bessel functions.** It was shown in §4.4 of Chapter IV that the Bessel functions $J_n(z)$ satisfy the differential equation

$$y''+\frac{1}{z}y'+\frac{z^2-n^2}{z^2}y=0 \tag{1}$$

and consequently they are cylindrical functions of $z$ with index $n$. Hence they are linear combinations of functions $H_n^{(1)}(z)$ and $H_n^{(2)}(z)$. We shall show that

$$J_n(z) = \frac{1}{2}[H_n^{(1)}(z) + H_n^{(2)}(z)]. \tag{2}$$

At first, we show that

$$J_0(x) = \frac{1}{2}[H_0^{(1)}(x) + H_0^{(2)}(x)]. \tag{3}$$

It was remarked in §4.4 of Chapter IV that a solution of the Bessel equation (1), bounded for $x = 0$, is unique to within a constant factor. Therefore to prove (3) it is sufficient to show that

$$\frac{1}{2}\lim_{x\to 0}[H_0^{(1)}(x) + H_0^{(2)}(x)] = J_0(0) = 1. \tag{4}$$

But, by (3) and (3′) of §3.4, we have

$$\frac{1}{2}[H_0^{(1)}(x) + H_0^{(2)}(x)] = \frac{1}{2\pi i}\int_{-\infty}^{\infty}[e^{ix\operatorname{ch}t} - e^{-ix\operatorname{ch}t}]\,dt$$

$$= \frac{2}{\pi}\int_0^{\infty}\sin(x\operatorname{ch}t)\,dt = \frac{2}{\pi}\int_x^{\infty}\frac{\sin y}{\sqrt{y^2-x^2}}\,dy. \tag{5}$$

As $\int_0^\infty (\sin y/y)\,dy = \pi/2$, equality (4) follows from (5). This proves (3).

Since the functions $J_n(x)$ and $\frac{1}{2}[H_n^{(1)}(x) + H_n^{(2)}(x)]$ satisfy the same recurrence relations (see formulas (9) and (9′) of subsection 2 and formulas (2) and (3) of §4.3, Chapter IV), it follows from (3) that

$$J_n(x) = \frac{1}{2}[H_n^{(1)}(x) + H_n^{(2)}(x)] \tag{6}$$

for all integral values of $n$.

Starting from formula (6), we define the Bessel function $J_\nu(z)$ for all values of $\nu$ by the formula

$$J_\nu(z) = \frac{1}{2}[H_\nu^{(1)}(z) + H_\nu^{(2)}(z)]. \tag{7}$$

This function is one of the particular solutions of equation (5) of subsection 3. We note another particular solution of this equation, called the *Neumann function*. It is defined by the equality

$$N_\nu(z)=\frac{1}{2i}[H_\nu^{(1)}(z)-H_\nu^{(2)}(z)]. \tag{8}$$

Thus we have

$$H_\nu^{(1)}(z)=J_\nu(z)+iN_\nu(z) \tag{9}$$

and

$$H_\nu^{(2)}(z)=J_\nu(z)-iN_\nu(z). \tag{9'}$$

## §5. FUNCTIONAL RELATIONS FOR THE HANKEL AND MACDONALD FUNCTIONS

1. **Introductory remarks.** The subsequent considerations are based on the following observation. If $Q_R(g_i)$, $i = 1, 2$, are integral operators with kernels $K(\lambda, \mu; R; g_i)$, then, because of

$$Q_R(g_1g_2)=Q_R(g_1)\,Q_R(g_2), \tag{1}$$

one has the equality

$$\int_{b-i\infty}^{b+i\infty} K(\lambda, \mu; R; g_1g_1)F(\mu)\,d\mu = \int_{b-i\infty}^{b+i\infty} K(\lambda, \nu; R; g_1)\,d\nu \int_{a-i\infty}^{a+i\infty} K(\nu, \mu; R; g_2)F(\mu)\,d\mu. \tag{2}$$

Therefore, provided that the integrals are absolutely convergent, we have

$$K(\lambda, \mu; R; g_1g_2) = \int_{a-i\infty}^{a+i\infty} K(\lambda, \nu; R; g_1)K(\nu, \mu; R; g_2)\,d\nu. \tag{3}$$

For different choices of the elements $g_1$ and $g_2$ of the group $MH(2)$, we obtain different relations for the Macdonald and Hankel functions from equality (3).

2. **Integral representations.** The following equality holds:

$$g(r, -r, 0)\,g(r, r, 0)=g(2r, 0, 0). \tag{1}$$

Owing to formula (3) of subsection 1, for $\operatorname{Re} R > 0$ this equality gives

$$K(\lambda, \mu; R; g(2r, 0, 0)) = \int_{a-i\infty}^{a+i\infty} K(\lambda, \nu; R; g(r, -r, 0))\,K(\nu, \mu; R; g(r, r, 0))\,d\nu. \tag{2}$$

On substituting in formula (2) the expressions obtained in §3.5 for the kernels, we get

$$K_{\lambda-\mu}(z)=\frac{1}{4\pi i}\int_{a-i\infty}^{a+i\infty}\Gamma(\lambda-\nu)\,\Gamma(\mu-\nu)\left(\frac{z}{2}\right)^{2\nu-\lambda-\mu}d\nu, \tag{3}$$

where $\operatorname{Re} z \equiv \operatorname{Re} 2Rr > 0$. Here, from the convergence conditions of the integrals expressing the kernels of operators $Q_R(g(r, -r, 0))$ and $Q_R(g(r, r, 0))$ (see §2.2), we have $\operatorname{Re}\lambda > a$, $\operatorname{Re}\mu > a$.

Replacing $\lambda - \mu$ by $\lambda$ and $\lambda - \nu$ by $\nu$ in (3), we get

$$K_{\lambda}(z)=\frac{1}{4\pi i}\int_{a-i\infty}^{a+i\infty}\Gamma(\nu)\,\Gamma(\nu-\lambda)\left(\frac{z}{2}\right)^{\lambda-2\nu}d\nu, \tag{3'}$$

where $\operatorname{Re} z > 0$, $\operatorname{Re}\nu > 0$, $\operatorname{Re}\nu > \operatorname{Re}\lambda$.

In order to obtain integral representations of the functions $H_\lambda^{(1)}(z)$ and $H_\lambda^{(2)}(z)$, we make use of relations (4) and (4') of §3.4:

$$H_\lambda^{(1)}(z)=-\frac{e^{-\frac{\lambda\pi i}{2}}}{2\pi^2}\int_{a-i\infty}^{a+i\infty}\Gamma(\nu)\,\Gamma(\nu-\lambda)\left[\frac{e^{-\frac{i\pi}{2}}z}{2}\right]^{\lambda-2\nu}d\nu, \tag{4}$$

where $\operatorname{Im} z > 0$, $a > 0$, $a > \operatorname{Re}\lambda$, and

$$H_\lambda^{(2)}(z)=\frac{e^{\frac{\lambda\pi i}{2}}}{2\pi^2}\int_{a-i\infty}^{a+i\infty}\Gamma(\nu)\,\Gamma(\nu-\lambda)\left[\frac{e^{\frac{i\pi}{2}}z}{2}\right]^{\lambda-2\nu}d\nu, \tag{4'}$$

where $\operatorname{Im} z < 0$, $a > 0$, $a > \operatorname{Re}\lambda$.

Formulas (4) and (4') remain valid for $\operatorname{Im} z = 0$, $z \neq 0$. Hence, for $x > 0$, we have

$$J_\lambda(x)=\frac{1}{2}[H_\lambda^{(1)}(x)+H_\lambda^{(2)}(x)]$$

$$=\frac{-i}{2\pi^2}\int_{a-i\infty}^{a+i\infty}\Gamma(\nu)\Gamma(\nu-\lambda)\sin(\nu-\lambda)\pi\left(\frac{x}{2}\right)^{\lambda-2\nu}d\nu$$

$$=\frac{1}{2\pi i}\int_{a-i\infty}^{a+i\infty}\frac{\Gamma(\nu)}{\Gamma(\lambda-\nu+1)}\left(\frac{x}{2}\right)^{\lambda-2\nu}d\nu, \qquad (5)$$

where $x>0$, $a>0$, $a>\operatorname{Re}\lambda$.

3. **Expansions in power series.** In formula (5) of subsection 2, let us add to the contour of integration a semicircle of infinitely large radius, lying in the left half plane. A simple estimate shows that the integral over this semicircle is equal to zero.

We calculate the integral thus obtained with the help of residues. According to §1.5, the function $\Gamma(x)$ has poles at the points $x=0,-1,-2,\cdots,-n,\cdots$, where

$$\operatorname*{Res}_{x=-n}\Gamma(x)=\frac{(-1)^n}{n!}.$$

Therefore we get

$$J_\lambda(x)=\sum_{n=0}^{\infty}\frac{(-1)^n}{n!\,\Gamma(\lambda+n+1)}\left(\frac{x}{2}\right)^{\lambda+2n}. \qquad (1)$$

In just the same way, from formulas (3′), (4) and (4′) of subsetion 2 we get [1])

$$H_\lambda^{(1,2)}(z)$$

$$=\pm\frac{1}{\pi i}\left[e^{\mp\lambda\pi i}\sum_{k=0}^{\infty}\frac{\Gamma(-k-\lambda)}{k!}\left(\frac{z}{2}\right)^{\lambda+2k}+\sum_{k=0}^{\infty}\frac{\Gamma(\lambda-k)}{k!}\left(\frac{z}{2}\right)^{-\lambda+2k}\right], \qquad (2)$$

$$K_\lambda(z)$$

$$=\frac{1}{2}\left[\sum_{k=0}^{\infty}\frac{(-1)^k\Gamma(-k-\lambda)}{k!}\left(\frac{z}{2}\right)^{\lambda+2k}+\sum_{k=0}^{\infty}\frac{(-1)^k\Gamma(\lambda-k)}{k!}\left(\frac{z}{2}\right)^{-\lambda+2k}\right]. \qquad (3)$$

Equality (2) gives

1) In formula (2) the upper sign relates to $H_\lambda^{(1)}(z)$, and the lower to $H_\lambda^{(2)}(z)$.

$$N_\lambda(z)=\frac{1}{2i}[H_\lambda^{(1)}(z)-H_\lambda^{(2)}(z)]$$

$$=-\frac{1}{\pi}\left[\cos\lambda\pi\sum_{k=0}^{\infty}\frac{\Gamma(-k-\lambda)}{k!}\left(\frac{z}{2}\right)^{\lambda+2k}+\sum_{k=0}^{\infty}\frac{\Gamma(\lambda-k)}{k!}\left(\frac{z}{2}\right)^{-\lambda+2k}\right]. \tag{4}$$

Formulas (1)–(4) show that the only singular point for cylindrical functions is the branch point $z = 0$.

Expressions (2)–(4) are valid only for nonintegral values of $\lambda$. If $\lambda = n$, an integer, then the integrand in integrals (3 ′)–(5) of subsection 2 has poles of the second order.

4. **Mellin transforms.** In formula (3 ′) of subsection 2 we replace $z$ by $2z$ and $2\nu$ by $\nu$, compare it with the inverse Mellin transform formula (see § 4.3 of Chapter II). We find that the function $\frac{1}{4}\Gamma(\nu/2)\Gamma(\nu/2-\lambda)$ is the Mellin transform for $x^{-\lambda}K_\lambda(2x)$, and therefore

$$\int_0^\infty x^{\nu-\lambda-1}K_\lambda(x)\,dx=2^{\nu-\lambda-2}\Gamma\left(\frac{\nu}{2}\right)\Gamma\left(\frac{\nu}{2}-\lambda\right). \tag{1}$$

Similarly, it follows from formula (5) of subsection 2 that

$$\int_0^\infty x^{\nu-\lambda-1}J_\lambda(x)\,dx=\frac{2^{\nu-\lambda-1}\Gamma\left(\frac{\nu}{2}\right)}{\Gamma\left(\lambda-\frac{\nu}{2}+1\right)}. \tag{2}$$

More general formulas are obtained in the following way. Let us consider the equality

$$g(r_1,\ r_1,\ 0)\,g(r_2,\ 0,\ 0)=g(r_1+r_2,\ r_1,\ 0).$$

If $r_1 > 0$, $r_2 > 0$, then

$$g(r_1+r_2,\ r_1,\ 0)=g(0,\ 0,\ \theta)\,g(r,\ 0,\ 0)\,g(0,\ 0,\ -\theta),$$

where

$$\operatorname{th}\theta=\frac{r_1}{r_1+r_2} \tag{3}$$

and

$$r^2=r_2^2+2r_1r_2 \tag{3 ′}$$

(see § 2.2).

Since to a hyperbolic rotation by the angle $\theta$ corresponds the operator of multiplication by $e^{\lambda\theta}$, for $\operatorname{Re} R > 0$ we get

$$e^{(\lambda-\mu)\theta} K(\lambda, \mu; R; g(r, 0, 0)) = \int_{a-i\infty}^{a+i\infty} K(\lambda, \nu; R; g(r_1, r_1, 0)) K(\nu, \mu; R; g(r_2, 0, 0))\, d\nu, \tag{4}$$

where $a > \operatorname{Re}\lambda$. In this formula we substitute the values of the kernels calculated in § 3.5, and replace $\lambda - \mu$ by $\lambda$, $\nu - \mu$ by $\nu$, $Rr_1$ by $z_1$ and $Rr_2$ by $z_2$. We obtain

$$e^{\lambda\theta} K_\lambda(z) = \frac{1}{2\pi i} \int_{a-i\infty}^{a+i\infty} \Gamma(\nu-\lambda) z_1^{\lambda-\nu} K_\nu(z_2)\, d\nu, \tag{5}$$

where $a > \operatorname{Re}\lambda$, $\operatorname{Re} z_1 > 0$, $\operatorname{Re} z_2 > 0$ and

$$\operatorname{th}\theta = \frac{z_1}{z_1+z_2}, \qquad -\frac{\pi}{2} < \operatorname{Im}\theta < \frac{\pi}{2}, \tag{6}$$

$$z^2 = z_2^2 + 2z_1 z_2, \qquad \operatorname{Re} z > 0 \tag{6$'$}$$

(these conditions determine $z$ and $\theta$ uniquely).

In particular, for $z_1 = z_2$ we get

$$3^{\frac{\lambda}{2}} K_\lambda(z\sqrt{3}) = \frac{1}{2\pi i} \int_{a-i\infty}^{a+i\infty} \Gamma(\nu-\lambda) z^{\lambda-\nu} K_\nu(z)\, d\nu. \tag{7}$$

Formula (5) remains valid for purely imaginary values of $z_1$ and $z_2$. Therefore, putting $z_1 = \pm ix_1$, $z_2 = \pm ix_2$ we find that

$$e^{\lambda\theta} H_\lambda^{(1,2)}(x) = \frac{1}{2\pi i} \int_{a-i\infty}^{a+i\infty} \Gamma(\nu-\lambda) x_1^{\lambda-\nu} H_\nu^{(1,2)}(x_2)\, d\nu, \tag{8}$$

where

$$\operatorname{th}\theta = \frac{x_1}{x_1+x_2}, \quad x^2 = x_2^2 + 2x_1x_2, \tag{9}$$

$$x_1 > 0, \quad x_2 > 0, \quad a > \operatorname{Re}\lambda, \quad -1 < a < 1, \quad -1 < \operatorname{Re}\lambda < 1.$$

Formula (8) can also be derived by considering the motion

$$g(r_1, r_1, 0)\, g(r_2, 0, 0), \quad r_1 > 0, \quad r_2 > 0,$$

for purely imaginary $R = \pm i\rho$, and utilizing the formulas of § 3.5.

Analogous formulas appear on considering the motion

$$g(r_1, r_1, 0)\, g(-r_2, 0, 0) = g(r_1 - r_2, r_1, 0),$$

where $r_1 > 0$, $r_2 > 0$. For $r_2 < 2r_1$ we have

$$g(r_1 - r_2,\ r_1,\ 0) = g(0,\ 0,\ \theta)\, g(0,\ r,\ 0)\, g(0,\ 0,\ -\theta),$$

where $\operatorname{th}\theta = (r_1 - r_2)/r_1$, $r^2 = 2r_1r_2 - r_2^2$. Putting $R = i\rho$ and utilizing the formulas of §3.5, we easily get

$$J_1 \equiv \frac{1}{2\pi i}\int_{a-i\infty}^{a+i\infty} \Gamma(\lambda+\nu)\, x_1^{-\nu} H_\nu^{(1)}(x_2)\, d\nu = \frac{2}{\pi i} e^{-\lambda\theta} x_1^\lambda K_\lambda(x), \tag{10}$$

where $x_1$, $x_2$, $x$, $\theta$ are connected by

$$\operatorname{th}\theta = \frac{x_1 - x_2}{x_1}, \quad x^2 = 2x_1x_2 - x_2^2. \tag{11}$$

But if $x_2 > 2x_1 > 0$, then we get

$$J_1 = e^{-\lambda\theta} x_1^\lambda H_{-\lambda}^{(1)}(x), \tag{12}$$

where

$$\operatorname{th}\theta = \frac{x_1}{x_1 - x_2}, \quad x^2 = -2x_1x_2 + x_2^2, \tag{13}$$

$x > 0$, and $a > -\operatorname{Re}\lambda$.

Lastly, for $x_2 = 2x_1$ we get

$$J_1 = \frac{\Gamma(-\lambda)}{\pi i} x_1^{2\lambda}, \tag{14}$$

where $\operatorname{Re}\lambda < 0$, $a > -\operatorname{Re}\lambda$.

Let us now consider the motion

$$g(r_1,\ r_1,\ 0)\, g(0,\ -r_2,\ 0) = g(r_1,\ r_1 - r_2,\ 0).$$

For $0 < x_2 < 2x_1$ and $a > -\operatorname{Re}\lambda$, $-1 < a < 1$, $-1 < \operatorname{Re}\lambda < 1$, we similarly find that

$$J_2 \equiv \frac{1}{2\pi i}\int_{a-i\infty}^{a+i\infty} \Gamma(\lambda+\nu)\, x_1^{-\nu} e^{-\nu\pi i} K_\nu(x_2)\, d\nu = -\frac{\pi i}{2} e^{-\lambda\theta} x_1^\lambda H_\lambda^{(2)}(x), \tag{15}$$

where

$$\operatorname{th}\theta = \frac{x_1 - x_2}{x_1}, \quad x^2 = 2x_1x_2 - x_2^2, \quad x > 0. \tag{16}$$

For $0 < 2x_1 < x_2$, $a > -\operatorname{Re}\lambda$, $-1 < a < 1$, $-1 < \operatorname{Re}\lambda < 1$, we have

$$J_2 = e^{-\lambda(\theta - \pi i)} x_1^\lambda K_\lambda(x), \tag{17}$$

where

$$\operatorname{th}\theta=\frac{x_1}{x_1-x_2},\qquad x^2=-2x_1x_2+x_2^2,\qquad x>0. \tag{18}$$

Finally, for $x_2 = 2x_1$ we get

$$J_2=\frac{\Gamma(-\lambda)\,e^{\lambda\pi i}}{2}\,x_1^{2\lambda}, \tag{19}$$

where $\operatorname{Re}\lambda < 0$, $a > -\operatorname{Re}\lambda$, $-1 < a < 1$.

5. **Mellin transforms** (continued). Let us compare formula (5) of subsection 4 with the inversion formula for the Mellin transform (see §4.3 of Chapter II). We see that $\Gamma(\nu-\lambda)K_\nu(x_2)$ is the Mellin transform for $x_1^{-\lambda}e^{\lambda\theta}K_\lambda(x)$ (as a function of $x_1$), i.e.

$$\int_0^\infty x_1^{\nu-\lambda-1}e^{\lambda\theta}K_\lambda(x)\,dx_1=\Gamma(\nu-\lambda)\,K_\nu(x_2), \tag{1}$$

where $x_1$, $x_2$, $\theta$ and $x$ are connected by relations similar to (6)–(6′) of subsection 4, and $\operatorname{Re}\nu > \operatorname{Re}\lambda$.

In just the same way, starting from formulas (10) and (12), we get

$$\int_0^{\frac{1}{2}x_2} e^{-\lambda\theta}x_1^{\lambda+\nu-1}H_{-\lambda}^{(1)}(x)\,dx_1+\frac{2}{\pi i}\int_{\frac{1}{2}x_2}^\infty e^{-\lambda\hat\theta}x_1^{\lambda+\nu-1}K_\nu(\hat x)\,dx$$
$$=\Gamma(\lambda+\nu)\,H_\nu^{(1)}(x_2), \tag{2}$$

where

$$\operatorname{th}\theta=\frac{x_1}{x_1-x_2},\qquad x^2=-2x_1x_2+x_2^2,\qquad x>0, \tag{3}$$

$$\operatorname{th}\hat\theta=\frac{x_1-x_2}{x_1},\qquad \hat x^2=2x_1x_2-x_2^2,\qquad \hat x>0, \tag{3′}$$

and $\operatorname{Re}\nu > -\operatorname{Re}\lambda$, $-1 < \operatorname{Re}\nu < 1$, $-1 < \operatorname{Re}\lambda < 1$.

Formulas (15) and (17) lead to

$$\int_0^{\frac{1}{2}x_2} e^{-\lambda(\theta-\pi i)} x_1^{\lambda+\nu-1} K_\lambda(x)\,dx_1 - \frac{\pi i}{2}\int_{\frac{1}{2}x_2}^{\infty} e^{-\lambda\hat\theta} x_1^{\lambda+\nu-1} H_\lambda^{(2)}(\hat x)\,dx_1 = \Gamma(\lambda+\nu)\,e^{-\nu\pi i}K_{-\nu}(x_2), \tag{4}$$

where

$$\operatorname{th}\theta = \frac{x_1}{x_1-x_2},\quad x^2 = -2x_1x_2+x_2^2,\qquad x>0, \tag{5}$$

$$\operatorname{th}\hat\theta = \frac{x_1-x_2}{x_1},\quad \hat x^2 = 2x_1x_2-x_2^2,\qquad \hat x>0 \tag{5'}$$

and $\operatorname{Re}\nu > -\operatorname{Re}\lambda$, $-1<\operatorname{Re}\lambda<1$, $-1<\operatorname{Re}\nu<1$.

Further, in formula (5) of subsection 4, let us add to the contour of integration a semicircle of infinitely large radius, lying in the left half plane. Calculating the integral with the help of residues, we get the equality

$$\sum_{n=0}^{\infty}\frac{(-1)^n}{n!} z_1^n K_{\lambda-n}(z_2) = e^{\lambda\theta}K_\lambda(z), \tag{6}$$

where $\operatorname{Re} z_1>0$, $\operatorname{Re} z_2>0$ and $z_1, z_2, \theta, z$ are connected by formulas (6) and (6′) of subsection 4.

Similarly, formulas (10) and (12) lead to

$$\sum_{n=0}^{\infty}\frac{(-1)^n}{n!} x_1^n H_{-\lambda-n}^{(1)}(x_2) = \begin{cases} \frac{2}{\pi i} e^{-\lambda\hat\theta} K_\lambda(\hat x), & \text{if } 0<x_2<2x_1,\\ e^{-\lambda\theta} H_{-\lambda}^{(1)}(x), & \text{if } 0<2x_1<x_2.\end{cases} \tag{7}$$

The values of $\theta, x, \hat\theta, \hat x$ are given by formulas (3) and (3′).

Lastly,

$$\sum_{n=0}^{\infty}\frac{1}{n!} x_1^n K_{\lambda+n}(x_2) = \begin{cases} -\frac{\pi i}{2} e^{-\lambda\hat\theta} H_\lambda^{(2)}(\hat x), & \text{if } 0<x_2<2x_1,\\ e^{-\lambda(\theta-\pi i)} K_\lambda(x), & \text{if } 0<2x_1<x_2.\end{cases} \tag{8}$$

6. **Composition theorems.** We shall now derive formulas analogous to the composition formula for Bessel functions (see §4.1 of Chapter IV). To that end, we consider the motion

$$g = g(r_1,\ 0,\ 0)\,g(0,\ 0,\ \theta)\,g(r_2,\ 0,\ 0),$$

where $r_1 > 0$, $r_2 > 0$. This motion can be presented in the form

$$g = g(0, 0, \varphi)\, g(r, 0, 0)\, g(0, 0, \theta - \varphi),$$

where

$$\operatorname{th}\varphi = \frac{r_2 \operatorname{sh}\theta}{r_2 \operatorname{ch}\theta + r_1}, \qquad r^2 = r_1^2 + 2r_1 r_2 \operatorname{ch}\theta + r_2^2, \qquad r > 0.$$

Since to a hyperbolic rotation by an angle $\theta$ there corresponds the operator of multiplication by $e^{\lambda\theta}$, we have

$$\int_{a-i\infty}^{a+i\infty} K(\lambda, \nu; R; g(r_1, 0, 0))\, K(\nu, \mu; R; g(r_2(0, 0))\, e^{(\nu-\mu)\theta}\, d\nu = e^{(\lambda-\mu)\varphi} K(\lambda, \mu; R; g(r, 0, 0)). \tag{1}$$

We substitute the values of the kernels in this formula. Then, for $\operatorname{Re} z_1 > 0$, $\operatorname{Re} z_2 > 0$,

$$\int_{a-i\infty}^{a+i\infty} K_{\lambda-\nu}(z_1)\, K_\nu(z_2)\, e^{\nu\theta} d\nu = \pi i e^{\lambda\varphi} K_\lambda(z), \tag{2}$$

where

$$\operatorname{th}\varphi = \frac{z_2 \operatorname{sh}\theta}{z_2 \operatorname{ch}\theta + z_1}, \qquad -\frac{\pi}{2} < \operatorname{Im}\varphi < \frac{\pi}{2}, \tag{3}$$

$$z^2 = z_1^2 + 2z_1 z_2 \operatorname{ch}\theta + z_2^2, \qquad \operatorname{Re} z > 0. \tag{3'}$$

For purely imaginary values of $z_1$ and $z_2$, $z_1 = \pm ix_1$, $z_2 = \pm ix_2$, $x_1 > 0$, $x_2 > 0$, we get

$$\int_{a-i\infty}^{a+i\infty} H_{\lambda-\nu}^{(1,2)}(x_1)\, H_\nu^{(1,2)}(x_2)\, e^{\nu\theta}\, d\nu = \pm 2 e^{\lambda\varphi} H_\lambda^{(1,2)}(x), \tag{4}$$

where $-1 < a < 1$, $-1 < \operatorname{Re}\lambda < 1$, $a - 1 < \operatorname{Re}(\lambda - \nu) < a + 1$,

$$\operatorname{th}\varphi = \frac{x_2 \operatorname{sh}\theta}{x_2 \operatorname{ch}\theta + x_1}, \tag{5}$$

$$x^2 = x_1^2 + 2x_1 x_2 \operatorname{ch}\theta + x_2^2, \quad x > 0, \tag{5'}$$

and where the upper sign relates to the function $H_\lambda^{(1)}(x)$, and the lower to $H_\lambda^{(2)}(x)$.

By considering other combinations of motions (for example $g(-r_1, 0, 0) g(0, 0, \theta)\, g(r_2, 0, 0)$, $g(r_1, 0, 0)\, g(0, 0, \theta)\, g(0, r_2, 0)$ etc.), we obtain a number of new formulas, similar to the ones established above. Omitting the details, we mention the final results.

For $x_1 > x_2 > 0$ and $x_2/x_1 < e^\theta < x_1/x_2$, we have

$$J_1 \equiv \int_{a-i\infty}^{a+i\infty} e^{-\nu\theta} H^{(1)}_{\nu+\mu}(x_1)\, H^{(2)}_{\nu}(x_2)\, d\nu = -2e^{\mu\varphi} H^{(1)}_{\mu}(x), \tag{6}$$

where $-1 < \operatorname{Re}\mu < 1$, $-1 < a < 1$, $-a-1 < \operatorname{Re}\mu < 1-a$, and

$$\operatorname{th}\varphi = \frac{x_2 \operatorname{sh}\theta}{x_2 \operatorname{ch}\theta - x_1}, \tag{7}$$

$$x^2 = x_1^2 + x_2^2 - 2x_1x_2 \operatorname{ch}\theta, \quad x > 0. \tag{7'}$$

For $x_1 > x_2 > 0$ and $e^{\theta} > x_1/x_2$,

$$J_1 = \frac{4i}{\pi} e^{\mu\varphi} K_{\mu}(x), \tag{8}$$

where $-1 < a < 1$, $-a-1 < \operatorname{Re}\mu < 1-a$ and

$$\operatorname{th}\varphi = \frac{x_2 \operatorname{ch}\theta - x_1}{x_2 \operatorname{sh}\theta}, \tag{9}$$

$$x^2 = 2x_1x_2 \operatorname{ch}\theta - x_1^2 - x_2^2, \quad x > 0. \tag{9'}$$

For $e^{\theta} < x_2/x_1$, $-1 < a < 1$, $-a-1 < \operatorname{Re}\mu < 1-a$,

$$J_1 = \frac{4i}{\pi} e^{\mu(\varphi-\pi i)} K_{\mu}(x) \tag{10}$$

where $\phi$ and $x$ are given by (9) and (9').

For $e^{\theta} = x_1/x_2$, $-1 < a < 1$, $-a-1 < \operatorname{Re}\mu < 1-a$, $\operatorname{Re}\mu > 0$

$$J_1 = \frac{2i}{\pi} \Gamma(\mu) \left(\frac{2x_1}{x_1^2 - x_2^2}\right)^{\mu}. \tag{11}$$

For $e^{\theta} = x_2/x_1$, $-1 < a < 1$, $-a-1 < \operatorname{Re}\mu < 1-a$, $\operatorname{Re}\mu < 0$

$$J_1 = \frac{2ie^{-\mu\pi i}}{\pi} \Gamma(-\mu) \left(\frac{x_1^2 - x_2^2}{2x_1}\right)^{\mu}. \tag{12}$$

Further, for $x_1 > x_2 > 0$ and $x_2/x_1 < e^{\theta} < x_1/x_2$

$$J_2 \equiv \int_{a-i\infty}^{a+i\infty} e^{-\nu(\theta-\pi i)} K_{\nu-\mu}(x_1)\, K_{\nu}(x_2)\, d\nu = \pi i e^{-\mu\varphi} K_{\mu}(x), \tag{13}$$

where $x$ and $\phi$ are given by (7) and (7').

For $x_1 > x_2 > 0$, $e^{\theta} > x_1/x_2$, $-1 < \operatorname{Re}\mu < 1$

$$J_2 = \frac{\pi^2}{2} e^{-\mu\varphi} H^{(2)}_{-\mu}(x). \tag{14}$$

For $x_1 > x_2 > 0$, $e^\theta < x_2/x_1$, $-1 < \operatorname{Re}\mu < 1$

$$J_2 = -\frac{\pi^2}{2} e^{-\mu\varphi} H_\mu^{(1)}(x), \tag{15}$$

where $x$ and $\phi$ are given by (9) and (9′).

For $e^\theta = x_1/x_2$, $\operatorname{Re}\mu > 0$

$$J_2 = \frac{\pi i}{2} \Gamma(-\mu) \left(\frac{x_1^2 - x_2^2}{2x_1}\right)^\mu. \tag{16}$$

For $e^\theta = x_2/x_1$, $\operatorname{Re}\mu > 0$

$$J_2 = \frac{\pi i}{2} \Gamma(\mu) \left(\frac{2x_1}{x_1^2 - x_2^2}\right)^\mu. \tag{17}$$

For $e^\theta > 0$ and $e^\theta < x_2/x_1$, $a - 1 < \operatorname{Re}\mu < a + 1$

$$J_3 \equiv \int_{a-i\infty}^{a+i\infty} e^{-\nu\theta} H_{\mu-\nu}^{(1)}(x_1) K_\nu(x_2)\, d\nu = 2e^{-\mu\varphi} K_\mu(x), \tag{18}$$

where

$$\operatorname{th}\varphi = \frac{x_1 + x_2 \operatorname{sh}\theta}{x_2 \operatorname{ch}\theta}, \tag{19}$$

$$x^2 = x_2^2 - 2x_1 x_2 \operatorname{sh}\theta - x_1^2, \quad x > 0. \tag{19′}$$

For $x_1 > 0$, $x_2 > 0$, $e^\theta > x_2/x_1$, $a - 1 < \operatorname{Re}\mu < a + 1$, $-1 < \operatorname{Re}\mu < 1$

$$J_3 = \pi i e^{-\mu\varphi} H_\mu^{(1)}(x), \tag{20}$$

where

$$\operatorname{th}\varphi = \frac{x_2 \operatorname{ch}\theta}{x_1 + x_2 \operatorname{sh}\theta}, \tag{21}$$

$$x^2 = x_1^2 + 2x_1 x_2 \operatorname{sh}\theta - x_2^2. \tag{21′}$$

For $e^\theta = x_2/x_1$, $a - 1 < \operatorname{Re}\mu < a + 1$, $\operatorname{Re}\mu > 0$

$$J_3 = \Gamma(\mu) \left(\frac{2x_1}{x_1^2 + x_2^2}\right)^\mu. \tag{22}$$

For $x_1 > 0$, $x_2 > 0$, $e^\theta > x_1/x_2$, $a - 1 < \operatorname{Re}\mu < a + 1$

$$J_4 \equiv \int_{a-i\infty}^{a+i\infty} e^{\nu\theta} H_{\mu-\nu}^{(2)}(x_1) K_\nu(x_2)\, d\nu = -2e^{\mu\varphi} K_\mu(x), \tag{23}$$

where

$$\operatorname{th}\varphi = \frac{x_2 \operatorname{sh}\theta - x_1}{x_2 \operatorname{ch}\theta}, \tag{24}$$

$$x^2 = x_2^2 + 2x_1x_2 \operatorname{sh}\theta - x_1^2, \qquad x > 0. \tag{24'}$$

For $x_1 > 0$, $x_2 > 0$, $e^\theta < x_1/x_2$, $a - 1 < \operatorname{Re}\mu < a + 1$, $-1 < \operatorname{Re}\mu < 1$

$$J_4 = \pi i e^{\mu\varphi} H_\mu^{(2)}(x), \tag{25}$$

where

$$\operatorname{th}\varphi = \frac{x_2 \operatorname{ch}\theta}{x_2 \operatorname{sh}\theta - x_1}, \tag{26}$$

$$x^2 = x_1^2 - 2x_1x_2 \operatorname{sh}\theta - x_2^2, \quad x > 0. \tag{26'}$$

For $e^\theta = x_1/x_2$, $a - 1 < \operatorname{Re}\mu < a + 1$, $\operatorname{Re}\mu > 0$

$$J_4 = -\Gamma(\mu)\left(\frac{2x_1}{x_1^2 + x_2^2}\right)^\mu. \tag{27}$$

7. **Multiplication theorems.** Let us apply the Fourier inversion formula to equality (2) of subsection 6. We obtain

$$\frac{1}{2}\int_{-\infty}^{\infty} e^{\lambda\varphi - \nu\theta} K_\lambda(z)\, d\theta = K_{\lambda-\nu}(z_1) K_\nu(z_2), \tag{1}$$

where $\phi$, $z$, $\theta$, $z_1$, $z_2$ are connected by formulas (3) and (3′) of subsection 6.

In particular, for $\lambda = 0$ we get the equality

$$\frac{1}{2}\int_{-\infty}^{\infty} e^{-\nu\theta} K_0\left(\sqrt{z_1^2 + 2z_1z_2 \operatorname{ch}\theta + z_2^2}\right) d\theta = K_\nu(z_1) K_\nu(z_2). \tag{2}$$

We may recall that in these formulas $\operatorname{Re} z_1 > 0$, $\operatorname{Re} z_2 > 0$ and $\operatorname{Re} z > 0$, where $z = \sqrt{z_1^2 + 2z_1z_2 \operatorname{ch}\theta + z_2^2}$.

Similarly, formula (4) of subsection 6 yields

$$\pm\frac{1}{\pi i}\int_{-\infty}^{\infty} e^{\lambda\varphi - \nu\theta} H_\lambda^{(1,2)}(x)\, d\theta = H_{\lambda-\nu}^{(1,2)}(x_1) H_\nu^{(1,2)}(x_2), \tag{3}$$

where $\phi$, $x$, $\theta$, $x_1$, $x_2$ are connected by formulas (5) and (5′) of subsection 6,

$x_1 > 0$, $x_2 > 0$, $x > 0$. The upper sign in this equality relates to $H_\nu^{(1)}(x)$, and the lower, to $H_\nu^{(2)}(x)$.

8. **Reciprocal integral transformations.** Since $Q_R(g)Q_R(g^{-1}) = E$, the integral transformations

$$\Phi(\lambda) = \int_{a-i\infty}^{a+i\infty} K(\lambda,\ \mu;\ R;\ g) F(\mu)\, d\mu \tag{1}$$

and

$$F(\lambda) = \int_{a-i\infty}^{a+i\infty} K(\lambda,\ \mu;\ R;\ g^{-1}) \Phi(\mu)\, d\mu \tag{2}$$

are reciprocal. But the operators $Q_R(g)$ and $Q_R(g^{-1})$ are simultaneously integral only in the unitary case (see §3.3), i.e. for $R = \rho i$. Utilizing the results of §3.3, we get the following assertions.

The integral transformations

$$\Phi(\lambda) = -\frac{1}{2} \int_{a-i\infty}^{a+i\infty} e^{\frac{(\lambda-\mu)\pi i}{2}} H_{\mu-\lambda}^{(2)}(x) F(\mu)\, d\mu \tag{3}$$

and

$$F(\lambda) = \frac{1}{2} \int_{a-i\infty}^{a+i\infty} e^{\frac{(\mu-\lambda)\pi i}{2}} H_{\mu-\lambda}^{(1)}(x) \Phi(\mu)\, d\mu, \tag{4}$$

where $-1 < \operatorname{Re}(\lambda - \mu) < 1$, are reciprocal for all $x > 0$.

The integral transformations

$$\Phi(\lambda) = \frac{1}{\pi i} \int_{a-i\infty}^{a+i\infty} e^{\frac{(\lambda-\mu)\pi i}{2}} K_{\mu-\lambda}(x) F(\mu)\, d\mu \tag{5}$$

and

$$F(\lambda) = \frac{1}{\pi i} \int_{a-i\infty}^{a+i\infty} e^{\frac{(\mu-\lambda)\pi i}{2}} K_{\lambda-\mu}(x) \Phi(\mu)\, d\mu, \tag{6}$$

where $-1 < \operatorname{Re}(\lambda - \mu) < 1$, are reciprocal for all $x > 0$.

Lastly, it follows from formulas (17) and (19) of §3.2 that the integral transformations

$$\Phi(\lambda)=\frac{1}{2\pi i}\int_{a-i\infty}^{a+i\infty}\Gamma(\mu-\lambda)(ix)^{\lambda-\mu}F(\mu)\,d\mu, \tag{7}$$

where $a > \operatorname{Re}\lambda$, $x > 0$, and

$$F(\lambda)=\frac{1}{2\pi i}\int_{b-i\infty}^{b+i\infty}\Gamma(\mu-\lambda)(-ix)^{\lambda-\mu}\Phi(\mu)\,d\mu, \tag{8}$$

where $b > \operatorname{Re}\lambda$, $x > 0$, are reciprocal. These transformations establish a connection between the functions $F(\lambda)$ and $\Phi(\lambda)$ in the half plane $\operatorname{Re}\lambda < a$.

Similarly, it follows from formulas (16) and (18) of §3.2 that the integral transformations

$$\Phi(\lambda)=\frac{1}{2\pi i}\int_{a-i\infty}^{a+i\infty}\Gamma(\lambda-\mu)(ix)^{\mu-\lambda}F(\mu)\,d\mu, \tag{9}$$

where $a < \operatorname{Re}\lambda$, $x > 0$, and

$$F(\lambda)=\frac{1}{2\pi i}\int_{b-i\infty}^{b+i\infty}\Gamma(\lambda-\mu)(-ix)^{\mu-\lambda}F(\mu)\,d\mu, \tag{10}$$

where $b > \operatorname{Re}\lambda$, $x > 0$, are reciprocal.

## §6. DECOMPOSITION OF THE QUASI-REGULAR REPRESENTATION OF THE GROUP $MH(2)$

1. The quasi-regular representation of $MH(2)$. We shall denote by $\mathfrak{L}^2$ the space of functions $f(x)$ on the pseudo-euclidean plane for which

$$\|f\|^2\equiv\int|f(\mathbf{x})|^2\,d\mathbf{x}<+\infty, \tag{1}$$

where $d\mathbf{x} = dx\,dy$ for $\mathbf{x} = (x, y)$. Putting

$$L(g)f(\mathbf{x})=f(g^{-1}\mathbf{x}), \tag{2}$$

where $g \in MH(2)$, we get the quasi-regular representation of $MH(2)$. This representation is unitary relative to the norm (1).

In order to decompose the representation $L(g)$ into irreducible components, we use the Fourier transform of the function $f(\mathbf{x})$, writing it in the form

$$F(\mathbf{y})=\int f(\mathbf{x})\,e^{i[\mathbf{x},\,\mathbf{y}]}\,d\mathbf{x}. \tag{3}$$

It is easy to show that operators of the quasi-regular representation $L(g)$ then turn into operators of the form

$$\hat{L}(g) F(\mathbf{y}) = e^{i[\mathbf{b}, \mathbf{y}]} F(\mathbf{y}_{-\alpha}), \tag{4}$$

where $g = g(\mathbf{b}, \alpha)$ and $\mathbf{y}_{-\alpha}$ is the point into which the point $\mathbf{y}$ is carried by the hyperbolic rotation

$$y_1' = y_1 \operatorname{ch} \alpha - y_2 \operatorname{sh} \alpha,$$
$$y_2' = -y_1 \operatorname{sh} \alpha + y_2 \operatorname{ch} \alpha.$$

The representation $\hat{L}(g)$ is decomposed into irreducible components in just the same way as it was done in §5 of Chapter IV for the group of euclidean motions Only instead of spaces of functions on circles centered at the origin, we consider spaces of functions on the hyperbolas $[\mathbf{y}, \mathbf{y}] = C$. Namely, we denote by $\mathfrak{H}_R^{(1)}$ the space of functions

$$F_R^{(1)}(\psi) = F(R \operatorname{ch} \psi, R \operatorname{sh} \psi), \qquad -\infty < R < \infty, \ R \neq 0,$$

such that

$$\int_{-\infty}^{\infty} |F(R \operatorname{ch} \psi, R \operatorname{sh} \psi)|^2 \, d\psi < +\infty,$$

and by $\mathfrak{H}_R^{(2)}$ we denote the space of functions of the form

$$F_R^{(2)}(\psi) = F(R \operatorname{sh} \psi, R \operatorname{ch} \psi), \quad -\infty < R < \infty, \quad R \neq 0. \tag{5}$$

One can easily see that all these spaces are invariant for the operators (4). In accordance with this, we get the decomposition of $\hat{L}(g)$ into irreducible representations of the form

$$\hat{L}_R^{(1)}(g) F_R^{(1)}(\psi) = e^{iR(b_1 \operatorname{ch} \psi - b_2 \operatorname{sh} \psi)} F_R^{(1)}(\psi - \alpha), \tag{6}$$

$$\hat{L}_R^{(2)}(g) F_R^{(2)}(\psi) = e^{iR(b_1 \operatorname{sh} \psi - b_2 \operatorname{ch} \psi)} F_R^{(2)}(\psi - \alpha), \tag{7}$$

where $R$ traverses the rays $(0, \infty)$ and $(-\infty, 0)$, and

$$\mathbf{y} = g(\mathbf{b}, \alpha) = g(b_1, b_2, \alpha).$$

We have

$$\hat{L}(g) = \int_{-\infty}^{\infty} L_R^{(1)}(g) R \, dR + \int_{-\infty}^{\infty} L_R^{(2)}(g) R \, dR. \tag{8}$$

Indeed.

$$\|F\|^2=\int|F(y)|^2\,dy$$
$$=\int_{-\infty}^{\infty} R\,dR\int_{-\infty}^{\infty}|F(R\operatorname{ch}\psi,\ R\operatorname{sh}\psi)|^2\,d\psi$$
$$+\int_{-\infty}^{\infty} R\,dR\int_{-\infty}^{\infty}|F(R\operatorname{sh}\psi,\ R\operatorname{ch}\psi)|^2\,d\psi. \tag{9}$$

If $\hat{L}(g)$ is constructed not in the space of square-integrable functions, but in the space of generalized functions, then, besides components (6) and (7), the decomposition contains other components, corresponding to the spaces of functions on the rays $(y, y)$, $(y, -y)$, $(-y, y)$, $(-y, -y)$ at the point $N(0, 0)$. These components are reducible representations. Their decomposition into irreducible ones is effected with the help of the Mellin transform.

2. **Integral transformations.** Let $f(\mathbf{x})$ be an infinitely differential finite function and $F(\mathbf{y})$ its Fourier transform:

$$F(\mathbf{y})=\int f(\mathbf{x})\,e^{i[\mathbf{x},\,\mathbf{y}]}\,d\mathbf{x}. \tag{1}$$

We put

$$f_1(r,\ \varphi)=f(r\operatorname{ch}\varphi,\ r\operatorname{sh}\varphi),$$
$$f_2(r,\ \varphi)=f(r\operatorname{sh}\varphi,\ r\operatorname{ch}\varphi),$$
$$f_3(r,\ \varphi)=f(-r\operatorname{ch}\varphi,\ -r\operatorname{sh}\varphi),$$
$$f_4(r,\ \varphi)=f(-r\operatorname{sh}\varphi,\ -r\operatorname{ch}\varphi),$$

where $0<r<\infty$, $-\infty<\phi<\infty$.

The functions $F_k(R, \psi)$, $1\le k\le 4$ are defined by analogous formulas. Thus the functions $F_j(r, \phi)$ coincide with $f(x)$ in one of the quadrants and are equal to zero outside it. The functions $F_j(R, \psi)$ have a similar significance.

It follows from (1) that

$$F_j(R,\ \psi)=\sum_{k=1}^{4}\int_0^{\infty} r\,dr\int_{-\infty}^{\infty} f_k(r,\ \varphi)\,e^{iRra_{jk}(\psi-\varphi)}\,d\varphi, \tag{2}$$

where

$$a_{11}(\alpha)=-a_{13}(\alpha)=-a_{22}(\alpha)=a_{24}(\alpha)$$
$$=-a_{31}(\alpha)=a_{33}(\alpha)=a_{42}(\alpha)=-a_{44}(\alpha)=\operatorname{ch}\alpha$$

and

$$-a_{12}(\alpha)=a_{14}(\alpha)=a_{21}(\alpha)=-a_{23}(\alpha)=a_{32}(\alpha)$$
$$=-a_{34}(\alpha)=-a_{41}(\alpha)=a_{43}(\alpha)=\operatorname{sh}\alpha.$$

Similarly, it follows from the inversion formula for one Fourier transform

$$f(\mathbf{x})=\frac{1}{(2\pi)^2}\int F(y)\,e^{-i[\mathbf{x},\mathbf{y}]}\,dy \tag{3}$$

that

$$f_k(r,\varphi)=\frac{1}{(2\pi)^2}\sum_{j=1}^{4}\int_0^\infty r\,dr\int_{-\infty}^{\infty}F_j(R,\psi)\,e^{-iRra_{jk}(\psi-\varphi)}\,d\psi. \tag{4}$$

We have seen in §3.1 that to each value of $R$ and of the index $j$ there corresponds an irreducible representation of the group $MH(2)$, realized in the space of functions $F_j(R,\psi)$ (for a given $R$). Here, in order to obtain the representations operators in the form of integral operators, it is necessary to pass from the functions $F_j(R,\psi)$ to their Fourier transforms with respect to $\psi$:

$$\Phi_j(R,\lambda)=\int_{-\infty}^{\infty}F_j(R,\psi)\,e^{i\lambda\psi}\,d\psi. \tag{5}$$

Let us express $\Phi_j(R,\lambda)$ directly by means of the function $f(\mathbf{x})$. To do this we substitute in formula (5) the expression (2) for $F_j(R,\psi)$, and change the order of integration. Utilizing the results of §3.4, we obtain

$$\Phi_j(R,\lambda)=\sum_{k=1}^{4}\int_0^\infty r\,dr\int_{-\infty}^{\infty}f_k(r,\varphi)\,e^{i\lambda\varphi}A_{jk}(Rr,\lambda)\,d\varphi, \tag{6}$$

where

$$A_{11}(\rho,\lambda)=A_{24}(\rho,\lambda)=A_{33}(\rho,\lambda)=A_{42}(\rho,\lambda)=\pi i e^{\frac{\lambda\pi}{2}}H^{(1)}_{-i\lambda}(\rho), \tag{7'}$$

$$A_{13}(\rho,\lambda)=A_{22}(\rho,\lambda)=A_{31}(\rho,\lambda)=A_{44}(\rho,\lambda)=-\pi i e^{\frac{\lambda\pi}{2}}H^{(2)}_{i\lambda}(\rho), \tag{7''}$$

$$A_{14}(\rho,\lambda)=A_{21}(\rho,\lambda)=A_{32}(\rho,\lambda)=A_{43}(\rho,\lambda)=2e^{-\frac{\lambda\pi}{2}}K_{i\lambda}(\rho), \tag{7'''}$$

$$A_{12}(\rho,\lambda)=A_{23}(\rho,\lambda)=A_{34}(\rho,\lambda)=A_{41}(\rho,\lambda)=2e^{\frac{\lambda\pi}{2}}K_{i\lambda}(\rho). \tag{7''''}$$

An expression for $f_k(r,\phi)$ in terms of $\Phi_y(R,\lambda)$ is obtained in just the same way with the help of the inversion formulas (3) and

$$F_j(R, \psi) = \frac{1}{2\pi} \int_{-\infty}^{\infty} \Phi_j(R, \lambda) e^{-i\lambda\psi} d\lambda \tag{8}$$

for the Fourier transform (5). As a result, we get

$$f_k(r, \varphi) = \frac{1}{(2\pi)^2} \sum_{j=1}^{4} \int_0^{\infty} R\,dR \int_{-\infty}^{\infty} \Phi_j(R, \lambda) e^{-i\lambda\varphi} A_{kj}(Rr, \lambda)\, d\lambda, \tag{9}$$

where $A_{kj}(\rho, \lambda)$ are given by (7′)–(7″″).

Thus, we have obtained reciprocal integral transformations (6) and (9), connecting the functions $f_k(r, \phi)$, $1 \leq k \leq 4$ with functions $\Phi_j(R, \lambda)$, $1 \leq j \leq 4$. These transformations easily lead to transformations for functions of one variable. Namely, we introduce the function

$$P_k(r, \lambda) = \int_{-\infty}^{\infty} f_k(r, \varphi) e^{i\lambda\varphi} d\varphi \tag{10}$$

which is the Fourier transform for $f_k(r, \phi)$ with respect to $\phi$. It follows from (6) that

$$\Phi_j(R, \lambda) = \sum_{k=1}^{4} \int_0^{\infty} A_{jk}(Rr, \lambda) P_k(r, \lambda)\, r\, dr, \quad 1 \leqslant j \leqslant 4. \tag{11}$$

On the other hand, if one regards formula (9) as an inverse Fourier transform formula with respect to $\lambda$, then

$$P_k(r, \lambda) = \frac{1}{(2\pi)^2} \sum_{j=1}^{4} \int_0^{\infty} A_{kj}(Rr, \lambda) \Phi_j(R, \lambda)\, R\, dR, \quad 1 \leqslant k \leqslant 4. \tag{12}$$

We have thus proved that if

$$\Phi_j(R) = \sum_{k=1}^{4} \int_0^{\infty} A_{jk}(Rr, \lambda) P_k(r)\, r\, dr, \tag{13}$$

where $-\infty < \lambda < \infty$, $1 \leq j \leq 4$, then

$$P_k(r) = \frac{1}{(2\pi)^2} \sum_{j=1}^{4} \int_0^{\infty} A_{kj}(Rr, \lambda) \Phi_j(R)\, R\, dR, \tag{14}$$

where $1 \leq k \leq 4$. For $P_1(r) = P_2(r) = P_3(r) = P_4(r) = P(r)$, we get

$$\Phi_1(R) = \Phi_2(R) = \Phi_3(R) = \Phi_4(R) = \Phi(R).$$

In this case, it follows from (13) and (14) that if

$$\Phi(R) = \int_0^\infty A(Rr, \lambda) P(r)\, dr, \tag{15}$$

then

$$P(r) = \frac{1}{(2\pi)^2} \int_0^\infty A(Rr, \lambda) \Phi(R) R\, dR, \tag{16}$$

where

$$A(\rho, \lambda) = \pi i e^{\frac{\lambda\pi}{2}} [H_{-i\lambda}^{(1)}(\rho) - H_{i\lambda}^{(2)}(\rho)] + 4 \operatorname{ch} \frac{\lambda\pi}{2} K_{i\lambda}(\rho). \tag{17}$$

Utilizing the results of §3.4, it is easy to show that the kernel $A(\rho, \lambda)$ can be presented as follows:

$$A(\rho, \lambda) = \frac{\pi i}{\operatorname{sh} \frac{\lambda\pi}{2}} [J_{i\lambda}(\rho) + e^{\frac{\lambda\pi}{2}} J_{i\lambda}(i\rho) - J_{-i\lambda}(\rho) - e^{-\frac{\lambda\pi}{2}} J_{-i\lambda}(i\rho)]$$

$$= \frac{\pi i}{\operatorname{sh} \frac{\lambda\pi}{2}} [J_{i\lambda}(\rho) + I_{i\lambda}(\rho) - J_{-i\lambda}(\rho) - I_{-i\lambda}(\rho)].$$

Another special case of (13) and (14) is obtained if we take $P_1(r, \lambda) = P_3(r, \lambda) \equiv P(r, \lambda)$, $P_2(r) = P_4(r) = 0$. We find that if

$$\Phi_1(R, \lambda) = \pi i e^{\frac{\lambda\pi}{2}} \int_0^\infty [H_{-i\lambda}^{(1)}(Rr) - H_{i\lambda}^{(2)}(Rr)] P(r, \lambda) r\, dr, \tag{18}$$

$$\Phi_2(R, \lambda) = 4 \operatorname{ch} \frac{\lambda\pi}{2} \int_0^\infty K_{i\lambda}(Rr) P(r, \lambda) r\, dr, \tag{19}$$

then

$$P(r, \lambda) = \pi i e^{\frac{\lambda\pi}{2}} \int_0^\infty [H_{-i\lambda}^{(1)}(Rr) - H_{i\lambda}^{(2)}(Rr)] \Phi_1(R, \lambda) R\, dR$$

$$+ 4 \operatorname{ch} \frac{\lambda\pi}{2} \int_0^\infty K_{i\lambda}(Rr) \Phi_2(R, \lambda) R\, dR, \tag{20}$$

and

$$0=4\operatorname{ch}\frac{\lambda\pi}{2}\int_0^\infty K_{i\lambda}(Rr)\,\Phi_1(R,\lambda)\,R\,dR$$
$$+\pi ie^{\frac{\lambda\pi}{2}}\int_0^\infty\left[H_{-i\lambda}^{(1)}(Rr)-H_{i\lambda}^{(2)}(Rr)\right]\Phi_2(R,\lambda)\,R\,dR. \quad (21)$$

## CHAPTER VI

# REPRESENTATIONS OF THE GROUP $QU(2)$ OF UNIMODULAR QUASI-UNITARY MATRICES OF THE SECOND ORDER AND THE FUNCTIONS OF LEGENDRE AND JACOBI

In this chapter we shall consider representations of the group $QU(2)$ of unimodular quasi-unitary matrices, in many respects similar to the group $SU(2)$. However, unlike $SU(2)$, the group of quasi-unitary matrices is noncompact. Because of this it possesses a continuous series of unitary representations. The special functions to which a consideration of this group leads are allied to the Legendre and Jacobi polynomials, studied in Chapter III. We shall call them Legendre functions and Jacobi functions.

## §1. THE GROUP $QU(2)$

1. Description. We call matrices of the form $\begin{pmatrix} \alpha & \beta \\ \bar{\beta} & \bar{\alpha} \end{pmatrix}$, where $|\alpha|^2 - |\beta|^2 = 1$, quasi-unitary unimodular matrices of the second order.

It is easy to see that these matrices form a group. We denote it by $QU(2)$. Evidently, if a matrix $g$ belongs to the group $QU(2)$, then $gsg^* = s$, where $s = \begin{pmatrix} 1 & 0 \\ 0 & -1 \end{pmatrix}$, and $g^* = \begin{pmatrix} \bar{\alpha} & \beta \\ \bar{\beta} & \alpha \end{pmatrix}$ is the matrix Hermitian adjoint to $g$. Conversely, if $gsg^* = s$ and $\operatorname{Det} g = 1$, then $g$ belongs to the group $QU(2)$. Thus $QU(2)$ can be defined as the group of matrices $g$ such that $gsg^* = s$ and $\operatorname{Det} g = 1$.

We show that $QU(2)$ is isomorphic to the group $SL(2, R)$ of real unimodular matrices of the second order. Indeed, with each matrix $g = \begin{pmatrix} \alpha & \beta \\ \gamma & \delta \end{pmatrix}$ of $SL(2, R)$ we associate a matrix $h = t^{-1}gt$, where $t = \begin{pmatrix} 1 & i \\ i & 1 \end{pmatrix}$. Then

$$h=\begin{pmatrix} a & b \\ \bar{b} & \bar{a} \end{pmatrix}, \qquad |a|^2-|b|^2=1, \tag{1}$$

where

$$a=\frac{1}{2}[\alpha+\delta+i(\beta-\gamma)], \tag{2}$$

$$b=\frac{1}{2}[\beta+\gamma+i(\alpha-\delta)]. \tag{2'}$$

Therefore the matrix $h$ belongs to $QU(2)$.

This establishes a mapping of $SL(2, R)$ into $QU(2)$. A simple calculation shows that this mapping is an isomorphism. However, for $n > 2$ the groups $SL(n, R)$ and $QU(n)$ are no longer isomorphic.

We mention another realization of $QU(2)$ (more precisely, of its factor group with respect to the center).

In the three-dimensional linear space $E_3$ let us consider the indefinite quadratic form

$$[\mathbf{x}, \mathbf{x}] = x_1^2 - x_2^2 - x_3^2. \tag{3}$$

The equation $[\mathbf{x}, \mathbf{x}] = 0$ determines a cone in $E_3$ with vertex at the origin. It divides $E_3$ into three domains: the interior of the upper half of the cone, the interior of the lower half of the cone and the domain exterior to the cone.

We shall denote by $SH(3)$ the group of linear transformations of $E_3$, transforming each of these domains into itself. Such transformations are the analogues of rotations in three-dimensional space, with the sole difference that they act transitively not on spheres, but on hyperboloids and the cones $[\mathbf{x}, \mathbf{x}] = 0$. Therefore such transformations are called hyperbolic rotations.

The connection between the groups $QU(2)$ and $SH(3)$ is the same as the connection between $SU(2)$ and $SO(3)$. Namely, with each point $\mathbf{x}(x_1, x_2, x_3)$ of $E_3$ we associate a quasi-unitary matrix

$$h=\begin{pmatrix} x_1 & x_2+ix_3 \\ x_2-ix_3 & x_1 \end{pmatrix}, \tag{4}$$

the upper left-hand element of which is real.

If $g$ is any matrix of $QU(2)$, then $ghg^*$ is a quasi-unitary matrix, the upper left-hand element of which is real. Therefore $ghg^*$ has the form

$$ghg^*=\begin{pmatrix} y_1 & y_2+iy_3 \\ y_2-iy_3 & y_1 \end{pmatrix} \tag{5}$$

and to it corresponds a point $\mathbf{y}(y_1, y_2, y_3)$ of $E_3$. The transformation $g\mathbf{x} = \mathbf{y}$ is linear, since elements of the matrix $ghg^*$ are linearly expressed by the elements of $h$.

Thus, with each matrix $g$ of $QU(2)$, a linear transformation is associated in $E_3$. Since $g$ is unimodular,

$$\operatorname{Det} ghg^* = \operatorname{Det} h.$$

But if $h = \begin{pmatrix} x_1 & x_2 + ix_3 \\ x_2 - ix_3 & x_1 \end{pmatrix}$, then $\operatorname{Det} h = [\mathbf{x}, \mathbf{x}]$. Thus the linear transformation preserves the quadratic form $[\mathbf{x}, \mathbf{x}]$. It is easy to verify that if $x_1 > 0$, then $y_1 > 0$ also, and so the upper half of the cone $[\mathbf{x}, \mathbf{x}] = 0$ is transformed into the upper half. Consequently the linear transformation $\mathbf{y} = g\mathbf{x}$, corresponding to the matrix $g$ of $QU(2)$, belongs to $SH(3)$.

We have established a mapping of $QU(2)$ into $SH(3)$. This mapping is homomorphic; moreover, the homomorphism kernel is the center of $QU(2)$, consisting of the matrices $e = \begin{pmatrix} 1 & 0 \\ 0 & 1 \end{pmatrix}$ and $-e = \begin{pmatrix} -1 & 0 \\ 0 & -1 \end{pmatrix}$.

We may also note the realizations of $QU(2)$ and $SL(2, R)$ as groups of linear fractional transformations of the complex plane. With each second order matrix $g = \begin{pmatrix} \alpha & \beta \\ \gamma & \delta \end{pmatrix}$ we associate a linear fractional transformation

$$w = gz = \frac{\alpha z + \beta}{\gamma z + \delta}.$$

A simple calculation shows that

$$(g_1 g_2) z = g_1 (g_2 z)$$

and so the group of linear-fractional transformations is a homomorphic image of the group $SL(2, C)$ of complex unimodular matrices of the second order. The kernel of this homomorphism consists of the matrices $e = \begin{pmatrix} 1 & 0 \\ 0 & 1 \end{pmatrix}$ and $-e = \begin{pmatrix} -1 & 0 \\ 0 & -1 \end{pmatrix}$

Under this homomorphism, to matrices of the form $\begin{pmatrix} \alpha & \beta \\ \bar{\beta} & \bar{\alpha} \end{pmatrix}$, $|\alpha|^2 - |\beta|^2 = 1$, i.e. to elements of the group $QU(2)$, correspond linear fractional transformations which transform the unit circle and its complement into themselves. To matrices of the subgroup $SL(2, R)$, however, correspond linear fractional transformations which transform the upper and the lower half planes into themselves. The isomorphism between $QU(2)$ and $SL(2, R)$ is linked with the existence of the linear fractional transformation $w = (z - i)/(z + i)$, which maps the upper half plane into the unit circle.

2. **Subgroups of $SL(2, R)$.** The representations of the groups $SU(2)$, $M(2)$ and $MH(2)$, considered in previous chapters, possessed the following important property: the representation operators corresponding to elements of certain subgroups were given in a suitable realization by diagonal matrices. In $SU(2)$ such such a subgroup was the subgroup of matrices of the form $\begin{pmatrix} e^{it/2} & 0 \\ 0 & e^{-it/2} \end{pmatrix}$, in $M(2)$ it was the subgroup of rotations around the origin, and in $MH(2)$ it was the subgroup of hyperbolic rotations. Due to such a choice of the realization we succeeded in expressing the matrix elements of representations as a product of three functions, each of which depended only on one variable.

In group $SL(2, R)$ there are three one-parametric subgroups such that any other one-parametric subgroup is conjugate to one of them. These subgroups are the subgroup $SO(2)$ of orthogonal real matrices $\begin{pmatrix} \cos\frac{t}{2} & -\sin\frac{t}{2} \\ \sin\frac{t}{2} & \cos\frac{t}{2} \end{pmatrix}$, the subgroup $SH(2)$ of diagonal matrices $\begin{pmatrix} e^{-t/2} & 0 \\ 0 & e^{t/2} \end{pmatrix}$ and the subgroup $Z$ of triangular matrices of the form $\begin{pmatrix} 1 & 0 \\ t & 1 \end{pmatrix}$.

To each of these subgroups corresponds a realization of representations of $SL(2, R)$, under which the elements of that subgroup are expressed by diagonal matrices (or operators of multiplication by a function – "continual diagonal matrices").

In this chapter we shall consider a realization under which elements of the compact subgroup $SO(2)$ are expressed by diagonal matrices. For studying this realization, it is more convenient to consider the group $QU(2)$ of matrices of the form $\begin{bmatrix} \alpha & \beta \\ \bar{\beta} & \bar{\alpha} \end{bmatrix}$, $|\alpha^2| - |\beta^2| = 1$. Under the isomorphism of $SL(2, R)$ and $QU(2)$, established above, to the subgroup $SO(2)$ corresponds the subgroup of diagonal matrices of the form $\begin{pmatrix} e^{it/2} & 0 \\ 0 & e^{-it/2} \end{pmatrix}$ in $QU(2)$. Thus, in this chapter we shall study representations of $QU(2)$ under which elements of a diagonal subgroup are expressed by diagonal matrices. Under such a realization, the matrix elements are expressed by means of functions similar to the Jacobi and Legendre functions studied in Chapter III.

3. Parameterizations of $QU(2)$. Every matrix $g = \begin{Vmatrix} \alpha & \beta \\ \bar{\beta} & \bar{\alpha} \end{Vmatrix}$ of $QU(2)$ can be specified by two complex numbers $\alpha$ and $\beta$, such that $|\alpha|^2 - |\beta|^2 = 1$. In many cases it is more convenient to replace these parameters by Euler angles. We may recall that, according to § 1.4 of Chapter III, every complex unitary matrix $g = \begin{Vmatrix} \alpha & \beta \\ \gamma & \delta \end{Vmatrix}$ is specified by three complex numbers $\phi, \theta, \psi$, namely the Euler angles of this matrix. The matrix with Euler angles $\phi, \theta, \psi$ has the form

$$g = \begin{pmatrix} \cos\frac{\theta}{2} e^{\frac{i(\varphi+\psi)}{2}} & i\sin\frac{\theta}{2} e^{\frac{i(\varphi-\psi)}{2}} \\ i\sin\frac{\theta}{2} e^{\frac{i(\psi-\varphi)}{2}} & \cos\frac{\theta}{2} e^{-\frac{i(\varphi+\psi)}{2}} \end{pmatrix}. \tag{1}$$

Let us clarify the restrictions imposed on the Euler angles if a matrix $g = \begin{Vmatrix} \alpha & \beta \\ \gamma & \delta \end{Vmatrix}$ is to belong to $QU(2)$. For this, the equalities $\delta = \bar{\alpha}$ and $\gamma = \bar{\beta}$ must be satisfied, i.e. one must have

$$\cos\frac{\theta}{2} e^{-\frac{i(\varphi+\psi)}{2}} = \overline{\cos\frac{\theta}{2}} e^{-\frac{i(\bar{\varphi}+\bar{\psi})}{2}} \tag{2}$$

and

$$\sin\frac{\theta}{2} e^{\frac{i(\psi-\varphi)}{2}} = -\overline{\sin\frac{\theta}{2}} e^{\frac{i(\bar{\psi}-\bar{\varphi})}{2}}. \tag{3}$$

We rewrite these equalities as

$$\cos\frac{\theta}{2} = \overline{\cos\frac{\theta}{2}} e^{\frac{i(\varphi-\bar{\varphi}+\psi-\bar{\psi})}{2}} \tag{2'}$$

and

$$\sin\frac{\theta}{2} = -\overline{\sin\frac{\theta}{2}} e^{\frac{i(\bar{\psi}-\psi+\varphi-\bar{\varphi})}{2}}. \tag{3'}$$

But $(i/2)(\phi - \bar{\phi} + \psi - \bar{\psi})$ and $(i/2)(\bar{\psi} - \psi + \phi - \bar{\phi})$ are real numbers. Therefore, if $g = g(\phi, \theta, \psi)$ belongs to $QU(2)$, then $\cos\theta/2$ is real and $\sin\theta/2$ purely imaginary. Consequently, the Euler angle $\theta$ is purely imaginary and so $\tau = \theta i$ is real.

Further, it follows from equalities (2) and (3) that $\phi + \psi$ and $\phi - \psi$ are real, and so the Euler angles $\phi$ and $\psi$ are real. Thus the Euler angles of a matrix

$g = \begin{bmatrix} \alpha & \beta \\ \bar\beta & \bar\alpha \end{bmatrix}$ of $QU(2)$ have the form $\phi, -\tau i, \psi$, where $\phi, \tau, \psi$ are real numbers. Taking into account the restrictions on the Euler angles of matrices of the group $SL(2, C)$, we find that the parameters $\phi, \tau, \psi$ vary as follows:

$$\left.\begin{aligned} 0 \leqslant \varphi \leqslant 2\pi, \\ 0 < \tau < \infty, \\ -2\pi \leqslant \psi < 2\pi. \end{aligned}\right\} \tag{4}$$

It is easy to show that, to different points of the domain given by (4), there correspond different matrices $g$ of the group $QU(2)$. Moreover, the matrices, written in terms of the parameters $\phi, \tau, \psi$, have the form

$$g = \begin{pmatrix} \operatorname{ch}\frac{\tau}{2} e^{\frac{i(\varphi+\psi)}{2}} & \operatorname{sh}\frac{\tau}{2} e^{\frac{i(\varphi-\psi)}{2}} \\ \operatorname{sh}\frac{\tau}{2} e^{\frac{i(\psi-\varphi)}{2}} & \operatorname{ch}\frac{\tau}{2} e^{-\frac{i(\varphi+\psi)}{2}} \end{pmatrix}. \tag{5}$$

Thus $QU(2)$ is also one of the real forms of the group $SL(2, C)$.

In the sequel it will be more convenient to utilize in the group $QU(2)$, instead of the Euler angles $\phi, \theta, \psi$, the corresponding real parameters $\phi, \tau, \psi$. In this chapter by Euler angles we shall understand the real numbers $\phi, \tau, \psi$. Let us find the rule of transformation of these parameters for multiplication of the elements of $QU(2)$. Let $g_1 = (0, \tau_1, 0)$, $g_2 = (\phi_2, \tau_2, 0)$ and $g_1 g_2 = (\phi, \tau, \psi)$. Utilizing formulas (2)–(2″) of §1.2, Chapter III, we get

$$\operatorname{ch}\tau = \operatorname{ch}\tau_1 \operatorname{ch}\tau_2 + \operatorname{sh}\tau_1 \operatorname{sh}\tau_2 \cos\varphi_2, \tag{6}$$

$$e^{i\varphi} = \frac{\operatorname{sh}\tau_1 \operatorname{ch}\tau_2 + \operatorname{ch}\tau_1 \operatorname{sh}\tau_2 \cos\varphi_2 + i \operatorname{sh}\tau_2 \sin\varphi_2}{\operatorname{sh}\tau}, \tag{6'}$$

$$e^{\frac{i(\varphi+\psi)}{2}} = \frac{\operatorname{ch}\frac{\tau_1}{2} \operatorname{ch}\frac{\tau_2}{2} e^{\frac{i\varphi_2}{2}} + \operatorname{sh}\frac{\tau_1}{2} \operatorname{sh}\frac{\tau_2}{2} e^{-\frac{i\varphi_2}{2}}}{\operatorname{ch}\frac{\tau}{2}}. \tag{6''}$$

These formulas yield

$$\operatorname{tg}\varphi = \frac{\operatorname{sh}\tau_2 \sin\varphi_2}{\operatorname{ch}\tau_1 \operatorname{sh}\tau_2 \cos\varphi_2 + \operatorname{sh}\tau_1 \operatorname{ch}\tau_2}, \tag{7'}$$

$$\operatorname{tg}\psi = \frac{\operatorname{sh}\tau_1 \sin\varphi_2}{\operatorname{sh}\tau_1 \operatorname{ch}\tau_2 \cos\varphi_2 + \operatorname{ch}\tau_1 \operatorname{sh}\tau_2}. \tag{7''}$$

The general case is easily reduced to the one analyzed above (compare §1.2 of Chapter III).

A simple calculation shows that the element $g(\phi, \tau, \psi)$ is inverse to the element $g(\pi - \psi, \tau, -\pi - \phi)$.

4. Invariant integration. Since $QU(2)$ is locally compact, on it there exists an invariant measure $dg$, i.e. a measure such that for any finite continuous function $f(g)$ we have

$$\int f(g)\,dg = \int f(gg_0)\,dg.$$

Repeating word for word the arguments of §6.1 of Chapter III, we see that the expression of $dg$ in Euler parameters has the form

$$dg = \operatorname{sh}\tau \, d\tau \, d\phi \, d\psi. \tag{1}$$

It is easy to verify that it is invariant not only for the transformations $g \to gg_0$, but also for the transformations $g \to g_0 g$ and $g \to g^{-1}$ Thus invariant integration on $QU(2)$ is given as follows:

$$\int f(g)\,dg = \int_0^\infty \int_0^{2\pi} \int_{-2\pi}^{2\pi} f(\varphi, \tau, \psi) \operatorname{sh}\tau \, d\psi \, d\varphi \, d\tau. \tag{2}$$

This can also be written in the form

$$\int f(g)\,dg = \int f(\alpha, \beta)\,\delta(|\alpha|^2 - |\beta|^2 - 1)\,d\alpha_1\,d\alpha_2\,d\beta_1\,d\beta_2, \tag{2'}$$

where $\alpha = \alpha_1 + i\alpha_2$, $\beta = \beta_1 + i\beta_2$, $g = \begin{pmatrix} \alpha & \beta \\ \bar\beta & \bar\alpha \end{pmatrix}$.

5. The Lie algebra. Let us determine the Lie algebra of $QU(2)$. We consider three one-parametric subgroups of this group. The subgroup $\Omega_1$ consists of matrices of the form

$$\omega_1(t) = \begin{pmatrix} \operatorname{ch}\frac{t}{2} & \operatorname{sh}\frac{t}{2} \\ \operatorname{sh}\frac{t}{2} & \operatorname{ch}\frac{t}{2} \end{pmatrix}, \tag{1}$$

the subgroup $\Omega_2$, of matrices of the form

$$\omega_2(t) = \begin{pmatrix} \operatorname{ch}\frac{t}{2} & i\operatorname{sh}\frac{t}{2} \\ -i\operatorname{sh}\frac{t}{2} & \operatorname{ch}\frac{t}{2} \end{pmatrix} \tag{1'}$$

and the subgroup $\Omega_3$, of matrices of the form

$$\omega_3(t)=\begin{pmatrix} e^{\frac{it}{2}} & 0 \\ 0 & e^{-\frac{it}{2}} \end{pmatrix}. \qquad (1'')$$

Tangent matrices to these subgroups have the following form:

$$a_1=\frac{1}{2}\begin{pmatrix} 0 & 1 \\ 1 & 0 \end{pmatrix}, \qquad (2)$$

$$a_2=\frac{i}{2}\begin{pmatrix} 0 & 1 \\ -1 & 0 \end{pmatrix}, \qquad (2')$$

$$a_3=\frac{i}{2}\begin{pmatrix} 1 & 0 \\ 0 & -1 \end{pmatrix}. \qquad (2'')$$

As these matrices are linearly independent, they form a basis of the Lie algebra of $QU(2)$.

The commutation relations for $a_1, a_2, a_3$ are

$$[a_1, a_2]=-a_3, \qquad (3)$$

$$[a_2, a_3]=a_1, \qquad (3')$$

$$[a_3, a_1]=a_2. \qquad (3'')$$

They differ from the commutation relations for $SU(2)$ only in the sign of $a_3$ in formula (3) (see §1.3 of Chapter III).

## §2. IRREDUCIBLE REPRESENTATIONS OF $QU(2)$

In this section we shall construct a representation series $T_\chi(g)$ of the group $QU(2)$. These representations are specified by a complex number $l$ and a number $\epsilon$, taking the values 0 and $\frac{1}{2}$, and are constructed in the space of homogeneous functions $f(z)$ of a given parity. We shall show that the representations $T_\chi(g)$ are reducible if and only if $2l$ and $2\epsilon$ are integers of the same parity, and we shall also indicate the conditions under which these representations are equivalent or unitary.

1. The space $\mathfrak{D}_\chi$. We denote by $\chi$ the set $\chi=(l, \epsilon)$ of complex numbers $l$ and numbers $\epsilon$ taking the values 0 and $\frac{1}{2}$. With each such set we associate a space $\mathfrak{D}_\chi$ of functions $\phi(z)$ of the complex variable $z = x + iy$, such that:

1) the functions $\phi(z)$ are infinitely differentiable with respect to $x$ and $y$ at all points $z = x + iy$, except the point $z = 0$;

2) for any positive number $a$ one has the equality

$$\phi(az) = a^{2l}\phi(z); \tag{1}$$

3) the functions $\phi(z)$ have given parity: for $\epsilon = 0$ they are even, and for $\epsilon = \frac{1}{2}$ they are odd:

$$\phi(-z) = (-1)^{2\epsilon}\phi(z). \tag{2}$$

If $\Gamma$ is some curve on the complex plane, which intersects any straight line passing through the point $z = 0$ at one and only one point, then a function $\phi(z)$ of the space $\mathfrak{D}_\chi$ is uniquely determined by its values on this curve. Namely, if $z$ is a point of the complex plane and $z_0$ is the point of intersection of the curve $\Gamma$ and the line joining this point to the origin, then

$$\varphi(z) = \left|\frac{z}{z_0}\right|^{2(l-\epsilon)}\left(\frac{z}{z_0}\right)^{2\epsilon}\varphi(z_0). \tag{3}$$

Therefore $\mathfrak{D}_\chi$ can be regarded as a space of functions on $\Gamma$.

If $\Gamma$ intersects the straight lines emanating from the origin at several points, then $\mathfrak{D}_\chi$ is realized as the space of functions given on $\Gamma$ and satisfying certain additional conditions, arising from the homogeneity and parity of $\phi(z)$.

For example, if $\Gamma$ is the unit circle, then for $\epsilon = 0$ the space $\mathfrak{D}_\chi$ is realized as the space of infinitely differentiable even functions on the circle, and for $\epsilon = \frac{1}{2}$ as the space of infinitely differentiable odd functions on the circle.[1]

It will be convenient for us to realize the space $\mathfrak{D}_\chi$ in another way on the circle. Namely, for $\epsilon = 0$ with each function $\phi(z)$ we associate a function $f(e^{i\theta})$, defined by

$$f(e^{i\theta}) = \varphi(e^{\frac{i\theta}{2}}). \tag{4}$$

Since $\phi(z)$ is even, the function $f(e^{i\theta})$ is uniquely defined. If $\epsilon = \frac{1}{2}$, we take

$$f(e^{i\theta}) = e^{\frac{i\theta}{2}}\varphi(e^{\frac{i\theta}{2}}). \tag{5}$$

This function is uniquely defined, since $\phi(z)$ is odd. In this way, for any $\chi = (l, \epsilon)$ the space $\mathfrak{D}_\chi$ can be realized as the space $\mathfrak{D}$ of infinitely differentiable functions on the circle.

---

[1] A function given on the circle is said to be even if it takes equal values at diametrically opposite points, and odd if its values at diametrically opposite points differ only in sign.

2. **The representation** $T_\chi(g)$. With each element $g = \begin{pmatrix} \alpha & \beta \\ \bar\beta & \bar\alpha \end{pmatrix}$ of $QU(2)$ we associate an operator $T_\chi(g)$ in $\mathfrak{D}_\chi$, defined by the formula

$$T_\chi(g)\phi(z) = \phi(\alpha z + \overline{\beta z}). \tag{1}$$

Evidently, the function $T_\chi(g)\phi(z)$ has the same degree of homogeneity $2l$ and parity $2\epsilon$ as $\phi(z)$, and therefore $T_\chi(g)$ transforms functions of $\mathfrak{D}_\chi$ into functions of the same space. Further, a simple calculation shows that

$$T_\chi(g_1)T_\chi(g_2) = T_\chi(g_1g_2), \tag{2}$$

and so $T_\chi(g)$ is a representation of $QU(2)$.

Let us find the expression for operators of the representation $T_\chi(g)$ under the realization of $\mathfrak{D}_\chi$ as the space $\mathfrak{D}$ of infinitely differentiable functions on the circle. For $\epsilon = 0$ we have taken $f(e^{i\theta}) = \phi(e^{i\theta/2})$. Hence

$$T_\chi(g)f(e^{i\theta}) = T_\chi(g)\varphi(e^{\frac{i\theta}{2}}) = \varphi(\alpha e^{\frac{i\theta}{2}} + \bar\beta e^{-\frac{i\theta}{2}})$$
$$= \left|\alpha e^{\frac{i\theta}{2}} + \bar\beta e^{-\frac{i\theta}{2}}\right|^{2l} \varphi\left(\frac{\alpha e^{\frac{i\theta}{2}} + \bar\beta e^{-\frac{i\theta}{2}}}{\left|\alpha e^{\frac{i\theta}{2}} + \bar\beta e^{-\frac{i\theta}{2}}\right|}\right).$$

We denote $(\alpha e^{i\theta/2} + \bar\beta e^{-i\theta/2})/|\alpha e^{i\theta/2} + \bar\beta e^{-i\theta/2}|$ by $e^{i\psi/2}$. Then

$$e^{i\psi} = \frac{\alpha e^{i\theta} + \bar\beta}{\beta e^{i\theta} + \bar\alpha}.$$

Since, in addition,

$$|\alpha e^{\frac{i\theta}{2}} + \bar\beta e^{-\frac{i\theta}{2}}| = |\beta e^{i\theta} + \bar\alpha|,$$

it follows that

$$T_\chi(g)f(e^{i\theta}) = |\beta e^{i\theta} + \bar\alpha|^{2l}\varphi\left(e^{\frac{i\psi}{2}}\right) = |\beta e^{i\theta} + \bar\alpha|^{2l} f(e^{i\psi})$$
$$= |\beta e^{i\theta} + \bar\alpha|^{2l} f\left(\frac{\alpha e^{i\theta} + \bar\beta}{\beta e^{i\theta} + \bar\alpha}\right).$$

Thus we have proved that for $\chi = (l, 0)$ the representation $T_\chi(g)$ is realized in the space $\mathfrak{D}$ of infinitely differentiable functions on the circle, and is given by the formula

$$T_\chi(g)f(e^{i\theta}) = |\beta e^{i\theta} + \bar{\alpha}|^{2l} f\left(\frac{\alpha e^{i\theta} + \bar{\beta}}{\beta e^{i\theta} + \bar{\alpha}}\right). \tag{3}$$

One similarly proves that, for $\chi = (l, \frac{1}{2})$ the operators $T_\chi(g)$ are given by the formula

$$T_\chi(g)f(e^{i\theta}) = |\beta e^{i\theta} + \bar{\alpha}|^{2l-1}(\beta e^{i\theta} + \bar{\alpha}) f\left(\frac{\alpha e^{i\theta} + \bar{\beta}}{\beta e^{i\theta} + \bar{\alpha}}\right). \tag{4}$$

Formulas (3) and (4) can be replaced by a single formula:

$$T_\chi(g)f(e^{i\theta}) = (\beta e^{i\theta} + \bar{\alpha})^{l+\varepsilon}(\bar{\beta} e^{-i\theta} + \alpha)^{l-\varepsilon} f\left(\frac{\alpha e^{i\theta} + \bar{\beta}}{\beta e^{i\theta} + \bar{\alpha}}\right). \tag{5}$$

**3. Infinitesimal operators.** We calculate the infinitesimal operator $A_1$ of the representation $T_\chi(g)$, corresponding to the one-parametric subgroup $\Omega_1$ of matrices of the form

$$\omega_1(t) = \begin{pmatrix} \operatorname{ch}\frac{t}{2} & \operatorname{sh}\frac{t}{2} \\ \operatorname{sh}\frac{t}{2} & \operatorname{ch}\frac{t}{2} \end{pmatrix}. \tag{1}$$

To these matrices correspond the operators

$$T_\chi(\omega_1(t))f(e^{i\theta}) = \left(\operatorname{sh}\frac{t}{2}e^{i\theta} + \operatorname{ch}\frac{t}{2}\right)^{l+\varepsilon}\left(\operatorname{sh}\frac{t}{2}e^{-i\theta} + \operatorname{ch}\frac{t}{2}\right)^{l-\varepsilon} \times f\left(\frac{\operatorname{ch}\frac{t}{2}e^{i\theta} + \operatorname{sh}\frac{t}{2}}{\operatorname{sh}\frac{t}{2}e^{i\theta} + \operatorname{ch}\frac{t}{2}}\right). \tag{2}$$

Therefore

$$A_1 = \left.\frac{dT_\chi(\omega_1(t))}{dt}\right|_{t=0} = \frac{1}{2}\left[(l+\varepsilon)e^{i\theta} + (l-\varepsilon)e^{-i\theta} - 2\sin\theta\frac{d}{d\theta}\right]. \tag{3}$$

In just the same way, to the subgroup $\Omega_2$ of matrices of the form

$$\omega_2(t) = \begin{pmatrix} \operatorname{ch}\frac{t}{2} & i\operatorname{sh}\frac{t}{2} \\ -i\operatorname{sh}\frac{t}{2} & \operatorname{ch}\frac{t}{2} \end{pmatrix} \tag{4}$$

corresponds the infinitesimal operator

$$A_2 = \frac{i}{2}\left[(l+\varepsilon)e^{i\theta} - (l-\varepsilon)e^{-i\theta} + 2l\cos\theta\frac{d}{d\theta}\right]. \tag{5}$$

And to the subgroup $\Omega_3$ of diagonal matrices

$$\omega_3(t) = \begin{pmatrix} e^{\frac{it}{2}} & 0 \\ 0 & e^{-\frac{it}{2}} \end{pmatrix} \tag{6}$$

corresponds the infinitesimal operator

$$A_3 = -i\varepsilon + \frac{d}{d\theta}. \tag{7}$$

Instead of $A_1, A_2, A_3$, it is more convenient to use their linear combinations $H_+, H_-, H_3$:

$$H_+ = -A_1 - iA_2, \tag{8}$$

$$H_- = -A_1 + iA_2, \tag{8'}$$

$$H_3 = iA_3. \tag{8''}$$

From (3), (5) and (7) it follows that

$$H_+ = ie^{-i\theta}\frac{d}{d\theta} - (l-\varepsilon)e^{-i\theta}, \tag{9}$$

$$H_- = -ie^{i\theta}\frac{d}{d\theta} - (l+\varepsilon)e^{i\theta}, \tag{9'}$$

$$H_3 = \varepsilon + i\frac{d}{d\theta}. \tag{9''}$$

In $\mathfrak{D}$ there exists a basis $\{e^{-ik\theta}\}$, consisting of the eigenfunctions of $H_3$:

$$H_3 e^{-ik\theta} = (\varepsilon + k)e^{-ik\theta}. \tag{10}$$

Let us clarify how $H_+$ and $H_-$ act on the functions of this basis. Because of (9) and (9′) we have

$$H_+ e^{-ik\theta} = (k - l + \varepsilon)e^{-i(k+1)\theta} \tag{11}$$

and

$$H_- e^{-ik\theta} = -(k + l + \varepsilon)e^{-i(k-1)\theta}. \tag{12}$$

Thus $H_+$ transforms the function $e^{-ik\theta}$ corresponding to the eigenvalue $\epsilon + k$ of $H_3$ into the function for which the eigenvalue is greater by one. Similarly, an application of $H_-$ decreases the corresponding eigenvalue by one.

4. Irreducibility. We shall now prove that the representations $T_\chi(g)$, $\chi = (l, \epsilon)$ of $QU(2)$, constructed in subsection 2, are, in general, irreducible. The exception is constituted by those values of $\chi = (l, \epsilon)$ for which $2l$ is an integer of the same

parity as $2\epsilon$. In that case $l+\epsilon$ and $l-\epsilon$ are integers. Such values of $\chi$ will be called integral. So, we shall prove that for nonintegral values of $\chi$ the representation $T_\chi(g)$ is irreducible.

Let us begin by restricting $T_\chi(g)$ to the subgroup $\Omega_3$ consisting of diagonal matrices of the form $\begin{pmatrix} e^{\frac{it}{2}} & 0 \\ 0 & e^{-\frac{it}{2}} \end{pmatrix}$. We obtain the representation

$$T_\chi(g)F(e^{i\theta}) = e^{-\frac{i\epsilon t}{2}}F(e^{i(\theta+t)}) \tag{1}$$

of $\Omega_3$. This representation is equivalent to the regular representation of $SO(2)$, isomorphic to $\Omega_3$. Therefore it can be decomposed as a direct sum of one-dimensional representations, which are realized in the subspaces $\mathfrak{H}_k$ of functions of the form $c_k e^{-ik\theta}$.

Therefore (see §3.3, Chapter I) any invariant subspace $\mathfrak{T}$ in $\mathfrak{D}$ can be decomposed as a direct sum of some of the subspaces $\mathfrak{H}_k$, and consequently either is null or contains one of the functions $e^{-ik\theta}$. But because of the invariance of $\mathfrak{T}$, together with any function $e^{-ik\theta}$ it contains all functions $H_+^m e^{-ik\theta}$, $H_-^m e^{-ik\theta}$ as well. Formulas (11) and (12) of subsection 3 show that

$$H_+^m e^{-ik\theta} = \alpha_{km} e^{-i(k+m)\theta}, \tag{2}$$

$$H_-^m e^{-ik\theta} = \beta_{km} e^{-i(k-m)\theta}, \tag{3}$$

where

$$\alpha_{km} = \prod_{n=k}^{k+m-1} (n - l + \varepsilon), \tag{4}$$

$$\beta_{km} = (-1)^m \prod_{n=k}^{k+m-1} (n + l + \varepsilon). \tag{5}$$

Obviously, if $\chi=(l,\epsilon)$ is not integral, then $\alpha_{km}$ and $\beta_{km}$ do not vanish. Therefore $\mathfrak{T}$ contains all functions $e^{-in\theta}$, and so coincides with $\mathfrak{D}$. The irreducibility of $T_\chi(g)$ is thus proved.

**5. Integral representations.** Let us now consider the case when $\chi=(l,\epsilon)$ is integral, i.e. the case when $l+\epsilon$ and $l-\epsilon$ are integers. We show that in this case $T_\chi(g)$ is reducible.

If $l+\epsilon$ and $l-\epsilon$ are integers, then the functions $e^{i(l+\epsilon)\theta}$ and $e^{-i(l-\epsilon)\theta}$ are uniquely defined. We denote by $\mathfrak{D}_l^+$ the subspace in $\mathfrak{D}$ consisting of functions of the form $e^{-i(l-\epsilon)\theta}f(e^{i\theta})$, where $f(z)$ is analytic inside the unit circle. By $\mathfrak{D}_l^-$ we denote the space of functions of the form $e^{i(l+\epsilon)\theta}f(e^{i\theta})$, where $f(z)$ is analytic outside the unit circle (and at the point at infinity). Functions of $\mathfrak{D}_l^+$ can be expressed in the series

$$f(e^{i\theta}) = e^{-i(l-\epsilon)\theta} \sum_{n=-\infty}^{0} a_n e^{-in\theta}, \tag{1}$$

and functions of $\mathfrak{D}_l^-$ in the series

$$f(e^{i\theta}) = e^{i(l+\epsilon)\theta} \sum_{n=0}^{\infty} a_n e^{-in\theta}. \tag{2}$$

We show that the subspaces $\mathfrak{D}_l^+$ and $\mathfrak{D}_l^-$ in $\mathfrak{D}$ are invariant for the operators $T_\chi(g)$, $\chi = (l, \epsilon)$. Indeed,

$$T_\chi(g)\,[e^{-i(l-\epsilon)\theta} f(e^{i\theta})] = e^{-i(l-\epsilon)\theta} (\beta e^{i\theta} + \bar{\alpha})^{2l} f\left(\frac{\alpha e^{i\theta} + \bar{\beta}}{\beta e^{i\theta} + \bar{\alpha}}\right). \tag{3}$$

Since $2l$ is an integer and $|\alpha| > |\beta|$, therefore $(\beta z + \bar{\alpha})^{2l}$ is an analytic function within the unit circle. In addition, since the transformation $w = (\alpha z + \bar{\beta})/(\beta z + \bar{\alpha})$ carries the interior of the unit circle into itself, and $f(z)$ is analytic inside this circle, therefore the function $f((\alpha z + \bar{\beta})/(\beta z + \bar{\alpha}))$ is also analytic within this circle. Hence the function $(\beta z + \bar{\alpha})^{2l} f((\alpha z + \bar{\beta})/(\beta z + \bar{\alpha}))$ is analytic within the unit circle, and consequently

$$T_\chi(g)\,[e^{-i(l-\epsilon)\theta} f(e^{i\theta})]$$

belongs to $\mathfrak{D}_l^+$. By the same token, the invariance of this subspace is proved. The invariance of $\mathfrak{D}_l^-$ is proved in exactly the same way.

If $l - \epsilon < 0$, then $\mathfrak{D}_l^-$ and $\mathfrak{D}_l^+$ have null intersection. In this case the invariant subspaces in $\mathfrak{D}$ are $\mathfrak{D}_l^-$, $\mathfrak{D}_l^+$ and $\mathfrak{D}_l^- + \mathfrak{D}_l^+$. We shall denote by $T_\chi^-(g)$, $T_\chi^+(g)$ and $T_\chi^0(g)$ the representations induced by a representation in subspaces $\mathfrak{D}_l^-$ and $\mathfrak{D}_l^+$ and in the factor space $\mathfrak{D}_l^0 = \mathfrak{D}/(\mathfrak{D}_l^+ + \mathfrak{D}_l^-)$.

But if $l - \epsilon > 0$, then $\mathfrak{D}_l^-$ and $\mathfrak{D}_l^+$ have a nonnull intersection $\mathfrak{D}_l^0 = \mathfrak{D}_l^- \cap \mathfrak{D}_l^+$. In this case we shall denote by $T_\chi^-(g)$, $T_\chi^+(g)$ and $T_\chi^0(g)$ the representations induced by $T_\chi(g)$ in the factor spaces $\mathfrak{D}_l^-/\mathfrak{D}_l^0$ and $\mathfrak{D}_l^+/\mathfrak{D}_l^0$ and in the subspace $\mathfrak{D}_l^0$.

Thus, generally speaking, the three representations $T_l^-(g)$, $T_l^+(g)$ and $T_l^0(g)$ correspond to a representation $T_\chi(g)$ in the integral case. As the exception we have the case $l = -\frac{1}{2}$, $\epsilon = \frac{1}{2}$, when we have only two representations, since in this case $\mathfrak{D} = \mathfrak{D}_l^- + \mathfrak{D}_l^+$ and so $T_l^0(g)$ is the identity representation.

In just the same way as in the nonintegral case, one proves that $T_l^-(g)$, $T_l^+(g)$ and $T_l^0(g)$ are irreducible.

6. Conditions for equivalence. Here we shall formulate the conditions under which two representations $T_{\chi_1}(g)$ and $T_{\chi_2}(g)$ of the group $QU(2)$ are equivalent. We shall only give an outline of the proof. A detailed derivation of these conditions, as well as of the results of the following subsection, can be found in [15] (§4.5 of Chapter VII).

We shall call the representation $T_{\chi_1}(g)$ partially equivalent to the representation $T_{\chi_2}(g)$ if in $\mathfrak{D}$ there is a nonnull operator $Q$ such that

$$QT_{\chi_1}(g) = T_{\chi_2}(g)Q. \tag{1}$$

If $Q$ has a continuous inverse $Q^{-1}$, then partially equivalent representations are equivalent.

It follows from (1) that the infinitesimal operators $A_k^{\chi_1}$ and $A_k^{\chi_2}$ satisfy the equality

$$QA_k^{\chi_1} = A_k^{\chi_2}Q, \qquad k = 1,\ 2,\ 3. \tag{2}$$

We take $k = 3$, and observe that, in the basis $\{e^{-in\theta}\}$, the operator $A_3^{\chi}$ has a diagonal matrix, where all the diagonal elements are different from each other in the nonintegral case. Hence the matrix of $Q$ is diagonal too. In addition, we find that for the existence of $Q$ one must have $\epsilon_1 = \epsilon_2$, where $\chi_1 = (l_1, \epsilon_1)$ and $\chi_2 = (l_2, \epsilon_2)$.

As condition (2) is satisfied for $k = 1, 2$, we have

$$QH_+^{\chi_1} = H_+^{\chi_2}Q \quad \text{and} \quad QH_-^{\chi_1} = H_-^{\chi_2}Q.$$

Utilizing the form of $H_+$ and $H_-$ (see (9) and (9′) of subsection 3), we find that the diagonal elements $q_{nn}$ of the operator $Q$ must satisfy the conditions

$$(n - l_1 + \varepsilon)\, q_{n+1,\, n+1} = (n - l_2 + \varepsilon)\, q_{nn} \tag{3}$$

and

$$(-n - l_2 - \varepsilon - 1)\, q_{n+1,\, n+1} = (-n - l_1 - \varepsilon - 1)\, q_{nn}. \tag{4}$$

A comparison of (3) and (4) leads to the conclusion that if $q_{nn} \neq 0$, then either $l_1 = l_2$ or $l_1 = -l_2 - 1$.

Thus, if $T_{\chi_1}(g)$, $\chi_1 = (l_1, \epsilon_1)$ and $T_{\chi_2}(g)$, $\chi_2 = (l_2, \epsilon_2)$ are partially equivalent, then $\epsilon_1 = \epsilon_2$, and either $l_1 = l_2$ $(x_1 = x_2)$ or $l_1 = -l_2 - 1 = l$ $(x_1 = -x_2)$

In the first case $T_{\chi_1}(g)$ and $T_{\chi_2}(g)$ coincide. In the second case, relation (3) gives

$$q_{nn}=\frac{\Gamma(l-n-\varepsilon+1)}{\Gamma(-n-l-\varepsilon)}.$$

Thus, in the case when $l_1=-l_2-1\equiv l$ the matrix of $Q$ in the basis $\{e^{-in\theta}\}$ is diagonal with numbers $\Gamma(l-n-\epsilon+1)/\Gamma(-n-l-\epsilon)$ on its principal diagonal.

Clearly, in the nonintegral case the operator $Q^{-1}$ exists and is continuous (its matrix is diagonal with the numbers $\Gamma(-n-l-\epsilon)/\Gamma(l-n-\epsilon+1)$ on its principal diagonal). Thus, if $\chi_1=-\chi_2$ and $T_{\chi_1}(g)$ is not integral, then $T_{\chi_1}(g)$ and $T_{\chi_2}(g)$ are equivalent.

The situation is more complicated in the integral case. Let $\chi_1=(l,\epsilon)$ be integral, where $l-\epsilon<0$. Then $T_\chi(g)$ is reducible. It induces the irreducible representations $T_l^+(g)$, $T_l^-(g)$ and $T_l^0(g)$ in the spaces $\mathfrak{D}_l^+$, $\mathfrak{D}_l^-$ and $\mathfrak{D}_l^0\equiv\mathfrak{D}/(\mathfrak{D}_l^++\mathfrak{D}_l^-)$ respectively.

Since $\chi_2=(-l-1,\epsilon)$, the representation $T_{\chi_2}(g)$ is also reducible. It induces the irreducible representations $T^+_{-l-1}(g)$, $T^-_{-l-1}(g)$ and $T^0_{-l-1}(g)$ in the spaces

$$\mathfrak{D}^+_{-l-1}/\mathfrak{D}^0_{-l-1},\quad \mathfrak{D}^-_{-l-1}/\mathfrak{D}^0_{-l-1}\quad\text{and}\quad \mathfrak{D}^0_{-l-1}\equiv\mathfrak{D}^+_{-l-1}\cap\mathfrak{D}^-_{-l-1}.$$

In just the same way as in the nonintegral case, one proves that $T_l^+(g)$, $T_l^-(g)$ and $T_l^0(g)$ are equivalent to $T^+_{-l-1}(g)$, $T^-_{-l-1}(g)$ and $T^0_{-l-1}(g)$ respectively.

It can be shown that there are no other equivalences between representations, except those indicated here.

7. Conditions for being unitary. Before clarifying for which values of $\chi=(l,\epsilon)$ the representation $T_\chi(g)$ is unitary, let us determine when, in the space $\mathfrak{D}$, a Hermitian form $H(f_1,f_2)$ is invariant relative to $T_\chi(g)$, i.e.

$$H(f_1,f_2)=H(T_\chi(g)f_1,\ T_\chi(g)f_2). \tag{1}$$

It follows from (1) that for any infinitesimal operator $A^\chi$ of $T_\chi(g)$ we have

$$(A^\chi f_1,f_2)+(f_1,A^\chi f_2)=0. \tag{2}$$

In order to see this, it is sufficient to write $g=g(t)$ in (1), differentiate with respect to $t$ and put $t=0$.

The arguments now proceed in a way similar to the preceding section. We thus arrive at the following result.

*The representation* $T_\chi(g)$, $\chi = (l, \epsilon)$, *possesses an invariant Hermitian form* $H(f_1, f_2)$ *if either* $l = \bar{l}$, *or* $l = -\bar{l} - 1$. *In the first case* $l$ *is real, and in the second it has the form* $l = -1/2 + i\rho$, *where* $\rho$ *is real.*

The Hermitian form, invariant relative to $T_\chi(g)$, $\chi = (-\frac{1}{2} + i\rho, \epsilon)$, has the form

$$(f_1, f_2) = \sum_{n=-\infty}^{\infty} a_n \bar{b}_n. \tag{3}$$

Here

$$f_1(e^{i\theta}) = \sum_{n=-\infty}^{\infty} a_n e^{-in\theta}, \quad f_2(e^{i\theta}) = \sum_{n=-\infty}^{\infty} b_n e^{-in\theta}.$$

This form can be written as an integral:

$$(f_1, f_2) = \frac{1}{2\pi} \int_0^{2\pi} f_1(e^{i\theta}) \overline{f_2(e^{i\theta})}\, d\theta. \tag{4}$$

Similarly, for real $l$ the Hermitian form invariant relative to $T_\chi(g)$, $\chi = (l, \epsilon)$ has the form

$$H(f_1, f_2) = \sum_{n=-\infty}^{\infty} \frac{\Gamma(l - n - \varepsilon + 1)}{\Gamma(-n - l - \varepsilon)} a_n \bar{b}_n \tag{5}$$

or

$$H(f_1, f_2) = C \int_0^{2\pi} \int_0^{2\pi} e^{-i\varepsilon(\psi - \varphi)} \left| \sin \frac{\psi - \varphi}{2} \right|^{-2l-2} \times \left[ \operatorname{sign} \sin \frac{\psi - \varphi}{2} \right]^{2\varepsilon} f_1(e^{i\psi}) \overline{f_2(e^{i\varphi})}\, d\psi\, d\varphi, \tag{6}$$

using integrals, where

$$C = \frac{\Gamma(l - \varepsilon + 1)\Gamma(-l + \varepsilon) e^{i\varepsilon\pi}}{2^{2l+4}\pi^2 \Gamma(-2l - 1)}.$$

It is now easy to determine the values of $\chi = (l, \epsilon)$ for which the invariant Hermitian form $H(f_1, f_2)$ is positive definite, i.e. for which the representation $T_\chi(g)$ is unitary.

Obviously, for any function $f(e^{i\theta})$ of $\mathfrak{D}$ we have

$$(f, f) = \sum_{n=-\infty}^{\infty} |a_n|^2 \geq 0.$$

Therefore all representations $T_\chi(g)$, $\chi = (-\frac{1}{2} + i\rho, \epsilon)$, are unitary. We shall call them unitary representations of the fundamental series. Moreover $T_\chi(g)$,

$\chi = (-\tfrac{1}{2} + i\rho, 0)$, are called unitary representations of the first fundamental series, and $T_\chi(g)$, $\chi = (-\tfrac{1}{2} + i\rho, \tfrac{1}{2})$, are the unitary representations of the second fundamental series.

Let us now consider the case when $\chi = (l, \epsilon)$, where $l$ is real. In order that the invariant Hermitian form be positive definite, it is necessary and sufficient that, for all $n$, one should have

$$\frac{\Gamma(l-n-\varepsilon+1)}{\Gamma(-n-l-\varepsilon)} > 0.$$

Since

$$\frac{\Gamma(l-n-\varepsilon+1)}{\Gamma(-n-l-\varepsilon)} = \frac{(l-n-\varepsilon)\Gamma(l-n-\varepsilon)}{(-n-l-\varepsilon-1)\Gamma(-n-l-\varepsilon-1)},$$

it follows that, for all $n$, we have

$$\frac{l-n-\varepsilon}{-n-l-\varepsilon-1} > 0. \tag{7}$$

In particular, for $n = -1$ we get $(l + 1 - \epsilon)/(l + \epsilon) < 0$. Since $\epsilon = 0$ or $\tfrac{1}{2}$, we have $l + 1 - \epsilon \geq l + \epsilon$. Hence it follows from (7) that $l + 1 - \epsilon > 0$, $l + \epsilon < 0$. This gives $\epsilon = 0$ and $-1 < l < 0$. Thus, if $\chi = (l, \epsilon)$, where $l$ is real, then the representation $T_\chi(g)$ is unitary if and only if $\epsilon = 0$ and $-1 < l < 0$. The corresponding representations $T_\chi(g)$ are called unitary representations of the complementary series.

Finally, unitary representations of $QU(2)$ arise in the integral case, i.e. in the case when $l + \epsilon$ and $l - \epsilon$ are integers. Let $l - \epsilon < 0$. Then in $\mathfrak{D}$ there are two nonintersecting invariant subspaces $\mathfrak{D}_l^+$ and $\mathfrak{D}_l^-$ (see subsection 5). The invariant Hermitian form in $\mathfrak{D}_l^+$ is given by

$$H_l^+(f_1, f_2) = \sum_{n=-\infty}^{l-\varepsilon} \frac{\Gamma(l-n-\varepsilon+1)}{\Gamma(-n-l-\varepsilon)} a_n \bar{b}_n, \tag{8}$$

and in $\mathfrak{D}_l^-$, by

$$H_l^-(f_1, f_2) = \sum_{n=-l-\varepsilon}^{\infty} \frac{\Gamma(1+n+l+\varepsilon)}{\Gamma(n-l+\varepsilon)} a_n \bar{b}_n. \tag{9}$$

These forms are positive definite. Hence, for integral $\chi = (l, \epsilon)$, $l - \epsilon < 0$, the representations $T_l^+(g)$ and $T_l^-(g)$ are unitary. The representations $T_{-l-1}^+$ and $T_{-l-1}^-$, which are equivalent to them and act in the factor spaces $\mathfrak{D}_{-l-1}^+/\mathfrak{D}_{-l-1}^0$ and $\mathfrak{D}_{-l-1}^-/\mathfrak{D}_{-l-1}^0$ respectively, where $\mathfrak{D}_{-l-1}^0 = \mathfrak{D}_{-l-1}^+ \cap \mathfrak{D}_{-l-1}^-$, are also unitary. For these representations invariant scalar products are given by the equalities

$$H^{\pm}_{-l-1}(f_1, f_2) = \sum_{n=-\infty}^{-l-\varepsilon-1} \frac{\Gamma(l-n-\varepsilon+1)}{\Gamma(-n-l-\varepsilon)} a_n \bar{b}_n \tag{10}$$

and

$$H^{-}_{-l-1}(f_1, f_2) = \sum_{n=l-\varepsilon+1}^{\infty} \frac{\Gamma(1+n+l-\varepsilon)}{\Gamma(n-l+\varepsilon)} a_n \bar{b}_n. \tag{11}$$

Forms (8) and (9) can also be presented as integrals. Namely,

$$H^{+}_{l}(f_1, f_2) = \frac{1}{\pi\Gamma(-2l-1)} \int\int_{|z|<1} (1-|z|^2)^{-2l-2} |z|^{2l-2\varepsilon} f_1(z) \overline{f_2(z)} dx\, dy, \tag{12}$$

where $z = x + iy$. Form (9) has a similar expression.

We may also note that $QU(2)$ does not have finite-dimensional irreducible unitary representations (apart from the identity representation). If $\chi = (l, \epsilon)$ is integral and $l - \epsilon < 0$, then the invariant Hermitian form for the finite-dimensional representation $T^0_\chi(g)$ is given by the equality

$$H^0_l(f_1, f_2) = \sum_{n=l-\varepsilon+1}^{-l-\varepsilon-1} \frac{(-1)^n}{\Gamma(n-l+\varepsilon)\Gamma(-n-l-\varepsilon)} a_n \bar{b}_n.$$

It is evident that this form is not positive definite.

8. **Unitary adjoint representations.** We shall call the representations $T(g)$ and $Q(g)$ unitary adjoint relative to some Hermitian form $H(f_1, f_2)$ if

$$H(T(g)f_1, f_2) = H(f_1, Q(g^{-1})f_2). \tag{1}$$

We now show that $T_\chi(g)$, $\chi = (l, \epsilon)$ and $T_{-\bar{\chi}}(g)$, $-\bar{\chi} = (-\bar{l} - 1, \epsilon)$ are unitary adjoint relative to the Hermitian form

$$(f_1, f_2) = \sum_{n=-\infty}^{\infty} a_n \bar{b}_n. \tag{2}$$

Indeed, we have

$$(f_1, f_2) = \langle f_1, e^{2i\varepsilon\theta} \bar{f}_2 \rangle,$$

where

$$\langle f_1, f_2 \rangle = \sum_{n=-\infty}^{\infty} a_n b_{-2\varepsilon-n}.$$

But it was shown in subsection 5 that

$$\langle T_\chi(g) f_1, f_2 \rangle = \langle f_1, T_{-\chi}(g^{-1}) f_2 \rangle,$$

where $-\chi = (-l - 1, \epsilon)$. Therefore

$$(T_\chi(g)f_1, f_2) = \langle f_1, T_{-\chi}(g^{-1}) e^{2il\theta}\overline{f_2}\rangle = (f_1, T_{-\overline{\chi}}(g^{-1})f_2).$$

This proves that $T_\chi(g)$ and $T_{-\overline{\chi}}(g)$ are unitary adjoint relative to the Hermitian form $(f_1, f_2)$. In particular, if $\chi = (-\frac{1}{2} + i\rho, \epsilon)$, then $-\overline{\chi} = \chi$, and we get an assertion, proved earlier, about the invariance of the form $(f_1, f_2)$ relative to the representations $T_\chi(g)$, $\chi = (-\frac{1}{2} + i\rho, \epsilon)$:

$$(T_\chi(g)f_1, T_\chi(g)f_2) = (f_1, T_\chi(g^{-1})T_\chi(g)f_2) = (f_1, f_2).$$

## §3. MATRIX ELEMENTS OF $T_\chi(g)$

1. **Calculation of matrix elements.** In this section we shall calculate the matrix elements of the representations $T_\chi(g)$ of $QU(2)$. In the space $\mathfrak{D}$ of the representation $T_\chi(g)$ we choose a basis, consisting of the functions $\{e^{-in\theta}\}$. Since

$$T_\chi(g)f(e^{i\theta}) = (\beta e^{i\theta} + \overline{\alpha})^{l+\epsilon}(\overline{\beta}e^{-i\theta} + \alpha)^{l-\epsilon} f\left(\frac{\alpha e^{i\theta} + \overline{\beta}}{\beta e^{i\theta} + \overline{\alpha}}\right) \tag{1}$$

(see formula (5), §2.2), we have

$$T_\chi(g)e^{-in\theta} = (\beta e^{i\theta} + \overline{\alpha})^{l+n+\epsilon}(\overline{\beta}e^{-i\theta} + \alpha)^{l-n-\epsilon}e^{-in\theta}. \tag{2}$$

The matrix elements $t^\chi_{mn}(g)$ of the representation $T_\chi(g)$ in the basis $\{e^{-in\theta}\}$ are Fourier coefficients in the expression of $T_\chi(g)e^{-in\theta}$ according to the system $\{e^{-im\theta}\}$:

$$T_\chi(g)e^{-in\theta} = \sum_{m=-\infty}^{\infty} t^\chi_{mn}(g)e^{-im\theta}. \tag{3}$$

The Fourier coefficient formula gives

$$t^\chi_{mn}(g) = \frac{1}{2\pi}\int_0^{2\pi}(\beta e^{i\theta} + \overline{\alpha})^{l+n+\epsilon}(\overline{\beta}e^{-i\theta} + \alpha)^{l-n-\epsilon}e^{i(m-n)\theta}\,d\theta. \tag{4}$$

We have obtained an integral representation for the matrix elements. With the help of the substitution $e^{i\theta} = z$ we get a representation of $t^\chi_{mn}(g)$ as an integral over the circle $\Gamma$: $|z| = 1$:

$$t^\chi_{mn}(g) = \frac{1}{2\pi i}\oint_\Gamma(\beta z + \overline{\alpha})^{l+n+\epsilon}(\alpha z + \overline{\beta})^{l-n-\epsilon}z^{m-l+\epsilon-1}\,dz. \tag{5}$$

The integrand in this formula has, within the contour $\Gamma$, two branch points $z = 0$ and $z = -\overline{\beta}/\alpha$, with exponents $m - l + \epsilon - 1$ and $l - n - \epsilon$ respectively. Since the sum of these indices is equal to the integer $m - n - 1$, the integrand is uniquely defined on $\Gamma$.

Let us expand the matrix elements in powers of $\alpha, \beta, \bar{\alpha}, \bar{\beta}$. Since $|\alpha| > |\beta|$, we have

$$(\bar{\alpha}+\beta e^{i\theta})^{l+n+\varepsilon} = \bar{\alpha}^{l+n+\varepsilon} \sum_{k=0}^{\infty} \frac{\Gamma(l+n+\varepsilon+1)}{\Gamma(k+1)\Gamma(l+n-k+\varepsilon+1)} \left(\frac{\beta e^{i\theta}}{\bar{\alpha}}\right)^k .$$

Similarly,

$$(\alpha+\bar{\beta} e^{-i\theta})^{l-n-\varepsilon} = \alpha^{l-n-\varepsilon} \sum_{k=0}^{\infty} \frac{\Gamma(l-n-\varepsilon+1)}{\Gamma(k+1)\Gamma(l-n-k-\varepsilon+1)} \left(\frac{\bar{\beta} e^{-i\theta}}{\alpha}\right)^k .$$

Multiplying these expressions side by side and substituting the resulting series in (4), we get

$$t^{\chi}_{mn}(g) = \Gamma(l+n+\varepsilon+1)\Gamma(l-n-\varepsilon+1)\alpha^{l-n-\varepsilon}\bar{\alpha}^{l+m+\varepsilon}\beta^{n-m} \times \sum_{s=\max(0,\, m-n)}^{\infty} \frac{|\beta/\alpha|^{2s}}{\Gamma(s+1)\Gamma(l-n-s-\varepsilon+1)\Gamma(n-m+s+1)\Gamma(l+m-s+\varepsilon+1)}. \tag{6}$$

2. Expression in terms of Euler angles. The basis $\{e^{-im\theta}\}$, chosen by us, is convenient for the space $\mathfrak{D}$ inasmuch as the matrices of the operators $T_\chi(h)$, where

$$h = \begin{pmatrix} e^{\frac{it}{2}} & 0 \\ 0 & e^{-\frac{it}{2}} \end{pmatrix} \tag{1}$$

are diagonal matrices, have a very simple form in it. Formula (1) of subsection 1 shows that the operator $T_\chi(h)$ has the form

$$T_\chi(h) f(e^{i\theta}) = e^{-i\epsilon t} f(e^{i(\theta+t)}) \tag{2}$$

and consequently

$$T_\chi(h) e^{-im\theta} = e^{-i(m+\epsilon)t} e^{-im\theta}. \tag{3}$$

Therefore the matrix of $T_\chi(h)$ in the basis $\{e^{-im\theta}\}$ is an infinite diagonal matrix, with the numbers $e^{-i(m+\epsilon)t}$ on its principal diagonal.

It was shown in §1.3 that any element $g$ of $QU(2)$ can be presented in the form

$$g = \begin{pmatrix} e^{\frac{i\varphi}{2}} & 0 \\ 0 & e^{-\frac{i\varphi}{2}} \end{pmatrix} \begin{pmatrix} \operatorname{ch}\frac{\tau}{2} & \operatorname{sh}\frac{\tau}{2} \\ \operatorname{sh}\frac{\tau}{2} & \operatorname{ch}\frac{\tau}{2} \end{pmatrix} \begin{pmatrix} e^{\frac{i\psi}{2}} & 0 \\ 0 & e^{-\frac{i\psi}{2}} \end{pmatrix}, \tag{4}$$

where $\phi, \tau, \psi$ are the Euler angles of $g$. As the matrix of $T_\chi(g)$ for diagonal matrices has already been found, we have to find the matrix of $T_\chi(g_\tau)$, corresponding to the element

$$g_\tau=\begin{pmatrix}\operatorname{ch}\frac{\tau}{2} & \operatorname{sh}\frac{\tau}{2}\\ \operatorname{sh}\frac{\tau}{2} & \operatorname{ch}\frac{\tau}{2}\end{pmatrix}$$

of $QU(2)$.

If in formula (4) of subsection 1 we substitute

$$\alpha=\operatorname{ch}\frac{\tau}{2},\quad \beta=\operatorname{sh}\frac{\tau}{2},$$

we obtain

$$t^\chi_{mn}(g_\tau)=\frac{1}{2\pi}\int_0^{2\pi}\left(\operatorname{ch}\frac{\tau}{2}+\operatorname{sh}\frac{\tau}{2}e^{i\theta}\right)^{l+n+\epsilon}\left(\operatorname{ch}\frac{\tau}{2}+\operatorname{sh}\frac{\tau}{2}e^{-i\theta}\right)^{l-n-\epsilon}e^{i(m-n)\theta}\,d\theta. \tag{5}$$

We introduce the function $\mathfrak{P}^l_{mn}(\operatorname{ch}\tau)$ by putting

$$\mathfrak{P}^l_{mn}(\operatorname{ch}\tau)=\frac{1}{2\pi}\int_0^{2\pi}\left(\operatorname{ch}\frac{\tau}{2}+\operatorname{sh}\frac{\tau}{2}e^{i\theta}\right)^{l+n}\left(\operatorname{ch}\frac{\tau}{2}+\operatorname{sh}\frac{\tau}{2}e^{-i\theta}\right)^{l-n}e^{i(m-n)\theta}\,d\theta, \tag{6}$$

where $0\le\tau<\infty$, $l$ is any complex number, and $m$, $n$ are either both integers or both semi-integers. Comparing (5) and (6), we get an expression of $t^\chi_{mn}(g_\tau)$ by means of $\mathfrak{P}^l_{mn}(\operatorname{ch}\tau)$:

$$t^\chi_{mn}(g_\tau)=\mathfrak{P}^l_{m'n'}(\operatorname{ch}\tau),\quad \chi=(l,\epsilon), \tag{7}$$

where $m'=m+\epsilon$, $n'=n+\epsilon$.

It is now easy to express $t^\chi_{mn}(g)$ in terms of the Euler angles of $g$. It follows from the factorization (4) that

$$T_\chi(g)=T_\chi(h_\varphi)\,T_\chi(g_\tau)\,T_\chi(h_\psi), \tag{8}$$

where $h_\phi$ and $h_\psi$ denote diagonal matrices, and $g_\tau$ is the matrix

$$\begin{pmatrix}\operatorname{ch}\frac{\tau}{2} & \operatorname{sh}\frac{\tau}{2}\\ \operatorname{sh}\frac{\tau}{2} & \operatorname{ch}\frac{\tau}{2}\end{pmatrix}.$$

We have seen that $T_\chi(h_\phi)$ and $T_\chi(h_\psi)$ are diagonal matrices, with the elements $e^{-i\phi(m+\epsilon)}\equiv e^{-im'\phi}$ and $e^{-i\psi(n+\epsilon)}\equiv e^{-in'\psi}$, respectively, on their principal diagonals. The elements of $T_\chi(g_\tau)$ are equal to $\mathfrak{P}^l_{m'n'}(\operatorname{ch}\tau)$. Therefore the elements of $T_\chi(g)$ have the form

$$t^{\chi}_{mn}(\varphi, \tau, \psi) = e^{-i(m'\varphi + n'\psi)}\mathfrak{P}^{l}_{m'n'}(\operatorname{ch}\tau), \tag{9}$$

where $\chi = (l, \epsilon)$, $m' = m + \epsilon$, $n' = n + \epsilon$. We have thus expressed $t^{\chi}_{mn}(\phi, \tau, \psi)$ in terms of Euler angles.

3. Various expressions for the functions $\mathfrak{P}^{l}_{mn}(z)$. We have proved that the matrix elements of $T_\chi(g)$ are expressed by means of the exponential function and the functions $\mathfrak{P}^{l}_{mn}(z)$. We pass on to a study of the functions $\mathfrak{P}^{l}_{mn}(z)$. As these functions play the same role for $QU(2)$ as the functions $P^{l}_{mn}(z)$ do for $SU(2)$, we shall call $\mathfrak{P}^{l}_{mn}(z)$ the Jacobi function of $z$.

From the expressions for the matrix elements $t^{\chi}_{mn}(g)$, obtained in subsection 1, follow corresponding expressions for functions $\mathfrak{P}^{l}_{mn}(\operatorname{ch}\tau)$. Namely, putting $\alpha = \operatorname{ch}\tau/2$, $\beta = \operatorname{sh}\tau/2$ in formula (5) of subsection 1, we get

$$\mathfrak{P}^{l}_{mn}(\operatorname{ch}\tau) = \frac{1}{2\pi i}\oint_{\Gamma}\left(\operatorname{ch}\frac{\tau}{2} + z\operatorname{sh}\frac{\tau}{2}\right)^{l+n}\left(z\operatorname{ch}\frac{\tau}{2} + \operatorname{sh}\frac{\tau}{2}\right)^{l-n} z^{m-l-1}\,dz, \tag{1}$$

where $\Gamma$ is the circle $|z| = 1$.

Similarly, putting $\alpha = \operatorname{ch}\tau/2$, $\beta = \operatorname{sh}\tau/2$ in formula (6) of subsection 1, we find that

$$\mathfrak{P}^{l}_{mn}(\operatorname{ch}\tau) = \Gamma(l+n+1)\,\Gamma(l-n+1)\left(\operatorname{ch}\frac{\tau}{2}\right)^{2l}\left(\operatorname{th}\frac{\tau}{2}\right)^{n-m} \times \sum_{s=\max(0,\, m-n)}^{\infty} \frac{\operatorname{th}^{2s}\frac{\tau}{2}}{\Gamma(n-m+s+1)\,\Gamma(l+m-s+1)\,\Gamma(s+1)\,\Gamma(l-n-s+1)}. \tag{2}$$

It follows from (2) that $\mathfrak{P}^{l}_{mn}(1) = \delta_{mn}$, where $\delta_{mn}$ is the Kronecker symbol. This is easily obtained directly: for $\tau = 0$ the matrix $g_\tau$ is the identity matrix, and so $T_\chi(g_\tau)$ is the identity operator.

Let us derive further integral representations for $\mathfrak{P}^{l}_{mn}(\operatorname{ch}\tau)$. We rewrite formula (1) as

$$\mathfrak{P}^{l}_{mn}(\operatorname{ch}\tau) = \frac{1}{2\pi i}\oint_{\Gamma}\left(\operatorname{ch}\tau + \frac{z^2+1}{2z}\operatorname{sh}\tau\right)^{l-n}\left(\operatorname{ch}\frac{\tau}{2} + z\operatorname{sh}\frac{\tau}{2}\right)^{2n} z^{m-n-1}\,dz. \tag{3}$$

Here, by virtue of Cauchy's theorem, we can assume that $\Gamma$ is the circle $|z| = a$, where $1 < a < \operatorname{cth}\tau/2$.

If in (3) we make the substitution

$$w = \operatorname{ch}\tau + \frac{z^2+1}{2z}\operatorname{sh}\tau,$$

we obtain

$$\mathfrak{P}^l_{mn}(\operatorname{ch}\tau) = \frac{1}{2\pi i}\oint_{\Gamma'} w^{l-n}\left(\operatorname{ch}\frac{\tau}{2} + z\operatorname{sh}\frac{\tau}{2}\right)^{2n}\frac{z^{m-n}dw}{\sqrt{w^2 - 2w\operatorname{ch}\tau + 1}}. \tag{4}$$

Here

$$z = \frac{w - \operatorname{ch}\tau \pm \sqrt{w^2 - 2w\operatorname{ch}\tau + 1}}{\operatorname{sh}\tau}, \tag{5}$$

where the radical is chosen such that $1 \leq |z| \leq \operatorname{cth}\tau/2$. $\Gamma'$ (Figure 4) denotes a contour encircling the segment $[e^{-\tau}, e^{\tau}]$ counterclockwise and intersecting the real axis in the intervals $(0, e^{-\tau})$ and $(e^{\tau}, \infty)$. The integrand is uniquely defined on this contour if the radical in formula (5) is chosen as indicated above, and by $w^{l-n}$ is understood the expression $\exp[(l-n)\ln w]$, where $\ln w$ is the principal value of the logarithm.

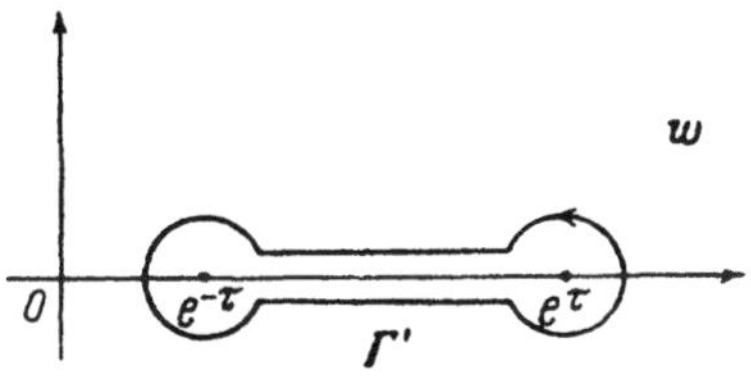


Figure 4.

The substitution $w = e^t$ transforms the expression obtained above into

$$\mathfrak{P}^l_{mn}(\operatorname{ch}\tau) = \frac{1}{2\pi i}\oint_{\Gamma''} e^{\left(l-n+\frac{1}{2}\right)t}\left(\operatorname{ch}\frac{\tau}{2} + z\operatorname{sh}\frac{\tau}{2}\right)^{2n}\frac{z^{m-n}dt}{\sqrt{2(\operatorname{ch}t - \operatorname{ch}\tau)}}. \tag{6}$$

Here $\Gamma''$ (Figure 5) is a contour, lying in the strip $-\pi < \operatorname{Im} t < \pi$ and encircling the segment $[-\tau, \tau]$ counterclockwise. The value of $z$ is obtained from (5) by substituting $w = e^t$.

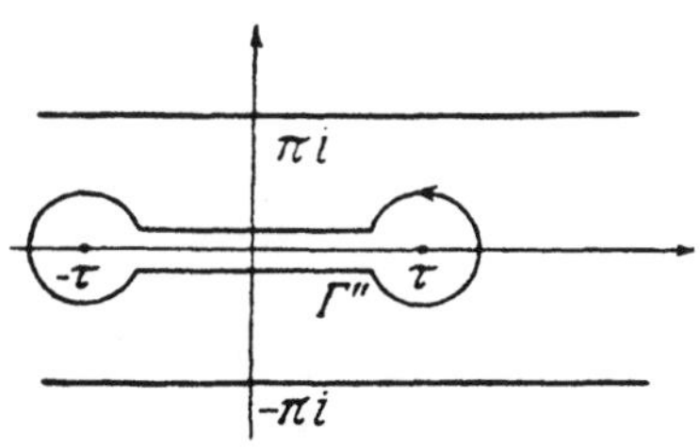


Figure 5.

In formula (3) $\Gamma$ can also be taken to be a circle of radius $a$, where $\operatorname{th}\tau/2 < a < 1$. We then obtain an expression of the form (4), but the contour $\Gamma'$ runs clockwise, and the radical in (5) is chosen such that $\operatorname{th}\tau/2 < |z| < 1$.

If in the formulas obtained above we contract the contours of integration to the segments which they enclose, we get expressions for $\mathfrak{P}^l_{mn}(\operatorname{ch}\tau)$ in the form of ordinary integrals. For example, we get

$$\mathfrak{P}^l_{mn}(\operatorname{ch}\tau)=\frac{1}{2\pi}\int_{-\tau}^{\tau}\frac{e^{\left(l-n+\frac{1}{2}\right)t}}{\sqrt{2(\operatorname{ch}\tau-\operatorname{ch}t)}}$$
$$\times\left[z_+^{m-n}\left(\operatorname{ch}\frac{\tau}{2}+z_+\operatorname{sh}\frac{\tau}{2}\right)^{2n}+z_-^{m-n}\left(\operatorname{ch}\frac{\tau}{2}+z_-\operatorname{sh}\frac{\tau}{2}\right)^{2n}\right]dt. \tag{7}$$

Here

$$z_\pm=\frac{e^t-\operatorname{ch}\tau\pm ie^{\frac{t}{2}}\sqrt{2(\operatorname{ch}\tau-\operatorname{ch}t)}}{\operatorname{sh}\tau}. \tag{8}$$

This expression simplifies if $n = m$ or $n = 0$. For $n = m$ we easily get

$$\mathfrak{P}^l_{nn}(\operatorname{ch}\tau)=\frac{1}{\pi}\int_0^{\tau}\frac{\operatorname{ch}\left(l-n+\frac{1}{2}\right)t\cos 2n\alpha}{\sqrt{\operatorname{ch}^2\frac{\tau}{2}-\operatorname{ch}^2\frac{t}{2}}}\,dt, \tag{9}$$

where

$$\cos\alpha=\frac{\operatorname{ch}\frac{t}{2}}{\operatorname{ch}\frac{\tau}{2}}. \tag{10}$$

This formula can be written in the form

$$\mathfrak{P}^l_{nn}(\operatorname{ch}\tau)=\frac{1}{\pi}\int_0^{\tau}\frac{\operatorname{ch}\left(l-n+\frac{1}{2}\right)t\,T_{2n}\left(\dfrac{\operatorname{ch}\frac{t}{2}}{\operatorname{ch}\frac{\tau}{2}}\right)dt}{\sqrt{\operatorname{ch}^2\frac{\tau}{2}-\operatorname{ch}^2\frac{t}{2}}}, \tag{11}$$

where $T_n(x)$ is the Čebyšev polynomial:

$$T_n(x) = \cos(n \arccos x).$$

If $n = 0$, we get

$$\mathfrak{P}^l_{m0}(\operatorname{ch}\tau)=\frac{1}{2\pi}\int_{-\tau}^{\tau}\frac{e^{\left(l+\frac{1}{2}\right)t}(z_+^m+z_-^m)}{\sqrt{2(\operatorname{ch}\tau-\operatorname{ch}t)}}\,dt. \tag{12}$$

This expression can be rewritten as follows:

$$\mathfrak{P}^l_{m0}(\operatorname{ch}\tau)=\frac{1}{2\pi}\int_{-\tau}^{\tau}\frac{e^{\left(l+\frac{1}{2}\right)t}\,T_m\left(\frac{e^t-\operatorname{ch}\tau}{\operatorname{sh}\tau}\right)dt}{\sqrt{\operatorname{ch}^2\frac{\tau}{2}-\operatorname{ch}^2\frac{t}{2}}}. \tag{12'}$$

In particular, for $m = 0$ we get

$$\mathfrak{P}^l_{00}(\operatorname{ch}\tau)=\frac{1}{\pi}\int_0^{\tau}\frac{\operatorname{ch}\left(l+\frac{1}{2}\right)t\,dt}{\sqrt{\operatorname{ch}^2\frac{\tau}{2}-\operatorname{ch}^2\frac{t}{2}}}. \tag{13}$$

Similar formulas are obtained if in (3) one chooses the contour $\Gamma$ depicted in Figure 6. Under the transformation $w = \operatorname{ch}\tau + ((z^2+1)/2z)\operatorname{sh}\tau$ it becomes the contour $\Gamma'$, depicted in Figure 7. If one denotes by $\rho$ the radius of the circle

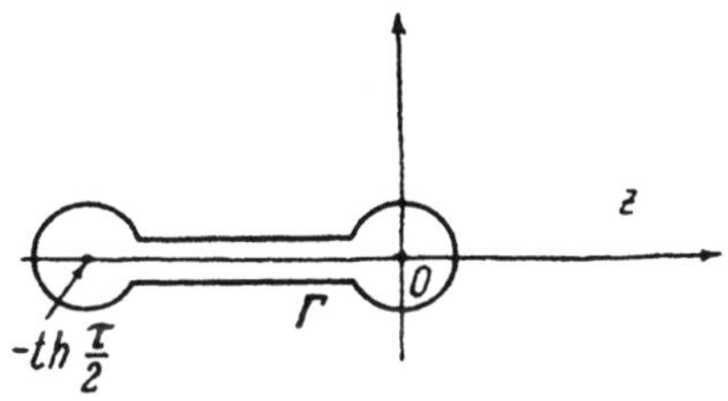


Figure 6.

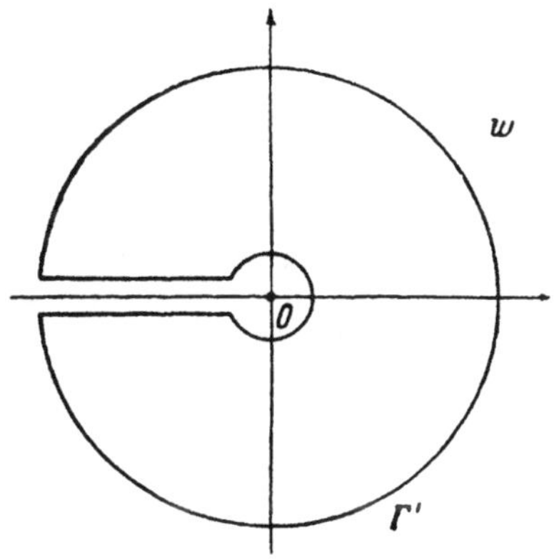


Figure 7.

with center at the point $O$ in Figure 7, then for $\rho \to 0$ the integrand in formula (4) is $O(\rho^{m-l+1})$, and the radius of the large circle is $O(\rho^{-1})$. Thus for $\operatorname{Re} l < m$, the integral over the large circle tends to zero when $\rho \to 0$. Similarly we establish that the integral over the small circle tends to zero when $\operatorname{Re} l > n-1$.

Thus, if $n-1 < \operatorname{Re} l < m$, we get

$$\mathfrak{P}^l_{mn}(\operatorname{ch}\tau)=\frac{1}{2\pi i}\oint_{\Gamma'} w^{l-n}\left(\operatorname{ch}\frac{\tau}{2}+z\operatorname{sh}\frac{\tau}{2}\right)^{2n}\frac{z^{m-n}dw}{\sqrt{w^2-2w\operatorname{ch}\tau+1}}, \tag{14}$$

where the contour $\Gamma'$ encircles the negative semiaxis clockwise. The radical is chosen such that the value of $z$ determined by formula (5) satisfies the inequalities $\operatorname{th}\tau/2 < |z| < 1$.

Let us contract the contour to the negative semiaxis; we obtain

$$\mathfrak{P}^l_{mn}(\operatorname{ch}\tau)=\frac{\sin(l-n)\pi}{\pi}\int_0^\infty w^{l-n}\left(\operatorname{ch}\frac{\tau}{2}+z\operatorname{sh}\frac{\tau}{2}\right)^{2n}\frac{z^{m-n}dw}{\sqrt{w^2+2w\operatorname{ch}\tau+1}},\tag{15}$$

where

$$z=\frac{-w-\operatorname{ch}\tau+\sqrt{w^2+2w\operatorname{ch}\tau+1}}{\operatorname{sh}\tau}.\tag{16}$$

For $n=0$ we get

$$\mathfrak{P}^l_{m0}(\operatorname{ch}\tau)=\frac{\sin l\pi}{\pi}\int_0^\infty\frac{w^lz^mdw}{\sqrt{w^2+2w\operatorname{ch}\tau+1}}.\tag{17}$$

In particular,

$$\mathfrak{P}^l_{00}(\operatorname{ch}\tau)=\frac{\sin l\pi}{\pi}\int_0^\infty\frac{w^ldw}{\sqrt{w^2+2w\operatorname{ch}\tau+1}}$$

$$=\frac{\sqrt{2}\sin l\pi}{\pi}\int_0^\infty\frac{\operatorname{ch}\left(l+\frac{1}{2}\right)t\,dt}{\sqrt{\operatorname{ch}t+\operatorname{ch}\tau}}.\tag{18}$$

**4. Zonal spherical functions of $T_\chi(g)$ and the Legendre functions.** Let $\epsilon=0$, i.e. $\chi=(l,0)$. Then in $\mathfrak{D}$ there is a function invariant for all operators $T_\chi(h)$, where $h$ is a diagonal matrix of $QU(2)$:

$$h=\begin{pmatrix}e^{\frac{it}{2}} & 0\\ 0 & e^{-\frac{it}{2}}\end{pmatrix}.$$

Namely, the basis element 1 is such a function. The matrix element $t^\chi_{00}(g)$ corresponding to such a basis element has been called (Chapter I, §2.5) the zonal spherical function of $T_\chi(g)$ for the subgroup $\Omega$ of diagonal matrices.

Thus for $\epsilon=0$ the representation $T_\chi(g)$ has a zonal spherical function, satisfying the functional equation (see Chapter I, §2.5).

$$t^\chi_{00}(h_1gh_2)=t^\chi_{00}(g),\tag{1}$$

where $h_1,h_2\in\Omega$. Hence $t^\chi_{00}(g)$ depends only on the Euler angle $\tau$ of $g$. This can, however, also be seen from formula (9) of subsection 2. Putting $\epsilon=m=n=0$ in this formula, we get

$$t^\chi_{00}(g)=\mathfrak{P}^l_{00}(\operatorname{ch}\tau).\tag{2}$$

We shall call $\mathfrak{P}^l_{00}(\operatorname{ch}\tau)$ the Legendre function with index $l$, and denote it by $\mathfrak{P}_l(\operatorname{ch}\tau)$. Thus,

$$\mathfrak{P}_l(\operatorname{ch}\tau)=t^\chi_{00}(0,\tau,0)=\mathfrak{P}^l_{00}(\operatorname{ch}\tau),\tag{3}$$

where $\chi = (l, 0)$. Putting $m = n = 0$ in formula (6) of subsection 2 and formulas (1) and (2) of subsection 3, we find the following expressions for Legendre functions:

$$\mathfrak{P}_l(\operatorname{ch}\tau) = \frac{1}{2\pi}\int_0^{2\pi}\left(\operatorname{ch}\frac{\tau}{2} + \operatorname{sh}\frac{\tau}{2}e^{i\theta}\right)^l\left(\operatorname{ch}\frac{\tau}{2} + \operatorname{sh}\frac{\tau}{2}e^{-i\theta}\right)^l d\theta$$

$$= \frac{1}{2\pi}\int_0^{2\pi}(\operatorname{ch}\tau + \operatorname{sh}\tau\cos\theta)^l\,d\theta, \tag{4}$$

$$\mathfrak{P}_l(\operatorname{ch}\tau) = \frac{1}{2\pi i}\int_\Gamma\left(\operatorname{ch}\frac{\tau}{2} + z\operatorname{sh}\frac{\tau}{2}\right)^l\left(z\operatorname{ch}\frac{\tau}{2} + \operatorname{sh}\frac{\tau}{2}\right)^l z^{-l-1}dz$$

$$= \frac{1}{2\pi i}\int_\Gamma\left(\operatorname{ch}\tau + \frac{z^2+1}{2z}\operatorname{sh}\tau\right)^l\frac{dz}{z}, \tag{5}$$

$$\mathfrak{P}_l(\operatorname{ch}\tau) = \frac{1}{\pi}\int_0^\tau\frac{\operatorname{ch}\left(l+\frac{1}{2}\right)t\,dt}{\sqrt{\operatorname{ch}^2\frac{\tau}{2} - \operatorname{ch}^2\frac{t}{2}}} = \frac{\sqrt{2}\sin\pi l}{\pi}\int_0^\infty\frac{\operatorname{ch}\left(l+\frac{1}{2}\right)t\,dt}{\sqrt{\operatorname{ch}t + \operatorname{ch}\tau}}, \tag{6}$$

$$\mathfrak{P}_l(\operatorname{ch}\tau) = \frac{\Gamma^2(l+1)}{2^l}(1+\operatorname{ch}\tau)^l\sum_{s=0}^\infty\frac{1}{\Gamma^2(s+1)\Gamma^2(l-s+1)}\left(\frac{\operatorname{ch}\tau-1}{\operatorname{ch}\tau+1}\right)^s. \tag{7}$$

It is evident from (4) that if $l$ is an integer, then the Legendre function $\mathfrak{P}_l(z)$ coincides with the Legendre polynomial $P_l(z)$:

$$\mathfrak{P}_l(z) = P_l(z). \tag{8}$$

The Legendre functions $\mathfrak{P}_{-1/2+i\rho}(\operatorname{ch}\tau)$, corresponding to the fundamental series of irreducible unitary representations of $QU(2)$, are called conical functions.

5. **Associated Legendre functions.** Let us now consider the associated Legendre functions of $T_\chi(g)$, $\chi = (l, 0)$, i.e. the matrix elements $t^\chi_{m0}(g)$ which are in the same column as the zonal spherical function $t^\chi_{00}(g)$. By formula (5) of subsection 2, these elements have the form

$$t^\chi_{m0}(g) = e^{-im\varphi}\mathfrak{P}^l_{m0}(\operatorname{ch}\tau). \tag{1}$$

It is obvious from (1) that they do not depend on the Euler angle $\psi$, i.e. they are constant on left cosets with respect to the subgroup $\Omega$ of diagonal matrices

$$t^\chi_{m0}(gh) = t^\chi_{m0}(g), \qquad h \in \Omega. \tag{2}$$

We define the associated Legendre functions $\mathfrak{P}^m_l(z)$ ($l$ complex, $m$ integral) by the formula

$$\mathfrak{P}_l^m(z) = \frac{\Gamma(l+m+1)}{2\pi\Gamma(l+1)} \int_0^{2\pi} (z + \sqrt{z^2-1}\cos\theta)^l e^{im\theta} d\theta. \tag{3}$$

This formula defines $\mathfrak{P}_l^m(z)$ uniquely in the complex plane, cut along the segment $[-1, 1]$. Since the mapping $z = \operatorname{ch}\tau$ carries the strip $0 < \operatorname{Im}\tau < \pi$ into the $z$-plane, cut along the segment $[-1, 1]$, the function $\mathfrak{P}_l^m(\operatorname{ch}\tau)$ is uniquely defined in this strip.

By formula (6) of subsection 2 we have

$$\mathfrak{P}_{m0}^l(\operatorname{ch}\tau) = \frac{1}{2\pi} \int_0^{2\pi} (\operatorname{ch}\tau + \operatorname{sh}\tau\cos\theta)^l e^{im\theta} d\theta. \tag{4}$$

Comparing (3) and (4), we get

$$\mathfrak{P}_l^m(\operatorname{ch}\tau) = \frac{\Gamma(l+m+1)}{\Gamma(l+1)} \mathfrak{P}_{m0}^l(\operatorname{ch}\tau), \tag{5}$$

and hence for $g = g(\phi, \tau, \psi)$ we have

$$t_{m0}^\chi(g) = \frac{\Gamma(l+1)}{\Gamma(l+m+1)} e^{-im\varphi} \mathfrak{P}_l^m(\operatorname{ch}\tau). \tag{6}$$

Formulas (1) and (2) of subsection 3 lead to the following expressions for associated Legendre functions:

$$\begin{aligned}\mathfrak{P}_l^m(\operatorname{ch}\tau) &= \frac{1}{2\pi i}\frac{\Gamma(l+m+1)}{\Gamma(l+1)} \int_\Gamma \left(\operatorname{ch}\frac{\tau}{2} + z\operatorname{sh}\frac{\tau}{2}\right)^l \left(z\operatorname{ch}\frac{\tau}{2} + \operatorname{sh}\frac{\tau}{2}\right)^l z^{m-l-1} dz \\ &= \frac{1}{2\pi i}\frac{\Gamma(l+m+1)}{\Gamma(l+1)} \int_\Gamma \left(\operatorname{ch}\tau + \frac{z^2+1}{2z}\operatorname{sh}\tau\right)^l z^{m-1} dz,\end{aligned} \tag{7}$$

$$\mathfrak{P}_l^m(\operatorname{ch}\tau) = \Gamma(l+m+1)\Gamma(l+1)\operatorname{ch}^{2l}\frac{\tau}{2} \times \sum_{s=\max(0,m)}^{\infty} \frac{\operatorname{th}^{2s-m}\frac{\tau}{2}}{\Gamma(s-m+1)\Gamma(l+m-s+1)\Gamma(s+1)\Gamma(l-s+1)}. \tag{8}$$

If $l$ is a negative integer and $|m| \le l$, then, from formulas (13) of §3.9 of Chapter III and formula (7), we get

$$\mathfrak{P}_l^m(z) = P_l^m(z). \tag{9}$$

6. Symmetry relations for the functions $\mathfrak{P}_{mn}^l(\operatorname{ch}\tau)$. Like the Jacobi polynomials $P_{mn}^l(\cos\theta)$, the Jacobi functions $\mathfrak{P}_{mn}^l(\operatorname{ch}\tau)$ satisfy certain symmetry relations involving the indices $l$, $m$, $n$. We shall first show that these functions do

not change on changing the signs of $m$ and $n$, i.e. we have

$$\mathfrak{P}^l_{mn}(\operatorname{ch}\tau)=\mathfrak{P}^l_{-m,\,-n}(\operatorname{ch}\tau). \tag{1}$$

We consider the operator $S$ transforming $f(e^{i\theta})$ into $e^{2i\epsilon\theta}f(e^{-i\theta})$. A direct calculation shows that $S$ is permutable with the operators $T_\chi(g_\tau)$, where

$$g_\tau=\begin{pmatrix}\operatorname{ch}\frac{\tau}{2} & \operatorname{sh}\frac{\tau}{2}\\ \operatorname{sh}\frac{\tau}{2} & \operatorname{ch}\frac{\tau}{2}\end{pmatrix},$$

i.e.

$$ST_\chi(g_\tau)=T_\chi(g_\tau)S. \tag{2}$$

This can also be seen from the following group-theoretic considerations. The representations $T_\chi(g)$ can be extended to the group of quasi-unitary matrices whose determinant is equal to $\pm 1$, without changing the formulas which give the representations. Moreover, $S = T_\chi(s)$, where $s = \begin{pmatrix}0 & 1\\ 1 & 0\end{pmatrix}$. But

$$sg_\tau=g_\tau s=\begin{pmatrix}\operatorname{sh}\frac{\tau}{2} & \operatorname{ch}\frac{\tau}{2}\\ \operatorname{ch}\frac{\tau}{2} & \operatorname{sh}\frac{\tau}{2}\end{pmatrix},$$

and therefore

$$T_\chi(s)\,T_\chi(g_\tau)=T_\chi(g_\tau)\,T_\chi(s),$$

i.e.

$$ST_\chi(g_\tau)=T_\chi(g_\tau)S.$$

The matrix elements $s_{mn}$ of $S$ in the basis $\{e^{-im\theta}\}$ are equal to zero if $m+n\neq -2\epsilon$, and are equal to unity if $m+n=-2\epsilon$. Multiplying the matrices in (2) and comparing the corresponding elements on the left and right, we find that

$$t^\chi_{-2\epsilon-m,\,n}(g_\tau)=t^\chi_{m,\,-2\epsilon-n}(g_\tau).$$

Since $t^\chi_{mn}(g_\tau)=\mathfrak{P}^l_{m'n'}(\operatorname{ch}\tau)$, where $m'=m+\epsilon$, $n'=n+\epsilon$, we get

$$\mathfrak{P}^l_{-m-\epsilon,\,n+\epsilon}(\operatorname{ch}\tau)=\mathfrak{P}^l_{m+\epsilon,\,-n-\epsilon}(\operatorname{ch}\tau).$$

In view of the arbitrariness of $m$, $n$, $\epsilon$, $l$, we get

$$\mathfrak{P}^l_{mn}(\operatorname{ch}\tau)=\mathfrak{P}^l_{-m,\,-n}(\operatorname{ch}\tau)$$

for all complex values of $l$ and all integral or semi-integral values of $m$ and $n$.

The symmetry relations (1) and formula (5) of subsection 5 lead to the following symmetry relations for associated Legendre functions:

$$\mathfrak{P}_l^{-m}(\operatorname{ch}\tau)=\frac{\Gamma(l-m+1)}{\Gamma(l+m+1)}\mathfrak{P}_l^m(\operatorname{ch}\tau), \tag{1'}$$

where $m$ is an integer.

In order to obtain the next symmetry relation, we observe that, on replacing $m$ by $-n$ and $n$ by $-m$, expression (2) of subsection 3 is multiplied by

$$\frac{\Gamma(l+m+1)\Gamma(l-m+1)}{\Gamma(l+n+1)\Gamma(l-n+1)}.$$

It follows that

$$\mathfrak{P}^l_{-n,\,-m}(\operatorname{ch}\tau)=\frac{\Gamma(l+m+1)\Gamma(l-m+1)}{\Gamma(l+n+1)\Gamma(l-n+1)}\mathfrak{P}^l_{mn}(\operatorname{ch}\tau) \tag{3}$$

or, owing to relation (1), that

$$\mathfrak{P}^l_{nm}(\operatorname{ch}\tau)=\frac{\Gamma(l+m+1)\Gamma(l-m+1)}{\Gamma(l+n+1)\Gamma(l-n+1)}\mathfrak{P}^l_{mn}(\operatorname{ch}\tau). \tag{3'}$$

We now pass on to relations connecting Jacobi functions with different values of $l$. We have proved in §2.6 that if $\chi$ is not integral, then the representations $T_\chi(g)$, $\chi=(l,\epsilon)$ and $T_{-\chi}(g)$, $-\chi=(-l-1,\epsilon)$ are equivalent. Namely, they are connected by the relation

$$QT_\chi(g)=T_{-\chi}(g)Q, \tag{4}$$

where $Q$ is an operator whose matrix in the basis $\{e^{-in\theta}\}$ is diagonal and has the following elements on its principal diagonal:

$$q_{nn}=\frac{\Gamma(l-n-\varepsilon+1)}{\Gamma(-l-n-\varepsilon)}. \tag{5}$$

Applying (4) to the element

$$g_\tau=\begin{pmatrix}\operatorname{ch}\frac{\tau}{2} & \operatorname{sh}\frac{\tau}{2}\\ \operatorname{sh}\frac{\tau}{2} & \operatorname{ch}\frac{\tau}{2}\end{pmatrix}$$

of $QU(2)$, and taking into consideration that

$$t^\chi_{mn}(g_\tau)=\mathfrak{P}^l_{m+\epsilon,\,n+\epsilon}(\operatorname{ch}\tau),$$

we obtain

$$\frac{\Gamma(l-m-\varepsilon+1)}{\Gamma(-l-m-\varepsilon)}\mathfrak{P}^l_{m+\varepsilon,\,n+\varepsilon}(\operatorname{ch}\tau)$$
$$=\frac{\Gamma(l-n-\varepsilon+1)}{\Gamma(-l-n-\varepsilon)}\mathfrak{P}^{-l-1}_{m+\varepsilon,\,n+\varepsilon}(\operatorname{ch}\tau). \tag{6}$$

Replacing $m+\epsilon$ by $m$ and $n+\epsilon$ by $n$, and utilizing the relation

$\Gamma(x)\Gamma(1-x)=\pi/\sin\pi x$, we rewrite equality (6) as

$$\mathfrak{P}^{-l-1}_{mn}(\operatorname{ch}\tau)=(-1)^{m-n}\frac{\Gamma(l-m+1)\Gamma(l+m+1)}{\Gamma(l-n+1)\Gamma(l+n+1)}\mathfrak{P}^{l}_{mn}(\operatorname{ch}\tau). \tag{7}$$

Hence, because of (3′), we get

$$\mathfrak{P}^{-l-1}_{mn}(\operatorname{ch}\tau)=(-1)^{m-n}\mathfrak{P}^{l}_{nm}(\operatorname{ch}\tau). \tag{7′}$$

On putting $n=0$ in (7) and employing relation (5) of subsection 5, we obtain

$$\mathfrak{P}^{m}_{l}(\operatorname{ch}\tau)=\mathfrak{P}^{m}_{-l-1}(\operatorname{ch}\tau). \tag{8}$$

In particular,

$$\mathfrak{P}_{l}(\operatorname{ch}\tau)=\mathfrak{P}_{-l-1}(\operatorname{ch}\tau). \tag{9}$$

The following relation for functions $\mathfrak{P}^{l}_{mn}(\operatorname{ch}\tau)$ follows from the unitary-adjointness of representations $T_\chi(g)$ and $T_{-\overline{\chi}}(g)$, $-\overline{\chi}=(-\overline{l}-1,\epsilon)$ (see §2.8):

$$(T_\chi(g)e^{-in\theta},\ e^{-im\theta})=(e^{-in\theta},\ T_{-\overline{\chi}}(g^{-1})e^{-im\theta})$$
$$=\overline{(T_{-\overline{\chi}}(g^{-1})e^{-im\theta},\ e^{-in\theta})}.$$

In other words,

$$t^{\chi}_{mn}(g)=\overline{t^{-\overline{\chi}}_{nm}(g^{-1})}.$$

Let us take

$$g=g_\tau\equiv\begin{pmatrix}\operatorname{ch}\frac{\tau}{2} & \operatorname{sh}\frac{\tau}{2}\\ \operatorname{sh}\frac{\tau}{2} & \operatorname{ch}\frac{\tau}{2}\end{pmatrix}.$$

Since the Euler angles of $g_\tau^{-1}$ are equal to $\pi,\tau,-\pi$, we get

$$\mathfrak{P}^{l}_{m'n'}(\operatorname{ch}\tau)=e^{i\pi(m'-n')}\overline{\mathfrak{P}^{-\overline{l}-1}_{n'm'}(\operatorname{ch}\tau)}$$

or

$$\overline{\mathfrak{P}^{l}_{mn}(\operatorname{ch}\tau)}=(-1)^{n-m}\mathfrak{P}^{-\overline{l}-1}_{nm}(\operatorname{ch}\tau)=\mathfrak{P}^{\overline{l}}_{mn}(\operatorname{ch}\tau). \tag{10}$$

Thus we have proved that

$$\overline{\mathfrak{P}^{l}_{mn}(\operatorname{ch}\tau)}=\mathfrak{P}^{\overline{l}}_{mn}(\operatorname{ch}\tau). \tag{10′}$$

In particular, it follows that $\mathfrak{P}^{l}_{mn}(\operatorname{ch}\tau)$, $0\le\tau<\infty$, with real index $l$ takes real values. We show that the same property is possessed by the functions

$$f(\tau)\equiv\frac{\Gamma\left(\frac{1}{2}+i\rho-m\right)}{\Gamma\left(\frac{1}{2}+i\rho-n\right)}\mathfrak{P}^{-\frac{1}{2}+i\rho}_{mn}(\operatorname{ch}\tau).$$

Indeed, it follows from (10) and (3′) that

$$\overline{\mathfrak{P}_{mn}^{-\frac{1}{2}+i\rho}(\operatorname{ch}\tau)}=(-1)^{n-m}\mathfrak{P}_{nm}^{-\frac{1}{2}+i\rho}(\operatorname{ch}\tau)$$

$$=(-1)^{n-m}\frac{\Gamma\left(\frac{1}{2}+i\rho+m\right)\Gamma\left(\frac{1}{2}+i\rho-m\right)}{\Gamma\left(\frac{1}{2}+i\rho+n\right)\Gamma\left(\frac{1}{2}+i\rho-n\right)}\mathfrak{P}_{mn}^{-\frac{1}{2}+i\rho}(\operatorname{ch}\tau). \tag{11}$$

Since

$$\frac{\Gamma\left(\frac{1}{2}-i\rho-m\right)\Gamma\left(\frac{1}{2}+i\rho+m\right)}{\Gamma\left(\frac{1}{2}-i\rho-n\right)\Gamma\left(\frac{1}{2}+i\rho+n\right)}=\frac{\sin\pi\left(\frac{1}{2}+i\rho+n\right)}{\sin\pi\left(\frac{1}{2}+i\rho+m\right)}=(-1)^{m-n},$$

it follows from (11) that $\overline{f(\tau)} = f(\tau)$. This proves that $f(\tau)$ is real. Hence the functions $\mathfrak{P}_{nn}^{-1/2+i\rho}(\operatorname{ch}\tau)$ take only real values for $0 \leq \tau < \infty$ and, in particular, $\mathfrak{P}_{-1/2+i\rho}(\operatorname{ch}\tau)$ is a real function.

7. The functions $\mathfrak{P}_{mn}^{l}(z)$ in the integral case. Let $\chi = (l, \epsilon)$ be integral, i.e. let $l+\epsilon$ and $l-\epsilon$ be integers. As shown in §2.5, in this case $T_\chi(g)$ is reducible.

If $l-\epsilon \leq 0$, then for $n \geq -l-\epsilon$ the function $T_\chi(g)e^{-in\theta}$ belongs to $\mathfrak{D}_l^-$, and can therefore be decomposed into the functions $e^{-im\theta}$, $m \geq -l-\epsilon$. Hence for $l-\epsilon \leq 0$ and $n \geq -l-\epsilon$, $m < -l-\epsilon$ we have $\mathfrak{P}_{mn}^{l}(z) = 0$. In just the same way one proves that, for $l-\epsilon \leq 0$ and $n \leq l-\epsilon$, $m > l-\epsilon$, we have $\mathfrak{P}_{mn}^{l}(z) = 0$.

Similarly, if $l-\epsilon \geq 0$, then $\mathfrak{P}_{mn}^{l}(z) \equiv 0$, for $n \geq +l-\epsilon$, $m < -l-\epsilon$ or for $n \leq -l-\epsilon$, $m > l-\epsilon$.

In the integral case the functions $\mathfrak{P}_{mn}^{l}(\operatorname{ch}\tau)$ are polynomials of $\operatorname{th}\tau/2$ and $\operatorname{sh}\tau/2$. Let us consider, for example, the case when $l$, $m$ and $n$ are simultaneously integers or semi-integers, where $m, n > l > 0$. Then the integrand in formula (1) of §3.3 has only one singular point $z = -\operatorname{th}\tau/2$ within the contour of integration. Therefore, owing to the residue theorem, we have

$$\mathfrak{P}_{mn}^{l}(\operatorname{ch}\tau)=\frac{1}{\Gamma(n-l)\operatorname{ch}^{n-l}\frac{\tau}{2}}\frac{d^{n-l-1}}{dz^{n-l-1}}\left[z^{m-l-1}\left(\operatorname{ch}\frac{\tau}{2}+z\operatorname{sh}\frac{\tau}{2}\right)^{l+n}\right],$$

where after differentiation one must substitute $z = -\operatorname{th}\tau/2$. Applying Leibnitz's theorem, we get

$$\mathfrak{P}^l_{mn}(\operatorname{ch}\tau)=(-1)^{m-n}\Gamma(l+n+1)\Gamma(m-l)\operatorname{sh}^{m-n}\frac{\tau}{2}\operatorname{ch}^{-m-n}\frac{\tau}{2}$$
$$\times\sum_{s=M}^{N}\frac{(-1)^s\operatorname{sh}^{2s}\frac{\tau}{2}}{s!\,\Gamma(l+n-s+1)\Gamma(n-l-s)\Gamma(m-n+s+1)},$$

where $M=\max(0, n-m)$, $N=\min(n-l-1, n+l)$.

Formulas for $\mathfrak{P}^l_{mn}(\operatorname{ch}\tau)$ in the other cases when $\chi=(l,\epsilon)$ is integral are easily obtained with the help of the symmetry relations of subsection 6.

We may note that, in the integral case, for $l-\epsilon<0$ the functions $\{e^{-in\theta}\}$, $-\infty<n\leq l-\epsilon$, form an orthogonal but not a normalized basis relative to the invari ant scalar product in $\mathfrak{D}^+_l$ (see §2.7). The orthonormal basis consists of the functions

$$\left[\frac{\Gamma(-n-l-\varepsilon)}{\Gamma(l-n-\varepsilon+1)}\right]^{\frac{1}{2}}e^{-in\theta},\qquad -\infty<n\leqslant l-\varepsilon.$$

Therefore the elements of the unitary matrix of $T^+_l(g)$ have the form

$$t^{\chi,+}_{mn}(g)=\left[\frac{\Gamma(l-m'+1)\Gamma(-l-n')}{\Gamma(l-n'+1)\Gamma(-l-m')}\right]^{\frac{1}{2}}e^{-i(m'\varphi+n'\psi)}\mathfrak{P}^l_{m'n'}(\operatorname{ch}\tau),$$

where $m'=m+\epsilon$, $n'=n+\epsilon$, $-\infty<m,\ n\leq l-\epsilon$. We note that

$$t^{\chi,+}_{nn}(g)=e^{-in'(\varphi+\psi)}\mathfrak{P}^l_{n'n'}(\operatorname{ch}\tau).$$

Similarly, the elements of the unitary matrix of $T^-_l(g)$ have the form

$$t^{\chi,-}_{mn}(g)=\left[\frac{\Gamma(-l-m')\Gamma(l-n'+1)}{\Gamma(-l-n')\Gamma(l-m'+1)}\right]^{\frac{1}{2}}e^{-i(m'\varphi+n'\psi)}\mathfrak{P}^l_{m'n'}(\operatorname{ch}\tau),$$

where $-l-\epsilon\leq m,\ n<\infty$.

## §4. FUNCTIONAL RELATIONS FOR $\mathfrak{P}^l_{mn}(\operatorname{ch}\tau)$

**1. The composition theorem.** The composition theorem for the functions $\mathfrak{P}^l_{mn}(\operatorname{ch}\tau)$ resembles the composition theorem for the functions $P^l_{mn}(\cos\theta)$, proved in §4.1 of Chapter III. To derive the composition theorem we use the relation

$$T_\chi(g_1g_2)=T_\chi(g_1)\,T_\chi(g_2), \tag{1}$$

which implies that

$$t^\chi_{mn}(g_1g_2)=\sum_{k=-\infty}^{\infty}t^\chi_{mk}(g_1)\,t^\chi_{kn}(g_2). \tag{2}$$

We apply (2) to the matrices $g_1 = g(0, \tau_1, 0)$ and $g_2 = g(\phi_2, \tau_2, 0)$. For these matrices, $t^{\chi}_{mk}(g_1) = \mathfrak{P}^l_{m'k'}(\operatorname{ch}\tau_1)$ and $t^{\chi}_{km}(g_2) = e^{-ik'\phi_2}\mathfrak{P}^l_{k'n'}(\operatorname{ch}\tau_2)$, where $k' = k + \epsilon$, $m' = m + \epsilon$, $n' = n + \epsilon$ (see §3.2). The matrix element $t^{\chi}_{mn}(g_1 g_2)$ has the form

$$t^{\chi}_{mn}(g_1g_2) = e^{-i(m'\varphi + n'\psi)}\mathfrak{P}^l_{m'n'}(\operatorname{ch}\tau), \tag{3}$$

where $\phi, \tau, \psi$ are the parameters of the matrix $g_1 g_2$. According to §1.3, these parameters are expressed in terms of $\tau_1, \phi_2, \tau_2$ by the formulas

$$\operatorname{ch}\tau = \operatorname{ch}\tau_1 \operatorname{ch}\tau_2 + \operatorname{sh}\tau_1 \operatorname{sh}\tau_2 \cos\varphi_2, \tag{4}$$

$$e^{i\varphi} = \frac{\operatorname{sh}\tau_1 \operatorname{ch}\tau_2 + \operatorname{ch}\tau_1 \operatorname{sh}\tau_2 \cos\varphi_2 + i \operatorname{sh}\tau_2 \sin\varphi_2}{\operatorname{sh}\tau}, \tag{4'}$$

$$e^{\frac{i(\varphi+\psi)}{2}} = \frac{\operatorname{ch}\frac{\tau_1}{2}\operatorname{ch}\frac{\tau_2}{2}e^{\frac{i\varphi_2}{2}} + \operatorname{sh}\frac{\tau_1}{2}\operatorname{sh}\frac{\tau_2}{2}e^{-\frac{i\varphi_2}{2}}}{\operatorname{ch}\frac{\tau}{2}}, \tag{4''}$$

where

$$\left.\begin{array}{c} 0 \leqslant \varphi < 2\pi, \\ 0 \leqslant \tau < \infty, \\ -2\pi \leqslant \psi < 2\pi. \end{array}\right\} \tag{5}$$

Substituting the values of $t^{\chi}_{mk}(g_1)$, $t^{\chi}_{kn}(g_2)$ and $t^{\chi}_{mn}(g_1 g_2)$ in (2) and replacing $m', k', n'$ by $m, k, n$, we get

$$e^{-i(m\varphi + n\psi)}\mathfrak{P}^l_{mn}(\operatorname{ch}\tau) = \sum_{k=-\infty}^{\infty} e^{-ik\varphi_2}\mathfrak{P}^l_{mk}(\operatorname{ch}\tau_1)\mathfrak{P}^l_{kn}(\operatorname{ch}\tau_2), \tag{6}$$

where the parameters $\tau_1, \phi_2, \tau_2, \phi, \tau, \psi$ are connected by the relations (4)–(4″). Moreover, if $m$ and $n$ are integral, $k$ runs through integral values, and if $m$ and $n$ are semi-integral, $k$ runs through semi-integral values. This is the composition theorem for functions $\mathfrak{P}^l_{mn}(\operatorname{ch}\tau)$.

As in the case of functions $P^l_{mn}(\cos\theta)$, one establishes the following special cases of formula (6).

If $\phi_2 = 0$, then $\tau = \tau_1 + \tau_2$, $\phi = \psi = 0$ and therefore

$$\mathfrak{P}^l_{mn}(\operatorname{ch}(\tau_1+\tau_2)) = \sum_{k=-\infty}^{\infty} \mathfrak{P}^l_{mk}(\operatorname{ch}\tau_1)\mathfrak{P}^l_{kn}(\operatorname{ch}\tau_2). \tag{7}$$

For $\phi_2 = \pi$, $\tau_1 \geq \tau_2$ we have $\tau = \tau_1 - \tau_2$, $\phi = 0$, $\psi = \pi$, and therefore

$$\mathfrak{P}^l_{mn}(\operatorname{ch}(\tau_1-\tau_2))=\sum_{k=-\infty}^{\infty}(-1)^{n-k}\mathfrak{P}^l_{mk}(\operatorname{ch}\tau_1)\mathfrak{P}^l_{kn}(\operatorname{ch}\tau_2). \tag{8}$$

In particular, for $\tau_1 = \tau_2 = \tau$ we have

$$\sum_{k=-\infty}^{\infty}(-1)^{n-k}\mathfrak{P}^l_{mk}(\operatorname{ch}\tau)\mathfrak{P}^l_{kn}(\operatorname{ch}\tau)=\mathfrak{P}^l_{mn}(1)=\delta_{mn}. \tag{9}$$

Utilizing formula (7′) of §3.6, we can rewrite (8) as

$$\mathfrak{P}^l_{mn}(\operatorname{ch}(\tau_1-\tau_2))=\sum_{k=-\infty}^{\infty}\mathfrak{P}^l_{mk}(\operatorname{ch}\tau_1)\mathfrak{P}^{-l-1}_{nk}(\operatorname{ch}\tau_2). \tag{10}$$

If $l = -\frac{1}{2} + i\rho$, then $-l - 1 = \bar{l}$, and by formula (10) of §3.6 we get

$$\mathfrak{P}^{-\frac{1}{2}+i\rho}_{mn}(\operatorname{ch}(\tau_1-\tau_2))=\sum_{k=-\infty}^{\infty}\mathfrak{P}^{-\frac{1}{2}+i\rho}_{mk}(\operatorname{ch}\tau_1)\overline{\mathfrak{P}^{-\frac{1}{2}+i\rho}_{nk}(\operatorname{ch}\tau_2)}. \tag{11}$$

In particular, for $\tau_1 = \tau_2 = \tau$ we have

$$\sum_{k=-\infty}^{\infty}\mathfrak{P}^{-\frac{1}{2}+i\rho}_{mk}(\operatorname{ch}\tau)\overline{\mathfrak{P}^{-\frac{1}{2}+i\rho}_{nk}(\operatorname{ch}\tau)}=\delta_{mn}. \tag{12}$$

This equality expresses the fact that representations of the fundamental series are unitary.

We may note the special case of the composition theorem for $\phi_2 = \pi/2$. In this case we get

$$e^{-i(m\varphi+n\psi)}\mathfrak{P}^l_{mn}(\operatorname{ch}\tau)=\sum_{k=-\infty}^{\infty}e^{-\frac{ik\pi}{2}}\mathfrak{P}^l_{mk}(\operatorname{ch}\tau_1)\mathfrak{P}^l_{kn}(\operatorname{ch}\tau_2), \tag{13}$$

where

$$\operatorname{ch}\tau=\operatorname{ch}\tau_1\operatorname{ch}\tau_2, \tag{14}$$

$$e^{i\varphi}=\frac{\operatorname{sh}\tau_1\operatorname{ch}\tau_2+i\operatorname{sh}\tau_2}{\operatorname{sh}\tau}, \tag{14′}$$

$$e^{\frac{i(\varphi+\psi)}{2}}=\frac{\sqrt{2}}{2}\,\frac{\operatorname{ch}\frac{\tau_1+\tau_2}{2}+i\operatorname{ch}\frac{\tau_1-\tau_2}{2}}{\operatorname{ch}\frac{\tau}{2}}. \tag{14″}$$

2. The integral case. The composition theorem for functions $\mathfrak{P}^l_{mn}(\operatorname{ch}\tau)$ assumes a slightly simpler form if $\chi = (l, \epsilon)$ is integral, i.e. if $2l$ has the same parity as $2\epsilon$. Then, as shown in §3.7, $\mathfrak{P}^l_{mn}(z) \equiv 0$ in the following cases:

a) $l - \epsilon < 0$ and $n > -l - \epsilon$, $m < -l - \epsilon$ or $n \leq l - \epsilon$, $m > l - \epsilon$;

b) $l-\epsilon \geq 0$ and $n \geq l-\epsilon$, $m < -l-\epsilon$ or $n \leq -l-\epsilon$, $m > l-\epsilon$.

Therefore if $l-\epsilon \leq 0$, then for $m, n \geq -l-\epsilon$ the summation in formula (6) above extends from $-l-\epsilon$ to $\infty$; for $m \geq -l-\epsilon$, $l-\epsilon \leq n \leq l+\epsilon$ it extends from $l-\epsilon$ to $\infty$; for $l-\epsilon \leq m$, $n \leq l+\epsilon$, from $l-\epsilon$ to $l+\epsilon$; for $m \leq l-\epsilon$, $l-\epsilon \leq n \leq l+\epsilon$, from $-\infty$ to $-l-\epsilon$; and for $m, n \leq l-\epsilon$, from $-\infty$ to $l-\epsilon$.

But if $l-\epsilon \geq 0$, then for $m, n \geq l-\epsilon$ the summation in formula (6) extends from $l-\epsilon$ to $\infty$; for $-l-\epsilon \leq m \leq l-\epsilon$, $n \geq l-\epsilon$ it extends from $-l-\epsilon$ to $\infty$; for $-l-\epsilon \leq m$, $n \leq l-\epsilon$, from $-l-\epsilon$ to $l-\epsilon$; for $-l-\epsilon \leq m \leq l-\epsilon$, $n \leq l-\epsilon$, from $\infty$ to $l-\epsilon$; and for $m, n \leq -l-\epsilon$, from $-\infty$ to $-l-\epsilon$.

**3. Composition theorems for Legendre functions.** We have defined Legendre functions and associated Legendre functions by the formulas

$$\mathfrak{P}_l(\operatorname{ch}\tau) = \mathfrak{P}^l_{00}(\operatorname{ch}\tau) \tag{1}$$

and

$$\mathfrak{P}^m_l(\operatorname{ch}\tau) = \frac{\Gamma(l+m+1)}{\Gamma(l+1)}\mathfrak{P}^l_{m0}(\operatorname{ch}\tau) = \frac{\Gamma(l+1)}{\Gamma(l-m+1)}\mathfrak{P}^l_{0m}(\operatorname{ch}\tau) \tag{2}$$

(see §§3.4 and 3.5). We put $m = n = 0$ in formula (6) above and use expressions (1) and (2). We get

$$\mathfrak{P}_l(\operatorname{ch}\tau) = \sum_{k=-\infty}^{\infty} \frac{\Gamma(l-k+1)}{\Gamma(l+k+1)} e^{-ik\varphi_2}\mathfrak{P}^k_l(\operatorname{ch}\tau_1)\,\mathfrak{P}^k_l(\operatorname{ch}\tau_2), \tag{3}$$

where

$$\operatorname{ch}\tau = \operatorname{ch}\tau_1\operatorname{ch}\tau_2 + \operatorname{sh}\tau_1\operatorname{sh}\tau_2\cos\varphi_2. \tag{4}$$

Utilizing formula (1′) of §3.6, we rewrite this equality as

$$\mathfrak{P}_l(\operatorname{ch}\tau) = \sum_{k=-\infty}^{\infty} e^{-ik\varphi}\mathfrak{P}^k_l(\operatorname{ch}\tau_1)\,\mathfrak{P}^{-k}_l(\operatorname{ch}\tau_2). \tag{5}$$

The composition formula for associated Legendre functions is obtained by putting $n = 0$ in equality (6) of subsection 1. We get

$$e^{im\varphi}\mathfrak{P}^m_l(\operatorname{ch}\tau) = \sum_{k=-\infty}^{\infty} \frac{\Gamma(l+m+1)}{\Gamma(l+k+1)} e^{ik\varphi_2}\mathfrak{P}^l_{mk}(\operatorname{ch}\tau_1)\,\mathfrak{P}^k_l(\operatorname{ch}\tau_2), \tag{6}$$

where the numbers $\phi$, $\tau$, $\phi_2$, $\tau_1$, $\tau_2$ are connected by the relations (4)–(4″) of subsection 1.

**4. The multiplication formula.** Let us multiply both sides of formula (6) of subsection 1 by $e^{ik\phi_2}$, and integrate with respect to $\phi_2$ from 0 to $2\pi$. We obtain

$$\mathfrak{P}^l_{mk}(\text{ch}\,\tau_1)\,\mathfrak{P}^l_{kn}(\text{ch}\,\tau_2)=\frac{1}{2\pi}\int_0^{2\pi} e^{i(k\varphi_2-m\varphi-n\psi)}\,\mathfrak{P}^l_{mn}(\text{ch}\,\tau)\,d\varphi_2. \tag{1}$$

The parameters $\phi$, $\psi$, $\phi_2$, $\tau_1$, $\tau_2$, $\tau$ in this formula are connected by relations (4)–(4″) of subsection 1.

Putting $m = n = 0$ in formula (1) and utilizing relation (1′) of §3.6, we get

$$\mathfrak{P}^k_l(\text{ch}\,\tau_1)\,\mathfrak{P}^{-k}_l(\text{ch}\,\tau_2)=\frac{1}{2\pi}\int_0^{2\pi} e^{ik\varphi_2}\mathfrak{P}_l(\text{ch}\,\tau_1\,\text{ch}\,\tau_2+\text{sh}\,\tau_1\,\text{sh}\,\tau_2\cos\varphi_2)\,d\varphi_2. \tag{2}$$

In particular,

$$\mathfrak{P}_l(\text{ch}\,\tau_1)\,\mathfrak{P}_l(\text{ch}\,\tau_2)=\frac{1}{2\pi}\int_0^{2\pi} \mathfrak{P}_l(\text{ch}\,\tau_1\,\text{ch}\,\tau_2+\text{sh}\,\tau_1\,\text{sh}\,\tau_2\cos\varphi_2)\,d\varphi_2. \tag{3}$$

These formulas have a simple geometrical significance, similar to the geometrical significance of formulas (2′)–(3) of §4.3 of Chapter III. Indeed, we consider the upper nappe of the two-sheeted hyperboloid

$$-x_1^2-x_2^2+x_3^2=1, \qquad x_3>0.$$

We introduce a metric on this nappe by taking as the distance between the points $M(x_1, x_2, x_3)$ and $N(y_1, y_2, y_3)$ the number $\tau$, where

$$\text{ch}\,\tau=-x_1y_1-x_2y_2+x_3y_3. \tag{4}$$

Finally, we introduce parameters $\tau$, $\phi$ on the hyperboloid by putting

$$x_1=\text{sh}\,\tau\sin\varphi,$$
$$x_2=\text{sh}\,\tau\cos\varphi,$$
$$x_3=\text{ch}\,\tau.$$

The functions $\mathfrak{P}^k_l(\text{ch}\,\tau)$ can be regarded as functions on the hyperboloid.

We consider the circle with center at the point $M(\tau_1, 0)$ and radius $\tau_2$. Points $N$ of this circle are specified by a number $\phi_2$, i.e. the angle between the geodesic lines joining $M$ with $N$ and with the point $O(0, 0, 1)$. It is not difficult to show that for the point $N$, corresponding to a given value of $\phi_2$, the coordinate $\tau$ is such that $\text{ch}\,\tau = \text{ch}\,\tau_1\,\text{ch}\,\tau_2 + \text{sh}\,\tau_1\,\text{sh}\,\tau_2\cos\phi_2$. Hence formula (3) signifies that $\mathfrak{P}_l(\text{ch}\,\tau_1)\mathfrak{P}_l(\text{ch}\,\tau_2)$ is the mean value of the function $\mathfrak{P}_l(\text{ch}\,\tau)$ on this circle. A similar interpretation holds for formula (2), which gives the value of Fourier coefficients for $\mathfrak{P}_l(\text{ch}\,\tau)$, regarded as a function of $\phi_2$.

Let us now rewrite formula (2) as

$$\mathfrak{P}_l^k(\operatorname{ch}\tau_1)\mathfrak{P}_l^{-k}(\operatorname{ch}\tau_2)$$
$$=\frac{1}{\pi}\int_0^\pi \mathfrak{P}_l(\operatorname{ch}\tau_1\operatorname{ch}\tau_2+\operatorname{sh}\tau_1\operatorname{sh}\tau_2\cos\varphi_2)\cos k\varphi_2\,d\varphi_2 \tag{5}$$

and make the change of variable

$$\operatorname{ch}\tau_1\operatorname{ch}\tau_2+\operatorname{sh}\tau_1\operatorname{sh}\tau_2\cos\varphi_2=\operatorname{ch}\tau. \tag{6}$$

We obtain

$$\mathfrak{P}_l^k(\operatorname{ch}\tau_1)\mathfrak{P}_l^{-k}(\operatorname{ch}\tau_2)$$
$$=\frac{1}{\pi}\int_{|\tau_1-\tau_2|}^{\tau_1+\tau_2}\frac{\mathfrak{P}_l(\operatorname{ch}\tau)\,T_k\left[\dfrac{\operatorname{ch}\tau-\operatorname{ch}\tau_1\operatorname{ch}\tau_2}{\operatorname{sh}\tau_1\operatorname{sh}\tau_2}\right]\operatorname{sh}\tau\,d\tau}{\sqrt{[\operatorname{ch}(\tau_1+\tau_2)-\operatorname{ch}\tau]\,[\operatorname{ch}\tau-\operatorname{ch}(\tau_1-\tau_2)]}}. \tag{7}$$

Here, in accordance with §4.3 of Chapter III, we have put

$$T_k(x)=\cos(k\arccos x).$$

In particular, for $k = 0$ we get

$$\mathfrak{P}_l(\operatorname{ch}\tau_1)\mathfrak{P}_l(\operatorname{ch}\tau_2)$$
$$=\frac{1}{\pi}\int_{|\tau_1-\tau_2|}^{\tau_1+\tau_2}\frac{\mathfrak{P}_l(\operatorname{ch}\tau)\operatorname{sh}\tau\,d\tau}{\sqrt{[\operatorname{ch}(\tau_1+\tau_2)-\operatorname{ch}\tau]\,[\operatorname{ch}\tau-\operatorname{ch}(\tau_1-\tau_2)]}}. \tag{8}$$

The expression in the denominator of (7) and (8) is the normalized area of a triangle on the hyperboloid $-x_1^2-x_2^2+x_3^2=1$ with sides $\tau_1, \tau_2, \tau$ (in the metric mentioned above).

5. **Recurrence formulas.** Recurrence formulas for the functions $\mathfrak{P}^l_{mn}(z)$ are derived from the composition theorem in exactly the same way as the recurrence formulas for the functions $\mathfrak{P}^l_{mn}(z)$ (see §4.4 of Chapter III). Therefore we shall omit the derivation, and only state the final results.

One has the equalities

$$\sqrt{z^2-1}\,\frac{d\mathfrak{P}^l_{mn}(z)}{dz}=\frac{l+n}{2}\mathfrak{P}^l_{m,n-1}(z)+\frac{l-n}{2}\mathfrak{P}^l_{m,n+1}(z) \tag{1}$$

and

$$\frac{m-nz}{\sqrt{z^2-1}}\mathfrak{P}^l_{mn}(z)=-\frac{l+n}{2}\mathfrak{P}^l_{m,n-1}(z)+\frac{l-n}{2}\mathfrak{P}^l_{m,n+1}(z). \tag{2}$$

These equalities are obtained by differentiating formulas (7) and (13) of subsection 1 with respect to $\tau_2$ and then putting $\tau_2 = 0$. They yield

$$\sqrt{z^2-1}\,\frac{d\mathfrak{P}^l_{mn}(z)}{dz}+\frac{m-nz}{\sqrt{z^2-1}}\,\mathfrak{P}^l_{mn}(z)=(l-n)\,\mathfrak{P}^l_{m,\,n+1}(z) \tag{3}$$

and

$$\sqrt{z^2-1}\,\frac{d\mathfrak{P}^l_{mn}(z)}{dz}+\frac{nz-m}{\sqrt{z^2-1}}\,\mathfrak{P}^l_{mn}(z)=(l+n)\,\mathfrak{P}^l_{m,\,n-1}(z). \tag{4}$$

Hence, utilizing the symmetery relation (3′) of § 3.6, we get

$$\sqrt{z^2-1}\,\frac{d\mathfrak{P}^l_{mn}(z)}{dz}+\frac{n-mz}{\sqrt{z^2-1}}\,\mathfrak{P}^l_{mn}(z)=(l+m+1)\,\mathfrak{P}^l_{m+1,\,n}(z) \tag{5}$$

and

$$\sqrt{z^2-1}\,\frac{d\mathfrak{P}^l_{mn}(z)}{dz}+\frac{mz-n}{\sqrt{z^2-1}}\,\mathfrak{P}^l_{mn}(z)=(l-m+1)\,\mathfrak{P}^l_{m-1,\,n}(z). \tag{6}$$

Subtracting (4) from (3), we get a recurrence relation which connects three functions $\mathfrak{P}^l_{mn}(z)$ with contiguous indices:

$$(l-n)\,\mathfrak{P}^l_{m,\,n+1}(z)-(l+n)\,\mathfrak{P}^l_{m,\,n-1}(z)=2\,\frac{m-nz}{\sqrt{z^2-1}}\,\mathfrak{P}^l_{mn}(z). \tag{7}$$

Similarly, it follows from (5) and (6) that

$$(l+m+1)\,\mathfrak{P}^l_{m+1,\,n}(z)-(l-m+1)\,\mathfrak{P}^l_{m-1,\,n}(z)=2\,\frac{n-mz}{\sqrt{z^2-1}}\,\mathfrak{P}^l_{mn}(z). \tag{8}$$

Further, the differential equation satisfied by the functions $\mathfrak{P}^l_{mn}(z)$ is derived in just the same way as for the functions $P^l_{mn}(z)$ (see § 4.5 of Chapter III). It has the form

$$\left(\sqrt{z^2-1}\,\frac{d}{dz}+\frac{(n+1)z-m}{\sqrt{z^2-1}}\right)\left(\sqrt{z^2-1}\,\frac{d}{dz}+\frac{m-nz}{\sqrt{z^2-1}}\right)\mathfrak{P}^l_{mn}(z)$$
$$=(l-n)(l+n+1)\,\mathfrak{P}^l_{mn}(z), \tag{9}$$

i.e., in other words,

$$(z^2-1)\,\frac{d^2\mathfrak{P}^l_{mn}(z)}{dz^2}+2z\,\frac{d\mathfrak{P}^l_{mn}(z)}{dz}-\frac{m^2+n^2-2mnz}{z^2-1}\,\mathfrak{P}^l_{mn}(z)$$
$$=l(l+1)\,\mathfrak{P}^l_{mn}(z). \tag{9′}$$

We note that this equation is the same as equation (5) for the functions $P^l_{mn}(z)$ in § 4.5 of Chapter III.

From these relations, as a special case, one obtains relations for associated Legendre functions. Let us put $n=0$ in (3). Taking into consideration formula (5) of § 3.5, we get

$$\sqrt{z^2-1}\,\frac{d\mathfrak{P}_l^m(z)}{dz}-\frac{mz}{\sqrt{z^2-1}}\,\mathfrak{P}_l^m(z)=\mathfrak{P}_l^{m+1}(z). \qquad (10)$$

In just the same way (4) gives

$$\sqrt{z^2-1}\,\frac{d\mathfrak{P}_l^m(z)}{dz}+\frac{mz}{\sqrt{z^2-1}}\,\mathfrak{P}_l^m(z)=(l+m)(l-m+1)\,\mathfrak{P}_l^{m-1}(z). \qquad (11)$$

From equation (9′) we get the differential equation for associated Legendre functions:

$$(z^2-1)\frac{d^2\mathfrak{P}_l^m(z)}{dz^2}+2z\,\frac{d\mathfrak{P}_l^m(z)}{dz}-\frac{m^2}{z^2-1}\,\mathfrak{P}_l^m(z)=l(l+1)\,\mathfrak{P}_l^m(z). \qquad (12)$$

For Legendre functions one has the differential equation

$$(z^2-1)\frac{d^2\mathfrak{P}_l(z)}{dz^2}+2z\,\frac{d\mathfrak{P}_l(z)}{dz}=l(l+1)\,\mathfrak{P}_l(z). \qquad (13)$$

6. **The generating function.** Another method of deriving recurrence relations depends upon the generating function for $\mathfrak{P}^l_{mn}(\operatorname{ch}\tau)$. Namely, by formula (6) of §3.2, one has the equality

$$\left(\operatorname{ch}\frac{\tau}{2}+\operatorname{sh}\frac{\tau}{2}\,e^{i\theta}\right)^{l+n}\left(\operatorname{ch}\frac{\tau}{2}+\operatorname{sh}\frac{\tau}{2}\,e^{-i\theta}\right)^{l-n}e^{-in\theta}$$
$$=\sum_{m=-\infty}^{\infty}\mathfrak{P}^l_{mn}(\operatorname{ch}\tau)\,e^{-im\theta}. \qquad (1)$$

On replacing $e^{-i\theta}$ by $z$, we get

$$\Phi(z,\tau)\equiv\left(z\operatorname{ch}\frac{\tau}{2}+\operatorname{sh}\frac{\tau}{2}\right)^{l+n}\left(z\operatorname{sh}\frac{\tau}{2}+\operatorname{ch}\frac{\tau}{2}\right)^{l-n}$$
$$=\sum_{m=-\infty}^{\infty}\mathfrak{P}^l_{mn}(\operatorname{ch}\tau)\,z^{l+m}. \qquad (2)$$

This shows that $\Phi(z,\tau)$ is a generating function for $\mathfrak{P}^l_{mn}(\operatorname{ch}\tau)$.

Putting $z=1$ in (2), we get

$$\sum_{m=-\infty}^{\infty}\mathfrak{P}^l_{mn}(\operatorname{ch}\tau)=e^{l\tau}. \qquad (3)$$

But if we put $z=e^{\pi i}$, we get

$$\sum_{m=-\infty}^{\infty}(-1)^{m-n}\,\mathfrak{P}^l_{mn}(\operatorname{ch}\tau)=e^{-l\tau}. \qquad (4)$$

Lastly, putting $n = 0$ and $z = e^{\pi i/2}$, we have

$$\sum_{m=-\infty}^{\infty} i^m \mathfrak{P}^l_{m0}(\operatorname{ch}\tau) = \operatorname{ch}^l \tau. \tag{5}$$

Equality (5) can be rewritten as

$$\sum_{m=-\infty}^{\infty} \frac{i^m}{\Gamma(l+m+1)} \mathfrak{P}^m_l(\operatorname{ch}\tau) = \frac{\operatorname{ch}^l\tau}{\Gamma(l+1)}. \tag{5'}$$

Recurrence formulas for the functions $\mathfrak{P}^l_{mn}(\operatorname{ch}\tau)$ are derived from (2) in exactly the same way as for $P^l_{mn}(\cos\theta)$. Therefore we shall not derive the formulas already obtained above, but state some other relations connecting the functions $\mathfrak{P}^l_{mn}(\operatorname{ch}\tau)$ with different values of $l$.

We differentiate both sides of (2) with respect to $z$, apply expansion (2) (with $l$ replaced by $l - \frac{1}{2}$ and $n$ by $n \pm \frac{1}{2}$) to both the terms in the left-hand side, and compare the coefficients of like powers of $z$. We obtain

$$(l+n)\operatorname{ch}\frac{\tau}{2}\mathfrak{P}^{l-\frac{1}{2}}_{m-\frac{1}{2},\, n-\frac{1}{2}}(\operatorname{ch}\tau) + (l-n)\operatorname{sh}\frac{\tau}{2}\mathfrak{P}^{l-\frac{1}{2}}_{m-\frac{1}{2},\, n+\frac{1}{2}}(\operatorname{ch}\tau) = (l+m)\mathfrak{P}^l_{mn}(\operatorname{ch}\tau). \tag{6}$$

Further, multiplying both sides of (2) by $z \operatorname{ch}\tau/2 + \operatorname{sh}\tau/2$ and expanding the left-hand side, we obtain

$$\mathfrak{P}^{l+\frac{1}{2}}_{m+\frac{1}{2},\, n+\frac{1}{2}}(\operatorname{ch}\tau) = \operatorname{ch}\frac{\tau}{2}\mathfrak{P}^l_{mn}(\operatorname{ch}\tau) + \operatorname{sh}\frac{\tau}{2}\mathfrak{P}^l_{m+1,\, n}(\operatorname{ch}\tau). \tag{7}$$

Similarly, multiplication by $z \operatorname{sh}\tau/2 + \operatorname{ch}\tau/2$ gives

$$\mathfrak{P}^{l+\frac{1}{2}}_{m+\frac{1}{2},\, n-\frac{1}{2}}(\operatorname{ch}\tau) = \operatorname{sh}\frac{\tau}{2}\mathfrak{P}^l_{mn}(\operatorname{ch}\tau) + \operatorname{ch}\frac{\tau}{2}\mathfrak{P}^l_{m+1,\, n}(\operatorname{ch}\tau). \tag{8}$$

Equalities (7) and (8) immediately yield

$$\mathfrak{P}^l_{mn}(\operatorname{ch}\tau) = \operatorname{ch}\frac{\tau}{2}\mathfrak{P}^{l+\frac{1}{2}}_{m+\frac{1}{2},\, n+\frac{1}{2}}(\operatorname{ch}\tau) - \operatorname{sh}\frac{\tau}{2}\mathfrak{P}^{l+\frac{1}{2}}_{m+\frac{1}{2},\, n-\frac{1}{2}}(\operatorname{ch}\tau) \tag{9}$$

and

$$\mathfrak{P}^l_{mn}(\operatorname{ch}\tau) = -\operatorname{sh}\frac{\tau}{2}\mathfrak{P}^{l+\frac{1}{2}}_{m-\frac{1}{2},\, n+\frac{1}{2}}(\operatorname{ch}\tau) + \operatorname{ch}\frac{\tau}{2}\mathfrak{P}^{l+\frac{1}{2}}_{m-\frac{1}{2},\, n-\frac{1}{2}}(\operatorname{ch}\tau). \tag{10}$$

We may also mention the formula

$$\mathfrak{P}^{l+1}_{mn}(\operatorname{ch}\tau) = \frac{\operatorname{sh}\tau}{2}[\mathfrak{P}^l_{m-1,\, n}(\operatorname{ch}\tau) + \mathfrak{P}^l_{m+1,\, n}(\operatorname{ch}\tau)] + \operatorname{ch}\tau\,\mathfrak{P}^l_{mn}(\operatorname{ch}\tau), \tag{11}$$

which is obtained by multiplying both sides of (2) by the corresponding sides of the equality

$$\left(z \operatorname{ch}\frac{\tau}{2}+\operatorname{sh}\frac{\tau}{2}\right)\left(z \operatorname{sh}\frac{\tau}{2}+\operatorname{ch}\frac{\tau}{2}\right)=\frac{z^2+1}{2}\operatorname{sh}\tau+z\operatorname{ch}\tau.$$

Formula (11) can also be obtained by employing relations (7) and (8).

Similarly,

$$\mathfrak{P}^{l-1}_{mn}(\operatorname{ch}\tau)=-\frac{\operatorname{sh}\tau}{2}\left[\mathfrak{P}^{l}_{m,\,n+1}(\operatorname{ch}\tau)+\mathfrak{P}^{l}_{m,\,n+1}(\operatorname{ch}\tau)\right]+\operatorname{ch}\tau\,\mathfrak{P}^{l}_{mn}(\operatorname{ch}\tau). \tag{12}$$

Finally, we consider the two expansions

$$\left(z \operatorname{ch}\frac{\tau}{2}+\operatorname{sh}\frac{\tau}{2}\right)^{l_1+n_1}\left(z \operatorname{sh}\frac{\tau}{2}+\operatorname{ch}\frac{\tau}{2}\right)^{l_1-n_1} = \sum_{m_1=-\infty}^{\infty}\mathfrak{P}^{l_1}_{m_1n_1}(\operatorname{ch}\tau)\, z^{l_1+m_1}$$

and

$$\left(z \operatorname{ch}\frac{\tau}{2}+\operatorname{sh}\frac{\tau}{2}\right)^{l_2+n_2}\left(z \operatorname{sh}\frac{\tau}{2}+\operatorname{ch}\frac{\tau}{2}\right)^{l_2-n_2} = \sum_{m_2=-\infty}^{\infty}\mathfrak{P}^{l_2}_{m_2n_2}(\operatorname{ch}\tau)\, z^{l_2+m_2}.$$

We multiply these expansions side by side and apply (2) to the left-hand side. Comparing the coefficients of like powers of $z$, we find that

$$\mathfrak{P}^{l_1+l_2}_{m,\,n_1+n_2}(\operatorname{ch}\tau)=\sum_{m_1=-\infty}^{\infty}\mathfrak{P}^{l_1}_{m_1n_1}(\operatorname{ch}\tau)\,\mathfrak{P}^{l_2}_{m-m_1,\,n_2}(\operatorname{ch}\tau). \tag{13}$$

7. **A continual generating function.** Let us apply the Mellin transform inversion formula to formula (15) of §3.3. We obtain

$$F(w,\operatorname{ch}\tau)\equiv\frac{\left(\operatorname{ch}\frac{\tau}{2}+z\operatorname{sh}\frac{\tau}{2}\right)^{2n}z^{m-n}}{\sqrt{w^2+2w\operatorname{ch}\tau+1}} = -\frac{i}{2}\int_{a-i\infty}^{a+i\infty}\frac{P^{l+n}_{mn}(\operatorname{ch}\tau)\,\omega^{-l-1}\,dl}{\sin l\pi}, \tag{1}$$

where

$$z=\frac{-w-\operatorname{ch}\tau+\sqrt{w^2+2w\operatorname{ch}\tau+1}}{\operatorname{sh}\tau} \tag{2}$$

and $-1 < a < m-n$.

This equality shows that the function $F(w, t)$ can be regarded as the generating function for $\mathfrak{P}^{l}_{mn}(t)$ for fixed $m$ and $n$. However, in the right-hand side instead of a sum there appears an integral with respect to $l$. Therefore we shall call the function $F(w, t)$ the continual generating function for $\mathfrak{P}^l_{mn}(t)$ for fixed $m$ and $n$.

Formula (1) takes a particularly simple form for $m = n = 0$:

$$\frac{1}{\sqrt{w^2+2w\operatorname{ch}\tau+1}}=-\frac{i}{2}\int_{a-i\infty}^{a+i\infty}\frac{\mathfrak{P}_l(\operatorname{ch}\tau)\,w^{-l-1}}{\sin l\pi}\,dl, \tag{3}$$

where $-1 < a < 0$.

## §5. DECOMPOSITION OF THE REGULAR REPRESENTATION OF THE GROUP $QU(2)$

In this section we shall study the decomposition of the space $\mathfrak{H}$ of square-integrable functions $f(g)$, $g \in QU(2)$, into minimal subspaces invariant for left (or right) shifts.

1. **The regular representation of $QU(2)$.** The recurrence formulas for the functions $\mathfrak{P}^l_{mn}(\operatorname{ch}\tau)$ were derived from the composition theorem for these functions. As in the case of the group $SU(2)$, another method of deriving these formulas is linked with the considerating of infinitesimal operators of the regular representation of $QU(2)$.

According to the general definition (see §2.4 of Chapter I), a left regular representation of $QU(2)$ is constructed in the space $\mathfrak{H}$ of functions $f(g)$ on this group such that

$$\int |f(g)|^2\,dg<+\infty.$$

Operators of the left regular representation are given by the formula

$$L(g_0)f(g)=f(g_0^{-1}g). \tag{1}$$

Similarly, operators of the right regular representation are given by the formula

$$R(g_0)f(g)=f(gg_0). \tag{2}$$

Infinitesimal operators of the regular representation $L(g)$ for $QU(2)$ are sought in exactly the same way as for $SU(2)$ (see §4.6 of Chapter III). Therefore, without carrying out detailed calculations, we give the final results.

To the subgroup $\Omega_1$ of matrices

$$\omega_1(t)=\begin{pmatrix}\operatorname{ch}\frac{t}{2} & \operatorname{sh}\frac{t}{2}\\ \operatorname{sh}\frac{t}{2} & \operatorname{ch}\frac{t}{2}\end{pmatrix}$$

corresponds an infinitesimal operator $\hat{A}_1$ of the left regular representation, having the following form in terms of the Euler prarmeters:

$$\hat{A}_1 = \operatorname{cth} \tau \sin \varphi \frac{\partial}{\partial \varphi} - \cos \varphi \frac{\partial}{\partial \tau} - \frac{\sin \varphi}{\operatorname{sh} \tau} \frac{\partial}{\partial \psi}, \tag{3}$$

To the subgroup $\Omega_2$ of matrices

$$\omega_2(t) = \begin{pmatrix} \operatorname{ch} \frac{t}{2} & i \operatorname{sh} \frac{t}{2} \\ -i \operatorname{sh} \frac{t}{2} & \operatorname{ch} \frac{t}{2} \end{pmatrix}$$

corresponds the infinitesimal operator

$$\hat{A}_2 = -\operatorname{cth} \tau \cos \varphi \frac{\partial}{\partial \varphi} - \sin \varphi \frac{\partial}{\partial \tau} + \frac{\cos \varphi}{\operatorname{sh} \tau} \frac{\partial}{\partial \psi}. \tag{4}$$

Lastly, to the subgroup $\Omega_3$ of diagonal matrices

$$\omega_3(t) = \begin{pmatrix} e^{\frac{it}{2}} & 0 \\ 0 & e^{-\frac{it}{2}} \end{pmatrix}$$

corresponds the infinitesimal operator

$$\hat{A}_3 = -\frac{\partial}{\partial \varphi}. \tag{5}$$

Similarly, infinitesimal operators of the right regular representation, corresponding to the same subgroups, have the following forms in terms of Euler parameters:

$$\hat{B}_1 = \frac{\sin \psi}{\operatorname{sh} \tau} \frac{\partial}{\partial \varphi} + \cos \psi \frac{\partial}{\partial \tau} - \operatorname{cth} \tau \sin \psi \frac{\partial}{\partial \psi}, \tag{6}$$

$$\hat{B}_2 = \frac{\cos \psi}{\operatorname{sh} \tau} \frac{\partial}{\partial \varphi} - \sin \psi \frac{\partial}{\partial \tau} - \operatorname{cth} \tau \cos \psi \frac{\partial}{\partial \psi}, \tag{7}$$

$$\hat{B}_3 = \frac{\partial}{\partial \psi}. \tag{8}$$

As indicated above, it is more convenient to utilize, instead of the infinitesimal operators $\hat{A}_1, \hat{A}_2, \hat{A}_3$, their linear combinations $\hat{H}_+, \hat{H}_-, \hat{H}_3$ (see §2.3), which have the following expressions in parameters $\phi, \tau, \psi$:

$$\hat{H}_+ = -\hat{A}_1 - i\hat{A}_2 = e^{i\varphi}\left[i \operatorname{cth} \tau \frac{\partial}{\partial \varphi} + \frac{\partial}{\partial \tau} - \frac{i}{\operatorname{sh} \tau} \frac{\partial}{\partial \psi}\right], \tag{9}$$

$$\hat{H}_- = -\hat{A}_1 + i\hat{A}_2 = e^{-i\varphi}\left[-i \operatorname{cth} \tau \frac{\partial}{\partial \varphi} + \frac{\partial}{\partial \tau} + \frac{i}{\operatorname{sh} \tau} \frac{\partial}{\partial \psi}\right], \tag{10}$$

$$\hat{H}_3 = i\hat{A}_3 = -i \frac{\partial}{\partial \varphi}. \tag{11}$$

Similarly, for the operators $\hat{B}_1, \hat{B}_2, \hat{B}_3$ we have

$$\hat{K}_+ = -\hat{B}_1 - i\hat{B}_2 = e^{-i\psi}\left[-\frac{i}{\operatorname{sh}\tau}\frac{\partial}{\partial\varphi} - \frac{\partial}{\partial\tau} + i\operatorname{cth}\tau\frac{\partial}{\partial\psi}\right], \tag{12}$$

$$\hat{K}_- = -\hat{B}_1 + i\hat{B}_2 = e^{i\psi}\left[\frac{i}{\operatorname{sh}\tau}\frac{\partial}{\partial\varphi} - \frac{\partial}{\partial\tau} - i\operatorname{cth}\tau\frac{\partial}{\partial\psi}\right], \tag{13}$$

$$\hat{K}_3 = i\hat{B}_3 = i\frac{\partial}{\partial\psi}. \tag{14}$$

On $QU(2)$ there exists an analogue of the Laplace operator, namely the second order differential operator $\Delta$, permutable with left and right shifts. It is expressed in terms of infinitesimal operators by the formula

$$\Delta = -\hat{A}_1^2 - \hat{A}_2^2 + \hat{A}_3^2 = -\hat{B}_1^2 - \hat{B}_2^2 + \hat{B}_3^2. \tag{15}$$

We shall call $\Delta$ the Laplace operator on $QU(2)$. In Euler parameters it is expressed by the formula

$$\Delta = -\frac{1}{\operatorname{sh}\tau}\frac{\partial}{\partial\tau}\operatorname{sh}\tau\frac{\partial}{\partial\tau} - \frac{1}{\operatorname{sh}^2\tau}\left[\frac{\partial^2}{\partial\varphi^2} - 2\operatorname{ch}\tau\frac{\partial^2}{\partial\varphi\,\partial\psi} + \frac{\partial^2}{\partial\psi^2}\right]. \tag{16}$$

2. **Recurrence relations and infinitesimal operators.** Recurrence relations for functions $\mathfrak{P}^l_{mn}(\operatorname{ch}\tau)$ are derived with the help of infinitesimal operators in just the same way as for functions $P^l_{mn}(\cos\theta)$ (see Chapter III, §4.7). With each function $f(e^{i\theta})$ of the space $\mathfrak{D}$ we associate a function

$$F(g) = (T_\chi(g)f, e^{-im\theta}) = \frac{1}{2\pi}\int_0^{2\pi} T_\chi(g)f(e^{i\theta})\, e^{im\theta}\, d\theta \tag{1}$$

on $QU(2)$. It is an immediate consequence of formula (1) that to the function $T_\chi(g_0)f(e^{i\theta})$ corresponds the function $F(gg_0)$, i.e. the operators of the representation $T_\chi(g)$ are realized as right-shift operators on $QU(2)$.

We shall denote the space of functions of the form (1) by $\mathfrak{H}_{\chi m}$. It is clear that the matrix elements

$$t^\chi_{mn}(g) = (T_\chi(g)e^{-in\theta}, e^{-im\theta}), \qquad -\infty < n < \infty$$

form a basis in the space $\mathfrak{H}_{\chi m}$ corresponding to the basis $\{e^{-in\theta}\}$ in the space $\mathfrak{D}$. Hence for these matrix elements relations (11), (12) of §2.3 hold with $e^{-in\theta}$ replaced by $t^\chi_{mn}(g)$, and $H_+, H_-$ by $\hat{K}_+, \hat{K}_-$:

$$\hat{K}_+ t^\chi_{mn}(g) = (n' - l)\, t^\chi_{m,n+1}(g), \tag{2}$$

$$\hat{K}_- t^\chi_{mn}(g) = -(n' + l)\, t^\chi_{m,n-1}(g). \tag{3}$$

In formulas (2) and (3) let us replace the matrix elements $t^\chi_{mn}(g)$ by their expressions in terms of Euler parameters according to (9) (§3.2), and $\hat{K}_+$, $\hat{K}_-$ by their expressions (12) and (13) of subsection 1. After simple transformations we obtain the recurrence formulas (3) and (4) of §4.5. Similar recurrence relations are obtained by realizing the operators $T_\chi(g)$ as left-shift operators on $QU(2)$.

We pass on to the derivation of the differential equation for the functions $\mathfrak{P}^l_{mn}(\operatorname{ch}\tau)$. We show that an application of the Laplace operator

$$\Delta = -\hat{A}_1^2 - \hat{A}_2^2 + \hat{A}_3^2$$

to the matrix elements $t^\chi_{mn}(g)$ amounts to multiplication by $l(l+1)$. It has already been remarked that the matrix elements $t^\chi_{mn}(g)$, $-\infty < n < \infty$, form a canonical basis in the space $\mathfrak{H}_{\chi m}$. Therefore $\Delta t^\chi_{mn}(g)$ is the image of the function $(-A_1^2 - A_2^2 + A_3^2)e^{-in\theta}$ under the mapping (1) ($A_1$, $A_2$, $A_3$ are infinitesimal operators of $T_\chi(g)$).

Utilizing formulas (3), (5) and (7) of §2.3, we see that for all $n$

$$(-A_1^2 - A_2^2 + A_3^2)\,e^{-in\theta} = l(l+1)\,e^{-in\theta}.$$

Hence for all $n$ we have

$$\Delta t^\chi_{mn}(g) = l(l+1)\,t^\chi_{mn}(g).$$

In this equality we replace $t^\chi_{mn}(g)$ by its expression in terms of the Euler parameters, and $\Delta$ by its expression (16) of subsection 1. We easily obtain the required differential equation for functions $\mathfrak{P}^l_{mn}(\operatorname{ch}\tau)$ (see equation (9′) of §4.5).

3. **Expansion of functions on $QU(2)$.** In §4.4 of Chapter I we explained the connection between the decomposition of a regular representation into irreducible components and the expansion of functions on a group in the matrix elements of irreducible representations. In this subsection we shall construct an expansion of square-integrable functions $f(g)$ on $QU(2)$ in the matrix elements of $T_\chi(g)$. It turns out that the expansion contains only unitary representations of the first and second fundamental series and representations of the discrete series.

So, let $f(g) = f(\phi, \tau, \psi)$ be a function of the space $\mathfrak{H}$, i.e. a square-integrable function on $QU(2)$. It follows from formula (5) of §1.3 that $f(\phi, \tau, \psi)$ satisfies the equalities

$$f(\varphi, \tau, \psi) = f(\varphi + 4\pi, \tau, \psi) = f(\varphi, \tau, \psi + 4\pi) = f(\varphi + 2\pi, \tau, \psi + 2\pi). \tag{1}$$

Therefore $f(\phi, \tau, \psi)$ can be expanded in a double Fourier series in $\phi$ and $\psi$ of the form

$$f(\varphi, \tau, \psi) = \sum_{m, n=-\infty}^{\infty} \Phi_{mn}(\tau) e^{-i(m\varphi+n\psi)}, \tag{2}$$

where the summation is extended over all pairs $(m, n)$ such that $m$ and $n$ are simultaneously integers or semi-integers. The coefficients $\Phi_{mn}(\tau)$ are expressed by the formula

$$\Phi_{mn}(\tau) = \frac{1}{8\pi^2} \int_{-2\pi}^{2\pi} \int_{0}^{2\pi} f(\varphi, \tau, \psi) e^{i(m\varphi+n\psi)} \, d\varphi \, d\psi. \tag{3}$$

Moreover, one has the analogue of the Plancherel formula:

$$\frac{1}{8\pi^2} \int_{-2\pi}^{2\pi} \int_{0}^{2\pi} \int_{0}^{\infty} |f(\varphi, \tau, \psi)|^2 \operatorname{sh} \tau \, d\tau \, d\varphi \, d\psi = \sum_{m, n=-\infty}^{\infty} \int_{0}^{\infty} |\Phi_{mn}(\tau)|^2 \operatorname{sh} \tau \, d\tau. \tag{4}$$

It was shown in §3.2 that the matrix elements of $T_\chi(g)$ containing the factor $e^{-i(m\phi+n\psi)}$ have the form $e^{-i(m\phi+n\psi)}\mathfrak{P}^l_{mn}(\operatorname{ch}\tau)$. Therefore the problem of expanding functions on $QU(2)$ in the matrix elements of $T_\chi(g)$ amounts to the following.

One is given numbers $m$ and $n$, simultaneously integral or semi-integral, and a function $F(x)$ such that

$$\int_{1}^{\infty} |F(x)|^2 \, dx < +\infty.$$

It is required to expand $F(x)$ in functions $\mathfrak{P}^l_{mn}(x)$, where $l$ is the parameter over which the expansion is carried out, and $1 \leq x < \infty$.

This problem reduces to the problem of expansion in the eigenfunctions of a selfadjoint differential operator. Indeed, the functions $\mathfrak{P}^l_{mn}(x)$ satisfy the selfadjoint differential equation

$$\frac{d}{dx}(x^2-1)\frac{du}{dx} - \frac{m^2+n^2-2mnx}{x^2-1} u = \lambda u,$$

where $\lambda = l(l+1)$ (compare equation (9′) of §4.5). Moreover, they are continuous at the point $x = 1$ (corresponding to $\tau = 0$). In other words, the functions $\mathfrak{P}^l_{mn}(x)$ are eigenfunctions of the selfadjoint operator

$$\frac{d}{dx}(x^2-1)\frac{d}{dx} - \frac{m^2-2mnx+n^2}{x^2-1}, \qquad 1 \leqslant x < \infty, \tag{5}$$

continuous at the point $x = 1$.

Employing the usual techniques of expansion in eigenfunctions of selfadjoint operators (see for example [32], [38], [51]), we get the following result.

If $m$ and $n$ are integers, then any function $F(x)$ such that $\int_1^\infty |F(x)|^2 dx < +\infty$ has the expansion

$$F(x) = \int_0^\infty a(\rho)\,\mathfrak{P}_{mn}^{-\frac{1}{2}+i\rho}(x)\,\rho\,\operatorname{th}\pi\rho\,d\rho + \sum_{l=1}^{M}\left(l-\frac{1}{2}\right) b(l)\,\mathfrak{P}_{mn}^{l}(x), \tag{6}$$

where

$$M = \begin{cases} \min(|m|, |n|) & \text{for } mn \geqslant 0, \\ 0 & \text{for } mn < 0. \end{cases}$$

The summation extends over integral values of $l$. The coefficients $a(\rho)$ and $b(l)$ are expressed by the formulas

$$a(\rho) = \int_1^\infty F(x)\,\mathfrak{P}_{mn}^{-\frac{1}{2}-i\rho}(x)\,dx \equiv \int_1^\infty F(x)\,\overline{\mathfrak{P}_{mn}^{-\frac{1}{2}+i\rho}(x)}\,dx \tag{7}$$

and[1)]

$$b(l) = \int_1^\infty F(x)\,\mathfrak{P}_{mn}^{-l-1}(x)\,dx \equiv \frac{\Gamma(l+m+1)\,\Gamma(l-m+1)}{\Gamma(l+n+1)\,\Gamma(l-n+1)} \int_1^\infty F(x)\,\mathfrak{P}_{mn}^{l}(x)\,dx. \tag{8}$$

One has the following analogue of the Plancherel equality:

$$\int_1^\infty |F(x)|^2\,dx = \int_0^\infty |a(\rho)|^2 \rho\,\operatorname{th}\pi\rho\,d\rho + \sum_{l=1}^{M}\left(l-\frac{1}{2}\right)\frac{\Gamma(l+n+1)\,\Gamma(l-n+1)}{\Gamma(l+m+1)\,\Gamma(l-m+1)}\,|b(l)|^2. \tag{9}$$

Similarly, if $m$ and $n$ are semi-integers, then

1) Two of the $\Gamma$-factors become infinite and must be transformed by the formula $\Gamma(x)\Gamma(1-x) = \pi/\sin\pi x$.

$$F(x) = \int_0^\infty a(\rho)\, \mathfrak{P}_{mn}^{-\frac{1}{2}+i\rho}(x)\, \rho \operatorname{cth} \pi\rho \, d\rho + \frac{1}{4\pi^2} \sum_{l=\frac{1}{2}}^{M} \left(l - \frac{1}{2}\right) b(l)\, \mathfrak{P}_{mn}^{l}(x), \tag{10}$$

where

$$M = \begin{cases} \min(|m|, |n|) & \text{for } mn > 0, \\ 0 & \text{for } mn < 0. \end{cases}$$

The summation extends over semi-integral values of $l$. The coefficients $a(\rho)$ and $b(l)$ are expressed by the same formulas as in the case of integral $m$ and $n$. The Plancherel formula has the following form in this case:

$$\int_1^\infty |F(x)|^2 dx = \int_0^\infty |a(\rho)|^2 \rho \operatorname{cth} \pi\rho \, d\rho + \sum_{l=\frac{1}{2}}^{M} \left(l - \frac{1}{2}\right) \frac{\Gamma(l+n+1)\Gamma(l-n+1)}{\Gamma(l+m+1)\Gamma(l-m+1)} |b(l)|^2. \tag{11}$$

We observe that

$$\operatorname{cth} \pi\rho = \operatorname{th} \pi \left(\rho + \frac{i}{2}\right).$$

Therefore the terms containing integrals in formulas (6) and (9)–(11) can be written uniformly on replacing $\operatorname{th} \pi\rho$ and $\operatorname{cth} \pi\rho$ by $\operatorname{th} \pi(\rho + \epsilon i)$, where $\epsilon = 0$ in the integral case and $\epsilon = \frac{1}{2}$ in the semi-integral case.

Let us apply these results to the coefficients $\Phi_{mn}(\tau)$ of expansion (2). We obtain the following assertion.

Any function $f(g)$ on $QU(2)$ such that $\int |f(g)|^2 dg < +\infty$ can be expanded in the matrix elements

$$t_{mn}^{\chi}(g) = e^{-i(m'\varphi + n'\psi)} \mathfrak{P}_{m'n'}^{l}(\operatorname{ch} \tau),$$
$$\chi = (l, \varepsilon), \quad m' = m + \varepsilon, \quad n' = n + \varepsilon$$

of irreducible representations of this group. The expansion has the form

$$f(g)=\frac{1}{4\pi^2}\sum_{m,n,\epsilon}\left[\int_0^\infty a^\epsilon_{mn}(\rho)\,t_{mn}^{\left(-\frac{1}{2}+i\rho,\,\epsilon\right)}(g)\,\rho\,\mathrm{th}\,\pi(\rho+\epsilon l)\,d\rho\right.$$

$$\left.+\sum_{l=1-\epsilon}^{\infty}\left(l-\frac{1}{2}\right)b^\epsilon_{mn}(l)\,t^{(l,\,\epsilon)}_{mn}(g)\right], \tag{12}$$

where $\epsilon$, $m$, $n$, $l$ have the same meaning as above.

The values of the coefficients $a^\epsilon_{mn}(\rho)$ and $b^\epsilon_{mn}(l)$ are expressed by the formulas

$$a^\epsilon_{mn}(\rho)=\int f(g)\,\overline{t_{mn}^{\left(-\frac{1}{2}+i\rho,\,\epsilon\right)}(g)}\,dg \tag{13}$$

and

$$b^\epsilon_{mn}(l)=\frac{(-1)^{m-n}\,\Gamma(l+m+\epsilon+1)\,\Gamma(l-m-\epsilon+1)}{\Gamma(l+n+\epsilon+1)\,\Gamma(l-n-\epsilon+1)}\int f(g)\,\overline{t^{(l,\,\epsilon)}_{mn}(g)}\,dg, \tag{14}$$

where

$$dg=\mathrm{sh}\,\tau\,d\tau\,d\varphi\,d\psi.$$

Finally, one has the analogue of the Plancherel formula

$$\int|f(g)|^2\,dg=\frac{1}{4\pi^2}\sum_{m,n,\epsilon}\left[\int_0^\infty|a^\epsilon_{mn}(\rho)|^2\,\rho\,\mathrm{th}\,\pi(\rho+\epsilon l)\,d\rho\right.$$

$$\left.+(-1)^{m-n}\sum_{l=1-\epsilon}^{M}\frac{\Gamma(l+n+\epsilon+1)\,\Gamma(l-n-\epsilon+1)}{\Gamma(l+m+\epsilon+1)\,\Gamma(l-m-\epsilon+1)}\left(l-\frac{1}{2}\right)|b^\epsilon_{mn}(l)|^2\right]. \tag{15}$$

These formulas are simplified if, in place of the matrix elements $t^{(l,\,\epsilon)}_{mn}(g)$, one introduces the matrix elements corresponding to unitary representations of the discrete series, which we shall denote by $t^{(l,\,\epsilon,\,\pm)}_{mn}(g)$. Namely, we obtain

$$f(g)=\frac{1}{4\pi^2}\sum_{m,n,\epsilon}\left[\int_0^\infty a^\epsilon_{mn}(\rho)\,t_{mn}^{\left(-\frac{1}{2}+i\rho,\,\epsilon\right)}(g)\,\rho\,\mathrm{th}\,\pi(\rho+\epsilon l)\,d\rho\right.$$

$$\left.+\sum_{l=1-\epsilon}^{M}\left(l-\frac{1}{2}\right)b^{\epsilon,\,\pm}_{mn}(l)\,t^{(l,\,\epsilon,\,\pm)}_{mn}(g)\right], \tag{16}$$

where the sign + is chosen if $m$ and $n>0$, and the sign − if $m$ and $n<0$. Moreover, the expression for $a^\epsilon_{mn}(\rho)$ remains as before, and the coefficients $b^{\epsilon,\,\pm}_{mn}(l)$ have the form

$$b^{\epsilon,\,\pm}_{mn}(l)=\int f(g)\,\overline{t^{(l,\,\epsilon,\,\pm)}_{mn}(g)}\,dg. \tag{17}$$

Formula (15) now becomes

$$\int |f(g)|^2 dg = \frac{1}{4\pi^2} \sum_{m,n,\epsilon} \left[ \int_0^\infty |a^\epsilon_{mn}(\rho)|^2 \rho \operatorname{th} \pi(\rho + \epsilon i)\, d\rho + \sum_{l=1-\epsilon}^{M} \left(l - \frac{1}{2}\right) |b^{\epsilon,\pm}_{mn}(l)|^2 \right]. \quad (18)$$

Thus we see that in the expansion of square-integrable functions $f(g)$ on $QU(2)$ only the matrix elements of irreducible unitary representations of this group appear. Moreover, only representations of the first and second fundamental series appear together with representations of the discrete series. But representations of the complementary series are not involved in the expansion. In this the group $QU(2)$ differs from all the groups considered earlier, for which the expansion of square-integrable functions involved the matrix elements of all irreducible unitary representations.

Example. In formula (7) of §4.4 we put $l = -\frac{1}{2} + i\rho$. We obtain

$$\mathfrak{P}^k_{-\frac{1}{2}+i\rho}(\operatorname{ch} \tau_1)\, \mathfrak{P}^{-k}_{-\frac{1}{2}+i\rho}(\operatorname{ch} \tau_2) = \frac{1}{\pi} \int_{|\tau_1-\tau_2|}^{\tau_1+\tau_2} \frac{\mathfrak{P}_{-\frac{1}{2}+i\rho}(\operatorname{ch} \tau)\, T_k\left(\frac{\operatorname{ch} \tau - \operatorname{ch} \tau_1 \operatorname{ch} \tau_2}{\operatorname{sh} \tau_1 \operatorname{sh} \tau_2}\right)}{\sqrt{|\operatorname{ch}(\tau_1+\tau_2) - \operatorname{ch} \tau|\,|\operatorname{ch} \tau - \operatorname{ch}(\tau_1 - \tau_2)|}} \operatorname{sh} \tau\, d\tau, \quad (19)$$

where $T_k(x)$ are Čebyšev polynomials. Employing formulas (6) and (7), we deduce that

$$\int_0^\infty \mathfrak{P}_{-\frac{1}{2}+i\rho}(\operatorname{ch} \tau)\, \mathfrak{P}^k_{-\frac{1}{2}+i\rho}(\operatorname{ch} \tau_1)\, \mathfrak{P}^{-k}_{-\frac{1}{2}+i\rho}(\operatorname{ch} \tau_2)\, \rho \operatorname{th} \pi\rho\, d\rho = \begin{cases} \dfrac{4\pi T_k\left(\dfrac{\operatorname{ch} \tau - \operatorname{ch} \tau_1 \operatorname{ch} \tau_2}{\operatorname{sh} \tau_1 \operatorname{sh} \tau_2}\right)}{\sqrt{|\operatorname{ch}(\tau_1+\tau_2) - \operatorname{ch} \tau|\,|\operatorname{ch} \tau - \operatorname{ch}(\tau_1 - \tau_2)|}} & \text{for } |\tau_1 - \tau_2| < \tau < \tau_1 + \tau_2, \\ 0 & \text{otherwise.} \end{cases} \quad (20)$$

4. Decomposition of the regular representation of $QU(2)$ into irreducible components. The expansion of the functions $f(g)$ of the space $\mathfrak{H}$ has a simple group-theoretic significance: it is linked with the decomposition of the regular representation of $QU(2)$ into irreducible components. Namely, with each $\chi = (l, \epsilon)$

and each $m$ we associate a space $\mathfrak{H}_{\chi}^{m}$ of functions on $QU(2)$, having the form

$$F(g)=(T_{\chi}(g)\varphi, e^{-im\theta}),$$

where

$$\int_0^{2\pi} |\varphi(e^{i\theta})|^2 d\theta < +\infty.$$

The space of such functions is invariant for right shifts, and the restriction of the regular representation $R(g)$ to $\mathfrak{H}_{\chi}^{m}$ is equivalent to the representation $T_{\chi}(g)$. It is not difficult to show that the functions $f_m^{\chi}(g)$ of $\mathfrak{H}_{\chi}^{m}$ can be written in the form

$$f_m^{\chi}(g)=\sum_n a_{mn}(\chi)\, t_{mn}^{\chi}(g).$$

Utilizing this, we deduce from expansion (12) and the Plancherel formula (15) above that

$$\mathfrak{H}=\frac{1}{4\pi^2}\Big[\sum_{\varepsilon=0,\frac{1}{2}}\ \sum_{m=-\infty}^{\infty}\int_0^{\infty}\mathfrak{H}^{m}_{-\frac{1}{2}+i\rho,\varepsilon}\,\rho\,\mathrm{th}\,\pi(\rho+\varepsilon i)\,d\rho$$

$$+\sum_{l=1-\varepsilon}^{|m|}\Big(l-\frac{1}{2}\Big)\mathfrak{H}^{m}_{l,\varepsilon,\pm}\Big], \tag{1}$$

and

$$R(g)=\frac{1}{4\pi^2}\sum_{\varepsilon=0,\frac{1}{2}}\ \sum_{m=-\infty}^{\infty}\Big[\int_0^{\infty}T^{(m)}_{-\frac{1}{2}+i\rho,\varepsilon}(g)\,\rho\,\mathrm{th}\,\pi(\rho+\varepsilon i)\,d\rho$$

$$+\sum_{l=1-\varepsilon}^{|m|}\Big(l-\frac{1}{2}\Big)T^{(m)}_{l,\varepsilon,\pm}(g)\Big]. \tag{2}$$

Here the summation extends over integral values of $m$, and the $T^{(m)}_{-\frac{1}{2}+i\rho,\epsilon}(g)$ denote representations equivalent to $T_{\chi}(g)$, $\chi=(-\frac{1}{2}+i\rho,\epsilon)$. Similarly, the $T^{(m)}_{l,\epsilon,\pm}(g)$ are representations equivalent to the representations $T^{\pm}_{\chi}(g)$, $\chi=(l,\epsilon)$, of the discrete series.

The decomposition (2) can be written in a different way by changing the order of summation and integration. Let

$$X_{-\frac{1}{2}+i\rho,\varepsilon}(g)=\sum_{m=-\infty}^{\infty}T^{(m)}_{-\frac{1}{2}+i\rho,\varepsilon}(g), \tag{3}$$

$$X_{l,\varepsilon,+}=\sum_{m}T^{(m)}_{l,\varepsilon,+}(g) \tag{4}$$

and

$$\mathrm{X}_{l,\varepsilon,-}=\sum_m T^{(m)}_{l,\varepsilon,-}(g). \tag{5}$$

Then one has the equality

$$R(g)=\frac{1}{4\pi^2}\sum_{\varepsilon=0,\frac{1}{2}}\Big[\int_0^\infty \mathrm{X}_{-\frac{1}{2}+i\rho,\varepsilon}(g)\rho\,\mathrm{th}\,\pi(\rho+\varepsilon l)\,d\rho + \sum_{l=1-\varepsilon}^{\infty}\Big(l-\frac{1}{2}\Big)[\mathrm{X}_{l,\varepsilon,+}(g)+\mathrm{X}_{l,\varepsilon,-}(g)]\Big]. \tag{6}$$

It gives the decomposition of $R(g)$ into representations which are multiples of irreducible representations.

5. **Decomposition of induced representations of $QU(2)$.** We denote by $\mathfrak{H}_m$ the space of functions $f(g)$ on $QU(2)$, such that $\int|f(g)|^2dg<+\infty$ and $f(hg)=e^{-imt}f(g)$, where $h=\begin{pmatrix} e^{\frac{it}{2}} & 0 \\ 0 & e^{-\frac{it}{2}}\end{pmatrix}$. This space is invariant for right-shift operators, and the formula

$$R_m(g_0)f(g)=f(gg_0)$$

determines a representation of $QU(2)$ (the restriction of the right regular representation to $\mathfrak{H}_m$). According to §2.6 of Chapter I it is called the representation induced by the representation $e^{-imt}$ of the subgroup $\Omega$ of diagonal matrices.

It is clear from the definition of $\mathfrak{H}_m$ that it consists of functions of the form

$$f(\varphi,\tau,\psi)=F(\tau,\psi)e^{-im\varphi},$$

where

$$\int_0^{2\pi}\int_0^\infty|F(\tau,\psi)|^2\,\mathrm{sh}\,\tau\,d\tau\,d\psi<+\infty.$$

Therefore a decomposition of $\mathfrak{H}$ as the direct sum of its subspaces $\mathfrak{H}_m$ leads to an expansion of the functions $f(\phi,\tau,\psi)$ in a Fourier series in $\phi$:

$$f(\varphi,\tau,\psi)=\sum_m F_m(\tau,\psi)e^{-im\varphi},$$

where $m$ runs through integral or semi-integral values, and

$$F_m(\tau, \psi) = \frac{1}{4\pi} \int_{-2\pi}^{2\pi} f(\varphi, \tau, \psi) e^{im\varphi} d\varphi.$$

Employing decomposition (6) of subsection 4, we easily get the decomposition of induced representations into irreducible ones. Namely, if $m$ is an integer, then

$$R_m(g) = \frac{1}{4\pi^2} \left[ \int_0^\infty T_{-\frac{1}{2}+i\rho,\, 0}(g)\, \rho \operatorname{th} \pi\rho\, d\rho + \sum_{l=1}^{|m|} \left( l - \frac{1}{2} \right) [T^+_{l,\, 0}(g) + T^-_{l,\, 0}(g)] \right]. \tag{1}$$

And if $m$ is a semi-integer, then

$$R_m(g) = \frac{1}{4\pi^2} \left[ \int_0^\infty T_{-\frac{1}{2}+i\rho,\, \frac{1}{2}}(g)\, \rho \operatorname{cth} \pi\rho\, d\rho + \sum_{l=\frac{1}{2}}^{|m|} \left( l - \frac{1}{2} \right) [T^+_{l,\, \frac{1}{2}}(g) + T^-_{l,\, \frac{1}{2}}(g)] \right]. \tag{2}$$

Thus in the decomposition of induced representations each irreducible representation appears once at the most.

In particular, let us consider the representation $R_0(g)$. We shall call this representation quasi-regular. It is obvious from (1) that

$$R_0(g) = \frac{1}{4\pi^2} \int_0^\infty T_{-\frac{1}{2}+i\rho,\, 0}(g)\, \rho \operatorname{th} \pi\rho\, d\rho. \tag{3}$$

Thus the quasi-regular representation of $QU(2)$ is the continuous direct sum of irreducible unitary representations of the first fundamental series.

Utilizing the local isomorphism of $QU(2)$ and $SH(3)$, described in §1.1, decomposition (3) easily yields a decomposition of the representation

$$L(g) f(\xi) = f(g^{-1}\xi), \qquad [\xi, \xi] = -1,$$

of $SH(3)$. We shall not dwell on that here, since the corresponding analysis will be carried out for all the groups $SH(n)$ in Chapter X.

Let us also consider the connection between the decompositions of induced representations for different values of $m$. It follows from formula (12) of subsection 3 that the expansion of functions $f(g)$ of $\mathfrak{H}_m$ has the form

$$f(g)=\frac{1}{4\pi^2}\sum_{n,\varepsilon}\Big[\int_0^\infty a^\varepsilon_{mn}(\rho)\, t_{mn}^{\left(-\frac{1}{2}+i\rho,\,\varepsilon\right)}(g)\,\rho\,\mathrm{th}\,\pi(\rho+\varepsilon l)\,d\rho + \sum_{l=1-\varepsilon}^{M}\left(l-\frac{1}{2}\right) b^\varepsilon_{mn}(l)\,(t)^{(l,\,\varepsilon)}_{mn}(g)\Big]. \tag{4}$$

But it is easy to see that the left-hand infinitesimal operator $\hat{H}_+$ transforms functions of $\mathfrak{H}_m$ into functions of $\mathfrak{H}_{m+1}$, and that $\hat{H}_-$ transforms functions of $\mathfrak{H}_m$ into functions of $\mathfrak{H}_{m-1}$ (see subsection 1). Therefore, applying $\hat{H}_+$ to both sides of (4), we obtain the expansion of the function $\hat{H}_+ f(g)$, belonging to $\mathfrak{H}_{m+1}$. However, generally speaking, one does not thus obtain all the functions of $\mathfrak{H}_{m+1}$, because $\hat{H}_+$ is not one-to-one. We leave aside a detailed discussion of this matter.

6. Orthogonality relations for the functions $\mathfrak{P}^l_{mn}(x)$. The results of §4.3 of Chapter I easily lead to orthogonality relations for functions $\mathfrak{P}^l_{mn}(x)$. In the integral case we get

$$\int_1^\infty \mathfrak{P}^l_{mn}(x)\,\overline{\mathfrak{P}^{l_1}_{mn}(x)}\,dx=0, \tag{1}$$

if $l$, $l_1$, $m$, $n$ are simultaneously integers or semi-integers, where $l \neq l_1$. Further, if $1-\epsilon \le l \le M$, where

$$M=\begin{cases}\min(|m|,\ |n|) & \text{for } mn \geqslant 0,\\ 0 & \text{for } mn<0,\end{cases}$$

then

$$\int_1^\infty |\mathfrak{P}^l_{mn}(x)|^2\,dx=\frac{(-1)^{n-m}}{4\pi^2}\left(l-\frac{1}{2}\right)\frac{\Gamma(l+n+1)\Gamma(l-n+1)}{\Gamma(l+m+1)\Gamma(l-m+1)}. \tag{2}$$

One similarly establishes that

$$\int_1^\infty \mathfrak{P}_{mn}^{-\frac{1}{2}+i\rho}(x)\,\overline{\mathfrak{P}_{mn}^{-\frac{1}{2}+i\tau}(x)}\,dx=\frac{1}{4\pi^2}\,\rho\,\mathrm{th}\,\pi(\rho+\varepsilon l)\,\delta(\tau-\rho). \tag{3}$$

## CHAPTER VII

# REPRESENTATIONS OF THE GROUP OF REAL UNIMODULAR MATRICES AND THE HYPERGEOMETRIC FUNCTION

In this chapter we shall consider representations of the group of real unimodular matrices of the second order, under which, corresponding to diagonal matrices, we have the operator of multiplication by a function. Under these representations to matrices of the form $\begin{Vmatrix} \operatorname{ch} t & \operatorname{sh} t \\ \operatorname{sh} t & \operatorname{ch} t \end{Vmatrix}$ and $\begin{Vmatrix} \cos t & -\sin t \\ \sin t & \cos t \end{Vmatrix}$ correspond integral operators whose kernels are expressed by means of the hypergeometric function. On the basis of this connection we shall obtain a number of properties of the hypergeometric function.

## §1. THE HYPERGEOMETRIC FUNCTION

1. Definition. We shall define the hypergeometric function $F(\alpha, \beta; \gamma; z)$ with the help of an integral representation. We cut the complex plane along the ray $1 \leq z < \infty$ of the real axis. For any $t$, $0 \leq t \leq 1$, the function $(1 - tz)^{-\alpha}$ is uniquely defined in the cut plane. We choose that branch of this function which takes the value 1 at the point $z = 0$. Further, we choose those branches of the functions $t^{\beta - 1}$ and $(1 - t)^{\gamma - \beta - 1}$ which take the value 1 for $t = 1$ and $t = 0$ respectively. Then the function $t^{\beta - 1}(1 - t)^{\gamma - \beta - 1}(1 - tz)^{-\alpha}$ is uniquely defined in the cut plane.

Let

$$F(\alpha, \beta; \gamma; z) = \frac{\Gamma(\gamma)}{\Gamma(\beta)\Gamma(\gamma - \beta)} \int_0^1 t^{\beta - 1}(1 - t)^{\gamma - \beta - 1}(1 - tz)^{-\alpha}\, dt. \tag{1}$$

Under the assumptions made above, this integral converges in the domain $\operatorname{Re}\gamma > \operatorname{Re}\beta > 0$ and defines a function $F(\alpha, \beta; \gamma; z)$, analytically depending on $\alpha, \beta, \gamma$. We call it the hypergeometric function.

For those values of $\beta$, $\gamma$ which do not belong to the domain mentioned, we define the hypergeometric function by means of analytic continuation.

For values of $z$ not lying on the ray $1 \leq z < \infty$, the expression $1 - tz$ does not vanish for $0 \leq t \leq 1$. Therefore the function $F(\alpha, \beta; \gamma; z)$, given by the integral (1), analytically depends on $z$ in the complex plane, cut along the ray $1 \leq z < \infty$. We shall assume that in this plane we have $|\arg(1-z)| < \pi$.

For $|z| < 1$ the hypergeometric function can be expanded in a power series. In fact, for $|z| < 1$ and $0 \leq t \leq 1$, from the binomial theorem we have

$$(1-tz)^{-\alpha} = \sum_{k=0}^{\infty} \frac{\Gamma(\alpha+k)}{\Gamma(\alpha)\, k!} (tz)^k.$$

We substitute this expansion in the integral (1) and then integrate. Since, by formula (2) of §1.7, Chapter V, we have

$$\int_0^1 t^{\beta+k-1}(1-t)^{\gamma-\beta-1}\, dt = \frac{\Gamma(\beta+k)\Gamma(\gamma-\beta)}{\Gamma(\gamma+k)},$$

therefore

$$F(\alpha, \beta; \gamma; z) = \frac{\Gamma(\gamma)}{\Gamma(\alpha)\Gamma(\beta)} \sum_{k=0}^{\infty} \frac{\Gamma(\alpha+k)\Gamma(\beta+k)}{\Gamma(\gamma+k)\, k!} z^k. \tag{2}$$

This expansion defines $F(\alpha, \beta; \gamma; z)$ for $|z| < 1$ for all values of $\alpha$, $\beta$, $\gamma$, except when $\gamma = 0, -1, -2, \cdots, -n, \cdots$. If $\gamma = -n$ and neither $\alpha$ nor $\beta$ is equal to $-m$, where $m < n$ and $m = 0, 1, 2, \cdots$, then the function $F(\alpha, \beta; \gamma; z)$ has a simple pole. Let us find the residue of $F(\alpha, \beta; \gamma; z)$ for $\gamma = -n$. For this, we note that

$$\lim_{\gamma \to -n} \frac{1}{\Gamma(\gamma+k)} = \begin{cases} 0 & \text{for } k \leqslant n, \\ \dfrac{1}{\Gamma(k-n)} & \text{for } k > n. \end{cases}$$

Hence from formula (2) we have

$$\lim_{\gamma \to -n} \frac{F(\alpha, \beta; \gamma; z)}{\Gamma(\gamma)} = \frac{1}{\Gamma(\alpha)\Gamma(\beta)} \sum_{k=n+1}^{\infty} \frac{\Gamma(\alpha+k)\Gamma(\beta+k)}{\Gamma(k-n)\, k!} z^k$$
$$= \frac{\Gamma(\alpha+n+1)\Gamma(\beta+n+1)}{\Gamma(\alpha)\Gamma(\beta)(n+1)!} z^{n+1} F(\alpha+n+1, \beta+n+1; n+2; z).$$

Since

$$\operatorname*{Res}_{\gamma=-n} \Gamma(\gamma) = \frac{(-1)^n}{n!},$$

we get

$$\operatorname*{Res}_{\gamma=-n} F(\alpha, \beta; \gamma; z) = \frac{(-1)^n \Gamma(\alpha+n+1)\Gamma(\beta+n+1)}{n!\,(n+1)!\,\Gamma(\alpha)\Gamma(\beta)} z^{n+1} F(\alpha+n+1, \quad \beta+n+1; n+2; z). \tag{3}$$

We also note that

$$F(\alpha, \beta; \gamma; 1) = \frac{\Gamma(\gamma)}{\Gamma(\beta)\Gamma(\gamma-\beta)} \int_0^1 t^{\beta-1}(1-t)^{\gamma-\beta-\alpha-1}\,dt = \frac{\Gamma(\gamma)\Gamma(\gamma-\beta-\alpha)}{\Gamma(\gamma-\alpha)\Gamma(\gamma-\beta)}. \tag{4}$$

**2. Some relations.** We note some relations for the hypergeometric function which follow directly from its definition.

From the expansion (2) above it follows that the parameters $\alpha$ and $\beta$ appear symmetrically in the hypergeometric function:

$$F(\alpha, \beta; \gamma; z) = F(\beta, \alpha; \gamma; z). \tag{1}$$

For $|z| > 1$ equality (1) follows from the analyticity of $F(\alpha, \beta; \gamma; z)$ in $z$.

Further, differentiating expansion (2) and replacing $k$ by $k+1$, we get

$$\frac{dF(\alpha, \beta; \gamma; z)}{dz} = \frac{\Gamma(\gamma)}{\Gamma(\alpha)\Gamma(\beta)} \sum_{k=0}^{\infty} \frac{\Gamma(\alpha+k+1)\Gamma(\beta+k+1)}{(\gamma+k+1)\,k!} z^k.$$

Hence we have the equality

$$\frac{dF(\alpha, \beta; \gamma; z)}{dz} = \frac{\alpha\beta}{\gamma} F(\alpha+1, \beta+1; \gamma+1; z). \tag{2}$$

In order to obtain the next relation, we make the substitution $t = 1 - s$ in the integral (1) of subsection 1. We get

$$F(\alpha, \beta; \gamma; z) = \frac{\Gamma(\gamma)}{\Gamma(\beta)\Gamma(\gamma-\beta)} \int_0^1 s^{\gamma-\beta-1}(1-s)^{\beta-1}(1-z+sz)^{-\alpha}\,ds$$

$$= \frac{\Gamma(\gamma)}{\Gamma(\beta)\Gamma(\gamma-\beta)} (1-z)^{-\alpha} \int_0^1 s^{\gamma-\beta-1}(1-s)^{\beta-1}\left(1-\frac{sz}{z-1}\right)^{-\alpha} ds.$$

Because of equality (1) of subsection 1, the integral on the right hand side of this equality can be expressed by means of the hypergeometric function. It follows that

$$F(\alpha, \beta; \gamma; z) = (1-z)^{-\alpha} F\left(\alpha, \gamma-\beta; \gamma; \frac{z}{z-1}\right). \tag{3}$$

Using the symmetry of $F(\alpha, \beta; \gamma; z)$ with respect to $\alpha$ and $\beta$, we get

$$F(\alpha, \beta; \gamma; z) = (1-z)^{-\beta} F\left(\gamma-\alpha, \beta; \gamma; \frac{z}{z-1}\right). \tag{4}$$

From (3) and (4) it follows that

$$F(\alpha, \beta; \gamma; z) = (1-z)^{\gamma-\alpha-\beta} F(\gamma-\alpha, \gamma-\beta; \gamma; z). \tag{5}$$

3. **Some integrals involving the hypergeometric function.** Certain frequently encountered integrals are expressible in terms of the hypergeometric function. Let us make the substitution $t = 1/s$ in equality (1) of subsection 1. We get

$$F(\alpha, \beta; \gamma; z) = \frac{\Gamma(\gamma)}{\Gamma(\beta)\Gamma(\gamma-\beta)} \int_1^\infty s^{\alpha-\gamma}(s-1)^{\gamma-\beta-1}(s-z)^{-\alpha}\, ds. \tag{1}$$

Replacing $s$ by $s+1$ and $1-z$ by $u$ in this integral, we find

$$F(\alpha, \beta; \gamma; 1-u) = \frac{\Gamma(\gamma)}{\Gamma(\beta)\Gamma(\gamma-\beta)} \int_0^\infty s^{\gamma-\beta-1}(1+s)^{\alpha-\gamma}(s+u)^{-\alpha}\, ds. \tag{2}$$

Many other integrals reduce to integrals of the forms (1) of subsection 1, (1) and (2). For example, the substitution $(s-a)/(b-a) = t$ reduces the integral

$$J = \int_a^b (s-a)^{\beta-1}(b-s)^{\gamma-\beta-1}(z-s)^{-\alpha}\, ds$$

to the form (1) of subsection 1. We obtain

$$J = \frac{\Gamma(\beta)\Gamma(\gamma-\beta)}{\Gamma(\gamma)}(b-a)^{\gamma-1}(z-a)^{-\alpha} F\left(\alpha, \beta; \gamma; \frac{b-a}{z-a}\right). \tag{3}$$

Let us replace $\beta$ by $\gamma-\lambda$ in (2). Then

$$\Phi(\lambda) \equiv \frac{\Gamma(\lambda)\Gamma(\gamma-\lambda)}{\Gamma(\gamma)} F(\alpha, \gamma-\lambda; \gamma; 1-u)$$

emerges as the Mellin transform for $f(s) \equiv (1+s)^{\alpha-\gamma}(s+u)^{-\alpha}$. By the inversion formula for the Mellin transform, this leads to

$$\frac{1}{2\pi i}\int_{a-i\infty}^{a+i\infty} \Gamma(\lambda)\Gamma(\gamma-\lambda)F(\alpha, \gamma-\lambda; \gamma; 1-u)\, s^{-\lambda}\, d\lambda$$
$$= \Gamma(\gamma)(1+s)^{\alpha-\gamma}(s+u)^{-\alpha}, \tag{4}$$

where $0 < a < \operatorname{Re} \gamma$.

In exactly the same way, from formula (1) of subsection 1, defining $f(\alpha, \beta; \gamma, z)$ we get

$$\frac{1}{2\pi i} \int_{a-i\infty}^{a+i\infty} \frac{\Gamma(\lambda)}{\Gamma(\beta+\lambda)} F(\alpha, \beta; \beta+\lambda; z) s^{-\lambda} d\lambda = \frac{(1-s)^{\beta-1}(1-z+sz)^{-\alpha}}{\Gamma(\beta)}, \tag{5}$$

where $a > 0$, $s > 0$.

**4. Expression of Jacobi functions and polynomials by means of the hypergeometric function.** We show that the functions $P^l_{mn}(z)$ and $\mathfrak{P}^l_{mn}(z)$, introduced in Chapters III and VI, are expressible by means of the hypergeometric function. For this we use the series expansions of these functions.

Let $m \geq n$. On replacing the summation index $j$ by $m+s$ in formula (1) of §3.4 of Chapter III, we get

$$P^l_{mn}(z) = \frac{i^{m-n}}{2^m} \sqrt{\frac{(l-m)!\,(l-n)!}{(l+m)!\,(l+n)!}} (1+z)^{\frac{m+n}{2}} (1-z)^{\frac{m-n}{2}} \times \sum_{s=0}^{l-m} \frac{(-1)^s (l+m+s)!}{s!\,(m-n+s)!\,(l-m-s)!} \left(\frac{1-z}{2}\right)^s. \tag{1}$$

Comparing this formula with formula (2) of subsection 1, we see that

$$P^l_{mn}(z) = \frac{i^{m-n}}{2^m (m-n)!} \sqrt{\frac{(l-n)!\,(l+m)!}{(l-m)!\,(l+n)!}} \times (1+z)^{\frac{m+n}{2}} (1-z)^{\frac{m-n}{2}} F\left(l+m+1, m-l; m-n+1; \frac{1-z}{2}\right). \tag{2}$$

An analogous formula holds for $n \geq m$

$$P^l_{mn}(z) = \frac{i^{n-m}}{2^n (n-m)!} \sqrt{\frac{(l-m)!\,(l+n)!}{(l-n)!\,(l+m)!}} \times (1+z)^{\frac{m+n}{2}} (1-z)^{\frac{n-m}{2}} F\left(l+n+1, n-l; n-m+1; \frac{1-z}{2}\right). \tag{2'}$$

In exactly the same way, from formula (2) of §3.4 of Chapter III, it follows that for $m \geq n$

$$P^l_{mn}(z) = \frac{i^{m-n}}{2^l} \sqrt{\frac{(l+m)!\,(l-n)!}{(l-m)!\,(l+n)!}} \times \tag{3}$$

$$\times(1-z)^{\frac{m-n}{2}}(1+z)^{l-\frac{m-n}{2}}F\left(m-l,\ -n-l;\ m-n+1;\ \frac{z-1}{z+1}\right). \tag{3}$$

In particular, putting $n = 0$, we get

$$P_l^m(z)=\frac{(-1)^m(l+m)!}{2^m m!\ (l-m)!}(1-z^2)^{\frac{m}{2}}F\left(l+m+1,\ m-l;\ m+1;\ \frac{1-z}{2}\right)$$
$$=\frac{(-1)^m}{m!}\frac{(l+m)!}{(l-m)!}\left(\frac{1-z}{1+z}\right)^{\frac{m}{2}}F\left(-l,\ l+1;\ m+1;\ \frac{1-z}{2}\right). \tag{4}$$

And taking $m = n = 0$, we find

$$P_l(z)=F\left(l+1,\ -l;\ 1;\ \frac{1-z}{2}\right)=\left(\frac{1+z}{2}\right)^l F\left(-l,\ -l;\ 1;\ \frac{z-1}{z+1}\right). \tag{5}$$

Further, from formula (2) of §3.3 of Chapter VI it follows that for $m \leq n$

$$\mathfrak{P}_{mn}^l(z)=\frac{\Gamma(l+n+1)}{2^l\Gamma(l+m+1)(n-m)!}(z-1)^{\frac{n-m}{2}}(z+1)^{l-\frac{n-m}{2}}$$
$$\times F\left(-l-m,\ n-l;\ n-m+1;\ \frac{z-1}{z+1}\right)$$
$$=\frac{\Gamma(l+n+1)}{2^n\Gamma(l+m+1)(n-m)!}(z-1)^{\frac{n-m}{2}}(z+1)^{\frac{m+n}{2}}$$
$$\times F\left(n+l+1,\ n-l;\ n-m+1;\ \frac{1-z}{2}\right). \tag{6}$$

In particular, for $m = 0$ we get

$$\mathfrak{P}_l^n(z)=\frac{\Gamma(l+n+1)}{\Gamma(l+1)}\mathfrak{P}_{n0}^l(z)=\frac{\Gamma(l+1)}{\Gamma(l-n+1)}\mathfrak{P}_{0n}^l(z)$$
$$=\frac{\Gamma(l+n+1)}{2^n\Gamma(l-n+1)\,n!}(z^2-1)^{\frac{n}{2}}F\left(n+l+1,\ n-l;\ n+1;\ \frac{1-z}{2}\right)$$
$$=\frac{\Gamma(l+n+1)}{\Gamma(l-n+1)\,n!}\left(\frac{z-1}{z+1}\right)^{\frac{n}{2}}F\left(l+1,\ -l;\ n+1;\ \frac{1-z}{2}\right). \tag{7}$$

These expressions show that the functions $P_{mn}^l(z)$ and $\mathfrak{P}_{mn}^l(z)$, studied by us, reduce to the hypergeometric function. Moreover, we get hypergeometric functions with certain integrality conditions on the parameters $\alpha$, $\beta$, $\gamma$. For example, for $P_{mn}^l(z)$ we have $\alpha = l + m + 1$, $\beta = m - l$, $\gamma = m - n + 1$, and thus $\alpha$, $\beta$, $\gamma$ are integers.

In order to obtain the group-theoretic interpretation of the hypergeometric function in the general case, it is necessary to consider representations of the group $SL(2, R)$ of real unimodular matrices of the second order, under which operators of multiplication by a function correspond to diagonal matrices.

## §2. THE GROUP $SL(2, R)$ OF REAL UNIMODULAR MATRICES OF THE SECOND ORDER

1. Introductory remarks. In the preceding chapter we considered representations of the group $QU(2)$, under which diagonal matrices are represented by (infinite) diagonal matrices. As already indicated in §1.2 of that chapter, these representations correspond to representations of the group $SL(2, R)$ of real unimodular matrices of the second order, under which elements of the subgroup $SO(2)$ of matrices of the form $\begin{pmatrix} \cos t & -\sin t \\ \sin t & \cos t \end{pmatrix}$ are expressed by diagonal matrices (it will be recalled that the groups $QU(2)$ and $SL(2, R)$ are isomorphic).

In this chapter we shall consider a realization of the representations of $SL(2, R)$ under which the diagonal matrices $\begin{pmatrix} e^{-t} & 0 \\ 0 & e^{t} \end{pmatrix}$ are represented in a particularly simple manner. Namely, corresponding to these matrices we shall have operators of multiplication by a function. At the same time, to elements of the subgroup $SO(2)$, and also to elements of the form $\begin{pmatrix} \operatorname{ch} t & \operatorname{sh} t \\ \operatorname{sh} t & \operatorname{ch} t \end{pmatrix}$, there shall, under such a realization, correspond integral transforms whose kernel is expressed by means of the hypergeometric function. For triangular matrices the hypergeometric function degenerates in degree. Starting from this connection between representations of the group $SL(2, R)$ and the hypergeometric function, we shall establish various properties of the latter. In particular, we shall obtain continual analogues of the composition theorem for the hypergeometric function, and, by considering infinitesimal operators, also recurrence relations for it.

At the end of the chapter we consider another realization of representations of the group $SL(2, R)$, under which operators of multiplication by a function correspond to triangular matrices $\begin{pmatrix} 1 & 0 \\ t & 1 \end{pmatrix}$. We shall see that in this realization there arise integral transforms whose kernels are expressed by means of Hankel functions. Thus the consideration of such representations provides a new approach to the group-theoretic treatment of Hankel functions.

2. Parametrization. As parameters in the group $SL(2, R)$ one can choose three real numbers $\alpha, \gamma, \delta$. The value of $\beta$ is determined, for $\gamma \neq 0$, from the equality $\alpha\delta - \beta\gamma = 1$. At the points where $\gamma \neq 0$ one can take the three other elements of the matrix $g$ (all four elements $\alpha, \beta, \gamma, \delta$ cannot be zero, since $\alpha\delta - \beta\gamma = 1$).

In many cases it is convenient to have another parametrization of the group $SL(2, R)$, similar to the parametrization of $SU(2)$ with the help of Euler angles. It is based on the following assertion.

**Theorem 1.** *Any matrix $g$ of the group $SL(2, R)$, all of whose elements are nonzero, can be represented in the form*

$$g = d_1(-e)^{\epsilon_1} s^{\epsilon_2} p\, d_2, \tag{1}$$

*where $\epsilon_1, \epsilon_2 = 0, 1$, $d_1$ and $d_2$ are diagonal matrices with positive elements*

$$d_1 = \begin{pmatrix} e^{-\varphi} & 0 \\ 0 & e^{\varphi} \end{pmatrix}, \quad d_2 = \begin{pmatrix} e^{-\psi} & 0 \\ 0 & e^{\psi} \end{pmatrix}, \quad -e = \begin{pmatrix} -1 & 0 \\ 0 & -1 \end{pmatrix}, \quad s = \begin{pmatrix} 0 & 1 \\ -1 & 0 \end{pmatrix},$$

*and $p$ is a matrix of one of the following two types:*

a. $$p = \begin{pmatrix} \operatorname{ch}\theta & \operatorname{sh}\theta \\ \operatorname{sh}\theta & \operatorname{ch}\theta \end{pmatrix}, \quad -\infty < \theta < \infty, \tag{2}$$

b. $$p = \begin{pmatrix} \cos\theta & -\sin\theta \\ \sin\theta & \cos\theta \end{pmatrix}, \quad -\frac{\pi}{4} < \theta < \frac{\pi}{4}. \tag{3}$$

We first prove the following lemma.

**Lemma.** *Let $g = \begin{pmatrix} \alpha & \beta \\ \gamma & \delta \end{pmatrix}$ be a matrix of the group $SL(2, R)$ such that $|\alpha| = |\delta|$, $|\beta| = |\gamma|$, $|\alpha| \geq |\beta|$ and $\alpha > 0$. Then it is either a matrix of the form*

$$g = \begin{pmatrix} \operatorname{ch}\theta & \operatorname{sh}\theta \\ \operatorname{sh}\theta & \operatorname{ch}\theta \end{pmatrix}, \quad -\infty < \theta < \infty, \tag{4}$$

*or a matrix of the form*

$$g = \begin{pmatrix} \cos\theta & -\sin\theta \\ \sin\theta & \cos\theta \end{pmatrix}, \quad -\frac{\pi}{4} \leqslant \theta < \frac{\pi}{4}. \tag{5}$$

**Proof.** Since $\alpha\delta - \beta\gamma = 1$ and $|\alpha\delta| \geq |\beta\gamma|$, we have $\alpha\delta > 0$. Therefore $\alpha = \delta > 0$. Let $\alpha = \delta \geq 1$. Since $\alpha\delta - 1 = \beta\gamma$, we have $\beta\gamma > 0$ and as $|\beta| = |\gamma|$, we get $\beta = \gamma$. Putting $\beta = \operatorname{sh}\theta$, we get $\alpha = \operatorname{ch}\theta$; consequently $g$ has the form (4).

Now let $\alpha = \delta < 1$. Then from the equality $\alpha\delta - 1 = \beta\gamma$ we get $\beta\gamma < 0$ and hence $\gamma = -\beta$. For $\alpha = \cos\theta$, $\beta = -\sin\theta$, the matrix $g$ has the form (5). Here $-\pi/4 < \theta < \pi/4$, since $\alpha > 0$ and $|\alpha| > |\beta|$.

We now pass on to the proof of the theorem. We put

$$e^{\varphi} = \left|\frac{\gamma\delta}{\alpha\beta}\right|^{\frac{1}{2}}, \quad e^{\psi} = \left|\frac{\beta\delta}{\alpha\gamma}\right|^{\frac{1}{2}},$$

and denote by $g_1$ the matrix $g_1 = d_1^{-1} g d_2^{-1}$, where

$$d_1 = \begin{pmatrix} e^{-\varphi} & 0 \\ 0 & e^{\varphi} \end{pmatrix}, \quad d_2 = \begin{pmatrix} e^{-\psi} & 0 \\ 0 & e^{\psi} \end{pmatrix}.$$

Let $g_1 = \begin{pmatrix} \alpha_1 & \beta_1 \\ \gamma_1 & \delta_1 \end{pmatrix}$. Owing to the choice of $\phi$ and $\psi$, the equalities $|\alpha_1|$ $|\delta_1|$ and $|\beta_1| = |\gamma_1|$ are satisfied. Further, let us denote by $p$ the matrix $p = s^{-\epsilon_2}(-e)^{-\epsilon_1} g_1$, $p = \begin{pmatrix} \alpha_2 & \beta_2 \\ \gamma_2 & \delta_2 \end{pmatrix}$, where

a) $\varepsilon_1 = \varepsilon_2 = 0$, if $|\alpha_1| \geqslant |\beta_1|$, $\alpha_1 > 0$,
b) $\varepsilon_1 = 1$, $\varepsilon_2 = 0$, if $|\alpha_1| \geqslant |\beta_1|$, $\alpha_1 < 0$,
c) $\varepsilon_1 = 0$, $\varepsilon_2 = 1$, if $|\alpha_1| < |\beta_1|$, $\gamma_1 > 0$,
d) $\varepsilon_1 = \varepsilon_2 = 1$, if $|\alpha_1| < |\beta_1|$, $\gamma_1 < 0$.

Then we have the inequalities $|\alpha_2| = |\delta_2| > |\beta_2| = |\gamma_2|$, $\alpha_2 > 0$. Consequently $p$ has either the form (2) or the form (3). But then

$$g = d_1 g_1 d_2 = d_1 (-e)^{\varepsilon_1} s^{\varepsilon_2} p d_2.$$

From this theorem it follows that every matrix $g$ of the group $SL(2, R)$ is given by the numbers $\phi$, $\theta$, $\psi$ and the numbers $\epsilon_1$, $\epsilon_2$ taking the values 0 and 1. In addition, it must be indicated whether the matrix $p$ has the form (2) or (3). We thus obtain eight ranges in the group $SL(2, R)$, characterized by the values of $\epsilon_1$ and $\epsilon_2$ and by the type of the matrix $p$. In each of these ranges the matrix $g$ is uniquely determined by the values of $\phi$, $\theta$, $\psi$. Moreover, if $g$ has the form (2), then $\theta$ varies from $-\infty$ to $\infty$, and if $g$ is of type (3), then $\theta$ varies from $-\pi/4$ to $\pi/4$.

This parametrization of the group $SL(2, R)$ is closely connected with the parametrization of $SL(2, C)$ with the help of Euler angles, but differs slightly from it (because of the presence of a factor of the form $s^{\epsilon}$). Parametrization with the help of Euler angles is inconvenient inasmuch as in certain cases one has to express real matrices as a product of matrices with complex elements.

We now consider, in the group $SL(2, R)$, those matrices for which one of the matrix elements is equal to zero. As in Theorem 1, we can see that any matrix of such a form can be written as

$$g = d(-e)^{\epsilon_1} s^{\epsilon_2} p s^{\epsilon_3}, \tag{6}$$

where $d = \begin{bmatrix} e^{-\phi} & 0 \\ 0 & e^{\phi} \end{bmatrix}$, $\epsilon_1$, $\epsilon_2$ and $\epsilon_3$ take the values 0 and 1, $-e = \begin{bmatrix} -1 & 0 \\ 0 & -1 \end{bmatrix}$,

$s = \begin{bmatrix} 0 & 1 \\ -1 & 0 \end{bmatrix}$, and $p$ is a matrix of the form $p = \begin{bmatrix} 1 & 0 \\ x & 1 \end{bmatrix}$.

For example, if $g = \begin{bmatrix} 0 & \beta \\ \gamma & \delta \end{bmatrix}$, $\gamma > 0$, then

$$g = dps,$$

where

$$d = \begin{pmatrix} e^{-\varphi} & 0 \\ 0 & e^{\varphi} \end{pmatrix}, \quad e^{\varphi} = -\gamma, \quad p = \begin{pmatrix} 1 & 0 \\ -\frac{\delta}{\gamma} & 1 \end{pmatrix}.$$

3. **Lie algebra.** In the Lie algebra of $SL(2, R)$ we choose a basis consisting of tangent matrices to the one-parameter subgroups

$$\omega_+(t) = \begin{pmatrix} 1 & t \\ 0 & 1 \end{pmatrix}, \tag{1}$$

$$\omega_-(t) = \begin{pmatrix} 1 & 0 \\ t & 1 \end{pmatrix}, \tag{1'}$$

$$\omega_3(t) = \begin{pmatrix} e^{-t} & 0 \\ 0 & e^{t} \end{pmatrix}. \tag{1''}$$

These tangent matrices have, respectively, the forms

$$a_+ = \begin{pmatrix} 0 & 1 \\ 0 & 0 \end{pmatrix}, \quad a_- = \begin{pmatrix} 0 & 0 \\ 1 & 0 \end{pmatrix}, \quad a_3 = \begin{pmatrix} -1 & 0 \\ 0 & 1 \end{pmatrix}. \tag{2}$$

They satisfy the commutation relations

$$[a_+, a_-] = -a_3, \tag{3}$$

$$[a_+, a_3] = 2a_+, \tag{3'}$$

$$[a_-, a_3] = -2a_-. \tag{3''}$$

In place of $a_+$ and $a_-$ we can take their linear combinations

$$a_1 = a_+ + a_- = \begin{pmatrix} 0 & 1 \\ 1 & 0 \end{pmatrix} \tag{4}$$

and

$$a_2 = -a_+ + a_- = \begin{pmatrix} 0 & -1 \\ 1 & 0 \end{pmatrix}. \qquad (4')$$

They are tangent matrices to the subgroups

$$\mathfrak{Q}_1 : \omega_1(t) = \begin{pmatrix} \operatorname{ch} t & \operatorname{sh} t \\ \operatorname{sh} t & \operatorname{ch} t \end{pmatrix} \qquad (5)$$

and

$$\mathfrak{Q}_2 : \omega_2(t) = \begin{pmatrix} \cos t & -\sin t \\ \sin t & \cos t \end{pmatrix} \qquad (5')$$

respectively.

The commutation relations for $a_1$, $a_2$, $a_3$ have the form

$$[a_1, a_2] = -2a_3, \qquad (6)$$
$$[a_2, a_3] = -2a_1, \qquad (6')$$
$$[a_3, a_1] = 2a_2. \qquad (6'')$$

## §3. IRREDUCIBLE REPRESENTATIONS OF THE GROUP $SL(2, R)$

1. **Description.** Since the group $SL(2, R)$ is isomorphic to the group $QU(2)$, its irreducible representations are immediately obtained from the irreducible representations of $QU(2)$, described in §2 of Chapter VI. We recall that these representations are constructed in the space $\mathfrak{D}_\chi$, $\chi = (l, \epsilon)$, of functions $\Phi(z)$, homogeneous of degree $2l$ and with parity $2\epsilon$, and are given by the formula

$$T_\chi(h)\,\Phi(z) = \Phi(az + \bar{b}\bar{z}), \qquad (1)$$

where $h = \begin{pmatrix} a & b \\ \bar{a} & \bar{b} \end{pmatrix}$, $|a|^2 - |b|^2 = 1$.

Therefore, if $g = \begin{pmatrix} \alpha & \beta \\ \gamma & \delta \end{pmatrix}$ is a matrix of the group $SL(2, R)$, then the corresponding representation operator is given by formula (1), where $h = t^{-t}gt$, $t = \frac{\sqrt{2}}{2}\begin{pmatrix} 1 & i \\ i & 1 \end{pmatrix}$ (see §1.1 of Chapter VI). In the parameters $\alpha$, $\beta$, $\gamma$, $\delta$ the representation is written as follows:

$$T_\chi(g)\Phi(z)=\Phi\left[\left(\frac{\alpha+\delta}{2}+i\frac{\beta-\gamma}{2}\right)z+\left(\frac{\beta+\gamma}{2}+i\frac{\delta-\alpha}{2}\right)\bar{z}\right] \tag{2}$$

(see §1.1 of Chapter VI, formulas (2) and (2′)).

One obtains a more convenient notation for the representation if one passes from the complex number $z = x + iy$ to the pair of real numbers $u = x - y$, $v = x + y$, i.e. to

$$u=\frac{\bar{z}+iz}{1+i};\qquad v=\frac{z+i\bar{z}}{1+i}.$$

A simple calculation shows that corresponding to (2) one has the operator

$$T_\chi(g)\varphi(u, v)=\varphi(\alpha u+\gamma v,\ \beta u+\delta v) \tag{3}$$

in the space of functions

$$\varphi(u, v)\equiv\Phi(z). \tag{4}$$

The functions $\phi(u, v)$ are infinitely differentiable, are homogeneous of degree $2l$ and have parity $2\epsilon$.

Thus we have obtained the following realization of representations of the group $SL(2, R)$. Let $\chi = (l, \epsilon)$, where $l$ is a complex number, and $\epsilon = 0, 1/2$. We denote by $\mathfrak{D}_\chi$ the space of infinitely differentiable functions $\phi(x, y)$ of two real variables $x$ and $y$, such that, for any real number $a$,

$$\varphi(ax, ay)=|a|^{2l}(\operatorname{sign} a)^{2\epsilon}\varphi(x, y). \tag{5}$$

We define the operators $T_\chi(g)$ of representation of the group $SL(2, R)$ by the formula

$$T_\chi(g)\varphi(x, y)=\varphi(\alpha x+\gamma y,\ \beta x+\delta y), \tag{6}$$

where $\phi(x, y)\in\mathfrak{D}_\chi$ and $g=\begin{pmatrix}\alpha & \beta\\ \gamma & \delta\end{pmatrix}\in SL(2, R)$.

From this realization of the representations, we go over to the realization in the space of functions $f(x)$ given on the real axis. With each function $\phi(x, y)$ of the space $\mathfrak{D}_\chi$ we associate a function of one variable

$$f(x) = \phi(x, 1) \tag{7}$$

Because of the homogeneity relation (5), the function $\phi(x, y)$ is expressible in terms of $f(x)$ by the formula

$$\varphi(x, y)=|y|^{2l}(\operatorname{sign} y)^{2\epsilon}\varphi\left(\frac{x}{y}, 1\right)=|y|^{2l}(\operatorname{sign} y)^{2\epsilon} f\left(\frac{x}{y}\right). \tag{8}$$

Evidently, $f(x)$ is infinitely differentiable. In addition, putting $x = 1$ in formula (8) and granting the infinite differentiability of the function $\phi(1, y)$ with respect to $y$, we find that the function

$$\hat{f}(y)=|y|^{2l}(\operatorname{sign} y)^{2\epsilon} f\left(\frac{1}{y}\right) \tag{9}$$

is infinitely differentiable. It is easy to show that to any infinitely differentiable function $f(x)$, such that the function $\hat{f}(x)$ is also infinitely differentiable, corresponds a function $\phi(x, y)$ of the space $\mathfrak{D}_\chi$ according to formula (8).

Using (6) and (8), it is easy to see that in passing from $\phi(x, y)$ to $f(x)$ the representation operators $T_\chi(g)$ take the form

$$T_\chi(g) f(x)=|\beta x+\delta|^{2l} \operatorname{sign}^{2\epsilon}(\beta x+\delta) f\left(\frac{\alpha x+\gamma}{\beta x+\delta}\right). \tag{10}$$

2. **Another realization of the representations $T_\chi(g)$.** We pass on to another realization of the representations $T_\chi(g)$. For this, with each function $f(x)$ of $\mathfrak{D}_\chi$ we associate the function pair $(F_+(\lambda), F_-(\lambda))$, defined by

$$F_+(\lambda)=\int_0^\infty x^{\lambda-1} f(x)\, dx \equiv \int_{-\infty}^{\infty} x_+^{\lambda-1} f(x)\, dx \tag{1}$$

and

$$F_-(\lambda)=\int_0^\infty x^{\lambda-1} f(-x)\, dx \equiv \int_{-\infty}^{\infty} x_-^{\lambda-1} f(x)\, dx. \tag{2}$$

In other words, these functions are the Mellin transforms of the functions $f(x)$ and $f(-x)$, considered on the semiaxis $0 \leq x < \infty$.

We elucidate upon the values of $\psi$ for which the integrals (1) and (2) converge Let $f(x) \in \mathfrak{D}_\chi$. From (1) it follows that for $|x| \to \infty$ we have

$$|f(x)| \sim |\hat{f}(0)|\,|x|^{2l}. \tag{3}$$

Therefore integrals (1) and (2) are absolutely convergent at infinity, if $\operatorname{Re}(\lambda + 2l) < 0$. At zero these integrals converge for $\operatorname{Re} \lambda > 0$. Thus (1) and (2) are absolutely convergent in the strip $0 < \operatorname{Re} \lambda < -2 \operatorname{Re} l$. In this strip they define $F_+(\lambda)$

and $F_-(\lambda)$ as analytic functions of $\lambda$.

Further, from the results of §3.2 of Chapter II it is easily follows that if $f(x) \in \mathfrak{D}_\chi$ and $0 < a < -2\operatorname{Re} l$, then the functions $F_+(a+i\mu)$ and $F_-(a+i\mu)$ decrease rapidly for $|\mu| \to \infty$.

Outside the strip $0 < \operatorname{Re} \lambda < -2\operatorname{Re} l$ the functions $F_+(\lambda)$ and $F_-(\lambda)$ are defined by means of analytic continuation with respect to $\lambda$.

As shown in §4.3 of Chapter II, the function $f(x)$ is expressed in terms of the functions $F_+(\lambda)$ and $F_-(\lambda)$ by the following inversion formula:

$$f(x) = \begin{cases} \dfrac{1}{2\pi i} \displaystyle\int_{a-i\infty}^{a+i\infty} F_+(\lambda)\, x^{-\lambda}\, d\lambda, & x > 0, \\ \dfrac{1}{2\pi i} \displaystyle\int_{a-i\infty}^{a+i\infty} F_-(\lambda)\, (-x)^{-\lambda}\, d\lambda, & x < 0, \end{cases} \tag{4}$$

where $a$ is any point of the interval $0 < a < -2\operatorname{Re} l$. Moreover, one has the Plancherel equality

$$\int_{-\infty}^{\infty} |f(x)|^2\, dx = \frac{1}{2\pi i} \int_{a-i\infty}^{a+i\infty} F_+(1-\bar\lambda)\, \overline{F_+(\lambda)}\, d\lambda + \frac{1}{2\pi i} \int_{a-i\infty}^{a+i\infty} F_-(1-\bar\lambda)\, \overline{F_-(\lambda)}\, d\lambda. \tag{5}$$

For $\lambda = 1/2 + i\rho$ these formulas take a symmetric form:

$$\Phi_{\pm}(\rho) = \int_{-\infty}^{\infty} f(x)\, x_{\pm}^{-\frac{1}{2}+i\rho}\, dx, \tag{6}$$

$$f(x) = \begin{cases} \dfrac{1}{2\pi} \displaystyle\int_{-\infty}^{\infty} \Phi_+(\rho)\, x^{-\frac{1}{2}-i\rho}\, d\rho, & x > 0, \\ \dfrac{1}{2\pi} \displaystyle\int_{-\infty}^{\infty} \Phi_-(\rho)\, (-x)^{-\frac{1}{2}-i\rho}\, d\rho, & x < 0, \end{cases} \tag{7}$$

$$\int_{-\infty}^{\infty} |f(x)|^2\, dx = \frac{1}{2\pi} \int_{-\infty}^{\infty} |\Phi_+(\rho)|^2\, d\rho + \frac{1}{2\pi} \int_{-\infty}^{\infty} |\Phi_-(\rho)|^2\, d\rho,$$

where, for brevity, $\Phi_{\pm}(\rho) = F_{\pm}(1/2 + i\rho)$. Therefore we can extend the correspondence $f \to (F_+, F_-)$ to the correspondence $f \to (\Phi_+, \Phi_-)$, where $f, \Phi_+, \Phi_-$ are functions with square-integrable modulus.

We now find out how the representation operators $T_\chi(g)$ are expressed in the space of pairs $\mathbf{F} = (F_+, F_-)$. We denote the pair corresponding to the function $T_\chi(g)f(x)$ by $R_\chi(g)\mathbf{F} = (R_\chi(g)F_+, R_\chi(g)F_-)$.

Let $d = \begin{pmatrix} e^{-\phi} & 0 \\ 0 & e^{\phi} \end{pmatrix}$ be a diagonal matrix. Corresponding to it in the space of functions $f(x)$ we have the operator

$$T_\chi(d)f(x) = e^{2l\varphi} f(e^{-2\varphi}x). \tag{8}$$

Therefore

$$\begin{aligned} R_\chi(d)F_+(\lambda) &= e^{2l\varphi}\int_0^\infty f(e^{-2\varphi}x)\,x^{\lambda-1}\,dx \\ &= e^{2\varphi(\lambda+l)}\int_0^\infty f(x)\,x^{\lambda-1}\,dx = e^{2\varphi(\lambda+l)}F_+(\lambda). \end{aligned} \tag{9}$$

Similarly,

$$R_\chi(d)F_-(\lambda) = e^{2\varphi(\lambda+l)}F_-(\lambda) \tag{10}$$

and consequently

$$R_\chi(d)\mathbf{F}(\lambda) = e^{2\varphi(\lambda+l)}\mathbf{F}(\lambda). \tag{11}$$

Thus, to the operator $T_\chi(d)$, $d = \begin{pmatrix} e^{-\phi} & 0 \\ 0 & e^{\phi} \end{pmatrix}$ there corresponds, in the space of pairs $\mathbf{F} = (F_+, F_-)$, the operator of multiplication by the function $e^{2\phi(\lambda+l)}$. In other words, on going over to the pair space, the operators corresponding to the matrices $d$ take the diagonal form.

One similarly proves that to the matrix $s = \begin{pmatrix} 0 & 1 \\ -1 & 0 \end{pmatrix}$ corresponds the operator taking the pair $\mathbf{F}(\lambda) = (F_+(\lambda), F_-(\lambda))$ into the pair

$$R_\chi(s)\mathbf{F}(\lambda) = (F_-(-\lambda-2l),\ (-1)^{2\epsilon}F_+(-\lambda-2l)). \tag{12}$$

As for the matrix $-e = \begin{pmatrix} -1 & 0 \\ 0 & -1 \end{pmatrix}$, to it corresponds the operator taking the pair $\mathbf{F} = (F_+, F_-)$ into the pair

$$(-1)^{2\epsilon}\mathbf{F} = ((-1)^{2\epsilon}F_+, (-1)^{2\epsilon}F_-). \tag{13}$$

Lastly, to the matrix $t = \begin{pmatrix} 1 & 0 \\ 0 & -1 \end{pmatrix}$ corresponds the operator taking $\mathbf{F} = (F_+, F_-)$ into

$$R_\chi(t)\mathbf{F} = ((-1)^{2\epsilon}F_-, (-1)^{2\epsilon}F_+). \tag{14}$$

3. **Operators of the second realization of $T_\chi(g)$.** We now consider the matrix $g = \begin{pmatrix} \alpha & \beta \\ \gamma & \delta \end{pmatrix}$, all of whose elements are nonzero, and find the corresponding operator in the pair space. Since

$$T_\chi(g)f(x) = |\beta x + \delta|^{2l}\operatorname{sign}^{2\epsilon}(\beta x + \delta) f\left(\frac{\alpha x + \gamma}{\beta x + \delta}\right), \tag{1}$$

we have

$$R_\chi(g)F_+(\lambda) = \int_{-\infty}^{\infty} x_+^{\lambda-1}|\beta x + \delta|^{2l}\operatorname{sign}^{2\epsilon}(\beta x + \delta) f\left(\frac{\alpha x + \gamma}{\beta x + \delta}\right) dx \tag{2}$$

and

$$R_\chi(g)F_-(\lambda) = \int_{-\infty}^{\infty} x_-^{\lambda-1}|\beta x + \delta|^{2l}\operatorname{sign}^{2\epsilon}(\beta x + \delta) f\left(\frac{\alpha x + \gamma}{\beta x + \delta}\right) dx. \tag{3}$$

Let us make the substitution $(\alpha x + \gamma)/(\beta x + \delta) = y$ in formula (2). We get

$$R_\chi(g)F_+(\lambda) = \int_{-\infty}^{\infty} \left(\frac{\delta y - \gamma}{-\beta y + \alpha}\right)_+^{\lambda-1} |-\beta y + \alpha|^{-2l-2}\operatorname{sign}^{2\epsilon}(-\beta y + \alpha) f(y)\, dy. \tag{4}$$

Using the inversion formula (4) of subsection 2, this integral can be written as

$$R_\chi(g)F_+(\lambda) = \frac{1}{2\pi i}\int_0^{\infty} \left(\frac{\delta y - \gamma}{-\beta y + \alpha}\right)_+^{\lambda-1} |-\beta y + \alpha|^{-2l-2}$$

$$\times \operatorname{sign}^{2\epsilon}(-\beta y + \alpha) \int_{a-i\infty}^{a+i\infty} y^{-\mu}F_+(\mu)\, d\mu\, dy +$$

$$+\frac{1}{2\pi i}\int_{-\infty}^{0}\left(\frac{\delta y-\gamma}{-\beta y+\alpha}\right)_{+}^{\lambda-1}|-\beta y+\alpha|^{-2l-2}$$
$$\times \operatorname{sign}^{2\varepsilon}(-\beta y+\alpha)\int_{a-i\infty}^{a+i\infty} y^{-\mu}F_{-}(\mu)\,d\mu\,dy, \tag{5}$$

where $0 < a < -2\operatorname{Re} l$.

We show that under the conditions

$$\left.\begin{array}{c} 0<\operatorname{Re}\lambda<-2\operatorname{Re} l, \\ -1-2\operatorname{Re} l<\operatorname{Re}\mu<1 \end{array}\right\} \tag{6}$$

the integrals in (5) are absolutely convergent. Since the functions $F_{+}(\mu)$ and $F_{-}(\mu)$, $\mu = a + i\nu$, decrease rapidly for $|\nu| \to \infty$, it is sufficient to show the convergence of the integrals for the variable $y$. If all the elements of the matrix $g = \begin{pmatrix}\alpha & \beta \\ \gamma & \delta\end{pmatrix}$ are nonzero, then the points $y = 0$, $\alpha/\beta$, $\gamma/\delta$ and $\infty$ are the singular points of the integrand. At these points the integrand has, respectively, the orders

$$-\mu, \quad -\lambda-2l-1, \quad \lambda-1 \text{ and } -\mu-2l-2.$$

Hence the absolute convergence of the integrals in the domain (6) follows. The case when one of the elements of $g$ becomes zero will be considered below.

We shall assume that conditions (6) are satisfied. Then we can change the order of integration. We obtain

$$R_{\chi}(g)F_{+}(\lambda)=\int_{a-i\infty}^{a+i\infty} K_{++}(\lambda, \mu; \chi; g)F_{+}(\mu)\,d\mu$$
$$+\int_{a-i\infty}^{a+i\infty} K_{+-}(\lambda, \mu; \chi; g)F_{-}(\mu)\,d\mu, \tag{7}$$

where the kernels $K_{++}(\lambda, \mu; \chi; g)$ and $K_{+-}(\lambda, \mu; \chi; g)$ are given by

$$K_{++}(\lambda, \mu; \chi; g)$$
$$=\frac{1}{2\pi i}\int_{0}^{\infty} x^{\lambda-1}\left(\frac{\alpha x+\gamma}{\beta x+\delta}\right)_{+}^{-\mu}|\beta x+\delta|^{2l}\operatorname{sign}^{2\varepsilon}(\beta x+\delta)\,dx, \tag{8}$$

$$K_{+-}(\lambda, \mu; \chi; g)=\frac{1}{2\pi i}\int_{0}^{\infty} x^{\lambda-1}\left(\frac{\alpha x+\gamma}{\beta x+\delta}\right)_{-}^{-\mu}|\beta x+\delta|^{2l}\operatorname{sign}^{2\varepsilon}(\beta x+\delta)\,dx \tag{8'}$$

(in the integrals expressing the kernels we have made the substitution $y = (\alpha x + \gamma)/(\beta x + \delta)$).

Similarly, one proves that

$$R_\chi(g)F_-(\lambda) = \int_{a-i\infty}^{a+i\infty} K_{-+}(\lambda, \mu; \chi; g) F_+(\mu)\,d\mu + \int_{a-i\infty}^{a+i\infty} K_{--}(\lambda, \mu; \chi; g) F_-(\mu)\,d\mu, \tag{9}$$

where

$$K_{-+}(\lambda, \mu; \chi; g) = \frac{1}{2\pi i}\int_0^\infty x^{\lambda-1}\left(\frac{-\alpha x+\gamma}{-\beta x+\delta}\right)_+^{-\mu} |-\beta x+\delta|^{2l}\operatorname{sign}^{2\epsilon}(-\beta x+\delta)\,dx, \tag{10}$$

$$K_{--}(\lambda, \mu; \chi, g) = \frac{1}{2\pi i}\int_0^\infty x^{\lambda-1}\left(\frac{-\alpha x+\gamma}{-\beta x+\delta}\right)_-^{-\mu} |-\beta x+\delta|^{2l}\operatorname{sign}^{2\epsilon}(-\beta x+\delta)\,dx. \tag{10$'$}$$

Thus we have proved the following assertion.

*If* $0 < \operatorname{Re}\lambda < -2\operatorname{Re} l$, $-1-2\operatorname{Re} l < \operatorname{Re}\mu < 1$, *and if all the elements of the matrix* $g = \begin{pmatrix}\alpha & \beta\\ \gamma & \delta\end{pmatrix}$ *are nonzero, then on going over to the space of pairs* $\mathbf{F} = (F_+, F_-)$ *the representation operators* $T_\chi(g)$ *turn into the operators*

$$R_\chi(g)\mathbf{F}(\lambda) = \int_{a-i\infty}^{a+i\infty} \mathbf{K}(\lambda, \mu; \chi; g) F(\mu)\,d\mu. \tag{11}$$

*Here* $\mathbf{K}(\lambda, \mu; \chi; g)$ *denotes the matrix*

$$\mathbf{K} = \begin{pmatrix} K_{++} & K_{+-}\\ K_{-+} & K_{--}\end{pmatrix}, \tag{12}$$

*whose elements are given by formulas* (8), (8$'$), (10), (10$'$).

The matrix $\mathbf{K}$ will be called the kernel of the operator $R_\chi(g)$. This kernel is defined in the domain

$$0 < \operatorname{Re}\lambda < -2\operatorname{Re} l,$$
$$-1-2\operatorname{Re} l < \operatorname{Re}\mu < 1.$$

In the case when one of the elements of $g$ is equal to zero, the corresponding operator $R_\chi(g)$ is also given by a formula of the form (11). However, the convergence domain of integrals (8), (8$'$), (10) and (10$'$) varies. Namely, for $\alpha = 0$

these integrals converge in the domain

$$0<\operatorname{Re}\lambda;\quad \operatorname{Re}(\lambda+\mu)<-2\operatorname{Re}l<\operatorname{Re}\mu+1, \tag{13}$$

for $\beta = 0$, in the domain

$$0<\operatorname{Re}\lambda<\operatorname{Re}\mu<1, \tag{13'}$$

for $\gamma = 0$, in the domain

$$-2\operatorname{Re}l-1<\operatorname{Re}\mu<\operatorname{Re}\lambda<-2\operatorname{Re}l \tag{13''}$$

and for $\delta = 0$, in the domain

$$\operatorname{Re}\lambda<-2\operatorname{Re}l<\operatorname{Re}(\lambda+\mu);\qquad \operatorname{Re}\mu<1. \tag{13'''}$$

4. **Infinitesimal operators.** We calculate the infinitesimal operators of the representations of $SL(2, R)$ constructed above. The representation operators $T_\chi(g)$, corresponding to the elements

$$\omega_+(t)=\begin{pmatrix}1 & t\\ 0 & 1\end{pmatrix} \tag{1}$$

of the one-parameter subgroup $\Omega_+$, have the form

$$T_\chi(\omega_+(t))=|tx+1|^{2l}\operatorname{sign}^{2\epsilon}(tx+1)f\left(\frac{x}{tx+1}\right). \tag{2}$$

On differentiating both sides of this equality with respect to $t$ and putting $t = 0$, we obtain

$$A_+\equiv\left.\frac{dT_\chi(\omega_+(t))}{dt}\right|_{t=0}=2lx-x^2\frac{d}{dx}. \tag{3}$$

Similarly, the infinitesimal operator $A_-$ of this representation, corresponding to the subgroup $\Omega_-$ of matrices of the form $\begin{pmatrix}1 & 0\\ t & 1\end{pmatrix}$, is given by the formula

$$A_-=\frac{d}{dx}. \tag{4}$$

As for the subgroup $\Omega_3$ of matrices $\omega_3(t) = \begin{pmatrix}e^{-t} & 0\\ 0 & e^t\end{pmatrix}$, to it corresponds the operator

$$A_3=2l-2x\frac{d}{dx}. \tag{5}$$

Since the tangent matrix to the subgroup

$$\Omega_1: \begin{pmatrix} \operatorname{ch} t & \operatorname{sh} t \\ \operatorname{sh} t & \operatorname{ch} t \end{pmatrix}$$

has the form $a_1 = a_+ + a_-$, to the subgroup $\Omega_1$ corresponds the infinitesimal operator

$$A_1 = A_+ + A_- = 2lx + (1 - x^2)\frac{d}{dx}. \tag{6}$$

In exactly the same way, one proves that to the subgroup

$$\Omega_2: \begin{pmatrix} \cos t & -\sin t \\ \sin t & \cos t \end{pmatrix}$$

corresponds the infinitesimal operator

$$A_2 = -2lx + (1 + x^2)\frac{d}{dx}. \tag{7}$$

Let us now calculate infinitesimal operators for the representation $R_\chi(g)$. In order to find the operator corresponding to the subgroup $\Omega_+$, one must find the Mellin transforms of the functions $A_+f(x)$, $x > 0$ and $(A_+f)(-x)$, $x < 0$. The Mellin transform for $A_+f(x)$ has the form

$$\int_0^\infty x^{\lambda-1} A_+ f(x)\,dx = \int_0^\infty x^{\lambda-1}[2lxf(x) - x^2 f'(x)]\,dx$$

$$= 2l\int_0^\infty x^\lambda f(x)\,dx - \int_0^\infty x^{\lambda+1} f'(x)\,dx.$$

Integrating by parts and remembering that

$$\int_0^\infty x^{\lambda-1} f(x)\,dx = F_+(\lambda),$$

we get

$$\int_0^\infty x^{\lambda-1} A_+ f(x)\,dx = (2l + \lambda + 1)F_+(\lambda + 1).$$

Similarly we find

$$\int_0^\infty x^{\lambda-1}(A_+f)(-x)\,dx = -(2l + \lambda + 1)F_-(\lambda + 1).$$

Thus the infinitesimal operator of the representation $R_\chi(g)$ corresponding to $\Omega_+$ has the form

$$B_+(F_+(\lambda),\ F_-(\lambda)) = ((2l+\lambda+1)F_+(\lambda+1),\ -(2l+\lambda+1)F_-(\lambda+1)). \tag{8}$$

In the same way one proves that to the subgroup $\Omega_-$ corresponds the infinitesimal operator

$$B_-(F_+(\lambda),\ F_-(\lambda)) = (-(\lambda-1)F_+(\lambda-1),\ (\lambda-1)F_-(\lambda-1)), \tag{9}$$

and to the subgroup $\Omega_3$, the operator

$$B_3(F_+(\lambda),\ F_-(\lambda)) = ((2l+2\lambda)F_+(\lambda),\ (2l+2\lambda)F_-(\lambda)). \tag{10}$$

Infinitesimal operators corresponding to the subgroups $\Omega_1$ and $\Omega_2$ are linear combinations of the operators $B_+$ and $B_-$:

$$\begin{aligned} B_1(F_+(\lambda),\ F_-(\lambda)) = ((2l+\lambda+1)F_+(\lambda+1) - (\lambda-1)F_+(\lambda-1),& \\ -(2l+\lambda+1)F_-(\lambda+1) + (\lambda-1)F_-(\lambda-1)),& \end{aligned} \tag{11}$$

$$\begin{aligned} B_2(F_+(\lambda),\ F_-(\lambda)) = (-(2l+\lambda+1)F_+(\lambda+1) - (\lambda-1)F_+(\lambda-1),& \\ (2l+\lambda+1)F_-(\lambda+1) + (\lambda-1)F_-(\lambda-1)).& \end{aligned} \tag{12}$$

## §4. CALCULATION OF THE KERNELS OF $R_\chi(g)$

1. **Calculation of $K(\lambda, \mu; \chi; h)$ and $K(\lambda, \mu; \chi; u)$.** In §2.2 it was shown that any matrix $g$ of the group $SL(2, R)$ can be represented in the form of a product of diagonal matrices, the matrix $s = \begin{pmatrix} 0 & 1 \\ -1 & 0 \end{pmatrix}$ and a matrix having either the form

$$h = \begin{pmatrix} \operatorname{ch}\theta & \operatorname{sh}\theta \\ \operatorname{sh}\theta & \operatorname{ch}\theta \end{pmatrix}, \qquad -\infty < \theta < \infty, \tag{1}$$

or the form

$$u = \begin{pmatrix} \cos\theta & -\sin\theta \\ \sin\theta & \cos\theta \end{pmatrix}, \qquad -\frac{\pi}{4} < \theta < \frac{\pi}{4}. \tag{2}$$

For diagonal matrices, and also for the matrix $s$, we have already found the form of the representation operator. Therefore it is sufficient to calculate the kernels for the operators $R_\chi(h)$ and $R_\chi(u)$, where $h$ and $u$ are matrices having, respectively, the forms (1) and (2). We begin with matrices of the form

$$h = \begin{pmatrix} \operatorname{ch}\theta & \operatorname{sh}\theta \\ \operatorname{sh}\theta & \operatorname{ch}\theta \end{pmatrix}$$

for $\theta > 0$. From formula (8) of §3.3 it follows that

$$K_{++}(\lambda, \mu; \chi; h) = \frac{1}{2\pi i}\int_0^\infty x^{\lambda-1}\left(\frac{x \operatorname{ch}\theta + \operatorname{sh}\theta}{x \operatorname{sh}\theta + \operatorname{ch}\theta}\right)_+^{-\mu} |x \operatorname{sh}\theta + \operatorname{ch}\theta|^{2l} \operatorname{sign}^{2\epsilon}(x \operatorname{sh}\theta + \operatorname{ch}\theta)\,dx.$$

But for $\theta > 0$, $x > 0$ we have $x\operatorname{ch}\theta + \operatorname{sh}\theta > 0$ and $x\operatorname{sh}\theta + \operatorname{ch}\theta > 0$; hence

$$K_{++}(\lambda, \mu; \chi; h) = \frac{1}{2\pi i}\int_0^\infty x^{\lambda-1}(x \operatorname{ch}\theta + \operatorname{sh}\theta)^{-\mu}(x \operatorname{sh}\theta + \operatorname{ch}\theta)^{2l+\mu}dx. \tag{3}$$

We make the substitution $x = y\operatorname{th}\theta$ in this integral and utilize formula (2) of §1.3. Then for $\theta > 0$ we find that

$$K_{++}(\lambda, \mu; \chi; h) = \frac{1}{2\pi i}\frac{\Gamma(\lambda)\Gamma(-\lambda-2l)}{\Gamma(-2l)}\frac{\operatorname{sh}^{4l+\lambda+\mu}\theta}{\operatorname{ch}^{2l+\lambda+\mu}\theta}F\left(-2l-\lambda, -2l-\mu; -2l; -\frac{1}{\operatorname{sh}^2\theta}\right) \tag{4}$$

or, by formula (5) of §1.2,

$$K_{++}(\lambda, \mu; \chi; h) = \frac{1}{2\pi i}\frac{\Gamma(\lambda)\Gamma(-\lambda-2l)}{\Gamma(-2l)}\frac{\operatorname{ch}^{\lambda+\mu+2l}\theta}{\operatorname{sh}^{\lambda+\mu}\theta}F\left(\lambda, \mu; -2l; -\frac{1}{\operatorname{sh}^2\theta}\right). \tag{4'}$$

We note that integral (3) is absolutely convergent in the domain $0 < \operatorname{Re}\lambda < -2\operatorname{Re} l$.

The remaining elements of the matrix $\mathbf{K}(\lambda, \mu; \chi; h)$ are calculated in exactly the same way. We give the resulting values:

$$K_{+-}(\lambda, \mu; \chi; h) = 0, \tag{5}$$

$$\begin{aligned} K_{-+}(\lambda, \mu; \chi; h) &= \frac{1}{2\pi i}\operatorname{ch}^{\lambda+\mu+2l}\theta \\ &\times\left[\frac{\Gamma(\lambda)\Gamma(1-\mu)}{\Gamma(\lambda-\mu+1)}\operatorname{sh}^{\lambda-\mu}\theta F(\lambda, \lambda+2l+1; \lambda-\mu+1; -\operatorname{sh}^2\theta)\right. \\ &+(-1)^{2\epsilon}\frac{\Gamma(-\lambda-2l)\Gamma(\mu+2l+1)}{\Gamma(\mu-\lambda+1)}\operatorname{sh}^{\mu-\lambda}\theta \\ &\left.\times F(\mu, \mu+2l+1; \mu-\lambda+1; -\operatorname{sh}^2\theta)\right], \end{aligned} \tag{6}$$

where $0 < \operatorname{Re}\lambda < -2\operatorname{Re} l$, $-1-2\operatorname{Re} l < \operatorname{Re}\mu < 1$, and

$$\begin{aligned} K_{--}(\lambda, \mu; \chi; h) &= \frac{1}{2\pi i}\frac{\Gamma(1-\mu)\Gamma(\mu+2l+1)}{\Gamma(2l+2)} \\ &\times \operatorname{sh}^{-\lambda-\mu-4l-2}\theta \operatorname{ch}^{\lambda+\mu+2l}\theta F\left(\lambda+2l+1, \mu+2l+1; 2l+2; -\frac{1}{\operatorname{sh}^2\theta}\right), \end{aligned} \tag{7}$$

where $-1-2\mathrm{Re}\, l < \mathrm{Re}\, \mu < 1$.

From formulas (4)–(7) and relation (5) of §1.2 for the hypergeometric function, it follows that

$$K_{++}(\lambda, \mu; \chi; h) = K_{++}(-\lambda - 2l, -\mu - 2l; \chi; h) \tag{8}$$

and

$$K_{--}(\lambda, \mu; \chi; h) = K_{--}(-\lambda - 2l, -\mu - 2l; \chi; h), \tag{9}$$

and also that

$$K_{-+}(\lambda, \mu; \chi; h) = (-1)^{2\varepsilon} K_{-+}(-\lambda - 2l, -\mu - 2l; \chi; h). \tag{10}$$

We now calculate the kernel of the operator $R_\chi(h^{-1}), h^{-1} = \begin{pmatrix} \mathrm{ch}\,\theta & -\mathrm{sh}\,\theta \\ -\mathrm{sh}\,\theta & \mathrm{ch}\,\theta \end{pmatrix}, \theta > 0$

For this, we note that

$$h^{-1} = shs(-e),$$

where

$$s = \begin{pmatrix} 0 & 1 \\ -1 & 0 \end{pmatrix}, \quad -e = \begin{pmatrix} -1 & 0 \\ 0 & -1 \end{pmatrix} \text{ and } h = \begin{pmatrix} \mathrm{ch}\,\theta & \mathrm{sh}\,\theta \\ \mathrm{sh}\,\theta & \mathrm{ch}\,\theta \end{pmatrix}.$$

Therefore

$$R_\chi(h^{-1}) = R_\chi(s) R_\chi(h) R_\chi(s) R_\chi(-e).$$

The operators $R_\chi(s)$ and $R_\chi(-e)$ were calculated in §3.3. Using the description given there and relations (8)–(10), we easily establish that the matrix $\mathbf{K}(\lambda, \mu; \chi; h^{-1})$ is obtained from the matrix $\mathbf{K}(\lambda, \mu; \chi; h)$ by changing both the index signs. In other words,

$$\left.\begin{aligned} &K_{++}(\lambda, \mu; \chi; h^{-1}) = K_{--}(\lambda, \mu; \chi; h), \\ &K_{-+}(\lambda, \mu; \chi; h^{-1}) = 0 \end{aligned}\right\} \tag{11}$$

and so on.

Let us now pass on to the operator $R_\chi(u)$, where

$$u = \begin{pmatrix} \cos\theta & -\sin\theta \\ \sin\theta & \cos\theta \end{pmatrix}, \quad 0 < \theta < \frac{\pi}{4}. \tag{12}$$

Exactly as for the matrix $h$, we obtain

$$K_{++}(\lambda, \mu; \chi; u)=\frac{1}{2\pi i}\frac{\Gamma(\lambda)\Gamma(\mu+2l+1)}{\Gamma(\lambda+\mu+2l+1)}$$
$$\times\frac{\cos^{\lambda+\mu+2l}\theta}{\sin^{\lambda+\mu}\theta}F(\lambda, \mu; \lambda+\mu+2l+1; -\operatorname{ctg}^2\theta), \quad (13)$$

where $0 < \operatorname{Re}\lambda$ and $-1-2\operatorname{Re} l < \operatorname{Re}\mu$;

$$K_{+-}(\lambda, \mu; \chi; u)$$
$$=\frac{(-1)^{2\varepsilon}}{2\pi i}\frac{\Gamma(-\lambda-2l)\Gamma(\mu+2l+1)}{\Gamma(\mu-\lambda+1)}\sin^{\mu-\lambda}\theta\cos^{\lambda-\mu+2l}\theta$$
$$\times F(-\lambda-2l, \mu; \mu-\lambda+1; -\operatorname{tg}^2\theta), \quad (14)$$

where $\operatorname{Re}\lambda < -2\operatorname{Re} l$ and $-1-2\operatorname{Re} l < \operatorname{Re}\mu$;

$$K_{-+}(\lambda, \mu; \chi; u)=\frac{1}{2\pi i}\frac{\Gamma(\lambda)\Gamma(1-\mu)}{\Gamma(\lambda-\mu+1)}$$
$$\times\sin^{\lambda-\mu}\theta\cos^{\mu-\lambda+2l}\theta F(\lambda, -\mu-2l; \lambda-\mu+1; -\operatorname{tg}^2\theta), \quad (15)$$

where $0 < \operatorname{Re}\lambda$, $\operatorname{Re}\mu < 1$; and

$$K_{--}(\lambda, \mu; \chi; u)=\frac{1}{2\pi i}\frac{\Gamma(-\lambda-2l)\Gamma(1-\mu)}{\Gamma(1-\lambda-\mu-2l)}$$
$$\times\sin^{\lambda+\mu-2}\theta\cos^{-\lambda-\mu-2l}\theta F(1-\lambda, 1-\mu; 1-\lambda-\mu-2l; -\operatorname{ctg}^2\theta), \quad (16)$$

where $\operatorname{Re}\lambda < -2\operatorname{Re} l$ and $\operatorname{Re}\mu < 1$.

In this case also, in replacing $u$ by $u^{-1}$ both index signs are changed:

$$K_{++}(\lambda, \mu; \chi; u^{-1})=K_{--}(\lambda, \mu; \chi; u),$$
$$K_{+-}(\lambda, \mu; \chi; u^{-1})=K_{-+}(\lambda, \mu; \chi; u), \quad (17)$$

and so on. Here

$$u^{-1}=\begin{pmatrix}\cos\theta & \sin\theta\\ -\sin\theta & \cos\theta\end{pmatrix}, \qquad 0<\theta<\frac{\pi}{4}. \quad (18)$$

2. The case of triangular matrices. In the case when the matrix $g$ is triangular, the kernel of $R_\chi(g)$ has a simpler expression, since the hypergeometric function degenerates in degree. Suppose, for example, that $z=\begin{pmatrix}1 & 0\\ \gamma & 1\end{pmatrix}$, $\gamma > 0$. In this case, by formula (8) of §3.3, we have

$$K_{++}(\lambda, \mu; \chi; z)=\frac{1}{2\pi i}\int_0^\infty x^{\lambda-1}(x+\gamma)^{-\mu}\,dx. \quad (1)$$

This integral converges in the domain $0 < \operatorname{Re} \lambda < \operatorname{Re} \mu$. Using formula (7) of §1.6 Chapter V, we find that

$$K_{++}(\lambda, \mu; \chi; z) = \frac{1}{2\pi i} \frac{\Gamma(\lambda)\Gamma(\mu-\lambda)}{\Gamma(\mu)} \gamma^{\lambda-\mu}. \tag{2}$$

In exactly the same way, we get

$$K_{+-}(\lambda, \mu; \chi; z) = 0, \tag{3}$$

$$K_{-+}(\lambda, \mu; \chi; z) = \frac{1}{2\pi i} \frac{\Gamma(\lambda)\Gamma(1-\mu)}{\Gamma(\lambda-\mu+1)} \gamma^{\lambda-\mu}, \tag{4}$$

where $0 < \operatorname{Re} \lambda$ and $\operatorname{Re} \mu < 1$, and

$$K_{--}(\lambda, \mu; \chi; z) = \frac{1}{2\pi i} \frac{\Gamma(1-\mu)\Gamma(\mu-\lambda)}{\Gamma(1-\lambda)} \gamma^{\lambda-\mu}, \tag{5}$$

where $\operatorname{Re} \lambda < \operatorname{Re} \mu < 1$.

Here, too, for replacing $y$ by $-y$ one must change both the index signs:

$$K_{++}(\lambda, \mu; \chi; z) = K_{--}(\lambda, \mu; \chi; z^{-1}), \tag{6}$$

and so on.

For other triangular matrices the operator kernels are easily obtained from the kernel of $R_\chi(z)$. For example, let $\zeta = \begin{Vmatrix} 1 & \beta \\ 0 & 1 \end{Vmatrix}$. Then one has the equality

$$\zeta = sz^{-1}s(-e),$$

where $s = \begin{Vmatrix} 0 & 1 \\ -1 & 0 \end{Vmatrix}$ and $z^{-1} = \begin{Vmatrix} 1 & 0 \\ -\beta & 1 \end{Vmatrix}$. Therefore

$$R_\chi(\zeta) = R_\chi(s) R_\chi(z^{-1}) R_\chi(s) R_\chi(-e). \tag{7}$$

Using the expressions found above for $R_\chi(s)$ and $R_\chi(-e)$, and also for the kernel of $R_\chi(z)$, we find that for $\beta > 0$

$$K_{++}(\lambda, \mu; \chi; \zeta) = \frac{1}{2\pi i} \frac{\Gamma(\lambda-\mu)\Gamma(-\lambda-2l)}{\Gamma(-\mu-2l)} \beta^{\mu-\lambda}, \tag{8}$$

where $\operatorname{Re} \mu < \operatorname{Re} \lambda < -2\operatorname{Re} l$;

$$K_{+-}(\lambda, \mu; \chi; \zeta) = 0; \tag{9}$$

$$K_{-+}(\lambda, \mu; \chi; \zeta) = \frac{(-1)^{2\varepsilon}}{2\pi i} \frac{\Gamma(\mu+2l+1)\Gamma(-\lambda-2l)}{\Gamma(\mu-\lambda+1)} \beta^{\mu-\lambda}, \tag{10}$$

where $\operatorname{Re}\lambda < -2\operatorname{Re} l$, $-1-2\operatorname{Re} l < \operatorname{Re}\mu$; and

$$K_{--}(\lambda, \mu; \chi; \zeta) = \frac{1}{2\pi i} \frac{\Gamma(\lambda-\mu)\Gamma(\mu+2l+1)}{\Gamma(\lambda+2l+1)} \beta^{\mu-\lambda}, \tag{11}$$

where $-1-2\operatorname{Re} l < \operatorname{Re}\mu < \operatorname{Re}\lambda$.

Again, for passing from the matrix $\zeta$ to the matrix $\zeta^{-1} = \begin{pmatrix} 1 & -\beta \\ 0 & 1 \end{pmatrix}$ one must change the index signs:

$$K_{++}(\lambda, \mu; \chi; \zeta^{-1}) = K_{--}(\lambda, \mu; \chi; \zeta), \tag{12}$$

and so on.

For matrices of the form $\begin{pmatrix} \alpha & 1 \\ -1 & 0 \end{pmatrix}$ and $\begin{pmatrix} 0 & 1 \\ -1 & \delta \end{pmatrix}$ the kernels are calculated in exactly the same way. We leave the details to the reader.

3. **The general case.** The results obtained above suffice for calculating the kernel of $R_\chi(g)$ for any element $g$ of $SL(2, R)$. If all the elements of the matrix $g = \begin{pmatrix} \alpha & \beta \\ \gamma & \delta \end{pmatrix}$ are nonzero, then, as shown in §2.2, this matrix can be written in the form

$$g = d_1 (-e)^{\varepsilon_1} s^{\varepsilon_2} p\, d_2, \tag{1}$$

where $s = \begin{pmatrix} 0 & 1 \\ -1 & 0 \end{pmatrix}$, $-e = \begin{pmatrix} -1 & 0 \\ 0 & -1 \end{pmatrix}$, $d_1$ and $d_2$ are diagonal matrices, and $p$ is a matrix of one of the following two forms:

$$p = \begin{pmatrix} \operatorname{ch}\theta & \operatorname{sh}\theta \\ \operatorname{sh}\theta & \operatorname{ch}\theta \end{pmatrix}, \qquad -\infty < \theta < \infty, \tag{2}$$

$$p = \begin{pmatrix} \cos\theta & -\sin\theta \\ \sin\theta & \cos\theta \end{pmatrix}, \quad -\frac{\pi}{4} < \theta < \frac{\pi}{4}. \tag{3}$$

Since for all these matrices the operator kernels have already been calculated, we easily find the kernel of $R_\chi(g)$.

In the case when one of the elements of $g$ is equal to zero, one must use the factorization (6) of §2.2. Lastly, in the case when two elements of $g$ are equal to zero, one must use the results of §3.2.

4. **Some integral transforms connected with the hypergeometric function.** The

operators $T_\chi(g)$ and $T_\chi(g^{-1})$ are reciprocal: if $\mathbf{F} = T_\chi(g)\mathbf{f}$, then $\mathbf{f} = T_\chi(g^{-1})\mathbf{F}$. Therefore, from

$$\mathbf{F}(\lambda) = \int_{a-i\infty}^{a+i\infty} \mathbf{K}(\lambda, \mu; \chi; g)\,\mathbf{f}(\mu)\,d\mu, \tag{1}$$

where $\mathbf{f} = (f_+, f_-)$ and $\mathbf{F} = (F_+, F_-)$, it follows that

$$\mathbf{f}(\lambda) = \int_{a-i\infty}^{a+i\infty} \mathbf{K}(\lambda, \mu; \chi; g^{-1})\,\mathbf{F}(\mu)\,d\mu, \tag{2}$$

where $a$ satisfies the restrictions mentioned in §3.3. Choosing various elements $g$ of $SL(2, R)$, we obtain pairs of mutually reciprocal integral transforms connected with the hypergeometric function.

If we take $g = \begin{Vmatrix} \operatorname{ch}\theta & \operatorname{sh}\theta \\ \operatorname{sh}\theta & \operatorname{ch}\theta \end{Vmatrix}$, $\theta > 0$, then we get the following pair of integral transforms, connecting $\mathbf{F}$ and $\mathbf{f}$. Let

$$F_+(\lambda) = \frac{1}{2\pi i}\frac{\Gamma(\lambda)\Gamma(-\lambda-2l)}{\Gamma(-2l)}(\operatorname{sh}\theta)^{-\lambda}(\operatorname{ch}\theta)^{\lambda+2l} \times \int_{a-i\infty}^{a+i\infty} (\operatorname{cth}\theta)^{\mu} F\left(\lambda, \mu; -2l; -\frac{1}{\operatorname{sh}^2\theta}\right) f_+(\mu)\,d\mu, \tag{3}$$

$$\begin{aligned} F_-(\lambda) = {}& \frac{1}{2\pi i}\Gamma(\lambda)(\operatorname{sh}\theta)^{\lambda}(\operatorname{ch}\theta)^{\lambda+2l} \\ &\times \int_{a-i\infty}^{a+i\infty} \frac{\Gamma(1-\mu)}{\Gamma(\lambda-\mu+1)}(\operatorname{cth}\theta)^{\mu} F(\lambda, \lambda+2l+1; \lambda-\mu+1; -\operatorname{sh}^2\theta)\, f_+(\mu)\,d\mu \\ &+ \frac{(-1)^{\pm}\Gamma(-\lambda-2l)}{2\pi i}(\operatorname{sh}\theta)^{-\lambda}(\operatorname{ch}\theta)^{-\lambda-2l}\int_{a-i\infty}^{a+i\infty}\frac{\Gamma(\mu+2l+1)}{\Gamma(\mu-\lambda+1)} \\ &\times (\operatorname{th}\theta)^{\mu} F(1-\lambda, -\lambda-2l; \mu-\lambda+1; -\operatorname{sh}^2\theta)\, f_+(\mu)\,d\mu \\ &+ \frac{1}{2\pi i\,\Gamma(2l+2)}(\operatorname{sh}\theta)^{-\lambda-4l-2}(\operatorname{ch}\theta)^{\lambda+2l}\int_{a-i\infty}^{a+i\infty}\Gamma(1-\mu)\Gamma(\mu+2l+1)(\operatorname{cth}\theta)^{\mu} \\ &\times F\left(\lambda+2l+1, \mu+2l+1; 2l+2; -\frac{1}{\operatorname{sh}^2\theta}\right) f_-(\mu)\,d\mu, \end{aligned} \tag{4}$$

where $-1 - 2\operatorname{Re} l < a < 1$. Then

$$\begin{aligned} f_+(\lambda) = {}& \frac{1}{2\pi i\,\Gamma(2l+2)}(\operatorname{sh}\theta)^{-\lambda-4l-2}(\operatorname{ch}\theta)^{\lambda+2l}\int_{a-i\infty}^{a+i\infty}\Gamma(1-\mu)\Gamma(\mu+2l+1) \\ &\times (\operatorname{cth}\theta)^{\mu} F\left(\lambda+2l+1, \mu+2l+1; 2l+2; -\frac{1}{\operatorname{sh}^2\theta}\right) F_+(\mu)\,d\mu + \end{aligned}$$

$$
+\frac{1}{2\pi i}\Gamma(\lambda)(\operatorname{sh}\theta)^{\lambda}(\operatorname{ch}\theta)^{\lambda+2l}\int_{a-i\infty}^{a+i\infty}\frac{\Gamma(1-\mu)}{\Gamma(\lambda-\mu+1)}(\operatorname{cth}\theta)^{\mu}
$$

$$
\times F(\lambda,\lambda+2l+1;\lambda-\mu+1;-\operatorname{sh}^2\theta)F_{+}(\mu)\,d\mu
$$

$$
+\frac{(-1)^{2\varepsilon}\Gamma(-\lambda-2l)}{2\pi i}(\operatorname{sh}\theta)^{\lambda}(\operatorname{ch}\theta)^{\lambda-2l}\int_{a-i\infty}^{a+i\infty}\frac{\Gamma(\mu+2l+1)}{\Gamma(\mu-\lambda+1)}(\operatorname{th}\theta)^{\mu}
$$

$$
\times F(1-\lambda,-\lambda-2l;\mu-\lambda+1;-\operatorname{sh}^2\theta)F_{-}(\mu)\,d\mu \tag{5}
$$

and

$$
f_{-}(\lambda)=\frac{\Gamma(\lambda)\Gamma(-\lambda-2l)}{2\pi i\,\Gamma(-2l)}(\operatorname{sh}\theta)^{\lambda}(\operatorname{ch}\theta)^{\lambda+2l}
$$

$$
\times\int_{a-i\infty}^{a+i\infty}(\operatorname{cth}\theta)^{\mu}F\left(\lambda,\mu;-2l;-\frac{1}{\operatorname{sh}^2\theta}\right)F_{-}(\mu)\,d\mu, \tag{6}
$$

where $-1-2\operatorname{Re} l<a<1$.

Similarly, the case $g=\begin{pmatrix}\cos\theta & -\sin\theta\\ \sin\theta & \cos\theta\end{pmatrix}$, $0<\theta<\pi/4$, leads to the following transforms. Let

$$
F_{+}(\lambda)=\frac{\Gamma(\lambda)}{2\pi i}(\sin\theta)^{-\lambda}(\cos\theta)^{\lambda+2l}\int_{a-i\infty}^{a+i\infty}\frac{\Gamma(\mu+2l+1)}{\Gamma(\lambda+\mu+2l+1)}(\operatorname{ctg}\theta)^{\mu}
$$

$$
\times F(\lambda,\mu;\lambda+\mu+2l+1;-\operatorname{ctg}^2\theta)f_{+}(\mu)\,d\mu
$$

$$
+\frac{(-1)^{2\varepsilon}\Gamma(-\lambda-2l)}{2\pi i}(\sin\theta)^{-\lambda}(\cos\theta)^{\lambda+2l}\int_{a-i\infty}^{a+i\infty}\frac{\Gamma(\mu+2l+1)}{\Gamma(\mu-\lambda+1)}
$$

$$
\times(\operatorname{tg}\theta)^{\mu}F(-\lambda-2l,\mu;\mu-\lambda+1;-\operatorname{tg}^2\theta)f_{-}(\mu)\,d\mu \tag{7}
$$

and

$$
F_{-}(\lambda)=\frac{\Gamma(\lambda)}{2\pi i}(\sin\theta)^{\lambda}(\cos\theta)^{-\lambda+2l}\int_{a-i\infty}^{a+i\infty}\frac{\Gamma(1-\mu)}{\Gamma(\lambda-\mu+1)}(\operatorname{ctg}\theta)^{\mu}
$$

$$
\times F(\lambda,-\mu-2l;\lambda-\mu+1;-\operatorname{tg}^2\theta)f_{+}(\mu)\,d\mu
$$

$$
+\frac{\Gamma(-\lambda-2l)}{2\pi i}(\sin\theta)^{\lambda-2}(\cos\theta)^{-\lambda-2l}\int_{a-i\infty}^{a+i\infty}\frac{\Gamma(1-\mu)}{\Gamma(1-\lambda-\mu-2l)}(\operatorname{tg}\theta)^{\mu}
$$

$$
\times F(1-\lambda,1-\mu;1-\lambda-\mu-2l;-\operatorname{ctg}^2\theta)f_{-}(\mu)\,d\mu, \tag{8}
$$

where $-1-2\mathrm{Re}\, l < a < 1$. Then

$$f_+(\lambda)=\frac{\Gamma(-\lambda-2l)}{2\pi i}(\sin\theta)^{\lambda-2}(\cos\theta)^{-\lambda-2l}\int_{a-i\infty}^{a+i\infty}\frac{\Gamma(1-\mu)}{\Gamma(1-\lambda-\mu-2l)}(\mathrm{tg}\,\theta)^{\mu}$$
$$\times F(1-\lambda,\ 1-\mu;\ 1-\lambda-\mu-2l;\ -\mathrm{ctg}^2\theta)\,F_+(\mu)\,d\mu$$
$$+\frac{\Gamma(\lambda)}{2\pi i}(\sin\theta)^{\lambda}(\cos\theta)^{-\lambda+2l}\int_{a-i\infty}^{a+i\infty}\frac{\Gamma(1-\mu)}{\Gamma(\lambda-\mu+1)}(\mathrm{ctg}\,\theta)^{\mu}$$
$$\times F(\lambda,\ -\mu-2l;\ \lambda-\mu+1;\ -\mathrm{tg}^2\theta)\,F_-(\mu)\,d\mu \tag{9}$$

and

$$f_-(\lambda)=\frac{(-1)^{2\varepsilon}}{2\pi i}\Gamma(-\lambda-2l)(\sin\theta)^{-\lambda}(\cos\theta)^{\lambda+2l}\int_{a-i\infty}^{a+i\infty}\frac{\Gamma(\mu+2l+1)}{\Gamma(\mu-\lambda+1)}(\mathrm{tg}\,\theta)^{\mu}$$
$$\times F(-\lambda-2l,\ \mu,\ \mu-\lambda+1;\ -\mathrm{tg}^2\theta)\,F_+(\mu)\,d\mu$$
$$+\frac{\Gamma(\lambda)}{2\pi i}(\sin\theta)^{-\lambda}(\cos\theta)^{\lambda+2l}\int_{a-i\infty}^{a+i\infty}\frac{\Gamma(\mu+2l+1)}{\Gamma(\lambda+\mu+2l+1)}(\mathrm{ctg}\,\theta)^{\mu}$$
$$\times F(\lambda,\ \mu;\ \lambda+\mu+2l+1;\ -\mathrm{ctg}^2\theta)\,F_-(\mu)\,d\mu, \tag{10}$$

where $-1-2\mathrm{Re}\, l < a < 1$.

Finally, let us take $g=\begin{pmatrix}1 & 0\\ t & 1\end{pmatrix}$, $t>0$. Then we obtain the following pair of transforms. Let

$$F_+(\lambda)=\frac{\Gamma(\lambda)t^{\lambda}}{2\pi i}\int_{a-i\infty}^{a+i\infty}\frac{\Gamma(\mu-\lambda)}{\Gamma(\mu)}t^{-\mu}f_+(\mu)\,d\mu, \tag{11}$$

$$F_-(\lambda)=\frac{\Gamma(\lambda)t^{\lambda}}{2\pi i}\int_{a-i\infty}^{a+i\infty}\frac{\Gamma(1-\mu)}{\Gamma(\lambda-\mu+1)}t^{-\mu}f_+(\mu)\,d\mu$$
$$+\frac{t^{\lambda}}{2\pi i\,\Gamma(1-\lambda)}\int_{a-i\infty}^{a+i\infty}\Gamma(1-\mu)\Gamma(\mu-\lambda)t^{-\mu}f_-(\mu)\,d\mu, \tag{12}$$

where $0<\mathrm{Re}\,\lambda<a<1$. Then

$$f_+(\lambda)=\frac{t^{\lambda}}{2\pi i\,\Gamma(1-\lambda)}\int_{a-i\infty}^{a+i\infty}\Gamma(1-\mu)\Gamma(\mu-\lambda)t^{-\mu}F_+(\mu)\,d\mu$$
$$+\frac{\Gamma(\lambda)t^{\lambda}}{2\pi i}\int_{a-i\infty}^{a+i\infty}\frac{\Gamma(1-\mu)}{\Gamma(\lambda-\mu+1)}t^{-\mu}F_-(\mu)\,d\mu \tag{13}$$

and

$$f_-(\lambda)=\frac{\Gamma(\lambda)\,t^{\lambda}}{2\pi i}\int\limits_{a-i\infty}^{a+i\infty}\frac{\Gamma(\mu-\lambda)}{\Gamma(\mu)}t^{-\mu}F_-(\mu)\,d\mu, \tag{14}$$

where $0 < \operatorname{Re}\lambda < a < 1$.

## §5. RECURRENCE FORMULAS FOR THE HYPERGEOMETRIC FUNCTION. THE HYPERGEOMETRIC EQUATION

**1. Relations between infinitesimal operators and operators of the representation.** In order to derive recurrence formulas for the hypergeometric function, we first establish relations between operators of the representation $R_\chi(g)$ and infinitesimal operators of this representation.

We consider the matrix $g = h(\phi)\,\zeta(t)$, where

$$h(\varphi)=\begin{pmatrix}\operatorname{ch}\varphi & \operatorname{sh}\varphi\\ \operatorname{sh}\varphi & \operatorname{ch}\varphi\end{pmatrix},\qquad \zeta(t)=\begin{pmatrix}1 & t\\ 0 & 1\end{pmatrix}.$$

This matrix has the form

$$g=\begin{pmatrix}\operatorname{ch}\varphi & t\operatorname{ch}\varphi+\operatorname{sh}\varphi\\ \operatorname{sh}\varphi & t\operatorname{sh}\varphi+\operatorname{ch}\varphi\end{pmatrix}. \tag{1}$$

For $t > 0$, $\phi > 0$ all the elements of $g$ are positive, and so, by virtue of §2.2, it can be represented in the form

$$g\equiv h(\varphi)\,\zeta(t)=d(t_1)\,h(\theta)\,d(t_2). \tag{2}$$

Here $d(t_1)$ and $d(t_2)$ are diagonal matrices:

$$d(t)=\begin{pmatrix}e^{-t} & 0\\ 0 & e^{t}\end{pmatrix},\ \text{and}\ h(\theta)=\begin{pmatrix}\operatorname{ch}\theta & \operatorname{sh}\theta\\ \operatorname{sh}\theta & \operatorname{ch}\theta\end{pmatrix}.$$

The parameters $t_1$, $\theta$, $t_2$ are connected with the parameters $\phi$ and $t$ by the formulas

$$\operatorname{ch}2\theta=\operatorname{ch}2\varphi+t\operatorname{sh}2\varphi, \tag{3}$$

$$e^{-t_1-t_2}=\frac{\operatorname{ch}\varphi}{\operatorname{ch}\theta}, \tag{3'}$$

$$e^{t_1-t_2}=\frac{\operatorname{sh}\varphi}{\operatorname{sh}\theta}, \tag{3''}$$

directly derivable from equalities (1) and (2).

From (2) it follows that

$$R_\chi(h(\varphi))R_\chi(\zeta(t)) = R_\chi(d(t_1))R_\chi(h(\theta))R_\chi(d(t_2)). \tag{4}$$

We differentiate both sides of this with respect to $t$ and put $t = 0$; since $t_1$, $\theta$ and $t_2$ depend on $t$, we get

$$\begin{aligned} R_\chi(h(\varphi))\frac{dR_\chi(\zeta(t))}{dt}\bigg|_{t=0} = \bigg(&\frac{dR_\chi(d(t_1))}{dt_1}\frac{dt_1}{dt}R_\chi(h(\theta))R_\chi(d(t_2)) \\ &+ R_\chi(d(t_1))\frac{dR_\chi(h(\theta))}{d\theta}\frac{d\theta}{dt}R_\chi(d(t_2)) \\ &+ R_\chi(d(t_1))R_\chi(h(\theta))\frac{dR_\chi(d(t_2))}{dt}\frac{dt_2}{dt}\bigg)_{t=0}. \end{aligned} \tag{5}$$

But from formulas (3)–(3″) it follows that for $t = 0$ we have $t_1 = t_2 = 0$, $\theta = \phi$, and hence $R_\chi(d(t_1))$ and $R_\chi(d(t_2))$ are unit operators. Moreover, differentiating (3)–(3″) with respect to $t$ and putting $t = 0$, we get

$$\frac{d\theta}{dt}\bigg|_{t=0} = \frac{1}{2}, \quad \frac{dt_1}{dt}\bigg|_{t=0} = -\frac{1}{2\,\mathrm{sh}\,2\varphi}, \quad \frac{dt_2}{dt}\bigg|_{t=0} = \frac{\mathrm{ch}\,2\varphi}{2\,\mathrm{sh}\,2\varphi}. \tag{6}$$

Lastly, by the definition of infinitesimal operators (see §3.4) we have

$$\frac{dR_\chi(\zeta(t))}{dt}\bigg|_{t=0} = B_+, \quad \frac{dR_\chi(d(t))}{dt}\bigg|_{t=0} = B_3. \tag{7}$$

Substituting these expressions into formula (5), we find that

$$R_\chi(h)B_+ = -\frac{B_3}{2\,\mathrm{sh}\,2\varphi}R_\chi(h) + \frac{1}{2}\frac{dR_\chi(h)}{d\varphi} + \frac{1}{2}R_\chi(h)B_3\,\mathrm{cth}\,2\varphi, \tag{8}$$

where $h = h(\phi) = \begin{Vmatrix} \mathrm{ch}\,\phi & \mathrm{sh}\,\phi \\ \mathrm{sh}\,\phi & \mathrm{ch}\,\phi \end{Vmatrix}$.

Similarly one establishes the equality

$$R_\chi(h)B_- = \frac{B_3}{2\,\mathrm{sh}\,2\varphi}R_\chi(h) + \frac{1}{2}\frac{dR_\chi(h)}{d\varphi} - \frac{1}{2}R_\chi(h)B_3\,\mathrm{cth}\,2\varphi. \tag{9}$$

Subtracting (8) from (9), we get

$$R_\chi(h)(B_- - B_+) = \frac{B_3}{\mathrm{sh}\,2\varphi}R_\chi(h) - R_\chi B_3\,\mathrm{cth}\,2\varphi$$

or

$$R_\chi(h)B_2 = \frac{B_3}{\mathrm{sh}\,2\varphi}R_\chi(h) - R_\chi(h)B_3\,\mathrm{cth}\,2\varphi. \tag{10}$$

In the same way, considering the matrices

$$\begin{pmatrix}\cos\varphi & -\sin\varphi\\ \sin\varphi & \cos\varphi\end{pmatrix}\begin{pmatrix}1 & t\\ 0 & 1\end{pmatrix},\quad \begin{pmatrix}\cos\varphi & -\sin\varphi\\ \sin\varphi & \cos\varphi\end{pmatrix}\begin{pmatrix}1 & 0\\ t & 1\end{pmatrix},$$

we obtain

$$R_\chi(u)B_+ = \frac{B_3}{2\sin 2\varphi}R_\chi(u) - \frac{1}{2}\frac{dR_\chi(u)}{d\varphi} - \frac{1}{2}R_\chi(u)B_+\operatorname{ctg}2\varphi \tag{11}$$

and

$$R_\chi(u)B_- = \frac{B_3}{2\sin 2\varphi}R_\chi(u) + \frac{1}{2}\frac{dR_\chi(u)}{d\varphi} - \frac{1}{2}R_\chi(u)B_3\operatorname{ctg}2\varphi, \tag{12}$$

where $u(\phi) = \begin{pmatrix}\cos\phi & -\sin\phi\\ \sin\phi & \cos\phi\end{pmatrix}$. Hence

$$R_\chi(u)B_1 = \frac{B_3}{\sin 2\varphi}R_\chi(u) - R_\chi(u)B_3\operatorname{ctg}2\varphi. \tag{13}$$

2. **Recurrence formulas.** From the formulas of the preceding subsection we now derive recurrence relations for the hypergeometric function, i.e. relations connecting $F(\alpha, \beta; \gamma; x)$ with $F(\alpha \pm 1, \beta \pm 1; \gamma \pm 1; x)$. For this we replace the operators $R_\chi(h)$, $R_\chi(u)$ and $B_+$, $B_-$, $B_3$, appearing in these formulas, by their explicit expressions, and compare the kernels of the resulting operators on the left and on the right.

Let us find the kernel of the operator $R_\chi(h)B_+$. As shown in §3.4, the operator $B_+$ transforms the pair $\mathbf{F}(\mu) = (F_+(\mu), F_-(\mu))$ into the pair

$$B_+\mathbf{F}(\mu) = ((2l+\mu+1)F_+(\mu+1), -(2l+\mu+1)F_-(\mu+1)). \tag{1}$$

The operator $R_\chi(h)$ is an integral operator with the kernel

$$\mathbf{K}(\lambda, \mu; \chi; h) = \begin{pmatrix}K_{++}(\lambda, \mu) & 0\\ K_{-+}(\lambda, \mu) & K_{--}(\lambda, \mu)\end{pmatrix}$$

(for brevity we have omitted the arguments $\chi$ and $h$ on the right). Hence

$$R_\chi(h)B_+\mathbf{F} = (\hat{F}_+(\lambda), \hat{F}_-(\lambda)), \tag{2}$$

where

$$\hat{F}_+(\lambda) = \int_{a-i\infty}^{a+i\infty} K_{++}(\lambda, \mu)(2l+\mu+1)F_+(\mu+1)\,d\mu, \tag{3}$$

$$\hat{F}(\lambda) = -\int_{a-i\infty}^{a+i\infty} K_{+}(\lambda, \mu)(2l+\mu+1)F_{+}(\mu+1)\,d\mu$$
$$-\int_{a-i\infty}^{a+i\infty} K_{--}(\lambda, \mu)(2l+\mu+1)F_{-}(\mu+1)\,d\mu. \quad (4)$$

Replacing $\mu$ by $\mu - 1$, we see that the kernel of the operator $R_{\chi}(h)B_{+}$ has the form

$$\begin{pmatrix} (2l+\mu)K_{++}(\lambda, \mu-1) & 0 \\ -(2l+\mu)K_{-+}(\lambda, \mu-1) & -(2l+\mu)K_{--}(\lambda, \mu-1) \end{pmatrix}. \quad (5)$$

In order to find the kernel of the operator on the right hand side of equality (8) of subsection 1, it is sufficient to note that the operator $B_3$ leads to the multiplication of both the components of the pair $(F_{+}(\lambda), F_{-}(\lambda))$ by $2(\lambda + l)$. Therefore the kernel mentioned has the form

$$\frac{(\mu+l)\operatorname{ch} 2\varphi - (\lambda+l)}{\operatorname{sh} 2\varphi}\mathbf{K}(\lambda, \mu; \chi; h(\varphi)) + \frac{1}{2}\frac{d\mathbf{K}(\lambda, \mu; \chi; h(\varphi))}{d\varphi}, \quad (6)$$

where

$$\mathbf{K} = \begin{pmatrix} K_{++} & 0 \\ K_{-+} & K_{--} \end{pmatrix}.$$

From formulas (5) and (6) it follows that

$$(2l+\mu)K_{++}(\lambda, \mu-1)$$
$$= \frac{(\mu+l)\operatorname{ch} 2\varphi - (\lambda+l)}{\operatorname{sh} 2\varphi}K_{++}(\lambda, \mu) + \frac{1}{2}\frac{dK_{++}(\lambda, \mu)}{d\varphi}. \quad (7)$$

In this equality we substitute the expression for $K_{++}(\lambda, \mu; \chi; h)$, given by formula (4) of §4.1. On replacing $-1/\sin^2\phi$ by $z$, $\lambda$ by $\alpha$, $\mu$ by $\beta$ and $-2l$ by $\gamma$, we easily obtain the following relation for the hypergeometric function:

$$(\beta-\gamma)F(\alpha, \beta-1; \gamma; z) + (\gamma-\beta-\alpha z)F(\alpha, \beta; \gamma; z)$$
$$+ z(1-z)\frac{dF(\alpha, \beta; \gamma; z)}{dz} = 0. \quad (8)$$

Using formula (2) of §1.2, equality (8) can be rewritten as :

$$(\beta-\gamma)F(\alpha, \beta-1; \gamma; z) + (\gamma-\beta-\alpha z)F(\alpha, \beta; \gamma; z)$$
$$+ \frac{\alpha\beta}{\gamma}z(1-z)F(\alpha+1, \beta+1; \gamma+1; z) = 0. \quad (9)$$

Because of the symmetry of the hypergeometric function in the parameters $\alpha$ and $\beta$, this gives

$$(\alpha-\gamma)F(\alpha-1,\ \beta;\ \gamma;\ z)+(\gamma-\alpha-\beta z)F(\alpha,\ \beta;\ \gamma;\ z)$$
$$+\frac{\alpha\beta}{\gamma}z(1-z)F(\alpha+1;\ \beta+1;\ \gamma+1;\ z)=0. \qquad (10)$$

Another recurrence formula is obtained from equality (10) of subsection 1. One similarly establishes that the kernel of the operator $R_\chi(h)B_2$ has the form

$$-\begin{pmatrix}(2l+\mu)K_{++}(\lambda,\ \mu-1) & 0\\ -(2l+\mu)K_{-+}(\lambda,\ \mu-1) & -(2l+\mu)K_{--}(\lambda,\ \mu-1)\end{pmatrix}$$
$$-\begin{pmatrix}\mu K_{++}(\lambda,\ \mu+1) & 0\\ \mu K_{-+}(\lambda,\ \mu+1) & \mu K_{--}(\lambda,\ \mu+1)\end{pmatrix}. \qquad (11)$$

Substituting this expression into formula (10) of subsection 1 and then expressing $K_{++}(\lambda,\ \mu-1)$, $K_{++}(\lambda,\ \mu)$, $K_{++}(\lambda,\ \mu+1)$ according to formula (4′) of §4.1, we obtain the recurrence formula

$$(\alpha z-\beta z+2\beta-\gamma)F(\alpha,\ \beta;\ \gamma;\ z)+(\gamma-\beta)F(\alpha,\ \beta-1;\ \gamma;\ z)$$
$$+\beta(z-1)F(\alpha,\ \beta+1;\ \gamma;\ z)=0. \qquad (12)$$

Interchanging $\alpha$ and $\beta$, we have

$$(\beta z-\alpha z+2\alpha-\gamma)F(\alpha,\ \beta;\ \gamma;\ z)+(\gamma-\alpha)F(\alpha-1,\ \beta;\ \gamma;\ z)$$
$$+\alpha(z-1)F(\alpha+1,\ \beta;\ \gamma;\ z)=0. \qquad (13)$$

Analogous formulas are obtained by comparing the elements $K_{-+}$ in the kernels. After simple manipulations we obtain the following recurrence relations for the hypergeometric function:

$$(\alpha+\beta-\gamma)F(\alpha,\ \beta;\ \gamma;\ z)+\frac{(\gamma-\beta)(\gamma-\alpha)}{\gamma}F(\alpha,\ \beta;\ \gamma+1;\ z)$$
$$+\frac{\alpha\beta}{\gamma}(z-1)F(\alpha+1,\ \beta+1;\ \gamma+1;\ z)=0 \qquad (14)$$

and

$$\left[1-\frac{2\gamma-\alpha-\beta-1}{\gamma-1}z\right]F(\alpha,\ \beta;\ \gamma;\ z)+\frac{(\gamma-\alpha)(\gamma-\beta)}{\gamma(\gamma-1)}F(\alpha,\ \beta;\ \gamma+1;\ z)$$
$$+(z-1)F(\alpha,\ \beta;\ \gamma-1;\ z)=0. \qquad (15)$$

In the same way, one derives recurrence relations from formula (9) of subsection 1. Comparing the elements $K_{++}$ on the left and on the right, we get

$$F(\alpha,\ \beta;\ \gamma;\ z)-F(\alpha,\ \beta+1;\ \gamma;\ z)$$
$$+\frac{\alpha z}{\gamma}F(\alpha+1,\ \beta+1;\ \gamma+1;\ z)=0 \qquad (16)$$

and, by the symmetry in $\alpha$ and $\beta$,

$$F(\alpha, \beta; \gamma; z) - F(\alpha+1, \beta; \gamma; z) + \frac{\beta z}{\gamma} F(\alpha+1, \beta+1; \gamma+1; z) = 0. \tag{17}$$

Further, comparing the elements $K_{-+}$, we get

$$F(\alpha, \beta; \gamma; z) - F(\alpha, \beta; \gamma-1; z) + \frac{\alpha\beta z}{\gamma(\gamma-1)} F(\alpha+1, \beta+1; \gamma+1; z) = 0. \tag{18}$$

Using formula (12) of subsection 1 and comparing the elements $K_{++}$, we obtain

$$\gamma F(\alpha, \beta; \gamma; z) - (\gamma-\alpha) F(\alpha, \beta+1; \gamma+1; z) - \alpha(1-z) F(\alpha+1, \beta+1; \gamma+1; z) = 0. \tag{19}$$

Because of the symmetry in $\alpha$ and $\beta$, this gives

$$\gamma F(\alpha, \beta; \gamma; z) - (\gamma-\beta) F(\alpha+1, \beta; \gamma+1; z) - \beta(1-z) F(\alpha+1, \beta+1; \gamma+1; z) = 0. \tag{20}$$

Similarly, from formula (11) of subsection 1, comparing $K_{++}$ on both sides of the equality, we get

$$(\gamma-1) F(\alpha, \beta-1; \gamma-1; z) + (\alpha z-\gamma+1) F(\alpha, \beta; \gamma; z) - \frac{\alpha\beta}{\gamma} z(1-z) F(\alpha+1, \beta+1; \gamma+1; z) = 0. \tag{21}$$

By the symmetry in $\alpha$ and $\beta$, this gives

$$(\gamma-1) F(\alpha-1, \beta; \gamma-1; z) + (\beta z-\gamma+1) F(\alpha, \beta; \gamma; z) - \frac{\alpha\beta}{\gamma} z(1-z) F(\alpha+1, \beta+1; \gamma+1; z) = 0. \tag{22}$$

Finally, from formula (13) of subsection 1, comparing the values of $K_{++}$, we obtain the recurrence formula

$$\gamma(\gamma-1) F(\alpha, \beta-1; \gamma-1; z) + \gamma[(\alpha-\beta) z-\gamma+1] F(\alpha, \beta; \gamma; z) + \beta(\gamma-\alpha) z F(\alpha, \beta+1; \gamma+1; z) = 0, \tag{23}$$

and, by the symmetry in $\alpha$ and $\beta$,

$$\gamma(\gamma-1) F(\alpha-1, \beta; \gamma-1; z) + \gamma[(\beta-\alpha) z-\gamma+1] F(\alpha, \beta; \gamma; z) + \alpha(\gamma-\beta) z F(\alpha+1, \beta; \gamma+1; z) = 0. \tag{24}$$

These 15 recurrence formulas are basic. From them one can deduce a number of new formulas. For example, subtracting (9) from (10), we get

$$(\alpha-\gamma)F(\alpha-1, \beta; \gamma; z)+(\gamma-\beta)F(\alpha, \beta-1; \gamma; z) \\ +(\beta-\alpha)(1-z)F(\alpha, \beta; \gamma; z)=0. \tag{25}$$

Further, let us multiply (9) by $\alpha-\gamma$, (10) by $\gamma-\beta$ and (14) by $(\beta-\alpha)z$. Combining the relations obtained, we find

$$F(\alpha, \beta-1; \gamma; z)-F(\alpha-1, \beta; \gamma; z) \\ +\frac{\alpha-\beta}{\gamma}zF(\alpha, \beta; \gamma+1; z)=0. \tag{26}$$

Let us eliminate $F(\alpha+1, \beta+1; \gamma+1; z)$ from (19) and (20). We get

$$\gamma(\alpha-\beta)F(\alpha, \beta; \gamma; z)-\alpha(\gamma-\beta)F(\alpha+1, \beta; \gamma+1; z) \\ +\beta(\gamma-\alpha)F(\alpha, \beta+1; \gamma+1; z)=0. \tag{27}$$

However, if we eliminate $F(\alpha, \beta; \gamma; z)$ and replace $\alpha, \beta, \gamma$ by $\alpha-1, \beta-1, \gamma-1$, we get

$$(\alpha-\beta)(1-z)F(\alpha, \beta; \gamma; z)+(\beta-\gamma)F(\alpha, \beta-1; \gamma; z) \\ +(\gamma-\alpha)F(\alpha-1, \beta; \gamma; z)=0. \tag{28}$$

Similarly, from (16) and (17), we find that

$$(\beta-\alpha)F(\alpha, \beta; \gamma; z)+\alpha F(\alpha+1, \beta; \gamma; z) \\ -\beta F(\alpha, \beta+1; \gamma; z)=0. \tag{29}$$

From (21) and (22) it follows that

$$\gamma(\alpha-\beta)F(\alpha, \beta; \gamma; z)-\alpha(\gamma-\beta)F(\alpha+1, \beta; \gamma+1; z) \\ +\beta(\gamma-\alpha)F(\alpha, \beta+1; \gamma+1; z)=0 \tag{30}$$

and

$$(\alpha-\beta)(1-z)F(\alpha, \beta; \gamma; z)+(\beta-\gamma)F(\alpha, \beta-1; \gamma; z) \\ +(\gamma-\alpha)F(\alpha-1, \beta; \gamma; z)=0. \tag{31}$$

From (9) and (21) we obtain

$$\gamma F(\alpha, \beta; \gamma; z)-\beta F(\alpha, \beta+1; \gamma+1; z) \\ +(\beta-\gamma)F(\alpha, \beta; \gamma+1; z)=0, \tag{32}$$

and, by the symmetry in $\alpha$ and $\beta$,

$$\gamma F(\alpha, \beta; \gamma; z)-\alpha F(\alpha+1, \beta; \gamma+1; z) \\ +(\alpha-\gamma)F(\alpha, \beta; \gamma+1; z)=0. \tag{33}$$

From (23) and (21) we obtain

$$(\gamma-1)\gamma[F(\alpha, \beta-1; \gamma-1; z)-F(\alpha, \beta; \gamma; z)] + \alpha(\gamma-\beta) zF(\alpha+1, \beta; \gamma+1; z)=0, \tag{34}$$

and, by the symmetry in $\alpha$ and $\beta$,

$$(\gamma-1)\gamma[F(\alpha-1, \beta; \gamma-1; z)-F(\alpha, \beta; \gamma; z)] + \beta(\gamma-\alpha) zF(\alpha, \beta+1; \gamma+1; z)=0. \tag{35}$$

We also mention the equalities

$$\gamma[\alpha-(\gamma-\beta)z]F(\alpha, \beta; \gamma; z)-\gamma(1-z)F(\alpha, \beta+1; \gamma; z) + (\gamma-\alpha)(\gamma-\beta)z F(\alpha, \beta; \gamma+1; z)=0 \tag{36}$$

and

$$\gamma[\beta-(\gamma-\alpha)z]F(\alpha, \beta; \gamma; z)-\gamma(1-z)F(\alpha+1, \beta; \gamma; z) + (\gamma-\alpha)(\gamma-\beta) zF(\alpha, \beta; \gamma+1; z)=0, \tag{37}$$

ensuing from relations (14)–(16).

3. The hypergeometric equation. We now derive the second order differential equation satisfied by the hypergeometric function. For this, we note that by formula (8) of subsection 2 the operator

$$z(1-z)\frac{d}{dz}+(\gamma-\beta-\alpha z)$$

transforms $F(\alpha, \beta; \gamma; z)$ into $(\gamma-\beta)F(\alpha, \beta-1; \gamma; z)$. Using formulas (16) of subsection 2 and (2) of §2.1, one can similarly prove that the operator $zd/dz+\beta$ transforms $F(\alpha, \beta; \gamma; z)$ into $\beta F(\alpha, \beta+1; \gamma; z)$. Hence

$$\left(z\frac{d}{dz}+\beta-1\right)\left[z(1-z)\frac{d}{dz}+(\gamma-\beta-\alpha z)\right]F(\alpha, \beta; \gamma; z) = (\beta-1)(\gamma-\beta)F(\alpha, \beta; \gamma; z). \tag{1}$$

On simplification this becomes

$$\left[z(1-z)\frac{d^2}{dz^2}+(\gamma-(\alpha+\beta+1)z)\frac{d}{dz}-\alpha\beta\right]F(\alpha, \beta; \gamma; z)=0. \tag{2}$$

Thus the hypergeometric function is a particular solution of the equation

$$z(1-z)\frac{d^2y}{dz^2}+[\gamma-(\alpha+\beta+1)z]\frac{dy}{dz}-\alpha\beta y=0. \tag{3}$$

This equation is called the hypergeometric differential equation.

In order to find the second particular solution, we make the substitution $y=z^{1-\gamma}w$ in (3). We obtain the equation

$$z(1-z)\frac{d^2w}{dz^2}+[2-\gamma-(\alpha+\beta-2\gamma+1)z]\frac{dw}{dz}$$
$$+(\alpha-\gamma+1)(\beta-\gamma+1)w=0. \tag{4}$$

One of the solutions of (4) is

$$w=F(\alpha-\gamma+1,\ \beta-\gamma+1;\ 2-\gamma;\ z).$$

Therefore the second particular solution of (3) has the form

$$y=z^{1-\gamma}w=z^{1-\gamma}F(\alpha-\gamma+1,\ \beta-\gamma+1;\ 2-\gamma;\ z). \tag{5}$$

If $\gamma$ is not an integer, this solution is linearly independent of $F(\alpha, \beta; \gamma; z)$. The case when $\gamma$ is an integer is dealt with by using the limiting process.

## §6. INTEGRAL REPRESENTATIONS AND THE COMPOSITION FORMULA FOR THE HYPERGEOMETRIC FUNCTION

1. **Introductory remarks.** The results of this section are based on the following simple remark. Since $R_\chi(g)$ is a representation of the group $SL(2, R)$, one has the equality

$$R_\chi(g_1g_2)=R_\chi(g_1)R_\chi(g_2). \tag{1}$$

We write this equality by means of the kernels of the operators $R_\chi(g_1)$, $R_\chi(g_2)$, $R_\chi(g_1g_2)$. We find that for any pair $\mathbf{F}=(F_+, F_-)$ from the space $\mathfrak{D}_\chi$ one has the relation

$$\int_{b-i\infty}^{b+i\infty}\mathbf{K}(\lambda,\ \mu;\ \chi;\ g_1g_2)\mathbf{F}(\mu)\,d\mu$$
$$=\int_{a-i\infty}^{a+i\infty}\mathbf{K}(\lambda,\ \nu;\ \chi;\ g_1)\int_{b-i\infty}^{b+i\infty}\mathbf{K}(\nu,\ \mu;\ \chi;\ g_2)\mathbf{F}(\mu)\,d\mu\,d\nu. \tag{2}$$

Hence, if one can justify changing the order of integration, the kernel $\mathbf{K}(\lambda, \mu; \chi; g)$ must satisfy the relation

$$\mathbf{K}(\lambda,\ \mu;\ \chi;\ g_1g_2)=\int_{a-i\infty}^{a+i\infty}\mathbf{K}(\lambda,\ \nu;\ \chi;\ g_1)\mathbf{K}(\nu;\ \mu;\ \chi;\ g_2)\,d\nu, \tag{3}$$

which is the continual analogue of the composition theorem. The value of $a$ must be such that the integrals considered are absolutely convergent for $\nu = a + it$.

Choosing matrices $g_1$ and $g_2$ in a special way, we obtain various relations for the hypergeometric function. To begin with, we shall obtain integral representations for this function.

2. **Integral representations.** In order to obtain integral representations for

the hypergeometric function, we make use of the identity

$$h \equiv \begin{pmatrix} \operatorname{ch}\theta & \operatorname{sh}\theta \\ \operatorname{sh}\theta & \operatorname{ch}\theta \end{pmatrix} = \begin{pmatrix} 1 & \operatorname{th}\theta \\ 0 & 1 \end{pmatrix} \begin{pmatrix} \frac{1}{\operatorname{ch}\theta} & 0 \\ 0 & \operatorname{ch}\theta \end{pmatrix} \begin{pmatrix} 1 & 0 \\ \operatorname{th}\theta & 1 \end{pmatrix}. \tag{1}$$

Since, for $\theta > 0$, to the diagonal matrix $\begin{Bmatrix} 1/\operatorname{ch}\theta & 0 \\ 0 & \operatorname{ch}\theta \end{Bmatrix}$ corresponds the operator of multiplication by $(\operatorname{ch}\theta)^{2(l+\lambda)}$ (see §3.3), it follows from (1) that

$$\mathbf{K}(\lambda, \mu; \chi; h) = \int_{a-i\infty}^{a+i\infty} \mathbf{K}(\lambda, \nu; \chi; \zeta)(\operatorname{ch}\theta)^{2(\nu+l)} \mathbf{K}(\nu, \mu; \chi; z)\, d\nu, \tag{2}$$

where $\zeta = \begin{Bmatrix} 1 & \operatorname{th}\theta \\ 0 & 1 \end{Bmatrix}$, $z = \begin{Bmatrix} 1 & 0 \\ \operatorname{th}\theta & 1 \end{Bmatrix}$. The values of the parameters $\lambda$, $\mu$, $l$ and $a$ in (2) must be such that the corresponding integrals are absolutely convergent.

By §4.2 we have

$$K_{+-}(\lambda, \nu; \chi; \zeta) = K_{+-}(\nu, \mu; \chi; z) = 0. \tag{3}$$

Therefore, comparing the matrix elements on the left and on the right in formula (2), we find that

$$K_{++}(\lambda, \mu; \chi; h) = \int_{a-i\infty}^{a+i\infty} K_{++}(\lambda, \nu, \chi; \zeta) K_{++}(\nu, \mu; \chi; z) \operatorname{ch}^{2(\nu+l)}\theta\, d\nu. \tag{4}$$

But in §§4.1 and 4.2 we calculated $K_{++}(\lambda, \mu; \chi; h)$, $K_{++}(\lambda, \nu; \chi; \zeta)$ and $K_{++}(\nu, \mu; \chi; z)$. Substituting the values of these functions found there and cancelling the common factor, we get

$$\frac{\Gamma(\lambda)}{\Gamma(-2l)} \frac{\operatorname{sh}^{4l+\lambda+\mu}\theta}{\operatorname{ch}^{2l+\lambda+\mu}\theta} F\left(-2l-\lambda, -2l-\mu; -2l; -\frac{1}{\operatorname{sh}^2\theta}\right)$$
$$= \frac{1}{2\pi i} \int_{a-i\infty}^{a+i\infty} \frac{\Gamma(\lambda-\nu)\Gamma(\nu)\Gamma(\mu-\nu)}{\Gamma(-\nu-2l)\Gamma(\mu)} (\operatorname{th}\theta)^{2\nu-\lambda-\mu} \operatorname{ch}^{2(\nu+l)}\theta\, d\nu. \tag{5}$$

Applying formula (5) of §1.2 to the left-hand side and replacing $1/\operatorname{sh}^2\theta$ by $x$, $2l$ by $-\omega$, we easily obtain the equality

$$F(\lambda, \mu; \omega; -x) = \frac{\Gamma(\omega)}{2\pi i\, \Gamma(\nu)\Gamma(\mu)} \int_{a-i\infty}^{a+i\infty} \frac{\Gamma(\lambda-\nu)\Gamma(\mu-\nu)\Gamma(\nu)}{\Gamma(\omega-\nu)} x^{-\nu}\, d\nu. \tag{6}$$

We now specify the restrictions to be imposed upon the parameters $\lambda$, $\mu$, $l$ and $a$ for the absolute convergence of the corresponding integrals. As noted in §3.4, the integral defining $K_{++}(\lambda, \mu; \chi; h)$ converges for $0 < \operatorname{Re}\lambda < \operatorname{Re}\omega$ (we have replaced $2l$ by $-\omega$). The integrals defining $K_{++}(\lambda, \nu; \chi; \zeta)$ and $K_{++}(\nu, \mu; \chi; z)$ converge for $\operatorname{Re}\nu < \operatorname{Re}\lambda < \operatorname{Re}\omega$ and $0 < \operatorname{Re}\nu < \operatorname{Re}\mu$ respectively. Since $\operatorname{Re}\nu = a$, we obtain the following conditions for the validity of equality (6):

$$0 < a < \operatorname{Re}\lambda < \operatorname{Re}\omega \text{ and } a < \operatorname{Re}\mu.$$

It is easy to prove that under these conditions the integral (2) of subsection 1 is absolutely convergent.

Actually, the restriction $\operatorname{Re}\lambda < \operatorname{Re}\omega$ is unnecessary. The conditions $0 < a < \operatorname{Re}\lambda$, $a < \operatorname{Re}\mu$ signify that the path of integration separates the poles of the function $\Gamma(\nu)$ from the poles of $\Gamma(\lambda - \nu)$ and $\Gamma(\mu - \nu)$. This remark allows us to extend the domain of validity of (6). Namely, the integrand in this equality analytically depends on $\nu$ in the entire complex plane, excluding the poles of $\Gamma(\nu)$, $\Gamma(\lambda - \nu)$, $\Gamma(\mu - \nu)$. Therefore, the path of integration can be deformed in any way so long as its initial and terminal points are unchanged and it does not cross the poles mentioned. After this one can change the values of $\lambda$, $\mu$ and $\omega$. Consequently, we see that formula (6) is valid for all values of $\lambda$, $\mu$, $\omega$, $a$, provided only that the path of integration separates the poles of $\Gamma(\nu)$ from the poles of $\Gamma(\lambda - \nu)$ and $\Gamma(\mu - \nu)$.

One can similarly extend the domain of applicability of other formulas of this chapter. We shall not carry this out explicitly in every individual case.

Formulas expressing $K_{-+}(\lambda, \mu; \chi; h)$ and $K_{--}(\lambda, \mu; \chi; h)$ do not give new integral representations for the hypergeometric function, but again lead to formula (6). The equalities for $\mathrm{K}(\lambda, \mu; \chi; h^{-1})$ also lead to this same formula. In order to obtain other integral representations for the hypergeometric function, we consider the factorization

$$u \equiv \begin{pmatrix} \cos\theta & -\sin\theta \\ \sin\theta & \cos\theta \end{pmatrix} = \begin{pmatrix} 1 & -\operatorname{tg}\theta \\ 0 & 1 \end{pmatrix} \begin{pmatrix} \frac{1}{\cos\theta} & 0 \\ 0 & \cos\theta \end{pmatrix} \begin{pmatrix} 1 & 0 \\ \operatorname{tg}\theta & 1 \end{pmatrix}. \tag{7}$$

This gives

$$\mathrm{K}(\lambda, \mu; \chi; u) = \int_{a-i\infty}^{a+i\infty} \mathrm{K}(\lambda, \nu; \chi; \zeta) \cos^{2(\nu+l)}\theta \, \mathrm{K}(\nu, \mu; \chi; z)\, d\nu, \tag{8}$$

where $\zeta = \begin{pmatrix} 1 & -\operatorname{tg}\theta \\ 0 & 1 \end{pmatrix}$, $z = \begin{pmatrix} 1 & 0 \\ \operatorname{tg}\theta & 1 \end{pmatrix}$. By §4.2, for $\theta > 0$ we have

$$K_{-+}(\lambda, \nu; \chi; \zeta) = K_{+-}(\nu, \mu; \chi; z) = 0.$$

Therefore it follows from (8) that

$$K_{--}(\lambda, \mu; \chi; u) = \int_{a-i\infty}^{a+i\infty} K_{--}(\lambda, \nu; \chi; \zeta) \cos^{2(\nu+l)}\theta K_{--}(\nu, \mu; \chi; z)\, d\nu. \quad (9)$$

In this formula we substitute the values of $K_{--}(\lambda, \mu; \chi; u)$, $K_{--}(\lambda, \nu; \chi; \zeta)$ and $K_{--}(\nu, \mu; \chi; z)$ given by formulas (16) of §4.1 and (11) and (5) of §4.2. We obtain

$$\frac{\operatorname{tg}^{2(\lambda+\mu+2l)}\theta}{\Gamma(1-\lambda-\mu-2l)} F(-\lambda-2l, \ -\mu-2l; \ 1-\lambda-\mu-2l; \ -\operatorname{ctg}^2\theta)$$
$$= \frac{1}{2\pi i} \int_{a-i\infty}^{a+i\infty} \frac{\Gamma(\lambda-\nu)\Gamma(\mu-\nu)}{\Gamma(-2l-\nu)\Gamma(1-\nu)} \sin^{2\nu}\theta\, d\nu,$$

where $a < \operatorname{Re}\lambda < -2\operatorname{Re} l$, $a < \operatorname{Re}\mu < 1$.

To simplify this formula, we apply equality (5) of §1.2 to the left-hand side and put $\lambda = 1 - \lambda'$, $\mu = \mu' - \omega' + 1$, $2l = \lambda' - \mu' - 1$, $\nu = \mu' + \nu' - \omega' + 1$ $\sin^2\theta = 1 - x$. After simple manipulations we get

$$x^{\omega-1} F(\lambda, \mu; \omega; x) = \frac{\Gamma(\omega)}{2\pi i} \int_{a-i\infty}^{a+i\infty} \frac{\Gamma(\omega-\lambda-\mu-\nu)\Gamma(-\nu)}{\Gamma(\omega-\lambda-\nu)\Gamma(\omega-\mu-\nu)} (1-x)^{\nu}\, d\nu, \quad (10)$$

where $a < 0 < \operatorname{Re}(\omega-\mu)$, $a < \operatorname{Re}(\omega-\lambda-\mu) < \operatorname{Re}(\omega-\lambda)$ and $0 < x < 1$.

The integral in the right hand side of this equality vanishes for $x < 0$. This is most easily seen by calculating it with the help of residues.

Further, calculating $K_{++}(\lambda, \mu; \chi; u)$ and replacing $\lambda + \mu + 2l + 1$ by $\omega'$ and $\sin^2\theta/2$ by $x$, we get

$$\frac{\Gamma(\lambda)\Gamma(\mu)\Gamma(\omega-\lambda)\Gamma(\omega-\mu)}{\Gamma(\omega)} F\left(\lambda, \mu; \omega; \frac{x-1}{x}\right)$$
$$= \frac{1}{2\pi i} \int_{a-i\infty}^{a+i\infty} \Gamma(\nu)\Gamma(\lambda-\nu)\Gamma(\mu-\nu)\Gamma(\omega-\lambda-\mu+\nu) x^{\nu}\, d\nu$$
$$+ \frac{(-1)^{2\varepsilon}}{2\pi i} \Gamma(\mu)\Gamma(1-\mu)\Gamma(\omega-\mu)\Gamma(1+\mu-\omega) \times$$

$$\times \int_{a-i\infty}^{a+i\infty} \frac{\Gamma(\nu)\Gamma(\nu+\omega-\lambda-\mu)}{\Gamma(\nu-\lambda+1)\Gamma(\nu-\mu+1)} x^{\nu}\, d\nu, \tag{11}$$

where

$$0<a<\operatorname{Re}\mu, \quad \operatorname{Re}(\lambda+\mu-\omega)<a<\operatorname{Re}\lambda \text{ and } 0<x<1. \tag{12}$$

Adding and subtracting formulas corresponding to the values $\epsilon = 0$ and $\epsilon = 1/2$, we find that, with inequalities (12) satisfied, we have

$$\frac{1}{2\pi i}\int_{a-i\infty}^{a+i\infty} \Gamma(\nu)\Gamma(\lambda-\nu)\Gamma(\mu-\nu)\Gamma(\omega-\lambda-\mu+\nu)\, x^{\nu}\, d\nu$$
$$=\frac{\Gamma(\lambda)\Gamma(\mu)\Gamma(\omega-\lambda)\Gamma(\omega-\mu)}{\Gamma(\omega)} F\left(\lambda,\ \mu;\ \omega;\ \frac{x-1}{x}\right) \tag{13}$$

and

$$\frac{1}{2\pi i}\int_{a-i\infty}^{a+i\infty} \frac{\Gamma(\nu)\Gamma(\nu+\omega-\lambda-\mu)}{\Gamma(\nu-\lambda+1)\Gamma(\nu-\mu+1)} x^{\nu}\, d\nu=0. \tag{13}$$

To calculate the integral (13) for $x > 1$, we replace $x$ by $1/x'$, $\mu$ by $\omega-\mu'$ and $\nu$ by $\lambda-\nu'$. We get

$$\frac{1}{2\pi i}\int_{a-i\infty}^{a+i\infty} \Gamma(\nu)\Gamma(\lambda-\nu)\Gamma(\mu-\nu)\Gamma(\omega-\lambda-\mu+\nu)\, x^{\nu}\, d\nu$$
$$=\frac{\Gamma(\lambda)\Gamma(\mu)\Gamma(\omega-\lambda)\Gamma(\omega-\mu)}{\Gamma(\omega)} F\left(\lambda,\ \mu;\ \omega;\ \frac{1-x}{x}\right), \tag{14}$$

where

$$0<a<\operatorname{Re}\mu, \quad \operatorname{Re}(\lambda+\mu-\omega)<a<\operatorname{Re}\lambda \text{ and } x>1.$$

Lastly, calculating $K_{+-}(\lambda,\ \mu;\ \chi;\ u)$, we arrive at the formula

$$\frac{1}{2\pi i}\int_{a-i\infty}^{a+i\infty} \frac{\Gamma(\lambda-\nu)\Gamma(\nu)}{\Gamma(\omega-\nu)\Gamma(1-\mu+\nu)} x^{-\nu}\, d\nu$$
$$=\frac{\Gamma(\lambda)}{\Gamma(\omega)\Gamma(1-\mu)} F(\lambda,\ \mu;\ \omega;\ x), \tag{15}$$

where

$$\operatorname{Re}\mu-1<0<a<\operatorname{Re}\lambda<\operatorname{Re}\omega \text{ and } 0<x<1. \tag{16}$$

From this formula we find, as in the case of formula (13), that for $x > 1$

$$\frac{1}{2\pi i}\int_{a-i\infty}^{a+i\infty}\frac{\Gamma(\lambda-\nu)\Gamma(\nu)}{\Gamma(\omega-\nu)\Gamma(1-\mu+\nu)}x^{-\nu}\,d\nu$$
$$=\frac{\Gamma(\lambda)(x-1)^{-\lambda}}{\Gamma(\lambda-\mu+1)\Gamma(\omega-\lambda)}F\left(\lambda,\ \omega-\mu;\ \lambda-\mu+1;\ \frac{1}{1-x}\right), \tag{17}$$

where $\operatorname{Re}\mu - 1 < 0 < a < \operatorname{Re}\lambda < \operatorname{Re}\omega$.

3. **Mellin transform.** A number of formulas, having the form of the Mellin transform for $F(\alpha, \beta; \gamma; x)$, considered as a function of the parameters $\alpha$, $\beta$ and $\gamma$, arise from a study of products of the form $g = h_1 z$, where

$$h_1=\begin{pmatrix}\operatorname{ch}\varphi & \operatorname{sh}\varphi\\ \operatorname{sh}\varphi & \operatorname{ch}\varphi\end{pmatrix},\quad z=\begin{pmatrix}1 & 0\\ z & 1\end{pmatrix},\quad -\infty<z<\infty, \tag{1}$$

so that

$$g=\begin{pmatrix}\operatorname{ch}\varphi+z\operatorname{sh}\varphi & \operatorname{sh}\varphi\\ \operatorname{sh}\varphi+z\operatorname{ch}\varphi & \operatorname{ch}\varphi\end{pmatrix}. \tag{2}$$

Let us first consider the case $\phi > 0$, $z > 0$. In this case all the elements of $g$ are positive. Therefore, using the results of §2.2, one can write $g$ in the form $g = d_1 h d_2$, where

$$d_1=\begin{pmatrix}e^{-t_1} & 0\\ 0 & e^{t_1}\end{pmatrix},\quad d_2=\begin{pmatrix}e^{-t_2} & 0\\ 0 & e^{t_2}\end{pmatrix},\quad h=\begin{pmatrix}\operatorname{ch}\theta & \operatorname{sh}\theta\\ \operatorname{sh}\theta & \operatorname{ch}\theta\end{pmatrix},\quad \theta>0.$$

One can easily see that parameters $t_1$, $\theta$, $t_2$ are connected with $z$ and $\phi$ by the relations

$$\left.\begin{aligned}\operatorname{ch}2\theta&=\operatorname{ch}2\varphi+z\operatorname{sh}2\varphi,\\ e^{4t_1}&=\frac{z\operatorname{ch}\varphi+\operatorname{sh}\varphi}{z\operatorname{sh}\varphi+\operatorname{ch}\varphi}\operatorname{cth}\varphi,\\ e^{4t_2}&=\frac{1}{1+z^2+2z\operatorname{cth}2\varphi}.\end{aligned}\right\} \tag{3}$$

Since to the diagonal matrix $d = \begin{pmatrix}e^{-t} & 0\\ 0 & e^{t}\end{pmatrix}$ corresponds the operator of multiplication by $e^{2t(\lambda+l)}$, one has the equality

$$e^{2t_1(\lambda+l)+2t_2(\mu+l)}\mathrm{K}(\lambda,\ \mu;\ \chi;\ h) = \int_{a-i\infty}^{a+i\infty}\mathrm{K}(\lambda,\ \nu;\ \chi;\ h_1)\,\mathrm{K}(\nu,\ \mu;\ \chi;\ z)\,d\nu. \tag{4}$$

But for $\phi > 0$, $z > 0$ we have

$$K_{+-}(\lambda, \mu; \chi; z) = K_{+-}(\lambda, \mu; \chi; h_1) = 0,$$

and therefore

$$e^{2t_1(\lambda+l)+2t_2(\mu+l)} K_{++}(\lambda, \mu; \chi; h) = \int_{a-i\infty}^{a+i\infty} K_{++}(\lambda, \nu; \chi; h_1) K_{++}(\nu, \mu; \chi; z)\, d\nu. \quad (5)$$

In this formula we substitute the expressions for $K_{++}$ given by equalities (4), (4′) of §4.1 and (1) of §4.2. We easily get

$$\frac{1}{2\pi i}\int_{a-i\infty}^{a+i\infty} z^{\nu}\Gamma(\nu)\Gamma(\mu-\nu)\operatorname{cth}^{\lambda+\nu}\varphi\, F\left(\lambda, \nu; 2l; -\frac{1}{\operatorname{sh}^2\varphi}\right)d\nu$$
$$= z^{\mu}\Gamma(\mu)\, e^{2t_1(\lambda-l)+2t_2(\mu-l)}\left(\frac{\operatorname{ch}\varphi}{\operatorname{ch}\theta}\right)^{2l}\operatorname{cth}^{\lambda+\mu}\theta\, F\left(\lambda, \mu; 2l; -\frac{1}{\operatorname{sh}^2\theta}\right), \quad (6)$$

where the variables $z$, $\phi$, $t_1$, $t_2$, $\theta$ are connected by (3) and

$$0 < \operatorname{Re}\lambda < \operatorname{Re} 2l, \qquad 0 < a < \operatorname{Re}\mu.$$

We put $z = 1/x$ in (6) and compare the formula obtained with the inversion formula for the Mellin transform (see §4.2 of Chapter II). We see that the formula obtained gives the inverse Mellin transform for the function

$$\Phi(\nu) = \Gamma(\nu)\Gamma(\mu-\nu)\operatorname{cth}^{\lambda+\nu}\varphi F\left(\lambda, \nu; 2l; -\frac{1}{\operatorname{sh}^2\varphi}\right).$$

Hence

$$\int_0^{\infty} z^{\mu-\nu-1} e^{2t_1(\lambda-l)+2t_2(\mu-l)}\left(\frac{\operatorname{ch}\varphi}{\operatorname{ch}\theta}\right)^{2l}\operatorname{cth}^{\lambda+\mu}\theta F\left(\lambda, \mu; 2l; -\frac{1}{\operatorname{sh}^2\theta}\right)dz$$
$$= \Gamma^{-1}(\mu)\Gamma(\nu)\Gamma(\mu-\nu)\operatorname{cth}^{\lambda+\nu}\varphi F\left(\lambda, \nu; 2l; -\frac{1}{\operatorname{sh}^2\varphi}\right), \quad (7)$$

where the variables $z$, $\phi$, $t_1$, $t_2$, $\theta$ are connected by (3).

One can obtain another formula from equality (6) as follows. Let $0 < z < \operatorname{th}\phi$. One can show that for $\operatorname{Re}\nu \to +\infty$ the integrand in (6) tends to zero. Therefore one can augment the contour of integration by an infinitely large semicircle, lying to the right of the contour. The poles lying within the resulting contour are the poles of the function $\Gamma(\mu-\nu)$, i.e. for the values $\nu = \mu + k$, $k = 0, 1, 2, \cdots$. Moreover,

$$\operatorname*{Res}_{\nu=\mu+k}\Gamma(\mu-\nu) = \frac{(-1)^k}{k!}.$$

Calculating (6) by the residue formula, we find that for $0 < z < \operatorname{th} \phi$ we have

$$\sum_{k=0}^{\infty} (-1)^k \frac{\Gamma(\mu+k)}{k!} (z \operatorname{cth} \varphi)^k F\left(\lambda, \mu+k; 2l; -\frac{1}{\operatorname{sh}^2 \varphi}\right)$$
$$= \Gamma(\mu) e^{2t_1(\lambda-l)+2t_2(\mu-l)} \left(\frac{\operatorname{ch} \varphi}{\operatorname{ch} \theta}\right)^{2l} \left(\frac{\operatorname{cth} \theta}{\operatorname{cth} \varphi}\right)^{\lambda+\mu} F\left(\lambda, \mu; 2l; -\frac{1}{\operatorname{sh}^2 \theta}\right), \quad (8)$$

where the variables $z$, $\phi$, $t_1$, $t_2$, $\theta$ are connected by relations (3).

In exactly the same way, one proves that for $z > \operatorname{th} \phi$

$$\sum_{k=0}^{\infty} \frac{(-1)^k \Gamma(\mu+k)}{k!\; z^{\mu+k}} \operatorname{cth}^{\lambda-k} \varphi F\left(\lambda, -k; 2l; -\frac{1}{\operatorname{sh}^2 \varphi}\right)$$
$$= \Gamma(\mu) e^{2t_1(\lambda-l)+2t_2(\mu-l)} \left(\frac{\operatorname{ch} \varphi}{\operatorname{ch} \theta}\right)^{2l} \operatorname{cth}^{\lambda+\mu} \theta F\left(\lambda, \mu; 2l; -\frac{1}{\operatorname{sh}^2 \theta}\right). \quad (9)$$

Formulas (6)–(9) have been obtained by considering the matrix element $K_{++}(\lambda, \mu; \chi; g)$. The consideration of other matrix elements of $\mathbf{K}(\lambda, \mu; \chi; g)$ does not lead to new formulas.

Let us now study the case when $\phi > 0$, $z = -z_1 < 0$. In this case we obtain three different answers, depending on the values of $z_1$ and $\phi$. Let $0 < z_1 < \operatorname{th} \phi$. Then all the elements of the matrix (2) are positive. Hence $g = d_1 h d_2$, where $d_1$, $d_2$ and $h$ have the same meaning as above. Here

$$\left.\begin{aligned} \operatorname{ch} 2\theta &= \operatorname{ch} 2\varphi - z_1 \operatorname{sh} 2\varphi, \qquad \theta > 0, \\ e^{4t_1} &= \frac{z_1 \operatorname{ch} \varphi - \operatorname{sh} \varphi}{z_1 \operatorname{sh} \varphi - \operatorname{ch} \varphi} \operatorname{cth} \varphi, \\ e^{4t_2} &= \frac{1}{1 + z_1^2 - 2z_1 \operatorname{cth} 2\varphi}. \end{aligned}\right\} \quad (10)$$

Moreover, in this case we have

$$K_{-+}(\lambda, \mu; \chi; z) = K_{+-}(\lambda, \mu; \chi; h_1) = 0.$$

Calculating $K_{++}(\lambda, \mu; \chi; g)$, we obtain

$$\frac{1}{2\pi i} \int_{a-i\infty}^{a+i\infty} \frac{\Gamma(\mu-\nu)}{\Gamma(1-\nu)} (z_1 \operatorname{cth} \varphi)^{\lambda+\nu} F\left(\lambda, \nu; -2l; -\frac{1}{\operatorname{sh}^2 \varphi}\right) d\nu$$
$$= \frac{1}{\Gamma(1-\mu)} e^{2t_1(\lambda+l)+2t_2(\mu+l)} \left(\frac{\operatorname{ch} \theta}{\operatorname{ch} \varphi}\right)^{2l} (z_1 \operatorname{cth} \theta)^{\lambda+\mu} F\left(\lambda, \mu; -2l; -\frac{1}{\operatorname{sh}^2 \theta}\right), \quad (11)$$

where $a < \operatorname{Re}\mu < 1$ and $0 < \operatorname{Re}\lambda < -2\operatorname{Re} l$. Similarly, calculating $K_{+-}(\lambda, \mu; \chi; g)$, we see that

$$\frac{1}{2\pi i}\int_{a-i\infty}^{a+i\infty}\frac{\Gamma(\nu)}{\Gamma(\nu-\mu+1)}(z_1 \operatorname{cth}\varphi)^{\lambda+\nu} F\left(\lambda,\ \nu;\ -2l;\ -\frac{1}{\operatorname{sh}^2\varphi}\right)d\nu = 0, \tag{12}$$

where $\operatorname{Re}\mu < 0 < \operatorname{Re}\lambda < -2\operatorname{Re} l$, $a > 0$.

Further, calculating $K_{-+}(\lambda, \mu; \chi; g)$, we obtain

$$\begin{aligned}\frac{1}{2\pi i}\int_{a-i\infty}^{a+i\infty}&\frac{\Gamma(\mu-\nu)}{\Gamma(1-\nu)}(z_1 \operatorname{ch}\varphi)^{\lambda+\nu}\Big[\frac{\Gamma(\lambda)\Gamma(1-\nu)}{\Gamma(\lambda-\nu+1)}\operatorname{sh}^{\lambda-\nu}\varphi\\ &\times F(\lambda,\ \lambda+2l+1;\ \lambda-\nu+1; -\operatorname{sh}^2\varphi)\\ &+(-1)^{2\epsilon}\frac{\Gamma(-\lambda-2l)\Gamma(\nu+2l+1)}{\Gamma(\nu-\lambda+1)}\operatorname{sh}^{\nu-\lambda}\varphi\\ &\times F(\nu,\ \nu+2l+1;\ \nu-\lambda+1;\ -\operatorname{sh}^2\varphi)\Big]\,d\nu\\ =e^{2l_1(\lambda+l)+2l_2(\mu+l)}&(z_1\operatorname{ch}\theta)^{\lambda+\mu}\left(\frac{\operatorname{ch}\theta}{\operatorname{ch}\varphi}\right)^{2l}\\ \times\Big[&\frac{\Gamma(\lambda)}{\Gamma(\lambda-\mu+1)}\operatorname{sh}^{\lambda-\mu}\theta F(\lambda,\ \lambda+2l+1;\ \lambda-\mu+1;\ -\operatorname{sh}^2\theta)\\ &+(-1)^{2\epsilon}\frac{\Gamma(-\lambda-2l)\Gamma(\mu+2l+1)}{\Gamma(\mu-\lambda+1)\Gamma(1-\mu)}\operatorname{sh}^{\mu-\lambda}\theta\\ &\times F(\mu,\ \mu+2l+1;\mu-\lambda+1;\ -\operatorname{sh}^2\theta)\Big],\end{aligned}\tag{13}$$

where $0 < \operatorname{Re}\lambda < -2\operatorname{Re} l < a+1 < 1+\operatorname{Re}\mu < 2$.

In this equality we put $\epsilon = 0$ and $\epsilon = 1/2$, and take half the sum and half the difference of the resulting formulas. We get

$$\begin{aligned}\frac{1}{2\pi i}\int_{a-i\infty}^{a+i\infty}&\frac{\Gamma(\mu-\nu)}{\Gamma(\lambda-\nu+1)}(z_1\operatorname{ch}\varphi)^{\lambda+\mu}\operatorname{sh}^{\lambda-\nu}\varphi\\ &\times F(\lambda,\ \lambda+2l+1;\ \lambda-\nu+1;\ -\operatorname{sh}^2\varphi)\,d\nu\\ =&\frac{1}{\Gamma(\lambda-\mu+1)}e^{2l_1(\lambda+l)+2l_2(\mu+l)}\left(\frac{\operatorname{ch}\theta}{\operatorname{ch}\varphi}\right)^{2l}(z_1\operatorname{ch}\theta)^{\lambda+\mu}\operatorname{sh}^{\lambda-\mu}\theta\\ &\times F(\lambda,\ \lambda+2l+1;\ \lambda-\mu+1;\ -\operatorname{sh}^2\theta)\end{aligned}\tag{14}$$

$$\begin{aligned}\frac{1}{2\pi i}\int_{a-i\infty}^{a+i\infty}&\frac{\Gamma(\mu-\nu)\Gamma(\nu+2l+1)}{\Gamma(1-\nu)\Gamma(\nu-\lambda+1)}(z_1\operatorname{ch}\varphi)^{\lambda+\mu}\operatorname{sh}^{\nu-\lambda}\varphi\\ &\times F(\nu,\ \nu+2l+1;\ \nu-\lambda+1;\ -\operatorname{sh}^2\varphi)\,d\nu\\ =&\frac{\Gamma(\mu+2l+1)}{\Gamma(\mu-\lambda+1)\Gamma(1-\mu)}e^{2l_1(\lambda+l)+2l_2(\mu+l)}\left(\frac{\operatorname{ch}\theta}{\operatorname{ch}\varphi}\right)^{2l}(z_1\operatorname{ch}\theta)^{\lambda+\mu}\operatorname{sh}^{\mu-\lambda}\theta\\ &\times F(\mu,\ \mu+2l+1;\ \mu-\lambda+1;\ -\operatorname{sh}^2\theta).\end{aligned}\tag{15}$$

In exactly the same way, calculating $K_{--}(\lambda, \mu; \chi; g)$, we obtain

$$\frac{1}{2\pi i}\int_{a-i\infty}^{a+i\infty}\Gamma(\nu)\Gamma(1-\nu)(z_1 \operatorname{ch}\varphi)^{\lambda+\nu}\Big[\frac{\Gamma(\lambda)}{\Gamma(\lambda-\nu+1)\Gamma(\nu-\mu+1)}\operatorname{sh}^{\lambda-\nu}\varphi$$
$$\times F(\lambda, \lambda+2l+1; \lambda-\nu+1; -\operatorname{sh}^2\varphi)$$
$$+\frac{\Gamma(\mu-\nu)\Gamma(\nu+2l+1)}{\Gamma(\mu)\Gamma(1-\mu)\Gamma(2l+2)}\operatorname{sh}^{-\lambda-\nu-4l-2}\varphi$$
$$\times F\left(\lambda+2l+1; \nu+2l+1; \quad 2l+2; \quad -\frac{1}{\operatorname{sh}^2\varphi}\right)\Big]d\nu$$
$$=\frac{\Gamma(\mu+2l+1)}{\Gamma(2l+2)}e^{2t_1(\lambda+l)+2t_2(\mu+l)}\left(\frac{\operatorname{ch}\theta}{\operatorname{ch}\varphi}\right)^{2l}(z_1\operatorname{ch}\theta)^{\lambda+\mu}\operatorname{sh}^{-\lambda-\mu-4l-2}\theta$$
$$\times F\left(\lambda+2l+1, \mu+2l+1; 2l+2; -\frac{1}{\operatorname{sh}^2\theta}\right), \tag{16}$$

where $0 < a < \operatorname{Re}\mu < 1$ and $0 < \operatorname{Re}\lambda < -2\operatorname{Re} l < 1 + a < 2$, and

$$\frac{1}{2\pi i}\int_{a-i\infty}^{a+i\infty}\frac{\Gamma(\nu)\Gamma(\nu+2l+1)}{\Gamma(\nu-\lambda+1)\Gamma(\nu-\mu+1)}(z_1\operatorname{ch}\varphi)^{\lambda+\nu}\operatorname{sh}^{\nu-\lambda}\varphi$$
$$\times F(\nu, \nu+2l+1; \nu-\lambda+1; -\operatorname{sh}^2\varphi)\,d\nu=0 \tag{17}$$

under the same restrictions on the parameters.

We leave the discussion of the cases $\phi > 0$, $z = -z_1 < 0$, $\operatorname{th}\phi < z_1 < \operatorname{cth}\phi$ and $\phi > 0$, $z = -z_1 < 0$, $z_1 > \operatorname{cth}\phi$ to the reader. In these cases the left hand sides of formulas (11)–(17) remain the same, but the right hand sides are changed, as the matrix $g$ has another canonical form. For example, for $\operatorname{th}\phi < z_1 < \operatorname{cth}\phi$ we have $g = d_1 u d_2$, where $d_1$ and $d_2$ are diagonal matrices,

$$d_1=\begin{pmatrix} e^{-t_1} & 0 \\ 0 & e^{t_1}\end{pmatrix},\quad d_2=\begin{pmatrix} e^{-t_2} & 0 \\ 0 & e^{t_2}\end{pmatrix} \text{ and } u=\begin{pmatrix}\cos\theta & -\sin\theta \\ \sin\theta & \cos\theta\end{pmatrix}.$$

Here

$$\left.\begin{aligned} \cos 2\theta &= \operatorname{ch}2\varphi - z_1\operatorname{sh}2\varphi, \qquad \theta<0,\\ e^{4t_1} &= \frac{z_1\operatorname{ch}\varphi-\operatorname{sh}\varphi}{\operatorname{ch}\varphi-z_1\operatorname{sh}\varphi}\operatorname{cth}\varphi,\\ e^{4t_2} &= \frac{1}{2z_1\operatorname{cth}2\varphi-1-z_1^2}.\end{aligned}\right\} \tag{18}$$

But if $z_1 > \operatorname{cth}\phi$, then $g = d_1 h(-s) d_2$, where $d_1$ and $d_2$ have the same meaning as above, $h = \begin{pmatrix}\operatorname{ch}\theta & \operatorname{sh}\theta \\ \operatorname{sh}\theta & \operatorname{ch}\theta\end{pmatrix}$ and $s = \begin{pmatrix} 0 & 1 \\ -1 & 0\end{pmatrix}$. Here

$$\left.\begin{aligned} \cos 2\theta &= z_1 \operatorname{sh} 2\varphi - \operatorname{ch} 2\varphi, \qquad \theta > 0, \\ e^{4t_1} &= \frac{z_1 \operatorname{ch}\varphi - \operatorname{sh}\varphi}{z_1 \operatorname{sh}\varphi - \operatorname{ch}\varphi} \operatorname{cth}\varphi, \\ e^{4t_2} &= \frac{1}{1 + z_1^2 - 2z_1 \operatorname{cth} 2\varphi}. \end{aligned}\right\} \tag{19}$$

Finally, we mention that analogous formulas arise from the consideration of matrices of the form $g = uz$, where $u = \begin{pmatrix} \cos\theta & -\sin\theta \\ \sin\theta & \cos\theta \end{pmatrix}$, $z = \begin{pmatrix} 1 & 0 \\ z & 1 \end{pmatrix}$. Here too one has to consider various cases, depending on the values of $\theta$ and $z$. Regarding formulas of this type, one may refer to [122], §6.

4. **Composition theorems.** We now apply equality (3) of subsection 1 to the matrices

$$g_1 = \begin{pmatrix} \operatorname{ch}\theta_1 & \operatorname{sh}\theta_1 \\ \operatorname{sh}\theta_1 & \operatorname{ch}\theta_1 \end{pmatrix} \quad \text{and} \quad g_2 = \begin{pmatrix} \operatorname{ch}\theta_2 & \operatorname{sh}\theta_2 \\ \operatorname{sh}\theta_2 & \operatorname{ch}\theta_2 \end{pmatrix}.$$

We have

$$g_1 g_2 = \begin{pmatrix} \operatorname{ch}(\theta_1 + \theta_2) & \operatorname{sh}(\theta_1 + \theta_2) \\ \operatorname{sh}(\theta_1 + \theta_2) & \operatorname{ch}(\theta_1 + \theta_2) \end{pmatrix}.$$

Therefore

$$\mathrm{K}(\lambda, \mu; \chi; \theta_1 + \theta_2) = \int_{a - i\infty}^{a + i\infty} \mathrm{K}(\lambda, \nu; \chi; \theta_1) \mathrm{K}(\nu, \mu; \chi; \theta_2)\, d\nu, \tag{1}$$

where the value of $a$ is suitably chosen.

In (1) we compare the matrix elements on the left and on the right. Let $\theta_1 > 0$, $\theta_2 > 0$. Then

$$K_{+-}(\lambda, \nu; \chi; \theta_1) = K_{+-}(\nu, \mu; \chi; \theta_2) = K_{+-}(\lambda, \mu; \chi; \theta_1 + \theta_2) = 0.$$

Hence

$$K_{++}(\lambda, \mu; \chi; \theta_1 + \theta_2) = \int_{a - i\infty}^{a + i\infty} K_{++}(\lambda, \nu; \chi; \theta_1) K_{++}(\nu, \mu; \chi; \theta_2)\, d\nu. \tag{2}$$

In this equality we substitute the expression for $K_{++}(\lambda, \mu; \chi; h)$, given by formula (4) of §4.1. We easily get

$$\frac{1}{2\pi i}\int_{a-i\infty}^{a+i\infty}\Gamma(\nu)\Gamma(\omega-\nu)(\operatorname{cth}\theta_1\operatorname{cth}\theta_2)^{\nu}$$

$$\times F\left(\lambda,\ \nu;\ \omega;\ -\frac{1}{\operatorname{sh}^2\theta_1}\right)F\left(\nu,\ \mu;\ \omega;\ -\frac{1}{\operatorname{sh}^2\theta_2}\right)d\nu$$

$$=\Gamma(\omega)\frac{\operatorname{th}^{\lambda}\theta_1\operatorname{th}^{\mu}\theta_2}{\operatorname{th}^{\lambda+\mu}(\theta_1+\theta_2)}\left(\frac{\operatorname{ch}\theta_1\operatorname{ch}\theta_2}{\operatorname{ch}(\theta_1+\theta_2)}\right)^{\omega}F\left(\lambda,\ \mu;\ \omega;\ -\frac{1}{\operatorname{sh}^2(\theta_1+\theta_2)}\right), \quad (3)$$

where $0 < \operatorname{Re}\lambda < \operatorname{Re}\omega$, $0 < a < \operatorname{Re}\omega$. The consideration of other elements of the matrix $K(\lambda, \mu; \chi; h)$ does not lead to new relations for the hypergeometric function.

Let us now consider the case when $\theta_1 > 0$ and $\theta_2 \equiv -\theta_3 < 0$, $0 < \theta_3 < \theta_1$. In this case $\theta_1 + \theta_2 > 0$ and so

$$K_{+-}(\lambda,\ \nu;\ \chi;\ \theta_1)=K_{-+}(\nu,\ \mu;\ \chi;\ -\theta_3)=K_{+-}(\lambda,\ \mu;\ \chi;\ \theta_1-\theta_3)=0.$$

Thus equality (2) holds in this case too. The substitution of the values of $K_{++}$ now leads to the relation

$$J_1\equiv\frac{1}{2\pi i}\int_{a-i\infty}^{a+i\infty}\left(\frac{\operatorname{th}\theta_3}{\operatorname{th}\theta_1}\right)^{\nu}F\left(\lambda,\ \nu;\ \omega;\ -\frac{1}{\operatorname{sh}^2\theta_1}\right)$$

$$\times F\left(1-\nu,\ 1-\mu;\ 2-\omega;\ -\frac{1}{\operatorname{sh}^2\theta_3}\right)d\nu$$

$$=\frac{\Gamma(2-\omega)}{\Gamma(1-\mu)\Gamma(\mu-\omega+1)}\frac{\operatorname{th}^{\lambda}\theta_1}{\operatorname{th}^{\mu}\theta_3\operatorname{th}^{\lambda+\mu}(\theta_1-\theta_3)}\left(\frac{\operatorname{ch}\theta_1}{\operatorname{ch}\theta_3\operatorname{ch}(\theta_1-\theta_3)}\right)^{\omega}\operatorname{sh}^2\theta_3$$

$$\times F\left(\lambda,\ \mu;\ \omega;\ -\frac{1}{\operatorname{sh}^2(\theta_1-\theta_3)}\right), \quad (4)$$

where $0 < \operatorname{Re}\lambda < \operatorname{Re}\omega < \operatorname{Re}\mu + 1 < 2$. In exactly the same way, calculating the value of $K_{+-}(\lambda, \mu;\ \chi;\ \theta_1 - \theta_3)$, we obtain the equalities

$$J_2'\equiv\frac{1}{2\pi i}\int_{a-i\infty}^{a+i\infty}\frac{\Gamma(\nu)}{\Gamma(\nu-\mu+1)}\left(\frac{\operatorname{th}\theta_3}{\operatorname{th}\theta_1}\right)^{\nu}F\left(\lambda,\ \nu;\ \omega;\ -\frac{1}{\operatorname{sh}^2\theta_1}\right)$$

$$\times F(1-\mu,\ \omega-\mu;\ \nu-\mu+1;\ -\operatorname{sh}^2\theta_3)\,d\nu=0, \quad (5)$$

$$J_2''\equiv\frac{1}{2\pi i}\int_{a-i\infty}^{a+i\infty}\frac{\Gamma(\omega-\nu)}{\Gamma(\mu-\nu+1)}(\operatorname{cth}\theta_1\operatorname{cth}\theta_3)^{\nu}$$

$$\times F\left(\lambda,\ \nu;\ \omega;\ -\frac{1}{\operatorname{sh}^2\theta_1}\right)F(\mu,\ \mu-\omega+1;\ \mu-\nu+1;\ -\operatorname{sh}^2\theta_3)\,d\nu=0, \quad (6)$$

in which $0 < \operatorname{Re}\lambda < \operatorname{Re}\omega < \operatorname{Re}\mu + 1 < 2$, $0 < a < \operatorname{Re}\omega$ (for calculating the integrals, we take half the sum and half the difference of formulas corresponding to the values $\epsilon = 0$ and $\epsilon = 1/2$).

Similarly, calculating $K_{-+}(\lambda, \mu; \chi; \theta_1 - \theta_3)$, we obtain

$$J_3' = \frac{1}{2\pi i}\int_{a-i\infty}^{a+i\infty} \frac{\Gamma(1-\nu)}{\Gamma(\lambda-\nu+1)} (\operatorname{cth}\theta_1 \operatorname{th}\theta_3)$$

$$\times F(\lambda, \omega-\nu; \lambda-\nu+1; \operatorname{th}^2\theta_1) F\left(1-\nu, 1-\mu; 2-\omega; -\frac{1}{\operatorname{sh}^2\theta_3}\right) d\nu$$

$$= \frac{\Gamma(2-\omega)}{\Gamma(\mu-\omega+1)\Gamma(\lambda-\mu+1)} \frac{\operatorname{cth}^\lambda\theta_1 \operatorname{cth}^\mu\theta_3}{\operatorname{cth}^{\mu-\lambda}(\theta_1-\theta_3)} \left(\frac{\operatorname{ch}\theta_1}{\operatorname{ch}\theta_3 \operatorname{ch}(\theta_1-\theta_3)}\right)^\omega \operatorname{sh}^2\theta_3$$

$$\times F(\lambda, \omega-\mu, \lambda-\mu+1; \operatorname{th}^2(\theta_1-\theta_3)), \tag{7}$$

$$J_3'' \equiv \frac{1}{2\pi i}\int_{a-i\infty}^{a+i\infty} \frac{\Gamma(\nu-\omega+1)}{\Gamma(\nu-\lambda+1)} (\operatorname{th}\theta_1 \operatorname{th}\theta_3)^\nu$$

$$\times F(\nu, \omega-\lambda; \nu-\lambda+1; \operatorname{th}^2\theta_1) F\left(1-\nu, 1-\mu; 2-\omega; -\frac{1}{\operatorname{sh}^2\theta_3}\right) d\nu$$

$$= \frac{\Gamma(2-\omega)}{\Gamma(1-\mu)\Gamma(\mu-\lambda+1)} \frac{\operatorname{th}^{\mu-\lambda}(\theta_1-\theta_3)}{\operatorname{th}^\lambda\theta_1 \operatorname{th}^\mu\theta_3} \left(\frac{\operatorname{ch}\theta_1}{\operatorname{ch}\theta_3 \operatorname{ch}(\theta_1-\theta_3)}\right)^\omega \operatorname{sh}^2\theta_3$$

$$\times F(\mu, \omega-\lambda, \mu-\lambda+1; \operatorname{th}^2(\theta_1-\theta_3)), \tag{8}$$

where $0 < \operatorname{Re}\lambda < \operatorname{Re}\omega < \operatorname{Re}\mu + 1 < 2$, $0 < a < \operatorname{Re}\omega < a + 1 < 2$.

Finally, calculating $K_{--}(\lambda, \mu; \chi; \theta_1 - \theta_3)$, we obtain

$$J_4' \equiv \frac{1}{2\pi i}\int_{a-i\infty}^{a+i\infty} (\operatorname{th}\theta_1 \operatorname{cth}\theta_3)^\nu \left[\frac{\Gamma(\lambda)\Gamma(1-\mu)\Gamma(\nu)\Gamma(1-\nu)}{\Gamma(\lambda-\nu+1)\Gamma(\nu-\mu+1)} (\operatorname{cth}\theta_1 \operatorname{th}\theta_3)^{2\nu}\right.$$

$$\times F(\lambda, \omega-\nu; \lambda-\nu+1; \operatorname{th}^2\theta_1) F(\nu, \omega-\mu; \nu-\mu+1; \operatorname{th}^2\theta_3)$$

$$+ \frac{\Gamma(\omega-\lambda)\Gamma(\mu-\omega-1)\Gamma(\omega-\nu)\Gamma(\nu-\omega+1)}{\Gamma(\nu-\lambda+1)\Gamma(\mu-\nu+1)} \operatorname{cth}^{2\lambda}\theta_1 \operatorname{th}^{2\mu}\theta_3$$

$$\times F(\nu, \omega-\lambda, \nu-\lambda+1; \operatorname{th}^2\theta_1) F(\mu, \omega-\nu; \mu-\nu+1; \operatorname{th}^2\theta_3)$$

$$+ \frac{\Gamma(1-\mu)\Gamma(\mu-\omega+1)\Gamma(\nu)\Gamma(\omega-\nu)}{\Gamma(2-\omega)\Gamma(\omega)} \operatorname{sh}^{-2}\theta_1$$

$$\left.\times F\left(-\lambda-1, -\nu-1; 2-\omega; -\frac{1}{\operatorname{sh}^2\theta_1}\right) F\left(\nu, \mu; \omega; -\frac{1}{\operatorname{sh}^2\theta_3}\right)\right] d\nu$$

$$= \frac{\Gamma(\lambda)\Gamma(1-\mu)}{\Gamma(\lambda-\mu+1)} \frac{\operatorname{th}^{\lambda-\mu}(\theta_1-\theta_3)}{\operatorname{th}^\lambda\theta_1 \operatorname{cth}^\mu\theta_3} \left(\frac{\operatorname{ch}\theta_3}{\operatorname{ch}\theta_1 \operatorname{ch}(\theta_1-\theta_3)}\right)^\omega$$

$$\times F(\lambda, \omega-\mu; \lambda-\mu+1; \operatorname{th}^2(\theta_1-\theta_3)), \tag{9}$$

and

$$J_4'' = \frac{1}{2\pi i}\int_{a-i\infty}^{a+i\infty} \left[\frac{\Gamma(\lambda)\Gamma(1-\nu)\Gamma(\omega-\nu)}{\Gamma(\omega-\lambda)\Gamma(\lambda-\nu+1)\Gamma(\mu-\nu+1)} \operatorname{th}^{\lambda-\nu}\theta_1 \operatorname{th}^{\mu-\nu}\theta_3\right.$$

$$\times F(\lambda, \omega-\nu; \lambda-\nu+1; \operatorname{th}^2\theta_1) F(\mu, \omega-\nu; \mu-\nu+1; \operatorname{th}^2\theta_3) +$$

$$+\frac{\Gamma(1-\mu)\Gamma(\nu-\omega+1)\Gamma(\nu)}{\Gamma(\mu-\omega+1)\Gamma(\nu-\lambda+1)\Gamma(\nu-\mu+1)}\operatorname{th}^{\nu-\lambda}\theta_1\operatorname{th}^{\nu-\mu}\theta_3$$
$$\left.\times F(\nu,\ \omega-\lambda;\ \nu-\lambda+1;\ \operatorname{th}^2\theta_1)F(\nu,\ \omega-\mu;\ \nu-\mu+1;\ \operatorname{th}^2\theta_3)\right]d\nu$$
$$=\frac{1}{\Gamma(\mu-\lambda+1)}\operatorname{th}^{\mu-\lambda}(\theta_1-\theta_3)\left(\frac{\operatorname{ch}\theta_1\operatorname{ch}\theta_3}{\operatorname{ch}(\theta_1-\theta_3)}\right)^{\omega}$$
$$\times F(\mu,\ \omega-\lambda;\ \mu-\lambda+1;\ \operatorname{th}^2(\theta_1-\theta_3)), \tag{10}$$

where $0 < \operatorname{Re}\lambda < \operatorname{Re}\omega < \operatorname{Re}\mu + 1 < 2$, $0 < a < \operatorname{Re}\omega < a + 1 < 2$.

The case $\theta_1 > 0$, $\theta_2 < 0$, $|\theta_2| > \theta_1$ does not lead to new formulas for the hypergeometric function.

New formulas of the same type are obtained from the equality

$$\begin{pmatrix}\cos\theta_1 & -\sin\theta_1\\ \sin\theta_1 & \cos\theta_1\end{pmatrix}\begin{pmatrix}\cos\theta_2 & -\sin\theta_2\\ \sin\theta_2 & \cos\theta_2\end{pmatrix}=\begin{pmatrix}\cos(\theta_1+\theta_2) & -\sin(\theta_1+\theta_2)\\ \sin(\theta_1+\theta_2) & \cos(\theta_1+\theta_2)\end{pmatrix} \tag{11}$$

This gives

$$\mathrm{K}(\lambda,\ \mu;\ \chi;\ u(\theta_1+\theta_2))=\int_{a-i\infty}^{a+i\infty}\mathrm{K}(\lambda,\ \nu;\ \chi;\ u(\theta_1))\,\mathrm{K}(\nu,\ \mu;\ \chi;\ u(\theta_2))\,d\nu. \tag{12}$$

Let us consider the case when $0 < \theta_1 < \pi/2$, $0 < \theta_2 < \pi/2$ and $0 < \theta_1 + \theta_2 < \pi/2$. In place of the matrices $\mathrm{K}(\lambda,\ \nu;\ \chi;\ u(\theta_1))$ etc., we substitute their expressions obtained in §4.1 and compare the corresponding matrix elements in the left and right hand sides of the equality. We obtain the relations

$$J_5'\equiv\frac{1}{2\pi i}\int_{a-i\infty}^{a+i\infty}\frac{\Gamma(\nu+\omega+1)\Gamma(\nu)}{\Gamma(\lambda+\nu+\omega+1)\Gamma(\mu+\nu+\omega+1)}(\operatorname{ctg}\theta_1\operatorname{ctg}\theta_2)^{\nu}$$
$$\times F(\lambda,\ \nu;\ \lambda+\nu+\omega+1;\ -\operatorname{ctg}^2\theta_1)F(\nu,\ \mu;\ \mu+\nu+\omega+1;\ -\operatorname{ctg}^2\theta_2)\,d\nu$$
$$=\frac{1}{\Gamma(\lambda+\mu+\omega+1)}\frac{\operatorname{tg}^{\lambda}\theta_1\operatorname{tg}^{\mu}\theta_2}{\operatorname{tg}^{\lambda+\mu}(\theta_1+\theta_2)}\left(\frac{\cos(\theta_1+\theta_2)}{\cos\theta_1\cos\theta_2}\right)^{\omega}$$
$$\times F(\lambda,\ \mu;\ \lambda+\mu+\omega+1;\ -\operatorname{ctg}^2(\theta_1+\theta_2)), \tag{13}$$

$$J_5''=\frac{1}{2\pi i}\int_{a-i\infty}^{a+i\infty}\frac{\Gamma(\nu)\Gamma(\nu+\omega+1)}{\Gamma(\nu-\lambda+1)\Gamma(\nu-\mu+1)}(\operatorname{tg}\theta_1\operatorname{tg}\theta_2)^{\nu}$$
$$\times F(-\lambda-\omega,\ \nu;\ \nu-\lambda+1;\ -\operatorname{tg}^2\theta_1)$$
$$\times F(-\mu-\omega,\ \nu;\ \nu-\mu+1;\ -\operatorname{tg}^2\theta_2)\,d\nu=0, \tag{14}$$

where $0 < \operatorname{Re}\lambda < -\operatorname{Re}\omega < \operatorname{Re}\mu + 1 < 2$, $a > 0$, $a > -1 - \operatorname{Re}\omega$, and

$$J_6 \equiv \frac{1}{2\pi i}\int_{a-i\infty}^{a+i\infty} \Gamma(\nu)\Gamma(1-\nu)$$

$$\times\Big[\frac{\Gamma(\mu+\omega+1)}{\Gamma(1-\mu)\Gamma(\lambda-\nu+1)\Gamma(\nu+\mu+\omega+1)}(\operatorname{ctg}\theta_1)^{\nu+\omega}(\operatorname{ctg}\theta_2)^{\nu}$$

$$\times F(-\omega-\nu,\ \lambda;\ \lambda-\nu+1;\ -\operatorname{tg}^2\theta_1)$$

$$\times F(\nu,\ \mu;\ \nu+\mu+\omega+1;\ -\operatorname{ctg}^2\theta_2)$$

$$+\frac{\Gamma(-\lambda-\omega)}{\Gamma(\lambda)\Gamma(-\lambda-\omega-\nu+1)\Gamma(\nu-\mu+1)}(\operatorname{tg}\theta_1)^{\nu+\omega}(\operatorname{tg}\theta_2)^{\nu}$$

$$\times F(-\lambda-\omega,\ -\nu-\omega;\ -\lambda-\omega-\nu+1;\ -\operatorname{ctg}^2\theta_1)$$

$$\times F(-\mu-\omega,\ \nu;\ \nu-\mu+1;\ -\operatorname{tg}^2\theta_2)\Big]d\nu$$

$$=\frac{1}{\Gamma(\lambda-\mu+1)}\left(\frac{\cos(\theta_1+\theta_2)}{\sin\theta_1\cos\theta_2}\right)^{\omega}\frac{(\operatorname{ctg}\theta_1)^{\lambda}(\operatorname{tg}\theta_2)^{\mu}}{(\operatorname{tg}(\theta_1+\theta_2))^{\mu-\lambda}}$$

$$\times F(\lambda,\ -\mu-\omega;\ \lambda-\mu+1;\ -\operatorname{tg}^2(\theta_1+\theta_2)), \tag{15}$$

where $0 < \operatorname{Re}\lambda < -\operatorname{Re}\omega < \operatorname{Re}\mu + 1 < 2$ and $0 < a < 1$.

The calculation of remaining elements of $\mathbf{K}(\lambda, \mu;\ \chi;\ u(\theta_1+\theta_2))$ does not give new formulas for the hypergeometric function.

We leave to the reader the discussion of the case $0 < \theta_1 < \pi/2$, $0 < \theta_2 < \pi/2$, $\theta_1 + \theta_2 > \pi/2$, and also the case when one of the angles $\theta_1$ or $\theta_2$ is less than zero.

## §7. REPRESENTATIONS OF THE GROUP OF REAL MATRICES OF THE SECOND ORDER AND HANKEL FUNCTIONS

In this chapter we have considered realizations of representations of the group $SL(2, R)$, under which, corresponding to diagonal matrices, we had operators of multiplication by a function. Here we shall consider another realization of representations of this group, under which operators of multiplication by a function correspond to triangular matrices $t = \begin{pmatrix} 1 & 0 \\ t & 1 \end{pmatrix}$. We shall see that, here, to the matrix $s = \begin{pmatrix} 0 & 1 \\ -1 & 0 \end{pmatrix}$ corresponds an integral operator whose kernel is expressed by means of the Hankel function.

1. A new realization of the representations $T_\chi(g)$. We recall that the representations $T_\chi(g)$ are realized in the space $\mathfrak{D}_\chi$ of functions on the line, and are given by the formula

$$T_\chi(g)f(x)=\chi(\beta x+\delta)f\left(\frac{\alpha x+\gamma}{\beta x+\delta}\right), \tag{1}$$

where $\chi(x) = |x|^{2l}\text{sign}^{2\epsilon}x$, $g = \begin{pmatrix}\alpha & \beta\\ \gamma & \delta\end{pmatrix}$. In order to pass on to the new realization, under which to the matrices $t$ there correspond operators of multiplication by a function, we employ the Fourier transform in $x$ (and not the Mellin transform, as we did in previous sections). So we pass on from $f(x)$ to

$$F(\lambda)=\int_{-\infty}^{\infty} f(x)e^{i\lambda x}\,dx. \tag{2}$$

Since $|f(x)| \sim |x|^{2l}$ for $|x| \to \infty$, this integral is absolutely convergent if $\lambda$ is a real number and $\operatorname{Re} l < -1/2$.

This realization possesses the required property, i.e. to triangular matrices $t$ in it correspond operators of multiplication by a function. In fact, by virtue of (1) we have

$$T_\chi(t)f(x)=f(x+t).$$

But

$$\int_{-\infty}^{\infty} f(x+t)e^{i\lambda x}dx=e^{-i\lambda t}\int_{-\infty}^{\infty} f(x)e^{i\lambda x}dx=e^{-i\lambda t}F(\lambda).$$

This equality shows that, in the new realization, to the matrix $t$ corresponds the operator of multiplication by $e^{-i\lambda t}$:

$$Q_\chi(t)F(\lambda)=e^{-i\lambda t}F(\lambda). \tag{3}$$

We now elucidate upon what becomes of the operators $T_\chi(g)$ corresponding to other matrices of the group $SL(2, R)$. We first note that any matrix $g$ of $SL(2, R)$ can be represented in one of the following two forms:

$$g=t\delta \tag{4}$$

or

$$g=t_1\delta s t_2. \tag{5}$$

Here $t_1$ and $t_2$ are triangular matrices with unities in the principal diagonals: $t_1 = \begin{pmatrix}1 & 0\\ t_1 & 1\end{pmatrix}$, $t_2 = \begin{pmatrix}1 & 0\\ t_2 & 1\end{pmatrix}$; $\delta = \begin{pmatrix}e^{-t} & 0\\ 0 & e^{t}\end{pmatrix}$ is a diagonal matrix and $s = \begin{pmatrix}0 & 1\\ -1 & 0\end{pmatrix}$

In fact, if $g = \begin{pmatrix} \alpha & 0 \\ \gamma & 1/\alpha \end{pmatrix}$, then $g = \begin{pmatrix} 1 & 0 \\ \gamma/\alpha & 1 \end{pmatrix} \begin{pmatrix} \alpha & 0 \\ 0 & 1/\alpha \end{pmatrix}$ and consequently $g$ has the form (4). Now let $g = \begin{pmatrix} \alpha & \beta \\ \gamma & \delta \end{pmatrix}$, where $\beta \neq 0$. We consider the matrix

$$t_1^{-1} g t_2^{-1} = \begin{pmatrix} \alpha - \beta t_2 & \beta \\ \gamma - \alpha t_1 - t_2(\delta - \beta t_2) & \delta - \beta t_1 \end{pmatrix}.$$

Since $\beta \neq 0$, we can take $t_2 = \alpha/\beta$ and $t_1 = \delta/\beta$. With such a choice we have

$$t_1^{-1} g t_2^{-1} = \begin{pmatrix} 0 & \beta \\ -\frac{1}{\beta} & 0 \end{pmatrix} = \begin{pmatrix} \beta & 0 \\ 0 & \frac{1}{\beta} \end{pmatrix} \begin{pmatrix} 0 & 1 \\ -1 & 0 \end{pmatrix} = \delta s.$$

Hence $g = t_1 \delta s t_2$.

From this assertion it follows that it is sufficient to find operators corresponding to diagonal matrices $\delta$ and to the matrix $s$. We have

$$T_\chi(\delta) f(x) = e^{2lt} f(e^{-2t} x)$$

and

$$\int_{-\infty}^{\infty} f(e^{-2t}x) e^{i\lambda x}\, dx = e^{2t} \int_{-\infty}^{\infty} f(y) e^{ie^{2t}\lambda y}\, dy = e^{2t} F(e^{2t}\lambda).$$

Therefore, to the diagonal matrix $\delta$ there corresponds an operator of the form

$$Q_\chi(\delta) F(\lambda) = e^{2(l+1)t} F(e^{2t}\lambda), \quad \delta = \begin{pmatrix} e^{-t} & 0 \\ 0 & e^t \end{pmatrix}. \tag{6}$$

2. **Calculation of the kernel of $Q_\chi(s)$.** It remains for us to find the operator corresponding to the matrix $s$. But

$$T_\chi(s) f(x) = |x|^{2l} \operatorname{sign}^{2\epsilon} x f\left(-\frac{1}{x}\right).$$

Therefore

$$Q_\chi(s) F(\lambda) = \int_{-\infty}^{\infty} |x|^{2l} \operatorname{sign}^{2\epsilon} x f\left(-\frac{1}{x}\right) e^{i\lambda x}\, dx$$
$$= \int_{-\infty}^{\infty} |y|^{-2l-2} \operatorname{sign}^{2\epsilon}(-y) f(y) e^{-i\lambda/y}\, dy.$$

By virtue of the inversion formula for the Fourier transform, we get

$$Q_\chi(s)F(\lambda)=\frac{1}{2\pi}\int_{-\infty}^{\infty}|y|^{-2l-2}\operatorname{sign}^{2\varepsilon}(-y)e^{-i\lambda/y}\int_{-\infty}^{\infty}F(\mu)e^{-i\mu y}\,d\mu\,dy.$$

If $-1<\operatorname{Re} l<0$, then we can change the order of integration in this equality and get

$$Q_\chi(s)F(\lambda)=\frac{1}{2\pi}\int_{-\infty}^{\infty}F(\mu)\int_{-\infty}^{\infty}|y|^{-2l-2}\operatorname{sign}^{2\varepsilon}(-y)e^{-i(\mu y+\lambda y^{-1})}\,dy\,d\mu. \tag{1}$$

Thus we have proved that the operator $Q_\chi(s)$ is an integral operator with the kernel

$$K_s(\lambda,\mu;\chi)=\frac{1}{2\pi}\int_{-\infty}^{\infty}|y|^{-2l-2}\operatorname{sign}^{2\varepsilon}(-y)e^{-i(\mu y+\lambda y^{-1})}\,dy. \tag{2}$$

This kernel is easily expressed by means of the Hankel function. The result depends on the signs of $\lambda$ and $\mu$.

Let $\lambda>0$ and $\mu>0$. We put $\sqrt{\lambda\mu}=R$, $\sqrt{\mu/\lambda}=e^a$ and $y=e^t$ for $y>0$, $y=-e^t$ for $y<0$. Then

$$\begin{aligned}
&K_s(\lambda,\mu;\chi)\\
&=\frac{1}{2\pi}\int_{-\infty}^{\infty}e^{2iR\operatorname{ch}(t+a)-t(2l+1)}\,dt+\frac{(-1)^{2\varepsilon}}{2\pi}\int_{-\infty}^{\infty}e^{-2iR\operatorname{ch}(t+a)-t(2l+1)}\,dt\\
&=\frac{ie^{(2l+1)a}}{2}[e^{(2l+1)\pi i/2}H^{(1)}_{2l+1}(2R)-(-1)^{2\varepsilon}e^{-(2l+1)\pi i/2}H^{(2)}_{2l+1}(2R)]\\
&=\frac{i}{2}\left(\frac{\mu}{\lambda}\right)^{l+1/2}[e^{(2l+1)\pi i/2}H^{(1)}_{2l+1}(2\sqrt{\lambda\mu})-(-1)^{2\varepsilon}e^{-(2l+1)\pi i/2}H^{(2)}_{2l+1}(2\sqrt{\lambda\mu})].
\end{aligned} \tag{3}$$

Similarly, for $\lambda<0$, $\mu<0$ we get

$$\begin{aligned}K_s(\lambda,\mu;\chi)=\frac{i}{2}\left(\frac{\mu}{\lambda}\right)^{l+1/2}[&-e^{-(2l+1)\pi i/2}H^{(2)}_{2l+1}(2\sqrt{\lambda\mu})\\
&+(-1)^{2\varepsilon}e^{(2l+1)\pi i/2}H^{(1)}_{2l+1}(2\sqrt{\lambda\mu})].\end{aligned} \tag{4}$$

If $\lambda>0$, $\mu<0$, then

$$\begin{aligned}K_s(\lambda,\mu;\chi)&=\frac{i}{2}\left(-\frac{\mu}{\lambda}\right)^{l+1/2}\\
&\times[(-1)^{2\varepsilon}H^{(1)}_{2l+1}(i\sqrt{-\lambda\mu})-H^{(2)}_{2l+1}(-i\sqrt{-\lambda\mu})],\end{aligned} \tag{5}$$

and if $\lambda<0$, $\mu>0$, then

$$\begin{aligned}&K_s(\lambda,\mu;\chi)\\
&=\frac{i}{2}\left(-\frac{\mu}{\lambda}\right)^{l+1/2}[-(-1)^{2\varepsilon}H^{(2)}_{2l+1}(-i\sqrt{-\lambda\mu})+H^{(1)}_{2l+1}(i\sqrt{-\lambda\mu})].\end{aligned} \tag{6}$$

Thus the kernel of the operator $Q_\chi(s)$, $\chi = (l, \epsilon)$, has been calculated for $-1 < l < 0$. In particular, formulas (3)–(6) are valid for representations of the fundamental series, since $l = -1/2 + ip$ for them.

It can be shown that, for representations of the discrete series, the operator $Q_\chi(s)$ is expressible by means of Bessel functions $J_n(x)$ with integral index. Using the formulas obtained above, one can find various properties of Hankel functions in the same way as in Chapter V.

# CHAPTER VIII

## REPRESENTATIONS OF THE GROUP OF THIRD ORDER TRIANGULAR MATRICES AND THE WHITTAKER FUNCTIONS

In this chapter we shall study the properties of the Whittaker functions $M_{\lambda,\mu}(z)$ and $W_{\lambda,\mu}(z)$. It will be shown that these functions are connected with the representations of the group of third order triangular matrices in exactly the same way as the hypergeometric function is connected with the representations of the group of real second order unimodular matrices. Namely, we shall construct a realization of irreducible representations of this group, in which corresponding to diagonal matrices we have operators of multiplication by a function. In this realization, corresponding to other group elements we have integral operators whose kernels are expressed by means of the Whittaker functions. Using these, we obtain recurrence relations, differential equations, continual analogues of the composition theorem and other relations for the Whittaker functions. In addition, we shall consider the Laguerre polynomials, which are closely connected with the Whittaker functions.

### §1. WHITTAKER FUNCTIONS AND THE CONFLUENT HYPERGEOMETRIC FUNCTION

1. **Definition.** We define the Whittaker functions $M_{\lambda,\mu}(z)$, $W_{\lambda,\mu}(z)$ by the following integral representations:

$$M_{\lambda,\mu}(z)=\frac{\Gamma(2\mu+1)\,z^{\mu+\frac{1}{2}}e^{-\frac{z}{2}}}{\Gamma\left(\mu-\lambda+\frac{1}{2}\right)\Gamma\left(\mu+\lambda+\frac{1}{2}\right)}\int_0^1 u^{\mu-\lambda-\frac{1}{2}}(1-u)^{\mu+\lambda-\frac{1}{2}}e^{zu}\,du, \quad (1)$$

$$W_{\lambda,\mu}(z)=\frac{z^{\mu+\frac{1}{2}}e^{-\frac{z}{2}}}{\Gamma\left(\mu-\lambda+\frac{1}{2}\right)}\int_0^\infty e^{-zu}u^{\mu-\lambda-\frac{1}{2}}(1+u)^{\mu+\lambda-\frac{1}{2}}\,du. \quad (2)$$

The first of these integrals converges for $\operatorname{Re}(\mu-\lambda+\tfrac{1}{2})>0, \operatorname{Re}(\mu+\lambda+\tfrac{1}{2})>0$

and the second for $\operatorname{Re}(\mu - \lambda + \tfrac{1}{2}) > 0$, $\operatorname{Re} z > 0$.

The integrals appearing in (1) and (2) are analytic functions of $z$ in the convergence domain. It is easy to prove that $M_{\lambda,\mu}(z)$ and $W_{\lambda,\mu}(z)$ can be analytically continued on the entire complex plane, cut along the negative semiaxis (the cut is made for choosing a single-valued branch of the factor $z^{\mu+1/2}$).

We also note that $M_{\lambda,\mu}(z)$, considered as a function of the parameters, is analytic everywhere, except for the points $\mu = -\tfrac{1}{2}, -1, -3/2, \cdots$, at which the factor $\Gamma(2\mu+1)$ has simple poles. But $W_{\lambda,\mu}(z)$ is defined for all values of $\lambda$ and $\mu$.

Starting from the integral representation (1), we obtain a series expansion of $M_{\lambda,\mu}(z)$. For this we expand $e^{zu}$ in a power series and integrate the resulting series term by term. Using integral (2) of §1.7, Chapter V, we get

$$M_{\lambda,\mu}(z)=\frac{\Gamma(2\mu+1)\,z^{\mu+\frac{1}{2}}e^{-\frac{z}{2}}}{\Gamma\left(\mu-\lambda+\frac{1}{2}\right)}\sum_{k=0}^{\infty}\frac{\Gamma\left(\mu-\lambda+k+\frac{1}{2}\right)}{k!\,\Gamma(2\mu+k+1)}z^{k}. \tag{3}$$

2. The confluent hypergeometric function. We introduce the confluent hypergeometric series

$$1+\frac{\alpha}{\gamma}\frac{z}{1!}+\frac{\alpha(\alpha+1)}{\gamma(\gamma+1)}\frac{z^2}{2!}+\cdots \tag{1}$$

This series is obtained from the ordinary hypergeometric series

$$1+\frac{\alpha\beta}{\gamma}\frac{z}{1!}+\frac{\alpha(\alpha+1)\beta(\beta+1)}{\gamma(\gamma+1)}\frac{z^2}{2!}+\cdots$$

on replacing $z$ by $z/k$, $\beta$ by $k$ and taking the limit as $k \to \infty$ (it is easy to see the validity of taking the limit term by term). Clearly the series (1) converges for all values of $z$. We denote its sum by $\Phi(\alpha;\ \gamma;\ z)$ and call it the confluent hypergeometric function.

From equalities (3) of subsection 1 and from (1), it follows that

$$M_{\lambda,\mu}(z)=z^{\mu+\frac{1}{2}}e^{-\frac{z}{2}}\Phi\left(\mu-\lambda+\frac{1}{2};\ 2\mu+1;\ z\right). \tag{2}$$

If, in the differential equation

$$z(1-z)\frac{d^2u}{dz^2}+[\gamma-(\alpha+\beta+1)z]\frac{du}{dz}-\alpha\beta u=0,$$

satisfied by the hypergeometric function $F(\alpha,\ \beta;\ \gamma;\ z)$ (see §5.3 of Chapter VII), we replace $z$ by $z/k$, $\beta$ by $k$ and take the limit as $k \to \infty$, then we obtain the equation

$$z\frac{d^2u}{dz^2}+(\gamma-z)\frac{du}{dz}-\alpha u=0, \tag{3}$$

called *the confluent hypergeometric equation*. It is natural to expect that one of its solutions will be the confluent hypergeometric function $\Phi(\alpha;\ \gamma;\ z)$ which is obtained from $F(\alpha,\ \beta;\ \gamma;\ z)$ by the same limiting process:

$$\Phi(\alpha;\ \gamma;\ z)=\lim_{k\to\infty} F\left(\alpha,\ k;\ \gamma;\ \frac{z}{k}\right). \tag{4}$$

That this is so is easily seen by direct verification.

From equation (3) one can easily derive the equation

$$\frac{d^2v}{dz^2}+\left(-\frac{1}{4}+\frac{1}{z}+\frac{\frac{1}{4}-\mu^2}{z^2}\right)v=0 \tag{5}$$

satisfied by the function $M_{\lambda,\mu}(z)$ (see §3).

## §2. THE GROUP OF THIRD ORDER TRIANGULAR MATRICES AND ITS REPRESENTATIONS

1. **Lie algebra.** We study the representations of the group $G$ of third order triangular matrices having the form

$$g=\begin{pmatrix}1 & a & b\\ 0 & c & d\\ 0 & 0 & 1\end{pmatrix}.$$

Here $a$, $b$, $c$, $d$ are real numbers and $c>0$.

We compute the Lie algebra of this group. Since the group elements depend on four parameters, we must select four one-parameter subgroups in $G$. It is con venient to choose the following as these subgroups:

$$g_+(t)=\begin{pmatrix}1 & t & 0\\ 0 & 1 & 0\\ 0 & 0 & 1\end{pmatrix}, \tag{1}$$

$$g_-(t)=\begin{pmatrix}1 & 0 & 0\\ 0 & 1 & t\\ 0 & 0 & 1\end{pmatrix}, \tag{2}$$

$$z(t)=\begin{pmatrix}1 & 0 & t\\ 0 & 1 & 0\\ 0 & 0 & 1\end{pmatrix}, \tag{3}$$

$$\varepsilon(t)=\begin{pmatrix}1 & 0 & 0\\ 0 & e^t & 0\\ 0 & 0 & 1\end{pmatrix}. \tag{4}$$

The tangent matrices to these subgroups have the form

$$a_+=\begin{pmatrix}0&1&0\\0&0&0\\0&0&0\end{pmatrix},\qquad a_-=\begin{pmatrix}0&0&0\\0&0&1\\0&0&0\end{pmatrix},$$
$$c=\begin{pmatrix}0&0&1\\0&0&0\\0&0&0\end{pmatrix},\qquad b=\begin{pmatrix}0&0&0\\0&1&0\\0&0&0\end{pmatrix}.$$

Direct claculation shows that

$$[a_+, a_-]=c,\quad [a_+, b]=a_+,\quad [a_-, b]=-a_-,\qquad\qquad (5)$$
$$[a_+, c]=[a_-, c]=[b, c]=0.$$

We note that the one-parameter subgroup $z(t)$ is the center of $G$, i.e. its elements commute with all the elements of $G$:

$$z(t)g=gz(t).$$

Along with the subgroups $g_+(t)$ and $g_-(t)$, we shall consider the following one-parameter subgroups in $G$:

$$g_1(t)=\begin{pmatrix}1&t&\frac{t^2}{2}\\0&1&t\\0&0&1\end{pmatrix}\qquad\qquad (6)$$

and

$$g_2(t)=\begin{pmatrix}1&-t&-\frac{t^2}{2}\\0&1&t\\0&0&1\end{pmatrix}.\qquad\qquad (7)$$

The corresponding tangent matrices are

$$a_1=\begin{pmatrix}0&1&0\\0&0&1\\0&0&0\end{pmatrix}=a_++a_-$$

and

$$a_2=\begin{pmatrix}0&-1&0\\0&0&1\\0&0&0\end{pmatrix}=a_--a_+.$$

Commutation relations for $a_1$ and $a_2$ follow from (5).

2. **Factorization into one-parameter subgroups.** Let

$$g(a, b, d, \tau) = \begin{pmatrix} 1 & a & b \\ 0 & e^{\tau} & d \\ 0 & 0 & 1 \end{pmatrix}$$

be an element of $G$. If $d = 0$, then we have the easily verifiable equality

$$g(a, b, 0, \tau) = \epsilon(\tau) g_{+}(ae^{-\tau}) z(b). \tag{1}$$

Similarly, if $a = 0$, then

$$g(0, b, d, \tau) = \varepsilon(\tau) g_{-}(de^{-\tau}) z(b). \tag{2}$$

We now consider the case when $a \neq 0$ and $b \neq 0$. Then for $ad > 0$ we have the equality

$$g(a, b, d, \tau) = \varepsilon(\tau_1) g_1(r) \varepsilon(\tau - \tau_1) z(b), \tag{3}$$

where

$$\left.\begin{aligned} r^2 &= ade^{-\tau}, \quad \operatorname{sign} r = \operatorname{sign} d, \\ e^{\tau_1} &= \frac{d}{r} = \sqrt{\frac{d}{a}}\, e^{\frac{\tau}{2}}. \end{aligned}\right\} \tag{4}$$

But if $ad < 0$, then

$$g(a, b, d, \tau) = \varepsilon(\tau_1) g_2(r) \varepsilon(\tau - \tau_1) z(b), \tag{5}$$

where

$$\left.\begin{aligned} r^2 &= -ade^{-\tau}, \quad \operatorname{sign} r = \operatorname{sign} d, \\ e^{\tau_1} &= \frac{d}{r} = \sqrt{-\frac{d}{a}}\, e^{\frac{\tau}{2}}. \end{aligned}\right\} \tag{6}$$

Thus in every case an element $g$ of $G$ can be represented in the form

$$g(a, b, d, \tau) = \varepsilon(\tau_1) h(r) \varepsilon(\tau - \tau_1) z(b), \tag{7}$$

where $h(r)$ is an element of one of the following four forms: $g_{+}(r)$, $g_{-}(r)$, $g_1(r)$, $g_2(r)$. In the first two cases $\tau_1 = \tau$.

3. **Irreducible representations of $G_1$.** Irreducible representations of $G_1$ are constructed in the space $\mathfrak{D}$ of finite infinitely differentiable functions on the real line. Each representation is specified by a pair $\chi = (\sigma, \omega)$ of complex numbers. Namely, with each element $g = (a, b, d, \tau)$ of the group $G$ we associate the operator

$$T_{\chi}(g) f(x) = e^{\omega\tau + \sigma(dx+b)} f(e^{\tau} x + a) \tag{1}$$

in the space $\mathfrak{D}$.

If $\operatorname{Re}\sigma \neq 0$ then one cannot extend this representation to the space of square-integrable functions, because of the exponential growth of the factor $e^{\sigma(dx+b)}$.

It is easy to show that for $\sigma \neq 0$ the representation $T_\chi(g)$ is irreducible. We shall not go into the details.

We calculate the infinitesimal operators of the representation $T_\chi(g)$, corresponding to the subgroups $g_+(t)$, $g_-(t)$, $z(t)$, $\epsilon(t)$. From formula (1) it follows that

$$T_\chi[g_+(t)]f(x) = f(x+t).$$

Therefore, to the subgroup $g_+(t)$ corresponds the infinitesimal operator

$$A_+f(x) = f'(x). \tag{2}$$

In exactly the same way one proves that to the subgroup $g_-(t)$ corresponds the infinitesimal operator

$$A_-f(x) = \sigma x f(x), \tag{3}$$

and to $z(t)$, the operator

$$Zf(x) = \sigma f(x). \tag{4}$$

Lastly, to $\epsilon(t)$ there corresponds the infinitesimal operator

$$Ef(x) = \omega f(x) + xf'(x). \tag{5}$$

Since $g_1(t) = g_-(t)g_+(t)z(t^2/2)$, to the subgroup $g_1(t)$ corresponds the infinitesimal operator

$$A_1f(x) \equiv (A_+ + A_-)f(x) = \sigma xf(x) + f'(x). \tag{6}$$

Similarly, to $g_2(t)$ corresponds the operator

$$A_2f(x) \equiv (A_- - A_+)f(x) = \sigma xf(x) - f'(x). \tag{7}$$

We leave to the reader the verification of commutation relations for these operators.

4. Another realization of the representations $T_\chi(g)$. As in the case of the group $SL(2, R)$, we now pass on to another realization of the representations $T_\chi(g)$. Namely, we choose a realization in which corresponding to diagonal matrices $\epsilon(\tau)$ we would have operators of multiplication by a function.

For this, as in Chapter VII, with each function $f(x)$ of $\mathfrak{D}$ we associate a pair of functions $F_+(\lambda)$ and $F_-(\lambda)$:

$$\left.\begin{aligned} F_+(\lambda) &= \int_{-\infty}^{\infty} f(x)\, x_+^{\lambda-1}\, dx, \\ F_-(\lambda) &= \int_{-\infty}^{\infty} f(x)\, x_-^{\lambda-1}\, dx. \end{aligned}\right\} \tag{1}$$

These functions are given in the half plane $\operatorname{Re}\lambda > 0$. As noted in Chapter VII, we have the inversion formula

$$f(x) = \begin{cases} \dfrac{1}{2\pi i} \displaystyle\int_{\rho-i\infty}^{\rho+i\infty} F_+(\mu)\, x^{-\mu}\, d\mu, & x > 0, \\ \dfrac{1}{2\pi i} \displaystyle\int_{\rho-i\infty}^{\rho+i\infty} F_-(\mu)\, (-x)^{-\mu}\, d\mu, & x < 0. \end{cases} \tag{2}$$

We explain how operators of the representation $T_\chi(g)$ transform under this transformation. For this, we find the pair $F^{(g)} = (F_+^{(g)}, F_-^{(g)})$ corresponding to the function

$$f^{(g)}(x) \equiv T_\chi(g) f(x).$$

We have

$$F_\pm^{(g)}(\lambda) = \int_{-\infty}^{\infty} T_\chi(g) f(x)\, x_\pm^{\lambda-1}\, dx = e^{\omega\tau+\sigma b} \int_{-\infty}^{\infty} e^{d\sigma x} f(e^\tau x + a)\, x_\pm^{\lambda-1}\, dx.$$

The substitution $e^\tau x + a = y$ transforms this equality into the form

$$F_\pm^{(g)}(\lambda) = \exp\left[(\omega - \lambda)\tau + \sigma(b - a d e^{-\tau})\right] \times \int_{-\infty}^{\infty} \exp\left[\sigma d e^{-\tau} y\right] (y - a)_\pm^{\lambda-1} f(y)\, dy. \tag{3}$$

In order to obtain the required operators, we have to replace $f(y)$ in (3) according to the inversion formula (2). We get

$$\begin{aligned} F_\pm^{(g)}(\lambda) = \frac{1}{2\pi i} \exp\left[(\omega - \lambda)\tau + \sigma(b - a d e^{-\tau})\right] \\ \times \Big[\int_0^\infty \int_{\rho-i\infty}^{\rho+i\infty} \exp\left[\sigma d e^{-\tau} y\right] y^{-\mu} (y - a)_\pm^{\lambda-1} F_+(\mu)\, d\mu\, dy \\ + \int_0^\infty \int_{\rho-i\infty}^{\rho+i\infty} \exp\left[-\sigma d e^{-\tau} y\right] y^{-\mu} (-y - a)_\pm^{\lambda-1} F_-(\mu)\, d\mu\, dy\Big]. \end{aligned} \tag{4}$$

Formally changing the order of integration, we obtain

$$F_+^{(g)}(\lambda) = \int_{\rho-i\infty}^{\rho+i\infty} K_{++}(\lambda, \mu; \chi; g) F_+(\mu)\, d\mu + \int_{\rho-i\infty}^{\rho+i\infty} K_{+-}(\lambda, \mu; \chi; g) F_-(\mu)\, d\mu \quad (5)$$

and

$$F_-^{(g)}(\lambda) = \int_{\rho-i\infty}^{\rho+i\infty} K_{-+}(\lambda, \mu; \chi; g) F_+(\mu)\, d\mu + \int_{\rho-i\infty}^{\rho+i\infty} K_{--}(\lambda, \mu; \chi; g) F_-(\mu)\, d\mu. \quad (6)$$

Here

$$K_{++}(\lambda, \mu; \chi; g) = \frac{1}{2\pi i} \exp[(\omega-\lambda)\tau + \sigma(b - ade^{-\tau})] \int_0^\infty \exp[\sigma de^{-\tau} y] y^{-\mu} (y-a)_+^{\lambda-1} dy, \quad (7)$$

$$K_{+-}(\lambda, \mu; \chi; g) = \frac{1}{2\pi i} \exp[(\omega-\lambda)\tau + \sigma(b - ade^{-\tau})] \times \int_0^\infty \exp[-\sigma de^{-\tau} y] y^{-\mu} (-y-a)_+^{\lambda-1} dy. \quad (8)$$

$K_{-+}$ is obtained from $K_{++}$ on replacing $(y-a)_+^{\lambda-1}$ by $(y-a)_-^{\lambda-1}$, and $K_{--}$ is obtained from $K_{+-}$ on replacing $(-y-a)_+^{\lambda-1}$ by $(-y-a)_-^{\lambda-1}$.

However, in contrast to the case of the hypergeometric function, studied in Chapter VII, for $\operatorname{Re}\sigma \neq 0$ one cannot here indicate a common domain in which all these formulas have a meaning. For example, if $a > 0$, $d > 0$ and $\operatorname{Re}\sigma < 0$, then the integrals expressing $K_{++}$, $K_{+-}$ and $K_{-+}$ converge for $\operatorname{Re}\mu < 1$, $\operatorname{Re}\lambda > 0$.

But the integral which expresses $K_{--}$ diverges. Therefore, in this case, in place of formula (6) we must write

$$F_-^{(g)}(\lambda) = \int_{\rho-i\infty}^{\rho+i\infty} K_{-+}(\lambda, \mu; \chi; g) F_+(\mu)\, d\mu + \mathfrak{K}_{--} F_-(\lambda), \quad (9)$$

where $\mathfrak{K}_{--}$ is not an integral operator.

In the case when $\operatorname{Re}\sigma = 0$ there exists a common convergence domain of the integrals for $K_{++}$, $K_{+-}$, $K_{-+}$, $K_{--}$. Namely, these integrals converge in the domain $\operatorname{Re}\mu < 1$, $\operatorname{Re}\lambda > 0$, $\operatorname{Re}\mu > \operatorname{Re}\lambda$.

If $g = \epsilon(\tau)$ it is more convenient to calculate $F_+^{(g)}(\lambda)$ and $F_-^{(g)}(\lambda)$ directly. We have

$$T_\chi[\varepsilon(\tau)] f(x) = e^{\omega\tau} f(e^\tau x).$$

Therefore

$$F_{\pm}^{\epsilon(\tau)}(\lambda)=e^{\omega\tau}\int_{-\infty}^{\infty}f(e^{\tau}x)\,x_{\pm}^{\lambda-1}\,dx$$
$$=e^{(\omega-\lambda)\tau}\int_{-\infty}^{\infty}f(y)\,y_{\pm}^{\lambda-1}\,dy=e^{(\omega-\lambda)\tau}F_{\pm}(\lambda). \tag{10}$$

Thus, to the diagonal matrix $\epsilon(\tau)$ corresponds the operator of multiplication by the function $e^{(\omega-\lambda)\tau}$.

In exactly the same way one proves that to a matrix of the form $z(b)$ there corresponds the operator of multiplication by $e^{\sigma b}$.

Henceforth operators of the representation $T_\chi(g)$, on passing to the realization in the space of pairs $\mathbf{F}(\lambda)=(F_+(\lambda), F_-(\lambda))$, will be denoted by $R_\chi(g)$. In other words, we set

$$R_\chi(g)\,\mathbf{F}(\lambda)=\mathbf{F}^{(g)}(\lambda). \tag{11}$$

5. **Infinitesimal operators of the representation $R_\chi(g)$.** Let us now calculate the infinitesimal operators $\hat{A}_+$, $\hat{A}_-$, $\hat{Z}$ and $\hat{E}$ for the representation $R_\chi(g)$. In order to find the operator corresponding to the subgroup $g_+(t)$, it is necessary to find the Mellin transform of the functions $A_+f(x)=f'(x)$ and $A_+f(-x)=f'(-x)$. The Mellin transform for $A_+f(x)$ has the form

$$\hat{A}_+F_+(\lambda)=\int_0^\infty A_+f(x)\,x^{\lambda-1}\,dx=\int_0^\infty f'(x)\,x^{\lambda-1}\,dx$$
$$=-(\lambda-1)\int_0^\infty f(x)\,x^{\lambda-2}\,dx=-(\lambda-1)F_+(\lambda-1)$$

One similarly proves the equality

$$\hat{A}_+F_-(\lambda)=(\lambda-1)F_-(\lambda-1).$$

Thus the infinitesimal operator of the representation $R_\chi(g)$, corresponding to the subgroup $g_+(t)$, has the form

$$\hat{A}_+(F_+(\lambda), F_-(\lambda))=(\lambda-1)(-F_+(\lambda-1), F_-(\lambda-1)). \tag{1}$$

In exactly the same way one proves that to the subgroups $g_-(t)$, $z(t)$, $\epsilon(t)$ correspond the infinitesimal operators

$$\hat{A}_-(F_+(\lambda), F_-(\lambda))=\sigma(F_+(\lambda+1), -F_-(\lambda+1)), \tag{2}$$
$$\hat{Z}(F_+(\lambda), F_-(\lambda))=\sigma(F_+(\lambda), F_-(\lambda)), \tag{3}$$
$$\hat{E}(F_+(\lambda), F_-(\lambda))=(\omega-\lambda)(F_+(\lambda), F_-(\lambda)). \tag{4}$$

Infinitesimal operators corresponding to the subgroups $g_1(t)$ and $g_2(t)$ have the form

$$\acute{A}_1(F_+(\lambda), F_-(\lambda))$$
$$=(\sigma F_+(\lambda+1)-(\lambda-1)F_+(\lambda-1), -\sigma F_-(\lambda+1)+(\lambda-1)F_-(\lambda-1)), \quad (5)$$

$$\hat{A}_2(F_+(\lambda), F_-(\lambda))$$
$$=(\sigma F_+(\lambda+1)+(\lambda-1)F_+(\lambda-1), -\sigma F_-(\lambda+1)-(\lambda-1)F_-(\lambda-1)) \quad (6)$$

6. **Calculation of the kernels of representations.** In subsection 2 it was proved that any matrix $g$ of the group $G$ can be written in the form

$$g=\varepsilon(\tau_1)h(r)\varepsilon(\tau-\tau_1)z(b),$$

where $z(b)$ and $\epsilon(b)$ are given by inequalities (3) and (4) of subsection 1, and $h(r)$ is a matrix belonging to one of the subgroups $g_+(t)$, $g_-(t)$, $g_1(t)$, $g_2(t)$. To the matrix $z(b)$ corresponds, in the representation $R_\chi(g)$, the operator of multiplication by the function $e^{\sigma b}$, and to the matrix $\epsilon(\tau)$, the operator of multiplication by $e^{(\omega-\lambda)\tau}$ (see subsection 4). Therefore it is sufficient to calculate the kernels for elements of the subgroups $g_+(t)$, $g_-(t)$, $g_1(t)$, $g_2(t)$.

We begin with the subgroup $g_+(t)$. For the elements of this subgroup $b = d = \tau = 0$, and so

$$K_{++}(\lambda, \mu; \chi; g_+(t))=\frac{1}{2\pi i}\int_0^\infty y^{-\mu}(y-t)_+^{\lambda-1}\,dy.$$

If $t > 0$, then we have

$$K_{++}(\lambda, \mu; \chi; g_+(t))=\frac{1}{2\pi i}\int_t^\infty y^{-\mu}(y-t)^{\lambda-1}\,dy=\frac{1}{2\pi i}\frac{\Gamma(\mu-\lambda)\Gamma(\lambda)}{\Gamma(\mu)}t^{\lambda-\mu}. \quad (1)$$

This integral converges for $\operatorname{Re}\mu > \operatorname{Re}\lambda > 0$.

Similarly, one proves that for $t > 0$

$$K_{+-}(\lambda, \mu; \chi; g_+(t))=0, \quad (2)$$

$$K_{-+}(\lambda, \mu; \chi; g_+(t))=\frac{\Gamma(\lambda)\Gamma(1-\mu)}{\Gamma(\lambda-\mu+1)}t^{\lambda-\mu}, \quad (3)$$

where $\operatorname{Re}\mu < 1$, $\operatorname{Re}\lambda > 0$, and

$$K_{--}(\lambda, \mu; \chi; g_+(t)) = \frac{\Gamma(1-\mu)\Gamma(\mu-\lambda)}{\Gamma(1-\lambda)} t^{\lambda-\mu}, \tag{4}$$

where $\operatorname{Re}\lambda < \operatorname{Re}\mu < 1$.

We now consider operators corresponding to elements of the subgroup $g_-(t)$. Here we get

$$K_{++}(\lambda, \mu; \chi; g_-(t)) = \frac{1}{2\pi i}\int_0^\infty e^{\sigma t y} y^{\lambda-\mu-1}\, dy.$$

This integral converges if $\operatorname{Re}\sigma t < 0$, and it is equal to

$$K_{++}(\lambda, \mu; \chi; g_-(t)) = \frac{\Gamma(\lambda-\mu)(-\sigma t)^{\mu-\lambda}}{2\pi i}. \tag{5}$$

As above, we shall take $\operatorname{Re}\sigma < 0$. Then formula (5) holds for $t > 0$. In exactly the same way one proves that for $\operatorname{Re}\sigma < 0$, $t > 0$, we have

$$K_{+-}(\lambda, \mu; \chi; g_-(t)) = 0 \tag{6}$$

and

$$K_{-+}(\lambda, \mu; \chi; g_-(t)) = 0. \tag{7}$$

But the kernel $K_{--}(\lambda, \mu; \chi; g_-(t))$, for $\operatorname{Re}\sigma < 0$, $t > 0$, is expressed by a divergent integral.

The kernels considered so far lead only to the power function. But to elements of the subgroups $g_1(t)$ and $g_2(t)$ correspond kernels which are expressed by the Whittaker functions. For definiteness we shall take $\sigma = -1$.

From formula (7) of subsection 4 it follows that

$$K_{++}(\lambda, \mu; \chi; g_1(t)) = \frac{1}{2\pi i} e^{t^2/2}\int_t^\infty e^{-ty} y^{-\mu}(y-t)^{\lambda-1}\, dy.$$

Making the substitution $y - t = tx$ and using equality (2) of §1.1, we deduce that for $t > 0$

$$K_{++}(\lambda, \mu; \chi; g_1(t))$$
$$= \frac{1}{2\pi i} e^{-\frac{t^2}{2}} t^{\lambda-\mu}\int_0^\infty e^{-t^2 x}(x+1)^{-\mu} x^{\lambda-1}\, dx = \frac{\Gamma(\lambda)}{2\pi i t} W_{\frac{1-\lambda-\mu}{2}, \frac{\lambda-\mu}{2}}(t^2), \tag{8}$$

where $\operatorname{Re}\lambda > 0$.

One similarly proves that for $\sigma = -1$, $t > 0$

$$K_{+-}(\lambda, \mu; \chi; g_1(t)) = 0, \tag{9}$$

$$K_{-+}(\lambda; \mu; \chi; g_1(t)) = \frac{1}{2\pi i t} \frac{\Gamma(\lambda)\Gamma(1-\mu)}{\Gamma(\lambda-\mu+1)} M_{\frac{1-\lambda-\mu}{2}, \frac{\lambda-\mu}{2}}(t^2), \tag{10}$$

where $\operatorname{Re} \lambda > 0$, $\operatorname{Re} \mu < 1$. Lastly, $K_{--}(\lambda, \mu; \chi; g_1(t))$ is expressed by a divergent integral.

In exactly the same way we establish that for $\sigma = -1$, $t > 0$,

$$K_{++}(\lambda, \mu; \chi; g_2(t)) = \frac{\Gamma(1-\mu)}{2\pi i t} W_{\frac{\lambda+\mu-1}{2}, \frac{\lambda-\mu}{2}}(t^2), \tag{11}$$

where $\operatorname{Re} \mu < 1$. Further,

$$K_{+-}(\lambda, \mu; \chi; g_2(t)) = \frac{1}{2\pi i t} \frac{\Gamma(\lambda)\Gamma(1-\mu)}{\Gamma(\lambda-\mu+1)} M_{\frac{\lambda+\mu-1}{2}, \frac{\lambda-\mu}{2}}(t^2), \tag{12}$$

where $\operatorname{Re} \lambda > 0$ and $\operatorname{Re} \mu < 1$, and

$$K_{-+}(\lambda, \mu; \chi; g_2(t)) = 0. \tag{13}$$

Lastly, the integral expressing $K_{--}(\lambda, \mu; \chi; g_2(t))$ diverges for $\sigma = -1$, $t > 0$.

In all the cases considered, the passage from $t$ to $-t$ amounts to changing both the index signs:

$$K_{++}(\lambda, \mu; \chi; g_+(-t)) = K_{--}(\lambda, \mu; \chi; g_+(t)), \tag{14}$$

and so on.

## §3. FUNCTIONAL RELATIONS FOR WHITTAKER FUNCTIONS

**1. Relations between infinitesimal operators and operators of the representation.** The derivation of recurrence relations for Whittaker functions is based on certain relations between operators of the representation $R_\chi(g)$ and infinitesimal operators of this representation.

Let us consider the matrix

$$g_+(t)\, g_1(x) = \begin{pmatrix} 1 & t+x & tx+\frac{x^2}{2} \\ 0 & 1 & x \\ 0 & 0 & 1 \end{pmatrix}.$$

One can easily verify that

$$g_+(t)\, g_1(x) = \varepsilon(\tau)\, g_1(r)\, \varepsilon(-\tau)\, z(b), \tag{1}$$

where

$$r^2=x^2+tx,\quad e^{2\tau}=\frac{x}{t+x},\quad b=\frac{1}{2}tx. \tag{2}$$

From (2) it follows that for $t=0$ we have $r=x$, $\tau=b=0$ and

$$\left.\frac{dr}{dt}\right|_{t=0}=\frac{1}{2},\quad \left.\frac{d\tau}{dt}\right|_{t=0}=-\frac{1}{2x},\quad \left.\frac{db}{dt}\right|_{t=0}=\frac{x}{2}. \tag{3}$$

Since $R_\chi(g)$ is a representation of the group $G$, we have

$$R_\chi(g_+(t))\,R_\chi(g_1(x))=R_\chi(\varepsilon(\tau))\,R_\chi(g_1(r))\,R_\chi(\varepsilon(-\tau))\,R_\chi(z(b)). \tag{4}$$

We differentiate with respect to $t$ and put $t=0$.

Because of equalities (3) and formulas (1)–(4) of §2.5 we get

$$\hat{A}_+R_\chi(g_1(x))$$
$$=\frac{1}{2x}\left(R_\chi(g_1(x))\,\hat{E}-\hat{E}R_\chi(g_1(x))\right)+\frac{x}{2}R_\chi(g_1(x))\hat{Z}+\frac{1}{2}\frac{dR_\chi(g_1(x))}{dx}.$$

Substituting the expressions for $\hat{E}$ and $\hat{Z}$, we deduce that for $\sigma=-1$

$$\hat{A}_+R_\chi(g_1(x))=\frac{1}{2}\left(\frac{\lambda-\mu}{x}-x\right)R_\chi(g_1(x))+\frac{1}{2}\frac{dR_\chi(g_1(x))}{dx}. \tag{5}$$

In exactly the same way, by considering the product $g_-(t)g_1(x)$ we deduce that

$$\hat{A}_-R_\chi(g_1(x))=\frac{1}{2}\left(\frac{\mu-\lambda}{x}+x\right)R_\chi(g_1(x))+\frac{1}{2}\frac{dR_\chi(g_1(x))}{dx}. \tag{6}$$

From the product $g_+(t)g_2(x)$ we obtain

$$\hat{A}_+R_\chi(g_2(x))=\frac{1}{2}\left(\frac{\mu-\lambda}{x}-x\right)R_\chi(g_2(x))-\frac{1}{2}\frac{dR_\chi(g_2(x))}{dx}, \tag{7}$$

and from the product $g_-(t)g_2x$,

$$\hat{A}\;R_\chi(g_2(x))=\frac{1}{2}\left(\frac{\mu-\lambda}{x}-x\right)R_\chi(g_2(x))+\frac{1}{2}\frac{dR_\chi(g_2(x))}{dx}. \tag{8}$$

2. **Recurrence relations.** From the above formulas we now derive recurrence relations connecting the functions $W_{\lambda,\mu}(x)$ and $M_{\lambda,\mu}(x)$ with the functions $W_{\lambda\pm1/2,\mu\pm1/2}(x)$ and $M_{\lambda\pm1/2,\mu\pm1/2}(x)$. For this we replace the operators $\hat{A}_+$, $\hat{A}_-$, $R_\chi(g_1(x))$, $R_\chi(g_2(x))$, appearing in these formulas, by their explicit expressions, and then compare the kernels of the resulting operators on the left and on the right.

The operator $R_\chi(g_1(x))$ transforms the pair $\mathrm{F}(\lambda)=(F_+(\lambda),F_-(\lambda))$ into the pair $F^{(g_1)}(\lambda) = (F_+^{(g_1)}(\lambda), F_-^{(g_1)}(\lambda))$, where

$$F_+^{(g_1)}(\lambda) = \int_{\rho - i\infty}^{\rho + i\infty} K_{++}(\lambda, \mu; \chi; g_1(x)) F_+(\mu)\, d\mu$$

and $F_-^{(g_1)}(\lambda)$ has a similar expression. But the operator $\hat{A}_+$ transforms $F_+^{(g_1)}(\lambda)$ into $-(\lambda - 1)F_+^{(g_1)}(\lambda - 1)$ (see §2.5). Therefore

$$\hat{A}_+ F_+^{(g_1)}(\lambda) = (1 - \lambda) \int_{\rho - i\infty}^{\rho + i\infty} K_{++}(\lambda - 1, \mu; \chi; g_1(x)) F_+(\mu)\, d\mu.$$

In exactly the same way one establishes that the corresponding operator in the right hand side of the formula (5) of subsection 1 has the form

$$\frac{1}{2} \int_{\rho - i\infty}^{\rho + i\infty} \left[\left(\frac{\lambda - \mu}{x} - x\right) K_{++}(\lambda, \mu; \chi; g_1(x)) + \frac{dK_{++}(\lambda, \mu; \chi; g_1(x))}{dx}\right] F_+(\mu)\, d\mu.$$

Consequently,

$$\begin{aligned}(1 - \lambda) K_{++}(\lambda - 1, \mu; \chi; g_1(x)) \\ = \frac{1}{2}\left(\frac{\lambda - \mu}{x} - x\right) K_{++}(\lambda, \mu; \chi; g_1(x)) + \frac{1}{2} \frac{dK_{++}(\lambda, \mu; \chi; g_1(y))}{dx}.\end{aligned}$$

Substituting expression (8) of §2.6 for $K_{++}(g_1(t))$ in this equality, we derive the recurrence relation for the function $W_{\lambda,\mu}(x)$):

$$\frac{dW_{\lambda,\mu}(x)}{dx} = -\frac{1}{\sqrt{x}} W_{\lambda + 1/2, \mu - 1/2}(x) - \left(\frac{2\mu - 1}{2x} - \frac{1}{2}\right) W_{\lambda,\mu}(x) \tag{1}$$

(we replaced $(1 - \lambda - \mu)/2$ by $\lambda$, $(\lambda - \mu)/2$ by $\mu$ and $x^2$ by $x$).

In exactly the same way, by considering the element $K_{-+}$ we derive the recurrence relation for the function $M_{\lambda,\mu}(x)$):

$$\frac{dM_{\lambda,\mu}(x)}{dx} = \frac{2\mu}{\sqrt{x}} M_{\lambda + 1/2, \mu - 1/2}(x) - \left(\frac{2\mu - 1}{2x} - \frac{1}{2}\right) M_{\lambda,\mu}(x). \tag{2}$$

Similarly, equality (6) of subsection 1 leads to the recurrence relations

$$\frac{dW_{\lambda,\mu}(x)}{dx} = \frac{\lambda - \mu - \frac{1}{2}}{\sqrt{x}} W_{\lambda - 1/2, \mu + 1/2}(x) + \left(\frac{2\mu + 1}{2x} - \frac{1}{2}\right) W_{\lambda,\mu}(x) \tag{3}$$

and

$$\frac{dM_{\lambda,\mu}(x)}{dx} = \frac{\mu - \lambda + \frac{1}{2}}{(2\mu + 1)\sqrt{x}} M_{\lambda - 1/2, \mu + 1/2}(x) + \left(\frac{2\mu + 1}{2x} - \frac{1}{2}\right) M_{\lambda,\mu}(x). \tag{4}$$

From equality (7) of subsection 1 we obtain the relations

$$\frac{dW_{\lambda,\mu}(x)}{dx}=\frac{\lambda+\mu-\frac{1}{2}}{\sqrt{x}}W_{\lambda-1/2,\,\mu-1/2}(x)-\left(\frac{2\mu-1}{2x}+\frac{1}{2}\right)W_{\lambda,\mu}(x) \tag{5}$$

and

$$\frac{dM_{\lambda,\mu}(x)}{dx}=\frac{2\mu}{\sqrt{x}}M_{\lambda-1/2,\,\mu-1/2}(x)-\left(\frac{2\mu-1}{2x}+\frac{1}{2}\right)M_{\lambda,\mu}(x), \tag{6}$$

and from (8), the relations

$$\frac{dW_{\lambda,\mu}(x)}{dx}=-\frac{W_{\lambda+1/2,\,\mu+1/2}(x)}{\sqrt{x}}+\left(\frac{2\mu+1}{2x}+\frac{1}{2}\right)W_{\lambda,\mu}(x) \tag{7}$$

and

$$\frac{dM_{\lambda,\mu}(x)}{dx}$$
$$=-\frac{\left(\lambda+\mu+\frac{1}{2}\right)}{(2\mu+1)\sqrt{x}}M_{\lambda+1/2,\,\mu+1/2}(x)+\left(\frac{2\mu+1}{2x}+\frac{1}{2}\right)M_{\lambda,\mu}(x). \tag{8}$$

From the formulas obtained above, it is easy to deduce relations which do not contain the differentiation operator. Thus, from (1) and (3) we find

$$(2\mu-x)\,W_{\lambda,\mu}(x)+\left(\lambda-\mu-\frac{1}{2}\right)\sqrt{x}\,W_{\lambda-1/2,\,\mu+1/2}(x)$$
$$+\sqrt{x}\,W_{\lambda+1/2,\,\mu-1/2}(x)=0; \tag{9}$$

from (3) and (7) we find

$$W_{\lambda,\mu}(x)+\frac{1}{\sqrt{x}}\left[\left(\mu-\lambda+\frac{1}{2}\right)W_{\lambda-1/2,\,\mu+1/2}(x)\right.$$
$$\left.-W_{\lambda+1/2,\,\mu+1/2}(x)\right]=0, \tag{10}$$

and from (1) and (5) we get

$$W_{\lambda,\mu}(x)-\frac{1}{\sqrt{x}}\left[\left(\lambda+\mu-\frac{1}{2}\right)W_{\lambda-1/2,\,\mu-1/2}(x)\right.$$
$$\left.+W_{\lambda+1/2,\,\mu-1/2}(x)\right]=0. \tag{11}$$

Similarly, from (2) and (4) we have

$$(x-2\mu)\,M_{\lambda,\mu}(x)+2\mu\sqrt{x}\,M_{\lambda+1/2,\,\mu-1/2}(x)$$
$$-\frac{\left(\mu-\lambda+\frac{1}{2}\right)\sqrt{x}}{2\mu+1}M_{\lambda-1/2,\,\mu+1/2}(x)=0, \tag{12}$$

from (4) and (8) we find

$$M_{\lambda,\mu}(x)-\frac{1}{(2\mu+1)\sqrt{x}}\left[\left(\mu-\lambda+\frac{1}{2}\right)M_{\lambda-1/2,\,\mu+1/2}(x)\right.$$
$$\left.+\left(\lambda+\mu+\frac{1}{2}\right)M_{\lambda+1/2,\,\mu+1/2}(x)\right]=0, \tag{13}$$

and from (2) and (6) we have

$$M_{\lambda,\mu}(x)+\frac{2\mu}{\sqrt{x}}\left(M_{\lambda+1/2,\,\mu-1/2}(x)-M_{\lambda-1/2,\,\mu-1/2}(x)\right)=0. \tag{14}$$

3. **The Whittaker differential equation.** We now derive the differential equation satisfied by the Whittaker functions $M_{\lambda,\mu}(x)$ and $W_{\lambda,\mu}(x)$. For this we note that because of formula (1) of subsection 2 we have

$$\left(\sqrt{x}\frac{d}{dx}+\frac{2\mu-1}{2\sqrt{x}}-\frac{\sqrt{x}}{2}\right)W_{\lambda,\mu}(x)=-W_{\lambda+1/2,\,\mu-1/2}(x).$$

From formula (3) of subsection 2 we obtain

$$\left(\sqrt{x}\frac{d}{dx}-\frac{\mu}{\sqrt{x}}+\frac{\sqrt{x}}{2}\right)W_{\lambda+1/2,\,\mu-1/2}(x)=\left(\lambda-\mu+\frac{1}{2}\right)W_{\lambda,\mu}(x).$$

Therefore

$$\left(\sqrt{x}\frac{d}{dx}-\frac{\mu}{\sqrt{x}}+\frac{\sqrt{x}}{2}\right)\left(\sqrt{x}\frac{d}{dx}+\frac{2\mu-1}{2\sqrt{x}}-\frac{\sqrt{x}}{2}\right)W_{\lambda,\mu}(x)$$
$$=\left(\mu-\lambda-\frac{1}{2}\right)W_{\lambda,\mu}(x).$$

On simplification, we see that $W_{\lambda,\mu}(x)$ satisfies the differential equation

$$\frac{d^2y}{dx^2}+\left(-\frac{1}{4}+\frac{\lambda}{x}+\frac{\frac{1}{4}-\mu^2}{x^2}\right)y=0. \tag{1}$$

This equation is called the Whittaker differential equation. The same differential equation follows from equalities (5) and (7) of subsection 2.

In exactly the same way, from formulas (2) and (4) of subsection 2 it follows that $M_{\lambda,\mu}(x)$ is also a solution of equation (1). The same conclusion can be reached by making use of formulas (6) and (8).

Combining in different ways the recurrence relations obtained in subsection 2, we obtain relations connecting $M_{\lambda,\mu}(x)$ and $W_{\lambda,\mu}(x)$ with the functions $M_{\lambda\pm1,\mu}(x)$, $M_{\lambda,\mu\pm1}(x)$, $W_{\lambda\pm1,\mu}(x)$, $W_{\lambda,\mu\pm1}(x)$. For example, from relations (1) and (5) of subsection 2 it follows that

$$\left(\sqrt{x}\frac{d}{dx}+\frac{\mu-1}{\sqrt{x}}+\frac{\sqrt{x}}{2}\right)\left(\sqrt{x}\frac{d}{dx}+\frac{2\mu-1}{2\sqrt{x}}-\frac{\sqrt{x}}{2}\right)W_{\lambda,\mu}(x)$$
$$=-\left(\lambda+\mu-\frac{1}{2}\right)W_{\lambda,\mu-1}(x).$$

Subtracting from the simplified form of this relation the equality

$$x\frac{d^2W_{\lambda,\mu}}{dx^2}+\left(-\frac{x}{4}+\lambda+\frac{1-4\mu^2}{4x}\right)W_{\lambda,\mu}=0$$

(see formula (1)), we obtain

$$(2\mu-1)\frac{dW_{\lambda,\mu}(x)}{dx}+\left[\frac{(2\mu-1)^2}{2x}-\lambda\right]W_{\lambda,\mu}(x)$$
$$=-\left(\lambda+\mu-\frac{1}{2}\right)W_{\lambda,\mu-1}(x). \tag{2}$$

In the same way, from (1) and (7) of subsection 2 we get

$$x\frac{dW_{\lambda,\mu}(x)}{dx}+\left(\lambda-\frac{x}{2}\right)W_{\lambda,\mu}(x)=-W_{\lambda+1,\mu}(x), \tag{3}$$

from (3) and (5) we have

$$x\frac{dW_{\lambda,\mu}(x)}{dx}-\left(\lambda-\frac{x}{2}\right)W_{\lambda,\mu}(x)=\left[\left(\lambda-\frac{1}{2}\right)^2-\mu^2\right]W_{\lambda-1,\mu}(x), \tag{4}$$

and from (3) and (7) ,

$$(2\mu+1)\frac{dW_{\lambda,\mu}(x)}{dx}-\left[\frac{(2\mu+1)^2}{2x}-\lambda\right]W_{\lambda,\mu}(x)=\left(\lambda-\mu-\frac{1}{2}\right)W_{\lambda,\mu+1}(x). \tag{5}$$

Analogous relations for $M_{\lambda,\mu}(x)$ have the following form:

$$(2\mu-1)\frac{dM_{\lambda,\mu}(x)}{dx}+\left[\frac{(2\mu-1)^2}{2x}-\lambda\right]M_{\lambda,\mu}(x)=2\mu(2\mu-1)M_{\lambda,\mu-1}(x), \tag{6}$$

$$x\frac{dM_{\lambda,\mu}(x)}{dx}+\left(\lambda-\frac{x}{2}\right)M_{\lambda,\mu}(x)=\left(\lambda+\mu+\frac{1}{2}\right)M_{\lambda+1,\mu}(x), \tag{7}$$

$$x\frac{dM_{\lambda,\mu}(x)}{dx}-\left(\lambda-\frac{x}{2}\right)M_{\lambda,\mu}(x)=\left(\mu-\lambda+\frac{1}{2}\right)M_{\lambda-1,\mu}(x), \tag{8}$$

$$(2\mu+1)\frac{dM_{\lambda,\mu}(x)}{dx}-\left[\frac{(2\mu+1)^2}{2x}-\lambda\right]M_{\lambda,\mu}(x)$$
$$=\frac{\left(\mu+\frac{1}{2}\right)^2-\lambda^2}{2(\mu+1)(2\mu+1)}M_{\lambda,\mu+1}(x). \tag{9}$$

4. **Symmetry relations for the Whittaker functions.** We have proved that the Whittaker functions $M_{\lambda,\mu}(x)$ and $W_{\lambda,\mu}(x)$ are solutions of the linear second order differential equation

$$\frac{d^2y}{dx^2}+\left(-\frac{1}{4}+\frac{\lambda}{x}+\frac{\frac{1}{4}-\mu^2}{x^2}\right)y=0. \tag{1}$$

This equation is unchanged on replacing $\mu$ by $-\mu$, and also on replacing simultaneously $x$ by $-x$ and $\lambda$ by $-\lambda$.

Hence, along with functions $M_{\lambda,\mu}(x)$ and $W_{\lambda,\mu}(x)$, the functions

$$M_{\lambda,-\mu}(x),\quad M_{-\lambda,\mu}(-x),\quad M_{-\lambda,-\mu}(-x),$$
$$W_{\lambda,-\mu}(x),\quad W_{-\lambda,\mu}(-x),\quad W_{-\lambda,-\mu}(-x)$$

are also solutions of the Whittaker equation. Therefore all these functions must be linearly expressible by means of two linearly independent solutions (1). We choose $M_{\lambda,\mu}(x)$ and $M_{\lambda,-\mu}(x)$ as these two solutions (their linear independence is easily proved by calculating the Wronskian).

An expression for $M_{-\lambda,\mu}(-x)$ in terms of $M_{\lambda,\mu}(x)$ is immediately obtained from the integral representation (1) of §1.1. On replacing $\lambda$ by $-\lambda$ and $z$ by $-x$ in this formula and making the substitution $u = 1 - v$, we get

$$M_{-\lambda,\mu}(-x) = \frac{\Gamma(2\mu+1)(-x)^{\mu+1/2}e^{-x/2}}{\Gamma\left(\mu-\lambda+\frac{1}{2}\right)\Gamma\left(\mu+\lambda+\frac{1}{2}\right)}\int_0^1 u^{\mu-\lambda-1/2}(1-u)^{\mu+\lambda-1/2}e^{zu}\,du.$$

Comparing this formula with formula (1) of §1.1, we find that

$$(-x)^{-\mu-1/2}M_{-\lambda,\mu}(-x) = x^{-\mu-1/2}M_{\lambda,\mu}(x). \tag{2}$$

Similarly, $M_{-\lambda,-\mu}(-x)$ is expressible in terms of $M_{\lambda,-\mu}(x)$.

We shall now find a relation connecting $W_{\lambda,\mu}(x)$ with $M_{\lambda,\mu}(x)$ and $M_{\lambda,-\mu}(x)$. We shall seek it in the form

$$W_{\lambda,\mu}(x) = AM_{\lambda,\mu}(x) + BM_{\lambda,-\mu}(x).$$

Replacing the Whittaker functions in this equality by their integral representations (1) and (2) of §1.1, we get the relation

$$\begin{aligned}\int_0^\infty & e^{-zu}u^{\mu-\lambda-1/2}(1+u)^{\mu+\lambda-1/2}du \\ &= A\frac{\Gamma(2\mu+1)}{\Gamma\left(\mu+\lambda+\frac{1}{2}\right)}\int_0^1 u^{\mu-\lambda-1/2}(1-u)^{\mu+\lambda-1/2}e^{zu}\,du \\ &\quad + B\frac{\Gamma(1-2\mu)z^{-2\mu}\Gamma\left(\mu-\lambda+\frac{1}{2}\right)}{\Gamma\left(-\mu-\lambda+\frac{1}{2}\right)\Gamma\left(-\mu+\lambda+\frac{1}{2}\right)} \\ &\quad \times\int_0^1 u^{-\mu-\lambda-1/2}(1-u)^{-\mu+\lambda-1/2}e^{zu}\,du.\end{aligned} \tag{3}$$

The integrals appearing in this relation converge for $z = 0$ in the domain

$$\operatorname{Re}\left(\mu-\lambda+\frac{1}{2}\right)>0, \quad \operatorname{Re}\left(\mu+\lambda+\frac{1}{2}\right)>0, \ \operatorname{Re}\mu<0,$$
$$\operatorname{Re}\left(-\mu-\lambda+\frac{1}{2}\right)>0, \ \operatorname{Re}\left(-\mu+\lambda+\frac{1}{2}\right)>0.$$

In equality (3) we put $z = 0$ and make use of integrals (7) of subsection 6 and (2) of subsection 7 of §1, Chapter V. Since for $\operatorname{Re}\mu < 0$ the expression $z^{-2\mu}$ vanishes at the point $z = 0$, a simple calculation gives

$$A=\frac{\Gamma(-2\mu)}{\Gamma\left(\frac{1}{2}-\lambda-\mu\right)}.$$

In order to find the value of $B$, we make the substitution $zu = v$ in the integral on the left hand side of (3) and multiply both the sides by $z^{2\mu}$.

We get the relation

$$\int_0^\infty e^{-u}u^{\mu-\lambda-1/2}(z+u)^{\mu+\lambda-1/2}\,du$$
$$=A\frac{\Gamma(2\mu+1)}{\Gamma\left(\mu+\lambda+\frac{1}{2}\right)}z^{2\mu}\int_0^1 u^{\mu-\lambda-1/2}(1-u)^{\mu+\lambda-1/2}e^{zu}\,du$$
$$+B\frac{\Gamma(1-2\mu)\Gamma\left(\mu-\lambda+\frac{1}{2}\right)}{\Gamma\left(-\mu-\lambda+\frac{1}{2}\right)\Gamma\left(-\mu+\lambda+\frac{1}{2}\right)}\int_0^1 u^{-\mu-\lambda-1/2}(1-u)^{-\mu+\lambda-1/2}e^{zu}\,du.$$

For $z = 0$ the integrals converge in the domain

$$\operatorname{Re}\mu>0, \ \operatorname{Re}\left(\mu-\lambda+\frac{1}{2}\right)>0, \ \operatorname{Re}\left(\mu+\lambda+\frac{1}{2}\right)>0,$$
$$\operatorname{Re}\left(-\mu-\lambda+\frac{1}{2}\right)>0, \ \operatorname{Re}\left(-\mu+\lambda+\frac{1}{2}\right)>0.$$

Putting $z = 0$ we obtain

$$B=\frac{\Gamma(2\mu)}{\Gamma\left(\frac{1}{2}-\lambda+\mu\right)}.$$

Thus we have proved that

$$W_{\lambda,\mu}(z)=\frac{\Gamma(-2\mu)}{\Gamma\left(\frac{1}{2}-\lambda-\mu\right)}M_{\lambda,\mu}(z)+\frac{\Gamma(2\mu)}{\Gamma\left(\frac{1}{2}-\lambda+\mu\right)}M_{\lambda,-\mu}(z). \tag{4}$$

We note that the restrictions imposed on $\lambda$ and $\mu$ in calculating the coefficients $A$ and $B$ are removed by analytic continuation in $\lambda$ and $\mu$. Therefore formula (4) holds for all values of $\lambda$ and $\mu$.

Right side of expressions (4) is unchanged on replacing $\mu$ by $-\mu$. Hence we have the symmetry relation.

$$W_{\lambda,\mu}(z) = W_{\lambda,-\mu}(z).$$

We leave to the reader the derivation of the formula for $W_{-\lambda,\mu}(-z)$.

We have thus expressed the functions $W_{-\lambda,\mu}(-z)$, $W_{\lambda,\mu}(z)$ etc. by means of $M_{\lambda,\mu}(z)$ and $M_{\lambda,-\mu}(z)$.

## §4. INTEGRALS CONNECTED WITH THE WHITTAKER FUNCTIONS

1. The Mellin-Barnes representation. As in the case of the hypergeometric function, integrals connected with the Whittaker functions are derived from the relation

$$\mathbf{K}(\lambda, \mu; \chi; g_1g_2) = \int_{\rho-i\infty}^{\rho+i\infty} \mathbf{K}(\lambda, \nu; \chi; g_1)\mathbf{K}(\nu, \mu; \chi; g_2)\, d\nu, \tag{1}$$

which expresses the fact that

$$T_\chi(g_1g_2) = T_\chi(g_1)\,T_\chi(g_2). \tag{2}$$

However, one must bear in mind that in the matrix $\mathbf{K}(\lambda, \mu; \chi; g)$ certain elements are not defined here.

Choosing in different ways the elements $g_1$ and $g_2$ and comparing corresponding matrix elements on the left and on the right, we arrive at integral relations for the Whittaker functions. Thus, in order to obtain an integral representation of the Mellin-Barnes type, we shall start from the following obvious equality:

$$g_-(t)\,g_+(t) = g_1(t)\,z\left(-\frac{t^2}{2}\right). \tag{3}$$

For definiteness we shall take $\operatorname{Re}\sigma < 0$, As remarked in §2, in this case the value of $K_{++}(g_1(t))$ is defined. Since $K_{+-}(g_-(t)) = K_{-+}(g_-(t)) = 0$, it follows from (3) that

$$e^{-\sigma t^2/2}K_{++}(\lambda, \mu; \chi; g_1(t)) = \int_{\rho-i\infty}^{\rho+i\infty} K_{++}(\lambda, \nu; \chi; g_-(t))\,K_{++}(\nu, \mu; \chi; g_+(t))\, d\nu.$$

On substituting the values of $K_{++}(g_+(t))$, $K_{++}(g_-(t))$, $K_{++}(g_1(t))$ given by equalities (1), (5) and (8) of §2.6, and putting

$$-\sigma t^2 = z, \quad \frac{1-\lambda-\mu}{2} = \lambda', \quad \frac{\lambda-\mu}{2} = \mu',$$
$$\nu = \nu' - \lambda',$$

we get

$$W_{\lambda,\mu}(z) = \frac{e^{-z/2}}{2\pi i} \int_{\rho-i\infty}^{\rho+i\infty} \frac{\Gamma\left(\mu-\nu+\frac{1}{2}\right)}{\Gamma\left(\mu-\lambda+\frac{1}{2}\right)} \frac{\Gamma\left(\frac{1}{2}-\mu-\nu\right)}{\Gamma\left(\frac{1}{2}-\mu-\lambda\right)} \Gamma(\nu-\lambda) z^\nu \, d\nu. \tag{4}$$

From the convergence conditions of the integrals expressing the kernels $K_{++}(g_+(t))$ etc., arise the following restrictions on the parameters $\lambda$, $\mu$, $\nu$: $\operatorname{Re}(\mu + \frac{1}{2}) > \operatorname{Re}\nu$ and $\operatorname{Re}(\frac{1}{2} - \mu) > \operatorname{Re}\nu > \operatorname{Re}\lambda$.

Another expression for $W_{\lambda,\mu}(z)$ is obtained from the equality

$$g_+(t)\, g_-(t) = g_1(t)\, z\left(\frac{t^2}{2}\right). \tag{5}$$

This gives

$$K_{++}(\lambda,\ \mu;\ \chi;\ g_1(t))\, e^{\frac{\sigma t^2}{2}} = \int_{\rho-i\infty}^{\rho+i\infty} K_{++}(\lambda,\ \nu;\ \chi;\ g_+(t))\, K_{++}(\nu,\ \mu;\ \chi;\ g_-(t))\, d\nu.$$

Substituting the values of the kernels, we find

$$W_{\lambda,\mu}(z) = \frac{e^{z/2}}{2\pi i} \int_{\rho-i\infty}^{\rho+i\infty} \frac{\Gamma\left(\frac{1}{2}-\mu-\nu\right)\Gamma\left(\frac{1}{2}+\mu-\nu\right)}{\Gamma(1-\lambda-\nu)} z^\nu \, d\nu, \tag{6}$$

where

$$\operatorname{Re}(1-\lambda) > \operatorname{Re}\left(\frac{1}{2}-\mu\right) > \operatorname{Re}\nu, \quad \operatorname{Re}\left(\frac{1}{2}+\mu\right) > \operatorname{Re}\nu.$$

If we compare the expressions for $K_{-+}$, we get

$$K_{-+}(\lambda,\ \mu;\ \chi;\ g_1(t))\, e^{\frac{\sigma t^2}{2}} = \int_{\rho-i\infty}^{\rho+i\infty} K_{-+}(\lambda,\ \nu;\ \chi;\ g_+(t))\, K_{++}(\nu,\ \mu;\ \chi;\ g_-(t))\, d\nu.$$

Substituting the expressions for the kernels here, we find the following integral representation for the Whittaker function $M_{\lambda,\mu}(z)$:

$$M_{\lambda,\mu}(z) = \frac{\Gamma(2\mu+1)\, e^{z/2}}{2\pi i \Gamma\left(\lambda+\mu+\frac{1}{2}\right)} \int_{\rho-i\infty}^{\rho+i\infty} \frac{\Gamma\left(\mu-\nu+\frac{1}{2}\right)}{\Gamma\left(\mu+\nu+\frac{1}{2}\right)} \Gamma(\lambda+\nu)\, z^\nu \, d\nu, \tag{7}$$

where

$$-\operatorname{Re}\lambda<\operatorname{Re}\nu<\operatorname{Re}\left(\mu+\frac{1}{2}\right) \text{ и } \operatorname{Re}\left(\mu-\lambda+\frac{1}{2}\right)>0.$$

2. The Mellin transforms in parameters. From equality (1) of subsection 1 we can obtain a whole series of other integral relations for the Whittaker functions. In order to derive the Mellin transform formulas for the Whittaker functions in parameters, we consider the matrix

$$g_+(x)\,g_1(t)=\begin{pmatrix}1 & t+x & tx+\frac{t^2}{2}\\ 0 & 1 & t\\ 0 & 0 & 1\end{pmatrix}$$

and apply to it the factorizations (3) and (5) of §2.2.

If $t>0$, $x>0$ then this factorization has the form

$$g_+(x)\,g_1(t)=\varepsilon(\tau)\,g_1(r)\,\varepsilon(-\tau)\,z\left(\frac{tx}{2}\right), \tag{1}$$

where $g_+(x)$, $\epsilon(\tau)$, $g_1(t)$ are given by formulas (1), (4) and (6) of §2.1. Moreover,

$$z^2=t^2+tx,\quad e^{2\tau}=\frac{t}{t+x}.$$

Since to the matrix $\epsilon(\tau)$ there corresponds the operator of multiplication by $e^{(\omega-\lambda)\tau}$, and to the matrix $z(t)$ corresponds the operator of multiplication by $e^{\sigma t}$, it follows from formula (1) that

$$\int_{\rho-i\infty}^{\rho+i\infty}\mathbf{K}(\lambda,\ \nu;\ \chi;\ g_+(x))\,\mathbf{K}(\nu,\ \mu;\ \chi;\ g_1(t))\,d\nu$$
$$=e^{\frac{\sigma tx}{2}+\tau(\mu-\lambda)}\mathbf{K}(\lambda,\ \mu;\ \chi;\ g_1(r)).$$

We put $\sigma=-1$ here and compare the elements $K_{++}$ on the left and on the right. We get

$$\int_{\rho-i\infty}^{\rho+i\infty}K_{++}(\lambda,\ \nu;\ \chi;\ g_+(x))\,K_{++}(\nu,\ \mu;\ \chi;\ g_1(t))\,d\nu$$
$$=e^{-\frac{tx}{2}+\tau(\mu-\lambda)}K_{++}(\lambda,\ \mu;\ \chi;\ g_1(r)).$$

Substituting the expressions for $K_{++}(\lambda, \nu; \chi; g_+(x))$, etc., we arrive at the equality

$$\frac{1}{2\pi i}\int_{\rho-i\infty}^{\rho+i\infty}\Gamma(\nu)\,x^{-\nu}W_{\lambda-\nu/2,\ \mu+\nu/2}(t^2)\,d\nu$$
$$=e^{-tx/2}\left(1+\frac{x}{t}\right)^{\mu-1/2}W_{\lambda,\mu}(t^2+tx), \qquad (2)$$

where

$$\operatorname{Re}\nu>0>\operatorname{Re}\left(\lambda-\mu-\frac{1}{2}\right) \quad \text{и} \quad t>0,\quad x>0.$$

This formula can be regarded as the inverse Mellin transform for the function

$$F(\nu)=\Gamma(\nu)\,W_{\lambda-\nu/2,\ \mu+\nu/2}(t^2).$$

Therefore, by virtue of the inversion formula for the Mellin transform (see §4.2 of Chapter II), for $t > 0$, we have

$$\int_0^\infty x^{\nu-1}e^{-tx/2}\left(1+\frac{x}{t}\right)^{\mu-1/2}W_{\lambda,\mu}(t^2+tx)\,dx=\Gamma(\nu)\,W_{\lambda-\nu/2,\mu+\nu/2}(t^2).$$

In formula (1) we now take $x \equiv -y < 0$, $t > 0$. Then, comparing the elements $K_{-+}$, we similarly find that for $0 < y < t$:

$$J_1\equiv\frac{1}{2\pi i}\int_{\rho-i\infty}^{\rho+i\infty}\frac{\Gamma(\nu)}{\Gamma(2\mu+\nu+1)}\,y^{-\nu}M_{\lambda-\nu/2,\ \mu+\nu/2}(t^2)\,d\nu$$
$$=\frac{e^{ty/2}}{\Gamma(2\mu+1)}\left(1-\frac{y}{t}\right)^{\mu-1/2}M_{\lambda,\mu}(t^2-ty), \qquad (4)$$

where

$$\operatorname{Re}\nu>0>\operatorname{Re}\left(\lambda-\mu-\frac{1}{2}\right),\quad \operatorname{Re}\left(\lambda+\mu+\frac{1}{2}\right)>0;$$

for $y \geq t > 0$ we have $J_1 = 0$, where $\lambda$, $\mu$, $\nu$ satisfy the same inequalities.

By the inversion formula for the Mellin transform, we have

$$\frac{1}{\Gamma(2\mu+1)}\int_0^t y^{\nu-1}e^{ty/2}\left(1-\frac{y}{t}\right)^{\mu-1/2}M_{\lambda,\mu}(t^2-ty)\,dy$$
$$=\frac{\Gamma(\nu)}{\Gamma(2\mu+\nu+1)}\,M_{\lambda-\nu/2,\ \mu+\nu/2}(t^2).$$

Considering the product $g_+(x)g_2(t)$, we similarly arrive at the following relations.

If $0 < x < t$, then

$$J_2 \equiv \frac{1}{2\pi i} \int_{\rho - i\infty}^{\rho + i\infty} \frac{\Gamma(\nu)}{\Gamma\left(\nu + \mu + \lambda + \frac{1}{2}\right)} x^{-\nu} W_{\lambda + \nu/2,\, \mu + \nu/2}(t^2)\, d\nu$$
$$= \frac{e^{-xt/2}}{\Gamma\left(\mu + \lambda + \frac{1}{2}\right)} \left(1 - \frac{x}{t}\right)^{\mu - 1/2} W_{\lambda, \mu}(t^2 - tx), \tag{6}$$

where

$$\operatorname{Re} \nu > 0 > \operatorname{Re}\left(-\frac{1}{2} - \lambda - \mu\right) \text{ и } \operatorname{Re}\left(\mu - \lambda - \frac{1}{2}\right) < 0. \tag{7}$$

But if $x > t > 0$, then

$$J_2 = \frac{e^{-xt/2}}{\Gamma\left(\mu - \lambda + \frac{1}{2}\right)} \left(\frac{x}{t} - 1\right)^{\mu - 1/2} W_{-\lambda, \mu}(tx - t^2), \tag{8}$$

where $\lambda$, $\mu$, $\nu$ satisfy the same inequalities (7).

Further, for $0 < x < t$ we have

$$J_3 \equiv \frac{1}{2\pi i} \int_{\rho - i\infty}^{\rho + i\infty} \frac{\Gamma(\nu)}{\Gamma(\nu + 2\mu + 1)} x^{-\nu} M_{\lambda + \frac{\nu}{2},\, \mu + \frac{\nu}{2}}(t^2)\, d\nu$$
$$= \frac{e^{-xt/2}}{\Gamma(2\mu + 1)} \left(1 - \frac{x}{t}\right)^{\mu - 1/2} M_{\lambda, \mu}(t^2 - tx), \tag{9}$$

where $\lambda$, $\mu$, $\nu$ satisfy the same inequalities (7). But if $x > t > 0$ and $\lambda$, $\mu$, $\nu$ satisfy the inequalities (7), then $J_3 = 0$.

Lastly, for $0 < t < x$ we have

$$J_4 \equiv \frac{1}{2\pi i} \int_{\rho - i\infty}^{\rho + i\infty} \frac{\Gamma\left(\frac{1}{2} - \lambda - \mu - \nu\right)}{\Gamma(1 - \nu)} x^{-\nu} W_{\lambda + \frac{\nu}{2},\, \mu + \frac{\nu}{2}}(t^2)\, d\nu = 0, \tag{10}$$

where

$$\operatorname{Re}\left(\lambda + \mu + \nu - \frac{1}{2}\right) < 0, \quad \operatorname{Re}\left(\lambda + \mu + \frac{1}{2}\right) > 0, \quad \operatorname{Re}\left(\lambda - \mu - \frac{1}{2}\right) < 0.$$

But if $x > t > 0$, then

$$J_4 = \frac{e^{-xt/2}}{\Gamma(2\mu + 1)} \left(\frac{x}{t} - 1\right)^{\mu - 1/2} M_{-\lambda, \mu}(tx - t^2), \tag{11}$$

where $\lambda$, $\mu$, $\nu$ satisfy the inequalities (7).

Applying the inversion formula for the Mellin transform, from equality (9) and from the fact that $J_3 = 0$ for $x > t > 0$, we deduce the relation

$$\frac{1}{\Gamma(2\mu+1)} \int_0^t x^{\nu-1} e^{-xt/2} \left(1 - \frac{x}{t}\right)^{\mu-\frac{1}{2}} M_{\lambda,\mu}(t^2 - tx)\, dx = \frac{\Gamma(\nu)}{\Gamma(\nu+2\mu+1)} M_{\lambda+\nu/2,\,\mu+\nu/2}(t^2). \tag{12}$$

From formulas (6) and (8) one can similarly deduce that

$$\frac{1}{\Gamma(2\mu+1)} \int_t^\infty x^{\nu-1} e^{-xt/2} \left(\frac{x}{t} - 1\right)^{\mu-1/2} M_{-\lambda,\mu}(tx - t^2)\, dx = \frac{\Gamma\left(\frac{1}{2} - \lambda - \mu - \nu\right)}{\Gamma(1-\nu)} W_{\lambda+\nu/2,\,\mu+\nu/2}(t^2). \tag{13}$$

We also note the relation

$$\frac{1}{2\pi i} \int_{\rho-i\infty}^{\rho+i\infty} \Gamma(\nu)\Gamma\left(\frac{1}{2} - \lambda - \mu - \nu\right) x^{-\nu} W_{\lambda+\nu/2,\,\mu+\nu/2}(t^2)\, d\nu = \Gamma\left(\frac{1}{2} - \lambda - \mu\right) e^{\frac{xt}{2}} \left(1 + \frac{x}{t}\right)^{\mu-\frac{1}{2}} W_{\lambda,\mu}(tx + t^2), \tag{14}$$

where

$$0 < \operatorname{Re}\nu < \operatorname{Re}\left(\frac{1}{2} - \lambda - \mu\right) \text{ and } \operatorname{Re}(\lambda - \mu) < \frac{1}{2}.$$

This follows from a consideration of $g_+(-x)g_2(t)$ for $x > 0$, $t > 0$.

Owing to the Mellin inversion formula we have

$$\Gamma\left(\frac{1}{2} - \lambda - \mu\right) \int_0^\infty x^{\nu-1} e^{xt/2} \left(1 + \frac{x}{t}\right)^{\mu-\frac{1}{2}} W_{\lambda,\mu}(tx + t^2)\, dx = \Gamma(\nu)\Gamma\left(\frac{1}{2} - \lambda - \mu - \nu\right) W_{\lambda+\nu/2,\,\mu+\nu/2}(t^2). \tag{15}$$

**3. Continual composition theorems.** In order to obtain continual composition theorems for the Whittaker functions, we consider matrices of the form

$$g_i(t_1)\,\varepsilon(\tau)\,g_j(t_2), \tag{1}$$

where $i$ and $j$ are equal to 1 or 2, and the matrices $\epsilon(\tau)$, $g_1(t)$, $g_2(t)$ are given by formulas (4), (6) and (7) of §2.1. As proved in §2.2, the matrix (1) can be represented in the form

$$g_i(t_1)\,\varepsilon(\tau)\,g_j(t_2)=\varepsilon(\tau_1)\,g_k(r)\,\varepsilon(\tau-\tau_1)\,z(b), \tag{2}$$

where $z(b)$ is defined by formula (3) of §2.1 and $g_k(r)$ is a matrix of one of the four types

$$g_1(r),\ g_2(r),\ g_+(r)\ \text{ or }\ g_-(r).$$

We know that to the matrix $z(b)$ there corresponds the operator of multiplication by $e^{\sigma b}$, and to the matrix $\epsilon(\tau)$ corresponds the operator of multiplication by $e^{\tau(\omega-\lambda)}$. Therefore it follows from (2) that

$$\int_{\rho-i\infty}^{\rho+i\infty} \mathbf{K}(\lambda,\ \nu;\ \chi;\ g_i(t_1))\,\mathbf{K}(\nu,\ \mu;\ \chi;\ g_j(t_2))\,e^{\tau(\mu-\nu)}\,d\nu$$
$$=e^{\sigma b+\tau_1(\mu-\nu)}\,\mathbf{K}(\lambda,\ \mu;\ \chi;\ g_k(r)). \tag{3}$$

Relation (3) is the general form of the composition theorem for the Whittaker functions. In order to obtain particular formulas from it, one must consider different choices of the indices $i$, $j$ and of the values of $t_1$, $\tau$, $t_2$.

We consider in detail the case when $i=j=1$. If $t_1\equiv t>0$, $t_2\equiv s>0$, then

$$g_1(t)\,\varepsilon(\tau)\,g_1(s)=\varepsilon(\tau_1)\,g_1(r)\,\varepsilon(\tau-\tau_1)\,z(b),$$

where

$$b=ts\operatorname{sh}\tau,\quad r^2=t^2+2ts\operatorname{ch}\tau+s^2,\qquad e^{\tau_1}=\frac{t+se^{\tau}}{r}.$$

Using these formulas and calculating the value of $K_{++}$ in (3), we get

$$\frac{1}{2\pi i}\int_{\rho-i\infty}^{\rho+i\infty}\Gamma\left(\nu+\frac{1}{2}\right)e^{-\nu\tau}W_{\frac{\lambda-\mu-\nu}{2},\,\frac{\lambda-\mu+\nu}{2}}(t^2)\,W_{\frac{\lambda+\mu-\nu}{2},\,\frac{\lambda+\mu+\nu}{2}}(s^2)\,d\nu$$
$$=\frac{ts}{r}\exp\left[(\lambda+\mu)\tau-2\mu\tau_1-ts\operatorname{sh}\tau\right]W_{\lambda,\mu}(r^2), \tag{4}$$

where

$$\operatorname{Re}(\lambda-\mu)<\frac{1}{2}\ \text{and}\ \operatorname{Re}\nu>-\frac{1}{2}.$$

We now consider the case when $t_1\equiv t>0$ and $t_2\equiv -s<0$, where $t>s$. Then for $e^{\tau}<s/t$ we have

$$g_1(t)\,\varepsilon(\tau)\,g_1(-s)=\varepsilon(\tau_1)\,g_2(r)\,\varepsilon(\tau-\tau_1)\,z(b),$$

where

$$b=-ts\operatorname{sh}\tau,\quad r^2=2ts\operatorname{ch}\tau-t^2-s^2,\qquad e^{\tau_1}=\frac{t-se^{\tau}}{r}.$$

This leads us to the formula

$$J_5 \equiv \frac{1}{2\pi i} \int_{\rho - i\infty}^{\rho + i\infty} \frac{\Gamma\left(\nu + \frac{1}{2}\right)}{\Gamma(\nu + \lambda + \mu + 1)} e^{-\tau\nu} \times W_{\frac{\lambda-\mu-\nu}{2}, \frac{\lambda-\mu+\nu}{2}}(t^2) M_{\frac{\lambda+\mu-\nu}{2}, \frac{\lambda+\mu+\nu}{2}}(s^2)\, d\nu$$
$$= \frac{1}{\Gamma(2\mu+1)} \frac{ts}{r} \exp\left[(\lambda+\mu)\tau - 2\mu\tau_1 + ts \operatorname{sh} \tau\right] M_{-\lambda,\mu}(r^2), \tag{5}$$

where

$$\operatorname{Re}\left(\lambda - \mu - \frac{1}{2}\right) < 0, \quad \operatorname{Re}\left(\nu + \frac{1}{2}\right) > 0, \quad \operatorname{Re}\left(\lambda + \mu + \frac{1}{2}\right) > 0. \tag{6}$$

If $e^{\tau} > s/t$, then we have $J_5 = 0$, where $\lambda$, $\mu$, $\nu$ satisfy inequalities (6).

In the case $s > t > 0$ we can prove the following relations in exactly the same way.

If $e^{\tau} < t/s$, then $J_5$ is expressed by formula (5), and if $e^{\tau} > s/t$, then $J_5 = 0$ But in case $t/s < e^{\tau} < s/t$ we have

$$g_1(t)\,\varepsilon(\tau)\,g_1(-s) = \varepsilon(\tau_1)\,g_1(-r)\,\varepsilon(\tau - \tau_1)\,z(b),$$

where

$$b = -ts \operatorname{sh} \tau, \quad r^2 = s^2 + t^2 - 2ts \operatorname{ch} \tau, \quad e^{\tau_1} = \frac{se^{\tau} - t}{r}.$$

Moreover

$$J_5 = \frac{1}{\Gamma(2\mu+1)} \frac{ts}{r} \exp\left[(\lambda+\mu)\tau - 2\mu\tau_1 + ts \operatorname{sh} \tau\right] M_{\lambda,\mu}(r^2), \tag{7}$$

where $\lambda$, $\mu$, $\nu$ satisfy (6).

The case when $i \neq j$ leads to a number of new formulas of the same type as considered above. Let us first consider the product

$$g_1(t_1)\,\varepsilon(\tau)\,g_2(t_2).$$

If $t_1 \equiv t > 0$, $t_2 \equiv s > 0$ and $e^{\tau} < s/t$, then

$$g_1(t_1)\,\varepsilon(\tau)\,g_2(s) = \varepsilon(\tau_1)\,g_2(r)\,\varepsilon(\tau - \tau_1)\,z(b),$$

where

$$b = ts \operatorname{ch} \tau, \quad r^2 = s^2 - t^2 - 2ts \operatorname{sh} \tau, \quad e^{\tau_1} = \frac{t + e^{\tau} s}{r}.$$

In exactly the same way we get

$$J_6 \equiv \frac{1}{2\pi i} \int_{\rho - i\infty}^{\rho + i\infty} e^{-\tau\nu} W_{\frac{\lambda - \mu - \nu}{2}, \frac{\lambda - \mu + \nu}{2}}(t^2)\, W_{\frac{\nu - \lambda - \mu}{2}, \frac{\nu + \lambda + \mu}{2}}(s^2)\, d\nu$$
$$= \frac{1}{\Gamma\left(\mu - \lambda + \frac{1}{2}\right)} \frac{ts}{r} \exp\left[(\lambda + \mu)\tau - 2\mu\tau_1 - ts \operatorname{ch} \tau\right] W_{-\lambda, \mu}(r^2), \tag{8}$$

where

$$\operatorname{Re}\left(\lambda - \mu - \frac{1}{2}\right) < 0, \quad \operatorname{Re}\left(\lambda + \mu + \frac{1}{2}\right) > 0. \tag{9}$$

Further,

$$J_7 \equiv \frac{1}{2\pi i} \int_{\rho - i\infty}^{\rho + i\infty} \frac{\Gamma\left(\nu + \frac{1}{2}\right)}{\Gamma(\nu + \lambda + \mu + 1)} e^{-\tau\nu} W_{\frac{\lambda - \mu - \nu}{2}, \frac{\lambda - \mu + \nu}{2}}(t^2)$$
$$\times M_{\frac{\nu - \lambda - \mu}{2}, \frac{\nu + \lambda + \mu}{2}}(s^2)\, d\nu$$
$$= \frac{1}{\Gamma(2\mu + 1)} \frac{ts}{r} \exp\left[(\lambda + \mu)\tau - 2\mu\tau_1 - ts \operatorname{ch} \tau\right] M_{-\lambda, \mu}(r^2), \tag{10}$$

where

$$\operatorname{Re}\left(\lambda - \mu - \frac{1}{2}\right) < 0, \quad \operatorname{Re}\left(\lambda + \mu + \frac{1}{2}\right) > 0, \quad \operatorname{Re} \nu > -\frac{1}{2} \tag{11}$$

and, finally,

$$J_8 \equiv \frac{1}{2\pi i} \int_{\rho - i\infty}^{\rho + i\infty} \frac{\Gamma\left(\frac{1}{2} - \nu\right)}{\Gamma(\mu - \lambda - \nu + 1)} e^{-\tau\nu} M_{\frac{\lambda - \mu - \nu}{2}, \frac{\mu - \lambda - \nu}{2}}(t^2)$$
$$\times W_{\frac{-\lambda - \mu + \nu}{2}, \frac{-\lambda - \mu - \nu}{2}}(s^2)\, d\nu = 0, \tag{12}$$

where

$$\operatorname{Re}\left(\lambda - \mu - \frac{1}{2}\right) < 0, \quad \operatorname{Re}\left(\lambda + \mu + \frac{1}{2}\right) > 0, \quad \operatorname{Re} \nu < \frac{1}{2}. \tag{13}$$

But if $e^{\tau} > s/t$, then

$$b = ts \operatorname{ch} \tau, \quad r^2 = t^2 - s^2 + 2ts \operatorname{sh} \tau, \qquad e^{\tau_1} = \frac{t + e^{\tau} s}{r}$$

and

$$J_6 = \frac{1}{\Gamma\left(\frac{1}{2} + \lambda + \mu\right)} \frac{ts}{2} \exp\left[(\lambda + \mu)\tau - 2\mu\tau_1 - ts \operatorname{ch} \tau\right] W_{\lambda, \mu}(r^2), \tag{14}$$

$$J_7 = 0, \tag{15}$$

$$J_8 = \frac{1}{\Gamma(2\mu + 1)} \frac{ts}{r} \exp\left[(\lambda + \mu)\tau - 2\mu\tau_1 - ts \operatorname{ch} \tau\right] M_{\lambda, \mu}(r^2). \tag{16}$$

In [125] the reader can find yet other formulas, obtained by considering a product of the form

$$g_2(t_2)\, \varepsilon(\tau)\, g_1(t_1).$$

4. **Dual formulas.** The Fourier inversion formulas in complex form can be written as follows:

$$\Phi(\nu) = \int_{-\infty}^{\infty} f(x)\, e^{\nu x} dx,$$

where

$$f(x) = \frac{1}{2\pi i} \int_{\rho - i\infty}^{\rho + i\infty} \Phi(\nu)\, e^{-\nu x} d\nu$$

(see §4.1 of Chapter II). Using these, we get a number of new relations from the formulas of the preceding subsection.

Thus, from formula (4) of subsection 3 we obtain

$$\int_{-\infty}^{\infty} \exp\left[(\lambda + \mu + \nu)\tau - 2\mu\tau_1 - ts \operatorname{sh} \tau\right] r^{-1} W_{\lambda, \mu}(r^2)\, d\tau$$
$$= \Gamma\left(\nu + \frac{1}{2}\right) t^{-1} W_{\frac{\lambda - \mu - \nu}{2},\, \frac{\lambda - \mu + \nu}{2}}(t^2)\, s^{-1} W_{\frac{\lambda + \mu - \nu}{2},\, \frac{\lambda + \mu + \nu}{2}}(s^2), \tag{1}$$

where

$$r^2 = t^2 + 2ts \operatorname{ch} \tau + s^2, \qquad e^{\tau_1} = \frac{t + se^{\tau}}{r}.$$

Similarly, from formula (5) and the equality $J_5 = 0$ for $e^{\tau} > s/t$ for $t > s > 0$ we get

$$\frac{1}{\Gamma(2\mu+1)}\int\limits_{-\infty}^{\ln\frac{s}{t}} \exp\left[(\lambda+\mu+\nu)\tau - 2\mu\tau_1 + ts\operatorname{sh}\tau\right] r^{-1}M_{-\lambda,\mu}(r^2)\,d\tau$$
$$=\frac{\Gamma\left(\nu+\frac{1}{2}\right)}{\Gamma(\nu+\lambda+\mu+1)}t^{-1}W_{\frac{\lambda-\mu-\nu}{2},\ \frac{\lambda-\mu+\nu}{2}}(t^2)\,s^{-1}M_{\frac{\lambda+\mu-\nu}{2},\ \frac{\lambda+\mu+\nu}{2}}(s^2), \qquad (2)$$

where

$$r^2 = 2ts\operatorname{ch}\tau - t^2 - s^2, \qquad e^{\tau_1} = \frac{t - se^{\tau}}{r}.$$

Further, from formulas (8) and (11) it follows that

$$\frac{1}{\Gamma(2\mu+1)}\int\limits_{-\infty}^{\ln\frac{s}{t}} \exp\left[(\lambda+\mu+\nu)\tau - 2\mu\tau_1 - ts\operatorname{ch}\tau\right] r^{-1}M_{-\lambda,\mu}(r^2)\,d\tau$$
$$=\frac{\Gamma\left(\nu+\frac{1}{2}\right)}{\Gamma(\nu+\lambda+\mu+1)}t^{-1}W_{\frac{\lambda-\mu-\nu}{2},\ \frac{\lambda-\mu+\nu}{2}}(t^2)\,s^{-1}M_{\frac{\nu-\lambda-\mu}{2},\ \frac{\nu+\lambda+\mu}{2}}(s^2), \qquad (3)$$

where

$$r^2 = s^2 - t^2 - 2ts\operatorname{sh}\tau, \qquad e^{\tau_1} = \frac{t + e^{\tau}s}{r},$$

and from formulas (9) and (12) that

$$\frac{1}{\Gamma(2\mu+1)}\int\limits_{\ln\frac{s}{t}}^{\infty} \exp\left[(\lambda+\mu+\nu)\tau - 2\mu\tau_1 - ts\operatorname{ch}\tau\right] M_{\lambda,\mu}(r^2)\,d\tau$$
$$=\frac{\Gamma\left(\frac{1}{2}-\nu\right)}{\Gamma(\mu-\lambda-\nu+1)}t^{-1}M_{\frac{\lambda-\mu-\nu}{2},\ \frac{\mu-\lambda-\nu}{2}}(t^2)\,s^{-1}W_{\frac{-\lambda-\mu+\nu}{2},\ \frac{-\lambda-\mu-\nu}{2}}(s^2), \qquad (4)$$

where

$$r^2 = t^2 - s^2 + 2ts\operatorname{sh}\tau, \qquad e^{\tau_1} = \frac{t + e^{\tau}s}{r}.$$

5. **Degenerate cases of composition theorems.** In the case of certain relations between the numbers $t_1$, $r$, $t_2$, the matrix $g_i(t_1)\epsilon(r)g_j(t_2)$ is expressed not in terms of $g_1(r)$ or $g_2(r)$ but in terms of $g_+(r)$ or $g_-(r)$. In these cases the

corresponding integrals are expressed not in terms of the Whittaker functions but in terms of the power function. We state the corresponding formulas.

If $i = j = 1$, $t > s > 0$ and $e^{\tau} = s/t$, then

$$g_1(t)\,\varepsilon(\tau)\,g_1(-s) = \varepsilon(\tau)\,g_-(r)\,z(b),$$

where

$$r = \frac{t^2 - s^2}{s}, \qquad b = \frac{t^2 - s^2}{2}.$$

Since $K_{+-}[g_-(r)] = 0$, the integral $J_5$ of subsection 3 vanishes for $e^{\tau} = s/t$, $t > s > 0$.

If $s > t > 0$, then for $e^{\tau} = t/s$

$$g_1(t)\,\varepsilon(\tau)\,g_1(-s) = \varepsilon(\tau)\,g_+(-r)\,z(b),$$

where

$$r = \frac{s^2 - t^2}{s}, \qquad b = \frac{s^2 - t^2}{2}.$$

Hence, for $s > t > 0$, $e^{\tau} = t/s$, the integral $J_5$ takes the value

$$J_5 = \frac{1}{\Gamma(2\mu + 1)} t^{\lambda-\mu+1} s^{1-\lambda-\mu} (s^2 - t^2)^{2\mu} \exp\left[\frac{t^2 - s^2}{2}\right].$$

One similarly proves that if $s > t > 0$ and $e^{\tau} = s/t$, then $J_5 = 0$.

We shall mention one other result of the same type. If $t > 0$, $s > 0$, $e^{\tau} = s/t$, then the integral $J_6$ of subsection 3 takes the value

$$J_6 = \frac{\Gamma(2\mu)}{\Gamma\left(\frac{1}{2} - \lambda + \mu\right)\Gamma\left(\frac{1}{2} + \lambda + \mu\right)} t^{\mu-\lambda+1} \times s^{\lambda+\mu+1} (s^2 + t^2)^{-2\mu} \exp\left[-\frac{t^2 + s^2}{2}\right].$$

But the integrals $J_7$ and $J_8$ of subsection 3 vanish for $t > 0$, $s > 0$, $e^{\tau} = s/t$.

## §5. THE LAGUERRE POLYNOMIALS AND REPRESENTATIONS OF THE GROUP OF COMPLEX THIRD ORDER TRIANGULAR MATRICES

1. Definition of the Laguerre polynomials. If in the confluent hypergeometric function $\Phi(\alpha;\ \gamma;\ x)$ the parameter $\alpha$ is a negative integer or zero, then the series terminates. Thus $\Phi(-n;\ \gamma;\ x)$ is a polynomial. We consider the polynomials

$$L_n^\alpha(x)=\frac{\Gamma(n+\alpha+1)}{n!\,\Gamma(\alpha+1)}\Phi(-n;\ \alpha+1;\ x), \tag{1}$$

called the Laguerre polynomials.

Many properties of the Laguerre polynomials follow directly from the corresponding properties of the confluent hypergeometric function. Thus, from expansion (1) of §1.2 it follows that

$$L_n^\alpha(x)=\sum_{m=0}^{n}(-1)^m\frac{\Gamma(n+\alpha+1)}{(n-m)!\,\Gamma(\alpha+m+1)}\frac{x^m}{m!} \tag{2}$$

Further, the Laguerre polynomials are solutions of the linear second order differential equation

$$z\frac{d^2z}{du^2}+(\alpha-z+1)\frac{du}{dz}+nu=0, \tag{3}$$

which is immediately obtained from equation (3) of §1.2.

From (2) it easily follows that

$$L_n^\alpha(x)=\frac{1}{n!}e^x x^{-\alpha}\frac{d^n}{dx^n}(e^{-x}x^{n+\alpha}). \tag{4}$$

For a proof it is sufficient to use Leibnitz's formula. We now give another proof. For this we note that, because of formulas (1) of §1.1 and 2 of §1.2, we have the following integral representation of the confluent hypergeometric function $\Phi(\alpha;\ \gamma;\ x)$:

$$\Phi(\alpha;\ \gamma;\ x)=\frac{\Gamma(\gamma)e^x}{\Gamma(\alpha)\Gamma(\gamma-\alpha)}\int_0^1 e^{-ux}u^{\gamma-\alpha-1}(1-u)^{\alpha-1}du. \tag{5}$$

For $\alpha=-n$ this loses its meaning. But in the theory of generalized functions [1] it is proved that

$$\lim_{\alpha\to-n}\frac{(1-u)_+^{\alpha-1}}{\Gamma(\alpha)}=\delta^{(n)}(1-u).$$

Hence, passing to the limit in (5) as $\alpha\to-n$, we obtain

$$\Phi(-n;\ \gamma;\ x)=\frac{\Gamma(\gamma)\,e^x}{\Gamma(\gamma+n)}\frac{d^n}{du^n}(e^{-ux}u^{\gamma-\alpha-1})\Big|_{u=1}.$$

Taking (1) into consideration, we have

---

1) See [18], Chapter 1, §3.5.

$$L_n^\alpha(x) = \frac{1}{n!} e^x \frac{d^n}{du^n} [e^{-ux} u^{\gamma-\alpha-1}] \Big|_{u=1}.$$

The substitution $ux = y$ transforms this relation into the form (4).

In conclusion, we establish the orthogonality relation for the Laguerre poly nomials. It has form

$$\int_0^\infty x^\alpha e^{-x} L_m^\alpha(x) L_n^\alpha(x)\, dx = \begin{cases} 0, & \text{if} \quad m \neq n, \\ \dfrac{\Gamma(\alpha+n+1)}{n!}, & \text{if} \quad m = n. \end{cases}$$

The validity of (6) for $m < n$ is easily proved by substituting for $L_m^\alpha(x)$ and $L_n^\alpha(x)$ from the expression (4) and integrating by parts $(m+1)$ times. Since $L_m^\alpha(x)$ is a polynomial of degree $m$, it vanishes in the process.

In the case $m = n$ the proof is carried out in exactly the same way. After integrating by parts $n$ times, we obtain

$$\int_0^\infty x^\alpha e^{-x} [L_n^\alpha(x)]^2\, dx = \frac{1}{n!} \int_0^\infty e^{-x} x^{\alpha+n} dx.$$

An application of equality (1) of §1.4, Chapter V, leads to the required result.

In conclusion we note that, for $|z| < 1$,

$$e^{xz}(1+z)^\alpha = \sum_{n=0}^\infty L_n^{\alpha-n}(x) z^n, \tag{7}$$

as is immediately proved by expanding the functions in the left hand side in powers of $z$ and multiplying the resulting series.

2. The group of complex third order triangular matrices and the Laguerre polynomials. In order to derive the composition theorem for the Laguerre polynomials, we establish the connection between these polynomials and representa tions of the group $G_1$ of matrices of the form

$$\begin{pmatrix} 1 & a & b \\ 0 & c & d \\ 0 & 0 & 1 \end{pmatrix},$$

where $a, b, c, d$ are complex numbers and $c \neq 0$. The representations of this group include those which are "analytic extensions" of the representation $T_\chi(g)$ of the group $G$ of §2.3. They are given by the formula

$$T_\chi(g)f(z)=c^\omega e^{\sigma(dz+b)}f(cz+a),$$

where the operators $T_\chi(g)$ act in the space of integral analytic functions of exponential type.

As the natural basis in the space of integral functions, we have the monomial set $\{z^k\}$, $0 \leq k < \infty$. Let us calculate the matrix elements of the representation $T_\chi(g)$ in this basis. We have

$$T_\chi(g)(z^n)=c^\omega e^{\sigma(dz+b)}(cz+a)^n.$$

Hence, using equality (7) of subsection 1, we get

$$T_\chi(g)(z^n)=\sum_{k=0}^{\infty} c^{\omega+k}a^{n-k}e^{\sigma b}L_k^{n-k}\left(-\frac{\sigma ad}{c}\right)z^k.$$

Consequently the required matrix elements are expressed as follows:

$$t_{kn}^\chi(g)=c^{\omega+k}a^{n-k}e^{\sigma b}L_k^{n-k}\left(-\frac{\sigma ad}{c}\right). \tag{1}$$

Using the equality

$$T_\chi(g_1)T_\chi(g_2)=T_\chi(g_1g_2),$$

it is now easy to derive the composition formula for the Laguerre polynomials. Namely, taking $\sigma = -1$, setting

$$g_1=\begin{pmatrix}1 & a_1 & b_1\\ 0 & c_1 & d_1\\ 0 & 0 & 1\end{pmatrix}, \qquad g_2=\begin{pmatrix}1 & a_2 & b_2\\ 0 & c_2 & d_2\\ 0 & 0 & 1\end{pmatrix}$$

and introducing numbers $a$, $b$, $x$, $y$ by the relations

$$a=a_1, \quad b=\frac{a_2}{c_2}, \quad x=\frac{d_1}{c_1}, \quad y=d_2,$$

we get

$$L_k^\alpha[(a+b)(x+y)]=(a+b)^{-\alpha}e^{ay}\sum_{m=-k}^{\infty} a^m b^{\alpha-m}L_k^m(ax)L_{k+m}^{\alpha-m}(by), \tag{2}$$

where $|a| < |b|$ (we have replaced $\alpha$ by $\alpha + k$ and $m$ by $m + k$). In particular, putting

$$a=r_1e^{\tau/2}, \quad x=r_1e^{-\tau/2}, \quad b=r_2e^{-\tau/2}, \quad y=r_2e^{\tau/2},$$

where $r_1 > 0$, $r_2 > 0$, $-2\pi < \tau < 2\pi$, we obtain the relation

$$L_k^\alpha(r_1^2+r_2^2+2r_1r_2\operatorname{ch}\tau)$$
$$=\left(\frac{r_2}{r_1e^\tau+r_2}\right)^\alpha \exp(r_1r_2e^\tau)\sum_{m=-k}^{\infty}\left(\frac{r_1}{r_2}\right)^m e^{m\tau}L_k^m(r_1^2)\,L_{k+m}^{\alpha-m}(r_2^2), \tag{3}$$

where $|r_1e^\tau|<r_2$. For $\tau=i\varphi$ relation (3) takes the form

$$L_k^\alpha(r_1^2+r_2^2+2r_1r_2\cos\varphi)$$
$$=\left(\frac{r_2}{r_1e^{i\varphi}+r_2}\right)^\alpha e^{r_1r_2\cos\varphi}\left[\cos(r_1r_2\sin\varphi]+i\sin(r_1r_2\sin\varphi)\right]$$
$$\times\sum_{m=-k}^{\infty}\left(\frac{r_1}{r_2}\right)^m L_k^m(r_1^2)\,L_{k+m}^{\alpha-m}(r^2)\,e^{im\varphi}, \tag{4}$$

where $-\pi\le\phi<\pi$, $0<r_1<r_2$.

CHAPTER IX

# THE GROUP OF ROTATIONS OF $n$-DIMENSIONAL EUCLIDEAN SPACE AND GEGENBAUER FUNCTIONS

The classes of special functions considered so far (Legendre functions and polynomials, Bessel functions, the functions of Hankel, Macdonald and Whittaker) were connected with representations of elementary groups consisting of matrices of the second or third order. We now turn to a study of special functions connected with representations of more complicated groups. We begin with the group $SO(n)$ of rotations of $n$-dimensional euclidean space. The special functions connected with this group are called Gegenbauer polynomials.

## § 1. THE GROUP $SO(n)$

1. Spherical coordinates. We shall denote vectors of the $n$-dimensional real euclidean space $E_n$ by $\mathbf{x} = (x_1, \cdots, x_n)$, and their length by $\|\mathbf{x}\|$ or $r$. The totality of vectors of the form $(x_1, \cdots, x_k, 0, \cdots, 0)$ will be denoted by $E_k$, and the unit sphere in $E_n$ by $S^{n-1}$. Obviously $S^{n-1}$ can be identified with the totality of vectors of unit length. As a rule, we shall denote unit vectors by $\xi$, $\eta$.

We now introduce spherical coordinates $r, \theta_1, \cdots, \theta_{n-1}$ in the space $E_n$. They are related to the cartesian coordinates $x_1, \cdots, x_n$ by the formulas

$$\left.\begin{array}{l} x_1 = r \sin\theta_{n-1} \ldots \sin\theta_2 \sin\theta_1, \\ x_2 = r \sin\theta_{n-1} \ldots \sin\theta_2 \cos\theta_1, \\ \cdot\ \cdot\ \cdot\ \cdot\ \cdot\ \cdot\ \cdot\ \cdot\ \cdot\ \cdot\ \cdot\ \cdot\ \cdot \\ x_{n-1} = r \sin\theta_{n-1} \cos\theta_{n-2}, \\ x_n = r\cos\theta_{n-1}. \end{array}\right\} \tag{1}$$

The numbers $r, \theta_1, \cdots, \theta_{n-1}$ here vary as follows:

$$\left.\begin{array}{ll} 0 \leqslant r < \infty, & \\ 0 \leqslant \theta_1 < 2\pi, & \\ 0 \leqslant \theta_k < \pi, & k \neq 1. \end{array}\right\} \tag{2}$$

It is evident that for every point $\mathbf{x}(x_1, \cdots, x_n)$ of $E_n$ there is a system of parameters $r, \theta_1, \cdots, \theta_{n-1}$, related to it by formulas (1) and lying in the domain (2). Moreover, for almost all points $\mathbf{x}$ [1] such a system of parameters is uniquely determined.

The parameters $r, \theta_1, \cdots, \theta_{n-1}$ are expressed in terms of cartesian coordinates by the formulas

$$\cos\theta_k = \frac{x_{k+1}}{r_{k+1}}, \tag{3}$$

$$\sin\theta_k = \frac{r_k}{r_{k+1}}, \tag{4}$$

$$r = r_n, \tag{5}$$

where

$$r_k^2 = x_1^2 + \ldots + x_k^2. \tag{6}$$

For fixed $r$ the point $\mathbf{x}(r, \theta_1, \cdots, \theta_{n-1})$ runs through the sphere of radius $r$. Therefore the numbers $\theta_1, \cdots, \theta_{n-1}$ can be regarded as coordinates on the unit sphere $S^{n-1}$. We shall call them spherical coordinates.

Let us find an expression for measure on the sphere $S^{n-1}$, invariant for rotations of the sphere around the origin. We observe that the measure $d\mathbf{x} = dx_1 \cdots \cdots dx_n$ in euclidean space is invariant for these rotations, as is the generalized function $\delta(r-1)$. Therefore, if for an arbitrary continuous function $f(\xi)$ on $S^{n-1}$ we put

$$\int f(\xi)\,d\xi = A\int f(\mathbf{x})\,\delta(r-1)\,d\mathbf{x}, \tag{7}$$

where $A$ is any positive number, we obtain an invariant measure on $S^{n-1}$ (here $f(\mathbf{x})$ denotes an arbitrary continuous function in euclidean space which coincides with $f(\xi)$ on $S^{n-1}$).

In cartesian coordinates formula (7) takes the form

$$d\xi = \frac{A\,dx_1 \ldots dx_{n-1}}{|x_n|}. \tag{8}$$

We pass from cartesian coordinates to spherical coordinates $\theta_1, \cdots, \theta_{n-1}$. An elementary calculation shows that

$$\frac{dx_1 \ldots dx_{n-1}}{|x_n|} = \sin^{n-2}\theta_{n-1} \ldots \sin\theta_2\, d\theta_1 \ldots d\theta_{n-1}.$$

---

1) We say that a certain assertion is valid for almost all points of a manifold $\mathfrak{M}$ if the points where it is false constitute a manifold of lesser dimension.

Therefore

$$d\xi = A \sin^{n-2}\theta_{n-1} \ldots \sin \theta_2 \, d\theta_1 \ldots d\theta_{n-1}.$$

We choose the constant $A$ such that the measure of the whole sphere is equal to unity. Since

$$\int_0^\pi \sin^{k-1} \theta_k \, d\theta_k = \frac{\Gamma\left(\frac{k}{2}\right) \sqrt{\pi}}{\Gamma\left(\frac{k+1}{2}\right)} \text{ and } \int_0^{2\pi} d\theta_1 = 2\pi ,$$

we get $A = \Gamma(n/2)/2\pi^{n/2}$. Hence

$$d\xi = \frac{\Gamma\left(\frac{n}{2}\right)}{2\pi^{n/2}} \sin^{n-2} \theta_{n-1} \ldots \sin \theta_2 \, d\theta_1 \ldots d\theta_{n-1}. \tag{9}$$

2. Description of the group $SO(n)$. By rotations of the euclidean space $E_n$ we mean linear transformations $g$ of this space, which do not change its orientation and leave invariant the distance of points from the origin:

$$\| g\mathbf{x} \| = \| \mathbf{x} \|.$$

Rotations of the $n$-dimensional euclidean space $E_n$ form a group, which we shall denote by $SO(n)$.

We choose the orthonormal basis $\xi_1, \cdots, \xi_n$ in the space $E_n$. We shall denote by $SO(k)$ the subgroup of $SO(n)$, consisting of those rotations which leave the vectors $\xi_{k+1}, \cdots, \xi_n$ of this basis invariant. $S^{n-1}$ can be identified with the space of left cosets of $SO(n)$ with respect to the subgroup $SO(n-1)$. Indeed, the rotations of $SO(n-1)$ (and only they) leave $\xi_n$ invariant. In other words, this subgroup is the stationary subgroup of the point $\xi_n(0, 0, \cdots, 1)$ of the unit sphere. Since $SO(n)$ is a transitive group of transformations of $S^{n-1}$, it follows that $S^{n-1}$ can be identified with $SO(n)/SO(n-1)$. Under this identification, to a left coset consisting of elements of the form $\{gh\}$, $h \in SO(n-1)$, corresponds the vector $g\xi_n$. In the sequel we shall often denote $g\xi_n$ by $\xi_g$.

To each element $g_0$ of the group $SO(n)$ corresponds a transformation

$$\{gh\} \to \{g_0gh\}, \quad h \in SO(n-1), \tag{1}$$

in the factor space $SO(n)/SO(n-1)$. Moreover, one has the equality

$$\xi_{g_0g} = g_0\xi_g. \tag{2}$$

This equality means that to transformations (1) in the factor space there correspond rotations of the sphere.

3. Euler angles. We have seen in §1.6 of Chapter III that the Euler angles $\phi$, $\theta$, $\psi$ are highly convenient parameters on the group $SO(3)$ of rotations of three-dimensional euclidean space. We generalize this parameterization to the group $SO(n)$ for arbitrary $n$.

We denote by $g_{jk}(\alpha)$ a rotation by the angle $\alpha$ in the $(x_j, x_k)$-plane, oriented such that the rotation from the vector $\xi_j$ to the vector $\xi_k$ is assumed to be positive. For brevity, the rotation $g_{k+1,k}(\alpha)$ will be denoted by $g_k(\alpha)$. We show that any rotation $g$ of $SO(n)$ can be presented as a product of rotations $g_k(\alpha)$.

Theorem 1. *Any rotation* $g$ *of the group* $SO(n)$ *can be presented in the form*

$$g = g^{(n-1)} \ldots g^{(1)}, \tag{1}$$

*where*

$$g^{(k)} = g_1(\theta_1^k) \ldots g_k(\theta_k^k). \tag{2}$$

Proof. The theorem is evident for $n = 2$, since in this case $g$ is a rotation in the $(x_1, x_2)$-plane, $g = g_1(\theta)$. We assume that it has been proved for $SO(n-1)$, and prove it for $SO(n)$. Let $g$ be an element of $SO(n)$. We denote by $\theta_1^{n-1}, \ldots$ $\ldots, \theta_{n-1}^{n-1}$ the spherical coordinates of the vector $g\xi_n$ and consider the rotation

$$g^{(n-1)} = g_1(\theta_1^{n-1}) \ldots g_{n-1}(\theta_{n-1}^{n-1}). \tag{3}$$

Let us determine the vector into which $\xi_n$ is transformed by the rotation $g^{(n-1)}$. The rotation $g_{n-1}(\theta_{n-1}^{n-1})$ can be written down as follows:

$$\left.\begin{aligned} x'_{n-1} &= x_{n-1} \cos\theta_{n-1}^{n-1} + x_n \sin\theta_{n-1}^{n-1}, \\ x'_n &= -x_{n-1} \sin\theta_{n-1}^{n-1} + x_n \cos\theta_{n-1}^{n-1}. \end{aligned}\right\} \tag{4}$$

Therefore $\xi_n(0, 0, \cdots, 1)$ is transformed by this rotation into the vector $g_{n-1}(\theta_{n-1}^{n-1})\xi_n$ with the coordinates

$$(0, 0, \ldots, \sin\theta_{n-1}^{n-1}, \cos\theta_{n-1}^{n-1}).$$

Repeating this reasoning and taking into consideration formula (3), we see that $g^{(n-1)}\xi_n = g\xi_n$. Hence $[g^{(n-1)}]^{-1} g\xi_n = \xi_n$, and consequently the rotation $[g^{(n-1)}]^{-1} g$ belongs to the subgroup $SO(n-1)$. By the induction hypothesis we have

$$[g^{n-1}]^{-1} g = g^{(n-2)} \ldots g^{(1)},$$

where $g^{(k)}$ is a rotation of the form (2). Therefore

$$g = g^{(n-1)} \ldots g^{(1)}.$$

The theorem is proved.

It is easy to show that, for almost all elements $g$ of $SO(n)$, the form (1) is unique (the exceptions are the cases when one of the angles $\theta_k^k$, $2 \le k \le n-1$, become zero or $\pi$). We shall call the numbers

$$\theta_j^k, \quad 1 \leqslant k \leqslant n-1, \quad 1 \leqslant j \leqslant k, \tag{5}$$

the Euler angles of the rotation $g$. These numbers determine $g$ uniquely. Moreover, it can be seen from the construction that the Euler angles vary as follows:

$$\left.\begin{array}{l} 0 \leqslant \theta_1^k < 2\pi, \\ 0 \leqslant \theta_j^k < \pi, \quad j \neq 1. \end{array}\right\} \tag{6}$$

The geometrical significance of Euler angles is clear from the construction: the angles $\theta_1^k, \cdots, \theta_k^k$ are spherical coordinates of the vector

$$\eta_{k+1} = [g^{(n-1)} \dots g^{(k+1)}]^{-1} g\xi_{k+1}, \tag{7}$$

where the rotations $g^{(j)}$ are defined by (4).

We note that $g$ can be written not only in the form (1), but also in the form

$$g = g_{(1)} g_{(2)} \cdots g_{(n-1)}, \tag{8}$$

where

$$g_{(k)} = g_k(\theta_k^{n-1}) g_{k-1}(\theta_{k-1}^{n-2}) \dots g_1(\theta_1^{n-k}). \tag{9}$$

This follows from the fact that $g_j(\alpha)$ and $g_k(\beta)$ are permutable for $|j-k| > 1$.

4. Invariant integration. From the visualisation of Euler angles, indicated above, it is easy to obtain an expression for the invariant measure on $SO(n)$. Namely, we have

$$dg = A_n \prod_{k=1}^{n-1} \prod_{j=1}^{k} \sin^{j-1} \theta_j^k \, d\theta_j^k, \tag{1}$$

where

$$A_n = \prod_{k=1}^{n} \frac{\Gamma\left(\frac{k}{2}\right)}{2\pi^{k/2}}. \tag{2}$$

In fact, the rotation $g$ is uniquely determined by the vectors

$$\eta_{k+1} = [g^{(n-1)} \dots g^{(k+1)}]^{-1} g\xi_{k+1}, \qquad 1 \leqslant k \leqslant n-1,$$

lying on the spheres $S^k$, $1 \le k \le n-1$. One can easily see that on replacing $g$ by a rotation $gg_0$, the vectors $\eta_k$ are replaced by the vectors $g_0^{(k)}\eta_k$, where $g_0^{(k)}$ is some rotation of $S^k$ (depending, in general, not only on $g_0$, but also on $g$).

Since, under the rotations $g_0^{(k)}$, the euclidean measures $d\eta_{k+1}$ remain invariant on the spheres $S^k$, the measure

$$dg = d\eta_2 \ldots d\eta_n$$

on $SO(n)$ remains invariant on the replacement of $g$ by $gg_0$.

But the Euler angles $\theta_1^k, \cdots, \theta_k^k$ are spherical coordinates of the vector $\eta_{k+1}$. Taking into account the expression for euclidean measure on $S^k$ in spherical coordinates (see formula (9) of subsection 1) we arrive at the expression (1) for the invariant measure $dg$ on $SO(n)$.

It follows from equality (1) of subsection 3 that each element $g$ of $SO(n)$ is given by an element $h = g^{(n-2)} \cdots g^{(1)}$ of $SO(n-1)$ and a point $\eta$ on $S^{n-1}$ with the spherical coordinates $\theta_1^{n-1}, \cdots, \theta_{n-1}^{n-1}$. It follows from formula (1) that

$$dg = dh\, d\eta. \tag{3}$$

## §2. REPRESENTATIONS OF CLASS 1 OF THE GROUP $SO(n)$ AND HARMONIC POLYNOMIALS

1. The quasi-regular representation. We begin the study of representations of $SO(n)$ with a consideration of its quasi-regular representation. The group $SO(n)$ is a transitive group of transformations of the $n$-dimensional unit sphere $S^{n-1}$. We shall denote by $\mathfrak{L}^2(S^{n-1})$ the space of functions $f(\xi)$ on $S^{n-1}$ for which the integral

$$\|f\|^2 = \int_{S^{n-1}} |f(\xi)|^2 d\xi \tag{1}$$

converges ($d\xi$ is the normalized euclidean measure on $S^{n-1}$). To each element $g$ of $SO(n)$ corresponds an operator $L(g)$ in the space $\mathfrak{L}^2(S^{n-1})$, which transforms $f(\xi)$ into the function

$$L(g)f(\xi) = f(g^{-1}\xi). \tag{2}$$

Evidently, $L(g_1g_2) = L(g_1)L(g_2)$ and so $L(g)$ is a representation of $SO(n)$.

Since the euclidean measure $d\xi$ on $S^{n-1}$ is invariant for rotations, we have

$$\|L(g)f\|^2 = \int_{S^{n-1}} |f(g^{-1}\xi)|^2 d\xi = \int_{S^{n-1}} |f(\xi)|^2 d\xi = \|f\|^2. \tag{3}$$

Thus the representation $L(g)$ is unitary relative to the scalar product

$$(f_1, f_2) = \int_{S^{n-1}} f_1(\xi)\overline{f_2(\xi)}\, d\xi. \tag{4}$$

We shall call this unitary representation of $SO(n)$ the quasi-regular representation.

2. Representations in spaces of homogeneous polynomials. The quasi-regular representation of $SO(n)$ is reducible. Below we shall construct irreducible representations of this group, into which $L(g)$ is decomposed. We shall denote by $\mathfrak{R}^{nl}$ the space of homogeneous polynomials of degree $l$ in $n$ variables. Every polynomial $f(\mathbf{x})$ of $\mathfrak{R}^{nl}$ is uniquely determined by its values on $S^{n-1}$:

$$f(r\xi)=r^l f(\xi). \tag{1}$$

Therefore we can regard $\mathfrak{R}^{nl}$ as a subspace in $\mathfrak{L}^2(S^{n-1})$. It is evident that under the transformation

$$L(g)f(\mathbf{x})=f(g^{-1}\mathbf{x}) \tag{2}$$

homogeneous polynomials turn into homogeneous polynomials, and so the subspace $\mathfrak{R}^{nl}$ is invariant in $\mathfrak{L}^2(S^{n-1})$. The representation $L(g)$ thus gives a representation $L^{nl}(g)$ in the invariant subspace $\mathfrak{R}^{nl}$:

$$L^{nl}(g)f(\mathbf{x})=f(g^{-1}\mathbf{x}), \qquad f(\mathbf{x})\in\mathfrak{R}^{nl}. \tag{3}$$

However, the representations $L^{nl}(g)$ are reducible too. Indeed, under rotations $g$ of $SO(n)$, the distance $r^2=x_1^2+\cdots+x_n^2$ of points $\mathbf{x}$ of $E_n$ from the origin remains unchanged. Therefore the subspace $r^2\mathfrak{R}^{n,l-2}$ of polynomials of the form $r^2 f(\mathbf{x})$, where $f(\mathbf{x})\in\mathfrak{R}^{n,l-2}$, is invariant for the operators $L^{nl}(g)$.

We denote by $T^{nl}(g)$ the representation given by $L^{nl}(g)$ in the factor space $\mathfrak{A}^{nl}\equiv\mathfrak{R}^{nl}/r^2\mathfrak{R}^{n,l-2}$. We shall prove below that this representation is an irreducible unitary representation of $SO(n)$, and that the quasi-regular representation of this group is a direct sum of representations $T^{nl}(g)$:

$$L(g)=\sum_{l=0}^{\infty}T^{nl}(g). \tag{4}$$

Thus the representations $T^{nl}(g)$ may be characterised as the irreducible components of the quasi-regular representation of $SO(n)$.

We shall further prove that these representation admit another characteristic. Namely, they exhaust the irreducible representations $T(g)$ of $SO(n)$ in whose spaces there is a vector $\mathbf{f}_0$ invariant for all operators $T(h)$, $h\in SO(n-1)$:

$$T(h)\mathbf{f}_0=\mathbf{f}_0, \qquad h\in SO(n-1), \tag{5}$$

i.e. the representations of class 1 relative to the subgroup $SO(n-1)$. So, we shall prove below that the representations $T^{nl}(g)$ are irreducible representations of class 1 of the group $SO(n)$, and that every irreducible representation of class 1 of this group is equivalent to one of the representations $T^{nl}(g)$.

3. **Harmonic polynomials.** In order to prove the assertions formulated at the end of the preceding section, we construct another realization of $T^{nl}(g)$, namely, a realization in the space $\mathfrak{H}^{nl}$ of homogeneous harmonic polynomials of degree $l$ in $n$ variables.

A polynomial $f(\mathbf{x})$ is called harmonic if $\Delta f = 0$, where

$$\Delta = \frac{\partial^2}{\partial x_1^2} + \ldots + \frac{\partial^2}{\partial x_n^2} \tag{1}$$

is the Laplace operator. We shall later need some relations, connected with the Laplace operator.

We first show that if $f_1(\mathbf{x}) \in \mathfrak{R}^{n,l-2}$, then

$$\Delta(r^2 f_1) = (2n + 4l - 8) f_1 + r^2 \Delta f_1. \tag{2}$$

Indeed, direct calculation shows that

$$\Delta(r^2 f_1) = 2n f_1 + 4 \sum_{k=1}^{n} x_k \frac{\partial f_1}{\partial x_k} + r^2 \Delta f_1. \tag{3}$$

But as $f_1(\mathbf{x})$ is a homogeneous polynomial of degree $l - 2$, by Euler's formula we have

$$\sum_{k=1}^{n} x_k \frac{\partial f_1}{\partial x_k} = (l-2) f_1,$$

whence equality (2) follows.

By repeated application of formula (2) one can easily show that if $f_k(\mathbf{x}) \in \mathfrak{R}^{n,l-2k}$, then

$$\Delta(r^{2k} f_k) = 2k(n + 2l - 2k - 2) r^{2k-2} f_k + r^{2k} \Delta f_k. \tag{4}$$

Equality (2) leads to the following assertion.

*If $h(\mathbf{x})$ is a homogeneous harmonic polynomial of degree $l$, then*

$$x_j h(\mathbf{x}) - \frac{1}{n + 2l - 2} r^2 \frac{\partial h}{\partial x_j} \tag{5}$$

*is a homogeneous harmonic polynomial of degree $l + 1$.*

Indeed, as $\Delta h = 0$, we have

$$\Delta[x_j h(\mathbf{x})] = x_j \Delta h + 2 \frac{\partial h}{\partial x_j} = 2 \frac{\partial h}{\partial x_j}. \tag{6}$$

On the other hand, since $\partial h / \partial x_j \in \mathfrak{R}^{n,l-1}$, by formula (2) we have

$$\Delta\left(r^2\frac{\partial h}{\partial x_j}\right)=(2n+4l-4)\frac{\partial h}{\partial x_j}+r^2\Delta\left(\frac{\partial h}{\partial x_j}\right)=2(n+2l-2)\frac{\partial h}{\partial x_j}. \tag{7}$$

Comparing (6) and (7), we get

$$\Delta\left[x_j h(\mathbf{x})-\frac{1}{n+2l-2}r^2\frac{\partial h}{\partial x_j}\right]=0. \tag{8}$$

Finally, we note that if $h(\mathbf{x})$ is a harmonic polynomial, then for any $j$ the polynomial $\partial h/\partial x_j$ is also harmonic. Indeed,

$$\Delta\left(\frac{\partial h}{\partial x_j}\right)=\frac{\partial(\Delta h)}{\partial x_j}.$$

Therefore, if $\Delta h=0$, then $\Delta(\partial h/\partial x_j)=0$ too.

4. Invariance of the subspace $\mathfrak{H}^{nl}$. The space $\mathfrak{H}^{nl}$ of homogeneous harmonic polynomials of degree $l$ in $n$ variables is a subspace of the space $\mathfrak{R}^{nl}$ of all homogeneous polynomials of degree $l$ in $n$ variables. We show that this subspace is invariant for operators of the representation $L^{nl}(g)$. For this we use the following important property of the Laplace operator.

*The Laplace operator is permutable with the operators of motion in euclidean space: if $g$ is a motion, then,*

$$\Delta[f(g^{-1}\mathbf{x})]=(\Delta f)(g^{-1}\mathbf{x}). \tag{1}$$

Equality (1) is most simply proved by noting that

$$\Delta f(\mathbf{a})=\lim_{r\to 0}\frac{\Gamma\left(\frac{n+2}{2}\right)\left[f_r(\mathbf{a})-f(\mathbf{a})\right]}{\pi^{n/2}r^2}, \tag{2}$$

where $f_r(\mathbf{a})$ is the mean value of function $f(\mathbf{x})$ on the sphere $S^{n-1}(\mathbf{a}, r)$ of radius $r$ with center at point $\mathbf{a}$. Equality (2) is derived by expanding $f(\mathbf{x})$ in a Taylor series in the neighborhood of $\mathbf{a}$, and integrating this series term by term over $S^{n-1}(\mathbf{a}, r)$. Since the operator in the right hand side of (2) is permutable with motions of euclidean space, the Laplace operator is also permutable with these motions.

It follows from (2) that $\mathfrak{H}^{nl}$ is invariant for the operators of the representation

$$L^{nl}(g)f(\mathbf{x})=f(g^{-1}\mathbf{x}),\qquad f(\mathbf{x})\in\mathfrak{R}^{nl}. \tag{3}$$

Indeed, if $\Delta f(\mathbf{x})=0$, then because of formula (2) we have $\Delta[f(g^{-1}\mathbf{x})]=0$. Therefore $f(\mathbf{x})\in\mathfrak{H}^{nl}$ implies that

$$L^{nl}(g)f(\mathbf{x})=f(g^{-1}\mathbf{x})\in\mathfrak{H}^{nl}. \tag{4}$$

5. Harmonic projection of a polynomial. Representation in a space of harmonic polynomials. We now prove that the invariant subspace $\mathfrak{H}^{nl}$ is the complement in $\mathfrak{R}^{nl}$ of the invariant subspace $r^2\mathfrak{R}^{n,l-2}$. In other words, we prove the following assertion.

*The space $\mathfrak{R}^{nl}$ of homogeneous polynomials of degree $l$ in $n$ variables is the direct sum of the subspace $\mathfrak{H}^{nl}$ of homogeneous harmonic polynomials of degree $l$ in $n$ variables and the subspace $r^2\mathfrak{R}^{n,l-2}$ of polynomials of the form $r^2 f_1(\mathbf{x})$, $f_1(\mathbf{x}) \in \mathfrak{R}^{n,l-2}$:*

$$\mathfrak{R}^{nl} = \mathfrak{H}^{nl} + r^2\mathfrak{R}^{n,\,l-2}. \tag{1}$$

First, we prove that

$$\mathfrak{H}^{nl} \cap r^2\mathfrak{R}^{n,\,l-2} = 0, \tag{2}$$

i.e. if $f_1(\mathbf{x}) \in \mathfrak{R}^{n,l-2}$ and $f_1(\mathbf{x}) \neq 0$, then $\Delta(r^2 f_1) \neq 0$. Indeed, if $f(\mathbf{x}) \in r^2\mathfrak{R}^{n,l-2}$, then $f(\mathbf{x})$ can be presented in the form $f(\mathbf{x}) = r^{2k} f_k(\mathbf{x})$, where $f_k(\mathbf{x}) \in \mathfrak{R}^{n,l-2k}$, $0 < k \leq [l/2]$, and $f_k(\mathbf{x})$ is not divisible by $r^2$. Let $\Delta f = 0$. Then by formula (4) of subsection 3 we have

$$2k(n + 2l - 2k - 2)\, r^{2k-2} f_k + r^{2k} \Delta f_k = 0. \tag{3}$$

Since $k \geq 1$ and $l \geq 2k$, we have $k(n + 2l - 2k - 2) > 0$. Cancelling $r^{2k-2}$ out, we see that $f_k$ is divisible by $r^2$, contrary to our assumption. This proves that $\mathfrak{H}^{nl} \cap r^2\mathfrak{R}^{n,l-2} = 0$.

We now prove that $\mathfrak{H}^{nl}$ and $r^2\mathfrak{R}^{n,l-2}$ make up the whole space $\mathfrak{R}^{nl}$. To do this we calculate the dimensions of these subspaces. We denote the dimension of $\mathfrak{R}^{nl}$ by $r(n, l)$. Arranging the homogeneous polynomials in powers of $x_1$ we see that

$$r(n, l) = r(n-1, l) + r(n-1, l-1) + \ldots + r(n-1, 0). \tag{4}$$

Moreover, it is clear that

$$r(1, l) = 1. \tag{5}$$

Hence by induction we get

$$r(n, l) = \frac{(n+l-1)!}{(n-1)!\, l!}. \tag{6}$$

Since the dimension of $r^2\mathfrak{R}^{n,l-2}$ is equal to the dimension of $\mathfrak{R}^{n,l-2}$, and $\mathfrak{H}^{nl} \cap r^2\mathfrak{R}^{n,l-2} = 0$, the dimension $h(n, l)$ of $\mathfrak{H}^{nl}$ satisfies the inequality

$$h(n, l) \leqslant r(n, l) - r(n, l-2). \tag{7}$$

On the other hand, the polynomial $\Delta f$, $f \in \mathfrak{R}^{nl}$ belongs to $\mathfrak{R}^{n,l-2}$. Therefore the condition $\Delta f = 0$ imposes not more than $r(n, l-2)$ linear relations on the coefficients of $f(\mathbf{x})$. Hence the dimension of $\mathfrak{H}^{nl}$ is not less than $r(n, l) - r(n, l-2)$:

$$h(n, l) \geqslant r(n, l) - r(n, l-2). \tag{8}$$

It follows from (7) and (8) that

$$h(n, l) = r(n, l) - r(n, l-2), \tag{9}$$

or, in other words,

$$\dim \mathfrak{H}^{nl} = \dim \mathfrak{R}^{nl} - \dim \mathfrak{R}^{n,\, l-2} = \dim \mathfrak{R}^{nl} - \dim r^2 \mathfrak{R}^{n,\, l-2}. \tag{10}$$

It follows from (2) and (10) that the subspaces $\mathfrak{H}^{nl}$ and $r^2 \mathfrak{R}^{n,l-2}$ make up the whole space $\mathfrak{R}^{nl}$, and so equality (1) is proved.

Equality (1) shows that the representation $T^{nl}(g)$ of the group $SO(n)$ in the factor space $\mathfrak{R}^{nl}/r^2 \mathfrak{R}^{n,l-2}$ is equivalent to the restriction of $L(g)$ to the invariant subspace $\mathfrak{H}^{nl}$. We shall denote the latter representation by the same symbol $T^{nl}(g)$. However, where it will be necessary to distinguish between the realizations, we shall denote the realization in $\mathfrak{H}^{nl}$ by $H^{nl}(g)$.

We may note that in deriving formula (1) we have incidentally obtained the expression for the dimension $h(n, l)$ of the space $\mathfrak{H}^{nl}$ of homogeneous harmonic polynomials of degree $l$ in $n$ variables:

$$h(n, l) = r(n, l) - r(n, l-2) = \frac{(2l+n-2)(n+l-3)!}{(n-2)!\, l!}. \tag{11}$$

It follows from (1) that any homogeneous polynomial $f(\mathbf{x})$ of degree $l$ can be presented in the form

$$f(\mathbf{x}) = h_l(\mathbf{x}) + r^2 f_1(\mathbf{x}), \tag{12}$$

where $f_1(\mathbf{x}) \in \mathfrak{R}^{n,l-2}$ and $h_l(\mathbf{x}) \in \mathfrak{H}^{nl}$ is a harmonic polynomial of degree $l$. Henceforth we shall call $h_l(\mathbf{x})$ the harmonic projection of the homogeneous polynomial $f(\mathbf{x})$. We shall denote the harmonic projection of polynomial $f(\mathbf{x})$ by $Hf(\mathbf{x})$.

Let us find an explicit formula for harmonic projection. We shall seek the harmonic projection of $f(\mathbf{x})$ in the form

$$Hf(\mathbf{x}) = \sum_{k=0}^{[l/2]} a_k r^{2k} \Delta^k f(\mathbf{x}), \tag{13}$$

where $\alpha_0 = 1$. We apply the Laplace operator $\Delta$ to both sides of this equality, and take into consideration the relation

$$\Delta(r^{2k}\Delta^k f) = 2k(n+2l-2k-2)r^{2k-2}\Delta^k f + r^{2k}\Delta^{k+1} f$$

(see formula (4) of subsection 3). We obtain

$$\Delta Hf(\mathbf{x}) = \sum_{k=0}^{[l/2]} [\alpha_k + (2k+2)(n+2l-2k-4)\alpha_{k+1}] r^{2k}\Delta^{k+1} f. \quad (14)$$

Equality (14) shows that for the polynomial given by formula (13) to be harmonic, it is necessary and sufficient that coefficients $\alpha_k$ satisfy the recurrence relation

$$\alpha_k + (2k+2)(n+2l-2k-4)\alpha_{k+1} = 0.$$

Remembering that $\alpha_0 = 1$, we get

$$Hf(\mathbf{x}) = \sum_{k=0}^{[l/2]} \frac{(-1)^k r^{2k}\Delta^k f(\mathbf{x})}{2^k k!(n+2l-4)(n+2l-6)\dots(n+2l-2k-2)}. \quad (15)$$

Formula (15) gives the harmonic projection of the polynomial $f(\mathbf{x}) \in \mathfrak{R}^{nl}$.

6. The canonical decomposition of homogeneous polynomials. We have just proved that

$$\mathfrak{R}^{nl} = \mathfrak{H}^{nl} + r^2\mathfrak{R}^{n,\, l-2}. \quad (1)$$

Decomposing $\mathfrak{R}^{n,l-2}$ in the same way and continuing this process, we obtain

$$\mathfrak{R}^{nl} = \sum_{k=0}^{[l/2]} r^{2k}\mathfrak{H}^{n,\, l-2k}, \quad (2)$$

where $r^{2k}\mathfrak{H}^{n,l-2k}$ is a subspace in $\mathfrak{R}^{nl}$ consisting of polynomials of the form $r^{2k}h_{l-2k}(\mathbf{x})$, $h_{l-2k}(\mathbf{x}) \in \mathfrak{H}^{n,l-2k}$.

Thus we have proved that every homogeneous polynomial of degree $l$ in $n$ variables is uniquely decomposed as a sum of polynomials of the form $r^{2k}h_{l-2k}(\mathbf{x})$, where $h_{l-2k}(\mathbf{x}) \in \mathfrak{H}^{n,l-2k}$:

$$f(\mathbf{x}) = \sum_{k=0}^{[l/2]} r^{2k} h_{l-2k}(\mathbf{x}). \quad (3)$$

We shall call this the *canonical decomposition* of the polynomial $f(\mathbf{x})$.

It follows from (3) that, on the unit sphere,

$$f(\xi)=\sum_{k=0}^{[l/2]} h_{l-2k}(\xi). \tag{4}$$

In other words, the value of any homogeneous polynomial $f(\mathbf{x})$ of degree $l$ on the unit sphere coincides with the value on this sphere of the harmonic (nonhomogeneous) polynomial $h(\mathbf{x})=\sum_{k=0}^{[l/2]} h_{l-2k}(\mathbf{x})$. Hence in considering polynomials on the unit sphere it is sufficient to restrict oneself to harmonic polynomials.

Since $r^2=x_1^2+\cdots+x_n^2$ is unchanged by rotations $g$ of $SO(n)$, it follows from (3) that

$$L^{nl}(g)f(\mathbf{x})=f(g^{-1}\mathbf{x})=\sum_{k=0}^{[l/2]} r^{2k}h_{l-2k}(g^{-1}\mathbf{x}). \tag{5}$$

Since, for $h_{l-2k}(\mathbf{x})\in \mathfrak{H}^{n,l-2k}$, we have

$$H^{n,l-2k}(g)\,h_{l-2k}(\mathbf{x})=h_{l-2k}(g^{-1}\mathbf{x}), \tag{6}$$

it follows from (5) that

$$L^{nl}(g)=\sum_{k=0}^{[l/2]} H^{n,l-2k}(g). \tag{7}$$

7. Decomposition of the quasi-regular representation. We have constructed above unitary representations $T^{nl}(g)$ of the group $SO(n)$, which are of class 1. We show that the quasi-regular representation

$$L(g)f(\xi)=f(g^{-1}\xi) \tag{1}$$

of this group is a direct sum of representations $T^{nl}(g)$:

$$L(g)=\sum_{l=0}^{\infty} T^{nl}(g). \tag{2}$$

The representation $L(g)$ is constructed in the space $\mathfrak{L}^2(S^{n-1})$ of functions on the unit sphere, having square-integrable moduli. In this space we consider subspaces consisting of the values on $S^{n-1}$ of functions $h(\mathbf{x})$ of $\mathfrak{H}^{nl}$. We shall denote these subspaces too by $\mathfrak{H}^{nl}$. Evidently they are invariant for operators of the quasi-regular representation.

The restriction of the quasi-regular representation to $\mathfrak{H}^{nl}$ is the representation $T^{nl}(g)$, and for $l\neq m$ the representations $T^{nl}(g)$ and $T^{mn}(g)$ are nonequivalent. Hence the subspaces $\mathfrak{H}^{nl}$ and $\mathfrak{H}^{nm}$, $l\neq m$, are orthogonal to each other. In other words, if $h_1(\xi) \in \mathfrak{H}^{nl}$ and $h_2(\xi) \in \mathfrak{H}^{nm}$, then

$$\int h_1(\xi)\overline{h_2(\xi)}\,d\xi = 0. \tag{3}$$

Therefore, to prove the decomposition

$$\mathfrak{L}^2(S^{n-1}) = \sum_{l=0}^{\infty} \mathfrak{H}^{nl} \tag{4}$$

it is sufficient to show that any function $f(\xi)$ of $\mathfrak{L}^2(S^{n-1})$ can be approximated in the mean, with any degree of accuracy required, by sums of functions of the spaces $\mathfrak{H}^{nl}$, $l = 0, 1, \cdots$.

By Weierstrass's theorem any continuous function $f(\xi)$ on $S^{n-1}$ can be uniformly approximated, with any degree of accuracy required, by a function of the form $\phi(\xi)$, where $\phi(\mathbf{x})$ is a polynomial. Since every polynomial can be written as a sum of homogeneous polynomials, we make the following deduction.

*For any continuous function $f(\xi)$ on the unit sphere $S^{n-1}$ and any $\epsilon > 0$, there are polynomials $\phi_l(\mathbf{x})$ of the spaces $\mathfrak{R}^{nl}$, $l = 1, 2, \cdots, m$, such that for all points $\xi \in S^{n-1}$ we have*

$$\left| f(\xi) - \sum_{l=1}^{m} \varphi_l(\xi) \right| < \varepsilon.$$

But in subsection 6 it was shown that if $\phi_l(\mathbf{x}) \in \mathfrak{R}^{nl}$, then $\phi_l(\xi)$ can be presented in the form

$$\varphi_l(\xi) = \sum_{k=0}^{[l/2]} h_{l-2k}(\xi),$$

where $h_{l-2k}(\xi) \in \mathfrak{H}^{n,l-2k}$. Hence any continuous function $f(\xi)$ on $S^{n-1}$ can be uniformly approximated, with any degree of accuracy required, by a sum of functions belonging to the spaces $\mathfrak{H}^{nl}$. Consequently the subspaces $\mathfrak{H}^{nl}$ make up the whole space $\mathfrak{L}^2(S^{n-1})$. This proves the decomposition (4).

It follows from (4) that $L(g)$ is a direct sum of the representations $T^{nl}(g)$, where each of the representations $T^{nl}(g)$ appears only once in the decomposition of $L(g)$. Equality (2) is thus proved.

As we have used the realization of the representations $T^{nl}(g)$ in $\mathfrak{H}^{nl}$, formula (2) is more precisely written in the form

$$L(g) = \sum_{l=0}^{\infty} H^{nl}(g). \tag{2'}$$

It remains to show that the $T^{nl}(g)$ are irreducible. We shall prove this with the help of induction on $n$. Let us first consider the restriction of $T^{nl}(g)$ to $SO(n-1)$.

**8.** Decomposition of the restriction of $T^{nl}(g)$ to $SO(n-1)$. We restrict the representation $T^{nl}(g)$ of $SO(n)$ to the subgroup $SO(n-1)$. We obtain a representa tion $T^{nl}(h)$ of this subgroup. We shall show that $T^{nl}(h)$, $h \in SO(n-1)$ is a direct sum of the representations $T^{n-1,k}(h)$, $0 \leq k \leq l$:

$$T^{nl}(h) = \sum_{k=0}^{l} T^{n-1,k}(h), \qquad h \in SO(n-1). \tag{1}$$

First let us construct the subspaces $\mathfrak{A}^{nlk}$ in $\mathfrak{R}^{nl}/r^2\mathfrak{R}^{n,l-2}$ in which the $T^{n-1,k}(h)$ are realized. We denote by $x_n^{l-k}\mathfrak{H}^{n-1,k}$ the space of polynomials of the form $x_n^{l-k}h_k(\mathbf{x}')$, where $h_k(\mathbf{x}') \in \mathfrak{H}^{n-1,k}$, i.e. is a homogeneous harmonic polynomial of degree $k$ in $n-1$ variables. Obviously, $x_n^{l-k}\mathfrak{H}^{n-1,k} \in \mathfrak{R}^{nl}$, $0 \leq k \leq l$. As $x_n$ is unchanged by rotations $h$ of $SO(n-1)$, for $f(\mathbf{x}) \in x_n^{l-k}\mathfrak{H}^{n-1,k}$ we have

$$L^{nl}(h)f(\mathbf{x}) = f(h^{-1}\mathbf{x}) = x_n^{l-k}h_k(h^{-1}\mathbf{x}'). \tag{2}$$

Since $\mathfrak{H}^{n-1,k}$ is invariant for the rotation operators of $SO(n-1)$ (see subsection 4), we have $h_k(h^{-1}\mathbf{x}') \in \mathfrak{H}^{n-1,k}$, and consequently

$$L^{nl}(h)[x_n^{l-k}h_k(\mathbf{x}')] \in x_n^{l-k}\mathfrak{H}^{n-1,k}. \tag{3}$$

Thus we have proved that the operators $L^{nl}(h)$, $h \in SO(n-1)$ leave the subspaces $x_n^{l-k}\mathfrak{H}^{n-1,k}$ invariant. Moreover, the representation

$$T(h)[x_n^{l-k}h_k(\mathbf{x}')] = x_n^{l-k}h_k(h^{-1}\mathbf{x}') \tag{4}$$

is the restriction of $L^{nl}(h)$ to $x_n^{l-k}\mathfrak{H}^{n-1,k}$, and according to subsection 5 is equivalent to the representation $T^{n-1,k}(h)$, $h \in SO(n-1)$.

We denote by $\mathfrak{A}^{nlk}$ the image of $x_n^{l-k}\mathfrak{H}^{n-1,k}$ under a mapping of $\mathfrak{R}^{nl}$ onto $\mathfrak{R}^{nl}/r^2\mathfrak{R}^{n,l-2}$. Evidently, $\mathfrak{A}^{nlk}$ is invariant for the operators of $T^{nl}(h)$, $h \in SO(n-1)$, induced in $\mathfrak{R}^{nl}/r^2\mathfrak{R}^{n,l-2}$ by $L^{nl}(h)$. We show that $\mathfrak{R}^{nl}/r^2\mathfrak{R}^{n,l-2}$ is a direct sum of its subspaces $\mathfrak{A}^{nlk}$:

$$\mathfrak{R}^{nl}/r^2\mathfrak{R}^{n,l-2} = \sum_{k=0}^{l} \mathfrak{A}^{nlk}, \tag{5}$$

and that the restriction of $T^{nl}(h)$, $h \in SO(n-1)$ to $\mathfrak{A}^{nlk}$ is equivalent to $T^{n-1,k}(h)$. It is evident that these assertions will be proved if we show that $\mathfrak{R}^{nl}$

is a direct sum of the subspace $r^2\mathfrak{R}^{n,l-2}$ and the subspaces $x_n^{l-k}\mathfrak{H}^{n-1,k}$:

$$\mathfrak{R}^{nl}=r^2\mathfrak{R}^{n,\,l-2}+\sum_{k=0}^{l}x_n^{l-k}\mathfrak{H}^{n-1,\,k}. \tag{6}$$

In other words, we have to prove that any polynomial $f(\mathbf{x})$ of $\mathfrak{R}^{nl}$ can be uniquely presented in the form

$$f(\mathbf{x})=r^2 f_1(\mathbf{x})+\sum_{k=0}^{l}x_n^{l-k}h_k(\mathbf{x}'), \tag{7}$$

where $f_1(\mathbf{x})$ is a homogeneous polynomial of degree $l-2$ in $n$ variables, and $h_k(\mathbf{x}')$ is a homogeneous harmonic polynomial of degree $k$ in $n-1$ variables.

In order to prove equality (7), we first show that

$$r(n,\,l)=r(n,\,l-2)+\sum_{k=0}^{l}h(n-1,\,k), \tag{8}$$

where $r(n,\,l)$ is the dimension of $\mathfrak{R}^{nl}$, and $h(n-1,\,k)$ is the dimension of $\mathfrak{H}^{n-1,k}$. By formula (9) of subsection 4 we have

$$h(n-1,\,k)=r(n-1,\,k)-r(n-1,\,k-2).$$

Therefore

$$\sum_{k=0}^{l}h(n-1,\,k)=\sum_{k=0}^{l}[r(n-1,\,k)-r(n-1,\,k-2)]$$
$$=r(n-1,\,l)+r(n-1,\,l-1), \tag{9}$$

where we have put $r(n-1,\,-1)=r(n-2,\,-2)=0$. Employing formula (11) of subsection 5, we get

$$r(n,\,l-2)+\sum_{k=0}^{l}h(n-1,\,k)=r(n,\,l-2)+r(n-1,\,l)$$
$$+r(n-1,\,l-1)=\frac{(n+l-3)!}{(n-1)!\,(l-2)!}+\frac{(n+l-2)!}{(n-2)!\,l!}+\frac{(n+l-3)!}{(n-2)!\,(l-1)!}$$
$$=\frac{(n+l-1)!}{(n-1)!\,l!}=r(n,\,l). \tag{10}$$

Relation (8) is thus proved.

We now show that the subspaces $r^2\mathfrak{R}^{n,l-2}$ and $x_n^{l-k}\mathfrak{H}^{n-1,k}$, $0\le k\le l$, make up the whole space $\mathfrak{R}^{nl}$. Let $f(\mathbf{x})\in\mathfrak{R}^{nl}$. Then the polynomial $f(\mathbf{x})$ can be presented in the form

$$f(\mathbf{x})=r^2F(\mathbf{x})+x_n\varphi_1(\mathbf{x}')+\varphi_2(\mathbf{x}'), \tag{11}$$

where

$$r^2=x_1^2+\ldots+x_n^2,\quad \mathbf{x}'=(x_1,\,\ldots,\,x_{n-1}),$$
$$F(\mathbf{x})\in\mathfrak{R}^{n,\,l-2},\quad \varphi_1(\mathbf{x}')\in\mathfrak{R}^{n-1,\,l-1}\text{ and }\varphi_2(\mathbf{x}')\in\mathfrak{R}^{n-1,\,l}.$$

On applying the canonical decomposition to the polynomials $\phi_1(\mathbf{x}')$ and $\phi_2(\mathbf{x}')$, we obtain

$$f(\mathbf{x})=r^2F(\mathbf{x})+\sum_{k=0}^{\left[\frac{l-1}{2}\right]} x_n r_{n-1}^{2k}\hat{h}_{l-2k-1}(\mathbf{x}')+\sum_{k=0}^{[l/2]} r_{n-1}^{2k}\hat{h}_{l-2k}(\mathbf{x}'), \quad (12)$$

where $r_{n-1}^2=x_1^2+\cdots+x_{n-1}^2$ and $\hat{h}_s(\mathbf{x}')\in\mathfrak{H}^{n-1,s}$.

We replace $r_{n-1}^2$ by $r^2-x_n^2$ in (12) and simplify the right hand side, transferring all those terms which contain $r^2$ to the first summand. We obtain

$$f(x)=r^2f_1(\mathbf{x})+\sum_{k=0}^{l} x_n^{l-k}h_k(\mathbf{x}'), \quad (13)$$

where $f_1(\mathbf{x})\in\mathfrak{R}^{n,l-2}$, $h_k(\mathbf{x})\in\mathfrak{H}^{n-1,k}$($h_k(\mathbf{x}')$ differs only in sign from $\hat{h}_k(\mathbf{x}')$ in the decomposition (12)).

Thus we have proved that the subspaces $r^2\mathfrak{R}^{n,l-2}$ and $x_n^{l-k}\mathfrak{H}^{n-1,k}$, $0\le k\le l$, make up the whole space $\mathfrak{R}^{nl}$. Since it was shown earlier that the dimension of $\mathfrak{R}^{nl}$ is equal to the sum of the dimensions of these subspaces, the decomposition

$$\mathfrak{R}^{nl}=r^2\mathfrak{R}^{n,l-2}+\sum_{k=0}^{l} x_n^{l-k}\mathfrak{H}^{n-1,k},$$

is proved.

It was remarked above that this decomposition leads to the following assertions.

1) *The space* $\mathfrak{R}^{nl}/r^2\mathfrak{R}^{n,l-2}$ *is the direct sum of the subspaces* $\mathfrak{A}^{nlk}$ (*the image of* $x_n^{l-k}\mathfrak{H}^{n-1,k}$ *under a mapping of* $\mathfrak{R}^{nl}$ *onto* $\mathfrak{R}^{nl}/r^2\mathfrak{R}^{n,l-2}$):

$$\mathfrak{R}^{nl}/r^2\mathfrak{R}^{n,l-2}=\sum_{k=0}^{l}\mathfrak{A}^{nlk}.$$

2) *The restriction of* $T^{nl}(h)$, $h\in SO(n-1)$ *to* $\mathfrak{A}^{nlk}$ *is equivalent to* $T^{n-1,k}(h)$, $0\le k\le l$.

9. Infinitesimal operators of $T^{nl}(g)$. In the following subsection we shall prove the irreducibility of the representations $T^{nl}(g)$ of the group $SO(n)$. In this connection we shall find it convenient to deal not with the operators $T^{nl}(g)$

themselves, but with the infinitesimal operators.

We denote by $A_{jk}$ the infinitesimal operator of $T^{nl}(g)$, corresponding to the subgroup $\Omega_{jk}$ of rotations in the $(x_j, x_k)$-plane. We shall assume that $T^{nl}(g)$ is realized in the space $\mathfrak{H}^{nl}$ of homogeneous harmonic polynomials. In that case it is given by the same formula

$$T^{nl}(g)f(\mathbf{x})=f(g^{-1}\mathbf{x}), \tag{1}$$

as the representation $L(g)$ in $\mathfrak{L}^2(S^{n-1})$. Therefore the infinitesimal operators of these representations are also given by one and the same set of formulas.

We denote by $g_{jk}(\phi)$ a rotation by an angle $\phi$ in the $(x_j, x_k)$-plane. The operator $L[g_{jk}(\phi)]$ transforms $f(\mathbf{x}) = f(x_1, \cdots, x_n)$ into

$$\begin{aligned} L[g_{jk}(\varphi)]f(\mathbf{x})&=f[g_{jk}(-\varphi)\mathbf{x}] \\ &=f(x_1,\ldots, x_j\cos\varphi+x_k\sin\varphi,\ldots,-x_j\sin\varphi+x_k\cos\varphi,\ldots,x_n). \end{aligned} \tag{2}$$

Therefore the infinitesimal operator of $L(g)$ has the form

$$A_{jk}=\frac{dL[g_{jk}(\varphi)]}{d\varphi}\bigg|_{\varphi=0}=x_k\frac{\partial}{\partial x_j}-x_j\frac{\partial}{\partial x_k}. \tag{3}$$

The infinitesimal operators of the $L^{nl}(g)$ (in $\mathfrak{R}^{nl}$) and $T^{nl}(g)$ (in $\mathfrak{H}^{nl}$) have the same form.

It follows from (3) that $A_{jk} = -A_{kj}$. Further, from the same formula it follows that if $j < k$, $p < q$, then

$$[A_{jk}, A_{pq}]=\delta_{kp}A_{jq}, \tag{4}$$

where $[A_{jk}, A_{pq}] = A_{jk}A_{pq} - A_{pq}A_{jk}$ is the commutator of the operators $A_{jk}$ and $A_{pq}$, and $\delta_{kp}$ is the Kronecker symbol.

**10. Irreducibility of the representations $T^{nl}(g)$.** We can now pass on to the proof of irreduciblity of the representations $T^{nl}(g)$ of $SO(n)$ for $n \geq 3$.

We observe that for $n = 2$ the representation $T^{nl}(g)$ is reducible; it is the direct sum of two one-dimensional representations. Indeed, it follows from formula (11) of subsection 5 that $h(2, l) = 2$, i.e. $T^{2l}(g)$ is a two-dimensional representation. Since $SO(2)$ is commutative, its irreducible representations are one-dimensional (see §3.3 of Chapter I). Consequently, $T^{2l}(g)$ is the direct sum of two irreducible representations.

In order to obtain the decomposition of $T^{2l}(g)$ in an explicit form, we note that the polynomials $(x_2 + ix_1)^l$ and $(x_2 - ix_1)^l$ belong to $\mathfrak{H}^{2l}$. Moreover, a simple calculation shows that if $g$ is the rotation by an angle $\phi$ in the $(x_1, x_2)$-plane, then

$$T^{2l}(g)(x_2+ix_1)^l = e^{il\varphi}(x_2+ix_1)^l \tag{1}$$

and

$$T^{2l}(g)(x_2-ix_1)^l = e^{-il\varphi}(x_2-ix_1)^l. \tag{1'}$$

Thus $\mathfrak{H}^{2l}$ is the direct sum of two invariant subspaces, one of which is spanned by the function $(x_2+ix_1)^l$ and the function $(x_2-ix_1)^l$. We shall denote the corresponding representations of $SO(2)$ by $T_+^{2l}(g)$ and $T_-^{2l}(g)$.

We now give a proof of irreducibility of the representations $T^{nl}(g)$ for $n \geq 3$. The proof is carried out by induction on $n$. The case $n = 3$ was considered in Chapter III. Irreducible representations $T_l(g)$, $0 \leq l < \infty$, of $SO(3)$ were constructed there. It is not difficult to show that these representations are only formally different from the representations $T^{3l}(g)$, and so the representations $T^{3l}(g)$ of $SO(3)$ are irreducible.

Suppose that the representations $T^{n-1,k}(h)$ of $SO(n-1)$, $n \geq 4$ have already been proved to be irreducible. We prove that the representations $T^{nl}(g)$ of $SO(n)$ are then irreducible too. Indeed, we consider the restriction of $T^{nl}(g)$ to the subgroup $SO(n-1)$. We have proved in subsection 8 that this restriction can be decomposed as a direct sum of the representations $T^{n-1,k}(h)$, $0 \leq k \leq l$, $h \in SO(n-1)$, which are, by hypothesis, irreducible. But $T^{n-1,k}(h)$ and $T^{n-1,m}(h)$ are nonequivalent for $k \neq m$ (because the dimensions $h(n-1,k)$ and $h(n-1,m)$ of the spaces of these representations are obviously different). Since these representations are irreducible, any invariant subspace $\mathfrak{T}$ in $\mathfrak{R}^{nl}/r^2\mathfrak{R}^{n,l-2}$ must be a direct sum of the subspaces $\mathfrak{A}^{nlk}$ in which the representations $T^{n-1,k}(h)$ are realized. (We may recall that $\mathfrak{A}^{nlk}$ is the image of $x_n^{l-k}\mathfrak{H}^{n-1,k}$ under a mapping of $\mathfrak{R}^{nl}$ onto $\mathfrak{R}^{nl}/r^2\mathfrak{R}^{n,l-2}$.)

Thus

$$\mathfrak{T} = \mathfrak{A}^{nlk_1} + \ldots + \mathfrak{A}^{nlk_s}. \tag{2}$$

Hence if $\mathfrak{T} \neq \varnothing$, then it contains at least one of the subspaces $\mathfrak{A}^{nlm}$.

It remains for us to show that if the invariant subspace $\mathfrak{T}$ contains $\mathfrak{A}^{nlm}$, then it contains all $\mathfrak{A}^{nlk}$, $0 \leq k \leq l$.

We observe that the invariant subspace $\mathfrak{T}$ must be invariant for the infinitesimal operators $A_{jk}$. According to subsection 9 the operator $A_{jk}$ carries a coset with respect to $r^2\mathfrak{R}^{n,l-2}$, containing the function $f(\mathbf{x})$, into a coset containing the function

$$A_{jk}f(\mathbf{x}) = x_k \frac{\partial f}{\partial x_j} - x_j \frac{\partial f}{\partial x_k}. \tag{3}$$

But $\mathfrak{A}^{nlm}$ consists of cosets containing functions of the form $x_n^{l-m}h_m(\mathbf{x}')$, where $h_m(\mathbf{x}') \in \mathfrak{H}^{n-1,m}$, i.e. is a harmonic polynomial of degree $m$ in $n-1$ variables. The operator $A_{jn}$ carries the function $x_n^{l-m}h_m(\mathbf{x}')$ into

$$A_{jn}[x_n^{l-m}h_m(\mathbf{x}')] = x_n^{l-m+1}\frac{\partial h_m}{\partial x_j} - (l-m)x_n^{l-m-1}x_jh_m(\mathbf{x}'). \tag{4}$$

We prove that the expression on the right hand side of this equality can be written as a linear combination of functions of the subspaces $x_n^{l-m+1}\mathfrak{H}^{n-1,m-1}$, $x_n^{l-m-1}\mathfrak{H}^{n-1,m+1}$ and $r^2\mathfrak{R}^{nl}$.

Indeed, as shown in subsection 3, the polynomial $\partial h_m/\partial x_j$ is harmonic and therefore $\partial h_m/\partial x_j \in \mathfrak{H}^{n-1,m-1}$ Further, it was shown that

$$\hat{h}_{m+1}(\mathbf{x}') \equiv x_jh_m(\mathbf{x}') - \frac{1}{n+2m-3}r_{n-1}^2\frac{\partial h_m}{\partial x_j} \tag{5}$$

is a harmonic polynomial. Since the degree of this polynomial is equal to $m+1$, we have $\hat{h}_{m+1}(\mathbf{x}') \in \mathfrak{H}^{n-1,n+1}$. But $r_{n-1}^2 = r^2 - x_n^2$, and so

$$x_jh_m(\mathbf{x}') = \hat{h}_{m+1}(\mathbf{x}') - \frac{1}{n+2m-3}x_n^2\frac{\partial h_m}{\partial x_j} + \frac{1}{n+2m-3}r^2\frac{\partial h_m}{\partial x_j}. \tag{6}$$

Owing to formulas (4)–(6) we have

$$\begin{aligned} A_{jn}[x_n^{l-m}h_m(\mathbf{x}')] = {} & \frac{l+n+m-3}{n+2m-3}x_n^{l-m+1}\frac{\partial h_m}{\partial x_j} \\ & -(l-m)x_n^{l-m-1}\hat{h}_{m+1}(\mathbf{x}') - \frac{l-m}{n+2m-3}r^2x_n^{l-m-1}\frac{\partial h_m}{\partial x_j}. \end{aligned} \tag{7}$$

Since $\partial h/\partial x_j \in \mathfrak{H}^{n-1,m-1}$, $\hat{h}_{m+1}(\mathbf{x}') \in \mathfrak{H}^{n-1,m+1}$ and

$$x_n^{l-m-1}\frac{\partial h}{\partial x_j} \in \mathfrak{R}^{n,l-2},$$

it follows that (7) is the required expression of $A_{jn}[x_n^{l-m}h_m(\mathbf{x}')]$ as a linear combination of functions of the subspaces

$$x_n^{l-m+1}\mathfrak{H}^{n-1,m-1},\ x_n^{l-m-1}\mathfrak{H}^{n-1,m+1} \text{and } r^2\mathfrak{R}^{n,l-2}.$$

If $h_m(\mathbf{x}') \neq 0$, then for $m \neq 0$, $m \neq l$ for at least one $j$, $1 \leq j \leq n$, the first two terms in (7) are nonzero. For the first term this follows from the fact that for $m \neq 0$ the function $h_m(\mathbf{x}')$ is not constant, and so at least one of its partial derivatives is nonzero. For the second term this follows from the fact that $x_j h_m(\mathbf{x}')$ is not divisible by $r^2$ (see §2.5 and so $\hat{h}_{m+1}(\mathbf{x}') \neq 0$; the factor $l - m$ is obviously nonzero for $m \neq l$. In case $m = 0$ we have $\hat{h}_{m+1} \neq 0$, and for $m = l$ we have $\partial h_m / \partial x_j \neq 0$.

Thus we have proved that there exists a $j$, $1 \leq j \leq n$, for which

$$A_{jn}[x_n^{l-m} h_m(\mathbf{x}')] \in x_n^{l-m+1} \mathfrak{H}^{n-1,\, m-1} + x_n^{l-m-1} \mathfrak{H}^{n-1,\, m+1} + r^2 \mathfrak{R}^{n,\, l-2}, \tag{8}$$

where the components in the subspaces $x_n^{l-m+1} \mathfrak{H}^{n-1,m-1}$ and $x_n^{l-m-1} \mathfrak{H}^{n-1,m+1}$ are nonzero (except for the cases $m = 0$, $m = l$ indicated above). Hence the image of the polynomial $A_{jn}[x_n^{l-m} h_m(\mathbf{x}')]$ under the mapping of $\mathfrak{R}^{nl}$ onto $\mathfrak{R}^{nl}/r^2\mathfrak{R}^{n,l-2}$ lies in the sum of subspaces $\mathfrak{A}^{nl,m-1}$ and $\mathfrak{A}^{nl,m+1}$. As this image belongs to the invariant subspace $\mathfrak{T}$, by virtue of decomposition (2) $\mathfrak{T}$ must contain both $\mathfrak{A}^{nl,m-1}$ and $\mathfrak{A}^{nl,m+1}$ (only $\mathfrak{A}^{nl,m+1}$ for $m = 0$, and only $\mathfrak{A}^{nl,m-1}$ for $m = l$). Hence $\mathfrak{T}$ contains all the subspaces $\mathfrak{A}^{nlk}$, and so coincides with $\mathfrak{R}^{nl}/r^2\mathfrak{R}^{n,l-2}$ The irreducibility of $T^{nl}(g)$ is proved.

Thus we have proved that the $T^{nl}(g)$ are irreducible representations of $SO(n)$, $n \geq 3$. We now show that $T^{nl}(g)$ are representation of class 1, i.e. in the spaces of these representations there are vectors invariant for all operators $T^{nl}(h)$, $h \in SO(n-1)$. For this it is more convenient to use the realization of $T^{nl}(g)$ in the factor space $\mathfrak{R}^{nl}/r^2\mathfrak{R}^{n,l-2}$. We take the coset $x_n^l + r^2\mathfrak{R}^{n,l-2}$. If $h \in SO(n-1)$, then a rotation $h$ does not change the coordinate $x_n$, and so

$$T^{nl}(h)[x_n^l + r^2\mathfrak{R}^{n,\, l-2}] = x_n^l + r^2\mathfrak{R}^{n,\, l-2}. \tag{9}$$

It is not difficult to show that any element in $\mathfrak{R}^{nl}/r^2\mathfrak{R}^{n,l-2}$, invariant for operators $T^{nl}(h)$, $h \in SO(n-1)$, is proportional to $x_n^l + r^2\mathfrak{R}^{n,l-2}$.

Finally, we note that under the realization in $\mathfrak{H}^{nl}$ the representation $T^{nl}(g)$ is unitary relative to the scalar product

$$(h_1, h_2) = \int_{S^{n-1}} h_1(\xi)\overline{h_2(\xi)}\, d\xi. \tag{10}$$

Indeed, from the invariance of the euclidean measure $d\xi$ on the sphere $S^{n-1}$ it follows that

$$(T^{nl}(g)h_1,\ T^{nl}(g)h_2) = \int_{S^{n-1}} h_1(g^{-1}\xi)\overline{h_2(g^{-1}\xi)}\,d\xi$$
$$= \int_{S^{n-1}} h_1(\xi)\overline{h_2(\xi)}\,d\xi = (h_1,\ h_2). \quad (11)$$

11. Completeness of the system of representations $T^{nl}(g)$. We have constructed the system of representations $T^{nl}(g)$, $0 \le l < \infty$, of the group $SO(n)$. We shall now show that this system is complete in the set of representations of class 1 (relative to the subgroup $SO(n-1)$). In other words, we shall show that any representation $T(g)$ of class 1 relative to the subgroup $SO(n-1)$ is equivalent to one of the representations $T^{nl}(g)$.

Let $\mathbf{e}_1$ be a normalized invariant vector in the space $\mathfrak{A}$ of the representation $T(g)$. With each vector $\mathbf{f}$ of $\mathfrak{A}$ we associate a function on $SO(n)$, defined by the formula

$$f(g) = (T(g^{-1})\mathbf{f}, \mathbf{e}_1) \quad (1)$$

($(\mathbf{f}, \mathbf{e})$ is the invariant scalar product in $\mathfrak{A}$). It follows from the results of §2.5, Chapter I, that $f(g)$ is constant on left cosets with respect to the subgroup $SO(n-1)$. Therefore $f(g)$ can be regarded as a function on the homogeneous manifold $SO(n)/SO(n-1)$, i.e. on the sphere $S^{n-1}$. Namely, if $\xi = g\xi_n$, we put $\phi(\xi) = f(g)$.

On passing from vectors $\mathbf{f}$ to the corresponding functions $f(g)$, the representation $T(g)$ is given by the shift operators

$$Q(g)\varphi(\xi) = \varphi(g^{-1}\xi). \quad (2)$$

By the continuity of $T(g)$, the functions $\phi(\xi)$ are continuous and consequently square-integrable. But then $Q(g)$ is an irreducible component of the quasi-regular representation $L(g)$. Since $L(g)$ is decomposed as a direct sum of pairwise non-equivalent irreducible representations $H^{nl}(g)$, it follows from §3.3 of Chapter I that $T(g)$ is equivalent to one of the representations $H^{nl}(g)$ or, what is the same thing, to one of the representations $T^{nl}(g)$.

By the same token it is proved that every irreducible unitary representation $T(g)$ of $SO(n)$, which is of class 1, is equivalent to one of the representations $T^{nl}(g)$. Moreover, we have constructed a realization of $T(g)$ by shift operators in the space $\mathfrak{H}^{nl}$ of harmonic polynomials of degree $l$ in $n$ variables. Hence we get the following assertion.

*Let $T(g)$ be an irreducible unitary representation of class 1 of $SO(n)$ and let*

$\mathbf{e}_1$ *be a normalized vector of the space* $\mathfrak{A}$ *of this representation, invariant for all operators* $T(h)$, $h \in SO(n-1)$. *With each vector* $\mathbf{f}$ *of* $\mathfrak{A}$ *we associate a function* $\phi(\xi)$ *on the sphere* $S^{n-1}$:

$$\varphi(\xi) = (T(g^{-1})\mathbf{f}, \mathbf{e}_1), \tag{3}$$

*where* $g\xi_n = \xi$. *Then the function*

$$\Phi(\mathbf{x}) = r^l \varphi\left(\frac{\mathbf{x}}{r}\right) \tag{4}$$

*is a harmonic polynomial of degree* $l$ *in the variables* $x_1, \cdots, x_n$. *Any harmonic polynomial of degree* $l$ *in* $x_1, \cdots, x_n$ *can be presented in the form* (4).

We may also note that from this assertion it follows that $SO(n-1)$ is a massive subgroup in $SO(n)$. Indeed, any irreducible unitary representation of class 1 is equivalent to one of the representations $T^{nl}(g)$, and in the space $\mathfrak{A}^{nl} = \mathfrak{R}^{nl}/r^2\mathfrak{R}^{n,l-2}$ of this representation there is only one normalized vector invariant for all operators $T^{nl}(h)$, $h \in SO(n-1)$.

## §3. ZONAL SPHERICAL FUNCTIONS OF REPRESENTATIONS $T^{nl}(g)$ AND GEGENBAUER POLYNOMIALS

1. Description of zonal spherical functions. As indicated in §2.5 of Chapter I, the zonal spherical function of a representation $T(g)$ (relative to a subgroup $H$) is defined as the function

$$f(g) = (T(g)\mathbf{e}_1, \mathbf{e}_1),$$

where $\mathbf{e}_1$ is a vector in the space of the representation $T(g)$, invariant for all operators $T(h)$, $h \in H$.

Let us apply this definition to the subgroup $SO(n-1)$ of the group $SO(n)$ and the representation $T^{nl}(g)$. We obtain the function

$$f_{nl}(g) = (T^{nl}(g)\mathbf{e}_1, \mathbf{e}_1).$$

Corresponding to it we have the function

$$\varphi_{nl}(\xi) = (T^{nl}(g)\mathbf{e}_1, \mathbf{e}_1), \qquad \xi = g\xi_n, \tag{1}$$

on the sphere $S^{n-1}$, invariant for all transformations

$$L(h)\varphi(\xi) = \varphi(h^{-1}\xi),$$

where $h \in SO(n-1)$ (i.e. for rotations around the axis $Ox_n$). Consequently, the harmonic polynomial

$$\Phi_{nl}(\mathbf{x}) = r^l \varphi_{nl}\left(\frac{\mathbf{x}}{r}\right),$$

corresponding to the function $\phi_{nl}(\xi)$, must possess the same invariance.

As shown in §2.5, to each harmonic polynomial of the space $\mathfrak{H}^{nl}$ there corresponds, in a one-to-one manner, a coset of the space $\mathfrak{R}^{nl}$ with respect to the subspace $r^2\mathfrak{R}^{n,l-2}$. But it is easy to see that the only such coset invariant for rotations around the axis $Ox_n$ is (to within a constant factor) the coset $x_n^l + r^2\mathfrak{R}^{n,l-2}$. Therefore the harmonic polynomial corresponding to the zonal spherical function $\phi_{nl}(\xi)$ is (to within a constant factor) the harmonic projection of the function $x_n^l$. On calculating this projection by formula (15) of §2.5, we obtain

$$Hx_n^l = \sum_{k=0}^{[l/2]} \frac{(-1)^k l(l-1)\ldots(l-2k+1)\, r^{2k} x_n^{l-2k}}{2^k k!\,(n+2l-4)(n+2l-6)\ldots(n+2l-2k-2)}. \tag{2}$$

Thus we have proved that the zonal spherical function of $T^{nl}(g)$ is (to within a constant factor) the value on $S^{n-1}$ of the harmonic polynomial $Hx_n^l$:

$$\begin{aligned}\varphi_{nl}(\xi) &= a_{nl} H\xi_n^l \\ &= a_{nl} \sum_{k=0}^{[l/2]} \frac{(-1)^k l(l-1)\ldots(l-2k+1)}{2^k k!\,(n+2l-4)(n+2l-6)\ldots(n+2l-2k-2)}\, \xi_n^{l-2k}.\end{aligned} \tag{2'}$$

The expression for $Hx_n^l$ can be written more concisely with the help of polynomials of a special form, called Gegenbauer polynomials. These polynomials are defined by the formula

$$\begin{aligned}C_m^p(t) = \frac{2^m\Gamma(p+m)}{m!\,\Gamma(p)}\Big[t^m &- \frac{m(m-1)}{2^2(p+m-1)}t^{m-2} \\ &+ \frac{m(m-1)(m-2)(m-3)}{2^4\, 1\cdot 2(p+m-1)(p+m-2)}t^{m-4} + \ldots\Big].\end{aligned} \tag{3}$$

Comparing (2) and (3), we get

$$Hx_n^l = \frac{l!\,\Gamma\left(\frac{n-2}{2}\right)}{2^l\,\Gamma\left(l+\frac{n-2}{2}\right)}\, r^l C_l^{\frac{n-2}{2}}\left(\frac{x_n}{r}\right). \tag{4}$$

Equality (4) shows that $r^l C_l^{(n-2)/2}(x_n/r)$ is a harmonic polynomial belonging to the space $\mathfrak{H}^{nl}$.

It follows from (2′) and (3) that

$$\varphi_{nl}(\xi) = b_{nl} C_l^{\frac{n-2}{2}}(\xi_n). \tag{5}$$

To find the value of the coefficient $b_{nl}$, we put $\xi = \xi_n = (0, 0, \cdots, 0, 1)$ in

(5) Formula (1) gives $\phi_{nl}(\xi_n) = (\mathbf{e}_1, \mathbf{e}_1) = 1$. Therefore we have

$$1 = b_{nl} C_l^{\frac{n-2}{2}}(1).$$

Thus

$$\varphi_{nl}(\xi) = \frac{C_l^{\frac{n-2}{2}}(\xi_n)}{C_l^{\frac{n-2}{2}}(1)}. \tag{6}$$

2. The differential equation and recurrence relations for Gegenbauer polynomials. In order to derive the differential equation for Gegenbauer polynomials, we observe that $r^l C_l^{(n-2)/2}(x_n/r)$ is a harmonic polynomial, i.e.

$$\Delta\left[r^l C_l^{\frac{n-2}{2}}\left(\frac{x_n}{r}\right)\right] = 0. \tag{1}$$

Let us go over to spherical coordinates $r, \theta_1, \cdots, \theta_{n-1}$. Then the Laplace operator has the form

$$\Delta = r^{1-n}\frac{\partial}{\partial r}\left(r^{n-1}\frac{\partial}{\partial r}\right) + r^{-2}\sin^{2-n}\theta_{n-1}\frac{\partial}{\partial\theta_{n-1}}\left[\sin^{n-2}\theta_{n-1}\frac{\partial}{\partial\theta_{n-1}}\right] + \ldots, \tag{2}$$

where the dots stand for terms containing derivatives with respect to the $\theta_k$, $1 \le k \le n-2$. Since $x_n/r = \cos\theta_{n-1}$, equality (1) takes the form

$$\frac{\partial^2}{\partial\theta_{n-1}^2} C_l^{\frac{n-2}{2}}(\cos\theta_{n-1}) + (n-2)\operatorname{ctg}\theta_{n-1}\frac{\partial}{\partial\theta_{n-1}} C_l^{\frac{n-2}{2}}(\cos\theta_{n-1}) + l(n+l-2)C_l^{\frac{n-2}{2}}(\cos\theta_{n-1}) = 0.$$

Substituting $\cos\theta_{n-1} = t$, $n-2 = 2p$ in this equation, we obtain the following differential equation for Gegenbauer polynomials:

$$(1-t^2)\frac{d^2}{dt^2}C_l^p(t) - (2p+1)t\frac{d}{dt}C_l^p(t) + l(2p+l)C_l^p(t) = 0. \tag{3}$$

From this we can obtain a recurrence formula for Gegenbauer polynomials. We differentiate both sides of equality (3) of subsection 1. Comparing the resulting formula with the expression for $C_{m-1}^{p+1}(t)$, we get

$$\frac{d}{dt}C_m^p(t) = 2pC_{m-1}^{p+1}(t). \tag{4}$$

It follows from (3) and (4) that

$$4p(p+1)(1-t^2)C_{l-2}^{p+2}(t) - 2p(2p+1)tC_{l-1}^{p+1}(t) + l(2p+l)C_l^p(t) = 0. \tag{5}$$

We may also note that formula (4) yields the relation

$$\frac{d^k}{dt^k} C^p_m(t) = \frac{2^k \Gamma(p+k)}{\Gamma(p)} C^{p+k}_{m-k}(t). \tag{6}$$

We derive another recurrence formula for Gegenbauer polynomials by using equality (4) of subsection 1. We rewrite this equality as

$$x_n^l = \frac{l!\,\Gamma\left(\frac{n-2}{2}\right)}{2^l\,\Gamma\left(l+\frac{n-2}{2}\right)} r^l C_l^{\frac{n-2}{2}}\left(\frac{x_n}{r}\right) + r^2 f_1(\mathbf{x}), \tag{7}$$

where $f_1(\mathbf{x}) \in \mathfrak{R}^{n,l-2}$, and multiply both sides by $x_n$. Since $r^l C_l^{(n-2)/2}(x_n/r)$ is a harmonic polynomial of degree $l$, therefore, according to §2.3, the polynomial

$$x_n r^l C_l^{\frac{n-2}{2}}\left(\frac{x_n}{r}\right) - \frac{1}{n+2l-2} r^2 \frac{\partial\left[r^l C_l^{\frac{n-2}{2}}\left(\frac{x_n}{r}\right)\right]}{\partial x_n}$$

is harmonic. Consequently,

$$Hx_n^{l+1} = \frac{l!\,\Gamma\left(\frac{n-2}{2}\right)}{2^l\,\Gamma\left(l+\frac{n-2}{2}\right)}\left[x_n r^l C_l^{\frac{n-2}{2}}\left(\frac{x_n}{r}\right) - \frac{1}{n+2l-2} r^2 \frac{\partial\left[r^l C_l^{\frac{n-2}{2}}\left(\frac{x_n}{r}\right)\right]}{\partial x_n}\right].$$

But $r^2 = x_1^2 + \cdots + x_n^2$, and so $\partial r/\partial x_n = x_n/r$. Taking (4) into consideration, we easily get

$$Hx_n^{l+1} = \frac{l!\,\Gamma\left(\frac{n-2}{2}\right)}{2^{l+1}\Gamma\left(l+\frac{n}{2}\right)}\left[(n+l-2)\, x_n r^l C_l^{\frac{n-2}{2}}\left(\frac{x_n}{r}\right)\right.$$
$$\left. - (n-2)\, r^{l-1}(r^2 - x_n^2)\, C_{l-1}^{\frac{n}{2}}\left(\frac{x_n}{r}\right)\right]. \tag{8}$$

On the other hand, by formula (4) of subsection 1 we have

$$Hx_n^{l+1} = \frac{(l+1)!\,\Gamma\left(\frac{n-2}{2}\right)}{2^{l+1}\Gamma\left(l+\frac{n}{2}\right)} r^{l+1} C_{l+1}^{\frac{n-2}{2}}\left(\frac{x_n}{r}\right). \tag{9}$$

Comparing (8) and (9) after putting $x_n = t$, $r = 1$, $n - 2 = 2p$ in them, we obtain

$$(l+1)\, C^p_{l+1}(t) = (2p+l)\, t C^p_l(t) - 2p(1-t^2)\, C^{p+1}_{l-1}(t). \tag{10}$$

Let us eliminate the term containing $(1-t^2)C^{p+2}_{l-2}(t)$ from equality (5) and the equality obtained from (10) on replacing $p$ by $p+1$ and $l$ by $l-1$. We obtain the recurrence relation

$$(2p+l)C_l^p(t)=2p[C_l^{p+1}(t)-tC_{l-1}^{p+1}(t)]. \tag{11}$$

3. **Special cases and particular values of Gegenbauer polynomials.** From equality (3) of subsection 1, defining Gegenbauer polynomials, we have

$$C_0^p(t)=1. \tag{1}$$

We give the vlaues of Gegenbauer polynomials for $m = 1, 2, 3$:

$$C_1^p(t)=2pt, \tag{2}$$

$$C_2^p(t)=2p(p+1)\left[t^2-\frac{1}{2p+2}\right], \tag{3}$$

$$C_3^p(t)=\frac{4}{3}p(p+1)(p+2)\left[t^3-\frac{3}{2p+4}t\right]. \tag{4}$$

Further, it follows from equality (3) of subsection 1 that in the Gegenbauer polynomials $C_m^p(t)$ only those powers of $t$ appear whose indices have the same parity as $m$. Therefore we have

$$C_m^p(-t)=(-1)^m C_m^p(t) \tag{5}$$

$$C_{2m+1}^p(0)=0. \tag{6}$$

The value of $C_{2m}^p(0)$ is obtained from formula (3) of subsection 2:

$$C_{2m}^p(0)=\frac{(-1)^m\Gamma(p+m)}{\Gamma(p)\Gamma(m+1)}. \tag{7}$$

We shall need the value of $C_l^p(1)$ later. To find it, we use formula (10) of subsection 2. This formula gives

$$C_l^p(1)=\frac{2p+l-1}{l}C_{l-1}^p(1) \tag{8}$$

and consequently,

$$C_l^p(1)=\frac{\Gamma(2p+l)}{l!\,\Gamma(2p)}. \tag{9}$$

On substituting this expression in formula (6) of subsection 1 we obtain

$$\varphi_{nl}(\xi)=\frac{l!\,\Gamma(n-2)}{\Gamma(l+n-2)}C_l^{\frac{n-2}{2}}(\xi_n). \tag{10}$$

4. **Orthogonality relations for Gegenbauer polynomials.** This connection between Gegenbauer polynomials and zonal spherical functions leads to othogonality relationsfor Gegenbauer polynomials. Indeed, by formula (1) of subsection 1, for $\xi = g\xi_n$ we have

$$\varphi_{nl}(\xi)=(T^{nl}(g)\mathbf{e}_1, \mathbf{e}_1). \tag{1}$$

Thus,

$$\varphi_{nl}(\xi)=t_{11}^{nl}(g), \tag{1'}$$

where $\xi = g\xi_n$ and $t_{11}^{nl}(g)$ is a matrix element of the representation $T^{nl}(g)$ (corresponding to the basis vector $\mathbf{e}_1$). The orthogonality relations for matrix elements of irreducible unitary representations of compact groups (see §4.3, Chapter I) yield

$$\int t_{11}^{nl}(g)\overline{t_{11}^{nm}(g)}\,dg = \frac{\delta_{lm}}{h(n, l)}, \tag{2}$$

where $\delta_{lm}$ is the Kronecker symbol, and

$$h(n, l) = \frac{(2l+n-2)(n+l-3)!}{(n-2)!\,l!} \tag{3}$$

is the dimension of $T^{nl}(g)$ (see formula (11) of §2.5).

From formula (1') of this subsection and formula (10) of subsection 3 it can be seen that the functions $t_{11}^{nl}(g)$ depend only on the coordinate $\xi_n$ of the vector $\xi = g\xi_n$, or, what is the same thing, on the spherical coordinate $\theta_{n-1}$ of this vector. The coordinate $\theta_{n-1}$ is none other than the Euler angle $\theta_{n-1}^{n-1}$ of the rotation $g$ (see §1.3). Going over to Euler angles in equality (2) and utilizing formula (10) of subsection 3, we find that for $l \neq m$

$$\int_0^{\pi} C_l^p(\cos\theta)\, C_m^p(\cos\theta)\sin^{2p}\theta\, d\theta = 0,$$

where for brevity we have put $p = (n-2)/2$. In other words,

$$\int_{-1}^{1} C_l^p(x)\, C_m^p(x)(1-x^2)^{p-\frac{1}{2}}\, dx = 0. \tag{4}$$

In case $l = m$ formulas (1'), (2) and (3) give

$$\int_0^{\pi} [C_l^p(\cos\theta)]^2 \sin^{2p}\theta\, d\theta = \frac{\pi\Gamma(2p+l)}{2^{2p-1}l!\,(l+p)\,\Gamma^2(p)}$$

(we have utilized the duplication formula for the $\Gamma$ function). Thus,

$$\int_{-1}^{1} [C_l^p(x)]^2 (1-x^2)^{p-\frac{1}{2}}\, dx = \frac{\pi\Gamma(2p+l)}{2^{2p-1}\, l!\,(l+p)\,\Gamma^2(p)}. \tag{5}$$

It follows from (4) and (5) that the polynomials

$$2^{p-1}\,\Gamma(p)\left[\frac{2(l+p)\,l!}{\Gamma(2p+l)\pi}\right]^{\frac{1}{2}} C_l^p(x), \qquad l = 0, 1, \ldots, \tag{6}$$

form an othonormal system on the segment $[-1, 1]$ relative to the weight $\sigma(x) = (1-x^2)^{p-\frac{1}{2}}$.

We shall denote by $\mathfrak{L}_\sigma^2$ the space of functions $\phi(x)$ such that

$$\int_{-1}^{1} |\varphi(x)|^2 (1-x^2)^{p-\frac{1}{2}} dx < +\infty.$$

The system of functions (6) is complete in this space. Indeed, from the results of §4.5, Chapter I it follows that any function $f(g)$ on $SO(n)$, having a square-integrable modulus and such that

$$f(h_1 g h_2) = f(g), \qquad h_1,\ h_2 \in SO(n-1), \tag{7}$$

can be expanded in a mean-convergent Fourier series of zonal spherical functions of the representations $T^{nl}(g)$:

$$f(g) = \sum_{l=0}^{\infty} a_l t_{11}^{nl}(g). \tag{8}$$

Here

$$a_l = h(n, l) \int f(g) \overline{t_{11}^{nl}(g)}\, dg. \tag{9}$$

From the functional equation (7) it follows that $f(g)$ depends only on the Euler angle $\theta_{n-1}^{n-1}$ of the element $g$, $f(g) = \phi(\cos \theta_{n-1}^{n-1})$. Moreover, from the fact that

$$\int |f(g)|^2 dg < +\infty,$$

it follows that

$$\int_{-1}^{1} |\varphi(x)|^2 (1-x^2)^{n-\frac{1}{2}} dx < +\infty, \tag{10}$$

i.e. $\phi(x)$ belongs to $\mathfrak{L}_\sigma^2$.

If in (8) we replace $f(g)$ by $\phi(x)$, and the zonal spherical functions by their expressions in terms of Gegenbauer polynomials, we easily get

$$\varphi(x) = \sum_{l=0}^{\infty} b_l C_l^p(x), \tag{11}$$

where

$$b_l = \frac{2^{2p-1}\, l!\, (l+p)\, \Gamma^2(p)}{\pi \Gamma(2p+l)} \int_{-1}^{1} \varphi(x)\, C_l^p(x) (1-x^2)^{p-\frac{1}{2}} dx. \tag{12}$$

Series (11) converges in the mean relative to the weight $\sigma(x) = (1-x^2)^{p-1/2}$ and satisfies the Parseval equality

$$\int_{-1}^{1} |\varphi(x)|^2 (1-x^2)^{p-\frac{1}{2}} dx = \sum_{n=0}^{\infty} \frac{\pi\Gamma(2p+l)}{2^{2p-1} l! (l+p) \Gamma^2(p)} |b_l|^2. \tag{13}$$

5. Decomposition of the space of harmonic polynomials. Here we shall decompose the space $\mathfrak{H}^{nl}$ as a direct sum of subspaces $\mathfrak{H}^{nls}$, invariant for the operators $T^{nl}(h)$, where $h \in SO(n-1)$. It was shown in § 2.5 that $\mathfrak{H}^{nl}$ is isomorphic to the factor space of the space $\mathfrak{R}^{nl}$ of homogeneous polynomials of degree $l$ with respect to the subspace $r^2\mathfrak{R}^{n,l-2}$ of polynomials of the form $r^2 f_1(\mathbf{x})$, where $r^2 = x_1^2 + \cdots + x_n^2$, $f_1(\mathbf{x}) \in \mathfrak{R}^{n,l-2}$:

$$\mathfrak{R}^{nl} = \mathfrak{H}^{nl} + r^2\mathfrak{R}^{n,\,l-2}. \tag{1}$$

On the other hand, we have shown in § 2.8 that $\mathfrak{R}^{nl}$ is a direct sum of the subspace $r^2\mathfrak{R}^{n,l-2}$ and the subspaces $x_n^{l-s}\mathfrak{H}^{n-1,s}$, consisting of polynomials of the form $x_n^{l-s} h_s(\mathbf{x}')$, where $\mathbf{x}' = (x_1, \cdots, x_{n-1})$, $h_s(\mathbf{x}') \in \mathfrak{H}^{n-1,s}$:

$$\mathfrak{R}^{nl} = \sum_{s=0}^{l} x_n^{l-s}\mathfrak{H}^{n-1,\,s} + r^2\mathfrak{R}^{n,\,l-2}. \tag{2}$$

It follows from (1) and (2) that any homogeneous harmonic polynomial $h(\mathbf{x})$ of degree $l$ can be uniquely presented in the form

$$h(\mathbf{x}) = \sum_{s=0}^{l} x_n^{l-s} h_s(\mathbf{x}') + r^2 f_1(\mathbf{x}), \tag{3}$$

where $h_s(\mathbf{x}') \in \mathfrak{H}^{n-1,s}$ and $f_1(\mathbf{x}) \in \mathfrak{R}^{n,l-2}$. Applying the operator of harmonic projection to both sides of this equality, we see that

$$h(\mathbf{x}) = \sum_{s=0}^{l} H(x_n^{l-s} h_s(\mathbf{x}')). \tag{4}$$

For fixed $s$ the polynomials $H(x_n^{l-s} h_s(\mathbf{x}'))$ form a subspace in $\mathfrak{H}^{nl}$. We denote this subspace by $\mathfrak{H}^{nls}$:

$$\mathfrak{H}^{nls} = H(x_n^{l-s}\mathfrak{H}^{n-1,\,s}). \tag{5}$$

Since every polynomial $h(\mathbf{x})$ belonging to $\mathfrak{H}^{nl}$ can be uniquely written in the form (3), it follows that $\mathfrak{H}^{nl}$ is the direct sum of the $\mathfrak{H}^{nls}$, $0 \le s \le l$:

$$\mathfrak{H}^{nl} = \sum_{s=0}^{l} \mathfrak{H}^{nls}. \tag{6}$$

We know that the subspaces $x_n^{l-s}\mathfrak{H}^{n-1,s}$ are invariant for the operator $L^{nl}(h)f(\mathbf{x}) = f(h^{-1}\mathbf{x})$, $h \in SO(n-1)$ (see § 2.8). Hence it easily follows that the $\mathfrak{H}^{nls}$ (i.e. the harmonic projections of the spaces $x_n^{l-s}\mathfrak{H}^{n-1,s}$) are invariant for the operators $T^{nl}(h)$, $h \in SO(n-1)$.

The subspaces $\mathfrak{H}^{nls}$ can be directly determined, without resorting explicitly to harmonic projection. For this we calculate the harmonic projection of the polynomial $x_n^{l-s} h_s(\mathbf{x}')$. Employing formula (15) of § 2.5, we get

$$H(x_n^{l-s}h_s(\mathbf{x}')) = \sum_{k=0}^{[l/2]} \frac{(-1)^k r^{2k} \Delta^k (x_n^{l-s} h_s(\mathbf{x}'))}{2^k k!\,(n+2l-4)(n+2l-6)\dots(n+2l-2k-2)}. \tag{7}$$

Since $h_s(\mathbf{x}')$ is a harmonic polynomial, for $p \geq 1$ we have $(\Delta')^p h_s(\mathbf{x}') = 0$, where

$$\Delta' = \frac{\partial^2}{\partial x_1^2} + \dots + \frac{\partial^2}{\partial x_{n-1}^2}$$

is the Laplace operator in $(n-1)$-dimensional space. But $\Delta = \Delta' + \partial^2/\partial x_n^2$, and so

$$\Delta^k [x_n^{l-s} h_s(\mathbf{x}')] = (l-s)\dots(l-s-2k+1)\, x_n^{l-s-2k} h_s(\mathbf{x}').$$

Thus,

$$H(x_n^{l-s}h_s(\mathbf{x}')) = h_s(\mathbf{x}') \sum_{k=0}^{\left[\frac{l-s}{2}\right]} \frac{(-1)^k (l-s)(l-s-1)\dots(l-s-2k+1)}{2^k k!\,(n+2l-4)\dots(n+2l-2k-2)} r^{2k} x_n^{l-s-2k}. \tag{8}$$

Taking equality (3) of subsection 1 into consideration, we can rewrite formula (8) as follows:

$$H(x_n^{l-s}h_s(\mathbf{x}')) = \frac{(l-s)!\,\Gamma\left(\frac{n-2}{2}+s\right)}{2^{l-s}\Gamma\left(\frac{n-2}{2}+l\right)} r^{l-s} C_{l-s}^{\frac{n-2}{2}+s}\left(\frac{x_n}{r}\right) h_s(\mathbf{x}'). \tag{9}$$

Hence we have proved that $\mathfrak{H}^{nls}$ is the space of functions of the form

$$r^{l-s} C_{l-s}^{\frac{n-2}{2}+s}\left(\frac{x_n}{r}\right) h_s(\mathbf{x}'), \tag{10}$$

where $h_s(\mathbf{x}') \in \mathfrak{H}^{n-1,s}$.

6. Construction of a canonical basis. We now pass on to the basic aim of this section—the construction of a basis in the space $\mathfrak{H}^{nl}$ of homogeneous

harmonic polynomials of degree $l$ in $n$ variables. From the results of the preceding subsection it follows that any polynomial $h(\mathbf{x})$ of $\mathfrak{H}^{nl}$ can be uniquely presented in the form

$$h(\mathbf{x}) = \sum_{s=0}^{l} r^{l-s} C_{l-s}^{\frac{n-2}{2}+s}\left(\frac{x_n}{r}\right) h_s(\mathbf{x}'), \tag{1}$$

where $h_s(\mathbf{x}') \in \mathfrak{H}^{n-1,s}$.

Since the functions $h_s(\mathbf{x}')$ belong to the spaces $\mathfrak{H}^{n-1,s}$, they admit a similar decomposition. We continue this process and take into consideration the fact that any harmonic polynomial of degree $s$ in two variables $x_1$ and $x_2$ is a linear combination of the polynomials $(x_2 + ix_1)^s$ and $(x_2 - ix_1)^s$. We thus get the following result.

*Any polynomial in $\mathfrak{H}^{nl}$ can be uniquely presented as a linear combination of functions of the form*

$$\Xi_K^l(\mathbf{x}) = A_K^l \prod_{j=0}^{n-3} r_{n-j}^{k_j - k_{j+1}} C_{k_j - k_{j+1}}^{\frac{n-j-2}{2}+k_{j+1}}\left(\frac{x_{n-j}}{r_{n-j}}\right)(x_2 \pm ix_1)^{k_{n-2}}, \tag{2}$$

*where $r_{n-j}^2 = x_1^2 + \cdots + x_{n-j}^2$ and $l \equiv k_0 \geq k_1 \geq \cdots \geq k_{n-2} \geq 0$. The symbol $K$ here denotes the integer sequence $(k_1, \cdots, \pm k_{n-2})$.*

The normalizing factor $A_K^l$ is chosen such that

$$\int_{S^{n-1}} |\Xi_K^l(\xi)|^2 \, d\xi = 1, \tag{3}$$

where $d\xi$ is the normalized euclidean measure on the unit sphere $S^{n-1}$. In order to calculate this coefficient in an explicit form, we write the functions $\Xi_K^l(\xi)$ in spherical coordinates. Since $x_{n-j}/r_{n-j} = \cos\theta_{n-j-1}$ and $r_{n-j-1}/r_{n-j} = \sin\theta_{n-j-1}$, we have

$$\Xi_K^l(\mathbf{x}) = A_K^l r^l \prod_{j=0}^{n-3} C_{k_j - k_{j+1}}^{\frac{n-j-2}{2}+k_{j+1}}(\cos\theta_{n-j-1}) \sin^{k_{j+1}}\theta_{n-j-1} e^{\pm ik_{n-2}\theta_1}. \tag{4}$$

On substituting this expression in (3) and taking into consideration the fact that

$$d\xi = \frac{\Gamma\left(\frac{n}{2}\right)}{2\pi^{\frac{n}{2}}} \sin^{n-2}\theta_{n-1} \ldots \sin\theta_2 \, d\theta_1 \ldots d\theta_{n-1},$$

we obtain

$$\frac{1}{(A_K^l)^2}=\frac{\Gamma\left(\frac{n}{2}\right)}{\pi^{\frac{n-2}{2}}}\prod_{j=0}^{n-3}\int_0^\pi [C_{k_j-k_{j+1}}^{\frac{n-j-2}{2}+k_{j+1}}(\cos\theta)]^2\sin^{2k_{j+1}+n-j-2}\theta d\theta.$$

Taking into account the normalization relations for Gegenbauer polynomials (see formula (5) of subsection 4), we easily get

$$(A_K^l)^2=\frac{1}{\Gamma\left(\frac{n}{2}\right)}\times\prod_{j=0}^{n-3}\frac{2^{2k_{j+1}+n-j-4}(k_j-k_{j+1})!\,(n-j+2k_j-2)\,\Gamma^2\left(\frac{n-j-2}{2}+k_{j+1}\right)}{\sqrt{\pi}\,\Gamma(k_j+k_{j+1}+n-j-2)}. \tag{5}$$

In particular, if $O=(0,\cdots,0)$, then

$$A_O^l=\sqrt{\frac{l!\,\Gamma(n-2)(2l+n-2)}{\Gamma(n+l-2)(n-2)}}, \tag{6}$$

and so

$$\Xi_O^l(\mathbf{x})=\sqrt{\frac{l!\,\Gamma(n-2)(2l+n-2)}{\Gamma(n+l-2)(n-2)}}\,r^lC_l^{\frac{n-2}{2}}\left(\frac{x_n}{r}\right). \tag{7}$$

We show, finally, that the basis $\Xi_K^l(\mathbf{x})$ in the space $\mathfrak{H}^{nl}$ is orthogonal relative to the scalar product

$$(h_1, h_2)=\int_{S^{n-1}}h_1(\xi)\overline{h_2(\xi)}\,d\xi \tag{8}$$

in this space. In other words, if the sequences $K=(k_1,\cdots,\pm k_{n-2})$ and $M=(m_1,\cdots,\pm m_{n-2})$ do not coincide, then

$$(\Xi_K^l,\ \Xi_M^l)\equiv\int_{S^{n-1}}\Xi_K^l(\xi)\overline{\Xi_M^l(\xi)}\,d\xi=0. \tag{9}$$

Indeed, in (9) we replace the functions $\Xi_K^l(\mathbf{x})$ and $\Xi_M^l(\mathbf{x})$ by their expressions according to formula (4) and take into consideration the expression for invariant measure on the sphere $S^{n-1}$. Then the integral (9) becomes (to within a constant factor) a product of integrals of the form

$$\int_0^{2\pi}e^{i(\pm k_{n-2}\mp m_{n-2})\theta}d\theta \tag{10}$$

and

$$\int_0^\pi C_{k_j-k_{j+1}}^{\frac{n-j-2}{2}+k_{j+1}}(\cos\theta)\,C_{m_j-m_{j+1}}^{\frac{n-j-2}{2}+m_{j+1}}(\cos\theta)\,\sin^{k_{j+1}+m_{j+1}+n-j-2}\theta\,d\theta, \tag{11}$$

$$j=0,\ 1,\ \ldots,\ n-3.$$

If $k_{j+1} = m_{j+1}$ but $k_j \neq m_j$, then, by virtue of the orthogonality relations for Gegenbauer polynomials, integral (11) is equal to zero. Therefore, if the sequences $K = (k_1, \cdots, \pm k_{n-2})$ and $M = (m_1, \cdots, \pm m_{n-2})$ are different, then at least one of the factors in the expression for $(\Xi^l_K, \Xi^l_M)$ is equal to zero (for $\pm k_{n-2} \neq \pm m_{n-2}$, a factor of the form (10) becomes zero).

In this way we have constructed an orthonormal basis $\Xi^l_K(\mathbf{x})$ in the space $\mathfrak{H}^{nl}$ of homogeneous harmonic polynomials of degree $l$ in $n$ variables. We shall call this the *canonical basis* in $\mathfrak{H}^{nl}$. Sometimes it is convenient to assume that this basis is ordered. Namely, we shall assume that $\Xi^l_K(\mathbf{x})$, $K = (k_0, \cdots, \pm k_{n-2})$, precedes $\Xi^l_M(\mathbf{x})$, $M = (m_0, \cdots, \pm m_{n-2})$, if there is a $j$ such that $k_i = m_i$, $0 \leq i \leq j$ and $k_{j+1} < m_{j+1}$ (or $\pm k_{n-2} < \pm m_{n-2}$).

It is clear from the construction of the basis $\Xi^l_K(\mathbf{x})$ that for $k_1 = s$ the function $\Xi^l_K(\mathbf{x})$ belongs to the subspace $\mathfrak{H}^{nls}$.

7. Expansion of functions on the $n$-dimensional sphere. For fixed $l$ the functions $\Xi^l_K(\xi)$, $K = (k_1, \cdots, \pm k_{n-2})$ form an orthonormal basis in $\mathfrak{H}^{nl}$. But in §2.7 it was shown that the space $\mathfrak{L}^2(S^{n-1})$ of functions on $S^{n-1}$, having square-integrable moduli, is the direct sum of the subspaces $\mathfrak{H}^{nl}$, $l = 0, 1, \cdots$

$$\mathfrak{L}^2(S^{n-1}) = \sum_{l=0}^{\infty} \mathfrak{H}^{nl}. \tag{1}$$

Hence the functions $\Xi^l_K(\xi)$, where $K$ runs through all possible integer sequences $K = (k_1, \cdots, \pm k_{n-2})$ such that $l \geq k_1 \geq \cdots \geq k_{n-2} \geq 0$, form an orthonormal basis in $\mathfrak{L}^2(S^{n-1})$.

Thus we have proved that any function $f(\xi)$ of $\mathfrak{L}^2(S^{n-1})$ can be expanded in a mean-convergent series of the form

$$f(\xi) = \sum_K a^l_K \Xi^l_K(\xi), \tag{2}$$

where

$$a^l_K = \int_{S^{n-1}} f(\xi) \overline{\Xi^l_K(\xi)} d\xi, \tag{3}$$

($d\xi$ is the normalized euclidean measure on sphere $S^{n-1}$). As one can see from the results of subsection 6, this expansion is connected with the decomposition of the quasi-regular representation of $SO(n)$ into irreducible representations. Namely, the terms of expansion (2), for which $l$ has a fixed value, belong to $\mathfrak{H}^{nl}$, and consequently this group of terms is transformed according to the irreducible representation $T^{nl}(g)$ of $SO(n)$.

## §4. MATRIX ELEMENTS OF THE NULL COLUMN

In the preceding section we construced a canonical basis in the space $\mathfrak{H}^{nl}$ of homogeneous harmonic polynomials of degree $l$ in $n$ variables. By the canonical matrix of the representation

$$T^{nl}(g)f(\mathbf{x})=f(g^{-1}\mathbf{x}), \qquad f(\mathbf{x})\in\mathfrak{H}^{nl},$$

we shall mean the matrix of this representation in the canonical basis. In this section we consider elements of the "null column" of a canonical matrix, corresponding to the basis element

$$\Xi^l_O(\mathbf{x}), \qquad O=(0, \ldots, 0)$$

and express them by means of Gegenbauer polynomials. This will lead to new properties of these polynomials.

1. Elements of the "null column" of a canonical matrix. We shall number the rows and columns of the canonical matrix of the representation $T^{nl}(g)$ by the symbols $K$, $M$, $K=(k_1, \cdots, \pm k_{n-2})$, $M=(m_1, \cdots, \pm m_{n-2})$. Thus elements of the canonical matrix of the operator $T^{nl}(g)$ are denoted by $t^{nl}_{KM}(g)$. These matrix elements are coefficients in the expansion of the function $T^{nl}(g)\Xi^l_M(\mathbf{x})$ in the functions $\Xi^l_K(\mathbf{x})$.

We denote by $O$ the sequence $(0, \cdots, 0)$. We show that the matrix elements $t^{nl}_{KO}(g)$ and $t^{nl}_{OK}(g)$ are very simply expressed by means of Gegenbauer polynomials.

Since $t^{nl}_{KM}(g)$ are matrix elements of $T^{nl}(g)$ in the canonical basis $\Xi^l_K(\mathbf{x})$, for any $g$ we have the equality

$$\Xi^l_M(g^{-1}\mathbf{x})=T^{nl}(g)\,\Xi^l_M(\mathbf{x})=\sum_K t^{nl}_{KM}(g)\,\Xi^l_K(\mathbf{x}). \tag{1}$$

Let us put $\mathbf{x}=\xi_n(0, 0, \cdots, 0, 1)$ in this equality. The values of $\Xi^l_K(\xi_n)$ are easily calculated by formula (4) of §3.6. Evidently for $K\neq 0$ the right hand side of that formula contains factors which are equal to zero, and so for $K\neq 0$ we have $\Xi^l_K(\xi_n)=0$. But if $K=0$, then by formula (7) of §3.6 we get

$$\Xi^l_O(\xi_n)=A^l_O C_l^{\frac{n-2}{2}}(1). \tag{2}$$

By formula (9) of §3.3 we have

$$C_l^{\frac{n-2}{2}}(1)=\frac{\Gamma(n+l-2)}{l!\,\Gamma(n-2)},$$

and by formula (6) of §3.6

$$A_O^l = \sqrt{\frac{l!\,\Gamma(n-2)(2l+n-2)}{\Gamma(n+l-2)(n-2)}}.$$

Substituting these values in formula (2), we get

$$\Xi_O^l(\xi_n) = \sqrt{\frac{\Gamma(l+n-2)(2l+n-2)}{l!\,\Gamma(n-1)}}.$$

Substituting the values of $\Xi_K^l(\xi_n)$ in formula (1) with $x = \xi_n$, we find that

$$\Xi_M^l(g^{-1}\xi_n) = \sqrt{\frac{\Gamma(l+n-2)(2l+n-2)}{l!\,\Gamma(n-1)}}\, t_{OM}^{nl}(g). \tag{3}$$

Since $T^{nl}(g)$ is unitary, $t_{OM}^{nl}(g) = \overline{t_{MO}^{nl}(g^{-1})}$. Hence it follows from (3) that

$$t_{MO}^{nl}(g) = \sqrt{\frac{l!\,\Gamma(n-1)}{\Gamma(l+n-2)(2l+n-2)}}\, \Xi_M^l(g\xi_n). \tag{4}$$

The value of the coefficient in (4) may also be obtained from group-theoretic considerations. Since $T^{nl}(g)$ is an irreducible representation of the group $SO(n)$, we have the equality

$$\int |t_{MO}^{nl}(g)|^2 dg = \frac{l!\,\Gamma(n-1)}{\Gamma(l+n-2)\,(2l+n-2)} \tag{5}$$

(see §4.3, Chapter I, and formula (11) of §2.5 of this chapter). On the other hand, it easily follows from equality (3) of §3.6 that

$$\int |\Xi_M^l(g\xi_n)|^2 dg = 1 \tag{6}$$

(by formula (3) of §1.4 the integration over $SO(n)$ reduces to integration over $SO(n-1)$ and over $S^{n-1}$; since for $h \in SO(n-1)$ we have $h\xi_n = \xi_n$, and on $S^{n-1}$ the functions $\Xi_M^l(\xi)$ are normalized, we get equality (6)). Comparing equalities (5) and (6), we get the value of the coefficient in formula (4).

From (4) one can easily derive an expression for the matrix elements $t_{MO}^{nl}(g)$ in terms of the Euler angles of the rotation $g$. Namely, it was shown in §1.3 that the Euler angles $\theta_1^{n-1}, \dots, \theta_{n-1}^{n-1}$ are equal to the spherical coordinates of the point $g\xi_n$ of the sphere $S^{n-1}$. An expression for $\Xi_M^l(\xi)$ in terms of the spherical coordinates of $\xi$ is given by formula (4) of §3.6. Making suitable substitutions, we find that if the Euler angles of $g$ are equal to $\theta_k^j$, then

$$t_{MO}^{nl}(g) = \sqrt{\frac{l!\,\Gamma(n-1)}{\Gamma(l+n-2)(2l+n-2)}}\, A_M^l$$
$$\times \prod_{j=0}^{n-3} C_{m_j - m_{j+1}}^{\frac{n-j-2}{2}+m_{j+1}}(\cos\theta_{n-j-1}^{n-1}) \sin^{m_{j+1}}\theta_{n-j-1}^{n-1}\, e^{\pm i m_{n-2}\theta_1^{n-1}}, \tag{7}$$

where $A^l_M$ is calculated according to formula (5) of §3.6 and

$$M = (m_1, \cdots, \pm m_{n-2}), \; m_0 \equiv l.$$

We note some special cases of this formula. If $M = O = (0, \cdots, 0)$ then all those factors in formula (7) for which $j > 0$ are equal to 1 ($C^p_0(\cos\theta) = 1$ and $\sin^0\theta = 1$). Therefore

$$t^{nl}_{OO}(g) = \frac{\Gamma(n-2)\, l!}{\Gamma(n+l-2)} C_l^{\frac{n-2}{2}}(\cos\theta^{n-1}_{n-1}). \tag{8}$$

Thus we see that the matrix element $t^{nl}_{00}(g)$ of the canonical matrix of $T^{nl}(g)$ depends only on the Euler angle $\theta^{n-1}_{n-1}$.

We also calculate the value of $t^{nl}_{MO}(g)$ for the case when $g = g_n(\phi)$ is a rotation by an angle $\phi$ in the $(x_n, x_{n-1})$-plane. In this case all Euler angles of $g$ are equal to zero, except for the angle $\theta^{n-1}_{n-1}$, which is equal to $\phi$. Hence it follows from formula (7) that even if one $m_j$, $j \geq 2$, is nonzero, $t^{nl}_{MO}[g_n(\phi)] = 0$. But if $M$ has the form $M = (m, 0, \cdots, 0)$, then, taking into account formulas (5) of subsection 6 and (9) of §3.3, we find from formula (7) that

$$t^{nl}_{MO}(g) = \frac{2^m \Gamma\left(\frac{n-2}{2}+m\right)}{\Gamma\left(\frac{n-2}{2}\right)} \times \sqrt{\frac{l!\,(l-m)!\,\Gamma(n-2)\,\Gamma(n+m-3)\,(n+2m-3)}{m!\,\Gamma(l+m+n-2)\,\Gamma(l+n-2)}} \times \sin^m\varphi\, C_{l-m}^{\frac{n-2}{2}+m}(\cos\varphi). \tag{9}$$

2. The composition theorem for Gegenbauer polynomials. From the definition of matrix elements if follows that for any element $g$ of the group $SO(n)$ we have

$$\Xi^l_O(g^{-1}\xi) = \sum_M t^{nl}_{MO}(g)\, \Xi^l_M(\xi). \tag{1}$$

Let us denote by $\xi^{(g)}_n$ the $n$th cartesian coordinate of the vector $g^{-1}\xi$. Using formula (7) of §3.6, we can rewrite equality (1) as follows:

$$C_l^{\frac{n-2}{2}}(\xi^{(g)}_n) = \sqrt{\frac{\Gamma(n+l-2)\,(n-2)}{l!\,\Gamma(n-2)\,(2l+n-2)}} \sum_M t^{nl}_{MO}(g)\, \Xi^l_M(\xi). \tag{2}$$

Since $t^{nl}_{MO}(g)$ and $\Xi^l_M(\xi)$ are expressible by means of Gegenbauer polynomials, equality (2) may be regarded as the general form of the composition theorem for these polynomials. However, formula (2) is not very convenient for applications, since the expressions for $t^{nl}_{MO}(g)$ and $\Xi^l_M(\xi)$ by means of Gegenbauer polynomials are rather complicated.

In order to obtain a simpler formula, we consider a special case of equality (2), choosing as $\xi$ the vector with spherical coordinates $0, 0, \cdots, \theta_{n-2}, \theta_{n-1}$, and as $g$ the rotation by the angle $\phi$ in the $(x_n, x_{n-1})$-plane, $g = g_n(\phi)$.

Calculating, by formula (4) of §3.6, the value of $\Xi^l_M(\xi)$ for the vector $\xi = \xi(0, \cdots, 0, \theta_{n-2}, \theta_{n-1})$, we get

$$\Xi^l_M(\xi) = 0,$$

if $m_k \neq 0$, $k \geq 2$, and

$$\Xi^l_M(\xi) = A^l_M \sin^m \theta_{n-1}\, C^{\frac{n-2}{2}+m}_{l-m}(\cos\theta_{n-1})\, C^{\frac{n-3}{2}}_m(\cos\theta_{n-2}),$$

if $M = (m, 0, \cdots, 0)$. Here

$$A^l_M = 2^{n+m-4}\Gamma\left(\frac{n-3}{2}\right)\Gamma\left(\frac{n-2}{2}+m\right)\sqrt{\frac{(l-m)!\,m!\,(n+2l-2)(n+2m-3)}{\pi\Gamma(l+m+n-2)\Gamma(m+n-3)(n-2)}}.$$

The value of $t^{nl}_{MO}[g_n(\phi)]$ is given by formula (9) of subsection 1.

It remains for us to calculate $\xi^{(g)}_n$. To do so, we observe that the cartesian coordinates of the point $\xi(0, \cdots, \theta_{n-2}, \theta_{n-1})$ have the form

$$\xi_k = 0, \qquad 1 \leqslant k \leqslant n-2,$$
$$\xi_{n-1} = \sin\theta_{n-1}\cos\theta_{n-2},$$
$$\xi_n = \cos\theta_{n-1}.$$

Therefore the rotation $g_n(\phi)^{-1}$ carries $\xi$ into the point $\xi^{(g)}$, the $n$th coordinate of which has the form

$$\xi_{-n} = \cos\varphi\cos\theta_{n-1} + \sin\varphi\sin\theta_{n-1}\cos\theta_{n-2}.$$

We substitute these expressions for $\xi_{-n}$, $\Xi^l_M(\xi)$ and $t^{nl}_{MO}[g_n(\phi)]$ in formula (2) and replace $\theta_{n-1}$ by $\theta$, $\theta_{n-2}$ by $\psi$ and $(n-2)/2$ by $p$. We obtain

$$C^p_l(\cos\theta\cos\varphi + \sin\theta\sin\varphi\cos\psi)$$
$$= \frac{\Gamma(2p-1)}{[\Gamma(p)]^2}\sum_{m=0}^{l}\frac{2^{2m}\Gamma^2(p+m)(l-m)!\,(2m+2p-1)}{\Gamma(l+m+2p)}$$
$$\times C^{p+m}_{l-m}(\cos\varphi)\sin^m\varphi\, C^{p+m}_{l-m}(\cos\theta)\sin^m\theta\, C^{p-\frac{1}{2}}_m(\cos\psi). \tag{3}$$

Formula (3) is usually called the composition formula for Gegenbauer polynomials.

Let us note some special cases of formula (3). If we put $\psi = 0$ in it, and take into consideration the fact that

$$C_m^{p-\frac{1}{2}}(1)=\frac{\Gamma(2p+m-1)}{m!\,\Gamma(2p-1)},$$

we obtain the equality

$$C_l^p[\cos(\theta-\varphi)] = \frac{1}{[\Gamma(p)]^2}\sum_{m=0}^{l}\frac{2^{2m}\Gamma^2(p+m)(l-m)!\,\Gamma(2p+m-1)(2m+2p-1)}{\Gamma(l+m+2p)\,m!} \times C_{l-m}^{p+m}(\cos\varphi)\sin^m\varphi\, C_{l-m}^{p+m}(\cos\theta)\sin^m\theta. \quad (4)$$

Similarly, on putting $\psi=\pi/2$ and taking into consideration the equalities $C_{2m+1}^p(0)=0$ and

$$C_{2m}^p(0)=\frac{(-1)^m\Gamma(p+m)}{\Gamma(p)\,m!}$$

(see §3.3), we obtain

$$C_l^p(\cos\theta\cos\varphi) = \frac{\Gamma(2p-1)}{[\Gamma(p)]^2}\sum_{m=0}^{[l/2]}\frac{(-1)^m 2^{4m}\Gamma^2(p+2m)(l-2m)!\,\Gamma(p+m-{}^1/_2)(4m+2p+1)}{\Gamma(l+2m+2p)\,\Gamma(p-{}^1/_2)\,m!} \times C_{l-2m}^{p+2m}(\cos\varphi)\sin^{2m}\varphi\, C_{l-2m}^{p+2m}(\cos\theta)\sin^{2m}\theta. \quad (5)$$

However, if we put $\phi=\theta=\pi/2$ and $l-m=2k$ in (3), we obtain an expression for $C_l^p(\cos\psi)$ in terms of the functions $C_k^{p-1/2}(\cos\psi)$:

$$C_l^p(\cos\psi)=\frac{\Gamma(2p-1)}{[\Gamma(p)]^2}\sum_{k=0}^{[l/2]}\frac{(2k)!\,(2l-4k+2p-1)}{\Gamma(2l-2k+2p)} \times\left[\frac{2^{l-2k}\Gamma(p+l-k)}{k!}\right]^2 C_{l-2k}^{p-\frac{1}{2}}(\cos\psi). \quad (6)$$

3. The multiplication formula for Gegenbauer polynomials. Let us multiply both sides of the composition formula (3) above by the expression $C_k^{p-1/2}(\cos\psi)\sin^{2p-1}\psi$, and integrate with respect to $\psi$ from 0 to $\pi$. By virtue of the orthogonality relations for Gegenbauer polynomials (§3.4) we obtain

$$\int_0^\pi C_l^p(\cos\theta\cos\varphi+\sin\theta\sin\varphi\cos\psi)\,C_k^{p-\frac{1}{2}}(\cos\psi)\sin^{2p-1}\psi\,d\psi$$
$$=\frac{2^{2k+2p-1}[\Gamma(p+k)]^2(l-k)!\,\Gamma(2p+k-1)}{k!\,\Gamma(2p-1)\,\Gamma(l+k+2p)}$$
$$\times C_{l-k}^{p+k}(\cos\varphi)\sin^k\varphi\, C_{l-k}^{p+k}(\cos\theta)\sin^k\theta. \qquad (1)$$

This formula is called the multiplication formula for Gegenbauer polynomials. In particular, for $k = 0$ we get

$$\int_0^\pi C_l^p(\cos\theta\cos\varphi+\sin\theta\sin\varphi\cos\psi)\sin^{2p-1}\psi\,d\psi$$
$$=\frac{2^{2p-1}\,l!\,[\Gamma(p)]^2}{\Gamma(2p+l)}C_l^p(\cos\varphi)\,C_l^p(\cos\theta). \qquad (2)$$

Equality (2) has a simple geometrical significance. In order to explain it, we observe that the zonal spherical functions $t_{00}^{nl}(g)$ may be regarded as functions on the sphere, which depend only on the polar angle $\theta$ (see subsection 2). Accordingly, we shall write

$$t_{00}^{nl}(g)=t_{00}^{nl}(\cos\theta).$$

By formula (8) of subsection 1 we have (for $n = 2p + 2$)

$$t_{00}^{nl}(\cos\theta)=\frac{\Gamma(2p)\,l!}{\Gamma(2p+l)}C_l^p(\cos\theta).$$

Hence (2) can be rewritten as follows:

$$\frac{\Gamma\left(p+\frac{1}{2}\right)}{\sqrt{\pi}\,\Gamma(p)}\int_0^\pi t_{00}^{nl}(\cos\gamma)\sin^{2p-1}\psi\,d\psi=t_{00}^{nl}(\cos\varphi)\,t_{00}^{nl}(\cos\theta), \qquad (3)$$

where $\cos\gamma=\cos\theta\cos\phi+\sin\theta\sin\phi\cos\psi$.

It is easy to verify that the expression on the left-hand side of (3) is none other than the mean value of the function $t_{00}^{nl}(\cos\gamma)$ on the $(n-2)$-dimensional sphere $S^{n-2}(P,\phi)$ with spherical radius $\phi$ and with center at a point $P$, the polar distance of which is equal to $\theta$ (i.e. the set of points $M$ of the sphere $S^{n-1}$, such that the arc $MP$ is equal to $\phi$).

Thus we have proved that the mean value of $t_{00}^{nl}(\cos\gamma)$ over the sphere $S^{n-2}(P,\phi)$ is equal to $t_{00}^{nl}(\cos\theta)\,t_{00}^{nl}(\cos\phi)$.

We now consider certain formulas which are obtained by transforming the multiplication formula (1). First, in formula (1) we make the substitution

$$\cos\theta\cos\varphi+\sin\theta\sin\varphi\cos\psi=\cos\gamma.$$

Since

$$\sin\psi = \frac{[\cos(\theta-\varphi)-\cos\gamma]^{\frac{1}{2}}[\cos\gamma-\cos(\theta+\varphi)]^{\frac{1}{2}}}{\sin\theta\sin\varphi}$$

and

$$d(\cos\psi) = \frac{d(\cos\gamma)}{\sin\theta\sin\varphi},$$

and $\gamma$ varies from $|\theta-\phi|$ to $\theta+\phi$ when $\psi$ varies from 0 to $\pi$, we get

$$\begin{aligned}\int_{|\theta-\varphi|}^{\theta+\varphi} & C_l^p(\cos\gamma)\, C_k^{p-\frac{1}{2}}\left(\frac{\cos\gamma-\cos\theta\cos\varphi}{\sin\theta\sin\varphi}\right)[\cos(\theta-\varphi)-\cos\gamma]^{p-1} \\ & \times[\cos\gamma-\cos(\theta+\varphi)]^{p-1}\sin\gamma\, d\gamma = \frac{2^{2k+2p-1}[\Gamma(p+k)]^2(l-k)!\,\Gamma(2p+k-1)}{k!\,\Gamma(2p-1)\,\Gamma(l+k+2p)} \\ & \times C_{l-k}^{p+k}(\cos\varphi)\sin^{2p+k-1}\varphi\, C_{l-k}^{p+k}(\cos\theta)\sin^{2p+k-1}\theta. \end{aligned} \tag{4}$$

In particular, for $k=0$, we have

$$\begin{aligned}\int_{|\theta-\varphi|}^{\theta+\varphi} & C_l^p(\cos\gamma)[\cos(\theta-\varphi)-\cos\gamma]^{p-1}[\cos\gamma-\cos(\theta+\varphi)]^{p-1}\sin\gamma\, d\gamma \\ & = \frac{2^{2p-1}l!\,[\Gamma(p)]^2}{\Gamma(2p+l)} C_l^p(\cos\varphi)\sin^{2p-1}\varphi\, C_l^p(\cos\theta)\sin^{2p-1}\theta. \end{aligned} \tag{5}$$

Next, applying the results of §3.4, we find that, for $|\theta-\phi|\le\gamma\le\theta+\phi$,

$$\begin{aligned}\sum_{l=0}^{\infty} & \frac{l!\,(l-k)!\,(l+p)}{\Gamma(2p+l)\,\Gamma(l+k+2p)} C_l^p(\cos\gamma)\, C_{l-k}^{p+k}(\cos\varphi)\, C_{l-k}^{p+k}(\cos\theta) \\ & = \frac{\pi k!\,\Gamma(2p-1)}{2^{2k+2p}[\Gamma(p)\,\Gamma(p+k)]^2\,\Gamma(2p+k-1)} \\ & \times\frac{\left[\sin\frac{\theta-\varphi+\gamma}{2}\sin\frac{\gamma-\theta+\varphi}{2}\sin\frac{\gamma+\theta+\varphi}{2}\sin\frac{\theta+\varphi-\gamma}{2}\right]^{p-1}}{\sin^{2p-1}\gamma\,\sin^{2p+k-1}\varphi\,\sin^{2p+k-1}\theta} \\ & \times C_k^{p-\frac{1}{2}}\left(\frac{\cos\gamma-\cos\theta\cos\varphi}{\sin\theta\sin\varphi}\right), \end{aligned} \tag{6}$$

and for the remaining values of $\gamma$ on the segment $[0, 2\pi]$ this sum is equal to zero. In particular, for $k=0$ and $|\theta-\phi|\le\gamma\le\theta+\phi$, we get

$$\begin{aligned}\sum_{l=0}^{\infty} & \frac{(l!)^2(l+p)}{[\Gamma(2p+l)]^2} C_l^p(\cos\gamma)\, C_l^p(\cos\varphi)\, C_l^p(\cos\theta) \\ & = \frac{\pi\Gamma(2p-1)\left[\sin\frac{\theta-\varphi+\gamma}{2}\sin\frac{\gamma-\theta+\varphi}{2}\sin\frac{\gamma+\theta+\varphi}{2}\sin\frac{\theta+\varphi-\gamma}{2}\right]^{p-1}}{2^{2p}[\Gamma(p)]^4\,\Gamma(2p-1)(\sin\gamma\sin\varphi\sin\theta)^{2p-1}}, \end{aligned} \tag{7}$$

and for the remaining values of $\gamma$ on the segment $[0, 2\pi]$ a sum of this form is equal to zero.

4. Realization of the representations $T^{nl}(g)$ in a space of functions of $n-1$ variables. In the following subsection we shall derive an integral representation for Gegenbauer polynomials. In order to obtain this representation in a sufficiently simple form we shall need another realization of the representation $T^{nl}(g)$.

In §2.2 we realized the representation $T^{nl}(g)$ of $SO(n)$ in the factor space $\mathfrak{R}^{nl}/r^2\mathfrak{R}^{n,l-2}$, where $\mathfrak{R}^{nl}$ is the space of homogeneous polynomials of degree $l$ in $n$ variables. We also showed that this factor space can be identified with the space $\mathfrak{H}^{nl}$ of homogeneous harmonic polynomials of degree $l$.

We now determine another realization of $T^{nl}(g)$. In every polynomial $f(\mathbf{x})$ of $\mathfrak{R}^{nl}$ we substitute $x_n = ir_{n-1}$, where $r_{n-1}^2 = x_1^2 + \cdots + x_{n-1}^2$. Obviously, this substitution reduces the polynomial $r^2 = x_1^2 + \cdots + x_{n-1}^2 + x_n^2$ to zero, and consequently the whole subspace $r^2\mathfrak{R}^{n,l-2}$ becomes zero as well. At the same time, none of those polynomials which do not belong to $r^2\mathfrak{R}^{n,l-2}$ vanishes under the transformation

$$Qf(x_1, \ldots, x_{n-1}, x_n) = f(x_1, \ldots, x_{n-1}, ir_{n-1}). \tag{1}$$

This can best be seen by writing $f(\mathbf{x})$ in the form

$$f(\mathbf{x}) = (x_1^2 + \ldots + x_n^2) f_1(\mathbf{x}) + x_n P_1(\mathbf{x}') + P_2(\mathbf{x}'), \tag{2}$$

where

$$f_1(\mathbf{x}) \in \mathfrak{R}^{n, l-2}, \quad P_1(\mathbf{x}') \in \mathfrak{R}^{n-1, l-1}, \quad P_2(\mathbf{x}') \in \mathfrak{R}^{n-1, l}.$$

Clearly, if at least one of the polynomials $P_1(\mathbf{x}')$ and $P_2(\mathbf{x}')$ is nonzero, then $Qf(\mathbf{x}) \neq 0$.

Thus the operator $Q$ establishes a one-to-one correspondence between the factor space $\mathfrak{R}^{nl}/r^2\mathfrak{R}^{n,l-2}$ and the space $\mathfrak{A}^{nl}$ of functions of the form $f(x_1, \cdots, x_{n-1}, ir_{n-1})$. Let us clarify the constituent functions of the space $\mathfrak{A}^{nl}$ For this we put $x_n = ir_{n-1}$ in (2). We obtain

$$Qf(\mathbf{x}) = ir_{n-1} P_1(\mathbf{x}') + P_2(\mathbf{x}'), \tag{3}$$

where $P_1(\mathbf{x}') \in \mathfrak{R}^{n-1,l-1}$, $P_2(\mathbf{x}') \in \mathfrak{R}^{n-1,l}$.

Conversely, any function of the form (3), where $P_1(\mathbf{x}') \in \mathfrak{R}^{n-1,l-1}$ and $P_2(\mathbf{x}') \in \mathfrak{R}^{n-1,l}$, belongs to the space $\mathfrak{A}^{nl}$; it is the transform of the polynomial

$x_n P_1(\mathbf{x}') + P_2(\mathbf{x}')$.

Hence the space $\mathfrak{A}^{nl}$ consists of functions of the form

$$ir_{n-1}P_1(\mathbf{x}') + P_2(\mathbf{x}'), \text{ where } P_1(\mathbf{x}') \in \mathfrak{R}^{n-1,\,l-1}, \quad P_2(\mathbf{x}') \in \mathfrak{R}^{n-1,\,l}.$$

Since $\mathfrak{A}^{nl}$ is in a natural way isomorphic to $\mathfrak{R}^{nl}/r^2\mathfrak{R}^{n,l-2}$, we can realize the representation $T^{nl}(g)$ in $\mathfrak{A}^{nl}$. Namely, we put

$$R^{nl}(g)F(\mathbf{x}') = QT^{nl}(g)Q^{-1}F(\mathbf{x}'), \tag{4}$$

where $F(\mathbf{x}') \in \mathfrak{A}^{nl}$, the operator $Q$ has the form indicated above, and $Q^{-1}$ is the operator transforming the function $F(\mathbf{x}')$ of the space $\mathfrak{A}^{nl}$ into the corresponding coset of $\mathfrak{R}^{nl}/r^2\mathfrak{R}^{n,l-2}$ (i.e. into the coset which contains, for example, the polynomial $x_n P_1(\mathbf{x}') + P_2(\mathbf{x}')$).

A simple calculation, the details of which we leave to the reader, shows that the operators of the represent ation $\mathfrak{R}^{nl}(g)$ have the following form: if $h \in SO(n-1)$, then

$$R^{nl}(h)F(\mathbf{x}') = F(h^{-1}\mathbf{x}'). \tag{5}$$

However, if $g = g_n(\phi)$ is a rotation by the angle $\phi$ in the $(x_n, x_{n-1})$-plane, then

$$\begin{aligned} R^{nl}[g_n(\varphi)]F(x_1, \ldots, x_{n-1}) &= F(x_1, \ldots, x_{n-2}, x_{n-1}\cos\varphi - ir_{n-1}\sin\varphi) \\ &\equiv (x_{n-1}\sin\varphi + ir_{n-1}\cos\varphi)P_1(x_1, \ldots, x_{n-2}, x_{n-1}\cos\varphi - ir_{n-1}\sin\varphi) \\ &\quad + P_2(x_1, \ldots, x_{n-2}, x_{n-1}\cos\varphi - ir_{n-1}\sin\varphi), \end{aligned} \tag{6}$$

where $F(\mathbf{x}') = ir_{n-1}P_1(\mathbf{x}') + P_2(\mathbf{x}')$.

5. Decomposition of $\mathfrak{A}^{nl}$. It was shown in § 2.8 that

$$\mathfrak{R}^{nl} = r^2\mathfrak{R}^{n,\,l-2} + \sum_{m=0}^{l} x_n^{l-m}\mathfrak{H}^{n-1,\,m}. \tag{1}$$

The mapping $Q$ carries the subspace $r^2\mathfrak{R}^{n,l-2}$ into zero, and the subspace $x_n^{l-m}\mathfrak{H}^{n-1,m}$ into the subspace $(ir_{n-1})^{l-m}\mathfrak{H}^{n-1,m}$. Hence $\mathfrak{A}^{nl}$ is a direct sum of the subspaces $(ir_{n-1})^{l-m}\mathfrak{H}^{n-1,m}$, consisting of functions of the form $(ir_{n-1})^{l-m}h_m(\mathbf{x})$, where $h_m(\mathbf{x}') \in \mathfrak{H}^{n-1,m}$:

$$\mathfrak{A}^{nl} = \sum_{m=0}^{l} (ir_{n-1})^{l-m}\mathfrak{H}^{n-1,\,m}. \tag{2}$$

As the subspaces $x_n^{l-m}\mathfrak{H}^{n-1,m}$ in $\mathfrak{R}^{nl}$ are invariant for the operators

$T^{nl}(h)$, $h \in SO(n-1)$ (see §2.8), the subspaces $(ir_{n-1})^{l-m}\mathfrak{H}^{n-1,m}$ in $\mathfrak{A}^{nl}$ are invariant for operators $R^{nl}(h)$, $h \in SO(n-1)$, as can also be shown directly by observing that

$$R^{nl}(h)[(ir_{n-1})^{l-m}h_m(\mathbf{x}')] = (ir_{n-1})^{l-m}h_m(h^{-1}\mathbf{x}').$$

Thus we have decomposed $\mathfrak{A}^{nl}$ as the direct sum of subspaces $(ir_{n-1})^{l-m}\mathfrak{H}^{n-1,m}$, invariant for the operators $R^{nl}(h)$, $h \in SO(n-1)$. It is easy to see that the restriction of the representation $R^{nl}(h)$ of the subgroup $SO(n-1)$ to the subspace $(ir_{n-1})^{l-m}\mathfrak{H}^{n-1,m}$ is isomorphic to the irreducible representation $T^{n-1,m}(h)$ of this subgroup.

6. An invariant scalar product in the space $\mathfrak{A}^{nl}$. Let us now construct a scalar porduct in $\mathfrak{A}^{nl}$ which will be invariant relative to the representation $R^{nl}(g)$ $g \in SO(n)$. $R^{nl}(g)$ is equivalent to the representation

$$H^{nl}(g)f(\mathbf{x}) = f(g^{-1}\mathbf{x}) \tag{1}$$

of $SO(n)$ in the space $\mathfrak{H}^{nl}$ of homogeneous harmonic polynomials. Let $F_1(\mathbf{x}')$ and $F_2(\mathbf{x}')$ be functions of $\mathfrak{A}^{nl}$. We take polynomials $f_1(\mathbf{x})$ and $f_2(\mathbf{x})$ in the space $\mathfrak{R}^{nl}$, such that $Qf_j(\mathbf{x}) = F_j(\mathbf{x}')$, $j = 1,2$. Then the invariant scalar product in $\mathfrak{A}^{nl}$ is defined by the formula

$$\langle F_1, F_2\rangle = \langle Qf_1, Qf_2\rangle = (Hf_1, Hf_2) = \int_{S^{n-1}} Hf_1(\xi)\overline{Hf_2(\xi)}\,d\xi \tag{2}$$

($Hf(\mathbf{x})$ is the harmonic projection of $f(\mathbf{x})$).

Our aim, however, is to obtain another expression for the scalar product $\langle F_1, F_2\rangle$. Namely, owing to equality (2) of subsection 5, any function $F(\mathbf{x}')$ on $\mathfrak{A}^{nl}$ can be presented in the form

$$F(\mathbf{x}') = \sum_{m=0}^{l} (ir_{n-1})^{l-m}h_m(\mathbf{x}'), \tag{3}$$

where $h_m(\mathbf{x}') \in \mathfrak{H}^{n-1,m}$. We want to express $\langle F_1, F_2\rangle$ by means of the functions $h_m^{(1)}(\mathbf{x}')$, and $h_m^{(2)}(\mathbf{x}')$, appearing in this decomposition.

First we consider the restriction of the representation $R^{nl}(g)$ to $SO(n-1)$. This restriction gives, in each of the subspaces $(ir_{n-1})^{l-m}\mathfrak{H}^{n-1,m}$, a representation equivalent to the irreducible representation $T^{n-1,m}(h)$ of $SO(n-1)$. But the invariant scalar product for $T^{n-1,m}(h)$ has the form

$$(\varphi_1, \varphi_2) = \int\limits_{S^{n-2}} \varphi_1(\xi') \overline{\varphi_2(\xi')} d\xi' \tag{4}$$

(see §2.8). Hence one easily finds that the required invariant scalar product in $\mathfrak{A}^{nl}$ has the form

$$(F_1, F_2) = \sum_{m=0}^{l} \alpha_m \int\limits_{S^{n-2}} h_m^{(1)}(\xi') \overline{h_m^{(2)}(\xi')} d\xi' \tag{5}$$

(see §3.3 of Chapter I).

It remains for us to determine the coefficients $\alpha_m$. We take some function $F(\mathbf{x}') = (ir_{n-1})^{l-m} h_m(\mathbf{x}')$, where $h_m(\mathbf{x}') \in \mathfrak{H}^{n-1,m}$. We have

$$\langle F, F \rangle = \alpha_m \int\limits_{S^{n-2}} |h_m(\xi')|^2 d\xi'. \tag{6}$$

But, by formula (1) of subsection 4, $F(\mathbf{x}') = Q[x_n^{l-m} h_m(\mathbf{x}')]$, and the harmonic projection of the polynomial $x_n^{l-m} h_m(\mathbf{x}')$ is calculated according to formula (9) of §3.5:

$$H[x_n^{l-m} h_m(\mathbf{x}')] = \frac{(l-m)!\,\Gamma\left(\frac{n-2}{2}+m\right)}{2^{l-m}\Gamma\left(\frac{n-2}{2}+l\right)} r^{l-m} C_{l-m}^{\frac{n-2}{2}+m}\left(\frac{x_n}{r}\right) h_m(\mathbf{x}'). \tag{7}$$

Utilizing formula (7), we get the equality

$$\alpha_m \int\limits_{S^{n-2}} |h_m(\xi')|^2 d\xi'$$
$$= \left[\frac{(l-m)!\,\Gamma\left(\frac{n-2}{2}+m\right)}{2^{l-m}\Gamma\left(\frac{n-2}{2}+l\right)}\right]^2 \int\limits_{S^{n-1}} |C_{l-m}^{\frac{n-2}{2}+m}(\xi_n) h_m(\mathbf{x}')|^2 d\xi, \tag{8}$$

where $\mathbf{x}' = (x_1, \cdots, x_{n-1})$, $\xi = (x_1, \cdots, x_{n-1}, \xi_n)$ and $\|\xi\| = 1$.

Normalizing $\mathbf{x}$, we obtain the vector

$$\xi' = \frac{\mathbf{x}'}{\sqrt{1-\xi_n^2}}.$$

By virtue of the homogeneity of $h_m(\mathbf{x}')$ we have

$$h_m(\mathbf{x}') = (1-\xi_n^2)^{\frac{m}{2}} h_m(\xi'). \tag{9}$$

Lastly, we take into consideration the fact that the normalized invariant measure $d\xi$ on $S^{n-1}$ is expressed in terms of the coordinate $\xi_n$ and the normalized invariant measure $d\xi'$ on $S^{n-2}$ by the formula

$$d\xi=\frac{\Gamma\left(\frac{n}{2}\right)}{\Gamma\left(\frac{n-1}{2}\right)\sqrt{\pi}}(1-\xi_n^2)^{\frac{n-3}{2}}\,d\xi_n\,d\xi'. \tag{10}$$

Hence the integral in the right hand side of (8) can be written in the form

$$\frac{\Gamma\left(\frac{n}{2}\right)}{\Gamma\left(\frac{n-1}{2}\right)\sqrt{\pi}}\int_{-1}^{1}\left[C_{l-m}^{\frac{n-2}{2}+m}(\xi_n)\right]^2(1-\xi_n^2)^{m+\frac{n-3}{2}}d\xi_n$$

$$\times\int_{S^{n-2}}|h_m(\xi')|^2\,d\xi'. \tag{11}$$

Because of the normalization relations for the Gegenbauer polynomials (formula (5), §3.4), formulas (8) and (11) give

$$\alpha_m=\frac{\Gamma\left(\frac{n}{2}\right)\sqrt{\pi}\,(l-m)!\,(n+l+m-3)!}{2^{2l+n-3}\,\Gamma\left(\frac{n-1}{2}\right)\Gamma\left(\frac{n}{2}+l\right)\Gamma\left(\frac{n-2}{2}+l\right)}. \tag{12}$$

We thus get the following expression for an invariant scalar product in $\mathfrak{A}^{nl}$:

$$\langle F_1,\,F_2\rangle=\frac{\Gamma\left(\frac{n}{2}\right)\Gamma\left(l+\frac{n-1}{2}\right)}{\Gamma\left(\frac{n-1}{2}\right)\Gamma\left(l+\frac{n}{2}\right)(2l+n-3)!}$$

$$\times\sum_{m=0}^{l}(l-m)!\,(n+l+m-3)!\int_{S^{n-2}}h_m^{(1)}(\xi')\overline{h_m^{(2)}(\xi')}\,d\xi', \tag{13}$$

where

$$F_j(\mathbf{x}')=\sum_{m=0}^{l}(ir_{n-1})^{l-m}h_m^{(j)}(\xi') \tag{14}$$

(use has been made here of the duplication formula for the $\Gamma$ function).

In particular, it follows from (13) that

$$\langle F,\,(ir_{n-1})^{l-m}\,h_m(\mathbf{x}')\rangle=\frac{\Gamma\left(\frac{n}{2}\right)\Gamma\left(l+\frac{n-1}{2}\right)i^{m-l}}{\Gamma\left(\frac{n-1}{2}\right)\Gamma\left(l+\frac{n}{2}\right)(2l+n-3)!}$$

$$\times(l-m)!\,(l+n+m-3)!\int_{S^{n-2}}F(\xi')\overline{h_m(\xi')}\,d\xi'. \tag{15}$$

This can be proved by taking into consideration formula (13) and the pairwise orthogonality of the subspaces $(ir_{n-1})^{l-m}\mathfrak{H}^{n-1,m}$.

Finally, in $\mathfrak{A}^{nl}$ we construct an orthonormal basis relative to the scalar product (13). It follows from equality (2) that for such a basis one can choose functions of the form

$$\hat{\Xi}_M^l(\mathbf{x}') = Q\Xi_M^l(\mathbf{x}), \tag{16}$$

where $\Xi_M^l(\mathbf{x})$ is the orthonormal basis, constructed in §3.6, of the space $\mathfrak{H}^{nl}$.

The functions $\hat{\Xi}_M^l(\mathbf{x}')$ can be easily written in an explicit form. For this let us recall that, by formula (9), §3.5, the functions

$$x_n^{l-m}h_m(\mathbf{x}') \text{ and } \frac{(l-m)!\,\Gamma\left(\frac{n-2}{2}+m\right)}{2^{l-m}\Gamma\left(\frac{n-2}{2}+l\right)} r^{l-m} C_{l-m}^{\frac{n-2}{2}+m}\left(\frac{x_n}{r}\right) h_m(\mathbf{x}')$$

lie in one and the same coset with respect to $r^2\mathfrak{R}^{n,l-2}$. Therefore the operator $Q$ transforms them into one and the same function. Consequently

$$Q\left[r^{l-m} C_{l-m}^{\frac{n-2}{2}+m}\left(\frac{x_n}{r}\right) h_m(\mathbf{x}')\right] = \frac{2^{l-m}\Gamma\left(\frac{n-2}{2}+l\right)}{(l-m)!\,\Gamma\left(\frac{n-2}{2}+m\right)} Q(x_n^{l-m}h_m(\mathbf{x}'))$$

$$= \frac{2^{l-m}\Gamma\left(\frac{n-2}{2}+l\right)}{(l-m)!\,\Gamma\left(\frac{n-2}{2}+m\right)} (ir_{n-1})^{l-m}h_m(\mathbf{x}'). \tag{17}$$

Applying (17) to formula (16), we find that, for $M=(m, m_2, \cdots, \pm m_{n-2})$ (see formula (2) of §3.6), we have

$$\hat{\Xi}_M^l(\mathbf{x}') = \frac{2^{l-m}\Gamma\left(\frac{n-2}{2}+l\right)}{(l-m)!\,\Gamma\left(\frac{n-2}{2}+m\right)} A_M^l (ir_{n-1})^{l-m}$$

$$\times \prod_{j=1}^{n-3} r_{n-j}^{m_j-m_{j+1}} C_{m_j-m_{j+1}}^{\frac{n-j-2}{2}+m_{j+1}}\left(\frac{x_{n-j}}{r_{n-j}}\right)(x_2 \pm ix_1)^{m_{n-2}}$$

$$= \frac{2^{l-1}\Gamma\left(\frac{n-2}{2}+l\right)}{\Gamma\left(\frac{n}{2}\right)}\left[\frac{(2l+n-2)\,\Gamma(n-1)}{\Gamma(l+n+m-2)\,(l-m)!}\right]^{\frac{1}{2}} (ir_{n-1})^{l-m}\,\Xi_{M_1}^m(\mathbf{x}'), \tag{18}$$

where $\mathbf{x}'=(x_1, \cdots, x_{n-1})$ and $M_1=(m_2, \cdots, \pm m_{n-2})$.

Evidently, $\hat{\Xi}_M^l(\mathbf{x}') \in (ir_{n-1})^{l-m}\mathfrak{H}^{n-1,m}$, and therefore

$$\langle F, \hat{\Xi}^l_M\rangle = \frac{[\pi\Gamma(n-1)\Gamma(l+n+m-2)(l-m)!(2l+n-2)]^{\frac{1}{2}}}{2^{l+n-2}\Gamma\left(\frac{n-1}{2}\right)\Gamma\left(l+\frac{n}{2}\right)} i^{m-l} \times \int_{S^{n-2}} F(\xi')\overline{\Xi^m_{M_1}(\xi')}\,d\xi'. \qquad (19)$$

7. **An integral representation for Gegenbauer polynomials.** We can now derive an integral representation for Gegenbauer polynomials. For this we use equality (6) of § 3.1 in the form

$$C^p_l(\cos\varphi) = \frac{\Gamma(2p+l)}{\Gamma(2p)\,l!} t^{nl}_{00}[g_n(\varphi)], \qquad (1)$$

where $n = 2p + 2$ and $g = g_n(\phi)$ is a rotation by the angle $\phi$ in the $(x_n, x_{n-1})$-plane. But, because the representation $T^{nl}(g)$ is unitary, we have

$$t^{nl}_{00}[g_n(\varphi)] = (T^{nl}[g_n(\varphi)]\,\Xi^l_0, \Xi^l_0). \qquad (2)$$

Further, it follows from equality (2) of subsection 6 that formula (2) can be rewritten as

$$t^{nl}_{00}[g_n(\varphi)] = \langle R^{nl}[g_n(\varphi)]\,\hat{\Xi}^l_0, \hat{\Xi}^l_0\rangle. \qquad (3)$$

Since $O = (0, \cdots, 0)$, we have $\Xi^0_{O_1}(\xi') = 1$.

Employing formula (19) above we get

$$t^{nl}_{00}[g_n(\varphi)] = \frac{[\pi\Gamma(2p+1)\Gamma(l+2p)\,l!\,(2l+2p)]^{\frac{1}{2}} i^{-l}}{2^{l+2p}\Gamma\left(p+\frac{1}{2}\right)\Gamma(l+p+1)} \times \int_{S^{n-2}} R^{nl}[g_n(\varphi)]\,\hat{\Xi}^l_0(\xi')\,d\xi'. \qquad (4)$$

Formula (18) yields

$$\hat{\Xi}^l_0(\mathbf{x}') = \frac{2^{l-1}\Gamma(p+l)}{\Gamma(p+1)}\left[\frac{(2l+2p)\Gamma(2p+1)}{\Gamma(l+2p)\,l!}\right]^{\frac{1}{2}}(ir_{n-1})^l,$$

and hence

$$R^{nl}[g_n(\varphi)]\,\hat{\Xi}^l_0(\mathbf{x}') = \frac{2^{l-1}\Gamma(p+l)}{\Gamma(p+1)}\left[\frac{(2l+2p)\Gamma(2p+1)}{\Gamma(l+2p)\,l!}\right]^{\frac{1}{2}}(x_{n-1}\sin\varphi + ir_{n-1}\cos\varphi)^l. \qquad (5)$$

We substitute this expression in (4) and go over to spherical coordinates. On the sphere $S^{n-2}$ we have $r_{n-1} = 1$ and $x_{n-1} = \cos\theta_{n-2}$, while

$$d\xi' = \frac{\Gamma\left(\frac{n-1}{2}\right)}{2\pi^{\frac{n-1}{2}}} \sin^{n-3}\theta_{n-2} \ldots \sin\theta_2\, d\theta_1 \ldots d\theta_{n-2}.$$

Therefore, taking (1) and (4) into account, we get

$$C_l^p(\cos\varphi) = \frac{\Gamma(2p+l)}{2^{2p-1}\Gamma^2(p)\, l!} \int_0^\pi (\cos\varphi - i\sin\varphi\cos\theta)^l \sin^{2p-1}\theta\, d\theta. \quad (6)$$

This integral representation of Gegenbauer polynomials is a special case of a more general integral formula. In order to derive the latter, we use the equality

$$t_{MO}^{nl}[g_n(\varphi)] = \langle \hat{T}^{nl}[g_n(\varphi)]\, \hat{\Xi}_O^l,\ \hat{\Xi}_M^l\rangle. \quad (7)$$

As shown in subsection 1, if $M = (m_1, \cdots, \pm m_{n-2})$ and $m_2 \neq 0$, then $t_{MO}^{nl}[g_n(\phi)]$ 0, Therefore we may assume that $M = (m, 0, \cdots, 0)$. Then, by formula (9) of subsection 1, we have

$$\begin{aligned} t_{MO}^{nl}&[g_n(\varphi)] \\ &= \frac{2^m\Gamma(p+m)}{\Gamma(p)} \sqrt{\frac{l!\,(l-m)!\,\Gamma(2p)\,\Gamma(2p+m-1)\,(2p+2m-1)}{m!\,\Gamma(2p+m+l)\,\Gamma(2p+l)}} \\ &\qquad \times \sin^m\varphi\, C_{l-m}^{p+m}(\cos\varphi). \end{aligned} \quad (8)$$

Comparing equalities (7) and (8), and utilizing equality (15) of subsection 6 for calculating the scalar product in (7), we get

$$\begin{aligned} \sin^m\varphi\, C_{l-m}^{p+m}(\cos\varphi) &= \frac{m!\,\Gamma\left(p-\frac{1}{2}\right)\Gamma(2p+m+l)\, i^m}{2^{m+1}\sqrt{\pi}\,\Gamma(p+m)\, l!\,\Gamma(2p+m-1)} \\ &\times \int_0^\pi (\cos\varphi - i\sin\varphi\cos\theta)^l\, C_m^{p-\frac{1}{2}}(\cos\theta) \sin^{2p-1}\theta\, d\theta. \end{aligned} \quad (9)$$

Similarly, the relation

$$t_{MO}^{nl}[g_n(\varphi)] = t_{OM}^{nl}[g_n(-\varphi)] = \langle \hat{T}^{nl}[g_n(-\varphi)]\, \hat{\Xi}_M^l,\ \hat{\Xi}_O^l\rangle$$

leads to

$$\begin{aligned} \sin^m\varphi\, C_{l-m}^{p+m}(\cos\varphi) &= \frac{m!\,\Gamma\left(p-\frac{1}{2}\right)\Gamma(2p+l)\, i^{-m}}{2^{m+1}\sqrt{\pi}\,\Gamma(p+m)\,(l-m)!\,\Gamma(2p+m-1)} \\ \times \int_0^\pi (\cos\varphi + i\cos\theta\sin\varphi)^l\, & C_m^{p-\frac{1}{2}}\left(\frac{\cos\theta\cos\varphi + i\sin\varphi}{\cos\varphi + i\cos\theta\sin\varphi}\right) \sin^{2p-1}\theta\, d\theta. \end{aligned} \quad (10)$$

In (10) we make the substitution

$$\frac{\cos\theta\cos\varphi + i\sin\varphi}{\cos\varphi + i\cos\theta\sin\varphi} = \cos\gamma.$$

We obtain

$$\sin^m\varphi\, C_{l-m}^{p+m}(\cos\varphi) = \frac{m!\Gamma\left(p-\frac{1}{2}\right)\Gamma(2p+l)\, i^{-m}}{2^{m+1}\sqrt{\pi}\,\Gamma(p+m)(l-m)!\Gamma(2p+m-1)}$$
$$\times\int_0^\pi (\cos\varphi - i\cos\gamma\sin\varphi)^{-l-2p} C_m^{p-\frac{1}{2}}(\cos\gamma)\sin^{2p-1}\gamma\, d\gamma. \qquad (11)$$

Hence for $m = 0$ we get

$$C_l^p(\cos\varphi) = \frac{\Gamma(2p+l)}{2^{2p-1}\Gamma^2(p)\, l!}\int_0^\pi (\cos\varphi - i\cos\varphi\sin\varphi)^{-l-2p}\sin^{2p-1}\gamma\, d\gamma. \qquad (12)$$

We may mention another special case of formula (9). For $\phi = \pi/2$, taking into consideration formula (7) of §3.3, we get

$$\int_0^\pi C_m^{p-\frac{1}{2}}(\cos\theta)\cos^{m+2k}\theta\sin^{2p-1}\theta\, d\theta$$
$$= \frac{\pi\Gamma(2k+m+1)\Gamma(2p+m-1)(2p-1)}{2^{2p+2k+m-1}m!k!\Gamma\left(p+\frac{1}{2}\right)\Gamma\left(p+m+k+\frac{1}{2}\right)}.$$

8. The connection between Gegenbauer polynomials and associated Legendre functions. The integral representation of Gegenbauer polynomials allows us to establish a connection between these polynomials and associated Legendre functions. On putting $p = ½$ in formula (6) of subsection 7, we get

$$C_l^{\frac{1}{2}}(\cos\varphi) = \frac{1}{\pi}\int_0^\pi (\cos\varphi - i\sin\varphi\cos\theta)^l\, d\theta. \qquad (1)$$

Comparing this with formula (16) of §3.9, Chapter III, we see that

$$C_l^{\frac{1}{2}}(\cos\varphi) = P_l(\cos\varphi). \qquad (2)$$

Thus the Legendre polynomials are a special case of Gegenbauer polynomials, corresponding to the value $p = ½$.

We now make use of formula (6) of §3.2. For positive integral $p$ this gives

$$C_{l-p}^{p+1/2}(t) = \frac{\sqrt{\pi}}{2^p\Gamma\left(p+\frac{1}{2}\right)}\frac{d^pC_l^{\frac{1}{2}}(t)}{dt^p} = \frac{\sqrt{\pi}}{2^p\Gamma\left(p+\frac{1}{2}\right)}\frac{d^pP_l(t)}{dt^p}. \qquad (3)$$

Since by formula (4) of §3.9, Chapter III we have

$$P_l(t) = \frac{1}{2^l l!} \frac{d^l (t^2-1)^l}{dt^l},$$

it follows that

$$C_{l-p}^{p+1/2}(t) = \frac{\sqrt{\pi}}{2^{p+l}\Gamma\left(p+\frac{1}{2}\right) l!} \frac{d^{p+l}(t^2-1)^l}{dt^{p+l}},$$

i.e.

$$C_l^q(t) = \frac{\sqrt{\pi}}{2^{2q+l}\Gamma(q)\Gamma\left(q+l+\frac{3}{2}\right)} \frac{d^{2q+l}(t^2-1)^{q+l+1/2}}{dt^{2q+l}}, \tag{4}$$

where $g$ is semi-integral, and $l$ integral, $g \geq ½$, $l \geq 0$.

Further, by formula (14) of §3.9, Chapter III, we have

$$P_l^p(t) = \frac{(-1)^p(1-t^2)^{p/2}}{2^l l!} \frac{d^{p+l}(t^2-1)^l}{dt^{p+l}}. \tag{5}$$

Comparing this with (4), we get

$$C_{l-p}^{p+1/2}(t) = \frac{(-1)^p\sqrt{\pi}(1-t^2)^{-p/2}}{2^p\Gamma\left(p+\frac{1}{2}\right)} P_l^p(t), \tag{6}$$

i e.

$$C_l^q(t) = \frac{(-1)^{q-1/2}\sqrt{\pi}(1-t^2)^{1/4-q/2}}{2^{q-1/2}\Gamma(q)} P_{l+q-1/2}^{q-1/2}(t)$$

$$= \frac{\sqrt{\pi}\,\Gamma(l+2q)(1-t^2)^{1/4-q/2}}{2^{q-1/2}\Gamma(q)\, l!} P_{l+q-1/2}^{1/2-q}(t). \tag{6'}$$

By (5) this equality can be rewritten as follows:

$$C_l^q(t) = \frac{(-1)^l\Gamma\left(q+\frac{1}{2}\right)\Gamma(l+2q)(1-t^2)^{1/2-q}}{2^l\Gamma(2q)\Gamma\left(l+q+\frac{1}{2}\right)l!} \frac{d^l(1-t^2)^{l+q-1/2}}{dt^l}. \tag{7}$$

This expression for $C_l^q(t)$ immediately leads to

$$\int_{-1}^{1}(1-t^2)^{\frac{p-1}{2}} F(t)\, C_l^{\frac{p}{2}}(t)\,dt$$

$$= \frac{1}{l!}\frac{p(p+1)\dots(p+l-1)}{(p+1)(p+3)\dots(p+2l-1)} \int_{-1}^{1}(1-t^2)^{l+\frac{p-1}{2}} \frac{d^l F}{dt^l}\,dt =$$

$$=\frac{\Gamma(p+l)\Gamma\left(\frac{p+1}{2}\right)}{2^l l!\,\Gamma(p)\Gamma\left(l+\frac{p+1}{2}\right)}\int_{-1}^{1}(1-t^2)^{l+\frac{p-1}{2}}\frac{d^lF}{dt^l}\,dt. \tag{8}$$

For a proof it is sufficient to replace $C_l^{p/2}(t)$ by its expression according to (7), and integrate by parts $l$ times. The integrated terms vanish on substituting the limits of integration.

Let us now consider the case when $p$ is an integer. We put $p = 1$ in equality (9) of subsection 7 and make the substitution $\cos\phi - i\sin\phi\cos\theta = z$. We get

$$C_l^1(\cos\varphi)=\frac{l+1}{2i\sin\varphi}\int_{e^{-i\varphi}}^{e^{i\varphi}} z^l\,dz=\frac{\sin(l+1)\varphi}{\sin\varphi}. \tag{9}$$

Employing the recurrence formula (6) of §3.2, we find

$$\begin{aligned}C_{l-p}^{p+1}(\cos\varphi)&=\frac{1}{2^p\Gamma(p+1)}\left(\frac{d}{d\cos\varphi}\right)^p\left[\frac{\sin(l+1)\varphi}{\sin\varphi}\right]\\&=\frac{1}{2^p\Gamma(p+1)(l+1)}\left(\frac{d}{d\cos\varphi}\right)^{p+1}(\cos(l+1)\varphi).\end{aligned} \tag{10}$$

Equality (10) can be rewritten as follows:

$$C_l^p(\cos\varphi)=\frac{1}{2^{p-1}\Gamma(p)(p+l)}\left(\frac{d}{d\cos\varphi}\right)^p[\cos(p+l)\varphi]. \tag{10'}$$

Here $p$ and $l$ are nonnegative integers.

For $p = 0$, Gegenbauer polynomials become zero. We now calculate $\lim_{p\to 0}\Gamma(p)C_l^p(\cos\phi)$. To do this we multiply by $\Gamma(p)$ both sides of the recurrence formula

$$(2p+l)C_l^p(\cos\varphi)=2p\,[\cos\varphi\, C_{l-1}^{p+1}(\cos\varphi)-C_{l-2}^{p+1}(\cos\varphi)]$$

(see §3.2) and take the limit for $p\to 0$. Using (9), we get the equality

$$\lim_{p\to 0}\Gamma(p)C_l^p(\cos\varphi)=\frac{2\cos l\varphi}{l}, \tag{11}$$

which can be rewritten as

$$\lim_{p\to 0}\frac{C_l^p(\cos\varphi)}{p}=\frac{2\cos l\varphi}{l}. \tag{11'}$$

9. **Some expansions in Gegenbauer polynomials.** Formula (9) of subsection 7 may be interpreted as the expression for Fourier coefficients in an expansion of the function $(\cos\phi - i\sin\phi\cos\theta)^l$ in polynomials $C_m^{p-1/2}(\cos\theta)$. Taking into

consideration formulas (11) and (12) of §3.4, we get

$$(\cos\varphi - i\sin\varphi\cos\theta)^l$$

$$= \frac{\Gamma(2p-1)\,l!}{\Gamma(p)}\sum_{m=0}^{l}\frac{(-2i)^m(2m+2p-1)\,\Gamma(p+m)}{\Gamma(2p+l+m)}\sin^m\varphi\, C_{l-m}^{p+m}(\cos\varphi)$$

$$\times C_m^{p-1/2}(\cos\theta). \tag{1}$$

We note some special cases of this formula. Putting $\phi = \pi/2$ in it, we get

$$\cos^l\theta = \frac{\sqrt{\pi}\,\Gamma(2p-1)\,l!}{2^{2p+l-1}\,\Gamma(p)}\sum_{k=0}^{[l/2]}\frac{(2p+2l-4k-1)}{k!\,\Gamma\left(p+l-k+\frac{1}{2}\right)}C_{l-2k}^{p-1/2}(\cos\theta). \tag{2}$$

Further, putting $\theta = 0$ or $\pi$, we find that

$$e^{\pm il\varphi} = \frac{l!}{\Gamma(p)}\sum_{m=0}^{l}\frac{(\pm 2i)^m(2m+2p-1)\,\Gamma(p+m)}{\Gamma(2p+l+m)\,m!}$$

$$\times\Gamma(2p+m-1)\sin^m\varphi\, C_{l-m}^{p+m}(\cos\varphi). \tag{3}$$

Finally, we multiply side by side the expansions (1) for $l = l_1$ and $l = l_2$, and apply the same expansion for $l = l_1 + l_2$ to the left hand side of the resulting equality. On multiplying both sides of the equality thus obtained by $\sin^{2p}\theta$ and integrating with respect to $\theta$ from $0$ to $\pi$, we find that, owing to the orthogonality relations for Gegenbauer polynomials, we have

$$C_{l_1+l_2}^{p}(\cos\varphi) = \frac{l_1!\; l_2!\;\Gamma(2p+l_1+l_2)}{(l_1+l_2)!\;\Gamma^2(p)}$$

$$\times\sum_{k=0}^{\min(l_1,\,l_2)}\frac{(-4)^k\,\Gamma^2(p+k)\,\Gamma(k+2p-1)}{k!\,\Gamma(2p+l_1+k)}\,\frac{(2k+2p-1)}{\Gamma(2p+l_2+k)}\sin^{2k}\varphi$$

$$\times C_{l_1-k}^{p+k}(\cos\varphi)\, C_{l_2-k}^{p+k}(\cos\varphi). \tag{4}$$

10. Other integral representations for Gegenbauer polynomials. For deriving the integral representation of Gegenbauer polynomials we have utilized the realization of the representations $T^{nl}(g)$ in the space $\mathfrak{A}^{nl}$. Other integral representations arise if these representations are realized in the space $\mathfrak{H}^{nl}$. Under such a realization we have

$$t_{00}^{nl}[g_n(\varphi)] = (T^{nl}[g_n(\varphi)]\,\Xi_0^l,\,\Xi_0^l) = \int_{S^{n-1}} T^{nl}[g_n(\varphi)]\,\Xi_0^l(\xi)\,\overline{\Xi_0^l(\xi)}\,d\xi. \tag{1}$$

But, by formula (7) of §3.6,

$$\Xi_0^l(\xi) = \sqrt{\frac{l!\Gamma(n-2)(2l+n-2)}{\Gamma(n+l-2)(n-2)}} C_l^{\frac{n-2}{2}}(\cos\theta_{n-1}),$$

and the operator $T^{nl}[g_n(\phi)]$ transforms the function $C_l^{(n-2)/2}(\cos\theta_{n-1})$ into

$$T^{nl}[g_n(\varphi)] C_l^{\frac{n-2}{2}}(\cos\theta_{n-1})$$

$$= C_l^{\frac{n-2}{2}}(\cos\theta_{n-1}\cos\varphi + \sin\theta_{n-1}\sin\varphi\cos\theta_{n-2}).$$

Therefore

$$t_{00}^{nl}[g_n(\varphi)] = \frac{l!\Gamma(n-2)(2l+n-2)}{\Gamma(n+l-2)(n-2)}$$

$$\times \int_{S^{n-1}} C_l^{\frac{n-2}{2}}(\cos\theta_{n-1}\cos\varphi + \sin\theta_{n-1}\sin\varphi\cos\theta_{n-2}) C_l^{\frac{n-2}{2}}(\cos\theta_{n-1})\, d\xi. \quad (2)$$

We go over to spherical coordinates and substitute the expression (1) of subsection 7 for $t_{00}^{nl}[g_n(\phi)]$. Replacing $n-2$ by $2p$, $\theta_{n-1}$ by $\theta$ and $\theta_{n-2}$ by $\psi$, we get

$$C_l^p(\cos\varphi) = \frac{l+p}{\pi} \int_0^\pi \int_0^\pi C_l^p(\cos\theta\cos\varphi + \sin\theta\sin\varphi\cos\psi)$$

$$\times C_l^p(\cos\theta)\sin^{2p}\theta\sin^{2p-1}\psi\, d\theta\, d\psi. \quad (3)$$

Formula (3) can also be seen to be valid on expressing $C_l^p(\cos\theta\cos\phi + \sin\theta\sin\phi\cos\psi)$ by the composition theorem and then integrating the resulting expression term by term.

Proceeding as in the case of (3) above, one derives the formula

$$\sin^m\varphi\, C_{l-m}^{p+m}(\cos\varphi)$$

$$= \frac{\Gamma(2p-1)(l+p)}{\pi\Gamma(2p+m-1)} \int_0^\pi \int_0^\pi C_l^p(\cos\theta\cos\varphi + \sin\theta\sin\varphi\cos\psi)$$

$$\times C_{l-m}^{p+m}(\cos\theta)\sin^{2p+m}\theta\, C_m^{p-\frac{1}{2}}(\cos\psi)\sin^{2p-1}\psi\, d\theta\, d\psi. \quad (4)$$

We leave it to the reader to obtain this from the equality

$$t_{MO}^{nl}[g_n(\varphi)] = (T^{nl}[g_n(\varphi)]\,\Xi_O^l,\ \Xi_M^l).$$

11. Certain integrals involving Gegenbauer polynomials. Some interesting integrals involving Gegenbauer polynomials are obtained if one calculates the zonal spherical function $t_{OO}^{nl}(g)$ not in spherical, but in bispherical coordinates. The latter are defined by the equalities

$$\left.\begin{aligned} x_1 &= \sin\alpha \sin\theta_{m-1} \ldots \sin\theta_1, \\ x_2 &= \sin\alpha \sin\theta_{m-1} \ldots \cos\theta_1, \\ &\cdots\cdots\cdots\cdots \\ x_m &= \sin\alpha\cos\theta_{m-1}, \\ x_{m+1} &= \cos\alpha \sin\psi_{n-m-1} \ \ldots\ \sin\psi_1, \\ &\cdots\cdots\cdots\cdots \\ x_n &= \cos\alpha\cos\psi_{n-m-1}, \end{aligned}\right\} \tag{1}$$

where

$$\left.\begin{aligned} &0 \leqslant \alpha \leqslant \frac{\pi}{2}, \\ &0 \leqslant \psi_k \leqslant \pi, \quad k \neq 1, \quad 0 \leqslant \psi_1 \leqslant 2\pi, \\ &0 \leqslant \theta_k \leqslant \pi, \quad k \neq 1, \quad 0 \leqslant \theta_1 \leqslant 2\pi. \end{aligned}\right\} \tag{2}$$

One can easily verify that, in bispherical coordinates, the normalized invariant measure on $S^{n-1}$ is expressed as follows:

$$d\xi = \frac{\Gamma\left(\frac{n}{2}\right)}{2\pi^{n/2}} \sin^{m-1}\alpha \cos^{n-m-1}\alpha \sin^{m-2}\theta_{m-1} \ldots \sin\theta_2 \\ \times \sin^{n-m-2}\psi_{n-m-1} \ldots \sin\psi_2 \, d\alpha\, d\theta_1 \ldots d\theta_{m-1}\, d\psi_1 \ldots d\psi_{n-m-1}. \tag{3}$$

We calculate integral (4) of subsection 7 in bispherical coordinates. Since on the sphere $S^{n-2}$ we have $r_{n-1} = 1$ and $x_{n-1} = \cos\alpha\cos\psi_{n-m-2}$, therefore, taking into account formulas (1), (4) and (5) of subsection 7 and equality (3), we get $(p = (n-2)/2)$

$$C_l^p(\cos\varphi) = \frac{\Gamma(2p+l)}{2^{2p-2}\Gamma(p)\Gamma\left(\frac{m}{2}\right)\Gamma\left(p-\frac{m}{2}\right)} \\ \times \int_0^{\frac{\pi}{2}}\int_0^{\pi} (\cos\varphi - i\sin\varphi\cos\alpha\cos\psi)^l \sin^{m-1}\alpha\cos^{2p-m}\alpha\sin^{2p-m-1}\psi\, d\psi\, d\alpha \tag{4}$$

(we have integrated here with respect to the coordinates $\theta_1, \cdots, \theta_{m-1}$ and $\psi_1, \cdots, \psi_{n-m-3}$).

We now integrate with respect to $\psi$. For this we multiply and divide the integrand by $(\cos^2\phi + \sin^2\phi\cos^2\alpha)^{l/2}$ and put

$$m = 2p - 2q, \quad \frac{\cos\varphi}{\sqrt{\cos^2\varphi + \sin^2\varphi\cos^2\alpha}} = \cos\beta,$$

$$\frac{\sin\varphi\cos\alpha}{\sqrt{\cos^2\varphi + \sin^2\varphi\cos^2\alpha}} = \sin\beta.$$

Employing formula (6) of subsection 7, we get

$$C_l^p(\cos\varphi) = \frac{2^{2q-2p+1} l!\, \Gamma(q)\Gamma(2p+l)}{\Gamma(p)\Gamma(p-q)\Gamma(2q+l)} \times \int_0^{\frac{\pi}{2}} C_l^q\left(\frac{\cos\varphi}{\sqrt{\cos^2\varphi + \sin^2\varphi\cos^2\alpha}}\right)(\cos^2\varphi + \sin^2\varphi\cos^2\alpha)^{\frac{l}{2}} \times (\sin\alpha)^{2p-2q-1}\cos^{2q}\alpha\, d\alpha. \tag{5}$$

We may mention a special case of formula (5). Since $\lim_{q\to 0}\Gamma(q)C_l^q(\cos\phi) = (2\cos l\phi)/l$, for $q = 0$ formula (5) takes the form

$$C_l^p(\cos\varphi) = \frac{2^{2-2p}\Gamma(2p+l)}{\Gamma^2(p)\Gamma(l+1)} \int_0^{\frac{\pi}{2}} T_l\left(\frac{\cos\varphi}{\cos^2\varphi + \sin^2\varphi\cos^2\alpha}\right) \times (\cos^2\varphi + \sin^2\varphi\cos^2\alpha)^{\frac{l}{2}}\cos^{2p-1}\alpha\, d\alpha. \tag{5'}$$

Here $T_l(x)$ is a Čebyšev polynomial:

$$T_l(x) = \cos(l \arccos x).$$

We note the following integral, which generalizes integral (5) of §8.5, Chapter III. If $l + m + n = 2g$, where $g$ is an integer, and if there exists an integral triangle with sides $l$, $m$, $n$, then

$$D(l, m, n, p) \equiv \int_{-1}^{1} C_l^p(x)\, C_m^p(x)\, C_n^p(x)(1 - x^2)^{p-\frac{1}{2}} dx$$

$$= \frac{2^{1-2p}\pi\Gamma(g+2p)\Gamma(g-l+p)\Gamma(g-m+p)\Gamma(g-n+p)}{[\Gamma(p)]^4\Gamma(g-l+1)\Gamma(g-m+1)\Gamma(g-n+1)\Gamma(g+p+1)}. \tag{6}$$

Otherwise integral (6) is equal to zero.

The vanishing of integral (6) in the case when $l + m + n$ is not an even number follows from the fact that the function $C_n^p(x)$ has the same parity as $n$, and $\int_{-1}^{1}\phi(x)$ is equal to zero if function $\phi(x)$ is odd. If a triangle with sides $l$, $m$, $n$ does not exist, then integral (6) is equal to zero because of the orthogonality relations for Gegenbauer polynomials.

We now consider the case when $l + m + n = 2g$, where $g$ is an integer and a triangle with sides $l$, $m$, $n$ exists. In this case we shall prove formula (6) with the help of induction on $l$. For $l = 0$ we have $m = n = g$, and so formula (6) becomes

$$D(0, n, n, p) = \int_{-1}^{1} [C_n^p(x)]^2 (1-x)^{p-\frac{1}{2}} dx = \frac{2^{1-2p}\pi\Gamma(n+2p)}{(n+p)[\Gamma(p)]^2\Gamma(n+1)}. \quad (6')$$

The validity of formula (6) in this case follows from equality (5) of §3.4.

Let us now assume that (6) is valid for all positive integers less than $l$. In (6) we make use of the recurrence formula (10) of §3.2:

$$C_l^p(x) = \frac{2(p+l-1)}{l} xC_{l-1}^p(x) - \frac{2p+l-2}{l} C_{l-2}^p(x)$$

first for replacing $C_l^p(x)$, and then for replacing $xC_n^p(x)$. We thus obtain

$$\begin{aligned} D(l, m, n, p) = {} & \frac{(p+l-1)(n+1)}{l(p+n)} D(l-1, m, n+1, p) \\ & + \frac{(p+l-1)(2p+n-1)}{l(p+n)} D(l-1, m, n-1, p) \\ & - \frac{2p+l-2}{l} D(l-2, m, n, p). \end{aligned}$$

In the right hand side here we now replace $D(l-1, m, n+1, p)$ etc. by their expressions according to (6) (which can be done by the induction hypothesis). After simple manipulations we obtain the expression (6) for $D(l, m, n, p)$ too.

Undoubtedly, formula (6) is connected with the Kronecker product of representations of the group $SO(n)$, and gives the expression for one of the Clebsch-Gordan coefficients for this product. It would be greatly interesting to construct a general theory of Kronecker products for representations of $SO(n)$.

Owing to the orthogonality relations for Gegenbauer functions, we get the following expansion from formula (6):

$$\begin{aligned} C_l^p(x) C_m^p(x) = & \sum_{n=|l-m|}^{l+m} \frac{(n+p)\Gamma(n+1)\Gamma(g+2p)}{[\Gamma(p)]^2 (g+p)!\Gamma(n+2p)} \\ & \times \frac{\Gamma(g-l+p)\Gamma(g-m+p)\Gamma(g-n+p)}{\Gamma(g-l+1)\Gamma(g-m+1)\Gamma(g-n+1)} C_n^p(x), \end{aligned} \quad (7)$$

where $l + m + n = 2g$, and summation extends over values of $n$ having the same parity as $l + m$.

12. The generating function for Gegenbauer polynomials. We have proved in subsection 8 that

$$C_{n-p}^{p+1/2}(t) = \frac{\sqrt{\pi}}{2^p\Gamma\left(p+\frac{1}{2}\right)} \frac{d^pP_n(t)}{dt^p}. \quad (1)$$

Utilizing this equality, one can easily find the generating function for Gegenbauer polynomials. Namely, in §5.3 of Chapter III it was shown that

$$(1-2th+h^2)^{-1/2}=\sum_{n=0}^{\infty} P_n(t)\,h^n.$$

We differentiate this equality $p$ times with respect to $t$, and then apply formula (1). We obtain

$$1\cdot 3\ldots(2p-1)(1-2th+h^2)^{-p-1/2}$$
$$=\sum_{n=p}^{\infty}\frac{d^pP_n(t)}{dt^p}\,h^{n-p}=\frac{2^p\,\Gamma\left(p+\frac{1}{2}\right)}{\sqrt{\pi}}\sum_{n=0}^{\infty} C_n^{p+1/2}(t)\,h^n.$$

But, because of the duplication formula for $\Gamma$ functions,

$$1\cdot 3\ldots(2p-1)=\frac{\Gamma(2p)}{2^{p-1}\Gamma(p)}=\frac{2^p\,\Gamma\left(p+\frac{1}{2}\right)}{\sqrt{\pi}},$$

and so, for positive integral $p$, we have

$$(1-2th+h^2)^{-p-1/2}=\sum_{n=0}^{\infty} C_n^{p+1/2}(t)\,h^{l}. \tag{2}$$

Equality (2) can serve for defining Gegenbauer polynomials with arbitrary index $\lambda$. Namely, by way of definition we put

$$(1-2th+h^2)^{\lambda}=\sum_{n=0}^{\infty} C_n^{\lambda}(t)\,h^n. \tag{3}$$

It can be shown that the recurrence relations, differential equation, integral representation etc, derived above, remain valid for Gegenbauer polynomials with arbitrary index $\lambda$.

It follows from (3) that

$$C_k^{\lambda+\mu}(t)=\sum_{n=0}^{k} C_n^{\lambda}(t)\,C_{k-n}^{\mu}(t). \tag{4}$$

To prove this it is sufficient to multiply (3) side by side with

$$(1-2th+h^2)^{-\mu}=\sum_{m=0}^{\infty} C_m^{\mu}(t)\,h^m,$$

and then compare the coefficients of like powers of $h$.

## §5. SPHERICAL FUNCTIONS AND THE LAPLACE OPERATOR. POLYSPHERICAL FUNCTIONS

In our account, the spherical functions appeared on choosing a basis in the spaces of irreducible representations of class 1 of the group $SO(n)$. Here we shall present another approach to spherical functions, connected with separation of variables for the Laplace operator in spherical coordinates. In addition, we shall construct other coordinate systems on the sphere, in which the Laplace operator admits separation of variables. This leads to a different choice of basis, related to polyspherical functions.

1. The Laplace operator on the sphere. We express the Laplace operator

$$\Delta = \frac{\partial^2}{\partial x_1^2} + \dots + \frac{\partial^2}{\partial x_n^2}$$

in spherical coordinates. It is known that if in an orthogonal system of curvilinear coordinates in $E_n$ the differential of arc length has the form

$$ds^2 = \sum_i A_i^2 (u_1, \dots, u_n)\, du_i^2, \tag{1}$$

then the Laplace operator is expressed in this coordinate system by the formula

$$\Delta = \frac{1}{A} \sum_{i=1}^{n} \frac{\partial}{\partial u_i} \frac{A}{A_i^2} \frac{\partial}{\partial u_i}, \tag{2}$$

where

$$A = \prod_{i=1}^{n} A_i. \tag{3}$$

Moreover, the invariant measure in the system of coordinates $u_1, \dots, u_n$ has the form

$$dx = A\, du_1 \dots du_n.$$

We apply this to the spherical coordinate system in the euclidean space $E_n$. It easily follows from formula (1) of §1.1 that in this system

$$ds^2 = dr^2 + r^2 d\theta_{n-1}^2 + r^2 \sin^2\theta_{n-1} d\theta_{n-2}^2 + \dots$$
$$\dots + r^2 \sin^2\theta_{n-1} \dots \sin^2\theta_2 d\theta_1^2. \tag{4}$$

Hence in spherical coordinates the Laplace operator has the form

$$\frac{1}{r^{n-1}} \frac{\partial}{\partial r} r^{n-1} \frac{\partial}{\partial r} + \frac{1}{r^2 \sin^{n-2}\theta_{n-1}} \frac{\partial}{\partial \theta_{n-1}} \sin^{n-2}\theta_{n-1} \frac{\partial}{\partial \theta_{n-1}}$$
$$+ \frac{1}{r^2 \sin^2\theta_{n-1} \sin^{n-3}\theta_{n-2}} \frac{\partial}{\partial \theta_{n-2}} \sin^{n-3}\theta_{n-2} \frac{\partial}{\partial \theta_{n-2}} + \dots$$
$$\dots + \frac{1}{r^2 \sin^2\theta_{n-1} \dots \sin^2\theta_2} \frac{\partial^2}{\partial \theta_1^2}. \tag{5}$$

If a function $f(\mathbf{x})$ in $E_n$ depends only on angular variables, then the Laplace operator on $f(\mathbf{x})$ is expressed by the formula $\Delta f(\mathbf{x}) = \Delta_0 f(\xi)$ on the unit sphere, where

$$\Delta_0 = \frac{1}{\sin^{n-2}\theta_{n-1}} \frac{\partial}{\partial\theta_{n-1}} \sin^{n-2}\theta_{n-1} \frac{\partial}{\partial\theta_{n-1}} + \ldots + \frac{1}{\sin^2\theta_{n-1} \ldots \sin^2\theta_2} \frac{\partial^2}{\partial\theta_1^2}. \quad (6)$$

The operator $\Delta_0$ is called the angular part of Laplace operator $\Delta$, or in other words the Laplace operator on the unit sphere $S^{n-1}$.

One can show that $\Delta_0$ is permutable with operators of the quasi-regular representation

$$L(g)f(\xi) = f(g^{-1}\xi)$$

(this follows from a more general assertion concerning the permutability of $\Delta$ with motions of $n$-dimensional euclidean space).

Since $\Delta_0$ is permutable with operators of the representation $L(g)$, the restriction of $\Delta_0$ to the space $\mathfrak{H}^{nl}$ of the irreducible component $T^{nl}(g)$ of the representation $L(g)$ is a multiple of the identity operator:

$$\Delta_0 f(\xi) = \lambda(n, l) f(\xi), \quad f(\xi) \in \mathfrak{H}^{nl}. \quad (7)$$

In order to calculate the coefficient $\lambda(n, l)$, we choose a function $\Xi_M^l(\xi)$ in $\mathfrak{H}^{nl}$, where $M = (l, l, \cdots, l)$. Owing to formula (4) of §3.6, the function $\Xi_M^l(\xi)$ has, in this case, the form

$$\Xi_M^l(\xi) = A_l \sin^l\theta_{n-1} \ldots \sin^l\theta_2 e^{il\theta_1}.$$

Substituting this expression into (7) in place of $f(\xi)$ and utilizing expression (6) for $\Delta_0$, we get $\lambda(n, l) = -l(l+n-2)$.

Thus all the functions $f(\xi)$ of the subspace $\mathfrak{H}^{nl}$ (i.e. all the values on $S^{n-1}$ of homogeneous harmonic polynomials of degree $l$) are eigenfunctions of the Laplace operator $\Delta_0$, corresponding to the eigenvalue $\lambda = -l(l+n-2)$.

One can arrive at this conclusion in a different way.

If $f(\xi) \in \mathfrak{H}^{nl}$, then $r^l f(\mathbf{x}/r)$ is a homogeneous harmonic polynomial of degree $l$. Thus,

$$\Delta r^l f\left(\frac{\mathbf{x}}{r}\right) = 0.$$

Going over to spherical coordinates and utilizing the formula

$$\Delta = \frac{1}{r^{n-1}} \frac{\partial}{\partial r} r^{n-1} \frac{\partial}{\partial r} + \frac{1}{r^2} \Delta_0$$

(see (5) and (6)), we get the required result.

2. Polyspherical coordinates. We have just shown that all the functions of the space $\mathfrak{H}^{nl}$ are eigenfunctions of the Laplace operator $\Delta_0$. The choice of the basis $\Xi_M^l(\xi)$ in $\mathfrak{H}^{nl}$ is linked with the fact that it consists of products of functions, each of which depends only on one variable. However, this choice is not unique; on the sphere there are other (types of) coordinates too, in which the Laplace operator $\Delta_0$ admits separation of variables. We shall describe one class of such coordinates, namely the polyspherical coordinates, as a special case of which we have the spherical coordinates.

Let us consider some tree with $n$ vertices. We shall call the root of the tree a vertex of zero rank; the vertices directly joined to the root, vertices of the first rank; and so on. With each vertex of the tree we associate a cartesian coordinate $x_k$, and we number these coordinates in the following way. We denote the coordinate of zero rank (i.e. corresponding to the root of the tree) by $x_0$, coordinates of the first rank by $x_{01}, \cdots, x_{0s}$, coordinates of the second rank, for which the corresponding vertices are joined to the coordinate $x_{0k}$, by $x_{0k1}, x_{0k2}, \cdots, x_{0km}$, and so on.

We arrange the coordinates in order by assuming that coordinates of a higher rank precede coordinates of a lower rank, while coordinates of equal rank are ordered lexicographically. For example, for the tree depicted in Figure 8, the coordinates are arranged in the following order:

$$x_{011},\ x_{012},\ x_{021},\ x_{01},\ x_{02},\ x_{03},\ x_0.$$

We shall call coordinates of the form $x_{0i_1\cdots i_s i_{s+1}\cdots i_m}$ subordinate to the coordinate $x_{0i_1\cdots i_s}$. Further, we shall say that the coordinate $x_{0j_1\cdots j_m}$ essentially precedes the coordinate $x_{0i_1\cdots i_s}$ if $m \geq s$ and $j_k = i_k$ for $1 \leq k \leq s-1$, and $j_s < i_s$. In this case we shall say that $x_{0i_1\cdots i_s}$ essentially follows $x_{0j_1\cdots j_m}$.

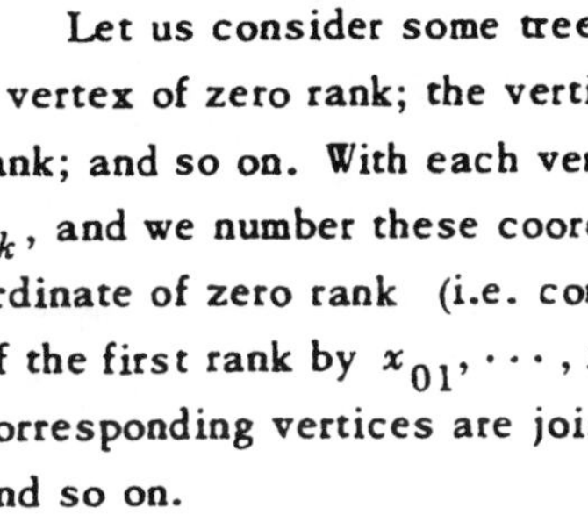


Figure 8.

We introduce a few more notations. The number of vertices essentially preceding $x_{0i_1\cdots i_m}$ is denoted by $p_{i_1\cdots i_m}$, and the number of vertices subordinate to the vertex $x_{0i_1\cdots i_m}$, by $q_{i_1\cdots i_m}$.

Let $x_{0i_1\cdots i_m}$ be one of the vertices of nonzero rank. With it we associate a rotation $g(\phi)$ by the angle $\phi = \phi_{i_1\cdots i_m}$ in the $(x_{0i_1\cdots i_{m-1}}, x_{0i_1\cdots i_m})$-plane

$$x_i' = x_i \cos\varphi - x_j \sin\varphi,$$
$$x_j' = x_i \sin\varphi + x_j \cos\varphi.$$

where, for conciseness, we have put $x_i = x_{0i_1\cdots i_{m-1}}$ and $x_j = x_{0i_1\cdots i_m}$.

We denote by $\mathbf{x}_0$ the point on the sphere corresponding to the coordinate axis $Ox_0$, and put

$$\mathbf{x} = \prod g(\varphi_{i_1\ldots i_m})\,\mathbf{x}_0, \tag{1}$$

where the product is extended over all rotations $g(\phi_{i_1\cdots i_m})$ corresponding to the vertices of the tree, the rotations being arranged according to the order of succession of the coordinates. For example, for Figure 8 we have

$$\mathbf{x} = g(\varphi_{11})\,g(\varphi_{12})\,g(\varphi_{21})\,g(\varphi_1)\,g(\varphi_2)\,g(\varphi_3)\,\mathbf{x}_0.$$

The coordinate $x_{0j_1\cdots j_s}$ can be expressed in terms of the angles $\phi_{i_1\cdots i_m}$ in the following way. We denote by $P_{j_1\cdots j_s}$ the set of coordinates subordinate to the coordinate $x_{0j_1\cdots j_s}$, by $R_{j_1\cdots j_s}$ the set of coordinates essentially following it, and by $S_{j_1\cdots j_s}$ the set of coordinates to which $x_{0j_1\cdots j_s}$ is subordinate. Then one has the equality

$$x_{0j_1\ldots j_s} = \prod_{P_{j_1\ldots j_s} \cup R_{j_1\ldots j_s}} \cos\varphi_{i_1\ldots i_m} \prod_{S_{j_1\ldots j_s} \cup x_{0j_1\ldots j_s}} \sin\varphi_{k_1\ldots k_l}, \tag{2}$$

easily derivable from formula (1).

Thus, to each set of angles $\{\phi_{i_1\cdots i_m}\}$ corresponds a point $\mathbf{x}$ of the unit sphere. Conversely, one can easily show that for any point $\mathbf{x}$ of sphere $S^{n-1}$ there is a set of angles $\{\phi_{i_1\cdots i_m}\}$, for which formula (1) is valid.

The angles $\{\phi_{i_1\cdots i_m}\}$ are, of course, not uniquely determined by the point $\mathbf{x}$ But if the conditions indicated below are imposed on them, then we shall have a one-to-one correspondance between almost every point $\mathbf{x}$ of the sphere $S^{n-1}$ and the angles $\{\phi_{i_1\cdots i_m}\}$. We shall call these angles the polyspherical coordinates of $\mathbf{x}$ (they depend on the choice of the tree and the correspondence between points of the tree and cartesian coordinates).

We now state the conditions imposed on the angles $\{\phi_{i_1\cdots i_m}\}$.

a) If $p_{i_1\cdots i_m} = 0$ and $q_{i_1\cdots i_m} = 0$, i.e. if the vertex $x_{0i_1\cdots i_m}$ has neither subordinate vertices nor essentially preceding ones, then

$$0 \leqslant \varphi_{i_1\ldots i_m} < 2\pi. \tag{3}$$

b) If $p_{i_1\ldots i_m} = 0$ and $q_{i_1\ldots i_m} \neq 0$, then

$$0 \leqslant \varphi_{i_1\ldots i_m} < \pi. \tag{3'}$$

c) If $p_{i_1\ldots i_m} \neq 0$ and $q_{i_1\ldots i_m} = 0$, then

$$-\frac{\pi}{2} \leqslant \varphi_{i_1\ldots i_m} < \frac{\pi}{2}. \tag{3''}$$

d) Lastly, if $p_{i_1\ldots i_m} \neq 0$ and $q_{i_1\ldots i_m} \neq 0$, then

$$0 \leqslant \varphi_{i_1\ldots i_m} < \frac{\pi}{2}. \tag{3'''}$$

The reader can easily verify that to almost every point of the sphere (i.e. with the exception of points constituting a set of smaller dimension) there exists one and only one set of numbers $\{\phi_{i_1\ldots i_m}\}$, lying between the limits indicated above.

From the definition of polyspherical coordinates it is easy to deduce their connection with cartesian coordinates. For example, for the tree depicted in Figure 8 we have

$$\begin{aligned}
x_0 &= \cos\varphi_3 \cos\varphi_2 \cos\varphi_1,\\
x_{03} &= \sin\varphi_3,\\
x_{02} &= \cos\varphi_3 \sin\varphi_2 \cos\varphi_{21},\\
x_{01} &= \cos\varphi_3 \cos\varphi_2 \sin\varphi_1 \cos\varphi_{12} \cos\varphi_{11},\\
x_{021} &= \cos\varphi_3 \sin\varphi_2 \sin\varphi_{21},\\
x_{012} &= \cos\varphi_3 \cos\varphi_2 \sin\varphi_1 \sin\varphi_{12},\\
x_{011} &= \cos\varphi_3 \cos\varphi_2 \sin\varphi_1 \cos\varphi_{12} \sin\varphi_{11}.
\end{aligned}$$

The ordinary spherical coordinates are a special case of the polyspherical ones. They correspond to the tree

$$x_0 \to x_{01} \to x_{011} \to \ldots \to x_{0111\ldots1}.$$

3. The differential of arc length and the Laplace operator in polyspherical coordinates. Utilizing expression (2) of subsection 2, connecting cartesian and polyspherical coordinates, we can easily obtain the form of the differential of arc length in polyspherical coordinates. It has the following form. With each coordinate $x_{0i_1\ldots i_m}$ we associate the expression

$$d\varphi_{i_1\ldots i_m}^2 + \sin^2\varphi_{i_1\ldots i_m}(\cdot) + \cos^2\varphi_{i_1\ldots i_m}(\cdot). \tag{1}$$

If a coordinate $x_{0i_1\cdots i_m+1}$ exists, then one must substitute this in the expression for the coordinate $x_{0i_1\ldots i_m+1}$ in the parentheses which have $\cos^2\phi_{i_1\cdots i_m+1}$ as coefficient. But if such a coordinate does not exist, then expression (1) is substituted in the expression for the coordinate $x_{0i_1\cdots i_{m-1}}$ in the parentheses which have $\sin^2\phi_{i_1\cdots i_{m-1}}$ as coefficient. After all the substitutions have been made, the rest of the empty brackets are taken to be equal to zero.

For example, for the tree in Figure 8 we have

$$ds^2 = d\varphi_3^2 + \cos^2\varphi_3\{d\varphi_2^2 + \sin^2\varphi_2 d\varphi_{21}^2 + \\ + \cos^2\varphi_2[d\varphi_1^2 + \sin^2\varphi_1(d\varphi_{12}^2 + \cos^2\varphi_{12}d\varphi_{11}^2)]\}.$$

Since the expression obtained on opening the brackets in $ds^2$ does not contain terms involving products of coordinates, the polyspherical coordinate system is orthogonal:

$$ds^2 = \sum A_{i_1\ldots i_m}^2 d\varphi_{i_1\ldots i_m}^2. \tag{2}$$

The value of $A_{i_1\cdots i_m}$ here is determined by the formula.

$$A_{i_1\ldots i_m}^2 = \prod_{R_{i_1\ldots i_m}} \cos^2\varphi_{j_1\ldots j_s} \prod_{S_{i_1\ldots i_m}} \sin^2\varphi_{k_1\ldots k_l}. \tag{3}$$

(The symbols $R_{i_1\cdots i_m}$ and $S_{i_1\cdots i_m}$ are defined in subsection 2.) It follows from (2) that, in polyspherical coordinates, the invariant measure on $S^{n-1}$ is given by

$$d\mathbf{x} = \prod A_{i_1\ldots i_m} d\varphi_{i_1\ldots i_m}, \tag{4}$$

where the product is extended to all the vertices of the tree with nonzero rank. This expression can be written as follows:

$$d\mathbf{x} = \prod \cos^{p_{i_1\ldots i_m}}\varphi_{i_1\ldots i_m} \sin^{q_{i_1\ldots i_m}}\varphi_{i_1\ldots i_m} d\varphi_{i_1\ldots i_m}, \tag{5}$$

where the product is extended to all the angles corresponding to a given tree (the symbols $p_{i_1\cdots i_m}$ and $q_{i_1\cdots i_m}$ are defined in subsection 2).

For example, for the tree in Figure 8 we have

$$d\mathbf{x} = \cos^5\varphi_3\cos^3\varphi_2 \sin\varphi_2 \sin^2\varphi_1 \cos\varphi_{12}\, d\varphi_{11}\, d\varphi_{12}\, d\varphi_{21}\, d\varphi_1\, d\varphi_2\, d\varphi_3.$$

Starting from the expression for $ds^2$, one can easily obtain a formula for the Laplace operator $\Delta_0$ on $S^{n-1}$ (see formula (2) of subsection 1).

For example, for the tree in Figure 8, we get

$$\Delta_0 = \frac{1}{\cos^5\varphi_3}\frac{\partial}{\partial\varphi_3}\cos^5\varphi_3\frac{\partial}{\partial\varphi_3} + \frac{1}{\cos^2\varphi_3\cos^3\varphi_2\sin\varphi_2}\frac{\partial}{\partial\varphi_2}\cos^3\varphi_2\sin\varphi_2\frac{\partial}{\partial\varphi_2}$$
$$+\frac{1}{\cos^2\varphi_3\sin^2\varphi_2}\frac{\partial^2}{\partial\varphi_{21}^2} + \frac{1}{\cos^2\varphi_3\cos^2\varphi_2\sin^2\varphi_1}\frac{\partial}{\partial\varphi_1}\sin^2\varphi_1\frac{\partial}{\partial\varphi_1}$$
$$+\frac{1}{\cos^2\varphi_3\cos^2\varphi_2\sin^2\varphi_1\cos\varphi_{12}}\frac{\partial}{\partial\varphi_{12}}\cos^2\varphi_{12}\frac{\partial}{\partial\varphi_{12}} + \frac{1}{\cos^2\varphi_3\cos^2\varphi_2\sin^2\varphi_1\cos^2\varphi_{12}}\frac{\partial^2}{\partial\varphi_{11}^2}.$$

4. **Eigenfunctions of the Laplace operator in polyspherical coordinates.** We pass on to the calculation, in polyspherical coordinates, of eigenfunctions of the Laplace operator on the sphere. This amounts to solving the differential equation

$$\frac{1}{\cos^p\varphi\sin^q\varphi}\frac{d}{d\varphi}\cos^p\varphi\sin^q\varphi\frac{du}{d\varphi} - \left[\frac{r(r+p-1)}{\cos^2\varphi} + \frac{s(s+q-1)}{\sin^2\varphi} - l(l+p+q)\right]u = 0. \qquad (1)$$

We make the substitution $u = \operatorname{tg}^s\phi\cos^l\phi v$ and then put $-\operatorname{tg}^2\phi = y$. Equation (1) is thus reduced to the hypergeometric equation, one of whose solutions is the function

$$v_1 = F\left(\frac{s-l+r}{2},\ \frac{s-l-r-p+1}{2};\ s+\frac{q+1}{2};\ y\right). \qquad (2)$$

Therefore one of the particular solutions of (1) has the form

$$u_1 = \operatorname{tg}^s\varphi\cos^l\varphi F\left(\frac{s-l+r}{2},\ \frac{s-l-r-p+1}{2};\ s+\frac{q+1}{2};\ -\operatorname{tg}^2\varphi\right). \qquad (3)$$

As we remarked in §5.3 of Chapter VII, the hypergeometric equation has, besides the function $F(\alpha, \beta; \gamma; x)$, another particular solution:

$$x^{1-\gamma}F(\alpha-\gamma+1,\ \beta-\gamma+1;\ 2-\gamma;\ x).$$

Hence we get the second particular solution of (1):

$$u_2 = \operatorname{ctg}^{q+s-1}\varphi\cos^l\varphi F\left(\frac{r-l-s-q+1}{2},\ 1-\frac{p+s+q+l+r}{2};\ -s-\frac{q-3}{2};\ -\operatorname{tg}^2\varphi\right). \qquad (4)$$

We now pass on to the separation of variables in the equation $\Delta_0 U + \lambda U = 0$. We substitute the expression for the Laplace operator $\Delta_0$ in polyspherical coordinates and write the required function as a product of functions, each of which depends only on one angle $\phi_{i_1\cdots i_m}$. Utilizing formula (3) of subsection 3, we get the following results (we omit the details of the straightforward but tedious derivation).

The function, which depends on the angle $\phi_{i_1\cdots i_m}$, satisfies the differential equation (1) with

$$\left.\begin{array}{lll} p=p_{i_1\ldots i_m}, & q=q_{i_1\ldots i_m}, & r=l_{i_1\ldots i_{m-1}}, \\ l=l_{i_1\ldots i_m}, & s=|l_{i_1\ldots i_m i_{m+1}}|, & \end{array}\right\} \qquad (5)$$

where $i_{m+1}$ has the largest possible value (in other words, the tree does not have a vertex $x_{0i_1\ldots i_{m+1}}+1$). If no vertices essentially precede the vertex $x_{0i_1\ldots i_m}$ (i.e. $p=0$), then $r=0$, and if none are subordinate to it, then $s=0$. The numbers $\{l_{i_1\ldots i_m}\}$ are parameters of variable separation.

Let us clarify the conditions on the separation parameters $\{l_{i_1\ldots i_m}\}$ under which equation (1) has a solution continuous on the whole sphere $S^{n-1}$. We first consider the case when

$$p_{i_1\ldots i_m}=q_{i_1\ldots i_m}=0.$$

Then equation (1) reduces to

$$\frac{d^2u}{d\varphi^2}+l^2u=0$$

and its solutions are given by the functions

$$u=e^{il\varphi} \text{ and } u=e^{-il\varphi}.$$

Since, in this case, $0\leq\phi<2\pi$ (see subsection 2), the parameter $l=l_{i_1\ldots i_m}$ takes arbitrary integral values, and to each value of $l$ corresponds a solution of equation (1).

If $p=0$ (and so $r=0$), while $q\neq 0$, then solutions (3) and (4) take the form

$$u_1=\operatorname{tg}^s\varphi\cos^l\varphi\, F\left(\frac{s-l}{2},\ \frac{s-l+1}{2};\ s+\frac{q+1}{2};\ -\operatorname{tg}^2\varphi\right) \qquad (6)$$

and

$$u_2=\operatorname{ctg}^{q+s-1}\varphi\cos^l\varphi \times F\left(\frac{-l-s-q+1}{2},\ 1-\frac{l+s+q}{2};\ -s-\frac{q-3}{2};\ -\operatorname{tg}^2\varphi\right). \qquad (7)$$

As the solution $u_1$ is single-valued and continuous at the point $\phi=0$, therefore $s$ is a nonnegative integer. For the continuity of $u_1$ at the point $\phi=\pi/2$ it is necessary that $l-s$ be a nonnegative integer too, i.e. $l\geq s$. If these conditions are satisfied, the function $u_1$ is continuous at the point $\pi/2$. Indeed, for $l\geq s$ one of the numbers $(s-l)/2$, $(s-l+1)/2$ is a nonpositive integer, and so the hypergeometric function in (6) degenerates into a polynomial in $\operatorname{tg}\phi$, whose degree does not exceed $l-s$. Since $\cos\phi$ appears in (6) with power $l-s$, this function is continuous for $\phi = \pi/2$.

Let us now consider the second solution (7). In order that it may be continuous at the point $\phi = 0$, it is necessary that $s \leq 1 - q$. Since we have $q > 0$, while $s = |l_{i_1 \cdots i_{m+1}}| \geq 0$, therefore $q = 1$ and $s = 0$. But in that case solution (6) coincides with (7).

Thus if $p = 0$ and $q \neq 0$, then the solutions of equation (1), continuous for $\phi = 0$, $\phi = \pi/2$, are given by the functions

$$\operatorname{tg}^s \cos^l \varphi F\left(\frac{s-l}{2}, \frac{s-l+1}{2}; s+\frac{q+1}{2}; -\operatorname{tg}^2 \varphi\right), \tag{8}$$

where $l = s, s+1, \cdots, s+k, \cdots$.

Let us now consider the case when $p \neq 0$ but $q = 0$, and therefore $s = 0$. In this case solutions (3) and (4) take the form

$$u_1 = \cos^l \varphi F\left(\frac{r-l}{2}, \frac{-p-l-r}{2}; \frac{1}{2}; -\operatorname{tg}^2 \varphi\right) \tag{9}$$

and

$$u_2 = \operatorname{tg} \varphi \cos^l \varphi F\left(\frac{r-l+1}{2}, 1-\frac{p+l+r}{2}; \frac{3}{2}; -\operatorname{tg}^2 \varphi\right). \tag{10}$$

These solutions are continuous and single-valued for $\phi = 0$. For $\phi = \pi/2$, for any integer $l \geq r$, only one of the two solutions (9) and (10) is continuous; if $l - r$ is even the first of these is continuous, and if $l - r$ is odd the second one is continuous. If $l - r = 2k$, then the continuous solution has the form

$$u_1 = \cos^{2k+r} \varphi F\left(-k, -k-\frac{p}{2}; \frac{1}{2}; -\operatorname{tg}^2 \varphi\right), \tag{11}$$

and if $l - r = 2k + 1$, then it has the form

$$u_2 = \sin \varphi \cos^{2k+r} \varphi F\left(-k, \frac{1-p}{2}-k-r; \frac{3}{2}; -\operatorname{tg}^2 \varphi\right). \tag{12}$$

Lastly, if $p \neq 0$ and $q \neq 0$, then continuous and single-valued solutions exist if and only if $l \geq s + r$, where $l - s - r$ is even: $l - s - r = 2k$. In that case they have the form

$$u = \sin^s \varphi \cos^{2k+r} \varphi F\left(-k, \frac{1-p}{2}-k-r; s+\frac{q+1}{2}; -\operatorname{tg}^2 \varphi\right). \tag{13}$$

Thus, we have proved that the number system $\Lambda = \{l_{i_1 \cdots i_m}\}$ is a system of separation parameters for equation (1) if and only if all the numbers $\{l_{i_1 \cdots i_m}\}$ are integral, where

a) $l_{i_1 \cdots i_m} \geq 0$, excluding the case when $p_{i_1 \cdots i_m} = q_{i_1 \cdots i_m} = 0$;

b)

$$l_{i_1\ldots i_m} \geqslant |l_{i_1\ldots i_m - 1}| + |l_{i_1\ldots i_m i_{m+1}}|. \tag{14}$$

Let us, finally, clarify the form taken by the eigenfunctions of the operator $\Delta_0$, corresponding to a given set $\Lambda = \{l_{i_1\ldots i_m}\}$ of separation parameters. We denote by $U_{i_1\ldots i_m}(\phi)$ a continuous and single-valued solution of equation (1) for the values of parameters $p, q, r, l, s$, given by equality (5). The eigenfunction of $\Delta_0$ corresponding to a given set $\{l_{i_1\ldots i_m}\}$ is given by

$$Y_\Lambda(\xi) = \prod U_{i_1\ldots i_m}(\varphi_{i_1\ldots i_m}), \tag{15}$$

where the product is extended to all vertices of nonzero rank. The functions $Y_\Lambda(\xi)$, corresponding to all admissible choices of $\{l_{i_1\ldots i_m}\}$, are called polyspherical functions corresponding to a given tree. In the case when the tree has the form depicted in Figure 8, polyspherical functions coincide with the ordinary spherical functions on $S^{n-1}$.

It is not difficult to show that the system of polyspherical functions corresponding to a given tree is orthogonal on the sphere $S^{n-1}$. This can be seen by utilizing the orthogonality relations for Jacobi polynomials and the expressions for solutions of equation (1).

## CHAPTER X

# REPRESENTATIONS OF THE GROUP OF HYPERBOLIC ROTATIONS OF $n$-DIMENSIONAL SPACE AND LEGENDRE FUNCTIONS

The connection between the theory of Legendre functions and the theory of group representations has already been analyzed in Chapters III and VI. In Chapter III it was shown that properties of the associated Legendre functions $P_l^m(x)$ with integral and semi-integral index $l$ can be derived by considering representations of the group $SU(2)$ of unitary unimodular matrices of the second order (i.e. representations of the group $SO(3)$ of rotations of three-dimensional euclidean space). In Chapter VI it was shown that many properties of the functions $\mathfrak{P}_l^m(x)$ with arbitrary index $l$ and integral $m$ are derived from the theory of representations of the group $QU(2)$ of quasi-unitary matrices. This group is locally isomorphic to the group $SH(2)$ of motions of the Lobačevskiĭ plane

Further properties of the functions $\mathfrak{P}_l^m(x)$ can be derived by studying representations of the group $SH(n)$ of motions of the $n$-dimensional Lobačevskiĭ space and that is what we shall do in this chapter.

## § 1. PSEUDO-EUCLIDEAN SPACE AND HYPERBOLIC ROTATIONS

1. Pseudo-euclidean space. By the pseudo-euclidean space of signature $(n-1, 1)$ we mean an $n$-dimensional real linear space $E_{n-1,1}$, in which the following bilinear form is given:

$$[\mathbf{x}, \mathbf{y}] = -x_1y_1 - \ldots - x_{n-1}y_{n-1} + x_ny_n.$$

This form is the analogue of the scalar product in euclidean space. With its help we can define distance between points in the space $E_{n-1,1}$ by putting

$$r^2(\mathbf{x}, \mathbf{y}) = [\mathbf{x}-\mathbf{y}, \mathbf{x}-\mathbf{y}].$$

In contrast to the case of euclidean space, the distance between points in pseudo-euclidean space can be either positive, zero, or purely imaginary.

The points lying within the cone $[\mathbf{x}, \mathbf{x}] = 0$ are at a positive distance from the origin $O$, and the points lying outside this cone are at a purely imaginary distance. The points on the cone itself are at zero distance from the origin. Points which are at a positive distance from the origin fall into two disconnected sets, namely points inside the upper half of the cone (i.e. such that $x_n > 0$), and points inside the lower half of the cone (i.e. such that $x_n < 0$). In this chapter we shall be mainly interested in those points of $E_{n-1,1}$ which are situated inside the upper half of the cone. For such points one has the inequalities $[\mathbf{x}, \mathbf{x}] > 0$, $x_n > 0$. We shall denote this domain by $\Omega_n$.

In $\Omega_n$ we introduce coordinates, similar to spherical coordinates in euclidean space. In analogy with § 1.1 of Chapter IX, we define these coordinates by

$$\left.\begin{aligned} x_1 &= r \operatorname{sh} \theta_{n-1} \sin \theta_{n-2} \ldots \sin \theta_2 \sin \theta_1, \\ x_2 &= r \operatorname{sh} \theta_{n-1} \sin \theta_{n-2} \ldots \sin \theta_2 \cos \theta_1, \\ &\cdots\cdots\cdots\cdots\cdots\cdots \\ x_{n-1} &= r \operatorname{sh} \theta_{n-1} \cos \theta_{n-2}, \\ x_n &= r \operatorname{ch} \theta_{n-1}, \end{aligned}\right\} \tag{1}$$

where $r^2 = [\mathbf{x}, \mathbf{x}]$.

Formula (1) establishes, to within sets of lower dimension, a one-to-one correspondence between $\Omega_n$ and the domain $\Omega'_n$:

$$\left.\begin{aligned} &0 \leqslant r < \infty, \\ &0 \leqslant \theta_1 < 2\pi, \\ &0 \leqslant \theta_k < \pi, \qquad 1 < k < n-1, \\ &0 \leqslant \theta_{n-1} < \infty \end{aligned}\right\} \tag{2}$$

of variables, $r, \theta_1, \cdots, \theta_{n-1}$. In other words, from the domains $\Omega_n$ and $\Omega'_n$ it is possible to remove sets of dimension less than $n$, such that equality (1) establishes a one-to-one correspondence between the remaining sets.

We now define the analogue of a euclidean sphere (with center at the origin) in pseudo-euclidean space. Namely, we call the set of points of pseudo-euclidean space for which $[\mathbf{x}, \mathbf{x}] = R^2$ a *pseudo-sphere of radius* $R$. Here we must distinguish between pseudo-spheres of positive, zero or purely imaginary radius. Pseudo-spheres of positive radius lie within the cone $[\mathbf{x}, \mathbf{x}] = 0$, those of purely imaginary radius lie outside this cone, while pseudo-spheres of zero radius coincide with the cone. As we have said above, we shall only consider those parts of pseudo-spheres of positive radius which lie in the upper half of the cone. From the point of view of euclidean geometry, they are upper halves of two-sheeted hyperboloids:

$[\mathbf{x}, \mathbf{x}] = R^2$, $x_n > 0$.

On a pseudo-sphere of positive radius the coordinate $r$ is constant. The remaining coordinates $\theta_1, \cdots, \theta_{n-1}$ give a coordinate system on the pseudo-sphere, which we shall call the *hyperbolic coordinate system*.

2. **The group** $SH(n)$. By *hyperbolic rotations* of pseudo-euclidean space (around the origin) we shall mean linear transformations which do not change the distance of points from the origin, preserve the orientation of the space and transform both halves of the cone $[\mathbf{x}, \mathbf{x}] > 0$ into themselves. Thus, hyperbolic rotations preserve the bilinear form $[\mathbf{x}, \mathbf{y}]$. Evidently the product of two hyperbolic rotations is again a hyperbolic rotation, as is the transformation inverse to a hyperbolic rotation. It follows that the hyperbolic rotations of $E_{n-1,1}$ form a group. We shall call it *the group of hyperbolic rotations* of $E_{n+1,1}$, and denote it by $SH(n)$.

From the invariance of the bilinear form $[\mathbf{x}, \mathbf{y}]$ it is easy to see that the matrix of a hyperbolic rotation is unimodular. Moreover, hyperbolic rotations leave the euclidean measure $d\mathbf{x} = dx_1 \cdots dx_n$ invariant in $E_{n-1,1}$.

In $SH(n)$ we consider the subgroup consisting of those hyperbolic rotations which do not change the coordinate $x_n$. Since hyperbolic rotations do not alter the expression $[\mathbf{x}, \mathbf{x}] = -x_1^2 - \cdots - x_{n-1}^2 + x_n^2$, these rotations leave $x_1^2 + \cdots + x_{n-1}^2$ invariant. In other words, to hyperbolic rotations which do not change the coordinate $x_n$, correspond euclidean rotations in the space $E_{n-1}: x_n = 0$. Therefore the subgroup of hyperbolic rotations which do not change the coordinate $x_n$ will be denoted by $SO(n-1)$. Similarly, the subgroup of hyperbolic rotations which do not change the coordinates $x_{n-s+1}, \cdots, x_n$ will be denoted by $SO(n-s)$.

We now show that the group $SH(n)$ acts transitively on the upper half of the hyperboloid $[\mathbf{x}, \mathbf{x}] = 1$. For this it is sufficient to show that any point $\xi(\xi_1, \cdots, \xi_n)$ of this hyperboloid can be carried into the point $\xi_n(0, \cdots, 0, 1)$ by a hyperbolic rotation. It is evident that, with the help of a euclidean rotation of $SO(n-1)$, the point $\xi$ can be carried into the point $\eta(0, \cdots, 0, \operatorname{sh}\alpha, \operatorname{ch}\alpha)$, where $\operatorname{ch}\alpha = \xi_n$. We then perform a hyperbolic rotation:

$$x'_{n-1} = x_{n-1}\operatorname{ch}\alpha - x_n\operatorname{sh}\alpha,$$
$$x'_n = -x_{n-1}\operatorname{sh}\alpha + x_n\operatorname{ch}\alpha$$

in the $(x_{n-1}, x_n)$-plane. Under this rotation the point $\eta$ goes into $\xi_n$. The transitivity of $SH(n)$ is thus proved.

The subgroup $SO(n-1)$ of euclidean rotations leaves the point $\xi_n = (0, \cdots, 0, 1)$ fixed, and, as one can easily see, it is the stationary subgroup of this point. It follows that the upper half of the hyperboloid $[\mathbf{x}, \mathbf{x}] = 1$, $x_n > 0$ is the space of left cosets of $SH(n)$ with respect to the subgroup $SO(n-1)$. Namely, to each left coset $\{gh\}$, $h \in SO(n-1)$ corresponds a point $\xi_g$ of the pseudo-sphere, into which the point $\xi_n$ is sent by hyperbolic rotations $gh$ of this coset. Moreover, to the transformation

$$\{gh\} \to \{g_0 gh\}, \qquad h \in SO(n-1),$$

in the factor-space $SH(n)/SO(n-1)$ corresponds the hyperbolic rotation $\xi_g \to \xi_{g_0 g}$ of the pseudo-sphere.

3. Lobačevskiĭ space. Let $\xi$ and $\eta$ be two points of the pseudo-sphere $[\mathbf{x}, \mathbf{x}] = 1$, $x_n > 0$. Then one has the inequality

$$[\xi, \eta] \geqslant 1, \tag{1}$$

where $[\xi, \eta] = 1$ if and only if $\xi = \eta$.

Indeed, it follows from $[\xi, \xi] = [\eta, \eta] = 1$ that

$$\xi_n = \sqrt{1 + \xi_1^2 + \ldots + \xi_{n-1}^2} \text{ and } \eta_n = \sqrt{1 + \eta_1^2 + \ldots + \eta_{n-1}^2}.$$

Therefore, by the Schwarz-Bunjakovskiĭ inequality,

$$1 + \xi_1\eta_1 + \ldots + \xi_{n-1}\eta_{n-1} \leqslant \sqrt{1 + \xi_1^2 + \ldots + \xi_{n-1}^2}\sqrt{1 + \eta_1^2 + \ldots + \eta_{n-1}^2} = \xi_n\eta_n. \tag{1'}$$

and consequently

$$[\xi, \eta] = -\xi_1\eta_1 - \ldots - \xi_{n-1}\eta_{n-1} + \xi_n\eta_n \geqslant 1.$$

In (1′) equality is attained only if $\xi_1 = \eta_1, \ldots, \xi_{n-1} = \eta_{n-1}$. But in that case $\xi_n = \eta_n$, and so $\xi = \eta$.

It follows from (1) that there exists a unique number $r \geq 0$ such that

$$\operatorname{ch} kr = [\xi, \eta] \tag{2}$$

($k$ being a fixed positive number). We shall call $r$ the distance between the points $\xi$ and $\eta$ of the pseudo-sphere.

It is easy to show that the pseudo-sphere is thus turned into a Riemannian space of constant negative curvature. This space is called the $(n-1)$*-dimensional Lobačevskiĭ space*, and is denoted by $\Lambda^{n-1}(R)$, where $R = 1/k$.

The group $SH(n)$ consists of those transformations of Lobačevskiĭ space which preserve the metric and orientation, i.e. is a group of motions of the $(n-1)$ dimensional Lobačevskiĭ space $\Lambda^{n-1}$ (just as the group $SO(n)$ is a group of rotations of the $n$-dimensional euclidean space $E_n$ and, at the same time, a group of motions of the $(n-1)$-dimensional sphere $S^{n-1}$).

Side by side with the model of Lobačevskiĭ space described above, other models of this space are sometimes found to be useful. In $E_{n-1,1}$ let us consider the totality of straight lines passing through the origin and lying inside the cone $[\mathbf{x}, \mathbf{x}] = 0$. We define the distance $r$ between such straight lines by the formula

$$\operatorname{ch} kr = \frac{[\mathbf{x}, \mathbf{y}]}{\sqrt{[\mathbf{x}, \mathbf{x}][\mathbf{y}, \mathbf{y}]}}, \tag{3}$$

where $\mathbf{x}$ and $\mathbf{y}$ are points lying on these lines. It is easy to see that the distance $r$ between the straight lines does not depend on the choice of points $\mathbf{x}$ and $\mathbf{y}$ on them. But if one chooses as these points the points of intersection of the straight lines with the hyperboloid $[\mathbf{x}, \mathbf{x}] = 1$, then equality (3) takes the form $\operatorname{ch} kr = [\mathbf{x}, \mathbf{y}]$, i.e. it coincides with (2). Hence the set of straight lines lying inside the cone $[\mathbf{x}, \mathbf{x}] = 1$, and with the metric given by formula (3), is another model of Lobačevskiĭ space.

Other models of Lobačevskiĭ space are obtained in the following way. Inside the upper half of the cone $[\mathbf{x}, \mathbf{x}] = 0$ we take any surface which intersects every ray emanating from the origin at one and only one point. With each straight line lying within the cone $[\mathbf{x}, \mathbf{x}] = 0$, we associate the point in which it intersects the surface taken by us. We thus obtain a model of Lobačevskiĭ space in the form of a set of points of a surface. If we take this surface to be the upper half of the hyperboloid $[\mathbf{x}, \mathbf{x}] = 0$, we obtain an earlier model of Lobačevskiĭ space.

We mention yet another model of Lobačevskiĭ space. Let us draw the plane $x_n = 1$. Obviously, this plane intersects the cone $[\mathbf{x}, \mathbf{x}] = 0$ along a sphere of radius 1. Lobačevskiĭ space can thus be interpreted as the interior of the $(n-1)$-dimensional ball of radius 1. The distance between points of this ball is defined by the formula

$$\operatorname{ch} kr = \frac{1 - (\mathbf{x}', \mathbf{y}')}{\sqrt{1 - (\mathbf{x}', \mathbf{x}')}\,\sqrt{1 - (\mathbf{y}', \mathbf{y}')}}, \tag{4}$$

where $\mathbf{x}' = (x_1, \ldots, x_{n-1})$, $\mathbf{y}' = (y_1, \ldots, y_{n-1})$, $(\mathbf{x}', \mathbf{y}') = x_1 y_1 + \cdots + x_{n-1} y_{n-1}$.

We now introduce the concept of the absolute of Lobačevskiĭ space. We take two straight lines through the origin, and make one of them approach the cone $[\mathbf{x}, \mathbf{x}] = 0$. Formula (3) shows that the distance between the straight lines will then tend to infinity. Thus straight lines lying on the cone $[\mathbf{x}, \mathbf{x}] = 0$ are the points at infinity of Lobačevskiĭ space. The set of all points at infinity is call *an absolute*. Hence, under the realization of Lobačevskiĭ space as a space of straight lines, the absolute consists of the set of rectilinear generators of the cone

$[\mathbf{x}, \mathbf{x}] = 0$. However, under the realization of Lobačevskiĭ space within the unit ball, the absolute is represented by the unit sphere.

**4. Euler angles in the group $SH(n)$.** We introduce parameters in $SH(n)$, analogous to Euler angles in the group $SO(n)$ of rotations of euclidean space. We shall first show that any element $g$ of $SH(n)$ can be written in the form

$$g = g_1(\theta_1^{n-1}) \ldots g_{n-1}(\theta_{n-1}^{n-1})\, h, \tag{1}$$

where $h \in SO(n-1)$, $g_k(\alpha)$ for $k < n-1$ denotes a rotation by the angle $\alpha$ in the $(x_{k+1}, x_k)$-plane, while $g_{n-1}(\alpha)$ denotes a hyperbolic rotation by the "angle" $\alpha$ in the $(x_n, x_{n-1})$-plane:

$$\left.\begin{aligned} x_{n-1}' &= x_{n-1} \operatorname{ch} \alpha + x_n \operatorname{sh} \alpha, \\ x_n' &= x_{n-1} \operatorname{sh} \alpha + x_n \operatorname{ch} \alpha. \end{aligned}\right\} \tag{2}$$

Indeed, let $g$ be an element of $SH(n)$. We denote by $\theta_1^{n-1}, \cdots, \theta_{n-1}^{n-1}$ the hyperbolic coordinates of the point $g\xi_n$, where $\xi_n = (0, \cdots, 0, 1)$. A simple calculation shows that $g\xi_n = g^{(n-1)}\xi_n$, where

$$g^{(n-1)} = g_1(\theta_1^{n-1}) \ldots g_{n-1}(\theta_{n-1}^{n-1}).$$

Therefore $[g^{(n-1)}]^{-1} g\xi_n = \xi_n$, and consequently the hyperbolic rotation $[g^{(n-1)}]^{-1} g = h$ belongs to $SO(n-1)$. Since

$$g = g^{(n-1)} h = g_1(\theta_1^{n-1}) \ldots g_{n-1}(\theta_{n-1}^{n-1})\, h,$$

our assertion is proved.

As shown in §1.3 of Chapter IX, a rotation $h$ of $SO(n-1)$ can be presented in the form

$$h = g^{(n-2)} \ldots g^{(1)},$$

where

$$g^{(k)} = g_1(\theta_1^k) \ldots g_k(\theta_k^k), \tag{3}$$

and $g_s(\alpha)$ is a rotation by angle $\alpha$ in the $(x_{s+1}, x_s)$-plane. Hence we get the following assertion.

*Any hyperbolic rotation $g$ of $SH(n)$ can be presented in the form*

$$g = g^{(n-1)} g^{(n-2)} \ldots g^{(1)}, \tag{4}$$

*where $g^{(k)}$ has the form (3) and $g_k(\alpha)$ for $k < n-1$ is a rotation by the angle $\alpha$ in the $(x_{k+1}, x_k)$-plane, and $g_{n-1}(\alpha)$ is the hyperbolic rotation by the "angle" $\alpha$ in the $(x_n, x_{n-1})$-plane.*

We shall call the numbers $\theta_j^k$, $1 \le k \le n-1$, $1 \le j \le k$, *the Euler angles of* $g$. These numbers vary as follows:

$$\left.\begin{array}{ll} 0 \leqslant \theta_1^k < 2\pi, & \\ 0 \leqslant \theta_{n-1}^{n-1} < \infty, & \\ 0 \leqslant \theta_j^k < \pi & \text{in all other cases.} \end{array}\right\} \tag{5}$$

By analogy with the group $SO(n)$, it is easy to show that the invariant measure $dg$ on $SH(n)$ is expressed, in terms of Euler angles, by the formula

$$dg = \operatorname{sh}^{n-2}\theta_{n-1}^{n-1}\sin^{n-3}\theta_{n-2}^{n-1}\dots\sin\theta_2^{n-1} \times \prod_{k=1}^{n-2}\prod_{j=1}^{k}\sin^{j-1}\theta_j^k \prod_{k=1}^{n-1}\prod_{j=1}^{k} d\theta_j^k. \tag{6}$$

## § 2. REPRESENTATIONS OF CLASS 1 OF THE GROUP $SH(n)$

A representation $T(g)$ of the group $SH(n)$ is called *a representation of class* 1 if in the space $\mathfrak{F}$ of this representation there is a vector $f_0$ which is invariant for all transformations $T(h)$, where $h$ is an element of $SO(n-1)$. In this section we shall describe these representations and study their properties.

1. Description of the representations $T^{n\sigma}(g)$. We shall denote by $\mathfrak{B}^{n\sigma}$ the space of functions $f(\mathbf{x})$ given on the upper half of the cone $[\mathbf{x}, \mathbf{x}] = 0$, $x_n > 0$, and such that

1) *the functions* $f(\mathbf{x})$ *are infinitely differentiable at every point of the upper half of the cone, and*

2) *the functions* $f(\mathbf{x})$ *are homogeneous of degree* $\sigma$:

$$f(a\mathbf{x}) = a^\sigma f(\mathbf{x}), \qquad a > 0. \tag{1}$$

Evidently, if a function $f(\mathbf{x})$ belongs to $\mathfrak{B}^{n\sigma}$ then the function $f(g^{-1}\mathbf{x})$, where $g \in SH(n)$, also belongs to this space. Therefore the equality

$$S^{n\sigma}(g) f(\mathbf{x}) = f(g^{-1}\mathbf{x}), \qquad g \in SH(n) \tag{2}$$

defines an operator in $\mathfrak{B}^{n\sigma}$. Since

$$S^{n\sigma}(g_1) S^{n\sigma}(g_2) f(\mathbf{x}) = S^{n\sigma}(g_1) f(g_2^{-1}\mathbf{x}) = f(g_2^{-1}g_1^{-1}\mathbf{x}) = f((g_1g_2)^{-1}\mathbf{x}) = S^{n\sigma}(g_1g_2) f(\mathbf{x}),$$

$S^{n\sigma}(g)$ is a representation of $SH(n)$.

The representation $S^{n\sigma}(g)$ can be realized as a representation of $SH(n)$ in the space of infinitely differentiable functions on the sphere. With each function $f(\mathbf{x})$ of the space $\mathfrak{B}^{n\sigma}$ we associate the function

$$F(\xi') = Qf(\xi') \equiv f(\xi_1, \ldots, \xi_{n-1}, 1)$$

on $S^{n-2}$, i.e. on the intersection of the cone $[\mathbf{x}, \mathbf{x}] = 0$ and the plane $x_n = 1$. It is evident that $f(\mathbf{x})$ is uniquely determined by $F(\xi')$. Namely, by virtue of the homogeneity of $f(\mathbf{x})$, we have

$$\begin{aligned} f(\mathbf{x}) &= f(x_1, \ldots, x_{n-1}, x_n) \\ &= x_n^\sigma f\left(\frac{x_1}{x_n}, \ldots, \frac{x_{n-1}}{x_n}, 1\right) = x_n^\sigma F\left(\frac{x_1}{x_n}, \ldots, \frac{x_{n-1}}{x_n}\right). \end{aligned} \tag{3}$$

Hence operators of the representation $S^{n\sigma}(g)$ correspond to operators of the representation

$$T^{n\sigma}(g) = QS^{n\sigma}(g)\, Q^{-1} \tag{4}$$

in the space $\mathfrak{D}$ of infinitely differentiable functions on the sphere $S^{n-2}$.

If $g$ belongs to $SO(n-1)$, then $T^{n\sigma}(g)$ can be easily written in an explicit form. Since a rotation $h$ of $SO(n-1)$ transforms the plane $x_n = 1$ into itself, and therefore also the sphere $S^{n-2}$ into itself, we have

$$T^{n\sigma}(h)\, F(\xi') = F(h^{-1}\xi'). \tag{5}$$

Thus the representation $T^{n\sigma}(h)$ of $SO(n-1)$ coincides with the quasi-regular representation of this group, considered in § 2.1 of Chapter IX.[1]

Let us now determine the form of the operator $T^{n\sigma}(g_{n-1}(\alpha))$, where $g_{n-1}(\alpha)$ is a hyperbolic rotation by the angle $\alpha$ in the $(x_n, x_{n-1})$-plane. We have

$$\begin{aligned} T^{n\sigma}[g_{n-1}(\alpha)]\, F(\xi_1, \ldots, \xi_{n-1}) &= QS^{n\sigma}[g_{n-1}(\alpha)] f(\xi_1, \ldots, \xi_{n-1}, \xi_n) \\ &= f(\xi_1, \ldots, \xi_{n-1}\operatorname{ch}\alpha - \operatorname{sh}\alpha, -\xi_{n-1}\operatorname{sh}\alpha + \operatorname{ch}\alpha). \end{aligned}$$

By virtue of the homogeneity of $f(\xi)$ and relation (3), this equality can be rewritten as follows:

$$\begin{aligned} T^{n\sigma}[g_{n-1}(\alpha)]\, F(\xi_1, \ldots, \xi_{n-1}) &= (\operatorname{ch}\alpha - \xi_{n-1}\operatorname{sh}\alpha)^\sigma \\ \times F\left(\frac{\xi_1}{\operatorname{ch}\alpha - \xi_{n-1}\operatorname{sh}\alpha}, \ldots, \right. & \left. \frac{\xi_{n-2}}{\operatorname{ch}\alpha - \xi_{n-1}\operatorname{sh}\alpha}, \frac{\xi_{n-1}\operatorname{ch}\alpha - \operatorname{sh}\alpha}{\operatorname{ch}\alpha - \xi_{n-1}\operatorname{sh}\alpha}\right). \end{aligned} \tag{6}$$

It is more convenient to write the operator $T^{n\sigma}[g_{n-1}(\alpha)]$ in spherical coordinates $\phi_1, \cdots, \phi_{n-2}$ on $S^{n-2}$. According to § 1.1, Chapter IX, these coordinates are related to cartesian coordinates $\xi_1, \cdots, \xi_{n-1}$ by the formulas

1) In § 2.1 of Chapter IX the quasi-regular representation was constructed in the completion of $\mathfrak{D}$ with respect to the norm

$$\|F\|^2 = \int_{S^{n-2}} |F(\xi')|^2\, d\xi'.$$

$$\cos\varphi_k=\frac{x_{k+1}}{r_{k+1}},\qquad \sin\varphi_k=\frac{r_k}{r_{k+1}},$$

where $r_k^2 = x_1^2+\cdots+x_k^2$. It follows that, under a simultaneous division of the coordinates $\xi_1,\cdots,\xi_{n-2}$ by $\operatorname{ch}\alpha-\xi_{n-1}\operatorname{sh}\alpha$, the values of the angles $\phi_1,\cdots,\phi_{n-3}$ are unchanged. The angle $\phi_{n-2}$ is transformed according to the formula

$$\cos\varphi'_{n-2}=\frac{\cos\varphi_{n-2}\operatorname{ch}\alpha-\operatorname{sh}\alpha}{\operatorname{ch}\alpha-\cos\varphi_{n-2}\operatorname{sh}\alpha}.$$

Consequently (6) can be rewritten as follows:

$$\begin{aligned}T^{n\sigma}[g_{n-1}(\alpha)]F_1(\cos\varphi_1,\ \ldots,\ \cos\varphi_{n-2})&\\ =(\operatorname{ch}\alpha-\cos\varphi_{n-2}\operatorname{sh}\alpha)^\sigma&\\ \times F_1\left(\cos\varphi_1,\ \ldots,\ \cos\varphi_{n-3},\ \frac{\cos\varphi_{n-2}\operatorname{ch}\alpha-\operatorname{sh}\alpha}{\operatorname{ch}\alpha-\cos\varphi_{n-2}\operatorname{sh}\alpha}\right).&\qquad(7)\end{aligned}$$

The representation $T^{n\sigma}(g)$ is a representation of class 1, i.e. in the space of this representation there is a function $F_0(\xi')$, invariant for all operators $T^{n\sigma}(h)$, $h\in SO(n-1)$. Such a function is given by

$$F_0(\xi')\equiv 1. \qquad (8)$$

2. **Adjoint representations.** Let us consider the representation $T^{n\sigma}(g)'$ adjoint to $T^{n\sigma}(g)$. By definition the adjoint representation $T^{n\sigma}(g)'$ is constructed in the space $\mathfrak{D}'$ of functionals for the space $\mathfrak{D}$, and is determined by the formula

$$(T^{n\sigma}(g)'\Phi,\ F)=(\Phi,\ T^{n\sigma}(g^{-1})F), \qquad (1)$$

where $\Phi\in\mathfrak{D}'$, $F\in\mathfrak{D}$.

We shall prove that $T^{n\sigma}(g)'$ is a continuation in $\mathfrak{D}'$ of the representation $T^{n,-n-\sigma+2}(g)$. To that end, with each function $\Phi(\xi')$ of $\mathfrak{D}$ we associate a functional

$$(\Phi,\ F)=\int_{S^{n-2}}\Phi(\xi')F(\xi')\,d\xi' \qquad (2)$$

in $\mathfrak{D}$, and show that

$$(T^{n,-n-\sigma+2}(g)\Phi,\ F)=(\Phi,\ T^{n\sigma}(g^{-1})F). \qquad (3)$$

Since every element $g$ of $SH(n)$ can be written in the form $g=h_1g_{n-1}(\alpha)h_2$, where $h_1,h_2\in SO(n-1)$, it is sufficient to give a proof for the cases when $g\in SO(n-1)$ and when $g=g_{n-1}(\alpha)$. In the first case our assertion is evident, since for $g\in SO(n-1)$ we have

$$
\begin{aligned}
(T^{n,-n-\sigma+2}(g)\Phi, F) &= \int_{S^{n-2}} T^{n,-n-\sigma+2}(g)\Phi(\xi')F(\xi')\,d\xi' \\
&= \int_{S^{n-2}} \Phi(g^{-1}\xi')F(\xi')\,d\xi' = \int_{S^{n-2}} \Phi(\xi')F(g\xi')\,d\xi' \\
&= \int_{S^{n-2}} \Phi(\xi')T^{n\sigma}(g^{-1})F(\xi')\,d\xi' = (\Phi, T^{n\sigma}(g^{-1})F).
\end{aligned}
$$

Now let $g = g_{n-1}(\alpha)$. In this case, by formula (6) of subsection 1 we have

$$
\begin{aligned}
&\int_{S^{n-2}} T^{n,-n-\sigma+2}(g)\Phi(\xi_1, \ldots, \xi_{n-1})F(\xi_1, \ldots, \xi_{n-1})\,d\xi' \\
&= \int_{S^{n-2}} (\operatorname{ch}\alpha - \xi_{n-1}\operatorname{sh}\alpha)^{-n-\sigma+2}\Phi\left(\frac{\xi_1}{\operatorname{ch}\alpha - \xi_{n-1}\operatorname{sh}\alpha}, \ldots, \frac{\xi_{n-2}}{\operatorname{ch}\alpha - \xi_{n-1}\operatorname{sh}\alpha},\right. \\
&\qquad\qquad \left.\frac{\xi_{n-1}\operatorname{ch}\alpha - \operatorname{sh}\alpha}{\operatorname{ch}\alpha - \xi_{n-1}\operatorname{sh}\alpha}\right) F(\xi_1, \ldots, \xi_{n-1})\,d\xi'. \qquad (4)
\end{aligned}
$$

Since $\xi_1^2 + \cdots + \xi_{n-1}^2 = 1$, we can choose $\xi_1, \cdots, \xi_{n-2}$ as the independent variables in this integral. In these variables, the invariant measure $d\xi'$ on $S^{n-2}$ is expressed by the formula

$$
d\xi' = \frac{\Gamma\left(\frac{n-1}{2}\right)}{2\pi^{\frac{n-1}{2}}} \frac{d\xi_1 \ldots d\xi_{n-2}}{\xi_{n-1}}. \qquad (5)
$$

In (4) we make the substitution

$$
\frac{\xi_k}{\operatorname{ch}\alpha - \xi_{n-1}\operatorname{sh}\alpha} = \eta_k, \qquad 1 \leqslant k \leqslant n-2.
$$

It is easy to show that

$$
\eta_{n-1} = \frac{\xi_{n-1}\operatorname{ch}\alpha - \operatorname{sh}\alpha}{\operatorname{ch}\alpha - \xi_{n-1}\operatorname{sh}\alpha}
$$

and

$$
d\xi' = (\operatorname{ch}\alpha + \eta_{n-1}\operatorname{sh}\alpha)^{n+2}\,d\eta'.
$$

Thus (4) takes the form

$$
\begin{aligned}
&\int_{S^{n-2}} (\operatorname{ch}\alpha + \eta_{n-1}\operatorname{sh}\alpha)^{\sigma}\Phi(\eta_1, \ldots, \eta_{n-1}) \\
&\times F\left(\frac{\eta_1}{\operatorname{ch}\alpha + \eta_{n-1}\operatorname{sh}\alpha}, \ldots, \frac{\eta_{n-2}}{\operatorname{ch}\alpha + \eta_{n-1}\operatorname{sh}\alpha}, \frac{\eta_{n-1}\operatorname{ch}\alpha + \operatorname{sh}\alpha}{\operatorname{ch}\alpha + \eta_{n-1}\operatorname{sh}\alpha}\right) d\eta' \\
&\qquad = \int_{S^{n-2}} \Phi(\eta')T^{n\sigma}[g_{n-1}(-\alpha)]F(\eta')\,d\eta'. \qquad (6)
\end{aligned}
$$

It follows from (4) and (6) that

$$
(T^{n,-n-\sigma+2}[g_{n-1}(\alpha)]\Phi, F) = (\Phi, T^{n\sigma}[g_{n-1}(-\alpha)]F).
$$

Thus (3) is proved for $g = g_{n-1}(\alpha)$ too. Consequently it is valid for all elements $g \in SH(n)$.

In exactly the same way one proves that $T^{n\sigma}(g)$ is Hermitian-adjoint to $T^{n, -n-\bar{\sigma}-2}(g)$. The only difference is that now with the function $\Phi(\xi')$ one associates the antilinear functional

$$\int_{S^{n-2}} \Phi(\xi')\overline{F(\xi')}\,d\xi'$$

in $\mathfrak{D}$.

**3. Irreducibility of the representations $T^{n\sigma}(g)$ for nonintegral $\sigma$.** We prove that if $\sigma$ is not an integer, then $T^{n\sigma}(g)$ is irreducible.

Let us first consider the restriction of $T^{n\sigma}(g)$ to $SO(n-1)$. We have already stated that this representation coincides with the quasi-regular representation of $SO(n-1)$. From the results of § 2.7 of Chapter IX it follows that $T^{n\sigma}(h)$, $h \in SO(n-1)$ is a direct sum of irreducible unitary representations $T^{n-1,k}(h)$ of $SO(n-1)$:

$$T^{n\sigma}(h) = \sum_{k=0}^{\infty} T^{n-1,k}(h), \qquad h \in SO(n-1). \tag{1}$$

The representations $T^{n-1,k}(h)$ are realized in the subspaces $\mathfrak{H}^{n-1,k}$ which consist of the values on $S^{n-2}$ of homogeneous harmonic polynomials of degree $k$ in $n-1$ variables. Moreover, for different $k$ the representations $T^{n-1,k}(h)$ are nonequivalent.

We denote by $\mathfrak{L}^2(S^{n-2})$ the space of those functions $F(\xi')$ on $S^{n-2}$ for which

$$\|F\|^2 = \int_{S^{n-2}} |F(\xi')|^2\,d\xi' < +\infty. \tag{2}$$

Let $\mathfrak{T}$ be an invariant subspace in the space $\mathfrak{L}^2(S^{n-2})$. From the decomposition (1) and the pairwise nonequivalence of the $T^{n-1,k}(h)$ it follows that $\mathfrak{T}$ is a direct sum of some of the subspaces $\mathfrak{H}^{n-1,k}$, $0 \le k < \infty$, and so contains at least one of these subspaces:

$$\mathfrak{T} = \sum_j \mathfrak{H}^{n-1,k_j}.$$

It remains for us to show that if the invariant subspace $\mathfrak{T}$ contains one of the subspaces $\mathfrak{H}^{n-1,k}$ and $\sigma$ is not an integer, then $\mathfrak{T}$ contains all the remaining subspaces $\mathfrak{H}^{n-1,k}$ too, and thus coincides with the subspace $\mathfrak{D}$.

From the invariance of $\mathfrak{T}$ for the operators $T^{n\sigma}(g)$ it follows that $\mathfrak{T}$ is invariant also for the infinitesimal operators of this representation. Let us calculate the infinitesimal operator corresponding to the subgroup $\Omega_n$ of hyperbolic rotations $g_{n-1}(\alpha)$ in the $(x_n, x_{n-1})$-plane. Employing formula (7) of subsection 1, we see that

$$AF_1 = \frac{dT^{n\sigma}[g_{n-1}(\alpha)]}{d\alpha}\Big|_{\alpha=0} F_1 = -\sigma\cos\varphi_{n-2}F_1 - \sin^2\varphi_{n-2}\frac{\partial F_1}{\partial\cos\varphi_{n-2}}.$$

Thus

$$A = -\sigma\cos\varphi_{n-2} - \sin^2\varphi_{n-2}\frac{\partial}{\partial\cos\varphi_{n-2}}. \tag{3}$$

We now assume that the invariant subspace $\mathfrak{T}$ contains the subspace $\mathfrak{H}^{n-1,k}$ According to §3.1, Chapter IX, one of the functions of $\mathfrak{H}^{n-1,k}$ is the function $C_k^{(n-3)/2}(\cos\phi_{n-2})$. Because of its invariance property, $\mathfrak{T}$ must also contain the function

$$AC_k^{\frac{n-3}{2}}(\cos\varphi_{n-2}) = -\sigma\cos\varphi_{n-2}C_k^{\frac{n-3}{2}}(\cos\varphi_{n-2}) - (n-3)\sin^2\varphi_{n-2}C_{k-1}^{\frac{n-1}{2}}(\cos\varphi_{n-2}). \tag{4}$$

Applying the recurrence relations

$$tC_n^p(t) = \frac{n+1}{2(p+n)}C_{n+1}^p(t) + \frac{2p+n-1}{2(p+n)}C_{n-1}^p(t)$$

and

$$(1-t^2)C_{n-1}^{p+1}(t) = \frac{2p+n}{2p}tC_n^p(t) - \frac{n+1}{2p}C_{n+1}^p(t)$$

for Gegenbauer polynomials (see §3.2 of Chapter IX), we see that $\mathfrak{T}$ contains the function[1)]

$$AC_k^{\frac{n-3}{2}}(\cos\varphi_{n-2}) = \frac{(k+1)(k-\sigma)}{2k+n-3}C_{k+1}^{\frac{n-3}{2}}(\cos\varphi_{n-2}) - \frac{(n+k-4)(n+k+\sigma-3)}{2k+n-3}C_{k-1}^{\frac{n-3}{2}}(\cos\varphi_{n-2}). \tag{5}$$

But $C_{k+1}^{(n-3)/2}(\cos\phi_{n-2}) \in \mathfrak{H}^{n-1,k+1}$ and $C_{k-1}^{(n-3)/2}(\cos\phi_{n-2}) \in \mathfrak{H}^{n-1,k-1}$. Moreover, if $\sigma$ is not an integer, then the coefficients of the Gegenbauer polynomials in (5) are nonzero.

It follows from (1) that if $\mathfrak{T}$ contains the sum of nonnull functions from the subspaces $\mathfrak{H}^{n-1,k-1}$ and $\mathfrak{H}^{n-1,k+1}$, then it contains these subspaces too.

Thus we have proved that if $\sigma$ is not an integer, the invariant subspace $\mathfrak{T}$, together with any given subspace $\mathfrak{H}^{n-1,k}$ , contain the subspaces $\mathfrak{H}^{n-1,k-1}$

1) For $k=0$ we take $C_{k-1}^{(n-3)/2}(\cos\phi_{n-2}) = 0$.

and $\mathfrak{H}^{n-1,k+1}$ (for $k = 0$ the subspace $\mathfrak{H}^{n-1,k+1}$ only). Therefore $\mathfrak{T}$ contains all subspaces $\mathfrak{H}^{n-1,m}$, and consequently it coincides with $\mathfrak{L}^2(S^{n-2})$. The irreducibility of $T^{n\sigma}(g)$ (and hence of $S^{n\sigma}(g)$ too) is thus proved.

**4. Reducibility of $T^{n\sigma}(g)$ for integral values of $\sigma$.** We now consider the case when $\sigma = k$, a nonnegative integer. We show that in this case $T^{nk}(g)$ is reducible Namely, an invariant subspace in $\mathfrak{L}^2(S^{n-2})$ is provided by the subspace $\mathfrak{A}^{n-1,k}$ of polynomials of degree $k$ in the variables $\xi_1, \cdots, \xi_{n-1}$. Indeed, it is evident that this subspace is invariant for the transformations $T^{nk}(h)\,F(\xi') = F(h^{-1}\xi')$, corresponding to elements $h \in SO(n-1)$. Let us now take the hyperbolic rotation in the $(x_n, x_{n-1})$-plane. If $F(\xi_1, \cdots, \xi_{n-1})$ is a polynomial of degree $k$ in $\xi_1, \cdots, \xi_{n-1}$, then

$$\begin{aligned} T^{nk}[g_{n-1}(\alpha)]\,F(\xi_1, \ldots, \xi_{n-1}) &= (\operatorname{ch}\alpha - \xi_{n-1}\operatorname{sh}\alpha)^k \\ \times F\left(\frac{\xi_1}{\operatorname{ch}\alpha - \xi_{n-1}\operatorname{sh}\alpha}, \ldots, \right. & \left. \frac{\xi_{n-2}}{\operatorname{ch}\alpha - \xi_{n-1}\operatorname{sh}\alpha}, \frac{\xi_{n-1}\operatorname{ch}\alpha - \operatorname{sh}\alpha}{\operatorname{ch}\alpha - \xi_{n-1}\operatorname{sh}\alpha}\right), \end{aligned} \tag{1}$$

is also a polynomial in $\xi_1, \cdots, \xi_{n-1}$ of the same degree.

The invariant subspace $\mathfrak{A}^{n-1,k}$ is a direct sum of subspaces $\mathfrak{H}^{n-1,j}$, $0 \le j \le k$. This follows from the decomposition of homogeneous polynomials, proved in §2.6 of Chapter IX, and from the fact than any polynomial of degree $k$ can be written as a sum of homogeneous polynomials of degree $j$, $0 \le j \le k$.

The representation $T^{nk}(g)$ is not completely reducible, since $\mathfrak{A}^{n-1,k}$ does not have an invariant complement. One can show that the restriction of $T^{nk}(g)$ to $\mathfrak{A}^{n-1,k}$ is irreducible. The representation induced by $T^{nk}(g)$ in the factor space $\mathfrak{D}/\mathfrak{A}^{n-1,k}$ is also irreducible.

The representation $T^{n\sigma}(g)$ is reducible in the case when $\sigma = -n-k+2$, where $k$ is a nonnegative integer. The invariant subspace in $\mathfrak{L}^2(S^{n-2})$, in this case, is provided by the orthogonal complement of $\mathfrak{A}^{n-1,k}$, i.e. by the space

$$\mathfrak{D}^{n-1,k} = \sum_{j=k+1}^{\infty} \mathfrak{H}^{n-1,j}. \tag{2}$$

Indeed, suppose that a function $\Phi(\xi')$ belongs to $\mathfrak{D}^{n-1,k}$, i.e. suppose that for any polynomial $F(\xi')$ of $\mathfrak{A}^{n-1,k}$ one has the equality

$$(\Phi, F) \equiv \int \Phi(\xi')\,\overline{F(\xi')}\,d\xi' = 0. \tag{3}$$

As remarked in subsection 2, the operators $T^{n,-n-k+2}(g)$ and $T^{nk}(g^{-1})$ are Hermitian-adjoint relative to the Hermitian functional $(\Phi, F)$. Therefore

$$(T^{n,-n-k+2}(g)\Phi, F) = (\Phi, T^{nk}(g^{-1})F). \tag{4}$$

But if $F(\xi') \in \mathfrak{A}^{n-1,k}$, then $T^{nk}(g^{-1})F(\xi') \in \mathfrak{A}^{n-1,k}$ as well, and so $(T^{n,-n-k+2}(g)\Phi, F) = 0$.

Hence, for all polynomials $F(\xi')$ of $\mathfrak{A}^{n-1,k}$, one has the equality

$$(T^{n,-n-k+2}(g)\Phi, F) = 0.$$

Consequently $T^{n,-n-k+2}(g)\Phi \in \mathfrak{D}^{n-1,k}$. The invariance of $\mathfrak{D}^{n-1,k}$ relative to $T^{n,-n-k+2}(g)$ is thus proved.

The representations induced by $T^{n,-n-k+2}$ in $\mathfrak{D}^{n-1,k}$ and $\mathfrak{D}/\mathfrak{D}^{n-1,k}$ are irreducible. The proof is similar to that in the case of nonintegral $\sigma$.

In the case $\sigma = -n-k+2$ the representation $T^{n\sigma}(g)$ is not completely reducible, since the subspace $\mathfrak{D}^{n-1,k}$ does not have an invariant complement.

5. **Conditions for $T^{n\sigma}(g)$ to be unitary.** Let us now ascertain for which values of $\sigma$ the representation $T^{n\sigma}(g)$ of $SH(n)$ is unitary, i.e. when does the space $\mathfrak{D}$ have a scalar product $(F_1, F_2)$, invariant relative to the operators of $T^{n\sigma}(g)$:

$$(F_1, F_2) = (T^{n\sigma}(g)F_1, T^{n\sigma}(g)F_2).$$

We shall first determine when an invariant Hermitian form can exist in $\mathfrak{D}$, and we shall then establish conditions on $\sigma$ under which this form is positive definite. We restrict $T^{n\sigma}(g)$ to $SO(n-1)$. By formula (1) of subsection 3 the representation $T^{n\sigma}(h)$, $h \in SO(n-1)$ is a direct sum of irreducible representations $T^{n-1,k}(h)$ of $SO(n-1)$, which are realized in the subspaces $\mathfrak{H}^{n-1,k}$. Moreover, we have

$$\mathfrak{L}^2(S^{n-2}) = \sum_{k=0}^{\infty} \mathfrak{H}^{n-1,k}.$$

Since an invariant scalar product in $\mathfrak{H}^{n-1,k}$ is defined by the formula

$$(F_1, F_2)_k = \int_{S^{n-2}} F_1(\xi')\overline{F_2(\xi')}\,d\xi', \quad F_1, F_2 \in \mathfrak{H}^{n-1,k},$$

the invariant scalar product $(F_1, F_2)$ must have the form

$$(F_1, F_2) = \sum_{k=0}^{\infty} \alpha_k \int_{S^{n-2}} F_1^{(k)}(\xi')\overline{F_2^{(k)}(\xi')}\,d\xi', \tag{1}$$

where $F_i = \sum F_i^{(k)}$, $i = 1, 2$, $F_i^{(k)} \in \mathfrak{H}^{n-1,k}$. The equation $(F_1, F_2) = \overline{(F_2, F_1)}$ implies that the coefficients $\alpha_k$ are real.

It remains for us to calculate the $\alpha_k$. From the invariance of the scalar product (1) it follows that any infinitesimal operator $A$ of the representation $T^{n\sigma}(g)$ satisfies the equality

$$(AF_1, F_2) = -(F_1, AF_2). \tag{2}$$

In particular, this equality must be satisfied by the infinitesimal operator

$$A = -\sigma \cos\varphi_{n-2} - \sin^2\varphi_{n-2}\frac{\partial}{\partial \cos\varphi_{n-2}},$$

corresponding to the one-parametric subgroup of hyperbolic rotations $g_{n-1}(\alpha)$ (see formula (3) of subsection 3).

Let us take

$$F_1 = C_k^{\frac{n-3}{2}}(\cos\varphi_{n-2})$$

and

$$F_2 = C_{k+1}^{\frac{n-3}{2}}(\cos\varphi_{n-2}).$$

We employ formula (4) of subsection 3 for calculating $AF_1$ and $AF_2$, and take into consideration the fact that $C_k^{(n-3)/2}(\cos\phi_{n-2})$ belongs to $\mathfrak{H}^{n-1,k}$. Substituting the values of $AF_1$ and $AF_2$ in (2) and taking (1) into account, we get

$$\frac{(k+1)(k-\sigma)\alpha_{k+1}}{2k+n-3}\int_0^\pi\left[C_{k+1}^{\frac{n-3}{2}}(\cos\varphi)\right]^2\sin^{n-3}\varphi\,d\varphi$$
$$=\frac{(n+k-3)(n+k+\bar{\sigma}-2)\alpha_k}{2k+n-1}\int_0^\pi\left[C_k^{\frac{n-3}{2}}(\cos\varphi)\right]^2\sin^{n-3}\varphi\,d\varphi.$$

On calculating the integrals in this equality (see § 3.4 of Chapter IX), we get the following recurrence relation for the $\alpha_k$:

$$(k-\sigma)\alpha_{k+1} = (n+k+\bar{\sigma}-2)\alpha_k. \tag{3}$$

As $\alpha_k$ is real, it follows that either $\alpha_{k+1} = \alpha_k$ or $\sigma$ is real. In the first case $k-\sigma = n+k+\bar{\sigma}-2$ and therefore

$$\sigma = -\frac{n-2}{2}+i\rho,$$

where $\rho$ is a real number. In the second case $\sigma = \bar{\sigma}$ and therefore

$$\alpha_{k+1} = \frac{n+k+\sigma-2}{k-\sigma}\alpha_k.$$

Putting $\alpha_0 = \Gamma(n+\sigma-2)/\Gamma(-\sigma)$ for simplicity, we find that

$$\alpha_k = \frac{\Gamma(n+k+\sigma-2)}{\Gamma(k-\sigma)}. \tag{4}$$

Thus we have proved that there are only two cases in which the space $\mathfrak{D}$ can have a Hermitian form $(F_1, F_2)$ invariant relative to the operators $T^{n\sigma}(g)$:

1) $\sigma$ has the form $\sigma = -(n-2)/2 + i\rho$, where $\rho$ is a real number. In this case

$$(F_1, F_2) = \sum_{k=0}^{\infty}\int_{S^{n-2}} F_1^{(k)}(\xi')\overline{F_2^{(k)}(\xi')}\,d\xi', \tag{5}$$

where $F_j = \sum_{k=0}^{\infty} F_j^{(k)}$, $F_j^{(k)} \in \mathfrak{H}^{n-1,k}$, $j = 1, 2$. Since the subspaces $\mathfrak{H}^{n-1,k}$ are pairwise orthogonal, equality (5) can be rewritten as follows:

$$(F_1, F_2) = \int F_1(\xi') \overline{F_2(\xi')}\, d\xi' \tag{5'}$$

2) $\sigma$ is real. In this case we have

$$(F_1, F_2) = \sum_{k=0}^{\infty} \frac{\Gamma(n+k+\sigma-2)}{\Gamma(k-\sigma)} \int_{S^{n-2}} F_1^{(k)}(\xi') \overline{F_2^{(k)}(\xi')}\, d\xi', \tag{6}$$

where $F_i^{(k)}(\xi')$ has the same meaning as above.

One can show that (6) may be written in the form

$$(F_1, F_2) = C_\sigma \int_{S^{n-2}} \int_{S^{n-2}} \frac{F_1(\xi') \overline{F_2(\eta')}}{(1-(\xi', \eta'))^{n-2-\sigma}}\, d\xi'\, d\eta', \tag{7}$$

where

$$(\xi', \eta') = \xi_1 \eta_1 + \ldots + \xi_{n-1} \eta_{n-1}$$

and

$$C_\sigma = \frac{\sqrt{\pi}\, \Gamma(\sigma)}{2^{\sigma-1} \Gamma\left(\frac{n-1}{2}\right) \Gamma\left(\sigma - \frac{n-2}{2}\right)}.$$

It is not difficult to show that the Hermitian forms (5′) and (6) are actually invariant relative to the representations $T^{n\sigma}(g)$, where $\sigma = -(n-2)/2 + i\rho$ for form (5′) and $\sigma = \sigma$ for form (6). We omit the details.

We shall now determine the values of $\sigma$ for which these invariant Hermitian forms are positive definite. For form (5′) the answer is evident; for any function $F$ we have $(F, F) \geq 0$. Thus, if $\sigma = -(n-2)/2 + i\rho$, where $\rho$ is real, then $T^{n\sigma}$ is unitary. These unitary representation are called *representations of the fundamental series of $SH(n)$*.

In the case of form (6) the inequality $(F, F) \geq 0$ holds for all functions $F$ if for all $k$ one has

$$\frac{\Gamma(n+k+\sigma-2)}{\Gamma(k-\sigma)} \geqslant 0. \tag{8}$$

It is evident that this occurs if $-n+2 < \sigma < 0$. Thus the representations $T^{n\sigma}(g)$ are unitary for real $\sigma$ if and only if $-n+2 < \sigma < 0$. These unitary representations are called *representations of the complementary series*.

Besides the fundamental and complementary series, there also exists a discrete series of unitary representations of $SH(n)$. This series corresponds to the values $\sigma = -n-k+2$, where $k$ is a nonnegative integer. We have shown in subsection 4 that in this case $\mathfrak{D}$ has a subspace invariant for the representation $T^{n\sigma}(g)$, namely

$$\mathfrak{D}^{n-1,k}=\sum_{j=k+1}^{\infty}\mathfrak{H}^{n-1,j}. \tag{9}$$

Formula (6) defines an invariant Hermitian form in $\mathfrak{D}^{n-1,k}$:

$$(F_1, F_2)=\sum_{j=k+1}^{\infty}\frac{\Gamma(j-k)}{\Gamma(j-\sigma)}\int_{S^{n-2}}F_1^{(j)}(\xi')\overline{F_2^{(j)}(\xi')}\,d\xi'. \tag{10}$$

This form is positive definite. Thus, for $\sigma=-n-k+2$ the representation $T^{n\sigma}(g)$ induces a unitary representation in $\mathfrak{D}^{n-1,k}$.

These unitary representations will be called *representations of the discrete series.*

For $\sigma=k$ we have a finite-dimensional invariant subspace

$$\mathfrak{A}^{n-1,k}=\sum_{j=0}^{k}\mathfrak{H}^{n-1,j}$$

in $\mathfrak{D}$. In this subspace there exists an invariant Hermitian form

$$\sum_{j=0}^{k}(-1)^j\Gamma(k-j+1)\Gamma(n+j+k-2)\int_{S^{n-2}}F_1^{(j)}(\xi')\overline{F_2^{(j)}(\xi')}\,d\xi'. \tag{11}$$

To prove this it is sufficient to use the recurrence relation (3) and put $\alpha_0=\Gamma(k+1)\Gamma(n+k-2)$. Form (11) is not positive definite. Thus $SH(n)$ does not have unitary finite-dimensional representations (except for the identity representation).

6. **Equivalence of the representations $T^{n\sigma}(g)$.** For certain values of $\sigma$ and $\tau$ the representations $T^{n\sigma}(g)$ and $T^{n\tau}(g)$ are equivalent. Repeating the arguments given in §2.6, Chapter VI, for $n=3$, we get the following result.

*If $\sigma$ is not an integer such that $\sigma\geq 0$ or $\sigma\leq -n+2$, then the representations $T^{n\sigma}(g)$ and $T^{n,-n-\sigma+2}(g)$ are equivalent to each other. If $\sigma=k\geq 0$ is an integer, then the representations induced by $T^{nk}(g)$ in $\mathfrak{A}^{n-1,k}$ and by $T^{n,-n-k-2}(g)$ in $\mathfrak{D}/\mathfrak{D}^{n-1,k}$ are equivalent. In addition, the representations induced by $T^{nk}(g)$ in $\mathfrak{D}/\mathfrak{A}^{n-1,k}$ and by $T^{n,-n-k+2}(g)$ in $\mathfrak{D}^{n-1,k}$ are equivalent.*

## §3. ZONAL AND ASSOCIATED SPHERICAL FUNCTIONS OF THE REPRESENTATIONS OF CLASS 1 OF GROUP $SH(n)$

In this section we shall study the matrix elements of the representations $T^{n\sigma}(g)$ of the group $SH(n)$, lying in the "zero column", i.e. corresponding to the function $F_0(\xi')\equiv 1$ invariant for the operators $T^{n\sigma}(h)$, $h\in SO(n-1)$. In particular, we shall study the zonal spherical function lying at the intersection of the zero column and the zero row. We shall see that these functions are expressed by means of the Legendre functions and associated Legendre functions studied in Chapter VI. With the help of this connection several new properties of Legendre functions, not revealed in Chapter VI, are established.

**1. Construction of a basis in $\mathfrak{D}$.** According to § 2.1 the representations $T^{n\sigma}(g)$ of $SH(n)$ are constructed in the space $\mathfrak{D}$ of infinitely differentiable functions on the sphere $S^{n-2}$. We introduce the scalar product

$$(F, F) = \int_{S^{n-2}} |F(\xi')|^2 d\xi' \tag{1}$$

in this space and complete it relative to the norm $\|F\|^2 = (F, F)$. We obtain the Hilbert space $\mathfrak{L}^2(S^{n-2})$. We shall assume that $T^{n\sigma}(g)$ is realized in this Hilbert space.

In this chapter we shall study the matrix elements of the representation $T^{n\sigma}(g)$. They are calculated in an orthonormal basis in the space $\mathfrak{L}^2(S^{n-2})$, which consists of functions of the form

$$\Xi_K(\xi') = A_K \prod_{j=0}^{n-4} C_{k_j - k_{j+1}}^{\frac{n-j-3}{2}} (\cos \varphi_{n-j-2}) \sin^{k_j+1} \varphi_{n-j-2} e^{\pm i k_{n-3} \varphi_1}, \tag{2}$$

where $\phi_1, \cdots, \phi_{n-2}$ are geographical coordinates on $S^{n-2}$,

$$K = (k_0, \ldots, \pm k_{n-3}),\ k_0 \geqslant k_1 \geqslant \ldots \geqslant k_{n-3} \geqslant 0$$

and

$$(A_K)^2 = \frac{1}{\Gamma\left(\frac{n-1}{2}\right)} \times \prod_{j=0}^{n-4} \frac{2^{2k_{j+1}+n-j-5} (k_j - k_{j+1})! (n-j+2k_j-3) \Gamma^2\left(\frac{n-j-3}{2} + k_{j+1}\right)}{\sqrt{\pi}\, \Gamma(k_j + k_{j+1} + n - j - 3)}. \tag{3}$$

This basis was considered by us in § 3.6 of Chapter IX.[1] It was shown there that in this basis the matrix of the quasi-regular representation of $SO(n-1)$ is block-diagonal; moreover, its principal diagonal contains the canonical matrices of irreducible unitary representations of the group.

One of the basis elements is the function $\Xi_O(\xi')$, $O = (0, \cdots, 0)$, identically

1) The notation has been slightly altered in comparison to Chapter IX, as regards the insertion of the number $l = k_0$ in the symbol $K$. Thus, for example, the basis element $\Xi_O(\xi')$ would be denoted by $\Xi_O^0(\xi') \equiv 1$ in Chapter IX.

equal to 1. Clearly this element is invariant for all rotations of $S^{n-2}$:

$$\Xi_O(h\xi') = \Xi_O(\xi'), \qquad h \in SO(n-1).$$

As shown in § 2.5 of Chapter I, it follows that the matrix elements $t^{n\sigma}_{KO}(g)$ satisfy the functional equation

$$t^{n\sigma}_{KO}(gh) = t^{n\sigma}_{KO}(g), \qquad h \in SO(n-1), \tag{4}$$

while the matrix elements $t^{n\sigma}_{OK}(g)$ satisfy the equation

$$t^{n\sigma}_{OK}(hg) = t^{n\sigma}_{OK}(g), \qquad h \in SO(n-1). \tag{5}$$

In particular, the matrix element $t^{n\sigma}_{OO}(g)$, i.e. the zonal spherical function, satisfies the functional equation

$$t^{n\sigma}_{OO}(h_1 g h_2) = t^{n\sigma}_{OO}(g), \qquad h_1, h_2 \in SO(n-1). \tag{6}$$

Since any element $g$ of $SH(n)$ can be written in the form $g = h_1 g_{n-1}(\theta^{n-1}_{n-1}) h_2$, where $h_1, h_2 \in SO(n-1)$ and $\theta^{n-1}_{n-1}$ is an Euler angle of $g$, we have

$$t^{n\sigma}_{OO}(g) = t^{n\sigma}_{OO}[g_{n-1}(\theta^{n-1}_{n-1})]. \tag{7}$$

Thus the matrix element $t^{n\sigma}_{OO}(g)$ depends only on the Euler angle $\theta^{n-1}_{n-1}$ of $g$.

**2. Integral representation of zonal and associated spherical functions.** In order to derive an integral representation of spherical functions, we use the formula

$$t^{n\sigma}_{KO}(g) = (T^{n\sigma}(g)\,\Xi_O, \Xi_K), \tag{1}$$

where $(F_1, F_2)$ is the scalar product in the space $\mathfrak{L}^2(S^{n-2})$:

$$(F_1, F_2) = \int_{S^{n-2}} F_1(\xi')\overline{F_2(\xi')}\, d\xi'. \tag{2}$$

Here $\Xi_O(\eta')$ denotes a function on $S^{n-2}$, identically equal to 1. We shall use the same symbol to denote a homogeneous function of degree $\sigma$ on the cone $[\xi, \xi] = 0$, equal to unity for $\xi_n = 1$. In other words, we shall take $\Xi_O(\xi) = \xi_n^\sigma$.

Let $\mathbf{x}_0$ denote the point with coordinates $(0, 0, \cdots, 1)$. Then $\Xi_O(\xi)$ can be written in the form

$$\Xi_O(\xi) = \xi_n^\sigma = [\mathbf{x}_0, \xi]^\sigma. \tag{3}$$

The operator $T^{n\sigma}(g)$ transforms $\Xi_O(\xi)$ into the function

$$T^{n\sigma}(g)\,\Xi_O(\xi) = \Xi_O(g^{-1}\xi) = [\mathbf{x}_0, g^{-1}\xi]^\sigma.$$

But the expression $[\mathbf{x}, \mathbf{y}]$ is invariant for shifts belonging to $SH(n)$:

$$[\mathbf{x}, \mathbf{y}] = [g\mathbf{x}, g\mathbf{y}],$$

and so

$$T^{n\sigma}(g)\,\Xi_O(\xi) = [g\mathbf{x}_0, \xi]^\sigma. \tag{4}$$

From (1), (2) and (4) it follows that

$$t_{KO}^{n\sigma}(g)=\int\limits_{S^{n-2}}[g\mathbf{x}_0,\ \boldsymbol{\xi}]^{\sigma}\,\overline{\Xi_K(\boldsymbol{\xi}')}\,d\boldsymbol{\xi}', \tag{5}$$

where $\xi=(\xi_1,\cdots,\xi_{n-1},1)$, $\xi'=(\xi_1,\cdots,\xi_{n-1})$.

In particular,

$$t_{OO}^{n\sigma}(g)=\int\limits_{S^{n-2}}[g\mathbf{x}_0,\ \boldsymbol{\xi}]^{\sigma}d\boldsymbol{\xi}. \tag{6}$$

Since the zonal spherical function depends only on the Euler angle $\theta_{n-1}^{n-1}$, it is sufficient to calculate it for the case when $g$ is a hyperbolic rotation by the angle $\theta$ in the $(x_{n-1},x_n)$-plane. This rotation carries $\mathbf{x}_0(0,\cdots,0,1)$ into $\mathbf{x}(0,\cdots,\operatorname{sh}\theta,\operatorname{ch}\theta)$. Therefore

$$[g\mathbf{x}_0,\ \boldsymbol{\xi}]=\operatorname{ch}\theta-\xi_{n-1}\operatorname{sh}\theta$$

and

$$t_{OO}^{n\sigma}(g_{n-1}(\theta))=\int\limits_{S^{n-2}}(\operatorname{ch}\theta-\xi_{n-1}\operatorname{sh}\theta)^{\sigma}\,d\boldsymbol{\xi}'.$$

Going over to spherical coordinates, we get

$$t_{OO}^{n\sigma}[g_{n-1}(\theta)]=\frac{\Gamma\left(\frac{n-1}{2}\right)}{\sqrt{\pi}\,\Gamma\left(\frac{n-2}{2}\right)}\int\limits_0^{\pi}(\operatorname{ch}\theta-\cos\varphi\operatorname{sh}\theta)^{\sigma}\sin^{n-3}\varphi\,d\varphi. \tag{7}$$

By the same token, we have obtained an integral representation of the zonal spherical function. Making the substitution

$$\cos\psi=\frac{\operatorname{ch}\theta\cos\varphi-\operatorname{sh}\theta}{\operatorname{ch}\theta-\cos\varphi\operatorname{sh}\theta},$$

we get

$$t_{OO}^{n\sigma}[g_{n-1}(\theta)]=\frac{\Gamma\left(\frac{n-1}{2}\right)}{\sqrt{\pi}\,\Gamma\left(\frac{n-2}{2}\right)}\int\limits_0^{\pi}(\operatorname{ch}\theta-\cos\varphi\operatorname{sh}\theta)^{-n-\sigma+2}\sin^{n-3}\varphi\,d\varphi. \tag{8}$$

The same result could be obtained by observing that the representations $T^{n\sigma}(g)$ and $T^{n,-n-\sigma+2}(g)$ are equivalent to each other.

3. **Expression of a zonal function by means of the hypergeometric function.** Let us take $\operatorname{ch}^{\sigma}\theta$ out of the integral sign in formula (7) above, expand $(1-\cos\phi\operatorname{th}\theta)^{\sigma}$ by the binomial theorem and integrate term by term. We obtain

$$t_{00}^{n\sigma}[g_{n-1}(\theta)] = \frac{\Gamma\left(\frac{n-1}{2}\right)\Gamma(\sigma+1)}{\sqrt{\pi}\,\Gamma\left(\frac{n-2}{2}\right)} \operatorname{ch}^\sigma\theta \sum_{k=0}^{\infty} \frac{(-1)^k \operatorname{th}^k\theta}{\Gamma(\sigma-k+1)\Gamma(k+1)} \int_0^\pi \cos^k\varphi \sin^{n-3}\varphi\, d\varphi.$$

Since

$$\int_0^\pi \cos^{2k}\varphi \sin^{n-3}\varphi\, d\varphi = \frac{\Gamma\left(k+\frac{1}{2}\right)\Gamma\left(\frac{n-2}{2}\right)}{\Gamma\left(k+\frac{n-1}{2}\right)}$$

and

$$\int_0^\pi \cos^{2k+1}\varphi \sin^{n-3}\varphi\, d\varphi = 0,$$

we get

$$t_{00}^{n\sigma}[g_{n-1}(\theta)] = \frac{\Gamma\left(\frac{n-1}{2}\right)\Gamma(\sigma+1)}{\sqrt{\pi}} \operatorname{ch}^\sigma\theta \sum_{k=0}^{\infty} \frac{\Gamma\left(k+\frac{1}{2}\right)\operatorname{th}^{2k}\theta}{\Gamma(\sigma-2k+1)\Gamma(2k+1)\Gamma\left(k+\frac{n-1}{2}\right)}. \tag{1}$$

But

$$\Gamma(2k+1) = \frac{2^{2k}}{\sqrt{\pi}}\Gamma\left(k+\frac{1}{2}\right)\Gamma(k+1),$$

$$\Gamma(\sigma+1) = -\frac{\pi}{\sin\sigma\pi\,\Gamma(-\sigma)}$$

and

$$\Gamma(\sigma-2k+1) = -\frac{2^{\sigma+1-2k}\pi\sqrt{\pi}}{\sin\sigma\pi\,\Gamma\left(k-\frac{\sigma}{2}\right)\Gamma\left(k-\frac{\sigma-1}{2}\right)}$$

(see §§ 1.6 and 1.7 of Chapter V). Therefore

$$t_{00}^{n\sigma}[g_{n-1}(\theta)] = \frac{\Gamma\left(\frac{n-1}{2}\right)\operatorname{ch}^\sigma\theta}{2^{\sigma+1}\sqrt{\pi}\,\Gamma(-\sigma)} \sum_{k=0}^{\infty} \frac{\Gamma\left(k-\frac{\sigma}{2}\right)\Gamma\left(k-\frac{\sigma-1}{2}\right)}{\Gamma\left(k+\frac{n-1}{2}\right)\Gamma(k+1)} \operatorname{th}^{2k}\theta. \tag{2}$$

Hence, by virtue of formula (4), § 1.7, Chapter V, and formula (2), § 1.1, Chapter VII, we get

$$t_{00}^{n\sigma}[g_{n-1}(\theta)] = \operatorname{ch}^\sigma\theta\, F\left(-\frac{\sigma}{2}, -\frac{\sigma-1}{2}; \frac{n-1}{2}; \operatorname{th}^2\theta\right). \tag{3}$$

We have thus obtained an expression for $t_{00}^{n\sigma}[g_{n-1}(\theta)]$ in terms of the hypergeometric function. Let us compare this expression with the expression for $\mathfrak{P}_l^m(z)$ in terms of the hypergeometric function (see § 1.4, Chapter VII). This gives

$$t_{OO}^{n\sigma}[g_{n-1}(\theta)]=\frac{2^{\frac{n-3}{2}}\Gamma\left(\frac{n-1}{2}\right)}{\operatorname{sh}^{\frac{n-3}{2}}\theta}\mathfrak{P}^{\frac{3-n}{2}}_{\sigma+\frac{n-3}{2}}(\operatorname{ch}\theta). \qquad (4)$$

4. **Calculation of associated spherical functions.** We now pass on to the calculation of associated spherical functions. We first consider the case when $g$ is a hyperbolic rotation by the angle $\theta$ in the $(x_{n-1}, x_n)$-plane. As we have seen above, in this case we have $[g\mathbf{x}_0, \xi] = \operatorname{ch}\theta - \xi_{n-1}\operatorname{sh}\theta$. Therefore by formula (1) of subsection 2 we get

$$t_{KO}^{n\sigma}(g_{n-1}(\theta))=\int_{S^{n-2}}(\operatorname{ch}\theta-\xi_{n-1}\operatorname{sh}\theta)^{\sigma}\,\overline{\Xi_K(\xi')}\,d\xi'. \qquad (1)$$

We go over to spherical coordinates and substitute expression (2) of subsection 1 for $\Xi_K(\xi')$.

From the orthogonality relations for Gegenbauer polynomials we find that $t_{KO}^{n\sigma}(g_{n-1}(\theta))$ is nonzero only if $k_1 = k_2 = \cdots = k_{n-3} = 0$.[1)] However, if $K = (k, 0, \cdots, 0)$ we get

$$t_{KO}^{n\sigma}(g_{n-1}(\theta))=\frac{\Gamma\left(\frac{n-3}{2}\right)}{2\sqrt{\pi}\,\Gamma\left(\frac{n-2}{2}\right)}\sqrt{\frac{k!\,(n+2k-3)\,\Gamma(n-2)}{\Gamma(n+k-3)}}$$

$$\times\int_0^{\pi}(\operatorname{ch}\theta-\cos\varphi\operatorname{sh}\theta)^{\sigma}C_k^{\frac{n-3}{2}}(\cos\varphi)\sin^{n-3}\varphi\,d\varphi$$

$$=\frac{\Gamma\left(\frac{n-3}{2}\right)}{2\sqrt{\pi}\,\Gamma\left(\frac{n-2}{2}\right)}\sqrt{\frac{k!\,(n+2k-3)\,\Gamma(n-2)}{\Gamma(n+k-3)}}$$

$$\times\int_{-1}^{1}(\operatorname{ch}\theta-x\operatorname{sh}\theta)^{\sigma}C_k^{\frac{n-3}{2}}(x)(1-x^2)^{\frac{n-4}{2}}dx. \qquad (2)$$

But, by formula (7), § 4.8, Chapter IX, we have

$$(1-x^2)^{\frac{n-4}{2}}C_k^{\frac{n-3}{2}}(x)=\frac{(-1)^k\,\Gamma(n+k-3)\,\Gamma\left(\frac{n-2}{2}\right)}{2^k\Gamma(n-3)\,\Gamma\left(\frac{n-2}{2}+k\right)k!}\frac{d^k}{dx^k}\left[(1-x^2)^{k+\frac{n-4}{2}}\right].$$

1) This also follows from Schur's lemma on taking into consideration the commutability of $T^{n\sigma}[g_{n-1}(\theta)]$ with all the operators $T^{n\sigma}(h)$, $h \in SO(n-2)$, and the form of the matrix $T^{n\sigma}(h)$.

We substitute this expression in (2) and integrate by parts $k$ times. We obtain

$$t_{KO}^{n\sigma}[g_{n-1}(\theta)] = \frac{(-1)^k \Gamma\left(\frac{n-3}{2}\right)\Gamma(\sigma+1)\operatorname{sh}^k\theta}{2^{k+1}\sqrt{\pi}\,\Gamma(n-3)\Gamma\left(\frac{n-2}{2}+k\right)\Gamma(\sigma-k+1)} \times \sqrt{\frac{(n+2k-3)\Gamma(n-2)\Gamma(n+k-3)}{k!}} \times \int_{-1}^{1}(\operatorname{ch}\theta - x\operatorname{sh}\theta)^{\sigma-k}(1-x^2)^{k+\frac{n-4}{2}}\,dx. \tag{3}$$

But, by formula (7) of subsection 2,

$$\int_{-1}^{1}(\operatorname{ch}\theta - x\operatorname{sh}\theta)^{\sigma-k}(1-x^2)^{k+\frac{n-4}{2}}\,dx = \frac{\sqrt{\pi}\,\Gamma\left(\frac{n-2}{2}+k\right)}{\Gamma\left(\frac{n-1}{2}+k\right)} t_{OO}^{n+2k,\sigma-k}[g_{n+2k-1}(\theta)],$$

and so, for $K = (k, 0, \cdots, 0)$, we get

$$t_{KO}^{n\sigma}[g_{n-1}(\theta)] = \frac{(-1)^k\Gamma\left(\frac{n-3}{2}\right)\Gamma(\sigma+1)}{2^{k+1}\Gamma(n-3)\Gamma\left(\frac{n-1}{2}+k\right)\Gamma(\sigma-k+1)} \times \sqrt{\frac{(n+2k-3)\Gamma(n-2)\Gamma(n+k-3)}{k!}}\operatorname{sh}^k\theta\, t_{OO}^{n+2k,\sigma-k}[g_{n+2k-1}(\theta)]. \tag{4}$$

We already know the expression for $t_{OO}^{n+2k,\sigma-k}[g_{n+2k-1}(\theta)]$ in terms of associated Legendre functions (see formula (4) of subsection 3). Therefore we have

$$t_{KO}^{n\sigma}[g_{n-1}(\theta)] = \frac{(-1)^k 2^{\frac{n-5}{2}}\Gamma\left(\frac{n-3}{2}\right)\Gamma(\sigma+1)}{\Gamma(n-3)\Gamma(\sigma-k+1)} \times \sqrt{\frac{(n+2k-3)\Gamma(n-2)\Gamma(n+k-3)}{k!}}\operatorname{sh}^{\frac{3-n}{2}}\theta\,\mathfrak{P}^{\frac{3-n}{2}-k}_{\sigma+\frac{n-3}{2}}(\operatorname{ch}\theta). \tag{5}$$

Let us now find an expression for matrix elements of the form $t_{OK}^{n\sigma}[g_{n-1}(\theta)]$. It was shown in §2.2 that the representations $T^{n\sigma}(g)$ and $T^{n,-n-\bar{\sigma}+2}(g)$ of $SH(n)$ are Hermitian-adjoint to each other. Hence

$$t_{OK}^{n\sigma}(g) = (T^{n\sigma}(g)\Xi_K, \Xi_O) = (\Xi_K, T^{n,-n-\bar{\sigma}+2}(g^{-1})\Xi_O) = (T^{n,-n-\sigma+2}(g^{-1})\Xi_O, \Xi_{\bar{K}}) = t_{\bar{K}O}^{n,-n-\sigma+2}(g^{-1}), \tag{6}$$

where $\bar{K} = (k, k_1, \cdots, \pm k_{n-3})$.

It follows from (2) that

$$t_{\bar{K}O}^{n,-n-\sigma+2}[g_{n-1}(-\theta)] = (-1)^k t_{\bar{K}O}^{n,-n-\sigma+2}[g_{n-1}(\theta)].$$

Thus

$$t^{n\sigma}_{OK}(g_{n-1}(\theta)) = (-1)^k t^{n,\,-n-\sigma-2}_{KO}(g_{n-1}(\theta)). \qquad (7)$$

We shall now calculate the associated spherical functions $t^{n\sigma}_{KO}(g)$ for an arbitrary element $g$ of $SH(n)$. Let the Euler angles of this element be equal to $\theta^j_i$. From equality (4) of subsection 1 it follows that $t^{n\sigma}_{KO}(g) = t^{n\sigma}_{KO}(g')$, where

$$g' = g_1(\theta^{n-1}_1)\ldots g_{n-1}(\theta^{n-1}_{n-1}).$$

Let

$$h' = g_1(\theta^{n-1}_1)\ldots g_{n-2}(\theta^{n-1}_{n-2}).$$

Then $g' = h' g_{n-1}(\theta^{n-1}_{n-1})$, and since

$$T^{n\sigma}(g') = T^{n\sigma}(h')\, T^{n\sigma}[g_{n-1}(\theta^{n-1}_{n-1})],$$

we have

$$t^{n\sigma}_{KO}(g') = \sum_M t^{n\sigma}_{KM}(h')\, t^{n\sigma}_{MO}[g_{n-1}(\theta^{n-1}_{n-1})]. \qquad (8)$$

The elements $t^{n\sigma}_{MO}[g_{n-1}(\theta^{n-1}_{n-1})]$ have been calculated above and have been shown to be nonzero only if $M = (m, 0, \cdots, 0)$. It remains for us to find $t^{n\sigma}_{KM}(h')$. But, as shown in § 2.1, the restriction of $T^{n\sigma}(g)$ to $SO(n-1)$ is a quasi-regular representation of this subgroup.

In the basis $\Xi_K(\xi')$ the matrix of a quasi-regular representation of $SO(n-1)$ is block-diagonal; moreover, the principal diagonal of this matrix contains the canonical matrices of the irreducible unitary representations $T^{n-1,k}(h)$ of $SO(n-1)$. Therefore the elements $t^{n\sigma}_{KM}(h')$, where $M = (m, 0, \cdots, 0)$ and $K = (k, k_1, \cdots, \pm k_{n-3})$, are equal to zero if $m \neq k$. But if $m = k$, then $t^{n\sigma}_{KM}(h') = t^{n-1,k}_{K'O'}(h')$, where $K' = (k_1, \cdots, \pm k_{n-3})$ and $O' = (0, \cdots, 0)$. Substituting this value in formula (8), we get

$$t^{n\sigma}_{KO}(g') = t^{n-1,\,k}_{K'O'}(h')\, t^{n\sigma}_{MO}[g^{n-1}(\theta^{n-1}_{n-1})],$$

where $M = (k, 0, \cdots, 0)$.

We replace the factors here by their expressions in terms of the associated Legendre functions and Gegenbauer polynomials (see formula (5) and formula (7) of § 4.1, Chapter IX). We thus obtain

$$t^{n\sigma}_{KO}(g) = t^{n\sigma}_{KO}(g') = B^{n\sigma}_{KO}\, \operatorname{sh}^{\frac{3-n}{2}} \theta^{n-1}_{n-1}\, \mathfrak{P}^{\frac{3-n}{2}-k_0}_{\sigma+\frac{n-3}{2}}(\operatorname{ch}\theta^{n-1}_{n-1})$$

$$\times \prod_{j=0}^{n-4} C^{\frac{n-j-3}{2}+k_{j+1}}_{k_j-k_{j+1}}(\cos\theta^{n-1}_{n-j-2})\, \sin^{k_{j+1}}\theta^{n-1}_{n-j-2}\, e^{\pm i k_{n-3}\theta^{n-1}_1}. \qquad (9)$$

where $k_0 = k$ and

$$B^{n\sigma}_{KO} = \frac{(-1)^{k_0} 2^{\frac{n-3}{2}} \Gamma(\sigma-1)}{\Gamma(\sigma-k_0+1)}$$

$$\times \left[\Gamma\left(\frac{n-1}{2}\right) \prod_{j=0}^{n-4} \frac{2^{2k_{j+1}+n-j-5}(k_j-k_{j+1})!}{\sqrt{\pi}\,\Gamma(k_j+k_{j+1}+n-j-3)}\right.$$

$$\left.\times (n-j+2k_j-3)\,\Gamma\left(\frac{n-j-3}{2}+k_{j+1}\right)\right]^{1/2}. \qquad (10)$$

5. **A composition theorem for Legendre functions.** Utilizing the connection between Legendre functions and elements of the zero column of the matrix $T^{n\sigma}(g)$, we derive a new composition theorem for Legendre functions.

Let us consider the element

$$g = g_{n-1}(\theta_1)\, g_{n-2}(\varphi_1)\, g_{n-1}(\theta_2) \qquad (1)$$

of $SH(n)$, where, in our usual notation, $g_{n-1}(\theta_1)$ and $g_{n-1}(\theta_2)$ are hyperbolic rotations in the $(x_n, x_{n-1})$-plane, while $g_{n-2}(\phi_1)$ is a rotation in the $(x_{n-1}, x_{n-2})$-plane. The rotation $g$ carries the point $M$ with coordinates $(0, 0, \cdots, 1)$ into a point $M'(x_1, \cdots, x_n)$, where

$$x_n = \operatorname{ch}\theta_1 \operatorname{ch}\theta_2 + \operatorname{sh}\theta_1 \operatorname{sh}\theta_2 \cos\varphi_1.$$

From the results of §1.4 it follows that the angle $\theta^{n-1}_{n-1} = \theta$ of the rotation $g$ is defined by the formula

$$\operatorname{ch}\theta = \operatorname{ch}\theta_1 \operatorname{ch}\theta_2 + \operatorname{sh}\theta_1 \operatorname{sh}\theta_2 \cos\varphi_1. \qquad (2)$$

It follows from (1) that

$$t^{n\sigma}_{OO}(g) = \sum_{K,M} t^{n\sigma}_{OK}[g_{n-1}(\theta_1)]\, t^{n\sigma}_{KM}[g_{n-2}(\varphi)]\, t^{n\sigma}_{MO}\, g_{n-1}(\theta_2)$$

$$= \sum_{K,M} (-1)^m t^{n\sigma}_{OK}[g_{n-1}(\theta_1)]\, t^{n\sigma}_{KM}[g_{n-2}(\varphi)]\, t^{n,-n-\sigma+2}_{O\overline{M}}[g_{n-1}(\theta_2)], \qquad (3)$$

where $\overline{M} = (m_0, m_1, \cdots, \mp m_{n-3})$.

As shown above, the elements $t^{n\sigma}_{OK}[g_{n-1}(\theta_1)]$ and $t^{n\sigma}_{O\overline{M}}[g_{n-1}(\theta_2)]$ are nonzero only if $K = (k, 0, \cdots, 0)$ and $M \equiv \overline{M} = (m, 0, \cdots, 0)$, while the element $t^{n\sigma}_{KM}[g_{n-2}(\phi)]$ is nonzero only if $k = m$. Therefore (3) takes the form

$$t^{n\sigma}_{OO}(g) = \sum_{k=0}^{\infty} (-1)^k t^{n\sigma}_{OK}[g_{n-1}(\theta_1)]$$

$$\times\, t^{n\sigma}_{KK}[g_{n-2}(\varphi_1)]\, t^{n,-n-\sigma+2}_{OK}[g_{n-1}(\theta_2)], \qquad (4)$$

where $K = (k, 0, \cdots, 0)$.

We replace the matrix elements in (4) by their expressions in terms of associated Legendre functions and Gegenbauer polynomials. Since

$$t^{n\sigma}_{KK}[g_{n-2}(\varphi_1)] = t^{n-1,\,k}_{O'O'}[g_{n-2}(\varphi_1)],$$

we get

$$\operatorname{sh}^{\frac{3-n}{2}}\theta\,\mathfrak{P}^{\frac{3-n}{2}}_{\sigma+\frac{n-3}{2}}(\operatorname{ch}\theta)$$

$$= 2^{\frac{n-5}{2}}\Gamma\left(\frac{n-3}{2}\right)\Gamma(\sigma+1)\Gamma(-n-\sigma+3)\sum_{k=0}^{\infty}(-1)^k\frac{(n+2k-3)}{\Gamma(\sigma-k+1)}$$

$$\times\Gamma(-n-\sigma-k+3)\operatorname{sh}^{\frac{3-n}{2}}\theta_1\operatorname{sh}^{\frac{3-n}{2}}\theta_2\,\mathfrak{P}^{\frac{3-n}{2}-k}_{\sigma+\frac{n-3}{2}}(\operatorname{ch}\theta_1)$$

$$\times\mathfrak{P}^{\frac{3-n}{2}-k}_{\sigma+\frac{n-3}{2}}(\operatorname{ch}\theta_2)\,C_k^{\frac{n-3}{2}}(\cos\varphi_1), \tag{5}$$

where

$$\operatorname{ch}\theta = \operatorname{ch}\theta_1\operatorname{ch}\theta_2 + \operatorname{sh}\theta_1\operatorname{sh}\theta_2\cos\varphi_1.$$

This formula is analogous to the composition formula for Gegenbauer polynomials proved in § 4.2 of Chapter IX.

**6. A multiplication theorem for Legendre functions.** Let us multiply both sides of equality (5) above by the function $\sin^{n-3}\phi_1 C_2^{(n-3)/2}(\cos\phi_1)$ and integrate with respect to $\phi_1$ from 0 to $\pi$. Taking into consideration the orthogonality relations for Gegenbauer functions, we obtain

$$\int_0^{\pi}\operatorname{sh}^{\frac{3-n}{2}}\theta\,\mathfrak{P}^{\frac{3-n}{2}}_{\sigma+\frac{n-3}{2}}(\operatorname{ch}\theta)\,C_m^{\frac{n-3}{2}}(\cos\varphi_1)\sin^{n-3}\varphi_1\,d\varphi_1$$

$$= \frac{(-1)^m\pi\,2^{\frac{5-n}{2}}\Gamma(\sigma+1)\Gamma(-\sigma-n+3)\Gamma(n+m-3)}{m!\,\Gamma(\sigma-m+1)\Gamma(-n-m-\sigma+3)\Gamma\left(\frac{n-3}{2}\right)}$$

$$\times\operatorname{sh}^{\frac{3-n}{2}}\theta_1\operatorname{sh}^{\frac{3-n}{2}}\theta_2\,\mathfrak{P}^{\frac{3-n}{2}-m}_{\sigma+\frac{n-3}{2}}(\operatorname{ch}\theta_1)\,\mathfrak{P}^{\frac{3-n}{2}-m}_{\sigma+\frac{n-3}{2}}(\operatorname{ch}\theta_2), \tag{1}$$

where

$$\operatorname{ch}\theta = \operatorname{ch}\theta_1\operatorname{ch}\theta_2 + \operatorname{sh}\theta_1\operatorname{sh}\theta_2\cos\varphi. \tag{2}$$

On replacing the variable $\phi_1$ by $\theta$, formula (1) takes the form

$$\int_{|\theta_1-\theta_2|}^{\theta_1+\theta_2} \operatorname{sh}^{\frac{5-n}{2}}\theta\, \mathfrak{P}^{\frac{3-n}{2}}_{\sigma+\frac{n-3}{2}}(\operatorname{ch}\theta)\, C_m^{\frac{n-3}{2}}\left(\frac{\operatorname{ch}\theta-\operatorname{ch}\theta_1\operatorname{ch}\theta_2}{\operatorname{sh}\theta_1\operatorname{sh}\theta_2}\right)$$

$$\times\{[\operatorname{ch}(\theta_1+\theta_2)-\operatorname{ch}\theta][\operatorname{ch}\theta-\operatorname{ch}(\theta_1-\theta_2)]\}^{\frac{n-4}{2}}\,d\theta$$

$$=\frac{(-1)^m\pi 2^{\frac{5-n}{2}}\Gamma(\sigma+1)\Gamma(-\sigma-n+3)\Gamma(n+m-3)}{m!\,\Gamma(\sigma-m+1)\Gamma(-n-m-\sigma+3)\Gamma\left(\frac{n-3}{2}\right)}$$

$$\times \operatorname{sh}^{\frac{n-3}{2}}\theta_1 \operatorname{sh}^{\frac{n-3}{2}}\theta_2\, \mathfrak{P}^{\frac{3-n}{2}-m}_{\sigma+\frac{n-3}{2}}(\operatorname{ch}\theta_1)\, \mathfrak{P}^{\frac{3-n}{2}-m}_{\sigma+\frac{n-3}{2}}(\operatorname{ch}\theta_2).$$

7. **A generating function for associated Legendre functions.** From formulas (2) and (5) of subsection 4 it follows that

$$\int_0^\pi(\operatorname{ch}\theta-\cos\varphi\operatorname{sh}\theta)^\sigma C_k^{\frac{n-3}{2}}(\cos\varphi)\sin^{n-3}\varphi\,d\varphi$$

$$=\frac{(-1)^k 2^{\frac{n-3}{2}}\sqrt{\pi}\,\Gamma\left(\frac{n-2}{2}\right)\Gamma(\sigma+1)\Gamma(n+k-3)}{k!\,\Gamma(n-3)\Gamma(\sigma-k+1)}$$

$$\times\operatorname{sh}^{\frac{3-n}{2}}\theta\,\mathfrak{P}^{\frac{3-n}{2}-k}_{\sigma+\frac{n-3}{2}}(\operatorname{ch}\theta). \tag{1}$$

Hence, owing to the orthogonality relations for Gegenbauer polynomials, we get

$$(\operatorname{ch}\theta-\cos\varphi\operatorname{sh}\theta)^\sigma=2^{\frac{n-5}{2}}\Gamma(\sigma+1)\Gamma\left(\frac{n-3}{2}\right)\operatorname{sh}^{\frac{3-n}{2}}\theta$$

$$\times\sum_{k=0}^\infty\frac{(-1)^k(n+2k-3)}{\Gamma(\sigma-k+1)}\mathfrak{P}^{\frac{3-n}{2}-k}_{\sigma+\frac{n-3}{2}}(\operatorname{ch}\theta)\,C_k^{\frac{n-3}{2}}(\cos\varphi). \tag{2}$$

Equality (2) may be regarded as a generating function for associated Legendre functions. With its help one can easily establish a number of new properties of these functions.

Let us put $\phi = 0$ or $\pi$ in equality (2). We get

$$e^{\pm\sigma\theta}=\frac{2^{\frac{n-5}{2}}\Gamma(\sigma+1)\Gamma\left(\frac{n-3}{2}\right)}{\Gamma(n-3)}\operatorname{sh}^{\frac{3-n}{2}}\theta$$

$$\times\sum_{k=0}^\infty\frac{(\pm1)^k(n+2k-3)\Gamma(n+k-3)}{k!\,\Gamma(\sigma-k+1)}\mathfrak{P}^{\frac{3-n}{2}-k}_{\sigma+\frac{n-3}{2}}(\operatorname{ch}\theta). \tag{3}$$

Formula (2) also yields

$$\Gamma(2p)\,\mathrm{sh}^p\theta\,\mathfrak{P}^{-p}_{\sigma+\tau+p}(\mathrm{ch}\,\theta) = 2^p\,\Gamma(p)\,\Gamma(\sigma+1)\,\Gamma(\tau+1) \times \sum_{k=0}^{\infty} \frac{(p+k)\,\Gamma(2p+k)}{k!\,\Gamma(\sigma-k+1)\,\Gamma(\tau-k+1)}\,\mathfrak{P}^{-p-k}_{\sigma+p}(\mathrm{ch}\,\theta)\,\mathfrak{P}^{-p-k}_{\tau+p}(\mathrm{ch}\,\theta). \quad (4)$$

## § 4. DECOMPOSITION OF REPRESENTATIONS OF THE GROUP $SH(n)$ AND THE MEHLER-FOCK TRANSFORMATION

**1. Introductory remarks.** In this section we shall decompose the quasi-regular representation of $SH(n)$. This representation is analogous to the quasi-regular representation of the group $SO(n)$ of rotations of $n$-dimensional euclidean space and is constructed in the space of functions on a sphere in pseudo-euclidean space. However, since pseudo-euclidean space has three types of spheres (of real, purely imaginary and zero curvature), there exist three types of quasi-regular representations of $SH(n)$. The first of these is realized in the space of functions on the upper half of the hyperboloid $[\mathbf{x}, \mathbf{x}] = 1$, the second is realized in the space of functions on the hyperboloid $[\mathbf{x}, \mathbf{x}] = -1$ and the third is realized in the space of functions of the upper half of the cone $[\xi, \xi] = 0$.

We shall restrict ourselves to the the representation connected with the two-sheeted hyperboloid $[\mathbf{x}, \mathbf{x}] = 1$, since the case of the hyperboloid of one sheet has not been fully studied so far. Decomposition of the representation connected with the hyperboloid $[\mathbf{x}, \mathbf{x}] = 1$ is carried out according to a general scheme, found by I. M. Gel'fand and M. I. Graev. The scheme consists of the following: first, with each function $f(\mathbf{x})$ on the hyperboloid $[\mathbf{x}, \mathbf{x}] = 1$ one associates a function $h(\xi)$ on the cone $[\xi, \xi] = 0$. The function $h(\xi)$ is decomposed into homogeneous components. To the quasi-regular representation correspond irreducible representations in the spaces of these homogeneous components.

The integral transformation which carries the function $f(\mathbf{x})$ on the hyperboloid $[\mathbf{x}, \mathbf{x}] = 1$ into the function $h(\xi)$ on the cone $[\xi, \xi] = 0$ is accomplished by integrating over certain manifolds on the hyperboloid, i.e. over the horispheres of Lobačevskiĭ space. The theory of these horispheres is explained, in the form we need, in [15], §1.4, Chapter V. There the following assertions are proved.

Suppose that Lobačevskiĭ space is realized on the upper half of the hyperboloid $[\mathbf{x}, \mathbf{x}] = 1$. Then its horispheres are sections of the hyperboloid by planes of the form $[\mathbf{x}, \xi] = 1$, where $\xi$ is a point of the upper half of the cone $[\xi, \xi] = 0$, $\xi_n > 0$. Thus there exists a one-to-one correspondance between the set of horispheres of Lobačevskiĭ space and points of the upper half of the cone $[\xi, \xi] = 0$,

$\xi_n > 0$.

2. **Invariant integration in Lobačevskiĭ space and on horispheres.** We define integration in Lobačevskiĭ space, invariant for shift operators, i.e. such that

$$\int f(\mathbf{x})\,d\mathbf{x} = \int f(g\mathbf{x})\,d\mathbf{x}. \tag{1}$$

We realize Lobačevskiĭ space on the hyperboloid $[\mathbf{x}, \mathbf{x}] = 1$.

The volume element $dv = dx_1 \cdots dx_n$ in the space $E_{n-1,1}$ is invariant for all linear transformations with determinant 1, including all hyperbolic rotations. In the domain $[\mathbf{x}, \mathbf{x}] > 0$ we introduce a new system of coordinates $x_1, \cdots, x_{n-1}, \rho$, where $\rho^2 = [\mathbf{x}, \mathbf{x}]$. In these variables the volume element $dv$ takes the form

$$dv = \frac{\rho\, d\rho\, dx_1 \ldots dx_{n-1}}{x_n} = \frac{\rho\, d\rho\, dx_1 \ldots dx_{n-1}}{\sqrt{1 + x_1^2 + \ldots + x_{n-1}^2}}.$$

Since hyperbolic rotations in $E_n$ preserve both $\rho$ and $dv$, they also preserve

$$d\mathbf{x} = \frac{dx_1 \ldots dx_{n-1}}{x_n} = \frac{dx_1 \ldots dx_{n-1}}{\sqrt{p^2 + x_1^2 + \ldots + x_{n-1}^2}}. \tag{2}$$

Formula (2) defines the invariant measure in Lobačevskiĭ space. It is analogous to the formula for invariant measure on a sphere established in § 1.1 of Chapter IX.

An invariant integral over the hyperboloid $[\mathbf{x}, \mathbf{x}] = 1$ can be written down with the help of the generalized function $\delta([\mathbf{x}, \mathbf{x}] - 1)$.[1)] Namely, it can be easily shown that

$$\int f(\mathbf{x})\,d\mathbf{x} = 2\int f(\mathbf{x})\,\delta([\mathbf{x}, \mathbf{x}] - 1)\,dv. \tag{3}$$

The invariance of the integral in the right hand side of (3) follows immediately from the invariance of the measure $dv$ and the function $[\mathbf{x}, \mathbf{x}]$ for hyperbolic rotations of $E_{n-1,1}$.

We shall also need the expression for the invariant measure in Lobačevskiĭ space in hyperbolic coordinates $\theta_1, \cdots, \theta_{n-1}$ (see § 1.1). Changing the variables according to formula (1) of § 1.1, we get

$$d\mathbf{x} = \operatorname{sh}^{n-2}\theta_{n-1} \sin^{n-3}\theta_{n-2} \ldots \sin\theta_2\, d\theta_1 \ldots d\theta_{n-1}. \tag{4}$$

In the same way one establishes that the invariant measure $d\xi$ on the cone $[\xi, \xi] = 0$ is given by the formula

$$d\xi = \frac{d\xi_1 \ldots d\xi_{n-1}}{\xi_n} = \frac{d\xi_1 \ldots d\xi_{n-1}}{\sqrt{\xi_1^2 + \ldots + \xi_{n-1}^2}}. \tag{5}$$

---

1) Concerning the generalized function $\delta(P)$ see [18], Chapter III, § 1.3, and in particular Russian p. 276. Of course, in the right hand side of (3) $f(\mathbf{x})$ stands for the continuous extension of a function, given on the hyperboloid $[\mathbf{x}, \mathbf{x}] = 1$, to the whole space $E_{n-1,1}$.

We shall now derive the formula for invariant integration over a horisphere. In subsection 1 it was remarked that, in the realization of Lobačevskiĭ space on the upper half of the hyperboloid $[\mathbf{x}, \mathbf{x}] = 1$, the equation of a horisphere has the form $[\mathbf{x}, \xi] = 1$, where $[\xi, \xi] = 0$.

We define the integral of a function $f(\mathbf{x})$ over the horisphere $\omega$: $[\mathbf{x}, \xi] = 1$ by the formula

$$\int_{\omega} f(\mathbf{x})\, d\omega = \int f(\mathbf{x})\, \delta([\mathbf{x}, \xi] - 1)\, d\mathbf{x}, \tag{6}$$

where $\mathbf{x}$ is a point of Lobačevskiĭ space, realized on the hyperboloid $[\mathbf{x}, \mathbf{x}] = 1$, and $d\mathbf{x}$ is the invariant measure in this space.

Since $d\mathbf{x}$ and $[\mathbf{x}, \xi]$ are invariant for a simultaneous shift of $\mathbf{x}$ and $\xi$, the measure defined in this way is invariant for a shift of the horisphere. It follows that if a motion $g$ of Lobačevskiĭ space carries the horisphere $\omega$ into the horisphere $g\omega$, then

$$\int_{\omega} f(g\mathbf{x})\, d\omega = \int_{g\omega} f(\mathbf{x})\, d\omega_g, \tag{7}$$

where $d\omega_g$ is the measure on $g\omega$. In particular, if the motion $g$ carries $\omega$ into itself, then

$$\int_{\omega} f(g\mathbf{x})\, d\omega = \int_{\omega} f(\mathbf{x})\, d\omega. \tag{8}$$

3. **The integral transformation of Gel'fand-Graev.** With each finite function $f(\mathbf{x})$ in Lobačevskiĭ space we associate its integrals over horispheres:

$$h(\omega) = \int_{\omega} f(\mathbf{x})\, d\omega, \tag{1}$$

where $d\omega$ is the measure on the horisphere $\omega$, defined above. With each finite function in Lobačevskiĭ space, one thus associates another function, $h(\omega)$, defined on the set of horispheres of this space. We shall call the transformation carrying $f(\mathbf{x})$ into $h(\omega)$, the integral transformation of Gel'fand-Graev.

It was shown in subsection 1 that the set of horispheres can be regarded as the set of points of the upper half of the cone $[\xi, \xi] = 0$ (to each point $\xi$ of this half corresponds a horisphere $[\mathbf{x}, \xi] = 1$). Therefore one can regard $h(\omega)$ as a function on the upper half of the cone, and write $h(\xi)$ in place of $h(\omega)$. In this way our integral transformation carries the space of functions given on the upper half of the hyperboloid $[\mathbf{x}, \mathbf{x}] = 1$, into the space of functions given on the upper half of the cone $[\xi, \xi] = 0$.

Recalling the definition of an integral over a horisphere under the realization of Lobačevskiĭ space on the upper half of the hyperboloid $[\mathbf{x}, \mathbf{x}] = 1$, one can write $h(\xi)$ in the form

$$h(\xi)=\int f(\mathbf{x})\,\delta([\mathbf{x},\ \xi]-1)\,d\mathbf{x}, \tag{2}$$

where $d\mathbf{x}$ is the invariant measure in Lobačevskiĭ space.

It can be proved (see [15], Russian p. 388) that if $f(\mathbf{x})$ is an infinitely differentiable finite function on the upper half of the hyperboloid $[\mathbf{x}, \mathbf{x}] = 1$, then the function $h(\xi)$ on the cone, defined by formula (2), is infinitely differentiable, finite and equal to zero in a certain neighborhood of the vertex of the cone.

In [15] (see §2.3, Chapter V) one can find an inversion formula for the integral transformation carrying $f(\mathbf{x})$ into $h(\xi)$. Namely, the following assertion is proved there.

*Let $f(\mathbf{x})$ be a finite function in $(n-1)$-dimensional Lobačevskiĭ space and let*

$$h(\xi)=\int f(\mathbf{x})\,\delta([\mathbf{x},\ \xi]-1)\,d\mathbf{x}. \tag{3}$$

*be its integrals over the horispheres $[\mathbf{x}, \xi] = 1$ of this space. If the dimension $n-1$ of the Lobačevskiĭ space is odd, $n = 2m + 2$, the inversion formula for the integral transformation (3) has the form*

$$f(\mathbf{a})=\frac{(-1)^m}{2(2\pi)^{2m}}\int h(\xi)\,\delta^{(2m)}([\mathbf{a},\ \xi]-1)\,d\xi. \tag{4}$$

*But if $n = 2m + 1$, then the inversion formula is written as follows:*

$$f(\mathbf{a})=\frac{(-1)^m\,\Gamma(2m)}{(2\pi)^{2m}}\int h(\xi)([\mathbf{a},\ \xi]-1)^{-2m}\,d\xi. \tag{5}$$

*Here the integral is understood in the sense of regularized value, i.e.*

$$\int h(\xi)([\mathbf{a},\ \xi]-1)^{-2m}\,d\xi=\int h(\xi)([\mathbf{a},\ \xi]-1)^{\mu}\,d\xi\,\big|_{\mu=-2m}. \tag{6}$$

4. **The quasi-regular representation of $SH(n)$.** We shall denote by $\mathfrak{K}$ the space of infinitely differentiable finite functions $f(\mathbf{x})$ on the upper half of the hyperboloid $[\mathbf{x}, \mathbf{x}] = 1$. With each element $g$ of $SH(n)$ we associate an operator $L(g)$ in $\mathfrak{K}$, given by the formula

$$L(g)f(\mathbf{x})=f(g^{-1}\mathbf{x}). \tag{1}$$

Clearly, $L(g_1)L(g_2) = L(g_1 g_2)$ and so $L(g)$ is a representation of $SH(n)$.

We complete $\mathfrak{K}$ with respect to the norm

$$\|f\|^2=\int_{[\mathbf{x},\ \mathbf{x}]=1}|f(\mathbf{x})|^2\,d\mathbf{x}. \tag{2}$$

We thus obtain a space $\mathfrak{H}$, consisting of all functions $f(\mathbf{x})$ on the hyperboloid $[\mathbf{x}, \mathbf{x}] = 1$, $x_n > 0$, such that

$$\int |f(\mathbf{x})|^2 \, d\mathbf{x} < +\infty.$$

The representation $L(g)$ can be extended to the space $\mathfrak{H}$. Moreover, from the invariance of the measure $d\mathbf{x}$ it follows that $L(g)$ is unitary relative to the norm $\|f\|$.

We decompose $L(g)$ into irreducible components. To that end, with each function $f(\mathbf{x})$ of $\mathfrak{R}$ we associate a function

$$h(\xi) = \int\limits_{[\mathbf{x}, \mathbf{x}]=1} f(\mathbf{x}) \, \delta([\mathbf{x}, \xi] - 1) \, d\mathbf{x}, \tag{3}$$

given on the upper half of the cone $[\xi, \xi] = 0$.

Let us find out how $L(g)$ is affected by this transformation. It can be seen from (3) that to the function $L(g)f(\mathbf{x}) = f(g^{-1}\mathbf{x})$ corresponds, on the cone, the function

$$\begin{aligned} h_g(\xi) &= \int f(g^{-1}\mathbf{x}) \, \delta([\mathbf{x}, \xi] - 1) \, d\mathbf{x} \\ &= \int f(\mathbf{x}) \, \delta([g\mathbf{x}, \xi] - 1) \, d\mathbf{x} = \int f(\mathbf{x}) \, \delta([\mathbf{x}, g^{-1}\xi] - 1) \, d\mathbf{x} = h(g^{-1}\xi). \end{aligned}$$

Thus, to the quasi-regular representation $L(g)$ corresponds the representation

$$T(g) \, h(\xi) = h(g^{-1}\xi) \tag{4}$$

in the space of functions on the upper half of the cone. The problem of decomposing $L(g)$ is thus reduced to the problem of decomposing $T(g)$. But the latter problem is considerably simpler and amounts to a decomposition of functions on the cone into homogeneous components.

Namely, with the function $h(\xi)$ we associate the function

$$\Phi(\xi, \sigma) = \int\limits_0^\infty h(t\xi) \, t^{-\sigma-1} \, dt. \tag{5}$$

For $a > 0$ we have

$$\Phi(a\xi, \sigma) = \int\limits_0^\infty h(at\xi) \, t^{-\sigma-1} \, dt = a^\sigma \int\limits_0^\infty h(t\xi) \, t^{-\sigma-1} \, dt = a^\sigma \, \Phi(\xi, \sigma).$$

and so the function $\Phi(\xi, \sigma)$ is a homogeneous function of $\xi$ of degree $\sigma$. Moreover, to the function $h(g^{-1}\xi)$ corresponds the function

$$T^{n\sigma}(g) \, \Phi(\xi, \sigma) = \Phi(g^{-1}\xi, \sigma). \tag{6}$$

But $T^{n\sigma}(g)$ is none other than the representation of $SH(n)$ in the space of homogeneous functions on the cone, constructed in § 2.1. These representations are irreducible (with the exception of the case when $\sigma$ or $-n-\sigma+2$ is a nonnegative integer). Therefore the representations $T^{n\sigma}(g)$ are the irreducible components of $T(g)$ (i.e. of $L(g)$).

We derive an explicit expression for the Fourier components $\Phi(\xi, \sigma)$ of $f(\mathbf{x})$. It follows from (3) and (5) that

$$\Phi(\xi, \sigma) = \int_0^\infty t^{-\sigma-1}\,dt \int_{[\mathbf{x},\mathbf{x}]=1} f(\mathbf{x})\,\delta(t[\mathbf{x}, \xi]-1)\,d\mathbf{x}$$

$$= \int_0^\infty t^{-\sigma-2}\,dt \int_{[\mathbf{x},\mathbf{x}]=1} f(\mathbf{x})\,\delta([\mathbf{x}, \xi]-t^{-1})\,d\mathbf{x}.$$

Changing the order of integration and replacing $t$ by $t^{-1}$, we get

$$\Phi(\xi, \sigma) = \int_{[\mathbf{x},\mathbf{x}]=1} f(\mathbf{x}) \left[\int_0^\infty t^{\sigma}\delta([\mathbf{x}, \xi]-t)\,dt\right] d\mathbf{x},$$

and therefore

$$\Phi(\xi, \sigma) = \int_{[\mathbf{x},\mathbf{x}]=1} f(\mathbf{x})\,[\mathbf{x}, \xi]^{\sigma}\,d\mathbf{x}. \tag{7}$$

We now derive an expression for $f(\mathbf{x})$ in terms of its Fourier components $\Phi(\xi, \sigma)$. For this we first express $h(\xi)$ by means of $\Phi(\xi, \sigma)$. From the inversion formula for the Mellin transformation (see §4.3 of Chapter II) it follows that

$$h(t\xi) = \frac{1}{2\pi i}\int_{a-i\infty}^{a+i\infty} t^{\sigma}\Phi(\xi, \sigma)\,d\sigma.$$

Putting $t=1$, we get

$$h(\xi) = \frac{1}{2\pi i}\int_{a-i\infty}^{a+i\infty} \Phi(\xi, \sigma)\,d\sigma. \tag{8}$$

We now utilize the inversion formula for the Gel'fand-Graev transformation (see subsection 3). We find that, for $n = 2m+2$,

$$f(\mathbf{x}) = \frac{(-1)^m}{2(2\pi)^{2m+1}i}\int_{[\xi,\xi]=0} \delta^{(2m)}([\mathbf{x}, \xi]-1)\int_{a-i\infty}^{a+i\infty} \Phi(\xi, \sigma)\,d\sigma\,d\xi$$

$$= \frac{(-1)^m}{2(2\pi)^{2m+1}i}\int_{a-i\infty}^{a+i\infty} d\sigma \int_{[\xi,\xi]=0} \Phi(\xi, \sigma)\,\delta^{(2m)}([\mathbf{x}, \xi]-1)\,d\xi. \tag{9}$$

Since $\Phi(\xi, \sigma)$ is homogeneous, it is uniquely determined by its values on any contour which intersects every generator of the cone $[\xi, \xi]=0$ once. As such a contour we choose the sphere $S^{n-2}$, i.e. the section of the cone by the plane $\xi_n = 1$. Points of $S^{n-2}$ will be denoted, as above, by $\xi'$. Clearly $\Phi(t\xi', \sigma) = t^{\sigma}\Phi(\xi', \sigma)$. The invariant measure $d\xi$ on the cone is related to the normalized euclidean measure $d\xi'$ on $S^{n-2}$ by the equality

$$d(t\xi') = \frac{2\pi^{\frac{n-1}{2}}}{\Gamma\left(\frac{n-1}{2}\right)} t^{n-3} dt\, d\xi'. \tag{10}$$

Hence for $n = 2m + 2$ we have

$$\begin{aligned} f(\mathbf{x}) &= \frac{(-1)^m}{2^{2m+1}\pi^{m+\frac{1}{2}}\Gamma\left(m+\frac{1}{2}\right) i} \\ &\times \int_{a-i\infty}^{a+i\infty} d\sigma \int_0^\infty t^{2m-1} \int_{S^{n-2}} \Phi(t\xi', \sigma)\, \delta^{(2m)}([\mathbf{x}, t\xi'] - 1)\, d\xi' dt \\ &= \frac{(-1)^m}{2^{2m+1}\pi^{m+\frac{1}{2}}\Gamma\left(m+\frac{1}{2}\right) i} \\ &\times \int_{a-i\infty}^{a+i\infty} d\sigma \int_{S^{n-2}} \Phi(\xi', \sigma) \int_0^\infty t^{\sigma-2}\delta^{(2m)}([\mathbf{x}, \xi'] - t^{-1})\, dt\, d\xi'. \end{aligned} \tag{11}$$

But

$$\begin{aligned} &\int_0^\infty t^{\sigma-2}\delta^{(2m)}([\mathbf{x}, \xi'] - t^{-1})\, dt \\ &\qquad = \int_0^\infty t^{-\sigma}\delta^{(2m)}([\mathbf{x}, \xi'] - t)\, dt = \frac{\Gamma(\sigma+2m)}{\Gamma(\sigma)} [\mathbf{x}, \xi']^{-\sigma-2m}. \end{aligned}$$

Therefore it follows from (11) that for $n = 2m + 2$

$$\begin{aligned} f(\mathbf{x}) &= \frac{(-1)^m}{2^{2m+1}\pi^{m+\frac{1}{2}}\Gamma\left(m+\frac{1}{2}\right) i} \\ &\times \int_{a-i\infty}^{a+i\infty} \frac{\Gamma(\sigma+2m)}{\Gamma(\sigma)} d\sigma \int_{S^{n-2}} \Phi(\xi', \sigma)\, [\mathbf{x}, \xi]^{-\sigma-2m}\, d\xi'. \end{aligned} \tag{12}$$

In just the same way, from formula (5) of subsection 3 we deduce that, for $n = 2m + 1$,

$$\begin{aligned} f(\mathbf{x}) &= \frac{(-1)^{m+1}}{2^{2m}\pi^m\Gamma(m)\, i} \int_{a-i\infty}^{a+i\infty} \frac{\Gamma(\sigma+2m-1)}{\Gamma(\sigma)} \operatorname{ctg} \pi\sigma\, d\sigma \\ &\times \int_{S^{n-2}} \Phi(\xi', \sigma)\, [\mathbf{x}, \xi']^{-\sigma-2m+1}\, d\xi', \end{aligned} \tag{13}$$

where $-2m + 1 < a < 1$.

In deducing formula (13) use is made of the equality

$$\int_0^\infty (t-1)^{-2m} t^{\sigma+2m-2}\,dt = -\frac{\pi \operatorname{ctg} \pi\sigma \Gamma(\sigma+2m-1)}{\Gamma(2m)\Gamma(\sigma)}, \tag{14}$$

where the integral is understood in the sense of regularized value. To derive this equality we observe that

$$\int_0^\infty |t-1|^\lambda t^{\sigma+2m-2}\,dt = \int_0^1 (1-t)^\lambda t^{\sigma+2m-2}\,dt + \int_1^\infty (t-1)^\lambda t^{\sigma+2m-2}\,dt.$$

The first integral converges for $\operatorname{Re}\lambda > -1$ and has the value

$$\frac{\Gamma(\lambda+1)\Gamma(\sigma+2m-1)}{\Gamma(\sigma+\lambda+2m)},$$

and the second converges for $\operatorname{Re}(\lambda+\sigma+2m) < 1$ and has the value

$$\frac{\Gamma(\lambda+1)\Gamma(-\sigma-\lambda-2m+1)}{\Gamma(-\sigma-2m+2)}.$$

Therefore the integral (14) is equal to the value, for $\lambda = -2m$, of the sum

$$\Gamma(\lambda+1)\left[\frac{\Gamma(\sigma+2m-1)}{\Gamma(\sigma+\lambda+2m)} + \frac{\Gamma(-\sigma-\lambda-2m+1)}{\Gamma(-\sigma-2m+2)}\right].$$

Calculating this value, we obtain (14). The condition on $a$ is linked with the existence of poles in the case of $(\Gamma(\sigma+2m-1)/\Gamma(\sigma)) \operatorname{ctg} \pi\sigma$.

Formulas (12) and (13) easily lead to an expression for $\|f\|^2$ in terms of the function $\Phi(\eta, \sigma)$. Namely, in the integral

$$\|f\|^2 = \int_{[\mathbf{x},\mathbf{x}]=1} f(\mathbf{x})\overline{f(\mathbf{x})}\,d\mathbf{x}$$

we replace $f(\mathbf{x})$ by its value according to formula (12) (or formula (13), if $n = 2m+1$). Utilizing formula (7), we thus find that, for $n = 2m+2$,

$$\|f\|^2 = \frac{(-1)^m}{2^{2m+1}\pi^{m+\frac{1}{2}}\Gamma\left(m+\frac{1}{2}\right)i} \int_{a-i\infty}^{a+i\infty} \frac{\Gamma(\sigma+2m)}{\Gamma(\sigma)}\,d\sigma \times \int_{S^{n-2}} \Phi(\xi', \sigma)\overline{\Phi(\xi', -\bar{\sigma}-2m)}\,d\xi', \tag{15}$$

while, for $n = 2m+1$,

$$\|f\|^2 = \frac{(-1)^{m+1}}{2^{2m}\pi^m\Gamma(m)i} \int_{a-i\infty}^{a+i\infty} \frac{\Gamma(\sigma+2m-1)}{\Gamma(\sigma)} \operatorname{ctg} \pi\sigma\,d\sigma \times \int_{S^{n-2}} \Phi(\xi', \sigma)\overline{\Phi(\xi', -\bar{\sigma}-2m+1)}\,d\xi', \tag{16}$$

where $-2m+1 < a < 1$.

Formulas (15) and (16) take a particularly simple form if we put $a = -(n-2)/2$.

Then $\sigma = -\bar{\sigma} - 2m$ (or $\sigma = -\bar{\sigma} - 2m + 1$), and for $n = 2m + 2$ we get

$$\|f\|^2 = \frac{(-1)^m}{2^{2m+1}\pi^{m+\frac{1}{2}}\Gamma\left(m+\frac{1}{2}\right)} \int_{-\infty}^{\infty} \frac{\Gamma(m+i\rho)}{\Gamma(-m+i\rho)} \times \int_{S^{n-2}} |\Phi(\xi', -m+i\rho)|^2 \, d\xi' \, d\rho, \tag{17}$$

while, for $n = 2m + 1$,

$$\|f\|^2 = \frac{(-1)^m i}{2^{2m}\pi^m\Gamma(m)} \int_{-\infty}^{\infty} \frac{\Gamma\left(m-\frac{1}{2}+i\rho\right)}{\Gamma\left(-m+\frac{1}{2}+i\rho\right)} \operatorname{th} \pi\rho \times \int_{S^{n-2}} \left|\Phi\left(\xi', -m+\frac{1}{2}+i\rho\right)\right|^2 d\xi' \, d\rho. \tag{18}$$

$\|f\|^2$ can be seen to be positive from the symmetry relation

$$\int_{S^{n-2}} \left|\Phi\left(\xi', -\frac{n-2}{2}+i\rho\right)\right|^2 d\xi' = \int_{S^{n-2}} \left|\Phi\left(\xi', -\frac{n-2}{2}-i\rho\right)\right|^2 d\xi', \tag{19}$$

satisfied by the functions $\Phi(\xi', \sigma)$. We shall not derive this relation, connected with the equivalence of the representations $T^{n\sigma}(g)$ and $T^{n,-n-\sigma+2}(g)$ of $SH(n)$.

**5. Integral transformations of functions on the hyperboloid.** In the preceding subsection we obtained an expression for the function $f(\mathbf{x})$, given on the hyperboloid $[\mathbf{x}, \mathbf{x}] = 1$, in terms of the function $\Phi(\xi', \sigma)$, given on the sphere $S^{n-2}$, i.e. the section of the cone $[\xi, \xi] = 0$ by the plane $\xi_n = 1$. But in § 3.7 of Chapter IX it was shown that every function on $S^{n-2}$ (with square-integrable modulus) can be expanded in a Fourier series in the system of functions $\Xi_K(\xi')$, where $K = (k_0, k_1, \ldots, \pm k_{n-3})$ (and here we include the weight $l = k_0$ in the symbol $K$). On applying this expansion to the function $\Phi(\xi', \sigma)$, we obtain

$$\Phi(\xi', \sigma) = \sum_K a_K(\sigma)\, \Xi_K(\xi'), \tag{1}$$

where

$$a_K(\sigma) = \int_{S^{n-2}} \Phi(\xi', \sigma)\, \overline{\Xi_K(\xi')}\, d\xi'. \tag{2}$$

Let us now find an expression for the coefficients $a_K(\sigma)$ in terms of $f(\mathbf{x})$. From formula (7) of subsection 4 and equality (2), we have

$$a_K(\sigma) = \int_{S^{n-2}} d\xi' \int_{[x,\,x]=1} f(x)\,[x, \xi]^{\sigma}\, \overline{\Xi_K(\xi')}\, dx$$

$$= \int_{[x,\,x]=1} f(x) \int_{S^{n-2}} [x, \xi]^{\sigma}\, \overline{\Xi_K(\xi')}\, d\xi'\, dx. \qquad (3)$$

Let $x = g_x x_0$, where $x_0 = (0, \cdots, 0, 1)$. It follows from formula (5) of § 3.2 that

$$a_K(\sigma) = \int_{[x,\,x]=1} f(x)\, t^{n\sigma}_{KO}(g_x)\, dx. \qquad (4)$$

We shall now find an expression for $f(x)$ in terms of the coefficients $a_K(\sigma)$. For this we substitute expansion (1) in the inversion formula (12) (or (13)) of subsection 1 and utilize formula (5) of § 3.2. Then for $n = 2m + 2$ we have

$$f(x) = \frac{(-1)^m}{2^{2m+1}\pi^{m+\frac{1}{2}}\Gamma\left(m+\frac{1}{2}\right)i} \sum_K \int_{a-i\infty}^{a+i\infty} \frac{\Gamma(\sigma+2m)}{\Gamma(\sigma)}\, a_K(\sigma)\, t^{n,\,-\sigma-2m}_{KO}(g_x)\, d\sigma. \qquad (5)$$

Similarly, for $n = 2m + 1$ we have

$$f(x) = \frac{(-1)^{m+1}}{2^{2m}\pi^m\Gamma(m)\, i}$$

$$\times \sum_K \int_{a-i\infty}^{a+i\infty} \frac{\Gamma(\sigma+2m-1)}{\Gamma(\sigma)} \operatorname{ctg} \pi\sigma\, a_K(\sigma)\, t^{n,\,-\sigma-2m+1}(g_x)\, d\sigma, \qquad (6)$$

where $-2m + 1 < a < 1$.

The functions $t^{n,\,-\sigma-2m}_{KO}(\xi)$, in which the expansions in (5) and (6) are carried out, are constant on left cosets with respect to $SO(n-1)$ (see § 3.1) and are therefore functions on the hyperboloid $[x, x] = 1$, i.e. the homogeneous space $SH(n)/SO(n-1)$. Let

$$\Theta^{n\sigma}_K(x) = t^{n\sigma}_{KO}(g_x), \qquad [x, x] = 1, \qquad (7)$$

where $g_x$ is the motion carrying the point $x_0(0, \cdots, 0, 1)$ into the point $x$. Then formulas (4)–(6) can be rewritten as follows:

$$a_K(\sigma) = \int_{[x,\,x]=1} f(x)\, \Theta^{n\sigma}_K(x)\, dx. \qquad (8)$$

If $n = 2m + 2$, then

$$f(\mathbf{x}) = \frac{(-1)^{m}}{2^{2m+1}\pi^{m+\frac{1}{2}}\Gamma\left(m+\frac{1}{2}\right)i} \times \sum_K \int_{a-i\infty}^{a+i\infty} \frac{\Gamma(\sigma+2m)}{\Gamma(\sigma)} a_K(\sigma)\, \Theta_K^{n,\,-\sigma-2m}(\mathbf{x})\, d\sigma, \tag{9}$$

while if $n = 2m + 1$, then

$$f(\mathbf{x}) = \frac{(-1)^{m+1}}{2^{2m}\pi^{m}\Gamma(m)i} \times \sum_K \int_{a-i\infty}^{a+i\infty} \frac{\Gamma(\sigma+2m-1)}{\Gamma(\sigma)} \operatorname{ctg} \pi\sigma\, a_K(\sigma)\, \Theta_K^{n,\,-\sigma-2m+1}(\mathbf{x})\, d\sigma, \tag{10}$$

where $-2m + 1 < a < 1$.

Formulas (9) and (10) take a particularly simple form in the case when the function $f(\mathbf{x})$ is invariant for rotations $h$ of the subgroup $SO(n-1)$, i.e. when $f(h\mathbf{x}) = f(\mathbf{x})$. In that case, the expansions contain only one term, corresponding to the value $K = 0$. We thus get the following formulas.

For $n = 2m + 2$ we have

$$f(\mathbf{x}) = \frac{(-1)^{m}}{2^{2m+1}\pi^{m+\frac{1}{2}}\Gamma\left(m+\frac{1}{2}\right)i} \times \int_{a-i\infty}^{a+i\infty} \frac{\Gamma(\sigma+2m)}{\Gamma(\sigma)} a(\sigma)\, t_{OO}^{n,\,-\sigma-2m}(g_{\mathbf{x}})\, d\sigma, \tag{11}$$

while, for $n = 2m + 1$,

$$f(\mathbf{x}) = \frac{(-1)^{m+1}}{2^{2m}\pi^{m}\Gamma(m)i} \times \int_{a-i\infty}^{a+i\infty} \frac{\Gamma(\sigma+2m-1)}{\Gamma(\sigma)} \operatorname{ctg} \pi\sigma\, a(\sigma)\, t_{OO}^{n,\,-\sigma-2m+1}(g_{\mathbf{x}})\, d\sigma, \tag{12}$$

where $-2m + 1 < a < 1$. The coefficients $a(\sigma)$ have the form

$$a(\sigma) = \int_{[\mathbf{x},\mathbf{x}]=1} f(\mathbf{x})\, t_{OO}^{n\sigma}(g_{\mathbf{x}})\, d\mathbf{x}. \tag{13}$$

An expression for the zonal spherical function $t_{OO}^{n\sigma}(g)$ in terms of Legendre functions is given by formula (4) of § 3.3. Applying this formula, we can rewrite (11) and (12) as follows.

If $n = 2m + 2$, then

$$f(\operatorname{ch}\theta) = \frac{(-1)^m}{2i\operatorname{sh}^{m-\frac{1}{2}}\theta} \int_{a-i\infty}^{a+i\infty} \frac{\Gamma(\sigma+2m)}{\Gamma(\sigma)} a(\sigma) \mathfrak{P}^{\frac{1}{2}-m}_{-\sigma-m-\frac{1}{2}}(\operatorname{ch}\theta)\, d\sigma, \tag{14}$$

while if $n = 2m + 1$, then

$$f(\operatorname{ch}\theta) = \frac{(-1)^{m+1}}{2i\operatorname{sh}^{m-1}\theta} \times \int_{a-i\infty}^{a+i\infty} \frac{\Gamma(\sigma+2m-1)}{\Gamma(\sigma)} \operatorname{ctg}\pi\sigma\, a(\sigma) \mathfrak{P}^{1-m}_{-\sigma-m}(\operatorname{ch}\theta)\, d\sigma, \tag{15}$$

where $-2m+1 < a < 1$. Moreover,

$$a(\sigma) = \int_0^\infty f(\operatorname{ch}\theta) \operatorname{sh}^{\frac{n-1}{2}}\theta\, \mathfrak{P}^{\frac{3-n}{2}}_{\sigma+\frac{n-3}{2}}(\operatorname{ch}\theta)\, d\theta. \tag{16}$$

Formulas (15) and (16) were established by Mehler and Fock for $n = 3$ and $\operatorname{Re}\sigma = -\frac{1}{2}$. Thus for $n = 3$ transformations (15) and (16) are called *the Mehler-Fock transformations.*

## §5. A LAPLACE OPERATOR ON THE HYPERBOLOID. POLYSPHERICAL AND HORISPHERICAL FUNCTIONS ON THE HYPERBOLOID

1. **A Laplace operator on the hyperboloid.** In pseudo-euclidean space $E_{n-1,1}$ a role analogous to the role of Laplace operator in euclidean space is played by the wave operator

$$\square = \frac{\partial^2}{\partial x_1^2} + \cdots + \frac{\partial^2}{\partial x_{n-1}^2} - \frac{\partial^2}{\partial x_n^2}.$$

Like the Laplace operator, it commutes with motions of pseudo-euclidean space.

We shall consider the operator $\square$ inside the cone $[\mathbf{x}, \mathbf{x}] > 0$, $x_n > 0$. In spherical coordinates (see §1.1) it is given by the formula

$$\square = -\frac{1}{r^{n-1}}\frac{\partial}{\partial r} r^{n-1}\frac{\partial}{\partial r} + \frac{1}{r^2\operatorname{sh}^{n-2}\theta_{n-1}}\frac{\partial}{\partial\theta_{n-1}}\operatorname{sh}^{n-2}\theta_{n-1}\frac{\partial}{\partial\theta_{n-1}} + \frac{1}{r^2\operatorname{sh}^2\theta_{n-1}\sin^{n-3}\theta_{n-2}}\frac{\partial}{\partial\theta_{n-2}}\sin^{n-3}\theta_{n-2}\frac{\partial}{\partial\theta_{n-2}} + \cdots + \frac{1}{r^2\operatorname{sh}^2\theta_{n-1}\sin^2\theta_{n-2}\cdots\sin^2\theta_2}\frac{\partial^2}{\partial\theta_1^2}.$$

If the function $f(\mathbf{x})$ does not depend on $r$, then on the unit hyperboloid $[\mathbf{x}, \mathbf{x}] = 1$, $x_n > 0$ we have $\square f(\mathbf{x}) = \square_0 f(\xi)$, where (in spherical coordinates)

$$\Box_0 = \frac{1}{\operatorname{sh}^{n-2}\theta_{n-1}}\frac{\partial}{\partial\theta_{n-1}}\operatorname{sh}^{n-2}\theta_{n-1}\frac{\partial}{\partial\theta_{n-1}} + \frac{1}{\operatorname{sh}^2\theta_{n-1}\sin^{n-3}\theta_{n-2}}\frac{\partial}{\partial\theta_{n-2}}\sin^{n-3}\theta_{n-2}\frac{\partial}{\partial\theta_{n-2}} + \ldots$$
$$\ldots + \frac{1}{\operatorname{sh}^2\theta_{n-1}\sin^2\theta_{n-2}\ldots\sin^2\theta_2}\frac{\partial^2}{\partial\theta_1^2}.$$

The operator $\Box_0$ is called the *Laplace operator on the hyperboloid* $[\mathbf{x}, \mathbf{x}] = 1$. Like the Laplace operator $\Delta_0$ on a sphere (see § 5.1, Chapter IX), the operator $\Box_0$ is permutable with hyperbolic rotations: if $g \in SH(n)$, then

$$L(g)\Box_0 = \Box_0 L(g),$$

where

$$L(g)f(\xi) = f(g^{-1}\xi).$$

We have seen in § 2 of Chapter IX that irreducible representations of class 1 of the group $SO(n)$ can be constructed in the spaces $\mathfrak{H}^{nl}$ of homogeneous harmonic polynomials. The representations $T^{n\sigma}(g)$ of $SH(n)$, constructed in § 2.1, can also be similarly realized. Namely, we denote by $\mathfrak{G}^{n\sigma}$ the space of homogeneous functions $f(\mathbf{x})$ of degree $\sigma$, $\mathbf{x} \in E_{n-1,1}$, such that $\Box f(\mathbf{x}) = 0$. Let $\mathfrak{H}_{n\sigma}$ denote the space of functions $f(\xi)$ on the hyperboloid $[\mathbf{x}, \mathbf{x}] = 1$, such that $r^{\sigma}f(\mathbf{x}/r) \in \mathfrak{G}_{n\sigma}$. It is not difficult to show that the space $\mathfrak{H}^{n\sigma}$ is invariant for the operators $L(g)$, $g \in SH(n)$. We thus get a representation $Q^{n\sigma}(g)$ of $SH(n)$, equivalent to the representation $T^{n\sigma}(g)$. One passes from $Q^{n\sigma}(g)$ to $T^{n\sigma}(g)$ by means of the Gel'fand-Graev transformation (see § 4.3).

2. Polyspherical coordinates on the hyperboloid $[\mathbf{x}, \mathbf{x}] = 1$. These coordinates are defined in just the same way as on a sphere (see § 5.2 of Chapter IX). The only difference consists of the following. With the root of the tree we associate the coordinate $x_0 \equiv x_n$, and with each of the vertices of first rank $x_{01}, \cdots, x_{0k}$ we associate a hyperbolic rotation in the $(x_{0m}, x_0)$-plane by the "angle" $\phi_m$:

$$\left.\begin{aligned} x_0' &= x_0 \operatorname{ch}\varphi_m + x_{0m}\operatorname{sh}\varphi_m, \\ x_{0m}' &= x_0 \operatorname{sh}\varphi_m + x_{0m}\operatorname{ch}\varphi_m. \end{aligned}\right\} \tag{1}$$

Therefore, in the expression for the differential of arc length on the hyperboloid, with vertices of the first rank we associate the expressions

$$d\varphi_m^2 + \operatorname{sh}^2\varphi_m(\cdot) + \operatorname{ch}^2\varphi_m(\cdot). \tag{2}$$

Corresponding changes are made in the expression for the Laplace operator and the invariant measure in polyspherical coordinates.

The parameters of variable separation on the hyperboloid are subjected to the same conditions as on the sphere, with the only difference that corresponding to vertices of the first rank we may have arbitrary complex numbers, and not solely integral ones. Thus, for vertices of the first rank, restrictions of the form (14) of §5.4, Chapter IX, are removed.

Under the separation of variables, to the vertex $x_{0m}$ of the first rank corresponds the differential equation

$$\frac{1}{\operatorname{ch}^p \varphi \operatorname{sh}^q \varphi} \frac{d}{d\varphi} \operatorname{ch}^p \varphi \operatorname{sh}^q \varphi \frac{du}{d\varphi} + \left[\frac{r(r+p-1)}{\operatorname{ch}^2 \varphi} - \frac{s(s+q-1)}{\operatorname{sh}^2 \varphi} - l(l+p+q)\right] u = 0. \tag{3}$$

Here

$$\left.\begin{array}{lll} \varphi = \varphi_m, & p = p_m, & q = q_m, \\ r = r_{m-1}, & l = l_m, & s = l_{mt}, \end{array}\right\} \tag{4}$$

where $t$ takes the largest possible value.

Solutions of this equation have the form

$$u_1 = \operatorname{th}^s \varphi \operatorname{ch}^l \varphi F\left(\frac{s-l+r}{2}, \frac{s-l-r-p+1}{2}; s+\frac{q+1}{2}; \operatorname{th}^2 \varphi\right) \tag{5}$$

and

$$u_2 = \operatorname{cth}^{q+s-1} \varphi \operatorname{ch}^l \varphi F\left(\frac{r-l-s-q+1}{2}, 1-\frac{p+s+q+l+r}{2}; -s-\frac{q-3}{2}; \operatorname{th}^2 \varphi\right). \tag{6}$$

Eigenfunctions of the Laplace operator on the hyperboloid are expressed by a formula analogous to formula (15) of §5.4, Chapter IX. Namely,

$$Y_\Lambda(\xi) = \Pi\, U_{i_1 \ldots i_s}(\varphi_{i_1 \ldots i_s}; \Lambda), \tag{7}$$

where for $s \geq 2$ the functions $U_{i_1 \ldots i_s}$ have the same meaning as in §5.4 of Chapter IX, while the functions $U_s(\phi_s; \Lambda)$ are solutions of equation (3) for the parameter values (4). In the case when the tree has the form depicted in Figure 8 (see Chapter IX, §5.4), the functions $Y_\Lambda(\xi)$ differ from the functions $t^{n\sigma}_{OK}(g)$ of §3.1 in a constant factor only.

3. **Horispherical coordinates on the hyperboloid.** We indicate another coordinate system on a two-sheeted hyperboloid, in which the Laplace operator admits a separation of variables. This system is constructed in the following way. We consider a tree containing a vertex $x_{011}$, but not containing $x_{012}$ (Figure 9). With all the vertices, except the vertex $x_{011}$, we

associate the same transformations as in subsection 2, i.e. euclidean rotations by the angle $\phi_{i_1 \cdots i_m}$ for the vertices whose rank is greater than unity, and hyperbolic rotations by the "angle" $\phi_k$ for the vertices $x_{0k}$. However, with the vertex $x_{011}$ we associate a birational transformation $h(t)$, which we shall call the horispherical rotation in the subspace $(x_0, x_{01}, x_{011})$ by the angle $t$:

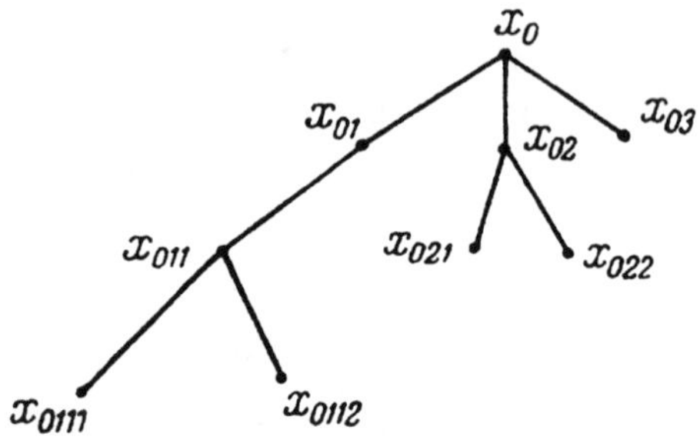


Figure 9.

$$\left.\begin{aligned} x_0' &= x_0 + \frac{2tx_{011} + t^2}{2(x_0 + x_{01})}, \\ x_{01}' &= x_{01} - \frac{2tx_{011} + t^2}{2(x_0 + x_{01})}, \\ x_{011}' &= x_{011} + t. \end{aligned}\right\} \tag{1}$$

We introduce a parameter $\phi_{11}$ by putting

$$t = \varphi_{11} e^{\varphi_1} \operatorname{ch} \varphi_2 \ldots \operatorname{ch} \varphi_k,$$

where $k$ is the number of vertices of the first rank, and we take the numbers $\{\phi_{i_1 \cdots i_m}\}$ as horispherical coordinates on the hyperboloid $[\mathbf{x}, \mathbf{x}] = 1$. Namely, the point with coordinates $\{\phi_{i_1 \cdots i_m}\}$ has the form

$$\mathbf{x} = \prod_{m \geq 3} g(\varphi_{i_1 \ldots i_m}) \, h(\varphi_{11} e^{\varphi_1} \operatorname{ch} \varphi_2 \ldots \operatorname{ch} \varphi_k) \prod{}' g(\varphi_{i_1, i_2}) \prod g(\varphi_i) \, \mathbf{x}_0. \tag{2}$$

Here $\Pi' g(\phi_{i_1 i_2})$ is extended to all vertices of the second rank, except $x_{011}$, while the remaining products are extended to all vertices of the corresponding rank; $\mathbf{x}_0$ is the point of the hyperboloid corresponding to the axis $Ox_0$.

From the definition of horispherical coordinates it follows, in particular, that

$$x_0 = \operatorname{ch} \varphi_2 \ldots \operatorname{ch} \varphi_k \left[ \operatorname{ch} \varphi_1 + \frac{\varphi_{11}^2}{2} e^{\varphi_1} \right],$$

$$x_{01} = \operatorname{ch} \varphi_2 \ldots \operatorname{ch} \varphi_k \left[ \operatorname{sh} \varphi_1 - \frac{\varphi_{11}^2}{2} e^{\varphi_1} \right],$$

$$x_{011} = e^{\varphi_1} \operatorname{ch} \varphi_2 \ldots \operatorname{ch} \varphi_k \, \varphi_{11} \cos \varphi_{111} \ldots \cos \varphi_{11m},$$

$$x_{011\,m} = e^{\varphi_1} \operatorname{ch} \varphi_2 \ldots \operatorname{ch} \varphi_k \, \varphi_{11} \sin \varphi_{11\,m},$$

where $k$ and $m$ have the largest values for the given tree.

A simple calculation shows that, in horispherical coordinates, the differential of arc length on the hyperboloid has the form

$$ds^2 = \sum B^2_{i_1 \ldots i_m} d\varphi^2_{i_1 \ldots i_m}, \tag{3}$$

where the coefficients $B_{i_1 \ldots i_m}$ are defined in the following way. If $(i_1, i_2) \neq (1, 1)$, then $B_{i_1 \ldots i_m} = A_{i_1 \ldots i_m}$, where $A_{i_1 \ldots i_m}$ is given by formula (3) of §5.3 of Chapter IX (with $\cos\phi_s$ and $\sin\phi_s$ replaced by $\operatorname{ch}\phi_s$ and $\operatorname{sh}\phi_s$, $1 \le \phi_s \le k$). Further

$$\left.\begin{aligned} B_1 &= \operatorname{ch}\varphi_2 \ldots \operatorname{ch}\varphi_k, \\ B_{11} &= e^{\varphi_1} \operatorname{ch}\varphi_2 \ldots \operatorname{ch}\varphi_k, \\ B_{11i_3 \ldots i_s} &= e^{\varphi_1} \operatorname{ch}\varphi_2 \ldots \operatorname{ch}\varphi_k \varphi_{11} A_{i_3 \ldots i_s}, \end{aligned}\right\} \tag{4}$$

where $A_{i_3 \ldots i_s}$ is defined by formula (3), §5.3, Chapter IX. For example,

$$B_{111} = e^{\varphi_1} \operatorname{ch}\varphi_2 \ldots \operatorname{ch}\varphi_k \varphi_{11} \cos\varphi_{112} \ldots \cos\varphi_{11m}.$$

The expression for $ds^2$ leads, in the usual way, to expressions for the invariant measure and Laplace operator on the hyperboloid. We shall not give these expressions.

**4. Variable separation in horispherical coordinates.** Separation parameters for the equation $\Box_0 u + \lambda u = 0$ in horispherical coordinates have the following form: to all the vertices whose rank is greater than 1 (except vertex $x_{011}$) correspond integers, while to vertices of the first rank and to the vertex $x_{011}$ correspond complex numbers. Moreover, the separation parameters $\{l_{i_1 \ldots i_m}\}$ must satisfy the same conditions as in §5.4 of Chapter IX.

After the separation of variables for all arguments, except $\phi_1$ and $\phi_{11}$, one obtains the same equations as in subsection 2. Hence it is only necessary to derive the equations for the arguments $\phi_1$ and $\phi_{11}$. For conciseness, we introduce the following notation: $\phi = \phi_{11}$, $\psi = \phi_1$, $q = q_{11}$ (the number of vertices subordinate to $x_{011}$), $l = l_{11}$, $s = l_{11m}$ (where $m$ takes the largest possible value), and $k = l_1$. Utilizing the expressions for the coefficients $B_{i_1 \ldots i_m}$, we get the following equation for $\phi_{11}$:

$$\varphi^{-q} \frac{d}{d\varphi} \varphi^q \frac{du}{d\varphi} - \left[\frac{s(s+q-1)}{\varphi^2} - l^2\right] u = 0. \tag{1}$$

The substitution $u = \phi^{(1-q)/2} v$ reduces it to the Bessel equation. Therefore the particular solutions of equation (1) have the form

$$u_1 = \varphi^{\frac{1-q}{2}} J_{s+\frac{q-1}{2}}(l\varphi) \tag{2}$$

and

$$u_2 = \varphi^{\frac{1-q}{2}} Y_{s+\frac{q-1}{2}}(l\varphi). \tag{3}$$

Only the first of these is continuous on the semiaxis $0 \leq \phi < \infty$.

Similarly, for $\phi_1$ we get the equation

$$e^{-(q+1)\psi} \frac{d}{d\psi} e^{(q+1)\psi} \frac{du}{d\psi} - [l^2 e^{-2\psi} + k(k+q+1)] u = 0. \tag{4}$$

The substitutions $e^{-\psi} = t$, $u = t^{(q+2)2} v$ reduce it to the modified Bessel equation

$$v'' + \frac{v'}{t} - \left[l^2 + \frac{\left(k+1+\frac{q}{2}\right)^2}{t^2}\right] v = 0. \tag{5}$$

Therefore the particular solutions of (4) have the form

$$u_1 = e^{\frac{q+2}{2}\psi} K_{k+\frac{q+2}{2}}\left(\frac{e^{-\psi}}{l}\right) \tag{6}$$

and

$$u_2 = e^{\frac{q+2}{2}\psi} I_{k+\frac{q+2}{2}}\left(\frac{e^{-\psi}}{l}\right). \tag{7}$$

Utilizing these results, one can easily write down the expression for eigenfunctions of the operator in question in horispherical coordinates (or horispherical functions, as they are briefly called). We leave this to the reader.

CHAPTER XI

# THE GROUP OF MOTIONS OF THE $n$-DIMENSIONAL EUCLIDEAN SPACE AND BESSEL FUNCTIONS

In Chapter IV several properties of Bessel functions were derived from the fact that, through these functions, one expresses matrix elements of irreducible representations of the group $M(2)$ of motions of the euclidean plane. In this chapter we shall consider representations of class 1 of the group $M(n)$ of motions of the $n$-dimensional euclidean space. We shall show that, with a suitable choice of basis in the representation space, certain matrix elements of the representation are expressed by means of Bessel functions. From this we shall derive new properties of Bessel functions.

## §1. THE GROUP $M(n)$

By a motion of the $n$-dimensional euclidean space $E_n$ we mean a nonhomogeneous linear transformation $g$ which preserves both the distance between points of this space and its orientation.

It is known that any motion $g$ in $E_n$ can be written in the form

$$\mathbf{x} \longrightarrow h\mathbf{x} + \mathbf{a}, \tag{1}$$

where $h$ is a rotation in the space $E_n$ (i.e. some element of the group $SO(n)$), and $\mathbf{a}$ is a vector of $E_n$. We shall therefore write

$$g = g(\mathbf{a}, h).$$

The motions of euclidean space form a group, which we denote by $M(n)$. The group $M(n)$ is isomorphic to the group of matrices of the $(n+1)$th order

$$\begin{pmatrix} h & \mathbf{a} \\ 0 & 1 \end{pmatrix}, \tag{2}$$

where $h$ is an orthogonal matrix, $\mathbf{a}$ is a column with $n$ elements, and $0$ is a row with $n$ zero elements.

Hence we easily get

$$g(\mathbf{a}_1, h_1)g(\mathbf{a}_2, h_2)=g(\mathbf{a}_1+h_1\mathbf{a}_2, h_1h_2). \tag{3}$$

In particular, we have the equality

$$g(\mathbf{a}, h)=g(\mathbf{a}, e)g(\mathbf{0}, h)=g(\mathbf{0}, h)g(h^{-1}\mathbf{a}, e), \tag{4}$$

where $e$ denotes unit rotation.

We introduce a parametrization in $M(n)$. Let $g = g(\mathbf{a}, h)$. The rotation $h$ is given by Euler angles $\theta_i^j$, $1 \leq j \leq n-1$, $1 \leq i \leq j$. It is convenient to specify the vector $\mathbf{a}$ by spherical coordinates $r, \phi_1, \cdots, \phi_{n-1}$ (see §1.1 of Chapter IX).

In the same way, the element $g$ of $M(n)$ is given by $n(n-1)/2$ angles $\theta_i^j$, $n-1$ angles $\phi_k$ and a positive number $r$. These numbers are subject to the following restrictions:

$$\begin{aligned}
&0 \leqslant \theta_1^j < 2\pi, \\
&0 \leqslant \theta_i^j < \pi, \qquad i > 1, \\
&0 \leqslant r < \infty, \\
&0 \leqslant \varphi_1 < 2\pi, \\
&0 \leqslant \varphi_k < \pi, \qquad k > 1.
\end{aligned}$$

As shown in §1.3 of Chapter IX, a vector $\mathbf{a}$ with spherical coordinates $r$, $\phi_1, \cdots, \phi_{n-1}$ can be written in the form

$$\mathbf{a}=g_1(\varphi_1)\ldots g_{n-1}(\varphi_{n-1})\mathbf{r}, \tag{5}$$

where $\mathbf{r}$ is a vector of the form $(0, \cdots, 0, r)$ and $g_k(\phi)$ is the rotation by an angle $\phi$ in the $(x_{k+1}, x_k)$-plane:

$$\begin{aligned}
x_k' &= x_k \cos\varphi + x_{k+1}\sin\varphi, \\
x_{k+1}' &= -x_k \sin\varphi + x_{k+1}\cos\varphi.
\end{aligned} \tag{6}$$

Hence an element $g(\mathbf{a}, h)$ of $M(n)$ can be written in the form

$$g(\mathbf{a}, h)=g(\mathbf{a}, e)g(\mathbf{0}, h)=g_1(\varphi_1)\ldots g_{n-1}(\varphi_{n-1})g(\mathbf{r}, e)g(\mathbf{0}, h). \tag{7}$$

It is easy to verify that an invariant measure in $M(n)$ is given by the formula

$$dg = dh\, d\mathbf{a}, \tag{8}$$

where $dh$ is the normalized invariant measure in the subgroup $SO(n)$, and $d\mathbf{a}$ is the euclidean measure in $E_n$.

## §2. IRREDUCIBLE REPRESENTATIONS OF CLASS 1 OF THE GROUP $M(n)$

1. Description of the representations $T_R(g)$. Irreducible representations of class 1 of the group $M(n)$ are constructed in the same way as the representations $T_R(g)$ of $M(2)$, with the only difference that the circle is replaced by the $(n-1)$-dimensional sphere.

Let $d\xi$ denote the normalized invariant measure on the sphere $S^{n-1}$:

$$d\xi = \frac{\Gamma\left(\frac{n}{2}\right) d\xi_1 \ldots d\xi_{n-1}}{2\pi^{n/2} |\xi_n|}. \tag{1}$$

The space of functions $f(\xi)$ on $S^{n-1}$ such that

$$\|f\|^2 = \int_{S^{n-1}} |f(\xi)|^2 d\xi,$$

is denoted by $\mathfrak{L}^2(S^{n-1})$. We shall construct representations of $M(n)$ in this space.

Let $R$ be a complex number. With each element $g = g(\mathbf{a}, h)$ of the group $M(n)$ we associate the operator

$$T_R(g) f(\xi) = e^{R(\mathbf{a}, \xi)} f(h^{-1}\xi) \tag{2}$$

in the space $\mathfrak{L}^2(S^{n-1})$. For $g_1 = g(\mathbf{a}_1, h_1)$, $g_2 = g(\mathbf{a}_2, h_2)$ we have

$$g_1 g_2 = g(\mathbf{a}_1 + h_1 \mathbf{a}_2, h_1 h_2)$$

and

$$T_R(g_1) T_R(g_2) f(\xi) = T_R(g_1) e^{R(\mathbf{a}_2, \xi)} f(h_2^{-1}\xi)$$
$$= e^{R(\mathbf{a}_1, \xi) + R(\mathbf{a}_2, h_1^{-1}\xi)} f(h_2^{-1} h_1^{-1}\xi) = e^{R(\mathbf{a}_1 + h_1\mathbf{a}_2, \xi)} f((h_1 h_2)^{-1}\xi).$$

Therefore

$$T_R(g_1) T_R(g_2) f(\xi) = T_R(g_1 g_2) f(\xi) \tag{3}$$

and consequently $T_R(g)$ is a representation of $M(n)$.

Corresponding to elements $h$ of the subgroup $SO(n-1)$ we have the operators

$$T_R(h) f(\xi) = f(h^{-1}\xi), \tag{4}$$

i.e. the operators of quasi-regular representation of the subgroup $SO(n-1)$.

Evidently, the function $f_0(\xi) \equiv 1$ is invariant for all operators $T_R(h)$, $h \in SO(n-1)$. Thus the representations $T_R(g)$ are representations of class 1 relative to the subgroup $SO(n-1)$.

Since the measure $d\xi$ is invariant for rotations of the sphere $S^{n-1}$, the representation $T_R(g)$ is unitary if and only if $R = \rho i$ is a purely imaginary number. In fact,

$$(T_R(g)f_1, T_R(g)f_2) = \int_{S^{n-1}} e^{(R+\bar{R})(\mathbf{a}, \xi)} f_1(h^{-1}\xi)\overline{f_2(h^{-1}\xi)}\,d\xi$$
$$= \int_{S^{n-1}} e^{(R+\bar{R})(\mathbf{a}, h\xi)} f_1(\xi)\overline{f_2(\xi)}\,d\xi.$$

Hence it is obvious that the equality

$$(f_1, f_2) = (T_R(g)f_1, T_R(g)f_2) \tag{5}$$

is satisfied for all $f_1$, $f_2$ and $g$ only if $R + \bar{R} = 0$. This means that $R$ is a purely imaginary number.

2. Irreducibility of the representations $T_R(g)$. The proof of irreducibility of the representations $T_R(g)$ for $R \neq 0$ is quite similar to the proof, given in §2.10, Chapter IX, of the irreducibility of the representations $T_l(g)$ of $SO(n)$. The only difference is that in the case of the group $M(n)$ it is necessary to take the infinitesimal operator corresponding to the subgroup of parallel translations along the axis $Ox_n$. Let $g_t = (\mathbf{a}, e)$, where $\mathbf{a} = (0, \cdots, 0, t)$. Then

$$T_R(g_t)f(\xi) = e^{Rt\xi_n} f(\xi).$$

Therefore the infinitesimal operator $A$, corresponding to the subgroup of elements of the form $g(\mathbf{a}, e)$, $\mathbf{a} = (0, \cdots, 0, t)$, is given by the formula

$$Af(\xi) = \left.\frac{dT_R(g_t)f(\xi)}{dt}\right|_{t=0} = R\xi_n f(\xi). \tag{1}$$

The rest of the proof is carried out in precisely the same way as in the case of $SO(n)$.

But if $R = 0$, then the representation $T_R(g)$ leads to the quasi-regular representation of the subgroup $SO(n)$:

$$T_0(g)f(\xi) = f(h^{-1}\xi), \quad g = g(\mathbf{a}, h),$$

and is, therefore, reducible.

## §3. ZONAL AND ASSOCIATED SPHERICAL FUNCTIONS OF REPRESENTATIONS OF CLASS 1 OF THE GROUP $M(n)$

In this section we shall study those matrix elements of the representations $T_R(g)$ which are in the zero column. We shall see that these elements are expressed by means of Bessel functions and Gegenbauer functions. This connection allows one to establish a number of new properties of Bessel functions.

1. A basis in the space $\mathfrak{L}^2(S^{n-1})$. As in Chapter X, we choose, as the basis in the space $\mathfrak{L}^2(S^{n-1})$, the functions $\Xi_M(\xi)$ [1] on the sphere $S^{n-1}$, described in §3.6 of Chapter IX, where $M = (m_0, \cdots, \pm m_{n-2})$ and $l = m_0 \geq m_1 \geq \cdots \cdots \geq m_{n-2} \geq 0$.

This basis is convenient because, in it, to elements $h$ of the subgroup $SO(n)$ there correspond block-diagonal matrices on whose principal diagonals we have canonical matrices of irreducible unitary representations $T_l(h)$ of the group $SO(n)$.

The function $\Xi_O(\xi)$ is identically equal to unity, and is consequently invariant for the operators $T_R(h)$, $h \in SO(n)$. We shall study the matrix elements $t^R_{MO}(g)$ (the associated spherical functions of the representation $T_R(g)$) and, in particular, the matrix element $t^R_{OO}(g)$ (the zonal spherical function of this representation). From the results of §2.5 of Chapter I it follows that, for $h_1, h_2 \in SO(n-1)$, we have

$$t^R_{MO}(gh_2) = t^R_{MO}(g)$$

and

$$t^R_{OO}(h_1 g h_2) = t^R_{OO}(g).$$

Any element $g$ of $M(n)$ can be represented in the form

$$g = g(\mathbf{a}, h) = g(\mathbf{a}, e)\, g(\mathbf{0}, h).$$

Therefore

$$t^R_{MO}(g) = t^R_{MO}[g(\mathbf{a}, e)]. \tag{1}$$

If $\mathbf{a} = h_1 \mathbf{a}_n$, where $\mathbf{a}_n = (0, \cdots, 0, r)$, then

$$g(\mathbf{a}, e) = g(\mathbf{0}, h_1)\, g(\mathbf{a}_n, e)\, g(\mathbf{0}, h_1^{-1})$$

and therefore

$$t^R_{OO}(g) = t^R_{OO}[g(\mathbf{a}_n, e)]. \tag{2}$$

---

1) It is convenient to insert the index $l$ in symbol $M$.

This equality shows that $t^R_{OO}(g)$, where $g = g(\mathbf{a}, h)$, depends only on the length of $\mathbf{a}$. But the matrix elements $t^R_{MO}(g)$ depend on the variables $\phi_1, \cdots, \phi_{n-1}, r$.

**2. Calculation of the zonal spherical functions.** As we have just shown, it is sufficient to calculate the zonal spherical functions $t^R_{OO}(g)$ for the element $g_r = g(\mathbf{a}_n, e)$, where $\mathbf{a}_n = (0, \cdots, 0, r)$. Corresponding to this element we have the operator transforming the function $\Xi_O(\xi) \equiv 1$ into the function

$$T_R(g_r)\,\Xi_O(\xi) = e^{Rr\xi_n}.$$

Therefore

$$t^R_{OO}(g_r) = (T^R(g_r)\Xi_O,\ \Xi_O) = (e^{Rr\xi_n},\ 1) = \int_{S^{n-1}} e^{Rr\xi_n}\,d\xi. \tag{1}$$

We go over to spherical coordinates. We get

$$t^R_{OO}(g_r) = \frac{\Gamma\left(\frac{n}{2}\right)}{\sqrt{\pi}\,\Gamma\left(\frac{n-1}{2}\right)} \int_0^\pi e^{Rr\cos\varphi} \sin^{n-2}\varphi\,d\varphi \tag{2}$$

(the normalizing factor preceding the integral is easily obtained from the fact that for $r = 0$ we have $g_r = e$, and $t^R_{OO}(e) = 1$).

We have obtained an integral representation for the zonal spherical function $t^R_{OO}(g_r)$. With the help of this representation it is easy to obtain an expression for $t^R_{OO}(g_r)$ in terms of Bessel functions. For this, we expand $e^{Rr\cos\phi}$ in powers of $\cos\phi$. Substituting this expansion in (2) and integrating term by term, we obtain

$$t^R_{OO}(g_r) = \frac{\Gamma\left(\frac{n}{2}\right)}{\sqrt{\pi}\,\Gamma\left(\frac{n-1}{2}\right)} \sum_{k=0}^\infty \frac{(Rr)^k}{k!} \int_0^\pi \cos^k\varphi \sin^{n-2}\varphi\,d\varphi.$$

But

$$\int_0^\pi \cos^{2m+1}\varphi \sin^{n-2}\varphi\,d\varphi = 0,$$

$$\int_0^\pi \cos^{2m}\varphi \sin^{n-2}\varphi\,d\varphi = \frac{\Gamma\left(\frac{n-1}{2}\right)\Gamma\left(m+\frac{1}{2}\right)}{\Gamma\left(m+\frac{n}{2}\right)}.$$

Therefore

$$t_{OO}^{R}(g_r)=\frac{\Gamma\left(\frac{n}{2}\right)}{\sqrt{\pi}}\sum_{m=0}^{\infty}\frac{\Gamma\left(m+\frac{1}{2}\right)(Rr)^{2m}}{\Gamma(2m+1)\Gamma\left(m+\frac{n}{2}\right)}.$$

Using the duplication formula for the $\Gamma$ function, we get

$$t_{OO}^{R}(g_r)=\Gamma\left(\frac{n}{2}\right)\sum_{m=0}^{\infty}\frac{1}{\Gamma(m+1)\Gamma\left(m+\frac{n}{2}\right)}\left(\frac{Rr}{2}\right)^{2m}. \tag{3}$$

Comparing this expansion with the expansion (3) of §3.3, Chapter IV, we obtain [1])

$$t_{OO}^{iR}(g_r)=\Gamma\left(\frac{n}{2}\right)\frac{J_{\frac{n-2}{2}}(Rr)}{\left(\frac{Rr}{2}\right)^{\frac{n-2}{2}}}. \tag{4}$$

We have thus established a connection between zonal spherical functions of the representations $T^R(g)$ of $M(n)$ and Bessel functions.

We note that formulas (1) and (4) lead to the equality

$$\int_{-1}^{1}e^{ixt}(1-t^2)^{\frac{n-1}{2}}dt=\frac{\sqrt{\pi}\,\Gamma\left(\frac{n+1}{2}\right)J_{\frac{n}{2}}(x)}{\left(\frac{x}{2}\right)^{\frac{n}{2}}} \tag{5}$$

(we have replaced $n$ by $n+2$).

3. **Associated spherical functions.** Let us now consider the associated spherical functions $t_{MO}^{R}(g)$ of the representations $T^R(g)$. As shown in subsection 1, these functions depend only on the parameters $\phi_1, \cdots, \phi_{n-1}, r$.

We first consider matrix elements of the form $t_{MO}^{R}(g_r)$, where $g_r = g(\mathbf{a}_n, e)$, $\mathbf{a}_n = (0, \cdots, 0, r)$. They are given by the formula

$$t_{MO}^{R}(g_r)=(T^R(g_r)\,\Xi_O,\ \Xi_M). \tag{1}$$

But, as indicated in subsection 2, the operator $T^R(g_r)$ transforms the function $\Xi_O\equiv 1$ into $e^{Rr\xi_n}$. Therefore

$$t_{MO}^{R}(g_r)=\int_{S^{n-1}}e^{Rr\xi_n}\,\overline{\Xi_M(\xi)}\,d\xi. \tag{2}$$

1) For convenience we have replaced $R$ by $iR$ here.

In this formula we substitute the value of $\Xi_M(\xi)$ (see formula (2) of §3.6, Chapter IX) and go over to spherical coordinates. From the orthogonality relations for Gegenbauer polynomials it follows that $t^R_{MO}(g_r)$ is nonzero only if $M$ has the form $M = (m, 0, \cdots, 0)$. For $M = (m, 0, \cdots, 0)$ we get [1)]

$$t^R_{MO}(g_r) = \frac{\Gamma\left(\frac{n}{2}\right)}{\sqrt{\pi}\,\Gamma\left(\frac{n-1}{2}\right)} \sqrt{\frac{m!\,\Gamma(n-2)(2m+n-2)}{\Gamma(n+m-2)(n-2)}} \times \int_{-1}^{1} e^{Rrx} C_m^{\frac{n-2}{2}}(x)(1-x^2)^{\frac{n-3}{2}} dx. \tag{3}$$

On applying formula (8) of §4.8, Chapter IX, we obtain

$$t^R_{MO}(g_r) = \frac{\Gamma\left(\frac{n}{2}\right)}{2^m \sqrt{\pi}\,\Gamma\left(m+\frac{n-1}{2}\right)} \sqrt{\frac{\Gamma(n+m-2)(2m+n-2)}{m!\,\Gamma(n-1)}} \times (Rr)^m \int_{-1}^{1} e^{Rrx}(1-x^2)^{m+\frac{n-3}{2}} dx. \tag{4}$$

Because of formula (5) of subsection 2, this equality can be rewritten as follows: if $M = (m, 0, \cdots, 0)$, then

$$t^{iR}_{MO}(g_r) = i^m \Gamma\left(\frac{n}{2}\right) \sqrt{\frac{\Gamma(n+m-2)(2m+n-2)}{m!\,\Gamma(n-1)}}\, \frac{J_{m+\frac{n-2}{2}}(Rr)}{\left(\frac{Rr}{2}\right)^{\frac{n-2}{2}}}, \tag{5}$$

where, we recall, $g_r$ is a (parallel) translation by $r$ along the axis $Ox_n$. It is easy to prove that, for $M = (m, 0, \cdots, 0)$, we have

$$t^R_{OM}(g_r) = t^R_{MO}(g_r). \tag{6}$$

We also note that formulas (3) and (5) lead to the equality

$$\int_{-1}^{1} e^{itx} C_m^{\frac{n-2}{2}}(x)(1-x^2)^{\frac{n-3}{2}} dx =$$

1) See formula (2) of §3.6, Chapter IX. It must be borne in mind that $\Xi_M(\xi)$, $M = (m, 0, \cdots, 0)$ was denoted by $\Xi^m_O(\xi)$ in Chapter IX.

$$= \frac{i^m \sqrt{\pi}\,\Gamma\left(\frac{n-1}{2}\right)\Gamma(n+m-2)}{m!\,\Gamma(n-2)} \frac{J_{m+\frac{n-2}{2}}(x)}{\left(\frac{x}{2}\right)^{\frac{n-2}{2}}}. \tag{7}$$

4. **A composition theorem for Bessel functions.** We now derive a new composition theorem for Bessel functions. Let us consider the element

$$g = g_{r_1} g_\phi g_{r_2}, \tag{1}$$

where $g_{r_1}$ and $g_{r_2}$ are parallel translations along the axis $Ox_n$ by segments of length $r_1$ and $r_2$ respectively, and $g_\phi$ is a rotation by angle $\phi$ in the $(x_n, x_{n-1})$ plane. Under this motion, the origin $O(0, \cdots, 0)$ goes into the point $M$, whose distance $r$ from $O$ is expressed by the formula

$$r = \sqrt{r_1^2 + r_2^2 + 2r_1 r_2 \cos\varphi}. \tag{2}$$

It follows from equality (1) that

$$T^{iR}(g) = T^{iR}(g_{r_1})\, T^{iR}(g_\varphi)\, T^{iR}(g_{r_2}), \tag{3}$$

so that

$$t_{OO}^{iR}(g) = \sum_{K,M} t_{OK}^{iR}(g_{r_1})\, t_{KM}^{iR}(g_\varphi)\, t_{MO}^{iR}(g_{r_2}). \tag{4}$$

As shown above, elements $t_{OK}^{iR}(g_{r_1})$ and $t_{MO}^{iR}(g_{r_2})$ are nonzero only if $K$ and $M$ have the forms

$$K = (k, 0, \ldots, 0), \qquad M = (m, 0, \ldots, 0).$$

Moreover, the element $t_{KM}^{iR}(g_\phi)$ is nonzero only if $k = m$. Therefore equality (3) takes the form

$$t_{OO}^{iR}(g) = \sum_{k=0}^{\infty} t_{OK}^{iR}(g_{r_1})\, t_{KK}^{iR}(g_\varphi)\, t_{KO}^{iR}(g_{r_2}), \tag{5}$$

where $K = (k, 0, \cdots, 0)$.

We have already found the values of $t_{OK}^{iR}(g_{r_1})$ and $t_{KO}^{iR}(g_{r_2})$ (see formulas (5) and (6) of subsection 3). Further (see subsection 1), the matrix of the operator $T^{iR}(g_\phi)$ is block-diagonal, and on its principal diagonal we have canonical matrices of irreducible representations of the group $SO(n)$. Therefore $t_{KK}^{iR}(g_\phi)$, where $K = (k, 0, \cdots, 0)$, is none other than the zonal spherical function of the representation $T^{nk}(g)$ of $SO(n)$.

Lastly, $t^{iR}_{00}(g)$, $g = g(\mathbf{a}, h)$ depends only on the length $r$ of the vector $\mathbf{a}$, given by formula (2). From these considerations we obtain

$$\frac{J_{\frac{n-2}{2}}(r)}{r^{\frac{n-2}{2}}} = 2^{\frac{n-2}{2}} \Gamma\left(\frac{n-2}{2}\right) \sum_{k=0}^{\infty} (-1)^k \left(k+\frac{n-2}{2}\right) \times \frac{J_{k+\frac{n-2}{2}}(r_1)}{r_1^{\frac{n-2}{2}}} \frac{J_{k+\frac{n-2}{2}}(r_2)}{r_2^{\frac{n-2}{2}}} C_k^{\frac{n-2}{2}}(\cos\varphi), \tag{6}$$

where

$$r = \sqrt{r_1^2 + r_1^2 + 2r_1 r_2 \cos\varphi}.$$

Formula (6) is analogous to the composition formula for Bessel functions, proved in §4.1 of Chapter IV.

5. **A multiplication theorem for Bessel functions.** Let us multiply both sides of equality (6) above by $C_m^{(n-2)/2}(\cos\phi)\sin^{n-2}\phi$ and integrate with respect to $\phi$ from 0 to $\pi$. Taking into consideration the orthogonality relations for Gegenbauer polynomials, we obtain

$$\int_0^{\pi} \frac{J_{\frac{n-2}{2}}(r)}{r^{\frac{n-2}{2}}} C_m^{\frac{n-2}{2}}(\cos\varphi)\sin^{n-2}\varphi\, d\varphi = \frac{(-1)^m \pi\Gamma(n+m-2)}{2^{\frac{n-4}{2}} m!\,\Gamma\left(\frac{n-2}{2}\right)} \frac{J_{m+\frac{n-2}{2}}(r_1)}{r_1^{\frac{n-2}{2}}} \frac{J_{m+\frac{n-2}{2}}(r_2)}{r_2^{\frac{n-2}{2}}}, \tag{1}$$

where

$$r = \sqrt{r_1^2 + r_2^2 + 2r_1 r_2 \cos\varphi}. \tag{2}$$

The replacement of $\phi$ by the variable $r$ transforms (1) into the form

$$\int_{|r_1-r_2|}^{r_1+r_2} \frac{J_{\frac{n-2}{2}}(r)}{r^{\frac{n-2}{2}}} C_m^{\frac{n-2}{2}}\left(\frac{r_1^2+r_2^2-r^2}{2r_1r_2}\right)$$

$$\times [(r_1+r_2+r)(r_1+r_2-r)(r+r_1-r_2)(r+r_2-r_1)]^{\frac{n-3}{2}} r\,dr$$

$$= \frac{\pi\Gamma(n+m-2)}{m!\,\Gamma\left(\frac{n-2}{2}\right)} (2r_1r_2)^{\frac{n-2}{2}} J_{m+\frac{n-2}{2}}(r_1) J_{m+\frac{n-2}{2}}(r_2). \tag{3}$$

In this formula we replace $r_1$, $r_2$, $r$ by $Rr_1$, $Rr_2$, $Rr$ respectively, where $R > 0$, and apply the Fourier-Bessel transform (see formula (3) of §5.2, Chapter IV). We get

$$\int_0^\infty J_{\frac{n-2}{2}}(Rr) J_{m+\frac{n-2}{2}}(Rr_1) J_{m+\frac{n-2}{2}}(Rr_2) R^{\frac{4-n}{2}} dR$$

$$= \begin{cases} \dfrac{m!\,\Gamma\left(\frac{n-2}{2}\right)}{4\pi\Gamma(n+m-2)S}\left(\dfrac{8S^2}{r_1r_2r}\right)^{\frac{n-2}{2}} C_m^{\frac{n-2}{2}}\left(\dfrac{r_1^2+r_2^2-r^2}{2r_1r_2}\right), \\ \qquad\qquad \text{if} \quad |r_1-r_2|<r<r_1+r_2; \\ 0 \qquad\qquad \text{otherwise}. \end{cases} \tag{4}$$

Here $S$ denotes the area of a triangle with sides $r_1$, $r_2$, $r$.

6. **A generating function for Bessel functions.** It follows from formulas (3) and (5) of subsection 3 that

$$\int_{-1}^{1} e^{itx} C_m^{\frac{n-2}{2}}(x)(1-x^2)^{\frac{n-3}{2}} dx = \frac{i^m \pi\Gamma(n+m-2)}{2^{\frac{n-4}{2}} m!\,\Gamma\left(\frac{n-2}{2}\right)} \frac{J_{m+\frac{n-2}{2}}(t)}{t^{\frac{n-2}{2}}}. \tag{1}$$

Hence, owing to the orthogonality relations for Gegenbauer polynomials (see Chapter IX, §3.4), we get

$$e^{itx} = \Gamma\left(\frac{n-2}{2}\right) \sum_{m=0}^{\infty} i^m \left(m+\frac{n-2}{2}\right) \frac{J_{m+\frac{n-2}{2}}(t)}{\left(\frac{t}{2}\right)^{\frac{n-2}{2}}} C_m^{\frac{n-2}{2}}(x). \tag{2}$$

Equality (2) can be regarded as a generating function for Bessel functions. With its help we can establish a number of new properties of these functions. Let us put $x = 1$ in (2). Since

$$C_m^{\frac{n-2}{2}}(1) = \frac{\Gamma(n+m-2)}{m!\,\Gamma(n-2)},$$

we get

$$e^{it} = \frac{\Gamma\left(\frac{n-2}{2}\right)}{\Gamma(n-2)} \sum_{m=0}^{\infty} \frac{i^m \Gamma(n+m-2)}{m!} \left(m + \frac{n-2}{2}\right) \frac{J_{m+\frac{n-2}{2}}(t)}{\left(\frac{t}{2}\right)^{\frac{n-2}{2}}}. \tag{3}$$

But if we put $x = 0$, we get

$$1 = \sum_{m=0}^{\infty} \frac{\Gamma\left(2m + \frac{n}{2}\right)}{2m!} \frac{J_{2m+\frac{n-2}{2}}(t)}{\left(\frac{t}{2}\right)^{\frac{n-2}{2}}}. \tag{4}$$

7. **Some integrals involving Bessel functions.** Let us calculate the zonal spherical function $t_{00}^{iR}(g_r)$, making use of bispherical coordinates (see Chapter IX, §4.11). Utilizing formulas (1) and (4) of subsection 2 and the expression for invariant measure in bispherical coordinates, we obtain

$$\frac{J_p(x)}{\left(\frac{x}{2}\right)^p} = \frac{2}{\sqrt{\pi}\,\Gamma\left(\frac{m}{2}\right)\Gamma\left(p - \frac{m-1}{2}\right)} \times \int_0^{\pi/2} \cos^{2p-m+1}\alpha \sin^{m-1}\alpha \, d\alpha \int_0^{\pi} e^{ix\cos\alpha\sin\varphi} \sin^{2p-m}\varphi \, d\varphi,$$

where $p = (n-2)/2$. Because of formula (5) of subsection 2, this equality can be rewritten as follows $(m = 2q+2)$:

$$\int_0^{\pi/2} J_{p-q}(x\cos\alpha) \cos^{p-q}\alpha \sin^{2q+1}\alpha \, d\alpha = 2^q \Gamma(q+1) \frac{J_p(x)}{x^{q+1}}. \tag{1}$$

Another integral is obtained in the following manner. We perform a (parallel) translation by a vector **a**, whose $n$th coordinate is equal to $r_1$, $m$th equal to $r_2$, and the others are equal to zero. This (parallel) translation $g(\mathbf{a})$ can be written in the form $g(\mathbf{a}) = h^{-1} g_r h$, where $h$ is the rotation transforming the vector $\mathbf{r} = (0, \cdots, 0, \sqrt{r_1^2 + r_2^2})$ into **a**. By virtue of the properties of a zonal spherical function, we have (for $R = i$)

$$t_{00}[g(\mathbf{a})] = t_{00}(g_r) = 2^p\Gamma(p+1)\frac{J_p(\sqrt{r_1^2+r_2^2})}{(r_1^2+r_2^2)^{p/2}}.$$

On the other hand, let us calculate $t_{00}(g(\mathbf{a}))$ in bispherical coordinates. Integrating with respect to $\theta_1, \cdots, \theta_{m-2}, \psi_1, \cdots, \psi_{n-m-2}$, we obtain

$$t_{00}(g_r) = \int_{S^{n-1}} e^{i(r_1\xi_n + r_2\xi_m)}\,d\xi$$

$$= \frac{2\Gamma(p+1)}{\pi\Gamma\left(\frac{m-1}{2}\right)\Gamma\left(p-\frac{m-1}{2}\right)}\int_0^{\pi/2}\int_0^{\pi}\int_0^{\pi} e^{i(r_1\cos\alpha\cos\psi + r_2\sin\alpha\cos\theta)}$$
$$\times \sin^{m-2}\theta\,\sin^{2p-m}\psi\,\cos^{2p-m+1}\alpha\,\sin^{m-1}\alpha\,d\theta\,d\psi\,d\alpha.$$

We integrate with respect to $\theta$ and $\psi$, taking equality (5) of subsection 2 into account. Putting $m = 2q$, we get

$$\int_0^{\pi/2} J_{p-q}(r_1\cos\alpha)\,J_{q-1}(r_2\sin\alpha)\cos^{p-q+1}\alpha\,\sin^q\alpha\,d\alpha$$
$$= r_1^{p-q}r_2^{q-1}(r_1^2+r_2^2)^{-\frac{p}{2}}J_p(\sqrt{r_1^2+r_2^2}). \quad (2)$$

## §4. LIMITING PROCESS FOR THE SPATIAL DIMENSION. HERMITE POLYNOMIALS

In previous chapters we have studied properties of Gegenbauer polynomials and Bessel functions, and their connection with representations of the groups $SO(n)$ and $M(n)$. Here we shall study the behavior of these functions for $n \to \infty$. We shall show that, under a suitable limiting process, Gegenbauer polynomials and Bessel functions tend to the Hermite polynomials. From this we shall derive a number of properties of Hermite polynomials.

1. **Hermite polynomials as a limit of Gegenbauer polynomials.** We shall start from the following integral representation of Gegenbauer polynomials:

$$C_l^p(x) = \frac{\Gamma(2p+l)\,\Gamma\left(p+\frac{1}{2}\right)}{\sqrt{\pi}\,l!\,\Gamma(2p)\,\Gamma(p)}\int_{-1}^{1}(x+i\sqrt{1-x^2}\,t)^l(1-t^2)^{p-1}\,dt \quad (1)$$

(see §4.7 of Chapter IX), where $p = (n-2)/2$. From the equalities

$$\int_{-1}^{1}(1-t^2)^{p-1}dt=\frac{\sqrt{\pi}\,\Gamma(p)}{\Gamma\left(p+\frac{1}{2}\right)}$$

and

$$\lim_{p\to\infty}(1-t^2)^{p-1}=0$$

for $-1\leq t\leq 1$, $t\neq 0$, it is clear that

$$\lim_{p\to\infty}\frac{\Gamma\left(p+\frac{1}{2}\right)}{\sqrt{\pi}\,\Gamma(p)}(1-t^2)^{p-1}=\delta(t).$$

Hence

$$\lim_{p\to\infty}\frac{l!\,\Gamma(2p)}{\Gamma(2p+l)}C_l^p(x)=\int_{-1}^{1}(x+i\sqrt{1-x^2}t)^l\delta(t)\,dt=x^l. \tag{2}$$

A more interesting limiting formula is obtained if in (1) we make the substitu tion $t=u/\sqrt{p}$ and replace $x$ by $x/\sqrt{p}$. We get

$$C_l^p\left(\frac{x}{\sqrt{p}}\right)=\frac{\Gamma(2p+l)\,\Gamma\left(p+\frac{1}{2}\right)}{\sqrt{\pi}\,l!\,\Gamma(2p)\,\Gamma(p)\,p^{\frac{l+1}{2}}}$$
$$\times\int_{-\sqrt{p}}^{\sqrt{p}}\left(x+i\sqrt{1-\frac{x^2}{p}}u\right)^l\left(1-\frac{u^2}{p}\right)^{p-1}du. \tag{3}$$

For $p\longrightarrow\infty$ the integral in the right-hand side tends to an integral of the form

$$\int_{-\infty}^{\infty}(x+iu)^le^{-u^2}\,du.$$

In order to study the behavior of the coefficient preceding the integral, when $p\longrightarrow\infty$, we utilize Stirling's formula

$$\Gamma(p)\sim\sqrt{2\pi}\,p^{p-\frac{1}{2}}e^{-p}. \tag{4}$$

Replacing the $\Gamma$-functions in (3) by Stirling's estimate, we obtain

$$\lim_{p\to\infty}p^{-\frac{l}{2}}C_l^p\left(\frac{x}{\sqrt{p}}\right)=\frac{2^l}{\sqrt{\pi}\,l!}\int_{-\infty}^{\infty}(x+iu)^le^{-u^2}\,du. \tag{5}$$

We now introduce the Hermite polynomials $H_l(x)$ by putting

$$H_l(x)=\frac{2^l}{\sqrt{\pi}}\int_{-\infty}^{\infty}(x+it)^l e^{-t^2}\,dt. \tag{6}$$

It follows from (5) that

$$\lim_{p\to\infty} l!\,p^{-\frac{l}{2}}C_l^p\left(\frac{x}{\sqrt{p}}\right)=H_l(x). \tag{7}$$

Thus, Hermite polynomials are obtained from Gegenbauer polynomials by making the spatial dimension tend to infinity.

Let us look a little more closely at the replacement of $t$ by $u/\sqrt{p}$ in (1). Gegenbauer polynomials appear on writing the matrix elements of representations of class 1 of the group $SO(n)$ in the form of integrals over the $(n-1)$-dimensional sphere (see §4.7 of Chapter IX). Moreover, the measure on the sphere is normalized by the condition that the measure of the whole sphere is equal to unity. In other words, we consider a sphere of radius such that the usual euclidean measure of this sphere is equal to unity. A simple calculation shows that this radius is given by

$$R_n=\left[\frac{\Gamma\left(\frac{n-1}{2}\right)}{2\pi^{\frac{n-1}{2}}}\right]^{\frac{1}{n-2}}$$

It follows from Stirling's formula that, for $n\longrightarrow\infty$,

$$R_n\sim\sqrt{\frac{n}{2\pi e}}.$$

In other words, for increasing $n$, the radius of a sphere of unit measure increases roughly like $\sqrt{n}$. This entails the replacement of $t$ by $u/\sqrt{p}$ in (1).

2. Some properties of Hermite polynomials. The connection between Hermite and Gegenbauer polynomials, established above, allows us to obtain various properties of Hermite polynomials by means of a limiting process. It must be borne in mind here that not only do the Gegenbauer polynomials themselves tend to Hermite polynomials, but their derivatives also tend to the corresponding derivatives of Hermite polynomials.

Let us first derive the differential equation for Hermite polynomials. It follows from equation (3) of §3.2, Chapter IX.

$$(1-x^2)\frac{d^2C_l^p(x)}{dx^2}-(2p+1)x\frac{dC_l^p(x)}{dx}+l(2p+l)C_l^p(x)=0. \tag{1}$$

We replace $x$ by $x/\sqrt{p}$, multiply both sides by $l!p^{-1-l/2}$ and take the limit as $p \longrightarrow \infty$. We obtain the following differential equation for Hermite polynomials:

$$\frac{d^2H_l(x)}{dx^2} - 2x\frac{dH_l(x)}{dx} + 2lH_l(x) = 0. \tag{2}$$

In exactly the same way, from the recurrence relations for Gegenbauer polynomials, it follows that

$$\frac{dH_l(x)}{dx} = 2lH_{l-1}(x) \tag{3}$$

and

$$H_{l+1}(x) - 2xH_l(x) + 2lH_{l-1}(x) = 0 \tag{4}$$

(compare with formula (2)).

Further, using the limiting process in formula (4) of §4.8, Chapter IX, we get

$$H_l(x) = (-1)^l e^{x^2}\frac{d^l}{dx^l}(e^{-x^2}). \tag{5}$$

Analogously, from (2) of §4.12, Chapter IX, which gives the generating function for the Gegenbauer polynomials, we obtain

$$e^{-t^2+2tx} = \sum_{k=0}^{\infty}\frac{t^k}{k!}H_k(x). \tag{6}$$

Let us now consider the composition theorem for Gegenbauer polynomials:

$$\begin{aligned} C_l^p(xy + \sqrt{1-x^2}\sqrt{1-y^2}t) &= \frac{\Gamma(2p-1)}{[\Gamma(p)]^2}\sum_{k=0}^{l}\frac{2^{2l}(l-k)![\Gamma(p+k)]^2(2p+2k-1)}{\Gamma(2p+l+k)} \\ &\times (1-x^2)^{\frac{k}{2}}(1-y^2)^{\frac{k}{2}}C_{l-k}^{p+k}(x)C_{l-k}^{p+k}(y)\,C_k^{p-\frac{1}{2}}(t). \end{aligned} \tag{7}$$

In this formula we replace $y$ by $y/\sqrt{p}$ and $t$ by $t/\sqrt{p}$, multiply both sides of the equality by $l!p^{-l/2}$ and take the limit as $p \longrightarrow \infty$. In the right-hand side, we multiply and divide the coefficients by

$$(l-k)!\,k!\,(p+k)^{-\frac{l-k}{2}}\left(p-\frac{1}{2}\right)^{-\frac{k}{2}}.$$

Utilizing formulas (2) and (7) of subsection 1 and equalities of the form

$$\frac{\Gamma(p+k)}{\Gamma(p)}=p(p+1)\ldots(p+k-1),$$

We easily obtain the composition formula for Hermite polynomials:

$$H_l(xy+\sqrt{1-x^2}t)=\sum_{k=0}^{l}\frac{l!}{k!\,(l-k)!}x^{l-k}(1-x^2)^{\frac{k}{2}}H_{l-k}(y)\,H_k(t). \tag{8}$$

In particular, for $x=\sqrt{2}/2$, we have

$$H_l\left(\frac{y+t}{\sqrt{2}}\right)=\sum_{k=0}^{l}\frac{l!}{2^{l/2}k!\,(l-k)!}H_{l-k}(y)\,H_k(t). \tag{9}$$

A more general composition formula is obtained by repeated application of formula (8):

$$H_l\left(\frac{(\mathbf{a},\,\mathbf{x})}{\sqrt{(\mathbf{a},\,\mathbf{a})}}\right)=\frac{1}{(\mathbf{a},\,\mathbf{a})^{l/2}}\sum_{k_1+\ldots+k_n=l}\frac{l!}{k_1!\ldots k_n!}\prod_{j=1}^{n}a_j^{k_j}H_{k_j}(x_j), \tag{10}$$

where $(\mathbf{a},\mathbf{x})=a_1x_1+\cdots+a_nx_n$ and $(\mathbf{a},\mathbf{a})=a_1^2+\cdots+a_n^2$.

We also mention the following formulas for Hermite polynomials. From the multiplication formula for Gegenbauer polynomials (see §4.3 of Chapter IX) it follows that, for $l\geq k$,

$$\int_{-\infty}^{\infty}H_l(xy+\sqrt{1-x^2}t)\,H_k(t)\,e^{-t^2}dt=\frac{2^k\sqrt{\pi}\,l!}{(l-k)!}x^{l-k}(1-x^2)^{\frac{k}{2}}H_{l-k}(y). \tag{11}$$

But if $l<k$ this integral is equal to zero. In particular, for $k=0$, we get

$$\int_{-\infty}^{\infty}H_l(xy+\sqrt{1-x^2}t)\,e^{-t^2}dt=\sqrt{\pi}x^lH_l(y). \tag{12}$$

Let us note the special case of this formula for $x=\sqrt{2}/2$:

$$\int_{-\infty}^{\infty}H_l\left(\frac{y+t}{\sqrt{2}}\right)e^{-t^2}dt=\frac{\sqrt{\pi}}{2^{l/2}}H_l(y). \tag{13}$$

Further, it follows from formula (6) of §4.11, Chapter IX, that if $l+m+n=2g$, where $g$ is a nonnegative integer, and if there exists a triangle with sides $l$, $m$, $n$. then

$$\int_{-\infty}^{\infty} H_l(x)\, H_m(x)\, H_n(x)\, e^{-x^2}\, dx = \frac{2^g \sqrt{\pi}\, l!\, m!\, n!}{(g-l)!\,(g-m)!\,(g-n)!}. \tag{14}$$

Otherwise this integral is equal to zero.

From formula (7) of §4.3, Chapter IX, we get

$$\sum_{n=0}^{\infty} \frac{x^n H_n(y)\, H_n(z)}{2^n n!} = (1-x^2)^{-\frac{1}{2}} \exp\left\{\frac{2xyz-(y^2-z^2)\,x^2}{1-x^2}\right\}. \tag{15}$$

We leave to the reader the details of the corresponding limiting processes. However, the expansion (15) can also be obtained by starting from formula (12), making the substitution $xy + \sqrt{1-x^2}t = u$ and utilizing the orthogonality relations for Hermite polynomials (see subsection 3 below). In the same way, from formula (11) we obtain

$$\sum_{n=0}^{\infty} \frac{x^n H_n(y)\, H_{n+k}(z)}{2^n n!}$$
$$= (1-x^2)^{-\frac{k+1}{2}} \exp\left[\frac{2xyz-(y^2+z^2)\,x^2}{1-x^2}\right] H_k\left(\frac{z-xy}{\sqrt{1-x^2}}\right). \tag{16}$$

3. Orthogonality relations for Hermite polynomials. From the orthogonality relations for Gegenbauer polynomials, we get the following orthogonality relations for Hermite polynomials:

$$\int_{-\infty}^{\infty} e^{-x^2}\, H_n(x)\, H_m(x)\, dx = \begin{cases} 0, & \text{if} \quad m \neq n, \\ 2^n n!\, \sqrt{\pi}, & \text{if} \quad m = n. \end{cases} \tag{1}$$

It follows that the functions

$$\frac{e^{\frac{x^2}{2}}}{2^{\frac{n}{2}} \sqrt{\pi^{\frac{1}{2}} n!}}\, H_n(x), \tag{2}$$

$n = 0, 1, \cdots$, form an orthonormal system in the space of square-integrable functions $f(x)$ on the $x$-axis. Moreover, by virtue of the completeness of the Gegenbauer polynomial system, the system of functions (2) is complete.

Hence any function on the $x$-axis such that

$$\|f\| = \int_{-\infty}^{\infty} |f(x)|^2\, dx < +\infty,$$

can be expanded in a mean-convergent series of the form

$$f(x) = \sum_{n=0}^{\infty} a_n e^{-\frac{x^2}{2}} H_n(x), \tag{3}$$

where

$$a_n = \frac{1}{2^n n! \sqrt{\pi}} \int_{-\infty}^{\infty} f(x) H_n(x) e^{-\frac{x^2}{2}}\, dx. \tag{4}$$

Here one has the Parseval equality

$$\int_{-\infty}^{\infty} |f(x)|^2\, dx = \sum_{n=0}^{\infty} 2^n n! \sqrt{\pi}\, |a_n|^2. \tag{5}$$

4. The Fourier transform of the functions $e^{-x^2/2} H_n(x)$. These functions possess the following remarkable invariance property relative to Fourier transformation:

$$\int_{-\infty}^{\infty} e^{ixy} e^{-\frac{x^2}{2}} H_n(x)\, dx = \sqrt{2\pi}\, i^n e^{-\frac{y^2}{2}} H_n(y). \tag{1}$$

In order to prove this property, we employ formula (6) of subsection 1. With the help of Cauchy's theorem we get

$$\int_{-\infty}^{\infty} e^{ixy} e^{-\frac{x^2}{2}} H_n(x)\, dx$$

$$= \frac{2^n e^{-\frac{y^2}{2}}}{\sqrt{\pi}} \int_{-\infty}^{\infty} \int_{-\infty}^{\infty} e^{-\frac{(x-iy)^2}{2}} (x+it)^n e^{-t^2}\, dt\, dx$$

$$= \frac{2^n i^n e^{-\frac{y^2}{2}}}{\sqrt{\pi}} \int_{-\infty-iy}^{\infty-iy} \int_{-\infty}^{\infty} e^{-\frac{x^2}{2}} (t+y-ix)^n e^{-t^2}\, dt\, dx$$

$$= \frac{2^{\frac{3}{2}n+\frac{1}{2}} i^n e^{-\frac{y^2}{2}}}{\sqrt{\pi}} \int_{-\infty}^{\infty} \int_{-\infty}^{\infty} e^{-x^2} \left(\frac{t+y}{\sqrt{2}} + ix\right)^n e^{-t^2}\, dt\, dx.$$

Again applying formula (6), we find that

$$\int_{-\infty}^{\infty} e^{ixy} e^{-\frac{x^2}{2}} H_n(x)\, dx$$
$$= 2^{\frac{n+1}{2}} i^n e^{-\frac{y^2}{2}} \int_{-\infty}^{\infty} H_n\left(\frac{t+y}{\sqrt{2}}\right) e^{-t^2} dt = \sqrt{2\pi}\, i^n H_n(y)\, e^{-\frac{y^2}{2}}$$

(see formula (14) of subsection 2). This proves equality (1).

Equality (1) gives a new proof of the inversion formula for the Fourier trans formation. Namely, it follows from the results of subsection 3 that any square-integrable function $f(x)$ can be expanded in a series of the form

$$f(x) = \sum_{n=0}^{\infty} a_n e^{-\frac{x^2}{2}} H_n(x).$$

From this expansion and formula (1) it follows that

$$F(y) = \int_{-\infty}^{\infty} e^{ixy} f(x)\, dx = \sqrt{2\pi} \sum_{n=0}^{\infty} a_n i^n e^{-\frac{y^2}{2}} H_n(y).$$

Applying formula (1) once again, we see that

$$\frac{1}{2\pi} \int_{-\infty}^{\infty} e^{-ixy} F(y) = \sum_{n=0}^{\infty} a_n e^{-\frac{x^2}{2}} H_n(x) = f(x).$$

By the same token, the inversion formula for Fourier transformation is proved. One similarly proves the Plancherel formula:

$$\int_{-\infty}^{\infty} |f(x)|^2\, dx = \frac{1}{2\pi} \int_{-\infty}^{\infty} |F(y)|^2\, dy = \sum_{n=0}^{\infty} 2^n n! \sqrt{\pi}\, |a_n|^2.$$

Let us now find the Fourier transform of the function $e^{-x^2} H_n(x)$. Employing formula (5) of subsection 2, we see that

$$\int_{-\infty}^{\infty} e^{ixy} e^{-x^2} H_n(x)\, dx = (-1)^n \int_{-\infty}^{\infty} e^{ixy} \frac{d^n}{dx^n}(e^{-x^2})\, dx.$$

Integrating $n$ times by parts and applying Cauchy's theorem, we get

$$\int_{-\infty}^{\infty} e^{ixy} e^{-x^2} H_n(x)\, dx = (iy)^n \int_{-\infty}^{\infty} e^{-x^2+ixy}\, dx =$$

$$=(iy)^n e^{-\frac{y^2}{4}} \int_{-\infty}^{\infty} e^{-\left(x-\frac{iy}{2}\right)^2} dx = (iy)^n e^{-\frac{y^2}{4}} \int_{-\infty}^{\infty} e^{-x^2} dx$$

$$= \sqrt{\pi}\,(iy)^n e^{-\frac{y^2}{4}}.$$

Hence, by virtue of the Fourier inversion formula, we get

$$\int_{-\infty}^{\infty} y^n e^{ixy} e^{-y^2} dy = \left(\frac{i}{2}\right)^n \sqrt{\pi}\, e^{-\frac{x^2}{4}} H_n(x).$$

Utilizing the orthogonality relations for Hermite polynomials, we deduce that

$$\sum_{n=0}^{\infty} \frac{(iy)^n}{2^n n!} H_n(x) = e^{\frac{y^2}{4} + ixy}$$

(see formula (6) of subsection 2).

5. **Limiting process for the dimension in the case of the group $M(n)$.** We now consider the limiting process for the dimension in the case of the group $M(n)$. As shown in §3.2, the zonal spherical functions for this group have the form

$$\frac{\Gamma(p+1) J_p(x)}{\left(\frac{x}{2}\right)^p} = \frac{\Gamma(p+1)}{\sqrt{\pi}\,\Gamma\left(p+\frac{1}{2}\right)} \int_{-1}^{1} e^{ixt} (1-t^2)^{p-\frac{1}{2}} dt,$$

where $p = (n-2)/2$. Now we make the substitution $t = u/\sqrt{p}$, replace $x$ by $x\sqrt{p}$ and take the limit as $p \longrightarrow \infty$. We get

$$\lim_{p\to\infty} \frac{\Gamma(p+1) J_p(2x\sqrt{p})}{(x\sqrt{p})^p} = \frac{1}{\sqrt{\pi}} \int_{-\infty}^{\infty} e^{2ixu-u^2} du = e^{-x^2}$$

(see formula (1) of subsection 4). An asymptotic expansion of the left-hand side of this equality for $p \longrightarrow \infty$ has been obtained in [128].

## REMARKS AND BIBLIOGRAPHICAL NOTES

### Chapter I

The theory of representations of linear transformation groups was originated by G. Frobenius [55]. Then Burnside and I. Schur found a substantially simpler approach by bringing the representation matrix itself into the foreground instead of its trace, namely the Frobenius character. The infinitesimal approach to representation theory was developed by É. Cartan [69]. The infinitesimal theory of Lie groups is expounded in the books of Pontrjagin [42], Helgason [56], Čebotarev [57], Chevalley [60], Eisenhart [61] and Kowalewski [78].

The general theory of continuous groups was, in the main, originated by L. S. Pontrjagin [42]. Invariant integration on Lie groups was introduced by Adolf Hurwitz in 1897 for proving theorems on invariants. In 1924 I. Schur [290] applied this process to representations of compact groups, in particular, the real orthogonal group. The existence of invariant measure on an arbitrary locally compact group was proved by Haar [252].

Induced representations of groups were utilized in a series of papers of I. M. Gel'fand and M. A. Naĭmark [17]. Concerning the general theory, see Mackey [277], Mautner [280] and Bruhat [240].

The theory of spherical functions on groups goes back to Cartan [244]; see Weyl [299] also. The subsequent development of the theory is given by Gel'fand, Naĭmark and Berezin [17], [105], [129], Vilenkin [116], Godement [251], Harish-Chandra [259], Mautner [281], Yoshizawa [303] and many other authors.

The results set forth here have, in the main, been obtained by I. Schur, F. Peter and H. Weyl. The fundamental result on completeness of the system of irreducible representations of a compact Lie group was proved by Peter and Weyl [284]. For infinite-dimensional representations, see Anna Hurevitsch [264]. These results were later carried over, in varying measure, to locally compact groups; see Rudin [90], Želobenko [178], Mackey [197,198], Naĭmark [204], Harish-Chandra [257–258] and Mautner [278]. Problems similar to those involved here were studied by M. G. Kreĭn [192,193].

Supplement to Chapter I. Concerning the topics touched upon here, see the books of I. M. Gel'fand, M. I. Graev and the author [14, 15], and the paper by M. A. Naĭmark and S. V. Fomin [209].

## Chapter II

The theory of trigonometrical series goes back to Bernoulli and Euler. These series were systematrically applied by Fourier and Cauchy in the 1820's for solving the problems of mathematical physics. Concerning the modern theory of trigonometrical series, see N. K. Bari, *Trigonometrical Series*, and A. Zygmund, *Trigonometrical Series*, volumes 1 and 2. The theory of Fourier integrals is expounded in the books of Bochner [5] and Titchmarsh [50]. Fourier transformation in the complex domain was first studied in detail by Wiener and Paley [12]. The proof of the inversion formula for Fourier transformation, given in the book, belongs to Gel'fand [132].

From the group-theoretic point of view, the theory of trigonometrical series and integrals is a special case of the theory of characters of commutative locally compact groups, constructed by L. S. Pontrjagin [42].

## Chapter III

The functions $P_n(x)$ were introduced almost simultaneously by Legendre and Laplace in connection with the study of spheroidal attraction. They were the first example of orthogonal polynomials. The Jacobi polynomials $P_k^{(\alpha,\beta)}(x)$ are a generalization of the functions $P_n(x)$. Concerning the general theory of orthogonal polynomials, see Jackson [22] and Szegö [46]. For a bibliography on these polynomials, see [92].

The classical theory of Legendre functions is expounded in the books of Hobson [19], Whittaker and Watson [53], Lense [79], Robin [89] and Wangerin [97, 98]. These functions are widely applied in classical and quantum physics. The study of the angular momentum operator in quantum mechanics led to a clarification of the connection between the polynomials of Legendre and Jacobi on the one hand, and representations of the group of rotations of three-dimensional euclidean space on the other. These matters are explained in the books of Bhagavantam and Venkatarayudu [2], Bauer [3], Beĭman [4], Van der Waerden [7], Wigner [11], Landau and Lifšic [30], Ljubarskiĭ [33], Heine [58], Hammermesch [72], Kahan [74], Racah [87] Weyl [99] and others.

A systematic construction of the theory of Legendre polynomials on the basis of the theory of group representations has been given by Gel'fand and Šapiro [155] see Gel'fand, Minlos and Šapiro [16] also. In these sources one can find an introduction to, and a detailed study of, the functions $P^l_{mn}(x)$, which are similar to Jacobi polynomials, yet somewhat different from them. The results of §§5.1 and 5.2 are due to the author [115,117].

The Clebsch-Gordan coefficients are employed for solving problems in spectroscopy; see Jucys, Levinsonas and Vanagas [63] and the books mentioned above. The derivation of properties of these coefficients, given here, is due basically to the author.

## Chapter IV

The function $J_0(x)$ was considered in 1738 by Daniel Bernoulli, while functions $J_n(x)$ with integral values of $n$ were considered in 1764 by Euler, and studied in detail by Bessel in 1824. Concerning the classical theory of Bessel functions, see Watson [8], Gray and Matthews [21], the handbook *Higher Transcendental Functions* [73] and Nielsen [83]. Applications to problems in mathematical physics are given in the books of Korenev [27], Lebedev [31], Morse and Feshbach [36] and Ufljand [54].

The author [113] has constructed certain sections of the theory of Bessel functions on the basis of the theory of representations of the group of euclidean plane motions. See Miller [81], Konstantinova and Sokolik [190] and Inonu and Wigner [265]. Representations of the universal covering group for $M(2)$ have been studied by Thoma [297]. For a generalization to the $\mathfrak{P}$-adic case, see Saito [289].

## Chapter V

Representations of the group of linear transformations of the straight line have been studied by Gel'fand and Naĭmark [145]; see Kohari [267] also. The functions $\Gamma(x)$ and $B(x, y)$ were introduced by Euler. The account of the theory of these functions, given here, belongs to the author. Various generalizations of the $\Gamma$-function have been considered by Godement [160], Gel'fand and Graev [143] and Gindikin [158].

The classical theory of the Macdonald and Hankel functions is presented in the treatise of Watson [8], already mentioned. The author has studied these functions on the basis of the theory of representations of the group of pseudo-euclidean plane motions. Some of the results (the composition theorem, for example) have been obtained independently by V. S. Ryko. It is possible that some of the formulas given here are new.

## Chapter VI

The group of real unitary matrices of the second order (i.e. the group of quasi-unitary matrices of the second order) was the first noncompact, noncommutative group the theory of whose representations was studied in detail (Bargmann [236]). Interest in the representations of this group and the closely related group of unitary complex matrices of the second order was evoked by the role of these groups in physics (they are isomorphic to the three-dimensional and four-dimensional Lorentz groups respectively). Representations of the group of complex second order matrices were first studied by Gel'fand and Naĭmark [144]. The results obtained here were used for a fundamental study of representations of semisimple groups (Gel'fand and Naĭmark [17, 147], Gel'fand and Graev [135, 136, 137, 138], Berezin [103, 104], Želobenko [176, 177], Ehrenpreis and Mautner [249], Harish-Chandra [253–259] and others). Many of the results of harmonic analysis on groups were first obtained on groups of the second order; moreover, all of these have not as yet been carried over to a more general case (Bogaevskiĭ [109], Gel'fand and Fomin [152], Želobenko [172–174], Naĭmark [201, 203, 207], Romm [220, 221], Kunze and Stein [269, 270] and Pukánszky [285, 287].

The exposition in this chapter basically follows Bargmann. Many of the topics are presented in the book of Gel'fand, Graev and the author [15], Chapter 7. Concerning other derivations of the Plancherel formula on the group $SL(2, R)$, see Gel'fand and Graev [136] and Pukánszky [287].

## Chapter VII

The hypergeometric function was introduced by Gauss. Concerning the classical theory of this function, see Kratzer and Franz [28], *Higher Transcendental Functions*, volume 1 [73], and Klein [77]. The account of the theory given here belongs to the author [122, 123]. Several formulas of this chapter are new.

## Chapter VIII

Concerning the theory of confluent hypergeometric functions and the closely related Whittaker functions, see Kratzer and Franz [28], Whittaker and Watson [53], Bailey [66], Buchholz [68], *Higher Transcendental Functions*, volume 1 [73], and Tricomi [94]. The theory of representations of the group of triangular third order matrices is a special case of the general theory of representations of solvable and nilpotent Lie groups (see Kirillov [188], Dixmier [167, 248]).

The connection between the theory of Whittaker functions and the theory of representations of the group of triangular matrices was established by the author [125]; see Miller [81] also. Several formulas of this chapter are new.

## Chapter IX

Schur [290], Cartan [242, 243], Weyl [300] and Brauer [238] have given the classification of representations of the orthogonal group and calculated the characters of these representations. Concerning these results, see Murnaghan [37, 82] and Weyl [10] also. Gel'fand and Cetlin [154] have calculated the matrix elements of irreducible infinitesimal representations of this group. The expression for the matrix elements of representations in terms of Euler parameters has been obtained by the author [114, 121] and Lambina [195].

The theory of spherical functions in the $n$-dimensional case is expounded in the books of Appell and Kampé de Fériet [65] and Nielsen [85]. In *Higher Transcendental Functions*, volume 2 [73], this theory is presented essentially on the basis of representation theory. The realization of representations given in our book belongs to the author, as does the theory of polyspherical functions. See Kirillov [186] also.

The connection between the theory of group representations and Gegenbauer polynomials was established by Cartan [244].

## Chapter X

Representations of the group of hyperbolic rotations of $n$-dimensional space and the related special functions were obtained, independently, by the author [114, 119, 121] and Takahashi [292]; see Hirai [262] also.

In the four-dimensional case these representations have been completely described by Gel'fand and Naĭmark [144, 16, 39, 201]. These representations, their matrix elements and related integral transformations have been studied in a number of papers (see, for example, Golodec [162], Dolginov, Toptygin and Moskalev [169–171], Želobenko [173, 174], Levinsonas and Jucys [196], Popov [218], Romm [220, 221], Šapiro [226, 226a], Èskin [228]. Takahashi [291] and Tamm [294].

The horisphere method for decomposing representations into irreducible components has been developed by Gel'fand and Graev [141, 142]. The derivation of the Fock-Mehler integral transformation on the basis of the horisphere method has been given by the author [119,121]. It can be shown that other integral transformations (of Kontorovich-Lebedev, Olevskiĭ [214] and others) are also obtained by this method. For the four-dimensional case this has been done by the author and Smorodinskiĭ [127]. Polyspherical and horispherical functions on the hyperboloid have been introduced by the author [124]. In the three-dimensional case various orthogonal coordinate systems, in which the Laplace operator admits separation of variables, have been studied by Olevskiĭ [215].

## Chapter XI

The study of properties of Bessel functions in connection with the theory of representations of the group of $n$-dimensional euclidean space motions has been carried out by the author [114,121] and independently by Orihara [283]. Ito [266] has described all irreducible unitary representations of this group. Bessel functions of matrix argument have been studied by Herz [261]. Concerning the theory of Hermite polynomials, see Szegö [46].

# BIBLIOGRAPHY

## Monographs and Textbooks


[1] N. I. Ahiezer and I. M. Glazman, *The theory of linear operators in Hilbert space*, GITTL, Moscow, 1950; English transl., Ungar, New York, 1961. MR 13, 358.

[2] S. Bhagavantum and T. Venkatarayudu, *Theory of groups and its application to physical problems*, 2nd ed., Andhra Univ., Waltair, India, 1951; Russian transl., IL, Moscow, 1959.

[3] E. Bauer, *Introduction à la théorie des groupes et à ses applications en physique quantique*, Ann. Inst. H. Poincaré 4 (1933–34), 1–170; reprint, Presses Univ. France, 1933; Russian transl., GITTL, Moscow, 1937.

[4] B. F. Beĭman, *Lectures on the application of group theory to nuclear spectroscopy*, Fizmatgiz, Moscow, 1961. (Russian)

[5] S. Bochner, *Lectures on Fourier integrals*, Ann. of Math. Studies, no. 42, Princeton Univ. Press, Princeton, N. J., 1959; Russian transl., Fizmatgiz, Moscow, 1962. MR 21 #5851.

[6] N. G. de Bruijn, *Asymptotic methods in analysis*, Bibliotheca Mathematica, vol. 4, North-Holland, Amsterdam, Noordhoff, Groningen and Interscience, New York, 1958; Russian transl., IL, Moscow, 1961. MR 20 #6003.

[7] B. L. van der Waerden, *Die gruppentheoretische Methode in der Quantenmechanik*, Die Grundlehren der math. Wissenschaften, Band 36, Springer, Berlin 1932; reprint, J. W. Edwards, Ann Arbor, Mich., 1944; Russian transl., DNVTU, Kharkov, 1938. MR 6, 145.

[8] G. N. Watson, *A treatise on the theory of Bessel functions*, 2nd ed., Cambridge Univ. Press, Cambridge and Macmillan, New York, 1944; Russian transl., Part I, IL, Moscow, 1947. MR 6, 64.

[9] A. Weil, *L'intégration dans les groupes topologiques et ses applications*, Actualités Sci. Indust., no. 869, Hermann, Paris, 1940; reprints, Princeton Univ. Press, Princeton, N. J., 1941 and J. W. Edwards, Ann Arbor, Mich., 1942; Russian transl., IL, Moscow, 1950. MR 3, 198.

[10] H. Weyl, *The classical groups. Their invariants and representations*, Princeton Univ. Press, Princeton, N. J., 1939; Russian transl., IL, Moscow, 1947, 1949. MR 1, 42.

[11] E. P. Wigner, *Gruppentheorie und ihre Anwendung auf die Quantenmechanik der Atomspectren*, J. W. Edwards, Ann Arbor, Mich., 1944; English transl., Pure and Appl. Phys., vol. 5., Academic Press, New York, 1959; Russian transl., IL, Moscow, 1961. MR 6, 39; MR 21 #5442.

[12] N. Wiener and R. Paley, *Fourier transforms in the complex domain*, Amer. Math. Soc. Colloq. Publ., vol. 19, Amer. Math. Soc., Providence, R.I., 1934; reprint, 1966; Russian transl., "Nauka", Moscow, 1964. MR 31 #5038.

[13] I. M. Gel'fand, *Lectures on linear algebra*, 2nd ed., GITTL, Moscow, 1951; English transl., Pure and Appl. Math., no. 9, Interscience, New York, 1961. MR 13, 99; MR 34 #4274.

[14] I. M. Gel'fand and N. Ja. Vilenkin, *Generalized functions*. Vol. 4: *Some applications of harmonic analysis*. Fizmatgiz, Moscow, 1961; English transl. Academic Press, New York, 1964. MR 26 #4173; MR 30 #4152.

[15] I. M. Gel'fand, M. I. Graev and N. Ja. Vilenkin, *Generalized functions*. Vol. 5 *Integral geometry and related problems in the theory of distributions*, Fizmatgiz, Moscow, 1962; English transl., Academic Press, New York, 1966. MR 28 #3324.

[16] I. M. Gel'fand, R. A. Minlos and Z. Ja Šapiro, *Representations of the rotation group and of the Lorentz group, and their applications*, Fizmatgiz, Moscow, 1958; English transl., Macmillan, New York. MR 22 #5694.

[17] I. M. Gel'fand and M. A. Naĭmark, *Unitary representations of the classical groups*, Trudy Mat. Inst. Steklov. 36 (1950); German transl., Akademie-Verlag, Berlin, 1957. MR 13, 722.

[18] I. M. Gel'fand and G. E. Šilov, *Generalized functions*. Vol. 1: *Generalized functions and operations on them*, 2nd ed., Fizmatgiz, Moscow, 1959; English transl. of 1st ed., Academic Press, New York, 1964. MR 20 #4182; MR 29 #3869.

[19] E. W. Hobson, *The theory of spherical and ellipsoidal harmonics*, Cambridge Univ. Press, Cambridge, 1931; 7th ed., Chelsea, New York, 1955; Russian transl., IL, Moscow, 1952. MR 16, 356.

[20] I. S. Gradšteĭn and I. M. Ryžik, *Tables of integrals, sums, series and products*, 4th ed., Fizmatgiz, Moscow, 1963; English transl., 4th ed., Academic Press, New York, 1965. MR 28 #5198; MR 33 #5952.

[21] A. Gray and G. B. Mathews, *A treatise on Bessel functions and their applications to physics*, 2nd ed., Macmillan, London, 1922; reprint, Dover, New York 1966; Russian transl., IL, Moscow, 1953. MR 33 #1506.

[22] D. Jackson, *Fourier series and orthogonal polynomials*, Carus Monograph Series, no. 6, Math. Assoc. of America, Buffalo, N. Y., 1941; Russian transl., IL, Moscow, 1948. MR 3, 230.

[23] M. A. Evgrafov, *Asymptotic estimates and entire functions*, GITTL, Moscow, 1957; 2nd rev. ed., Fizmatgiz, Moscow, 1962; English transl., Gordon and Breach, New York, 1961. MR 19, 642; MR 27 #4924.

[24] É. Cartan, *Leçons sur la théorie des spineurs*. I, II, Actualités Sci. Indust., nos. 643, 701, Hermann, Paris, 1938; English transl., *Theory of spinors*, MIT Press, Cambridge, Mass. and Hermann, Paris, 1967; Russian transl., IL, Moscow, 1947.

[25] ———, *Oeuvres complètes. Partie* I. *Groupes de Lie*. Vol. 1, "Selecta", Gauthier-Villars, Paris, 1939; complete, Gauthier-Villars, Paris, 1952; Russian transl., IL, Moscow, 1949. MR 14, 343.

[26] S. Kaczmarz and H. Steinhaus, *Theorie der Orthogonalreihen*, Monographs Mat., Bd. 6, Warsaw, 1935; reprint, Chelsea, New York, 1951; Russian transl. Fizmatgiz, Moscow, 1958. MR 20 #1148.

[27] B. G. Korenev, *Some problems in the theory of elasticity and heat conduction solved by Bessel functions*, Fizmatgiz, Moscow, 1960. (Russian) MR 22 #8780.

[28] A. Kratzer and W. Franz, *Transzendente Funktionen*, Mathematik und ihre Anwendungen in Physik und Technik, Reihe A, Bd. 28, Akademische Verlagsgesellschaft, Leipzig, 1960, Russian transl., IL, Moscow, 1963. MR 23 #A1843.

[29] R. Courant and D. Hilbert, *Methoden der mathematischen Physik*. Vol. 1, 2nd ed., Interscience, New York, 1943; Russian transl., 3rd ed., GITTL, Moscow, 1951; English transl., 2nd ed., Interscience, New York, 1953. MR 5, 97; MR 13, 800; MR 16, 426.

[30] L. D. Landau and E. M. Lifšic, *Mechanics*. 1, Part 1, 2nd ed., Nauka, Moscow, 1965. (Russian)

[31] N. N. Lebedev, *Special functions and their applications*, 2nd rev. ed., Fizmatgiz, Moscow, 1963; English transl., Prentice-Hall, Englewood Cliffs, N. J., 1965. MR **30** #4987; MR **30** #4988.

[32] B. M. Levitan, *Expansion in characteristic functions of differential equations of the second order*, GITTL, Moscow, 1950. (Russian) MR 12, 183.

[33] G. Ja. Ljubarskiĭ, *Group theory and its applications to physics*, Fizmatgiz, Moscow, 1957; English transl., Pergamon Press, New York, 1960. MR 21 #5441; MR 22 #7709.

[34] L. A. Ljusternik and V. I. Sobolev, *The elements of functional analysis*, GITTL, Moscow, 1951; 2nd rev. ed., "Nauka", Moscow, 1965; English transl., Hindustan, Delhi, Gordon & Breach, and Ungar, New York, 1961. MR 14, 54; MR 25 #5361; MR 25 #5362.

[35] A. I. Markuševič, *The theory of analytic functions*, GITTL, Moscow, 1950; English transl., Vols. I, II, Prentice-Hall, Englewood Cliffs, N. J., 1965; Vol. III, 1967. MR 12, 87; MR **30** #2125; MR 31 #5965.

[36] P. M. Morse and H. Feshbach, *Methods of theoretical physics*. Vols. 1, 2, McGraw-Hill, New York, 1953; Russian transl., IL, Moscow, 1958, 1960. MR 15, 583.

[37] F. D. Murnaghan, *Theory of representations of groups*, Johns Hopkins Press, Baltimore, Md., 1938; reprint, Dover, New York, 1963; Russian transl., IL, Moscow, 1950. MR 31 #258.

[38] M. A. Naĭmark, *Linear differential operators*, GITTL, Moscow, 1954; German transl., Akademie-Verlag, Berlin, 1960. MR 16, 702.

[39] ———, *Linear representations of the Lorentz group*, Fizmatgiz, Moscow, 1958; English transl., Macmillan, New York, 1964. MR 21 #4995; MR 30 #1211.

[40] ———, *Normed rings*, GITTL, Moscow, 1956; English transl., Noordhoff, Groningen, 1959; rev. ed., 1964. MR 19, 870; MR 22, 1824; MR **34** #4928.

[41] I. P. Natanson, *Constructive function theory*, GITTL, Moscow, 1949; English transl., Vol. I, Ungar, New York, 1964; Vols. II, III, 1965. MR **11**, 591; MR 33 #4529 a, b, c.

[42] L. S. Pontrjagin, *Continuous groups*, 2nd ed., GITTL, Moscow, 1954; English transl. of 1st ed., *Topological Groups*, 5th printing, Princeton Univ. Press, Princeton, N. J., 1958; of 2nd ed., Gordon & Breach, New York, 1966. MR 17, 171; MR 19, 867; MR 34 #1439.

[43] I. I. Pjateckiĭ-Šapiro, *Geometry of classical domains and theory of automorphic functions*, Fizmatgiz, Moscow, 1961; French transl., Travaux et Recherches Mathématique, no. 12, Dunod, Paris, 1966. MR 25 #231; MR 33 #5949.

[44] B. A. Rozenfel'd, *Non-euclidean geometries*, GITTL, Moscow, 1955. (Russian) MR 17, 293.

[45] M. Rose, *Multiple fields*, Wiley, New York and Chapman and Hall, London, 1955; Russian transl., IL, Moscow, 1957. MR 17, 326.

[46] G. Szegö, *Orthogonal polynomials*, Amer. Math. Soc. Colloq. Publ., vol. 23, Amer. Math. Soc., Providence, R. I., 1939; 2nd rev. ed., 1959; Russian transl., Fizmatgiz, Moscow, 1962. MR 1, 14; MR 21 #5029.

[47] Séminaire "Sophus Lie" de l'École Normale Supérieure 1954/1955, *Théorie des algèbres de Lie. Topologie des groupes de Lie*, Secrétariat Mathématique, Paris, 1955; Russian transl., IL, Moscow, 1962. MR 17, 384.

[48] I. N. Sneddon, *Fourier transforms*, McGraw-Hill, New York, 1951; Russian transl., IL, Moscow, 1955. MR 13, 29.

[49] N. Ja. Sonin, *Investigations of cylinder functions and special polynomials*, GITTL, Moscow, 1954. (Russian) MR 16, 552.

[50] E. Titchmarsh, *Introduction to the theory of Fourier integrals*, Clarendon Press, Oxford, 1937; Russian transl., 2nd ed., GITTL, Moscow, 1948.

[51] ———, *Eigenfunction expansions associated with second-order differential equations*. Vol. I, 2nd ed., Clarendon, Press, Oxford, 1962; Vol. II, 1958; Russian transl., IL, Moscow, 1960; 1961. MR 20 #1065; MR 22 #9662; MR 31 #426.

[52] C. J. Tranter, *Integral transforms in mathematical physics*, Methuen, London and Wiley, New York, 1951; Russian transl., GITTL, Moscow, 1956. MR 13, 127.

[53] E. T. Whittaker and G. N. Watson, *A course of modern analysis*. Vols. I, II, 4th ed. (reprint of 1927 ed.), Cambridge Univ. Press, New York, 1962; Russian transl., Fizmatgiz, Moscow, 1963. MR 31 #2375.

[54] Ja. S. Ufljand, *Integral transforms in problems of elasticity theory*, Idzat. Akad. Naud SSSR, Moscow, 1963. (Russian) MR 29 #6681.

[55] G. Frobenius, *Theory of characters and group representations*, Kharkov, 1937. (Russian)

[56] S. Helgason, *Differential geometry and symmetric spaces*, Pure and Appl. Math., vol. 12, Academic Press, New York, 1962; Russian transl., Izdat. "Mir", Moscow, 1964. MR 26 #2986.

[57] N. G. Čebotarev, *The theory of Lie groups*, GITTL, Moscow, 1940. (Russian)

[58] V. Heine, *Group theory in quantum mechanics: An introduction to its present usage*, Internat. Series of Monographs on Pure and Appl. Math., vol. 9, Pergamon Press, New York, 1960; Russian transl., IL, Moscow, 1963. MR 22 #4443.

[59] L. K. Hua, *Harmonic analysis of functions of several complex variables in the classical domains*, Science Press, Peking, 1958; Russian transl., IL, Moscow, 1959; English transl., Transl. Math. Monographs, vol. 6, Amer. Math. Soc., Providence, R.I., 1963. MR 23 #A3277; MR 30 #2162.

[60] C. Chevalley, (a) *Theory of Lie groups*. Vol. I, Princeton Math. Series, vol. 8 Princeton Univ. Press, Princeton, N.J., 1946; Russian transl., IL, Moscow, 1948.
(b) *Théorie des groupes de Lie*. Vols. II, III. Actualités Sci. Indust. nos. 1152, 1226, Hermann, Paris, 1951, 1955; Russian transl., IL, Moscow, 1958. MR 14, 448; MR 16, 901.

[61] L. P. Eisenhart, *Continuous groups of transformations*, Princeton Univ. Press, Princeton, N.J., 1933; reprint, Dover, New York, 1961; Russian transl., IL, Moscow, 1947. MR 23 #A1328.

[62] A. Erdélyi, *Asymptotic expansions*, Dover, New York, 1956; Russian transl., Fizmatgiz, Moscow, 1962. MR 17, 1202.

[63] A. Jucys, J. Levinsonas and V. Vanagas, *Mathematical techniques in the theory of angular momentum*, Akad. Nauk Litovsk. SSR, Inst. Fiz. i Mat. Publ. no. 3, Gosudarstv. Izdat. Politic. i Nauc. Lit. Litovsk. SSSR, Vilna, 1960. (Russian) MR 23 #B810.

[64] E. Jahnke, F. Emde and F.Lösch, *Tables of higher functions*, 6th ed., McGraw-Hill, New York, and Teubner, Leipzig, 1960; Russian transl., *Special Functions*, "Nauka", Moscow, 1964. MR 22 #5140.

[65] P. Appell and J. Kampé de Fériet, *Fonctions hypergéométriques et hypersphériques. Polynômes d'Hermite*, Gauthier-Villars, Paris, 1926.

[66] W. N. Bailey, *Generalized hypergeometric series*, Cambridge Univ. Press, Cambridge 1935; reprint, Cambridge Tracts in Math. and Math. Phys., no. 32, Stechert-Hafner, New York, 1964. MR 32 #2625.

[67] H. Boerner, *Darstellungen von Gruppen mit Berücksichtigung der Bedürfnisse der modernen Physik*, Die Grundlehren der math. Wissenschaften, Band 74, Springer-Verlag, Berlin, 1955. MR 17, 710.

[68] H. Buchholz, *Die konfluente hypergeometrische Funktion mit besonderer Berücksichtigung ihrer Anwendungen*. Band 2, Springer-Verlag, Berlin, 1953. MR 14, 978.

[69] É. Cartan, *Oeuvres complètes*. Partie 1, Gauthier-Villars, Paris, 1952. MR 14, 343.

[70] J. Dixmier, *Les $C^*$-algèbres et leurs representations*, Cahiers Scientifiques, Gauthier-Villars, Paris, 1964. MR 30 #1404:

[71] A. Erdélyi, W. Magnus, F. Oberhettinger and F. G. Tricomi, *Tables of integral transforms*. Vols. I, II, McGraw-Hill, New York, 1954. MR 15, 868; MR 16, 468.

[72] M. Hamermesh, *Group theory and its applications to physical problems*, Addison-Wesley, Reading, Mass., 1962. MR 25 #132.

[73] A. Erdélyi, W. Magnus, F. Oherhettinger, and F. G. Tricomi, *Higher transcendental functions*. Vols. I, II, III, McGraw-Hill, New York, 1953, 1955. MR 15, 419; MR 16, 586.

[74] Th. Kahan, *Théorie des groupes en physique classique et quantique*, Vol. 1, Dunod, Paris, 1960; English transl., American Elsevier, New York, 1966.

[75] J. Kampé de Fériet, *La fonction hypergéométrique de Gauss*, Mémorial des Sciences Mathématiques, Fasc. LXXXV, Gauthier-Villars, Paris, 1937.

[76] F. Klein, *Conférences sur les mathématiques* (Recueillies par A. Ziwet, traduites par L. Laugel), Chicago, 1893; Paris, 1898.

[77] F. Klein, *Vorlesungen über die hypergeometrische Funktion*, Berlin, 1933.

[78] G. Kowalewski, *Einführung in die Theorie der kontinuierlichen Gruppen*, Leipzig, 1931; Chelsea, New York, 1950. MR 12, 157.

[79] J. Lense, *Kugelfunktionen*, 2nd ed., Akademische Verlagsgesellschaft, Leipzig, 1954. MR 16, 470.

[80] D. E. Littlewood, *The theory of group characters and matrix representations of groups*, Oxford Univ. Press, New York, 1940. MR 2, 3.

[81] W. Miller, *On Lie algebras and some special functions of mathematical physics*, Mem. Amer. Math. Soc. 50 (1964). MR 30 #3246.

[82] F. D. Murnaghan, *The unitary and rotation groups*, Washington, 1962.

[83] N. Nielsen, *Handbuch der Theorie der Cylinderfunktionen*, Teubner, Leipzig, 1904.

[84] ———, *Handbuch der Theorie der Gamma-funktion*, Teubner, Leipzig, 1906.

[85] ———, *Théorie des functions métasphériques*, Gauthier-Villars, Paris, 1911.

[86] F. Oberhettinger, *Tabellen zur Fourier Transformationen*, Springer-Verlag, Berlin, 1957. MR 18, 481.

[87] G. Racah, *Group theory and spectroscopy*, Princeton lectures, Springer-Verlag, Berlin, 1951.

[88] E. D. Rainville, *Special functions*, Macmillan, New York, 1960. MR 21 #6447.

[89] L. Robin, *Fonctions sphériques de Legendre et fonctions sphéroidales*, Vols. I, II, Gauthier-Villars, Paris, 1957, 1958. MR 19, 954.

[90] W. Rudin, *Fourier analysis on groups*, Interscience Tracts in Pure and Appl. Math., no. 12, Interscience, New York, 1962. MR 27 #2808.

[91] F. W. Schäfke, *Einführung in die Theorie der speziellen Funktionen der mathematischen Physik*, Springer-Verlag, Berlin, 1963. MR 28 #252.

[92] J. B. Shohat, E. Hille and J. L. Walsh, *A bibliography on orthogonal polynomials*, Bulletin Nat. Research Council, no. 103, Nat. Research Council of Nat. Acad. Sci., Washington, D.C., 1940. MR 2, 197.

[93] L. J. Slater, *Confluent hypergeometric functions*, Cambridge Univ. Press, New York, 1960. MR 21 #5753.

[94] F. G. Tricomi, *Fonctions hypergéométriques confluentes*, Mémorial des Sciences Mathématiques, Fasc. CXL, Gauthier-Villars, Paris, 1960. MR 22 #11163.

[95] C. Truesdell, *An essay toward a unified theory of special functions based upon the functional equation* $\partial F(z, x)/\partial z = F(z, \alpha + 1)$, Ann of Math. Studies, no. 18, Princeton Univ. Press, Princeton, N. J., 1948. MR 9, 431.

[96] B. L. van der Waerden, *Gruppen von linearen Transformationen*, Chelsea, New York, 1948.

[97] A. Wangerin, *Theorie des Potentials und der Kugelfunktionen*, Bd. 1, 2, Leipzig, 1922.

[98] ———, *Theorie der Kugelfunktionen*, Encyclopädie, vol. II, Heft 1, 2, Teubner, Leipzig, pp. 695–759.

[99] H. Weyl, *Gruppentheorie und Quantenmechanik*, Leipzig, 1928.

[100] E. P. Wigner, *The application of group theory to the special functions of mathematical physics*, Princeton lectures, Parts I, II, Spring, 1955.

## Papers from journals

[101] E. L. Akim and A. A. Levin, *A generating function for the Clebsch-Gordan coefficients*, Dokl. Akad. Nauk SSSR 138 (1961), 503–505 = Soviet Math. Dokl. 2 (1961), 629–631. MR 25 #5133.

[102] Ju. M. Berezanskiĭ, *On certain normed rings constructed with orthogonal polynomials*, Ukrain. Mat. Ž. 3 (1951), 412–432. (Russian) MR 14, 884.

[103] F. A. Berezin, *Representation of complex semi-simple Lie groups in Banach space*, Dokl. Akad. Nauk SSSR 110 (1956), 897–900. (Russian) MR 19, 292.

[104] ———, *Laplace operators on semi-simple Lie groups*, Trudy Moskov. Mat. Obšč. 6 (1957), 371–463; English transl., Amer. Math. Soc. Transl. (2) 21 (1962), 239–339. MR 19, 867; MR 27 #245.

[105] F. A. Berezin and I. M. Gel'fand, *Some remarks on the theory of spherical functions on symmetric Riemannian manifolds*, Trudy Moskov. Mat. Obšč. 5 (1956), 311–351; English transl., Amer. Math. Soc. Transl., (2) 21 (1960) 193–238. (Russian) MR 19, 152; MR 27 #1910.

[106] F. A. Berezin, I. M. Gel'fand, M. I. Graev and M. A. Naĭmark, *Group representations*, Uspehi. Mat. Nauk 11 (1956), no. 6 (72), 13–40; English transl., Amer. Math. Soc. Transl. (2) 16 (1960), 325–353. MR 19, 662; MR 22 #8367.

[107] F. A. Berezin and F. I. Karpelevič, *Zonal spherical functions and Laplace operators on some symmetric spaces*, Dokl. Akad. Nauk SSSR 118 (1958), 9–12. (Russian) MR 20 #1722.

[108] A. N. Bogaevskiĭ, *Computation of zonal spherical functions*, Dokl. Akad. Nauk SSSR 129 (1959), 484–487. (Russian) MR 22 #5696.

[109] ———, *Harmonic functions on GL*(2), Dokl. Akad. Nauk SSSR 153 (1963), 751–753 = Soviet Math. Dokl. 4 (1963), 1734–1737. MR 29 #193.

[110] F. Bruhat, *Travaux de Harish-Chandra*, Séminaire Bourbaki: 9ième année, 1956/57, 2nd ed., exposé 143, Secrétariat mathématique, Paris, 1959; Russian transl., Matematika 6 (1962), 43–50.

[111] T. S. Bhanu Murti, *Plancherel's measure for the factor-space* $SL(n; R)/SO(n; R)$, Dokl. Akad. Nauk SSSR 133 (1960), 503–506 = Soviet Math. Dokl. 1 (1960), 860–861. MR 23 #A2481.

[112] ———, *The asymptotic behaviour of zonal spherical functions on the Siegel upper half-plane*, Dokl. Akad. Nauk SSSR 135 (1960), 1027–1030 = Soviet Math. Dokl. 1 (1960), 1325–1329. MR 23 #A967.

[113] N. Ja. Vilenkin, *Bessel functions and representations of the group of Euclidean motions*, Uspehi Mat. Nauk 11 (1956), no. 3 (60), 69–112. (Russian) MR 19, 153.

[114] ———, *The matrix elements of irreducible unitary representations of the group of real orthogonal matrices and group of Euclidean* $(n-1)$*-dimensional space motions*, Dokl. Akad. Nauk SSSR 113 (1957), 16–19. (Russian) MR 21 #3508.

[115] ———, *Deduction of certain properties of Jacobi polynomials from the theory of group representation*, Moskov. Gos. Ped. Inst. Učen. Zap. 108 (1957), 59–71. (Russian) MR 22 #5759.

[116] ———, *On the theory of associated spherical functions on Lie groups*, Mat. Sb. 42 (84) (1957), 485–496. (Russian) MR 20 #2634.

[117] ———, *On the generating function for Jacobi polynomials*, Uspehi Mat. Nauk 12 (1957), no. 6 (78), 137–142. (Russian) MR 20 #1804.

[118] ———, *Continuous analogue of the addition theorem for Jacobi polynomials*, Uspehi Mat. Nauk 13 (1958), no. 2 (80), 157–161. (Russian) MR 20 #1802.

[119] ———, *The matrix elements of irreducible unitary representations of a group of Lobachevsky space motions and the generalized Fock-Mehler transformation*, Dokl. Akad. Nauk SSSR 118 (1958). 219–222. (Russian) MR 21 #2877.

[120] ———, *Some relations for Gegenbauer functions*, Uspehi Mat. Nauk 13 (1958), no. 3 (81), 167–172. (Russian) MR 20 #2484.

[121] ———, *Special functions associated with class one representations of the motion group of spaces of constant curvature*, Trudy Moskov. Mat. Obšč. 12 (1963), 185–257 = Trans. Moscow Math. Soc. 1963, 209–290. MR 29 #191.

[122] ———, *The hypergeometric function and representations of the group of real second order matrices*, Mat. Sb. 64 (106) (1964), 497–520; English transl., Amer. Math. Soc. Transl. (2) 60 (1966), 159–184. MR 29 #3679.

[123] ———, *Functional composition theorems for the hypergeometric function*, Mat. Sb. 65 (107) (1964), 28–46. (Russian) MR 29 #3680.

[124] ———, *Polyspherical and horispherical functions*, Mat. Sb. 68 (110) (1965), 432–443. (Russian) MR 33 #672.

[125] ———, *Whittaker functions and representations of the group of third order traiangular matrices*, Mat. Sb. (to appear).

[126] N. Ja. Vilenkin, E. L. Akim and A. A. Levin, *Matrix elements of irreducible unitary representations of the group of Euclidean motions of three-dimensional space and their properties*, Dokl. Akad. Nauk SSSR 112 (1957), 987–989. (Russian) MR 20 #998.

[127] N. Ja. Vilenkin and Ja. A. Smorodinskiĭ, *Invariant expansions of relativistic amplitudes*, Ž. Èksper. Teoret. Fiz. 46 (1964), 1793–1808 = Soviet Physics JETP 19 (1964), 1209–1218. MR 31 #6572.

[128] N. Ja. Vilenkin and V. V. Cukerman, *On an asymptotic formula for a Bessel function*, Ž. Vyčisl. Mat. i Mat. Fiz. 4 (1964), 1097–1102. (Russian) MR 30 #1265.

[129] I. M. Gel'fand, *Spherical functions on symmetric Riemann spaces*, Dokl. Akad. Nauk SSSR 70 (1950), 5–8; English transl., Amer. Math. Soc. Transl. (2) 37 (1964), 39–43. MR 11, 498.

[130] ———, *The centre of an infinitesimal group ring*, Mat. Sb. 26 (68) (1950), 103–112. (Russian) MR 11, 498.

[131] ———, *The structure of a ring of rapidly decreasing functions on a Lie group*, Dokl. Akad. Nauk. SSSR 124 (1959), 19–21. (Russian) MR 22 #3987

[132] ———, *On the Fourier transformation formula*, Mat. Prosveščenie 5 (1960), 155–159. (Russian)

[133] ———, *On some problems of functional analysis*, Uspehi Mat. Nauk 11 (1956), no. 6 (72), 3–12; English transl., Amer. Math. Soc. Transl. (2) 16 (1960), 315–324. MR 19, 293; MR 22 #8366.

[134] ———, *Integral geometry and its relation to the theory of representations*, Uspehi Mat. Nauk 15 (1960), no. 2 (92), 155–164 = Russian Math. Surveys 15 (1960), no. 2, 143–151. MR 26 #1903.

[135] I. M. Gel'fand and M. I. Graev, *Unitary representations of the real unimodular group (principal nondegenerate series)*, Izv. Akad. Nauk SSSR Ser. Mat. 17 (1953), 189–248; English transl., Amer. Math. Soc. Transl. (2) 2 (1956), 147–205. MR 15, 199; MR 17, 876.

[136] ———, *Analogue of the Plancherel formula for the classical groups*, Trudy Moskov. Mat. Obšč. 4 (1955), 375–404; English transl., Amer. Math. Soc. Transl. (2) 9 (1958), 123–154. MR 17, 173; MR 19, 1181.

[137] ———, *Traces of unitary representations of the real unimodular group*, Dokl. Akad. Nauk SSSR 100 (1955), 1037–1040. (Russian) MR 16, 795.

[138] ———, *Fourier transforms of rapidly decreasing functions on complex semi-simple groups*, Dokl. Akad. Nauk SSSR 131 (1960), 496–499 = Soviet Math. Dokl. 1 (1960), 276–279. MR 22 #9876.

[139] ———, *Construction of irreducible representations of simple algebraic groups over a finite field*, Dokl. Akad. Nauk SSSR 147 (1962), 529–532 = Soviet Math. Dokl. 3 (1962), 1646–1648. MR 26 #6271.

[140] ———, *Categories of group representations and the classification problem of irreducible representations*, Dokl. Akad. Nauk SSSR 146 (1962), 757–760 = Soviet Math. Dokl. 3 (1962), 1378–1381. MR 26 #237.

[141] ———, *Geometry of homogeneous spaces, representations of groups in homogeneous spaces and related questions of integral geometry*. I, Trudy Moskov. Mat. Obšč. 8 (1959), 321–390; addendum, 9 (1960), 562; English transl., Amer. Math. Soc. Transl. (2) 37 (1964), 351–429. MR 23 #A4013.

[142] ———, *An application of the horisphere method to the spectral analysis of functions in the real and imaginary Lobatchevskiĭ spaces*, Trudy Moskov. Mat. Obšč. 11 (1962), 243–308. (Russian) MR 26 #6306.

[143] ———, *Representations of the group of second-order matrices with elements in a locally compact field and special functions on locally compact fields*, Uspehi Mat. Nauk 18 (1963), no. 4 (112), 29–99. (Russian) MR 27 #5864.

[144] I. M. Gel'fand and M. A. Naimark, *Unitary representations of the Lorentz group*, Izv. Akad. Nauk SSSR Ser. Mat. 11 (1947), 411–504. (Russian) MR 9, 495.

[145] ———, *Unitary representations of the group of linear transformations of the straight line*, Dokl. Akad. Nauk SSSR 55 (1947), 567–570. (Russian) MR 8, 563.

[146] ———, *Normed rings with involutions and their representations*, Izv. Akad. Nauk SSSR Ser. Mat. 12 (1948), 445–480. (Russian) MR 10, 199.

[147] ———, *Unitary representations of a unimodular group containing an identity representation of the unitary subgroup*, Trudy Moskov. Mat. Obšč. 1 (1952), 423–475; German transl., Appendix 1 to the German transl. of [17]. MR 14, 352.

[148] I. M. Gel'fand and I. I. Pjateckiĭ-Šapiro, *Theory of representations and the theory of automorphic functions*, Uspehi Mat. Nauk 14 (1959), no. 2 (86), 171–194; English transl., Amer. Math. Soc. Transl. (2) 26 (1963), 173–200. MR 22 #2912; MR 27 #1840.

[149] ———, *Unitary representations in uniform spaces with discrete stationary subgroups*, Dokl. Akad. Nauk SSSR 147 (1962), 17–20 = Soviet Math. Dokl. 3 (1962), 1528–1531. MR 26 #260.

[150] ———, *Unitary representations in the space $G/\Gamma$, where $G$ is a group of real matrices of order $n$ and $\Gamma$ is a subgroup of integral matrices*, Dokl. Akad. Nauk SSSR 147 (1962), 275–278 = Soviet Math. Dokl. 3 (1962), 1574–1577. MR 27 2586.

[151] ———, *Automorphic functions and the theory of representations*, Trudy Moskov. Mat. Obšč. 12 (1963), 389–412 = Trans. Moscow Math. Soc. 1963, 438–464. MR 28 #3115.

[152] I. M. Gel'fand and S. V. Fomin, *Geodesic flows on manifolds of constant negative curvature*, Uspehi Mat. Nauk 7 (1952), no. 1 (47), 118–137; English transl., Amer. Math. Soc. Transl. (2) 1 (1955), 49–55. MR 14, 660; MR 17, 514.

[153] I. M. Gel'fand and M. L. Cetlin, *Finite-dimensional representations of the group of unimodular matrices*, Dokl. Akad. Nauk SSSR 71 (1950), 825–828. (Russian) MR 12, 9.

[154] ———, *Finite-dimensional representations of the group of orthogonal matrices*, Dokl. Akad. Nauk SSSR 71 (1950), 1017–1020. (Russian) MR 11, 639.

[155] I. M. Gel'fand and Z. Ja. Šapiro, *Representations of the group of rotations of three-dimensional space and their applications*, Uspehi Mat. Nauk 7 (1952), no. 1 (47), 3–117; English transl., Amer. Math. Soc. Transl. (2) 2 (1956), 207–316. MR 13, 911; MR 17, 875.

[156] I. M. Gel'fand and A. M. Jaglom, *General Lorentz invariant equations and infinite-dimensional representations of the Lorentz group*, Ž. Èksper. Teoret. Fiz. 18 (1948), 703–733. (Russian) MR 10, 583.

[157] S. G. Gindikin, *The trace formula and zeta-function on certain symmetric spaces*, Moskov. Gos. Ped. Inst. Učen. Zap. **188** (1962), 23–53. (Russian)

[158] ———, *Analysis in homogeneous domains*, Uspehi Mat. Nauk **19** (1964), no. 4 (118), 3–92. (Russian) MR 30 #2167.

[159] S. G. Gindikin and F. I. Karpelevič, *Plancherel measure for symmetric Riemannian spaces of non-positive curvature*, Dokl. Akad. Nauk SSSR **145** (1962), 252–255 = Soviet Math Dokl. **3** (1962), 962–965. MR 27 #240.

[159a] ———, *A problem of integral geometry*, The collection "In memoriam N. B Čebotarev", Izdat. Kazan. Univ., Kazan, 1964, 30–43. (Russian) MR 33 #4875.

[160] R. Godement, *Où l'on généralise une intégral étudiée par C. L. Siegel et généralisant le fonction* Γ, Séminaire Henri Cartan, 10e année: 1957/58, Fonctions Automorphes, École Normale Supérieure, Paris, 1958; Russian transl., Matematika **4** (1960), 39–50.

[161] ———, *Fonctions holomorphes de carré sommable dans le demi-plan Siegel*, Séminaire Henri Cartan, 10e année: 1957/58, Fonctions Automorphes, École Normale Supérieure Paris, 1958; Russian transl., Matematika **4** (1960), 51–67.

[162] V. Ja. Golodec, *Matrix elements of irreducible unitary and spinor representations of the proper Lorentz group*, Vesci Akad. Navuk BSSR Ser. Fiz.-Tehn Navuk, 1961 19–28. (Russian) MR 27 #2584.

[163] M. I. Graev, *Unitary representations of real simple Lie groups*, Trudy Moskov. Mat. Obšč. 7 (1958), 335–389; English transl., Amer. Math. Soc. Transl. (2) 66 (1967), 1–62. MR 21 #3510.

[164] ———, *Irreducible unitary representations of certain classes of real simple Lie groups*, Dokl. Akad. Nauk SSSR **127** (1959), 13–16. (Russian) MR 22 #6867.

[165] M. I. Graev, F. I. Karpelevič and A. A. Kirillov, *The theory of representations of Lie groups*, Proc. Fourth All-Union Math. Congress, vol II: Sectional Lectures, 1961, "Nauka", Leningrad, 1964, pp. 275–281.

[166] J. Dixmier, *Quelques résultats d'Harish-Chandra*, Séminaire Bourbaki: 4ième année: 1951/52. Textes des Conferences, 2nd. ed., exposé 50, Secrétariat Mathématique, Paris, 1959; Russian transl., Matematika 6 (1962) 23–35.

[167] ———, *Sur les représentations unitaires des groupes de Lie nilpotents*. II, Bull. Soc. Math. France **85** (1957), 325–388; Russian transl., Matematika 5 (1961), 52–114. MR **20** #1928.

[168] ———, *Sur les représentations unitaires des groupes de Lie algébriques*, Ann. Inst. Fourier Grenoble 7 (1957), 315–328; Russian transl., Matematika 8 (1964), 69–79. MR **20** #5820.

[169] A. Z. Dolginov, *Relativistic spherical functions*, Ž. Èksper. Teoret. Fiz. **30** (1956), 746–755; supplement to **30**, no. 4, 6. (Russian) MR **18**, 176.

[170] A. Z. Dolginov and I. N. Toptygin, *Relativistic spherical functions*. II, Ž. Èksper. Teoret. Fiz. **37** (1959), 1441–1451 = Soviet Physics JETP **10** (1960), 1022–1028. MR **22** #6545.

[171] A. Z. Dolginov and A. N. Moskalev, *Relativistic spherical functions*. III, Ž. Èksper. Teoret. Fiz. 37 (1959), 1697–1707 = Soviet Physics JETP **10** (1960), 1202–1208. MR **22** #6546.

[172] D. P. Želobenko, *A description of a certain class of Lorentz group representations*, Dokl. Akad. Nauk SSSR **121** (1958), 586–589. (Russian) MR **21** #2920.

[173] ———, *Structure of the group ring of the Lorentz group*, Dokl. Akad. Nauk SSSR **126** (1959), 482–485. (Russian) MR **22** #3988.

[174] ———, *Linear representations of the Lorentz group*, Dokl. Akad. Nauk SSSR **126** (1959), 935–938. (Russian) MR **22** #906.

[175] ———, *Description of all irreducible representations of an arbitrary connected Lie group*, Dokl. Akad. Nauk SSSR **139** (1961), 1291–1294 = Soviet Math. Dokl. 2 (1961), 1076–1079. MR **24** #A1969.

[176] ———, *Classical groups. Spectral analysis of finite-dimensional representations*, Uspehi. Mat. Nauk **17** (1962), no. 1 (103), 27–120. (Russian) MR **25** #129.

[177] ———, *The theory of linear representations of complex and real Lie groups*, Trudy Moskov. Mat. Obšč. 12 (1963), 53–98 = Trans. Moscow Math. Soc. 1963, 57–110. MR **29** #2330.

[178] ———, *Harmonic analysis of functions on semisimple Lie groups*, I, Izv. Akad. Nauk SSSR Ser. Mat. 27 (1963), 1343–1394; English transl., Amer. Math. Soc. Transl. (2) **54** (1966), 177–230. MR **31** #2353.

[179] A. Selberg, *Harmonic analysis and discontinuous groups in weakly symmetric Riemann spaces with applications to Dirichlet series*, J. Indian Math Soc. 20 (1956), 47–87; Russian transl., Matematika 1 (1957), 3–28. MR 19, 531.

[180] F. I. Karpelevič, *Geodesics and harmonic functions on symmetric spaces*, Dokl. Akad. Nauk SSSR 124 (1959), 1199–1202. (Russian) MR 21 #3013.

[181] ———, *Horispherical radial parts of Laplace operators on symmetric spaces*, Dokl. Akad. Nauk SSSR 143 (1962), 1034–1037 = Soviet Math. Dokl. 3 (1962), 528–531. MR 25 #4032.

[182] P. Cartier, *On the character formula of H. Weyl*, Bull. Amer. Math. Soc. 67 (1961), 228–230; Russian transl., Matematika 6 (1962), 139–141. MR 26 #3828.

[183] ———, *Représentations des groupes de Lie (d'après Harish-Chandra)*, Séminaire Bourbaki, 6ième année: 1953/54. Textes des conferences, 2nd ed., exposé 96, Secrétariat Mathématique, Paris, 1959; Russian transl., Matematika 6 (1962), 33–41.

[184] G. I. Kac, *Linear representations of the unimodular group*, Nauk. Zap. Žitomir Ped. Inst. Fiz.-Matem. Ser. 3 (1956), 7–61. (Ukranian)

[185] ———, *Generalized functions on a locally compact group and decompositions of unitary representations*, Trudy Moskov. Mat. Obšč. 10 (1961), 3–40. (Russian) MR 27 #5863.

[186] A. A. Kirillov, *The representations of the group of rotations of an n-dimensional Euclidean space by spherical vector fields*, Dokl. Akad. Nauk SSSR 116 (1957), 538–541. (Russian) MR 20 #3930.

[187] ———, *On unitary representations of nilpotent Lie groups*, Dokl. Akad. Nauk SSSR 130 (1960), 966–968 = Soviet Math. Dokl. 1 (1960), 108–110. MR 24 #A3240.

[188] ———, *Unitary representations of nilpotent Lie groups*, Uspehi Mat. Nauk 17 (1962), no. 4 (106), 57–110. (Russian) MR 25 #5396.

[189] ———, *Infinite-dimensional representations of the complete matrix group*, Dokl. Akad. Nauk SSSR 144 (1962), 37–39 = Soviet Math. Dokl. 3 (1962), 652–655. MR 25 #3122.

[190] E. L. Konstantinova and G. A. Sokolik, *The two-dimensional Schrödinger equation and representations of the group of plane motions*, Ž. Èksper. Teoret. Fiz. 30 (1956), 430–431. (Russian)

[191] B. Kostant, *A formula for the multiplicity of a weight*, Trans. Amer. Math. Soc. 93 (1959), 53–79; Russian transl., Matematika 6 (1962), 133–152. MR 22 #80.

[192] M. G. Kreĭn, *A principle of duality for bicompact groups and quadratic block algebras*, Dokl. Akad. Nauk SSSR 69 (1949), 725–728. (Russian) MR 11, 491.

[193] ———, *Hermitian-positive kernels on homogeneous spaces*, I, II, Ukrain. Mat. Ž. 1 (1949), no. 4, 64–98; 2 (1950), no. 1, 10–59; English transl., Amer. Math. Soc. Transl. (2) 34 (1963), 69–108; 109–164. MR 14, 480; MR 12, 719.

[194] E. N. Lambina, *Matrix elements of irreducible unitary representations of the group $K_4$ of orthogonal matrices of Euclidean four space*, Dokl. Akad. Nauk BSSR 6 (1962), 613–615. (Russian) MR 26 #261.

[195] ———, *Matrix elements of irreducible unitary representations of the group $K_n$ of orthogonal matrices of n-dimensional Euclidean space*, Dokl. Akad. Nauk BSSR 9 (1965), 77–81. (Russian) MR 31 #2348.

[196] J. Levinsonas and A. P. Jucys, *Reduction of a direct product of representations of the proper homogeneous Lorentz group*, Trudy Akad. Nauk Litov. SSR Ser. B, 1958, no. 4(16), 3–16. (Russian)

[197] G. W. Mackey, *Functions on locally compact groups*, Bull. Amer. Math. Soc. 56 (1950), 385–412; Russian transl., Uspehi Mat. Nauk 8 (1953), no. 4 (56), 95–129. MR 12, 588; MR 15, 101.

[198] ———, *Infinite dimensional group representations*, Colloq. Lecture given at Stillwater, Okla., Sixty-Sixth Meeting of Amer. Math. Soc., Aug. 29–Sept. 1, 1961 (with added appendix and recent developments in Bull. Amer. Math. Soc. 69 (1963), 628–686); Russian transl., Matematika 6 (1962), 57–103. MR 27 #3745.

[199] V. B. Mandel'cveĭg, *Irreducible representations of the group $SU_3$*, Ž. Èksper Teoret. Fiz. 47 (1964), 1836–1846. = Soviet Physics JETP 20 (1965), 1237–1243.

[200] ———, *Decomposition of a representation of a reductive Lie algebra into representations of its regular reductive subalgebras of maximal rank*, Dokl. Akad. Nauk SSSR 162 (1965), 1245–1247 = Soviet Math. Dokl. 6 (1965), 851–853. MR 31 #5934.

[201] M. A. Naĭmark, *On irreducible linear representations of the proper Lorentz group*, Dokl. Akad. Nauk SSSR **97** (1954), 969–972. (Russian) MR **16**, 218.

[202] ———, *On the description of all unitary representations of the complex classical groups*. I, Mat. Sb. 35 (77) (1954), 317–356; II, ibid., 37 (79) (1955), 121–140; English transl., Amer. Math. Soc. Transl (2) **9** (1958), 155-193; 195–214. MR 16, 567; MR 17, 61; MR 19, 1181.

[203] ———, *Linear representations of the Lorentz group*, Uspehi Mat. Nauk 9 (1954), no. 4 (62), 19–93; English transl., Amer. Math. Soc. Transl. (2) 6 (1957), 379–458. MR 16, 566.

[204] ———, *A continuous analogue of Schur's lemma and its application to Plancherel's formula for the complex classical groups*, Izv. Akad. Nauk SSSR Ser. Mat. 20 (1956), 3–16; English transl., Amer. Math. Soc. Transl. (2) **9** (1958), 217–231. MR 17, 875; MR 19, 1181.

[205] ———, *On irreducible linear representations of the complete Lorentz group*, Dokl. Akad. Nauk SSSR 112 (1957), 583–586. (Russian) MR 20 #1240.

[206] ———, *On the resolution of irreducible representations of the principal series of a complex unimodular group of order n into representations of a second order complex unimodular group*, Dokl. Akad. Nauk SSSR 121 (1958) 590–593. (Russian) MR 20 #7229.

[207] ———, *Decomposition of a tensor product of irreducible representations of the proper Lorentz group into irreducible representations*. I, Trudy Moskov. Mat. Obšč. **8** (1959), 121–153; II, ibid., **9** (1960), 237–282; III, ibid, **10** (1961), 181–216; English transl., Amer. Math. Soc. Transl. (2) **36** (1964), 101–136; 137–187; 189–229. MR 22 #4966; MR 26 #6304a; MR 26 #6304b.

[208] ———, *Decomposition into factor representations of unitary representations of locally compact groups*, Sibirsk. Mat. Ž. 2 (1961), 88–99. (Russian) MR 24 #A187.

[209] M. A. Naĭmark and S. V. Fomin, *Continuous direct sums of Hilbert spaces and some of their applications*, Uspehi Mat. Nauk **10** (1955), no. 2 (64), 111–142; English transl., Amer. Math. Soc. Transl. (2) 5 (1957), 35–65. MR 17, 65; MR **18**, 748.

[210] M. N. Olevskiĭ, *Solution du problème de Cauchy et de certains problèmes limites pour l'équation des ondes, l'équation de la chaleur et l'équation de Laplace dans les espaces à courbure constante*, Dokl. Akad. Nauk SSSR **33** (1941), 282–286. MR 5, 98.

[211] ———, *On a generalization of Bessel functions*, Dokl. Akad. Nauk SSSR **40** (1943), 5–10. MR 6, 64.

[212] ———, *Quelques théorèmes de la moyenne dans les espaces à courbure constante*, Dokl. Akad. Nauk SSSR **45** (1944), 95–98. MR 6, 230.

[213] ———, *Solution du problème de Cauchy pour l'equation des ondes dans un espace à n dimensions à courbure constante*, Dokl. Akad. Nauk SSSR **46** (1945), 3–6. MR 6, 230.

[214] ———, *On the representation of an arbitrary function in the form of an integral with a kernel containing a hypergeometric function*, Dokl. Akad. Nauk SSSR **69** (1949), 11–14. (Russian) MR 11, 352.

[215] ———, *Triorthogonal systems in spaces of constant curvature, in which the equation $\Delta_2 u + \lambda u = 0$ allows a complete separation of variables*, Mat. Sb. **27** (69) (1950), 379–426. (Russian) MR 12, 415.

[216] ———, *The generalized wave equation, the equation of heat conduction, and special functions*, Trudy Moskov. Inst. Meh. i Èlektr. Sel'sk. Hozjaĭstv. **4** (1959), 129–136. (Russian)

[217] I. I. Pjateckiĭ-Šapiro, *Some questions in harmonic analysis in homogeneous cones*, Dokl. Akad. Nauk SSSR **116** (1957), 181–184. (Russian) MR 23 #A475.

[218] V. S. Popov, *On the theory of relativistic transformations of the wave functions and density matrix of particles with spin*, Ž. Èksper. Teoret. Fiz. **37** (1959), 1116–1126 = Soviet Physics JETP **10** (1960), 794–800. MR 22 #4463.

[219] V. S. Popov and È. I. Dolinskiĭ, *Group properties of complex angular momentum*, Ž. Èksper. Teoret. Fiz. **46** (1964), 1829–1841 = Soviet Physics JETP **19** (1964), 1232–1239. MR 32 #899.

[220] B. D. Romm, *Decomposition into irreducible representations of a tensor product of two irreducible representations of the real Lorentz group (the case of two discrete series)*, Izv. Akad. Nauk SSSR Ser. Mat. **28** (1964), 855–866. (Russian) MR 30 #196.

[221] ———, *Expansion into irreducible representations of the restrictions of representations of the principal series of the proper Lorentz group to the real Lorentz group*, Dokl. Akad. Nauk SSSR **152** (1963), 59–62 = Soviet Math. Dokl. **4** (1963), 1257–1260. MR 28 #161.

[221a] ———, *An analogue of Plancherel's formula for the real unimodular group of third order*, Dokl. Akad. Nauk SSSR **160** (1965), 1269–1270 = Soviet Math. Dokl. 6 (1965), 315–316. MR **30** #3942.

[222] V. B. Ryvkin, *Representation of the Clebsch-Gordan coefficients in the form of finite-difference analogues of Jacobi polynomials*, Dokl. Akad. Nauk BSSR **3** (1959), 183–185. (Russian)

[223] V. I. Semjanistyĭ, *Some integral transformations and integral geometry in an elliptic space*, Trudy Sem. Vektor. Tenzor. Anal. 12 (1963), 397–441. (Russian) MR 29 #3879.

[224] G. A. Sokolik, *Representations of the general Lorentz group and classification of relativistic equations*, Izv. Vysš. Učebn. Zaved. Fizika 5 (1962), 54–59. (Russian)

[225] ———, *A new class of representations of the full Lorentz group*, Ž. Èksper. Teoret. Fiz. **36** (1959), 1098–1102 = Soviet Physics JETP **9**, 781–783. MR **21** #7735.

[226] I. S. Šapiro, *Expansion of a wave function in irreducible representations of the Lorentz group*, Dokl. Akad. Nauk SSSR 106 (1956), 647–649. (Russian) MR 17, 1181.

[226a] ———, *Expansion of the scattering amplitude in relativistic spherical functions*, Ž. Èksper. Teoret. Fiz. **43** (1962), 1727–1730 = Soviet Physics JETP **16** (1963), 1219–1221 (See also Phys. Lett. 1 (1962), 253–255.). MR 27 #3279.

[227] Ju. M. Širokov, *A group-theoretical consideration of the basis of relativistic quantum mechanics*. I, II, III, Ž. Èksper. Teoret. Fiz. **33** (1957), 861–872, 1196–1207, 1208–1214 = Soviet Physics JETP **33** (6) (1958), 664–673, 919–928, 929–935. MR 21 #2480; MR 21 #2481; MR 21 #2482.

[228] L. D. Èskin, *Calculation of a spherical function of the group $D_2$*, Učen. Zap. Kazan. Univ. 115 (1955), no. 7, 19–24. (Russian)

[229] ———, *Remarks on the Laplace operators on a unimodular group*, Izv. Vysš. Učebn. Zaved. Matematika **1959**, no. 2 (9), 259–269. (Russian) MR 28 #163.

[230] ———, *On the theory of relativistic spherical functions*, Nauč. Dokl. Vysš. Skoly. Fiz.-Mat. Nauki 2 (1959), 95–97. (Russian)

[231] ———, *On the matrix elements of irreducible representations of the Lorentz group*, Izv. Vysš. Učebn. Zaved. Matematika 1961, no. 6 (25), 180–184. (Russian) MR 26 #5098.

[232] ———, *A ζ-function on the group of unitary matrices*, Dokl. Akad. Nauk SSSR 152 (1963), 1327–1328 = Soviet Math. Dokl. 4 (1963), 1579–1580. MR 27 #4893.

[233] ———, *The heat conduction equation on Lie groups*, The collection "In memoriam N. G. Čebotarev, 1894–1947", Izdat. Kazan. Univ., Kazan, 1964, pp. 113–132. (Russian) MR 34 #6353.

[234] A. M. Jaglom, *Certain types of random fields in n-dimensional space similar to stationary stochastic processes*, Teor. Verojatnost. i Primenen. 2 (1957), 292–338. (Russian) MR 20 #1353.

[235] ———, *Positive-definite functions and homogeneous random fields on groups and homogeneous spaces*, Dokl. Akad. Nauk SSSR 135 (1960), 1342–1345 = Soviet Math. Dokl. 1 (1960), 1402–1405. MR 23 #A2936.

[236] V. Bargmann, *Irreducible unitary representations of the Lorentz group*, Ann of Math. (2) 48 (1947), 568–640. MR 9, 133.

[237] L. C. Biedenharn, *On the representations of the semisimple Lie groups. I: The explicit construction of invariants for the unimodular unitary group in n dimensions*, J. Math. Phys. 4 (1963), 436–445. MR 26 #5097.

[238] R. Brauer, *Die stetigen Darstellungen der komplexen orthogonalen Gruppe*, S.-B. Preuss. Akad. Wissenschaften, Berlin, 1929, pp. 626–638.

[239] R. Brauer and H. Weyl, *Spinors in n dimensions*, Amer. J. Math. 57 (1935), 424–449.

[240] F. Bruhat, *Sur les représentations induites de groupes de Lie*, Bull. Soc. Math. France 84 (1956), 97–205. MR 18, 907.

[241] ———, *Sur les représentations des groupes classiques p-adiques*. I, II, Amer. J. Math. 83 (1961), 321–338, 343–368. MR 23 #A3184.

[242] É. Cartan, *Les groups projectifs, qui ne laissent invariante aucune multiplicité plane*, Bull. Soc. Math. France 41 (1913), 53–96.

[243] ———, *Les groupes projectifs continus réels, qui ne laissent invariante aucune miltiplicité plane*, J. Math. Pures Appl. 10 (1914), 149–186.

[244] ———, *Sur la determination d'un système orthogonal complet dans un espace de Riemann symmetrique clos*, Rend. Circ. Mat. Palermo 53 (1929), 217–252.

[245] J. Dieudonné, *Note sur les fonctions sphériques*, J. Math. Pures Appl. (9) 41 (1962), 233–240. MR 26 #3818.

[246] P. A. M. Dirac, *A remarkable representation of the* 3 + 2 *de Sitter group*. J. Math. Phys. 4 (1963), 901–909. MR **27** #3713.

[247] J. Dixmier, *Représentations intégrables du groupe de De Sitter* Bull. Soc. Math. France **89** (1961), 9–41. MR 25 #4031.

[248] ———, *Sur les représentations unitaires des groupes de Lie résolubles*, Math. J. Okayama Univ. 11 (1962), 1–18. MR 26 #5096.

[249] L. Ehrenpreis and F. J. Mautner, *Some properties of the Fourier transform on semi-simple Lie groups*. I, II, III, Ann. of Math. (2) 61 (1955), 406–439; Trans. Amer. Math. Soc. 84 (1957), 1–55; ibid., 90 (1959), 431–484. MR 16, 1017; MR 18, 745; MR 21 #1541.

[250] I. M. Gel'fand, *Automorphic functions and the theory of representations*, Proc. Internat. Congress Math. (Stockholm, 1962), pp. 74–85; Institute Mittag-Leffler, Djursholm, 1963. MR **28** #1.

[251] R. Godement, *A theory of spherical functions*. I, Trans. Amer. Math. Soc. 73 (1952), 496–556. MR 14, 620.

[252] A. Haar, *Der Massbegriff in der Theorie der kontinuierlichen Gruppen*, Ann. of Math. 34 (1933), 147–169.

[253] Harish-Chandra, *On some applications of the universal enveloping algebra of a semisimple Lie algebra*, Trans. Amer. Math. Soc. **70** (1951), 28–96. MR 13, 428.

[254] ———, *Plancherel formula for the* $2 \times 2$ *real unimodular group*, Proc. Nat. Acad. Sci. U.S.A. **38** (1952), 337–342. MR 13, 820.

[255] ———, *Representations of semisimple Lie groups*. I, Trans. Amer. Math. Soc. 75 (1953), 185–243; II, III, ibid., 76 (1954), 26–65; 234–253. MR 15, 100; MR 15, 398; MR 16, 11.

[256] ———, *Representations of semisimple Lie groups*. IV, Amer. J. Math. 77 (1955), 743–777; V, VI, ibid., 78 (1956), 1–41; 564–628. MR 17, 282; MR 18, 490.

[257] ———, *The Plancherel formula for complex semisimple Lie groups*, Trans Amer. Math. Soc. 76 (1954), 485–528. MR 16, 111.

[258] ———, *Fourier transforms on a semisimple Lie algebra*. I, II, Amer. J. Math. 79 (1957), 193–257, 653–686. MR **19**, 293; MR **20** #2396.

[259] ———, *Spherical functions on a semisimple Lie group*. I, II, Amer. J. Math. 80 (1958), 241–310, 553–613. MR **20** #925; MR 21 #92.

[260] R. N. D'Heedene, *Simultaneous invariance of generalized spherical harmonics under the operations of two rotation groups*, Quart. Appl. Math. 16 (1958), 188–192. MR **21** #498.

[261] C. S. Herz, *Bessel functions of matrix argument*, Ann. of Math. (2) 61 (1955), 474–523. MR **16**, 1107.

[262] T. Hirai, *On infinitesimal operators of irreducible representations of the Lorentz group of n-th order*, Proc. Japan Acad. **38** (1962), 83–87. MR 25 #2146.

[263] J. Horváth, *Singular integral operators and spherical harmonics*, Trans. Amer. Math. Soc. **82** (1956), 52–63. MR **19**, 270.

[264] A. Hurevitsch, *Unitary representation in Hilbert space of a compact topological group*, Mat. Sb. 13 (55) (1943), 79–86. MR 6, 147.

[265] E. Inönü and E. P. Wigner, *On the contraction of groups and their representations*, Proc. Nat. Acad. Sci. U.S.A. **39** (1953), 510–524. MR **14**, 1061.

[266] S. Itô, *Unitary representations of some linear groups*, I, Nagoya Math. J. 4 (1952), 1–3; II, ibid., 5 (1953), 79–96. MR 13, 722; MR **14**, 533.

[267] A. Kohari, *Harmonic analysis on the group of linear transformations of the straight line*, Proc. Japan Acad. **37** (1961), 250–254.

[268] B. Kostant, *Lie group representations on polynomial rings*, Amer. J. Math. 85 (1963), 327–404. MR **28** #1252.

[269] R. A. Kunze and E. M. Stein, *Analytic continuation of the principal series*, Bull. Amer. Math. Soc. 67 (1961), 593–596. MR **26** #6808.

[270] ———, *Uniformly bounded representations and harmonic analysis of the 2 × 2 real unimodular group*, I, Amer. J. Math. 82 (1960), 1–62; II, ibid., 83 (1961), 723–786. MR 29 #1287; MR 29 #1288.

[271] Y. Lehrer-Ilamed, *On the direct calculations of the representations of the three-dimensional pure rotation group*, Proc. Cambridge Philos. Soc. 60 (1964), 61–66. MR **28** #145.

[272] J. S. Lomont, *Decomposition of direct products of representations of the inhomogeneous Lorentz group*, J. Math. Phys. 1 (1960), 237–243. MR 22 #5695.

[273] H. Maass, *Spherical functions and quadratic forms*, J. Indian Math. Soc. 20 (1956), 117–162. MR 19, 252.

[274] ———, *Zetafunktionen mit Grössencharakteren und Kugelfunktionen*, Math Ann. 134 (1957), 1–32. MR 19, 838.

[275] ———, *Zur Theorie der Kugelfunktionen einer Matrixvariablen*, Math. Ann 135 (1958), 391–416. MR 22 #12249.

[276] ———, *Zur Theorie der harmonischen Formen*, Math. Ann. 137 (1959), 142–149. MR 22 #12250.

[277] G. Mackey, *Induced representations of locally compact groups*, I, Ann. of Math. (2) 55 (1952), 101–139; II, ibid., (2) 58 (1953), 193–221. MR 13, 434; MR 15, 101.

[278] F. I. Mautner, *Unitary representations of locally compact groups*. I, Ann. of Math. (2) 51 (1950), 1–25; II, ibid., (2) 52 (1950), 528–556. MR 11, 324; MR 12, 157.

[279] ———, *Fourier analysis and symmetric spaces*, Proc. Nat. Acad. Sci. U.S.A. 37 (1951), 529–533. MR 13, 434.

[280] ———, *Induced representations*, Amer. J. Math. 74 (1952), 737–758. MR 14, 134.

[281] ———, *Spherical functions over $\mathfrak{P}$-adic fields*. I, Amer. J. Math. 80 (1958) 441–457; II, ibid., 86 (1964), 171–200. MR 20 #82; MR 29.#3582.

[282] W. Miller, *Some applications of the representation theory of the Euclidean group in three-space*, Comm. Pure Appl. Math. 17 (1964), 527–540. MR 30 #1214.

[283] A. Orihara, *Bessel functions and the Euclidean motion group*, Tôhoku Math. J. (2) 13 (1961), 66–74. MR 29 #194.

[284] F. Peter and H. Weyl, *Die Vollständigkeit der primitiven Darstellungen einer geschlossenen kontinuierlichen Gruppe*, Math. Ann. 97 (1927), 735–755; Russian transl., Uspehi Mat. Nauk 2 (1936), 140–166.

[285] L. Pukánszky, *On the Kronecker products of irreducible representations of the* $2 \times 2$ *real unimodular group*. I, Trans. Amer. Math. Soc. 100 (1961), 116–152. MR 30 #3177.

[286] ———, *On the Kronecker product of irreducible unitary representations of the inhomogeneous Lorentz group*, J. Math. Mech. 10 (1961), 475–491. MR 25 #4034.

[287] ———, *The Plancherel formula for the universal covering group of SL(R, 2)*, Math. Ann. 156 (1964), 96–143. MR 30 #1215.

[288] A. Rubinowicz, *Über ein Additions-theorem für Laguerreschen Polynomen*, Zeeman Festschriften, 1935, 143–147.

[289] M. Saito, *Représentations unitaires du groupe des déplacements du plan* $\mathfrak{p}$*-adique*, Proc. Japan Acad. 39 (1963), 407–409. MR 29 #3577.

[290] I. Schur, *Neue Anwendungen der Integralrechunung auf Probleme der Invariantentheory*, S.-B. Preuss. Akad. Wiss. Berlin, 1924, 189–208, 297–321, 346–355.

[290a] M. Sugiura, *Representations of compact groups realized by spherical functions on symmetric spaces*, Proc. Japan Acad. 38 (1962), 111–113. MR 26 #258.

[291] R. Takahashi, *Sur les fonctions sphériques et la formule de Plancherel dans le groupe hyperbolique*, Japan J. Math. 13 (1961), 55–90. MR 27 #2809.

[292] ———, *Sur les représentations unitaires des groupes de Lorentz généralisés*, Bull. Soc. Math. France 91 (1963), 289–433. MR 31 #3544.

[293] T. Takahashi, *Generalized spherical harmonics as representation, matrix elements of rotation group*, J. Phys. Soc. Japan 7 (1952), 307–312. MR 14, 373.

[294] J. Tamm, *Die verallgemeinerten Kugelfunktionen und Wellenphysik*, Z. Physik. 71, 141–150.

[295] N. Tatsuuma, *Decomposition of representations of the three-dimensional Lorentz group*, Proc. Japan Acad. 38 (1962), 12–14. MR 25 #4035.

[296] ———, *Decomposition of Kronecker products of representations of the inhomogeneous Lorentz group*, Proc. Japan Acad. 38 (1962), 156–160. MR 25 #4036.

[297] E. Thoma, *Die unitären Darstellungen der universellen Überlagerungsgruppe der Bewegungsgruppe des $R^2$*, Math. Ann. 134 (1958), 428–459. MR 21 #821.

[298] L. H. Thomas, *On unitary representations of the group of De Sitter space*, Ann. of Math (2) 42 (1941), 113–126. MR 2, 216.

[299] H. Weyl, *Harmonics on homogeneous manifolds*, Ann. of Math. **35** (1934), 486–499.

[300] ———, *Theorie der Darstellung kontinuierlicher halbeinfacher Gruppen durch lineare Transformationen*, Math. Z. 23 (1925), 271–309; 24 (1925), 228–395.

[301] E. P. Wigner, *On unitary representations of the inhomogeneous Lorentz group*, Ann. of Math. 40 (1939), 149–204.

[302] S. Yamaguchi, *On certain zonal spherical functions*, Mem. Fac. Sci. Kyushu Univ. Ser. A 17 (1963), 131–134. MR **28** #5142.

[303] H. Yoshizawa, *Group representations and spherical functions*, Sûgaku 12 (1960/61), 21–37. (Japanese) MR 26 #3817.

## SUBJECT INDEX

# NOTATION INDEX

## Latin alphabet

## German alphabet

## Greek alphabet



## Special symbols

**Copying and reprinting.** Individual readers of this publication, and nonprofit libraries acting for them, are permitted to make fair use of the material, such as to copy a chapter for use in teaching or research. Permission is granted to quote brief passages from this publication in reviews, provided the customary acknowledgment of the source is given.

Republication, systematic copying, or multiple reproduction of any material in this publication (including abstracts) is permitted only under license from the American Mathematical Society. Requests for such permission should be addressed to the Assistant to the Publisher, American Mathematical Society, P. O. Box 6248, Providence, Rhode Island 02940-6248. Requests can also be made by e-mail to `reprint-permission@ams.org`.